WHY GALAXIES CARE ABOUT AGB STARS:
A CONTINUING CHALLENGE THROUGH COSMIC TIME

IAU SYMPOSIUM 343

COVER ILLUSTRATION: THE DETACHED GAS SHELL OF U ANTLIAE

Mass loss during the AGB phase is critical for the evolution of the star itself as well as for the chemical evolution of the hosting galaxy. During the last years, observations have revealed that its temporal evolution, the structure and dynamics of its extended atmosphere, and the chemistry related to molecule and dust formation are more complex than originally thought. As an example, the detached shell surrounding the carbon star U Antliae as observed by ALMA is shown.

Credit: ALMA (ESO/NAOJ/NRAO/F. Kerschbaum)

IAU SYMPOSIUM PROCEEDINGS SERIES

INTERNATIONAL ASTRONOMICAL UNION

UNION ASTRONOMIQUE INTERNATIONALE

WHY GALAXIES CARE ABOUT AGB STARS: A CONTINUING CHALLENGE THROUGH COSMIC TIME

PROCEEDINGS OF THE 343rd SYMPOSIUM OF THE INTERNATIONAL ASTRONOMICAL UNION HELD IN VIENNA, AUSTRIA 20–23 AUGUST, 2018

Edited by

FRANZ KERSCHBAUM
University of Vienna, Austria

MARTIN GROENEWEGEN
Royal Observatory of Belgium

and

HANS OLOFSSON
Chalmers University of Technology, Sweden

Editorial Assistant

Verena Baumgartner
University of Vienna, Austria

CAMBRIDGE UNIVERSITY PRESS
University Printing House, Cambridge CB2 8BS, United Kingdom
1 Liberty Plaza, Floor 20, New York, NY 10006, USA
10 Stamford Road, Oakleigh, Melbourne 3166, Australia

First published 2019

Printed in the UK by Bell & Bain, Glasgow, UK

Typeset in System LaTeX 2ε

A catalogue record for this book is available from the British Library Library of Congress Cataloguing in Publication data

ISBN 9781108471527 hardback
ISSN 1743-9213

Table of Contents

Nucleosynthesis, mixing, and rotation

Pulsation, dynamical atmospheres, and dust formation

Circumstellar envelopes of AGB stars and their progeny, planetary nebulae

Binarity, planets, and disks

AGB stars in the cosmic matter cycle

Resolved and unresolved AGB populations

Galaxy evolution, including the first AGB stars

Posters

Preface

Stars are the main components of galaxies, and the sites of the creation of most chemical elements in our universe. As such they are among the most important ingredients in our physical description of the Universe. Due to their high luminosity and production of heavy elements and cosmic dust, stars on the Asymptotic Giant Branch (AGB) play an important role in their capacity as both actors and probes. In addition, AGB stars are prime laboratories for studying complicated physics, such as hydro-dynamical instabilities, double-shell nuclear burning, and dynamical atmospheres where physical and chemical processes, active on different temporal and spatial scales, are at work simultaneously. Understanding these stars sheds light on different processes of great relevance to the understanding of stellar and galactic evolution in general.

Our IAU symposium follows in the tradition of great symposia like *Planetary Nebulae* (No. 155, Innsbruck, 1992), *The Carbon Star Phenomenon* (No. 177, Antalya, 1996), *Asymptotic Giant Branch Stars* (No. 191, Montpellier, 1998), and several other symposia related to variability research or the stellar content of galaxies which had AGB stars as important subtopics. In 2006, a new series of conferences under the main title *Why galaxies care about AGB stars* was initiated with the aim to bring two normally separated communities together, namely those studying the AGB proper and those interested in the evolution of galaxies. Already the first incarnation of the series held in Vienna, 2006 under the subtitle: *Their importance as actors and probes* saw a good acceptance by the community. It was followed by the 2010 conference *Shining examples and common inhabitants* and the 2014 one named *A closer look in space and time.*

The growing interest in these conferences triggered the formation of an international SOC comprising scientists covering a wide range of scientific backgrounds and working on all continents with the purpose of proposing an IAU symposium following this successful tradition for the IAU General Assembly to be held in Vienna in 2018. It should extend this interdisciplinary approach to an even wider community.

Following endorsement and sponsoring by IAU Commissions and Divisions and subsequent approval by the IAU Executive Committee in 2017, speakers were invited and the community at large informed. The overwhelming, positive response materialized into the impressive programme we now have.

It is a great pleasure to acknowledge the financial support of our sponsors, the active support of the members of the SOC and LOC in coping with the big challenges and realizing the numerous details always associated with such a symposium. Especially we would like to thank Verena Baumgartner for her invaluable work as editorial assistant for these proceedings.

Its is our hope that these proceedings will provide a useful resource for experts as well as newcomers to the field.

Franz Kerschbaum, Hans Olofsson, and Martin Groenewegen, editors of these proceedings.

Editors

Franz Kerschbaum
University of Vienna, Austria

Martin Groenewegen
Royal Observatory of Belgium

Hans Olofsson
Chalmers University of Technology, Sweden

Editorial Assistant

Verena Baumgartner
University of Vienna, Austria

Organising Committee

Scientific Organising Committee

SOC Chairs

Hans Olofsson	(SE)

SOC Co-Chairs

Franz Kerschbaum	(AT)
Paola Marigo	(IT)

SOC Members

Martha Boyer	(US)
Martin Groenewegen	(BE)
Susanne Höfner	(SE)
Atefeh Javadi	(IR)
Ciska Kemper	(TW)
Maria Lugaro	(HU)
Claudia Maraston	(UK)
Shazrene Mohamed	(ZA)
Keiichi Ohnaka	(CL)
Angela Speck	(US)
Hans Van Winckel	(BE)
Peter Wood	(AU)
Albert Zijlstra	(UK)

Acknowledgements

The symposium is sponsored and supported by the IAU Division G (Stars and Stellar Physics), Division H (Interstellar Matter and Local Universe), Division J (Galaxies and Cosmology) and by the IAU Commission G3 (Stellar Evolution).

Funding by the
International Astronomical Union,
University of Vienna,

and

Robert Wing Support Fund at Ohio State University,
are gratefully acknowledged.

CONFERENCE PHOTOGRAPH

Address by the Scientific Organizing Committee

Dear colleagues,

also from us a warm welcome to this IAU Symposium No. 343. We are delighted to see you all here in the beautiful city of Vienna during the 30th General Assembly of the International Astronomical Union. Vienna has already a long tradition in stellar astrophysics and hosted several successful symposia and conferences in the field and especially the very topic of our meeting, namely Asymptotic Giant Branch (AGB) stars. Many of us remember the AGB meetings in 2006, 2010 and 2014!

Symposium 343 builds a bridge between research on AGB stars themselves and its application to the modelling of stellar populations and the chemical evolution of galaxies and the Universe as a whole. Current developments and challenges seen from both domains are discussed to reach an understanding of possibilities, limitations, and needs in both areas, and hence to improve our understanding of the role of AGB stars in the context of galaxies over cosmic time. Despite the fact that major efforts have been carried out on both observational and theoretical grounds in recent years, our knowledge of AGB stars is still deficient due to uncertainties related to mass loss, convection, mixing, dredge-up efficiencies, and the role of binary interaction processes. These uncertainties in our understanding of AGB stars directly propagate into the field of extragalactic astronomy, where they affect critically the interpretation of galaxy properties, e.g. stellar masses, ages, and the chemical evolution. The complexity of the objects also makes it difficult for individual researchers to master all aspects of their role as galaxy inhabitants, a problem that the proposed symposium aims to illuminate and overcome.

New and upcoming major observational facilities like ALMA, Gaia, JWST, LSST, SKA, and the ELTs will provide exciting opportunities to tackle these challenges from the observational side, stretching from the detailed study of individual objects that are spatially resolved to AGB populations in distant galaxies. At the same time, thanks to high-performance computing, 3D modelling of stellar interiors is starting to become feasible, propelling us toward a better understanding of the uncertainties related to the physics of AGB stars. This makes it particularly important to outline a strategic programme of combined theoretical and observational activities at this time.

In order to cover the large breadth of the symposium it is divided into nine themes, each one covered by invited talks, oral contributions, and posters. The themes are:

- Stellar structure and evolution to, on, and past the AGB
- Nucleosynthesis, mixing and rotation
- Pulsation, dynamical atmospheres and dust formation
- Circumstellar envelopes of AGB stars and their progeny, planetary nebulae
- Binarity, planets and disks
- AGB stars in the cosmic matter cycle
- Resolved and unresolved AGB populations in stellar systems
- Galaxy evolution, including the first AGB stars.

There will also be two plenary talks, one on nucleosynthesis and one on new and future observational perspectives. These will be centred on AGB stars, but also give a broader perspective of interest for research on stars and galaxies in general.

We wish you all a very constructive and pleasant four working days here.

Hans Olofsson, Franz Kerschbaum, and Paola Marigo, co-chairs SOC
Vienna, August 20, 2018

In memoriam Thomas Posch (1974–2019)

The Department of Astrophysics of the University of Vienna and the Scientific Organizing Committee of IAU S343 mourns the loss of Thomas Posch, who passed away on the 4th of April 2019 after a long and difficult period of illness. Last year Thomas contributed in various forms at the 30th General Assembly of the International Union and until as recently as January this year he was still supervising students and holding tours of his home institute.

Thomas Posch was born on the 20th of February 1974 in Graz, Austria to Hildegard and Siegfried Posch. Between 1980 and 1984 he attended the primary school Volksschule Graben in the third district of Graz. He continued his education at the Bundesgymnasium Carnerigasse, where he chose to matriculate in the natural sciences branch of study. He graduated from the Gymnasium on the 1st of July 1992 with a high distinction. During his time at school he was the winner of the "Astronomy in Space" essay competition held by the European Space Agency (ESA). The following trips to the ESA/ESTEC institute in Noordwijk, the Netherlands and the Headquarters of the European Southern Observatory in Garching, Germany left a lasting impression on him. The 1990 Solar eclipse, which he viewed from Jansuu in southern Finland, also helped encourage his enthusiasm for astronomy and space research. It should also be mentioned that Thomas received the Bronze Medal at the 1991 Austrian Physics Olympics.

Thomas began his university education in 1992 in Graz by starting degrees in physics and astronomy. He spent an ERASMUS year at the Free University in Berlin, where he studied Physics and Philosophy. Upon returning to Austria, he moved to the University of Vienna, where he wrote his diploma thesis on the topic of "Circumstellar dust and the infrared-spectra of pulsating red giants" ("Zirkumstellarer Staub und die Infrarot-Spektren pulsierender Roter Riesen") under the supervision of Prof. Hans Michael Maitzen. He celebrated his graduation on the 9th of February 1999. Alongside taking part in many summer schools, Thomas visited the laboratory at the Department of Astrophysics at the University of Jena, Germany. This visit would prove to be of great importance for his future work in astrominerology.

His second great academic love, philosophy, and in particular the philosophy of nature, was the central topic of his next university endeavor between 1999 and 2002. Under the supervision of Prof. Friedrich Grimmlinger (Vienna) and Prof. Renate Wahsner (Berlin) Thomas delivered his dissertation titled "The Mechanics of Heat in the Jena System Design by Hegel from 1805/06" ("Die Mechanik der Wärme in Hegels Jenaer Systementwurf von 1805/06") which commented on the background to the development of the theories of thermodynamics between 1620 and 1840.

Thanks to a competitive research scholarship from the Austrian Academy of Science Thomas was able to submit his doctoral dissertation titled "Astromineralogy of Circumstellar Oxide Dust" in 2005. His supervisors for this project were Prof. Franz Kerschbaum (Vienna) und Prof. Thomas Henning (Heidelberg). The scientific core of the dissertation was the comparison of the signatures of cosmic dust, as observed by the ISO space telescope, with terrestrial analogues. The results of which were published in several international journals. His work has inspired several students and has fully established the field of astrominerology in Vienna.

As part of his military service in 2005, Thomas Posch conducted the first systematic study of the night sky brightness in Austria. This was an important base for his future work towards the preservation of the night sky.

After having worked as a guest scientist in Jena and after receiving a scholarship from the Max-Planck-Society, Thomas took up a position as Staff Scientist at the Department of Astronomy (now Astrophysics) at the University of Vienna in April 2006, almost exactly 13 years ago. Alongside his scientific endeavors, he also took on various responsibilities regarding the institute's library and historically important archives, as well as being engaged with public outreach activities. Of these his most noteworthy contributions include the expansion of the collection related to Maximilian Hell ("Schausammlung Maximilian Hell"), the screening and analysis of historical relevant archive material, and the supervision and coordination of large events such as the International Year of Astronomy 2009.

In 2011 years of fruitful astronomical research culminated in the comprehensive habilitation dissertation titled "Studies in Astromineralogy and Stellar Mass Loss", which allowed Thomas to teach at the level of docent. He distinguished himself both through his teaching and through his supervision of students. He brought together knowledge from many fields of research ranging from star formation and evolution, mineralogy, life, as well as historical topics and didactic methods in the natural sciences. A natural extension to his university courses were his vast, comprehensive, and popular public lectures, which made him into a well-known public educator both within and outside Austria. This was surely helped by his many contributions to radio and television programmes.

His national and international reputation led to a series of important roles in professional societies, as well as honors and accolades. Several positions worth mentioning include: chairman of the advisory board for "transdisciplinary science" at the Guardini foundation (since 2013), chairman of the working group for historical astronomy of the Astronomical Society (since 2014), contributing member of the ÖNORM committee on Light-Emission (between 2008 and 2012). In 2014 Thomas Posch also received the Galileo Award from the International Dark Sky Association.

During the last three years, even as the progressing illness curtailed his abilities, Thomas remained faithful to all areas of his work. He published papers, held lectures, taught, organized, and coordinated until the very end. The General Assembly of the International Astronomical Union held in the summer of 2018 in Vienna, with its multifaceted and diverse conference formats, was the last major event where Thomas played a leading role. Here he also chaired the annual meeting of the working group for the history of astronomy. As late as January 2019 Thomas was still supervising students and holding lectures. He passed away on the 4th of April 2019.

To honour the work of Thomas Posch the International Astronomical Union decided to name the main belt asteroid 2008 TP9 (328432) Thomasposch.

The following paragraphs are devoted to the major achievements from each of Thomas Posch's fields of research:

In the realms of astrophysics Thomas Posch concentrated mainly on astrominerology. His interest in explaining infrared emission bands from semi-periodic variable red giant stars as observed with the ISO satellite began during his time as a university student. Such bands are produced by mineral dust particles, primarily refractory oxides and silicates, condensing in stellar outflows. By working with Johann Dorschner and Thomas Henning at the Jena astrophysical laboratory Thomas was able to compare these observations with data from laboratory measurements of analogous terrestrial material, thereby explaining the nature of cosmic dust. Furthermore his collaboration with Johann Dorschner, Dirk Fabian, Harald Mutschke, Cornelia Jäger and Gabriele Born and their work on Aluminium oxides (Corundum, Spinel, Hibonite), Iron oxides, and Titanium oxides led to several important publications. In particular, the identifications of aluminium oxides with the stellar 13 μm band and of magnesium/iron oxides with the 19.5 μm band will remain related to his name for all time.

A related and also very important topic in Thomas Posch's work was the absorptivity of minerals in the wavelength range of stellar radiation. Again in the laboratory in Jena, Thomas investigated the influence which doping ions like Chromium and Iron had on the absorption spectrum of spinel. He was the first to calculate, using laboratory data, their temperature in the radiation field of a star. Together with Simon Zeidler he wrote a highly cited paper which also included absorption spectra of titanium oxides and silicates. Thomas demonstrated the need for, and the importance of, dust observations during the preparation of the far-infrared PACS spectrograph onboard the Herschel-Mission. Afterwards he provided comparative datasets for dust spectra in the corresponding spectral range. This included measurements on carbonates for temperatures down to 10 K in the Jena laboratory, and a spectroscopic study of hydrous silicates, which can provide an important hint to liquid water in planetary systems (with H. Mutschke). During the last years, he published together with S. Zeidler temperature-dependent optical constants of oxides and silicates, ultimately putting the assignment of aluminium oxide to the 13μm band on solid grounds.

From early on Thomas Posch was not only interested in the natural sciences, but rather also in the philosophical and epistemological questions connected to science. Quite often he would begin by using the historical approach. An example of this concerns the history of the Vienna institute during the Third Reich, which was then led by Bruno Thüring. Thüring was a long standing critic of Einstein and confidant of Wilhelm Führer. Führer, who was originally an astronomer, was Obersturmführer of the Waffen-SS and became a leading bureaucrat in the ministry of science of the Third Reich. Based on their preserved correspondence and published records, Thomas was able to build up a detailed picture of the times and also validate Thürings misuse of Hugo Dinglers philosophy.

The rare book collection of the University Observatory in Vienna dates back to the time of Peuerbach and Regiomontanus. Thomas Posch re-examined key works in this collection with an emphasis towards connections to Austria (e.g. by Kepler) and to astronomical phenomena. With these insights he composed literary works, and even theatrical pieces. His dramolett: "Kepler, Galilei and the Telescope" appropriately presented during the International Year of Astronomy 2009, told the story of the academic battle in the 17th century regarding the new emerging world view, the role that observations played, and the vanity of many of the central figures. Furthermore, his edition of Littrow's text detailing the history of astronomy ("Littrows Geschichte der Astronomie", together with Günter Bräuhofer and Karin Lackner, 2016) and the texts about Maximilian Hell's trip to Nordland based on excerpts from his on the diaries at the institute, would not have been possible without his extensive work with the archives.

Thomas Posch's research in philosophy also covered a wide range of topics. His lifelong interest in Hegel began during the time he wrote his dissertation under Friedrich Grimmlinger (Wien) and Renate Wahsner (Berlin). As part of the Society of System Philosophy (Gesellschaft für Systemphilosophie), Thomas was able to fruitfully introduce Hegel's thoughts into modern day philosophical and scientific discourse. His goal was to consistently achieve the problematization of the reductionistical point of view. By using this perspective he was able to foster the conversation between the natural sciences and theology. This was manifest in his contributions to present-day debates, in the already mentioned historical studies, and especially in his Monograph about Johannes Kepler ("Johannes Kepler. Die Entdeckung der Weltharmonie", WBG, 2017). Furthermore, he never shied away from questioning his supremely personal connection to faith and spirituality, especially given his occupation as a scientist. Given his background it is worth mentioning his well-grounded and thorough criticism of astrology, which he delivered through essays and lectures. Throughout his whole life it was important to Thomas to clearly separate science from pseudoscience and esotericism.

During the 1997 IAU General assembly in Kyoto, Japan, the problem of the rapid loss of dark night skies due to artificial light sources was raised. This motivated Franz Kerschbaum and soon thereafter also Thomas Posch to systematically quantify, and concurrently raise awareness of this ever growing problem. Simple and generally available measurement techniques needed to be developed. During Thomas's military service, he conducted the first study of the night sky brightness in Austria. Since then Thomas has accompanied many projects to modernize communal lighting as consultant. His edited-book "Das Ende der Nacht. Die globale Lichtverschmutzung und ihre Folgen" ("The End of Night. Global light pollution and its consequences") was released in 2009 and is currently in its second edition. It is seen as standard literature on the subject in German speaking countries and beyond. Especially in the last years of his life, one of Thomas' greatest concerns was to spread the message of the importance of preserving a naturally dark sky to both the general population and the relevant authorities. He was a consultant during the construction of a light-measurement-network in Upper Austria between 2014 and 2016. He was also one of the authors of the "Austrian guide to outdoor lighting" ("Österreichischer Leitfaden für Außenbeleuchtung"), which was signed by many provincial environment councilors. The data and results collected from the network were able to show which areas in the state of Upper Austria could still be classed as "Dark Sky Sites". Such places are eligible for an official certification as such. Through his role as a co-organizer of the Upper Austrian environment congress 2018 in Linz, Thomas was able to present this project to the public. It was enthusiastically received. Such protected area will serve as a role model for other regions in Austria. The implementation of such areas as well as passing on his knowledge to his doctoral student, Stefan Wallner, were top priorities for him until the very end. His work and the way he went about it will without a doubt be recognized well into the future.

Thomas's talent and desire to share the fascinating field of astronomy with the general public was already apparent during his time as a student. This was evident not only in his series of published popular scientific works, but also in his engagement with the public outreach activities at the university observatory in Vienna. An example of this engagement was the star counting initiative "How many stars can we still see", organized together with the Kuffner Observatory Association as part of "Science Week 2001". This initiative was the first Austrian-wide attempt to quantitatively determine the brightness of the night sky. His wide and varied knowledge and range of interests, as well as his welcoming and calm personality, meant that Thomas Posch was predestined for working with the wider public. A role which he gladly took on in 2006 as part of his position as staff scientist. Answering questions, contributing to popular science magazines and media outlets, reporting on the newest scientific results and insights (e.g. the yearly astronomical review on science.orf.at) were all part of Thomas's everyday activities.

One activity which Thomas Posch took on with great enthusiasm and competence, were the tours of the institute, and especially those to the large refractor telescope. During his time at the institute he gave more than 700 tours to guests of all ages and from all walks of life, from Austria and abroad. Given his broad knowledge of astronomy and astrophysics, as well as history and philosophy, these tours were a very special experience. Furthermore his excellent knowledge of foreign languages meant that he was able to hold tours not only in German, but in English, French, and Italian. His tours were always well received.

Naturally, Thomas also always played an important role in the organization of larger events, such as the "long night of research" ("Lange Nacht der Forschung") or the Children's University (Kinderuni). He also often took on the role of main organizer, e.g. for the Year of Astronomy 2009. His last great challenge was most certainly the 30th General Assembly of the IAU in Vienna in 2018, where he took on a series of official roles. Most notably those of the media spokesperson, and as a member of the editorial team for the daily conference newsletter. He was able to pass his enthusiasm, enjoyment, and wealth of experience in, and for, public outreach on to his students in many different ways, as well as always maintaining an emphasis on quality. A late example of this was his very active involvement with the course associated with the institute's mobile planetarium. Thomas Posch continued to maintain close contact to, and valued the contributions of, amateur astronomers as their contributions to public outreach was, and still is, invaluable. He also never stopped enjoying looking through his own not-so-small telescope.

Thomas's works as an author should also not be forgotten. His various pieces of short prose and somewhat more strongly condensed lyrical (if rhymeless) verses accompanied him throughout his life. The first collection of these works appeared in the Viennese "Edition Doppelpunkt" in 2001 with the title "Miniaturen. Ein bißchen Literatur" ("Miniatures - A little bit of Literature"). In the meantime he has written enough new texts that a second volume could be filled. In his pieces of Prose, Thomas used short stories to critically, yet amusingly, illuminate typical behaviours of modern-day humans. His lyrical verses described his personal, and not seldomly painful experiences in simple, yet deep and meaningful sentences. The literary styles in both verse and prose which he found are quite clearly his own creation, and very much worthy of attention and recognition.

An obituary by Franz Kerschbaum, with contributions from Josef Hron, Cornelia Jäger, Harald Mutschke, Johann Schelkshorn, Wilhelm Schwabe and Stefan Wallner. Translated into English by Kieran Leschinski.

Vienna, April 2019

Stellar structure and evolution to, on and past the AGB

Why Galaxies Care About AGB Stars:
A Continuing Challenge through Cosmic Time
Proceedings IAU Symposium No. 343, 2019
F. Kerschbaum, M. Groenewegen & H. Olofsson, eds.

doi:10.1017/S1743921318007019

AGB Stars: Remaining Problems

John Lattanzio

Monash Centre for Astrophysics
School of Physics and Astronomy, Monash University, 3800, Australia
email: `john.lattanzio@monash.edu`

Abstract. I present a subjective list of what I think are the most serious problems in the modelling of AGB stars. Because AGB stars represent the last phase of stellar evolution, they suffer from the accumulation of the effects of uncertainties in all the earlier phases. The complexity of AGB evolution adds further uncertainties specific to the evolutionary behaviour of those stars. Most of the problems are associated with mixing, specifically the boundaries of mixed regions. The nature of the "extra-mixing" remains a mystery, let alone how to model it reliably. Other problems are briefly mentioned and I finish with some hopes of making progress in the future.

Keywords. stars: AGB and post-AGB, stars: evolution, stars: interiors, convection, hydrodynamics

---

1. Introduction

It is useful to periodically review the state of knowledge of a field, with particular emphasis on the impediments to advances. Stellar modelling is no different, of course. If you ask almost anyone who works in the field of theoretical stellar physics "What are the biggest problems?" you will mostly hear "convection" or "turbulence" or "hydrodynamics". While that is largely true, it is still instructive to see how our ignorance of how to model these phenomena manifests itself in the theoretical models. These simple words actually cover a multitude of complex interactions. There are also other problems, of course, but I concentrate on mixing and related phenomena.

The AGB phase is the last nuclear phase of evolution of most stars, say those between about 1 and 8 $M_{\odot}$. Hence the models of AGB stars have accumulated all uncertainties from earlier evolutionary phases before we even start with AGB evolution. Uncertainty about convective overshoot on the main sequence affects AGB stars through the mass of their H-exhausted cores. Similarly for the core helium burning phase. The so-called "extra-mixing" on the red giant branch does not produce significant structural changes, but it can dramatically alter the surface composition, which is used as a diagnostic for events during the subsequent AGB evolution. Hence we start with a discussion of what I see as the main uncertainties prior to AGB evolution.

2. Previous Evolution

AGB stars began their lives with masses in the range of about 1–8 $M_{\odot}$. We do not discuss the super-AGB stars, which have been reviewed recently in Doherty *et al.* (2017) and Gil Pons *et al.* (2018). A note on nomenclature is appropriate, with reference to Figure 1. The Schwarzschild criterion gives us the position of neutral *buoyancy*, where the forces balance. A fluid element (if such exists in reality!) has a momentum that will carry it across this border. I shall refer to this as "overshoot", and this is shown schematically in the left panel of Figure 1. Further, at the maximum extent of this overshoot the fluid can "entrain" layers from the radiative region. In some cases, such

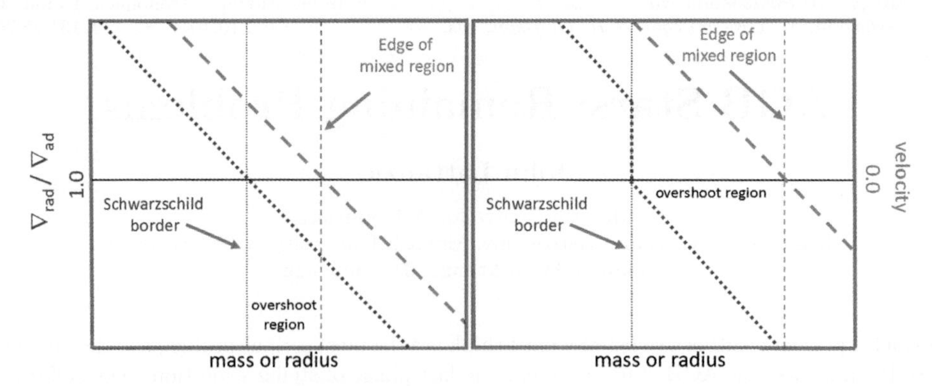

Figure 1. Schematic showing mixed regions in relation to the Schwarzschild criterion.

as core He burning and third dredge-up, there is a discontinuity in the acceleration (due to a discontinuity in composition) and although the Schwarzschild criterion does find a point of zero buoyancy, there is large positive buoyancy on the convective side, rather than a smooth decrease through zero at the border. This is shown in the right panel of Figure 1.

2.1. *Core Hydrogen Burning*

There is now much clear evidence for the existence of convective mixing beyond the Schwarzschild border for main sequence stars with convective cores. Indeed attempts have been made to quantify the required overshoot (Claret & Torres 2016, 2017). This is even more important for the higher masses. The size of the resultant H-exhausted core is critical for later evolution.

2.2. *First Dredge-Up*

The inward march of the convective envelope begins at the base of the giant branch and reaches its maximum extent essentially at the position of the bump in the luminosity function. The discrepancy between the predicted and observed position of the bump is well known now (e.g. see Angelou *et al.* 2015 and references within), and can be alleviated by the inclusion, again, of some overshoot beyond the Schwarzschild border. This affects the composition of the envelope at the end of the giant evolution.

2.3. *Extra-mixing on the Giant Branch*

For decades we have known that standard models do not match the observed abundances for stars beyond the first dredge-up, especially for lithium and carbon. Further, the models do not predict the observed variation with luminosity seen for some species, such as the carbon isotopic ratio. For a recent review see Karakas & Lattanzio (2014). This is very strong evidence for some form of "extra-mixing" connecting the convective envelope to the burning regions. This mixing is "extra" in the sense that it is not included in "standard" or "traditional" models, which only include mixing by convection.

Meridional circulation, caused by rotation, has long been suggested as a possible mechanism for the extra-mixing, but it appears that this phenomenon is not sufficient to explain the observations (Palacios *et al.* 2006). Many other mechanisms have been suggested, such as thermohaline mixing (Eggleton, Dearborn & Lattanzio 2006, 2008; Charbonnel & Zahn 2007) which has received a lot of attention in recent years. However, this mechanism is not free from criticisms, such as those arising from multi-dimensional hydrodynamical calculations (see Karakas & Lattanzio 2014 for details). Another of the regular suspects is magnetic fields (Busso *et al.* 2007), or a combination of meridional circulation and turbulent diffusion (Denissenkov & Tout 2000), and there have been many others. It is fair to say that none has overwhelming support and even though thermohaline mixing is perhaps the most favoured at present, the modelling of the process in 1D is still under debate, as discussed by Henkel, Karakas & Lattanzio (2017).

2.4. *Core Helium Burning*

Within the constraints of the tradition 1D formalism for convection, the discussion by Castellani, Giannone & Renzini (1971a,b) is still one of the best for explaining the problems that arise during core helium burning. The variation of opacity with temperature, density and composition contrives to produce a minimum in the ratio of the radiative to adiabatic gradients. When this minimum reaches unity then the convective core splits, and we face the problem of dealing with the outer region. This is a classic case of a region that is stable according to the Ledoux criterion but unstable according to the Schwarzschild criterion. Note that the Ledoux criterion includes the variation of composition in the stability condition. Various numerical schemes have been devised to handle this, but in all cases the details of how to handle the mixing are debated. This can produce significant changes in the size of the mixed region.

What is worse is the discovery of "core breathing pulses" where the convective core suddenly, and essentially instantaneously, grows into the semiconvective region. This results in an increase in the central helium content (e.g. see Castellani *et al.* 1985). Debate has raged over whether these are the result of a real instability or a numerical instability. While this behaviour shows many of the signs of a numerical instability, an analytic study by Sweigart & Demarque (1973) showed that there is a genuine physical basis for the instability, and indeed verified that it should only occur when the central He mass fraction reduces below about 0.12. Observations must be the ultimate arbiter, and traditionally this has been investigated by star counts to determine the relative timescales. The core-breathing pulses extend the lifetime of core helium burning because they mix extra helium into the core; they also simultaneously decrease the duration of the early AGB because the stars have removed some helium from the region beyond the core, which is usually burned during the early AGB. It seems that empirically, these pulses are minimal or not existent, as discussed in the recent work of Constantino *et al.* (2016).

The evolution during the core helium burning phase remains one of the last significantly uncertain areas in standard stellar evolution. The lifetime of this phase and, more importantly, the size of the helium exhausted core, are critically dependent on assumptions made for the calculation of mixing during core helium burning. This, in turn, affects the size of the hydrogen exhausted core because it determines how long the shell advances prior to exhaustion of the core helium supply. These are critical to later AGB evolution, with the thermal pulses beginning when the two active nuclear shells move close to each other and the thermal instability can develop. We desperately need more work in this area, such as has been performed by Constantino *et al.* (2015), Constantino *et al.* (2017), Spruit (2015).

3. AGB Evolution

So we come to AGB evolution itself. Again, the main uncertainties are associated with the convective phases, but this time we are concerned with those convective zones that develop during a pulse cycle.

3.1. *Depth of Third Dredge-Up*

The composition discontinuity that results during third dredge-up produces a discontinuity in the ratio of the gradients, as shown in the right panel of Figure 1 (only here the situation is reversed left-right, as the convection is in the outer regions of the star). Hence we have the usual uncertainties of how to calculate the extent of the mixing. Almost every person makes different assumptions for this phase, and hence the resultant depth of dredge-up is very dependent on how the calculations are performed (e.g. Frost & Lattanzio 1996; see also Kamath, Karakas & Wood 2012). There is very little reliability in the calculations for this phase, sadly.

3.2. *The Carbon Pocket*

AGB stars are powerhouses of s-process nucleosynthesis. It is well established that this is dependent upon a partially mixed region, at the bottom of the convective envelope, which mixes small amounts of protons into the carbon-rich intershell. These are then turned into ^{13}C by the first reaction in the CN cycle, and when the region later heats enough for α-capture we get neutrons produced by $^{13}C(\alpha,n)^{16}O$. The details of the formation of this so-called carbon pocket remain a mystery. How are the protons added? Is it simply partial mixing? Is it entrainment at the bottom of the convective envelope? Are magnetic fields implicated? Could it be shear instabilities from rotation? Gravity waves? In each case the proton profile is different, and there is an industry trying to work backwards from the resultant s-processing to determine constraints on the proton profile and hence mixing. The details remain unknown.

3.3. *Extra-mixing on the AGB*

Earlier we discussed extra-mixing on the first giant branch. There are similar reasons for thinking that some form of extra-mixing may occur during the AGB phase. The evidence appears to be contradictory. For example, Busso *et al.* (2010) argue for extra-mixing based on oxygen and aluminium isotopes in pre-solar grains as well as other evidence. In contrast Karakas, Campbell & Stancliffe (2010) argue that the C/O and carbon isotope ratios do not require extra-mixing. Interestingly, if one does not include extra-mixing on the first giant branch then the predictions for compositions on the AGB are incompatible with the models. But the real stars do have some extra-mixing on the first giant branch. If this is included then the models are more in accord on the AGB. In short, the situation is still uncertain.

3.4. *Overshoot inward from the Intershell Convective Zone*

For a convective core the only overshoot possible is outward, and for a convective envelope the only possible overshoot is inward. But for the intershell convective region that develops during a thermal pulse we have the possibility of overshoot at either (or both) edges of the convective region. It has been suggested by Herwig (2000) that overshoot inwards might mix carbon and, more importantly, oxygen into the intershell convective zone. This region usually has a negligible oxygen content, and yet there are indications that observations require some oxygen to be added to the envelope of AGB stars. In particular, this might explain the observed abundances in PG-1159 stars.

While one can easily apply an algorithm to force overshoot according to your favourite prescription, the true situation is far from understood. Lattanzio *et al.* (2017) attempted to model the proposed overshoot inward into the dense CO core, and found the extent of the overshoot to be negligible. At present there is no resolution to this conflict, except to allow that the mixing may exist, but may be produced by some mechanism other than traditional convective overshoot – something like a shear instability or another hydrodynamical instability.

4. Proton Ingestion Episodes

Low mass AGB stars, in particular, suffer "proton ingestion episodes" where a convective region, typically burning helium, makes contact with a hydrogen-rich region. These are sometimes called "convective-reactive" events (Jones *et al.* 2015), because the convective timescale is similar to the nuclear reaction timescale. I will use the term PIE for "proton ingestion episodes". These can occur during both core flashes and AGB thermal pulses, especially for the lower metallicities. Clearly the assumptions used in the mixing-length theory are inappropriate for such events, and indeed it seems very unlikely that the simple Schwarzschild (or Ledoux) criterion will be reliable either. Let me also warn against doing the mixing in such regions with a diffusion approximation, as is done almost universally in evolutionary calculations.

Interest in these events has grown following the recent work of Hempel *et al.* (2016). It seems that PIEs can produce a neutron density that is intermediate between that of the s- and r-processes (Campbell, Lugaro & Karakas 2010). The resultant neutron capture nucleosynthesis appears to produce an abundance pattern that matches the so-called CEMP-s/r stars.

5. Progress

After that depressing list of problems, let me end with something much more promising. I think that true progress is around the corner, finally. This progress answers to the names of "asteroseismology" and "hydrodynamical simulations". Neither of these will provide the full answer in the short term. But both are starting to yield true insights. Progress has been slow but the pace has accelerated in the last decade and there is room for optimism. Let us hope that at the next IAU General Assembly we can report some substantial progress on the problems outlined in this paper.

Acknowledgements

I would like to thank Peta Nocon for providing important and valuable distraction, both when requested and at less opportune times.

References

Angelou, G. C., *et al.* 2015, *MNRAS*, 450, 2423

Busso, M., Wasserburg, G. J., Nollett, K. M., & Callandra, A., 2007, *ApJ*, 671, 802

Busso, M., *et al.* 2010, *ApJ* (Letters), 717, L47

Campbell, S. W., Lugaro, M., & Karakas, A. I., 2010, *A&A*, 522, 6

Castellani, V., Chieffi, C., Pulone, L., & Tornambé, A., 1985, *A&A*, 296, 204

Castellani, V., Giannone, P., & Renzini, A., 1971a, *Ap&SS*, 10, 340

Castellani, V., Giannone, P., & Renzini, A., 1971b, *Ap&SS*, 10, 355

Charbonnel, C., & Zahn, J. P., 2007, *A&A*, 467, L15

Claret, A., & Torres, G., 2016, *A&A*, 592, A15

Claret, A., & Torres, G., 2017, *ApJ*, 849, 18

Constantino, T., Campbell, S. W., Christensen-Dalsgaard, J., & Lattanzio, J. C., 2015, *MNRAS*, 452, 123

Constantino, T., Campbell, S. W., Lattanzio, J. C., & van Duijneveldt, A., 2016, *MNRAS*, 456, 3866

Constantino, T., Campbell, S. W., & Lattanzio, J. C., 2017, *MNRAS*, 472, 4900

Denissenkov, P., & Tout, C. A., 2000, *MNRAS*, 316, 395

Doherty, C. L.. Gil-Pons, P., Siess, L., & Lattanzio, J. C., 2017, *PASA*, 34, 56

Eggleton, P. P., Dearborn, D. S., & Lattanzio, J. C., 2006, *Science*, 314, 1508
Eggleton, P. P., Dearborn, D. S., & Lattanzio, J. C., 2008, *ApJ*, 677, 581
Frost, C. A., & Lattanzio, J. C., 1996, *ApJ*, 473, 383
Gil-Pons, P., Doherty, C. L., Gutiérrez, J. L., Siess, L., Campbell, S. W., Lau, H. B., & Lattanzio, J. C., 2018, *PASA*, in press, arXiv:1810.00982
Hempel, M., Stancliffe, R. J., Lugaro, M., & Meyer, B. S., 2016, *ApJ*, 831, 171
Henkel, K., Karakas, A. I., & Lattanzio, J. C., 2017, *MNRAS*, 496, 4600
Herwig, F., 2000, *A&A*, 360, 952
Karakas, A. I., Campbell, S. W. & Stancliffe, R. J., 2010, *ApJ*, 713, 374
Kamath, D, Karakas, A. I., & Wood, P. R., 2012, *ApJ*, 746, 20
Karakas, A. I., & Lattanzio, J. C., 2014, *PASA*, 31,30
Lattanzio, J. C., Tout, C. A., Neumerzhitckii, E. V., Karakas, A. I., & Lesaffre, P., 2017, *MemSAIt*, 88, 248
Palacios, A., Charbonnel, C., Talon, S., & Siess, L., 2006, *A&A*, 453, 261
Jones, S., *et al.*, 2015, *MNRAS*, 455, 3848
Spruit, H., 2017, *A&A*, 582, 2
Sweigart, A. V., & Demarque, P. 1973, in IAU Colloq. 21, Astrophysics and Space Science Library, Variable Stars in Globular Clusters and in Related Systems, Vol. 36, ed. J. D. Fernie, D. Reidel Publishing Co, Dordrecht, 221

Discussion

QUESTION: Do you have a comment on late thermal pulse stars having $C/O < 1$ and high Ne abundances; neither of those match observations.

LATTANZIO: These are very complex stars, probably involving burning from a thermal pulse and mixing into the very thin envelope, and probably on similar timescales.

SAHAI: Has there been any progress in understanding J-type Carbon stars, where the $^{13}C/^{12}C$ is quite low ($\lesssim$5), although in general C stars have much higher ratios ($\gtrsim$30)(as per your talk)?

LATTANZIO: I am not aware of anything recent. The last I recall was that J-star properties were probably consistent with being evolved R stars; the latter being merged binaries.

QUESTION: What do you mean by an i-process?

LATTANZIO: Something between the r- and s-processes. These are two convenient extremes, but nature need not follow these two extremes, and it looks like there may well be something with a neutron density somewhere between the two.

Why Galaxies Care About AGB Stars:
A Continuing Challenge through Cosmic Time
Proceedings IAU Symposium No. 343, 2019
F. Kerschbaum, M. Groenewegen & H. Olofsson, eds.

doi:10.1017/S1743921318006658

3D modelling of AGB stars with CO5BOLD

Bernd Freytag, Susanne Höfner and Sofie Liljegren

Dept. of Astronomy & Space Physics, Uppsala University,
Box 516, SE-75120 Uppsala, Sweden
email: Bernd.Freytag@physics.uu.se

Abstract. Local three-dimensional radiation-hydrodynamics simulations of patches of the surfaces of solar-type stars, that are governed by small-scale granular convection, have helped analyzing and interpreting observations for decades. These models contributed considerably to the understanding of the atmospheres and indirectly also of the interiors and the active layers above the surface of these stars. Of great help was of course the availability of a close-by prototype of these stars – the sun.

In the case of an asymptotic-giant-branch (AGB) star, the convective cells have sizes comparable to the radius of the giant. Therefore, the extensions of the solar-type-star simulations to AGB stars have to be global and cover the entire object, including a large part of the convection zone, the molecule-formation layers in the inner atmosphere, and the dust-formation region in the outer atmosphere. Three-dimensional radiation-hydrodynamics simulations with CO5BOLD show how the interplay of large and small convection cells, waves, pulsations, and shocks, but also molecular and dust opacities of AGB stars create conditions very different from those in the solar atmosphere.

Recent CO5BOLD models account for frequency-dependent radiation transport and the formation of two independent dust species for an oxygen-rich composition. The drop of the comparably smooth temperature distribution below a threshold determines to onset of dust formation, further in, at higher temperatures, for aluminium oxides (Al_2O_3) than for silicates (Mg_2SiO_4). An uneven dust distribution is mostly caused by inhomogeneities in the density of the shocked gas.

Keywords. convection, hydrodynamics, radiative transfer, shock waves, waves, methods: numerical, stars: AGB and post-AGB, stars: atmospheres, stars: oscillations (including pulsations), stars: winds, outflows

---

1. Introduction

Even classical (quasi-)stationary one-dimensional models of stellar interiors and atmospheres have to take into account the effects of convection on the transport of energy and the mixing of material and of stellar winds onto mass loss. As time- and space-resolving observations of the accessible surface layers demonstrate, these complex, truly dynamical processes are accompanied by time-dependent small- and large-scale fluctuations in brightness, density, and velocity (see, e.g., Nordlund *et al.* 2009).

Detailed radiation hydrodynamics (RHD) simulations can help to qualitatively understand these processes and are the only way to quantitatively model dynamical layers in and around stars. The first "local" simulations of small representative patches of the solar surface, comprising a few granules and the photosphere above, were performed decades ago (Dravins *et al.* 1981, Nordlund & Stein 2001, Stein & Nordlund 2001). They have since been improved on in terms of algorithms, numerical resolution, extension, boundary conditions and in particular concerning microphysics (opacities, etc.). Grids of local RHD models produced by different groups with various codes are available for the atmospheres of a wide range of different types of stars (CO5BOLD: Ludwig *et al.* 2009, Tremblay *et al.* 2015; Stagger code: Magic *et al.* 2013, Trampedach *et al.* 2013; MURaM: Beeck *et al.* 2013).

In the meantime, these atmosphere models have been extended into the chromosphere and corona (with the Bifrost code: see, e.g., Gudiksen *et al.* 2011), while interior models, that exclude the surface layers, can simulate flows in the solar interior (with the ASH code: see, e.g., Clune *et al.* 1999).

Global RHD models with CO5BOLD, using a fixed cubic Cartesian grid and a prescribed gravitational potential, were presented by Freytag *et al.* (2002) for a red supergiant (RSG), by Freytag & Höfner (2008) for an AGB star and by Freytag *et al.* (2017) for a small grid of AGB stars. Ohlmann *et al.* (2017) used AREPO to perform hydrodynamics simulations without radiation transport of an AGB star with a "moving-mesh" accounting for self gravity.

2. Challenges for 3D modelling of AGB stars

Global models of the sun and sun-like stars are currently out of reach, due to the disparity in spatial scales (a few grid points are needed per photospheric pressure scale height of about 150 km, for the entire surface of the sun with a diameter of about 1 400 000 km) and time scales (numerical time steps of around a second for at least several rotation periods of about a month and better several magnetic cycles, each spanning 22 years). See Freytag *et al.* (2012) for a discussion.

However, due to the scaling of the pressure scale height $H_{\rm p}$ with the stellar radius R_* (ignoring the dependence on effective temperature, stellar mass, and composition),

$$H_{\rm p} \propto 1/g \propto R_*^2, \tag{2.1}$$

one derives

$$H_{\rm p}/R_* \propto R_*. \tag{2.2}$$

With the expectation that the typical horizontal size of surface granules is proportional to a local length scale – as, for example, the pressure scale height – Schwarzschild (1975) estimated that the surface of cool giants might be covered by a relatively small number of giant convection cells (already suggested by Stothers & Leung 1971). This picture is consistent with observations of surface inhomogeneities in red supergiants and AGB stars (e.g., Lim *et al.* 1998, Paladini *et al.* 2018). It brings "global" 3D RHD models covering the entire convective surface of cool giants into the realm of possibility, as several granules could already be resolved by early local RHD simulations.

The first global low-resolution model computed with CO5BOLD ("COnservative COde for the COmputation of COmpressible COnvection in a BOx of L Dimensions, $L = 2, 3$") was presented by Freytag *et al.* (2002). In spite of the similarities to the now commonplace local RHD simulations of the convective surface layers and the atmosphere of solar-type stars, global RHD models of cool (super-)giants face a number of additional difficulties, as outlined in the following.

The computational domain for local simulations can be extended to cover more granules or to reach deeper down into the interior or higher up into the upper atmosphere. Alternatively, the box can be shrunk to save grid points, i.e., CPU time and memory. Global simulations, in contrast, always have to include the entire star plus a bit of the environment, which, for a given number of grid points, limits the number of granules that can be resolved. It restricts current simulations therefore to the most "fluffy" stars with lowest surface gravity and/or low stellar mass (see Freytag *et al.* 2017). According to Eq. (2.2), stars with a smaller radius will require more grid points.

Local simulations not only cover most of the line-formation region in the optically thin atmosphere but also the optically thick top layers of the surface convection zone. Truly global simulations should ideally comprise these layers and include the stellar interior with the nuclear processes close to the core, the outer convective envelope with its violent dynamics due to convection and pulsations, the inner atmosphere with its network of small shocks, the outer atmosphere, where large-scale shock fronts create sufficient conditions for dust to form, and the wind-driving region further out. Current CO5BOLD models can neither resolve nor model the stellar core in detail, nor do they contain the wind-driving region or even the necessary physics to model radiative acceleration of matter.

A freshly started simulation should at least run long enough to cover the thermal relaxation phase (of typically a few years). For meaningful statistical results, averages have to be taken over several convective turnover times (of a few months for the small surface cells and several years for the global cells with downdrafts reaching deep into the interior) and pulsational periods (of about a year, see Table 1 in Freytag *et al.* 2017). This is still short compared to evolutionary time scales, but very long compared to the radiative time scales at the level of individual grid cells (of a few hundred seconds). It means that a typical global simulation requires several million radiation-transport sub steps, which is a much larger number than necessary for a local simulation of a sun-like star.

The steep sub-photospheric temperature step due to a peak in hydrogen opacity at low densities is accompanied by a drop in density. The resulting density inversion (dense above less dense material) contributes significantly to the driving of convective motions. However, the strong local gradients in temperature and density require a good numerical resolution – ideally more than in the current models – and pose high demands on the stability of the hydrodynamics and the radiation-transport solver. At higher densities, in local models, the problem is not quite as severe but still prevalent at higher effective temperatures (see, e.g., Mundprecht *et al.* 2013 and Vasilyev *et al.* 2017).

Due to the small ratio of radiation to gas pressure in local models of sun-like – and not too hot – stars, radiation pressure can usually be completely neglected. However, in AGB stars and in red supergiants, radiation pressure becomes significant in the interior and in the atmospheres. In the standard picture, radiative acceleration of dust grains is the driver for wind formation in AGB stars (see, e.g., Höfner & Olofsson 2018).

The formation of dust is a non-equilibrium process. This means that the properties of dust cannot just be inferred from temperature and pressure alone but require a detailed time-dependent treatment of formation, transport, and destruction of dust grains (see Gail & Sedlmayr 2013 for an overview and Freytag & Höfner 2008 about a way of treating dust in CO5BOLD). Most local solar-type models do not harbor conditions that allow dust to form, in contrast to models of brown dwarfs, that require a detailed treatment of dust (without radiative pressure onto grains but with gravitational settling of dust, see Freytag *et al.* 2010).

Despite recent advances in resolving surface structures on AGB stars (see Paladini *et al.* 2018 and Paladini, this volume), there are no observations of stars available that are even remotely close to the level of detail achievable by observations of the solar surface. This means that there is no sun-equivalent to check simulations against. A further complication for the comparison of numerical models and observations is the fact that parameters of AGB stars are known with much less accuracy than the parameters of the sun.

3. 3D models with CO5BOLD: current status

The CO5BOLD code (Freytag *et al.* 2012) is able to perform radiation-(magneto)-hydrodynamics simulations of the optically thick top layers of a stellar surface convection zone and the optically thin atmosphere above. Its bread-and-butter job is to compute local "box-in-a-star" models covering small patches on the surface of the sun (see, e.g., Wedemeyer *et al.* 2004), other roughly sun-like stars (see, e.g., Ludwig *et al.* 2009) and even brown dwarfs (see Freytag *et al.* 2010). Gravity is assumed is to be constant, pointing downward. The lateral boundary conditions are periodic, while energy enters at the lower boundary by radiation or matter transport, and radiative energy leaves through the upper boundary (see Freytag 2017).

However, for the global "box-in-a-star" models of AGB stars presented in this paper, an external central gravitational potential for a given stellar mass is assumed. It is smoothed in the center, because the small-scale flows in the tine stellar core cannot be resolved with the current grid (see also Freytag & Höfner 2008, and Chiavassa *et al.* 2009). A central energy source corresponding to the stellar luminosity is driving the convective flows in the outer stellar layers. All the outer boundaries are very similar to the upper boundary in the local models. They allow radiation to escape and the free flow of matter. More details about boundary conditions are given in Freytag (2017). Already early simulations containing only the topmost layers of the solar convection zone demonstrated that granulation is a genuine surface phenomenon and not caused by hot bubbles produced deep inside the star. Therefore, one can expect that the simplified treatment of the stellar core in CO5BOLD only has a minor impact on the properties of the surface convection. And for stellar pulsations the extended outer layers with relatively low sound speed (i.e., long sound-travel times) are more important than the small core region with high sound speed. Still, some downdrafts reach from the surface down into the core region. Their inner overturning behavior and their interaction with pulsations in the core region via a modulation of the convective over the pulsation cycle is somewhat affected by the details of the treatment of the core region.

The computational box, that covers about 2 stellar radii, limits studies to stellar pulsations (see Freytag *et al.* 2017), surface features (see Chiavassa *et al.* 2018) or near-surface features such as shocks (see Liljegren *et al.* 2018) and the formation of dust (see Freytag & Höfner 2008 and Höfner & Freytag, in prep.). The large wind-formation region, spanning several stellar radii, is not included in current models. The outer boundaries are implemented by filling a few layers of ghost cells with extrapolated values for density, internal energy, and velocity (see Freytag 2017). While the efficient radiation transport adjusts the temperature of the gas inside the computational box rapidly to its local equilibrium value, the gas density in the outer layers during long phases of material infall to some degree depends on the detailed settings for the outer boundary. A future enlargement of the computational domain and the inclusion of radiation pressure will get rid of these shortcomings of the current setup. The omission of radiation-pressure terms might also be responsible for the inability of current models to properly reproduce the observed large extension of supergiant atmospheres (see Arroyo-Torres *et al.* 2015).

The CO5BOLD code numerically integrates the Euler equations of hydrodynamics (HD) or, alternatively, the ideal magneto-hydrodynamics (MHD) equations explicitly in time, accounting for a tabulated equation of state and an external gravity field (Freytag *et al.* 2012 and Freytag 2013). The hydrodynamics module employs a Roe solver (Roe 1986) and is used for the examples below. In both cases, the fully compressible equations are used, so that sound waves and shocks can be modelled. However, these solvers are bound to the Courant-Friedrichs-Lewy condition, which imposes a limit to the numerical time step. It is not severe for the high-Mach-number flows with efficient radiative energy exchange in AGB models but becomes important for the low-Mach-number flows in brown dwarfs (see Freytag *et al.* 2010). The MHD solver is – in principle – capable of dealing with magnetic reconnection. However, such an event will only lead to a local heating of the gas and not to the acceleration of a small number of charged particles to very high (super-thermal) velocities. This would require a different kind of solver for a more complex set of plasmaphysics equations, necessary to adequately model thin plasmas far away from the stellar surface.

In the high-Reynolds-number flows in stellar atmospheres, kinetic energy is supposed to be generated on large granular scales, transferred via a turbulent cascade to smaller and smaller scales and finally dissipated on scales given by the mean-free-path length of particles. All these scales are impossible to resolve by current or foreseeable computers. However, the assumption, that the numerical viscosity inherent in all hydrodynamics schemes has a similar net effect, explains the success of RHD simulations of stellar surface convection with their relatively coarse grids (see Nordlund *et al.* 1997). Tests with – slightly – different numerical resolutions (see, e.g., Asplund *et al.* 1999, Freytag *et al.* 2017 or Collet *et al.* 2018) don't give reasons for major concerns. However, limitations due to the finite numerical resolution should also be kept in mind, for instance, when it comes to (radiative) shocks or small-scale magnetic phenomena.

The non-local transfer of radiative energy in optically thin or thick layers is based on detailed opacity tables merged, for example, from Phoenix (Hauschildt *et al.* 1997) and OPAL (Iglesias *et al.* 1992) data. The employed short-characteristics method is designed to be able to handle large variations of opacity and source function, among others by using a piecewise linear interpolation of the source function (see Freytag *et al.* 2012). Most previous global CO5BOLD models have used grey opacities only, adequate for the study of pulsations or (near-) surface features (see, e.g., Freytag *et al.* 2017, Chiavassa *et al.* 2018, Liljegren *et al.* 2018). However, for applications that require a more realistic photospheric temperature stratification, like spectrum synthesis (see Chiavassa *et al.* 2011) or dust formation (see the example below), the frequency dependence of the opacities is treated in an approximate way by an opacity-binning technique based on a method laid out by Nordlund (1982) and further refined later on (see Freytag *et al.* 2012 and references therein). For the sample model below, going from grey (single bin) to non-grey (simple 3-bin) opacities, increases the computational effort per time step by a factor 3 and reduces the time step by a factor 2. The increase of the total computational cost per simulation by a factor 6 makes it clear why grey models are still utilized in some cases. Currently used opacity tables treat scattering as true absorption and ignore dust completely. Radiation pressure is ignored. In other words, the microphysical processes necessary to drive a stellar wind are not completely implemented, yet.

The frequency resolution of the radiation-transport step in CO5BOLD is only designed to model the exchange of heat and is not fine enough to give a meaningful stellar spectrum. Instead, a detailed spectrum synthesis is performed in a post-processing step with OPTIM3D based on stored snapshots from CO5BOLD simulations (see Chiavassa *et al.* 2009).

One pillar of the concept of (magneto)hydrodynamics is the assumption that matter is locally in thermal equilibrium and that the local material properties can therefore be completely described by two state quantities, for example pressure and temperature or, typically in numerical simulations, gas density and internal energy. Further quantities (heat capacities, sound speed, etc.) can be derived from the equation of state, i.e., by interpolation in a table. It accounts for hydrogen and helium ionization, the formation of H_2 molecules, and a representative neutral metal. Other atoms or molecules contribute only relatively little to, e.g., the partial pressure and heat capacity and are therefore ignored in the equation of state, while they are accounted for in great detail in the opacity tables. Models of time-dependent non-equilibrium processes, for example, for carbon monoxide in the solar atmosphere (see Wedemeyer-Böhm *et al.* 2005), amorphous carbon dust around AGB stars (see Freytag & Höfner 2008) or forsterite dust clouds in the atmospheres of brown dwarfs (see Freytag *et al.* 2010) assume that the contributing species do not affect the equation of state but – possibly – the opacities.

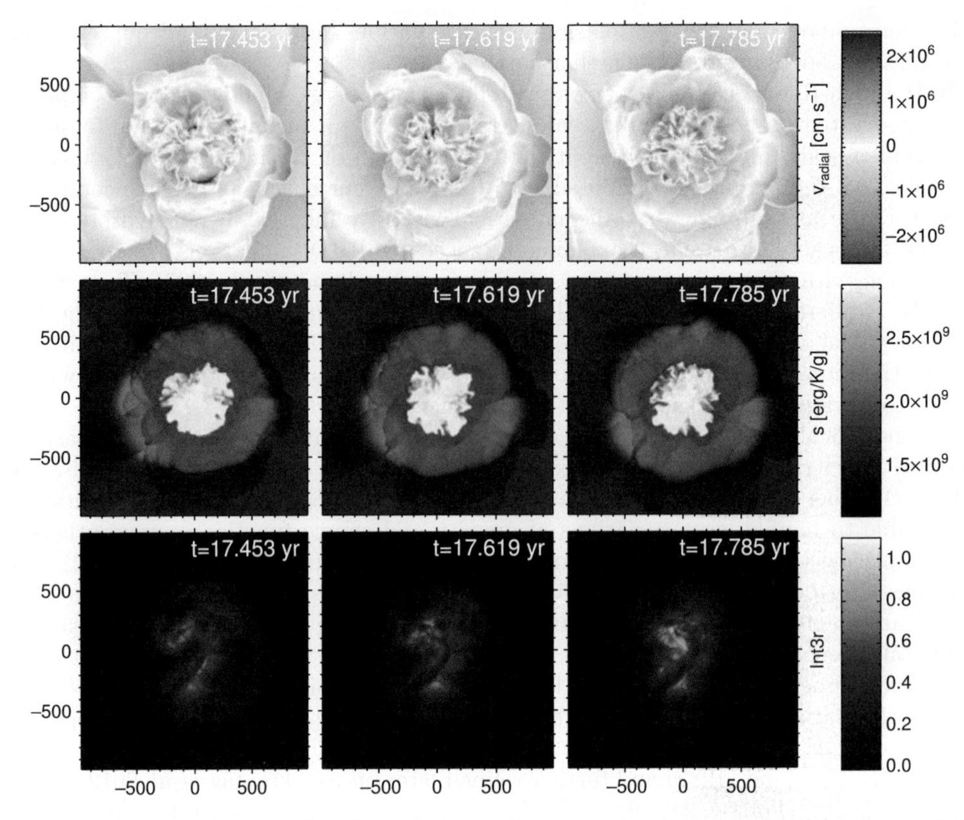

Figure 1. Time sequences of radial velocity and entropy for slices through the center of the model (st28gm06n038, rows 1–2), and the variation of relative surface intensity (bottom row). The snapshots are about 2 months apart (see the counter in the top of the panels).

4. Dust formation in 3D models

The latest 3D CO5BOLD model (st28gm06n038) of an AGB star, with $1\,M_\odot$, $7000\,L_\odot$ and solar composition, has a box size of $1970\,R_\odot$ with 401^3 grid points and uses a non-grey opacity table (see Figs. 1 and 2). It incorporates terms to describe the formation, transport, and destruction of aluminium-oxide (Al_2O_3) and silicate (Mg_2SiO_4) dust. For further details see Höfner & Freytag (in prep.).

The new extended high-resolution model confirms previous results about the dynamics in the convective envelopes and inner atmospheres of AGB stars (see Freytag & Höfner 2008, Freytag *et al.* 2017, Liljegren *et al.* 2018). Downdrafts reach from the surface of the convection zone into the core region of the model and outline a few global convection cells with lifetimes of several years. Surface cells on the other hand, driven by the narrow layer with strong superadiabaticity, flow on top of the global cells (cf. Fig. 1). Usually, they do not extend far below the surface. They have lifetimes of months. Particularly during merging events of small or large downdrafts, non-stationary convection excites acoustic waves. In the current models, only the fundamental radial mode shows up as distinct peak in a power spectrum (see Freytag *et al.* 2017). Waves with shorter wavelengths and periods do exist but are affected too much by changes in velocity field and sound speed of the background flow to achieve a lifetime that would cause a local peak in a power spectrum. When waves travel into the thin atmosphere with low sound speeds,

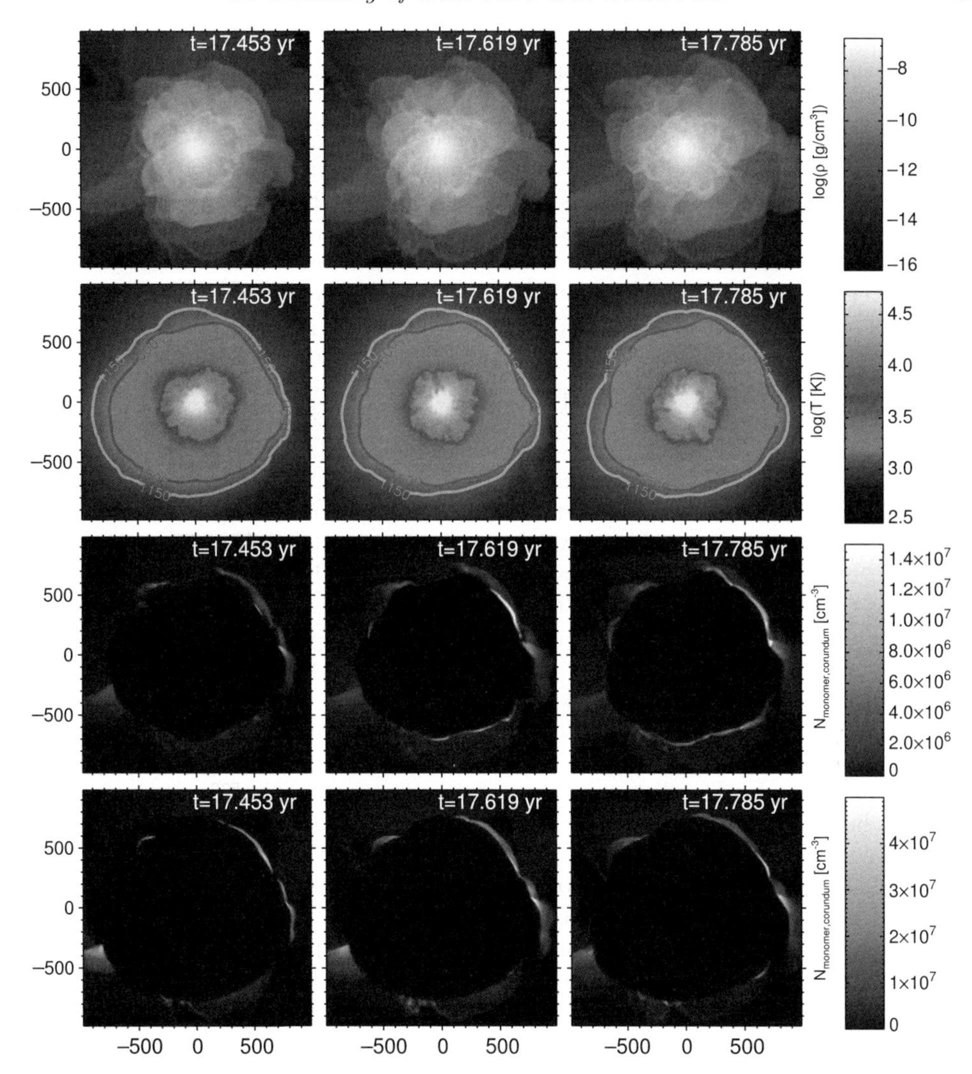

Figure 2. Time sequences of density, temperature, aluminium-oxide density and silicate density for slices through the center of the model (st28gm06n038). The snapshots are about 2 months apart (see the counter in the top of the panels).

they turn into shocks (see the velocity plots in Fig. 1 and the density plots in Fig. 2). Shorter-wavelength waves cause a complex small-scale network of shocks in the innermost atmosphere, while the fundamental pulsation mode causes a more or less spherical shock front, that is able to travel far out (see Liljegren *et al.* this volume).

New features in the current model generation are the source and sink terms for aluminium-oxide and silicate dust, based on a kinetic description of grain growth and thermal evaporation as discussed in Höfner *et al.* (2016). Once the temperatures are low enough, dust forms rather rapidly due to the high gas densities in the wake of shock fronts or at the bottom of a region with infalling material. While magnesium silicates can reach higher densities than aluminium oxides due to the higher amount of available magnesium compared to aluminium, the aluminium-oxides form further in, at higher temperatures, than silicates (compare the dust-density and the temperature plots in Fig. 2). The large density fluctuations of the shocked gas are reflected in the densities of the dust species (compare the dust-density and the gas density plots in Fig. 2).

5. Conclusions and outlook

The first global radiation hydrodynamics simulations with CO5BOLD of RSG stars (Freytag *et al.* 2002) and AGB stars (Freytag & Höfner 2008) were performed several years ago, but are much more demanding than local RHD models of sun-like stars or M-type main-sequence stars of the same effective temperature. That causes restrictions in numerical resolution, model extension, microphysics (e.g., opacity treatment) and number of models available. However, 3D effects can partly be incorporated in 1D atmosphere and wind models, e.g., by extracting a description of the surface velocity field from the 3D models and using this as an inner boundary condition for 1D models (see Freytag & Höfner 2008, Liljegren *et al.* 2018, and Liljegren, this volume). Grids of 1D models can easily cover large ranges of stellar parameters (see Bladh *et al.* 2015, Bladh, this volume and Eriksson, this volume).

The existing 3D models give interesting insights about the dynamics of the near-surface layers of cool giant and supergiant stars. For example, they confirmed previous ideas (e.g., of Stothers & Leung 1971 and Schwarzschild 1975) about the presence of giant convection cells but also showed that the contrast of surface features is much higher than on the sun. In addition, convective motions are much more violent, producing sound waves (as on the sun but with larger amplitudes), that turn into shocks as soon as they reach the thin atmosphere (and not higher up in the chromosphere as on the sun, see, e.g., Wedemeyer *et al.* 2004). The latest AGB model presented above demonstrates that the largest-scale shocks are able to lift high-density material into layers sufficiently cool for dust to form. The distance to the star is dictated by the temperature (i.e., the radiation field) with an inhomogeneous distribution caused by density fluctuations of the shocked gas. Aluminium oxides (Al_2O_3) form further in than silicates (Mg_2SiO_4) indicating that the combination of both plays a role for the generation of a stellar wind (see also Höfner *et al.* 2016 and references therein).

Future development will, on the one hand, focus on improving the treatment of the stellar interior to better model deep reaching convection and pulsations. This will be achieved by including terms for radiation pressure and by hierarchically refining the grid in the core region. On the other hand is the (more) refined treatment of dust (including detailed opacities and radiation pressure) crucial for the modelling of the stellar atmosphere and the wind-driving mechanism and for the computation of synthetic emergent spectra (the latter computed as a post-processing step). The inclusion of the wind-driving zone will require an enlargement of the computational box. Furthermore, first attempts are under way to include magnetic fields or stellar rotation.

Acknowledgements

This work has been supported by the Swedish Research Council (Vetenskapsrådet). The computations were performed on resources (rackham) provided by SNIC through Uppsala Multidisciplinary Center for Advanced Computational Science (UPPMAX) under Projects snic2017-1-41 and snic2018-3-74.

References

Arroyo-Torres, B., Wittkowski, M., Chiavassa, A., *et al.* 2015, *A&A*, 575, A50
Asplund, M., Nordlund, Å., Trampedach, R., & Stein, R. F. 1999, *A&A*, 346, L17
Beeck, B., Cameron, R. H., Reiners, A., & Schüssler, M. 2013, *A&A*, 558, A48
Bladh, S., Höfner, S., Aringer, B., & Eriksson, K. 2015, *A&A*, 575, A105
Chiavassa, A., Freytag, B., Masseron, T., & Plez, B. 2011, *A&A*, 535, A22

Chiavassa, A., Freytag, B., & Schultheis, M. 2018, *A&A*, 617, L1
Chiavassa, A., Plez, B., Josselin, E., & Freytag, B. 2009, *A&A*, 506, 1351
Clune, T. L., Elliott, J. R., Glatzmaier, G. L., Miesch, M. S., & Toomre, J. 1999, *Parallel Comput.*, 25, 361
Collet, R., Nordlund, Å., Asplund, M., Hayek, W., & Trampedach, R. 2018, *MNRAS*, 475, 3369
Dravins, D., Lindegren, L., & Nordlund, Å. 1981, *A&A*, 96, 345
Freytag, B. 2013, *MemSAItS*, 24, 26
Freytag, B. 2017, *MemSAIt*, 88, 12
Freytag, B., Allard, F., Ludwig, H.-G., Homeier, D., & Steffen, M. 2010, *A&A*, 513, A19
Freytag, B. & Höfner, S. 2008, *A&A*, 483, 571
Freytag, B., Liljegren, S., & Höfner, S. 2017, *A&A*, 600, A137
Freytag, B., Steffen, M., & Dorch, B. 2002, *AN*, 323, 213
Freytag, B., Steffen, M., Ludwig, H.-G., *et al.* 2012, *J.Comput.Phys.*, 231, 919
Gail, H.-P. & Sedlmayr, E. 2013, Physics and Chemistry of Circumstellar Dust Shells (Cambridge University Press)
Gudiksen, B. V., Carlsson, M., Hansteen, V. H., *et al.* 2011, *A&A*, 531, A154
Hauschildt, P. H., Baron, E., & Allard, F. 1997, *ApJ*, 483, 390
Höfner, S., Bladh, S., Aringer, B., & Ahuja, R. 2016, *A&A*, 594, A108
Höfner, S. & Olofsson, H. 2018, *A&AR*, 26, 1
Iglesias, C. A., Rogers, F. J., & Wilson, B. G. 1992, *ApJ*, 397, 717
Liljegren, S., Höfner, S., Freytag, B., & Bladh, S. 2018, arXiv:1808.05043
Lim, J., Carilli, C. L., White, S. M., Beasley, A. J., & Marson, R. G. 1998, *Nature*, 392, 575
Ludwig, H.-G., Caffau, E., Steffen, M., *et al.* 2009, *MemSAIt*, 80, 711
Magic, Z., Collet, R., Asplund, M., *et al.* 2013, *A&A*, 557, A26
Mundprecht, E., Muthsam, H. J., & Kupka, F. 2013, *MNRAS*, 435, 3191
Nordlund, Å. 1982, *A&A*, 107, 1
Nordlund, Å., Spruit, H. C., Ludwig, H.-G., & Trampedach, R. 1997, *A&A*, 328, 229
Nordlund, Å. & Stein, R. F. 2001, *ApJ*, 546, 576
Nordlund, Å., Stein, R. F., & Asplund, M. 2009, Living Reviews in Solar Physics, 6, 2
Ohlmann, S. T., Röpke, F. K., Pakmor, R., & Springel, V. 2017, *A&A*, 599, A5
Paladini, C., Baron, F., Jorissen, A., *et al.* 2018, *Nature*, 553, 310
Roe, P. 1986, *ARFM*, 18, 337
Schwarzschild, M. 1975, *ApJ*, 195, 137
Stein, R. F. & Nordlund, Å. 2001, *ApJ*, 546, 585
Stothers, R. & Leung, K.-C. 1971, *A&A*, 10, 290
Trampedach, R., Asplund, M., Collet, R., Nordlund, Å., & Stein, R. F. 2013, *ApJ*, 769, 18
Tremblay, P.-E., Ludwig, H.-G., Freytag, B., *et al.* 2015, *ApJ*, 799, 142
Vasilyev, V., Ludwig, H.-G., Freytag, B., Lemasle, B., & Marconi, M. 2017, *A&A*, 606, A140
Wedemeyer, S., Freytag, B., Steffen, M., Ludwig, H.-G., & Holweger, H. 2004, *A&A*, 414, 1121
Wedemeyer-Böhm, S., Kamp, I., Bruls, J., & Freytag, B. 2005, *A&A*, 438, 1043

Discussion

QUESTION: How do you define the radius of the star in your simulations?

FREYTAG: In contrast to, for example the stellar mass or the luminosity, the stellar radius and with it effective temperature and surface gravity are not well-defined quantities. A radius definition consistent with observations would require the generation of synthetic images in appropriate filter bands, the degradation of the images with the instrumental profile, and the application of a similar radius-measurement algorithm as used for observed images. To circumvent this, the radius is defined as the point where the luminosity (computed with $4\pi r^2 \sigma T^4$) from the averages of the temperature over spherical shells and time matches the stellar luminosity.

DECIN: What is the physical reason for shock temperatures to be around 2000 K in the 3D models and not reach values around 10 000 K?

FREYTAG: The peak temperatures, that shocks can reach, decreases with distance from the star. In the layers of the atmosphere that are relevant for dust formation, efficient non-grey radiative energy transfer smoothes small-scale temperature fluctuations on very short time scale of a few hundred seconds. With the current numerical mesh, the travel time of a shock front across a grid cell is so long, that the shock is essentially not adiabatic but isothermal. Simple grey radiation transport causes longer relaxation time scales, so that shocks are accompanied by a noticeable rise in temperature. However, they never reach 10000 K, even remotely. A proper modelling of radiative shocks would require very high numerical resolution and a non-LTE treatment of gas chemistry and radiation transport – completely out of reach for our 3D simulations.

DECIN: How is the dust formed, and in particular, the first seeds in your models?

FREYTAG: The multi-dimensional simulations can only afford a relatively simple dust model. Seeds are assumed to be always present, in a prescribed concentration. The amount of available monomers is computed from the gas and the dust density in a grid cell. The rate of grain growth or evaporation is computed via a number of of temperature- and density-dependent reaction rates.

QUESTION: How do you manage to resolve the surface structures of an AGB star with your current numerical grid? Do you have enough grid points to resolve, for example, a pressure scale height?

FREYTAG: With currently feasible grids of 400^3 points or so, we can only model stars with lowest surface gravities, i.e., largest pressure scale heights and largest granules relative to the star, at the tip of the asymptotic giant branch. Going down the AGB will require a significant refinement of the grid.

Why Galaxies Care About AGB Stars:
A Continuing Challenge through Cosmic Time
Proceedings IAU Symposium No. 343, 2019
F. Kerschbaum, M. Groenewegen & H. Olofsson, eds.

doi:10.1017/S1743921318005367

Magnetic fields of AGB and post-AGB stars

Wouter Vlemmings

Chalmers University of Technology, Department of Space, Earth and Environment,
Onsala Space Observatory, SE-43992 Onsala, Sweden
email: wouter.vlemmings@chalmers.se

Abstract. There is ample evidence for the presence of strong magnetic fields in the envelopes of (post-)Asymptotic Giant Branch (AGB) stars as well as supergiant stars. The origin and role of these fields are still unclear. This paper updates the current status of magnetic field observations around AGB and post-AGB stars, and describes their possible role during these stages of evolution. The discovery of magnetically aligned dust around a supergiant star is also highlighted. In our search for the origin of the magnetic fields, recent observations show the signatures of possible magnetic activity and rotation, indicating that the magnetic fields might be intrinsic to the AGB stars.

Keywords. magnetic fields, polarization, stars: AGB and post-AGB, supergiants, rotation, spots

1. Introduction

Magnetic fields are ubiquitous throughout the Universe and play an important role across a wide range of scales. Primordial magnetic fields could have played a role in the formation of the first stars just as magnetic fields in molecular clouds are an important ingredient in current star formation. Magnetic fields have also been detected in almost all stellar types and in almost all phases of stellar evolution (e.g. Berdyugina 2009), and have significant effects on stellar evolution through, e.g. their influence on the internal mixing. The magnetic field of stars can have either a dynamo origin, i.e. be generated by a dynamo process in the star itself (e.g. Charbonneau 2014), or can be the result of a remnant 'fossil' field, which are fields that originate from the star formation process (e.g. Braithwaite & Spruit 2004). The stellar magnetic field is affected by the changes of physical properties during stellar evolution and, because of flux conservation, becomes increasingly difficult to observe at the stellar surface when the star expands in the final phases of its life. However, in stellar end products, such as white dwarfs and neutron stars, magnetic fields are also shown to be significant.

The role of magnetic fields around AGB stars is not clear. In principle, they could help levitate material off the stellar surface through Alfvén waves (e.g. Falceta-Gonçalves & Jatenco-Pereira 2002), or through the creation of cool spots on the surface above where dust can form easier (Soker 1998). A specific model for the AGB star o Ceti (Mira A) has shown that a hybrid magnetohydrodynamic-dust-driven wind scenario can explain its mass loss (Thirumalai & Heyl 2013). In such a model, Alfvén waves add energy to lift material before dust forms and radiation pressure accelerates a wind. Magnetic fields also play an important role in the internal mixing required for s-process (slow) neutron capture reactions that define the stellar yields (e.g. Trippella *et al.* 2016).

After the AGB phase, the stellar envelopes undergo a major modification as they evolve to Planetary Nebulae (PNe). The standard assumption is that the initial slow

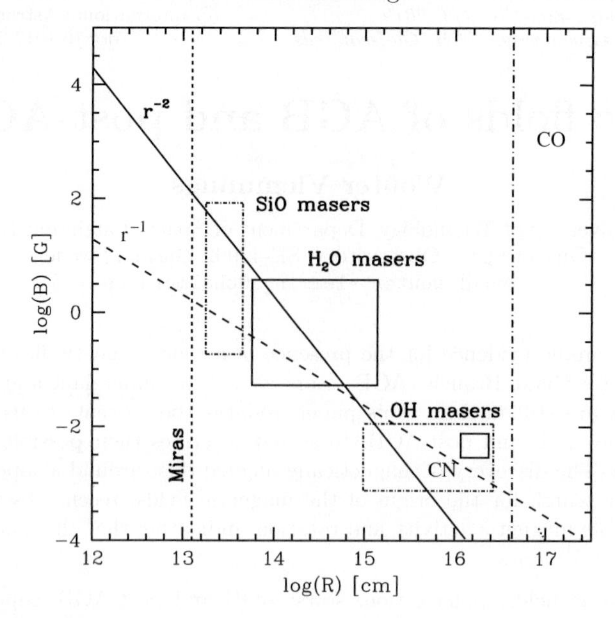

Figure 1. The circumstellar magnetic field strength vs. radius relation as indicated by current (maser) polarization observation. The boxes show the range of observed magnetic field strengths derived from the observations of SiO masers (Kemball *et al.* 2009, Herpin *et al.* 2006), H_2O masers (Vlemmings *et al.* 2002, Vlemmings *et al.* 2005, Leal-Ferreira *et al.* 2013), OH masers (Rudnitski *et al.* 2010, Gonidakis *et al.* 2014) and CN (Duthu *et al.* 2017). The thick solid and dashed lines indicate an r^{-2} solar-type and r^{-1} toroidal magnetic field configuration. The vertical dashed line indicates the stellar surface. Observations of the Goldreich-Kylafis effect in CO (e.g. Vlemmings *et al.* 2012) will uniquely probe the outer edge of the envelope (vertical dashed dotted line).

AGB mass loss quickly changes into a fast superwind, generating shocks and accelerating the surrounding envelope (Kwok *et al.* 1978). It is during this phase that the typically spherical circumstellar envelope (CSE) evolves into a PN. As the majority of pre-PNe are aspherical, an additional mechanism is needed to explain the departure from sphericity. This mechanism is still a matter of fierce debate. One possibility is that the interaction of the post-AGB star and a binary companion or massive planet supports a strong magnetic field that is capable of shaping the outflow (e.g. Nordhaus *et al.* 2007).

This paper expands on (and partly reproduces) the reviews presented in Vlemmings (2018) and Vlemmings (2014) and I refer interested readers to those review (and references therein) for further background.

2. Overview of magnetic field observations

2.1. *AGB stars*

Generally, AGB magnetic field measurements come from maser polarization observations (SiO, H_2O and OH). These have revealed a strong magnetic field throughout the CSE. Figure 1 shows the magnetic field strength in the regions of the CSE traced by the maser measurements throughout AGB envelopes. The field appears to vary between $B \propto R^{-2}$ (solar-type) and $B \propto R^{-1}$ (toroidal). Although the maser observations trace only oxygen-rich AGB stars, recent CN Zeeman splitting observations (Duthu *et al.* 2017) indicate that similar fields strengths are found around carbon-rich stars. The CSE magnetic fields are also consistent with the, thus far, only direct measurement of the

Table 1. Energy densities in AGB envelopes

		Photosphere	SiO	H_2O	OH	CO/CN
B	[G]	$\sim 1-10?$	~ 3.5	~ 0.3	~ 0.003	$\sim 0.003-0.008$
R	[AU]	-	~ 3	~ 25	~ 50	$\sim 50-100$
$V_{\rm exp}$	[km s^{-1}]	~ 20	~ 5	~ 8	~ 10	~ 10
$n_{\rm H_2}$	[cm^{-3}]	$\sim 10^{11}$	$\sim 10^{10}$	$\sim 10^{8}$	$\sim 10^{6}$	$\sim 10^{5}$
T	[K]	~ 2500	~ 1300	~ 500	~ 300	~ 150
$B^2/8\pi$	[dyne cm^{-2}]	$\mathbf{10^{-1.4,+0.6}}$	$\mathbf{10^{+0.1}}$	$\mathbf{10^{-2.4}}$	$10^{-6.4}$	$\mathbf{10^{-6.0,-6.4}}$
nKT	[dyne cm^{-2}]	$10^{-1.5}$	$10^{-2.7}$	$10^{-5.2}$	$10^{-7.4}$	$10^{-8.7}$
$\rho V_{\rm exp}^2$	[dyne cm^{-2}]	$10^{-0.3}$	$10^{-2.5}$	$10^{-4.1}$	$\mathbf{10^{-5.9}}$	$10^{-6.9}$
V_A	[km s^{-1}]	~ 20	~ 100	~ 300	~ 8	~ 8

Energy densities through AGB star CSEs (dominating energy densities are in bold face). From left to right the columns indicate the stellar photosphere, maser regions and the region probed by CO/CN, with increasing distance to the central star. The top rows are the typical magnetic field strength B, distance to the star R, expansion velocity $V_{\rm exp}$, hydrogen number density $n_{\rm H}$ and temperature T. The bottom rows are the magnetic, thermal and kinematic energy and a rough estimate of the Alvfén velocity V_A.

Zeeman effect on the surface of an AGB star, the Mira variable χ Cyg (Lèbre *et al.* 2014). In Table 1 an overview is given of the energy densities throughout the AGB CSEs.

The large-scale structure of the magnetic field is more difficult to infer, predominantly because the maser observations often probe only limited line-of-sights. Even though specifically OH observations seem to indicate a systematic field structure, it has often been suggested that there might not be a large-scale component to the field that would be necessary to shape the outflow (Soker 2002). Until recently the only tight shape constraints throughout the CSE had been determined for the field around the supergiant star VX Sgr, where maser observations spanning 3 orders of magnitude in distance are all consistent with a large-scale, possibly dipole-shaped, magnetic field (Vlemmings *et al.* 2005, Vlemmings *et al.* 2011).

Very recent ALMA observations have shown that it will soon be possible to finally overcome the problems with determining the circumstellar magnetic field structure. This involves observations aimed at measuring the Goldreich-Kylafis effect, which allows us to use the polarisation of non-maser molecular lines (in this case CO) to determine the magnetic field morphology in the more diffuse circumstellar gas. The first of these observations, for the post-AGB star OH 17.7-2.0, indicate that the magnetic field structure probed by the CO is consistent with that derived from OH maser observations (Fig. 2, Tafoya & Vlemmings, in prep.). This puts to rest the decades old question if maser magnetic field measurements can really be used to probe the large-scale fields. The second set of observations has given us the first velocity-resolved view of the large-scale magnetic field in the AGB star IRC+10216 (Fig. 3, Vlemmings *et al.*, in prep.).

2.2. *post-AGB stars*

Similar to the AGB stars, masers are the main source of magnetic field information of post-AGB and pre-PNe and even for some PNe. OH maser observations indicate magnetic field strengths similar to those of AGB stars (few mG) and a clear large-scale magnetic field structure (Bains *et al.* 2003, Gómez *et al.* 2016). Also dust polarization observations indicate a large-scale magnetic field (e.g. Sabin *et al.* 2015a).

Magnetic fields have also been detected around the so-called 'water-fountain' sources. These sources exhibit fast and highly collimated H_2O maser jets that often extend beyond even the regular OH maser shell. With the dynamical age of the jet of order 100 years, they potentially are the progenitors of the bipolar (pre-)PNe. Observations

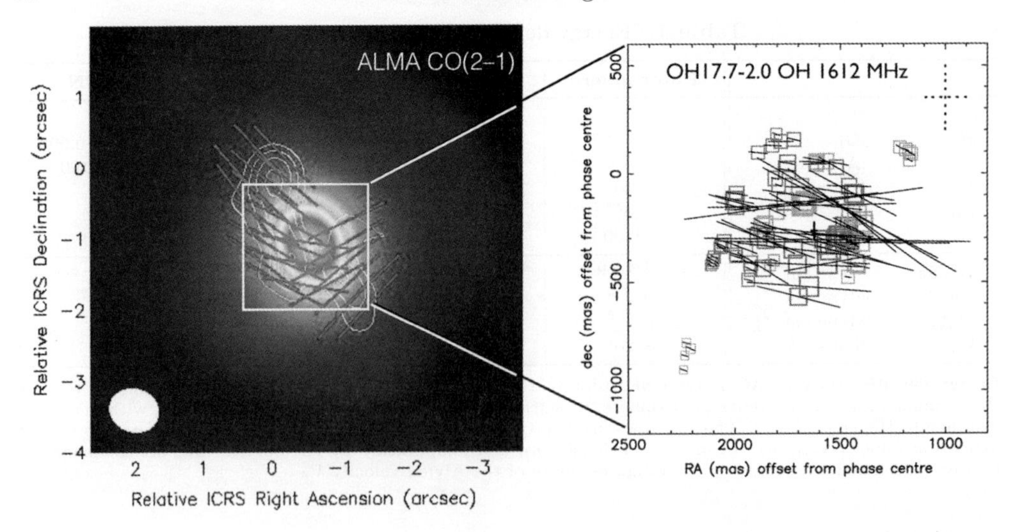

Figure 2. A comparison of the magnetic field determined using ALMA observations of the Goldreich-Kylafis effect on circumstellar CO (left, Tafoya & Vlemmings, in prep.) and MERLIN observations of OH masers (right, Bains *et al.* 2003) around the post-AGB star OH 17.7-2.0. These observations show that CO and OH trace the same large-scale magnetic field.

of the arch-type of the water-fountains, W43A, have revealed a strong toroidal magnetic field that is collimating the jet (Vlemmings *et al.* 2006). For another water-fountain source, IRAS 15445–5449, a synchrotron jet related to strong magnetic fields has been detected (Pérez-Sánchez *et al.* 2013). Similar, synchrotron emission has been found from what could be one of the youngest PNe (Suárez *et al.* 2015).

Finally, recently also surface fields have been measured for 2 post-AGB stars (Sabin *et al.* 2015b). These fields are consistent with the fields inferred from the CSE measurements.

2.3. *Supergiant stars*

Many maser observations show that strong magnetic fields are also present in the envelopes of Red Supergiant stars (e.g. Vlemmings *et al.* 2002; Herpin *et al.* 2006). The questions about local or large-scale fields, are the same as around AGB stars. As noted above, the supergiant VX Sgr is one of the first stars where a large-scale magnetic field, with a structure consistent throughout the envelope, was found. At (sub-)millimeter wavelengths it is now possible to simultaneously study the polarization of masers, regular molecular lines, and circumstellar dust using ALMA. Recent observations of VY CMa indicate magnetically aligned dust and consistent structures between the maser and non-maser molecular lines (Fig, 4. Vlemmings *et al.* 2017a). The observations indicate that magnetic fields could be involved in the mass loss of these massive stars.

3. Indirect tracers and origin of the magnetic field

The origin of AGB magnetic fields is unclear and might require an extra source of angular momentum to maintain a stellar dynamo. This however depends strongly on the magnetic coupling throughout the star itself. If a sufficiently strong magnetic field persist at the AGB stellar surface, it might be possible to detect signs of magnetic activity. Recently, it has been shown that the majority of the AGB stars are UV-emitters

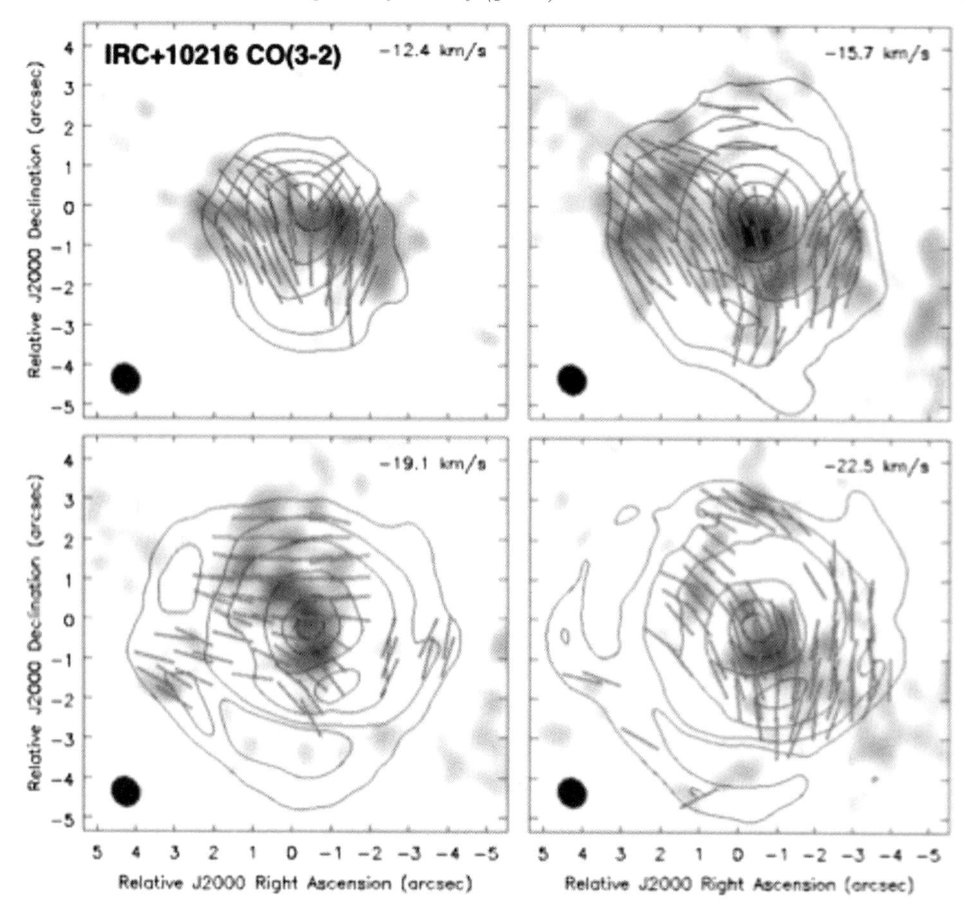

Figure 3. Four channel maps showing the Goldreich-Kylafis effect on the CO(3-2) line in the CSE of the AGB star IRC+10216. The ALMA observations for the first time clearly resolve the magnetic field structure throughout the CSE. The red vectors indicate the polarisation direction. The linearly polarised emission is shown in greyscale. The contours indicate the total intensity emission. A structure function analysis will be able to reveal the velocity resolved field strength and an initial analysis indicates a structured field with a strength >1 G at the stellar surface (Vlemmings *et al.*, in prep.).

(Montez *et al.* 2017) which could be a sign of (magnetic) activity. Similarly, recent observation of the surface of the AGB star W Hya show high brightness temperature hotspots (Fig 5, Vlemmings *et al.* 2017a). These spots can arise from strong shocks but could also point to magnetic activity.

As previously noted, the angular momentum imparted by a stellar (or sub-stellar) companion might be needed to maintain a stellar dynamo that can generate the observed magnetic fields. However, rotation is very difficult to measure for the extended AGB stars that are undergoing pulsations and show large convective cells. Only very recently has ALMA been able to measure the fast (~ 1 km s^{-1}) rotation of the AGB star R Dor (Fig. 6, Vlemmings 2018). As the rotation is almost two orders of magnitude larger than otherwise expected, it is a likely sign of interaction with an hitherto unknown companion. Unfortunately, no magnetic field observations exist yet for R Dor and it is thus not yet possible to establish a link between the generation of a magnetic field and the fast rotation.

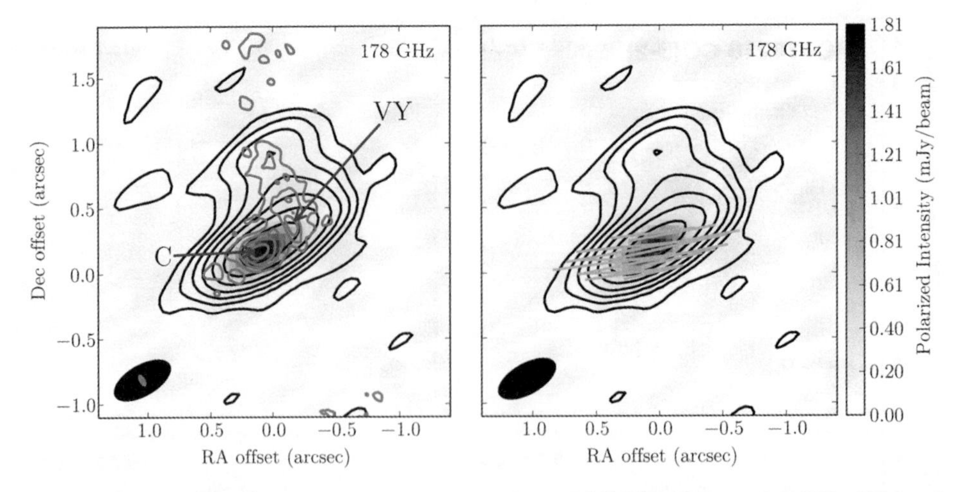

Figure 4. ALMA observations of the dust around the RSG VY CMa at 178 GHz (Vlemmings *et al.* (2017b)). Arrows indicate a strong dust clump (C) and the star (VY). The grey scale image is the linearly polarized intensity. The similarly spaced red contours (left) indicate the ALMA 658 GHz continuum from O'Gorman *et al.* (2015). The vectors (right) indicate the direction of the magnetic field traced by magnetically aligned dust grains.

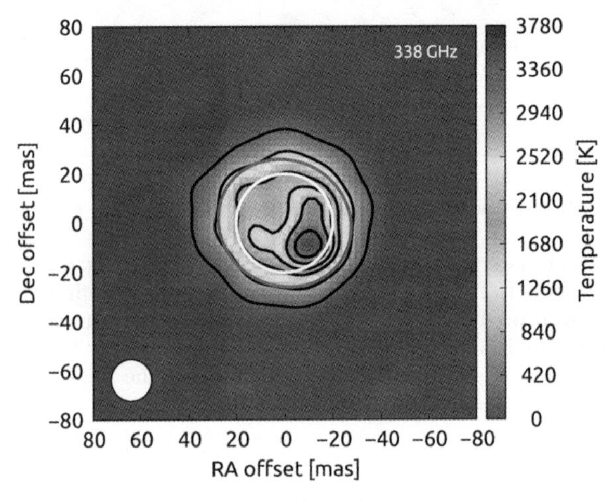

Figure 5. Brightness temperature map of the AGB star W Hya observed with ALMA at 338 GHz (Vlemmings *et al.* 2017a). The red ellipse indicates the size of the stellar disk at 338 GHz while the white circle indicates the size of the optical photosphere. The hotspot is unresolved and it brightness temperature in the map is a lower limit. From size measurements we can constrain the true brightness temperature to be $> 50\,000$ K, which could be a sign of shock interaction or magnetic activity.

4. Conclusions

Magnetic fields are ubiquitous around AGB and post-AGB stars, and several observations indicate a link between the magnetic field and the collimated outflows found in pre-PNe. Additionally, indirect observations of hotspots and UV-emission might point to magnetic activity on the surface of AGB stars. However, it is only now becoming possible to start probing the morphology of the magnetic field in AGB CSEs and to finally determine the role of magnetism around evolved stars.

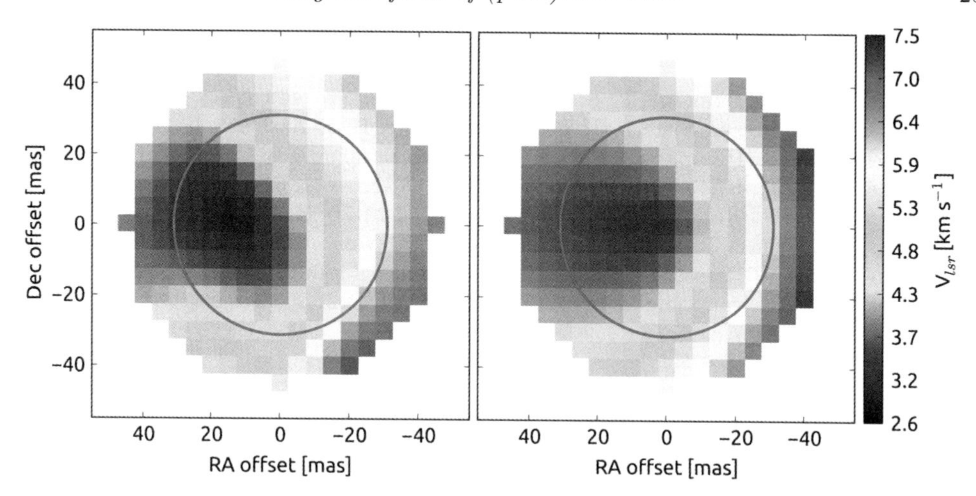

Figure 6. From Vlemmings *et al.* (2018) *(left)*: Center velocity of the SiO $v = 3, J = 5 - 4$ emission line indicating the fast rotation in the inner CSE of R Dor. The red ellipse indicates the measured size of the star at 214 GHz. *(right)*: The best fit model of solid-body rotation including a small expansion velocity component.

References

Bains, I., Gledhill, T. M., Yates, J. A., & Richards, A. M. S. 2003, *MNRAS*, 338, 287

Berdyugina, S. V. 2009, Cosmic Magnetic Fields: From Planets, to Stars and Galaxies, 259, 323

Braithwaite, J., & Spruit, H. C. 2004, *Nature*, 431, 819

Charbonneau, P. 2014, *ARAA*, 52, 251

Duthu, A., Herpin, F., Wiesemeyer, H., *et al.* 2017, *A&A*, 604, A12

Falceta-Gonçalves, D. & Jatenco-Pereira, V. 2002, *ApJ*, 576, 976

Gómez, J. F., Uscanga, L., Green, J. A., *et al.* 2016, *MNRAS*, 461, 3259

Gonidakis, I., Chapman, J. M., Deacon, R. M., & Green, A. J. 2014, *MNRAS*, 443, 3819

Herpin, F., Baudry, A., Thum, C., Morris, D., & Wiesemeyer, H. 2006, *A&A*, 450, 667

Kemball, A. J., Diamond, P. J., Gonidakis, I., *et al.* 2009, *ApJ*, 698, 1721

Kwok, S., Purton, C. R., & Fitzgerald, P. M. 1978, *ApJL*, 219, L125

Leal-Ferreira, M. L., Vlemmings, W. H. T., Kemball, A., & Amiri, N. 2013, *A&A*, 554, A134

Lèbre, A., Aurière, M., Fabas, N., *et al.* 2014, *A&A*, 561, A85

Montez, R., Jr., Ramstedt, S., Kastner, J. H., Vlemmings, W., & Sanchez, E. 2017, *ApJ*, 841, 33

Nordhaus, J., Blackman, E. G., & Frank, A. 2007, *MNRAS*, 376, 599

O'Gorman, E., Vlemmings, W., Richards, A. M. S., *et al.* 2015, *A&A*, 573, L1

Pérez-Sánchez, A. F., Vlemmings, W. H. T., Tafoya, D., & Chapman, J. M. 2013, *MNRAS*, 436, L79

Rudnitski, G. M., Pashchenko, M. I., & Colom, P. 2010, *Astronomy Reports*, 54, 400

Sabin, L., Hull, C. L. H., Plambeck, R. L., *et al.* 2015a, *MNRAS*, 449, 2368

Sabin, L., Wade, G. A., & Lèbre, A. 2015b, *MNRAS*, 446, 1988

Soker, N. 1998, *MNRAS*, 299, 1242

Soker, N. 2002, *MNRAS*, 336, 826

Suárez, O., Gómez, J. F., Bendjoya, P., *et al.* 2015, *ApJ*, 806, 105

Thirumalai, A., & Heyl, J. S. 2013, *MNRAS*, 430, 1359

Trippella, O., Busso, M., Palmerini, S., Maiorca, E., & Nucci, M. C. 2016, *ApJ*, 818, 125

Vlemmings, W. H. T. 2012, in IAU Symposium, Vol. 287, Cosmic Masers - from OH to H0, ed. R. S. Booth, W. H. T. Vlemmings, & E. M. L. Humphreys, 31–40

Vlemmings, W. H. T. 2014, in IAU Symposium, Vol. 302, Magnetic Fields throughout Stellar Evolution, ed. P. Petit, M. Jardine, & H. C. Spruit, 389–397

Vlemmings, W. H. T. 2018, Contributions of the Astronomical Observatory Skalnate Pleso, 48, 187
Vlemmings, W. H. T., Diamond, P. J., & Imai, H. 2006, *Nature*, 440, 58
Vlemmings, W. H. T., Diamond, P. J., & van Langevelde, H. J. 2002, *A&A*, 394, 589
Vlemmings, W. H. T., Humphreys, E. M. L., & Franco-Hernández, R. 2011, *ApJ*, 728, 149
Vlemmings, W. H. T., Khouri, T., De Beck, E., *et al.* 2018, *A&A*, 613, L4
Vlemmings, W. H. T., Khouri, T., Martí-Vidal, I., *et al.* 2017b, *A&A*, 603, A92
Vlemmings, W., Khouri, T., O'Gorman, E., *et al.* 2017a, *Nature Astronomy*, 1, 848
Vlemmings, W. H. T., Ramstedt, S., Rao, R., & Maercker, M. 2012, *A&A*, 540, L3
Vlemmings, W. H. T., van Langevelde, H. J., & Diamond, P. J. 2005, *A&A*, 434, 1029

Discussion

De Marco: There seem to be too many AGB stars with B-fields to be justified by a close-by companion. So, are you saying that there *must* be an alternative scenario to the binary scenario?

Vlemmings: Yes. Although the sample can still be considered small, magnetic fields appear to be present in all studied sources with extrapolated surface field strengths of a few Gauss. Certainly these sources do not all have close-by *stellar* companions.

Why Galaxies Care About AGB Stars: A Continuing Challenge through Cosmic Time
Proceedings IAU Symposium No. 343, 2019
F. Kerschbaum, M. Groenewegen & H. Olofsson, eds.

doi:10.1017/S1743921318007378

Constraining convection across the AGB with high-angular-resolution observations

Claudia Paladini[1], Fabien Baron[2], A. Jorissen[3], J.-B. Le Bouquin[4,5], B. Freytag[6], S. Van Eck[3], M. Wittkowski[7], J. Hron[8], A. Chiavassa[9], J.-P. Berger[10], C. Siopis[3], A. Mayer[8], G. Sadowski[3], K. Kravchenko[3], S. Shetye[3,11], F. Kerschbaum[8], J. Kluska[11] and S. Ramstedt[6]

[1]European Southern Observatory, Alonso de Cordova 3107, Vitacura, Santiago, Chile
email: cpaladin@eso.org

[2]Department of Physics and Astronomy, Georgia State University, PO Box 5060 Atlanta, Georgia 30302-5060, USA

[3]Institut dAstronomie et dAstrophysique, Université libre de Bruxelles, CP 226, 1050 Bruxelles, Belgium

[4]Institut de Planétologie et d'Astrophysique de Grenoble, CNRS, Univ. Grenoble Alpes, France

[5]University of Michigan, 1085 S. University Ave. 303B West Hall Ann Arbor, USA

[6]Department of Physics and Astronomy, Uppsala University, Box 516, 75120 Uppsala, Sweden

[7]European Southern Observatory, Karl-Schwarzschild-Strasse 2, 85748 Garching bei München, Germany

[8]Department of Astrophysics, University of Vienna, Türkenschanzstrasse 17, 1180 Vienna, Austria

[9]Laboratoire Lagrange, UMR 7293, Université de Nice Sophia-Antipolis, CNRS, Observatoire de la Côte dAzur, BP 4229, 06304 Nice Cedex 4, France

[10]Univ. Grenoble Alpes, CNRS, IPAG, 38000 Grenoble, France

[11]Institute for Astronomy, KU Leuven, Celestijnenlaan 200D B2401, B-3001 Leuven, Belgium

Abstract. We present very detailed images of the photosphere of an AGB star obtained with the PIONIER instrument, installed at the Very Large Telescope Interferometer (VLTI). The images show a well defined stellar disc populated by a few convective patterns. Thanks to the high precision of the observations we are able to derive the contrast and granulation horizontal scale of the convective pattern for the first time in a direct way. Such quantities are then compared with scaling relations between granule size, effective temperature, and surface gravity that are predicted by simulations of stellar surface convection.

Keywords. convection, stars: AGB and post-AGB, stars: imaging, stars: mass loss, techniques: high angular resolution, techniques: interferometric

---

1. Introduction

Convection plays a major role in many astrophysical processes, including energy transport, pulsation, dynamos and winds on evolved stars. Most of our knowledge about stellar convection comes from studying the Sun. Two millions of convective cells are observed on the surface of our star, each one with a size of about 2000 km. Following predictions dating back to the '70, the surface of evolved stars (e.g. the Asymptotic Giant Branch stars, AGB) is expected to be populated by a few large convective cells several tens of thousand times the size of the solar ones. Direct evidence of the presence of such structures has been observed for the red supergiant (RSG) star Betelgeuse with the

Hubble telescope by Gilliland & Dupree (1996), and later with the VLTI by Haubois *et al.* (2009). Asymmetric structures on the surface and in the very inner region of AGB stars have been detected by several authors (Ragland *et al.* 2006, Cruzalèbes *et al.* 2015). The properties of granulation (i.e., size of the cells and contrast) were usually derived via geometric modelling. This method however is degenerate and does not provide information about the physical origin of the convective cells. To observe directly convection on the surface of AGBs one had to wait till 2014, when Monnier *et al.* (2014) imaged for the first time the oxygen-rich mira R Car within the frame of the *Image Beauty Contest 2014*. Few years later, Wittkowski *et al.* (2017) reported the observations of the surface of the carbon semiregular variable R Scl. Observing directly and characterizing convection on the surface of AGB stars is challenging for several reasons. AGB stars are variable, and the observations need to be collected almost simultaneously. This is often not possible because of the limited amount of apertures of the interferometers, and scheduling requirements. AGB stars are smaller than RSGs, and the details of the surfaces can be retrieved only for stars within 600 pc. Most importantly the surface of AGB stars is obscured by dust and molecular opacity, which partially masks the photosphere of the star.

2. The target and the observations

π^1 Gruis is a semiregular variable (SRb) with a Period of 195 days and a parallax of 6.13±0.76 milliarcseconds. The star was observed with several ground and space facilities. It has an effective temperature ($T_{\rm eff}$) of $\sim$3100 K, a mass of 1.5–2 $M_\odot$, a luminosity (L) of $\log(L/L_\odot) \sim 3.86$, and it is close to the tip of the AGB with a mass-loss rate of $2.7{\times}10^{-6} M_\odot$ yr^{-1} (Mayer *et al.* 2014, and references therein).

We observed π^1 Gruis with the PIONIER instrument (Le Bouquin *et al.* 2011) in September 2014. PIONIER combines the light of 4 telescopes in a coherent way providing visibility and closure phase interferometric measurements. The observations were made in the H-band in low spectral resolution ($R \sim 35$). π^1 Gruis was observed for a total of 1.5 nights using the compact and medium array configuration of the VLTI. The baselines span between 8 and 90 m achieving and angular resolution ($\lambda/2B$) of $\sim$2 mas.

3. Method

We reconstructed 3 independent images, one per each spectral channel (1.625, 1.678, and 1.73 μm), using both the SQUEEZE (Baron *et al.* 2010) and MiRa (Thiebaut 2008). image reconstruction tools. The dusty envelope is transparent in the wavelength range covered by PIONIER. The molecular contribution is mainly due to CO and CN, and the opacity reaches a minimum in the third spectral channel. Both imaging algorithms give a consistent result showing a nearly spherical disc with structures of convective nature on the surface (Fig. 1).

Contrast and typical granulation size were derived from the images after correcting for the limb darkening effect. Such correction was executed using three different masks: (i) a square mask, (ii) a gaussian mask, (iii) an intensity profile from the MARCS model atmosphere (Van Eck *et al.* 2017) that best fits the spectral energy distribution of the star. The results presented here correspond to the ones obtained using the square mask. The other two methods produce similar results, and we refer the reader to Paladini *et al.* (2018) for a detailed discussion. The intensity contrast is estimated as $\Delta I_{\rm rms}/I_{\rm mean}$ following the definition from Wedemeyer *et al.* (2009). To derive the typical granulation size we estimated the spatial power-spectrum density (PSD) from the three images (Fig. 2).

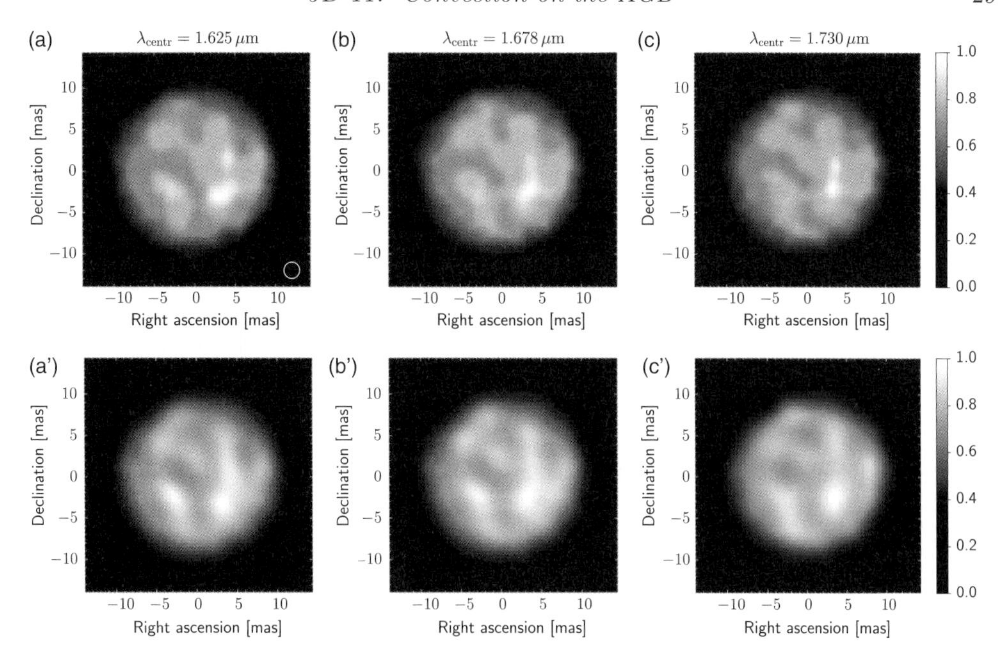

Figure 1. Images of the stellar surface of π^1 Gruis. The upper row shows the three spectral channels reconstructed using SQUEEZE and the smoothness regularizer function. The lower row shows the reconstruction obtained with MiRa and the total variation regularizer. (See also Paladini *et al.* 2018)

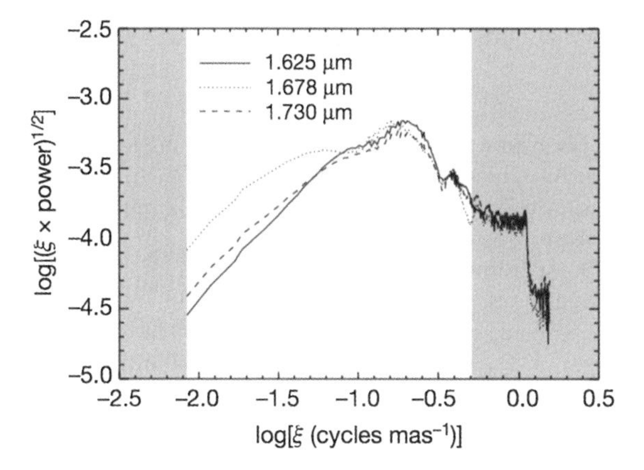

Figure 2. Spatial power spectrum density derived from the three PIONIER SQUEEZE images. The grey-shaded area to the left represents the size of the box of the image. The grey-shaded area to the right marks the angular resolution of the observations, e.g., the observations are not sensitive to surface structure at $\log \xi > -0.3$. The peak of the PSD gives information on the typical granulation size.

4. Results and discussion

The photosphere of π^1 Gruis has an intensity contrast of the order of 12%, which increases slightly towards shorter wavelengths. The systematic errors are very difficult to assess, and there are also no models to compare our results with. However the constrast so determined is in qualitative agreement with the contrast derived from bolometric intensity maps of RSGs (Freytag *et al.* 2017). The characteristic granulation size

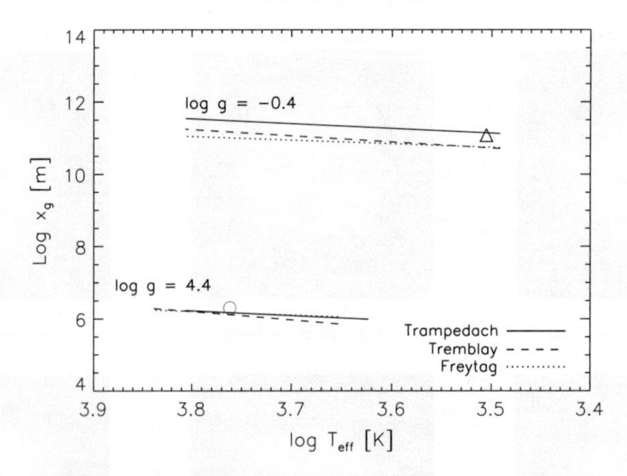

Figure 3. The characteristic granulation size of convection measured on the Sun (circle), and the one derived in this work for π^1 Gruis (triangle). The different lines represent the parametric formulas derived from theoretical model predictions of convection.

obtained is 5.3 ± 0.2 mas, corresponding to 1.2×10^{11} m. Such measurement was compared with parametric formulas relating the granulation size to the stellar parameters, and obtained from theoretical model predictions of stellar convection (Freytag *et al.* 1997, Trampedach *et al.* 2013, Tremblay *et al.* 2013). Figure 3 shows how our results are in agreement with the model predictions, which is remarkable considering the models do not cover the parameter space of π^1 Gruis. As a next step we have an ongoing monitoring program to derive the time-scale of convective granules on the surface of AGB stars. We also plan to image more AGB stars and repeat the same kind of analysis to help constraining the theory of convection on the AGB.

References

Baron, F., Monnier, J. D., Kloppenborg, & B. 2010, *SPIE*, 7734, 21
Cruzalèbes, P., Jorissen, A., Chiavassa, A., *et al.* 2015, *MNRAS*, 446, 3227
Freytag, B., Holweger, H., Steffen, M., & Ludwig, H.-G. 1997, in: F. Paresce (ed.), *Science with the VLT Interferometer*, (Springer), p. 316
Freytag, B., Liljegren, S., & Höfner, S. 2017, *A&A*, 600, 137
Gilliland, R. L., & Dupree, A. K. 1996, *ApJ*, 463, 29
Haubois, X., Perrin, G., Lacour, S., *et al.* 2009, *A&A*, 508, 923
Le Bouquin, J.-B., Berger, J.-P., Lazareff, B., *et al.* 2011, *A&A*, 535, 67
Mayer, A., Jorissen, A., Paladini, C., *et al.* 2014, *A&A*, 570, A113
Monnier, J. D., Berger, J.-P., Le Bouquin, J.-B., *et al.* 2014, *SPIE*, 9146, 1
Paladini, C., Baron, F., Jorissen, A., *et al.* 2018, *Nature*, 553, 310
Ragland, S., Traub, W. A., Berger, J.-P., *et al.* 2006, *ApJ*, 652, 650
Thiebaut, E. 2008, *SPIE*, 7013, 1
Trampedach, R., Asplund, M., Collet, R., Nordlund, Åke; S., & Robert, F. 2013 *ApJ*, 769, 18
Tremblay, P.-E., Ludwig, H.-G., Freytag, B., Steffen, M., & Caffau, E. 2013 *A&A*, 557, 7
Van Eck, S., Neyskens, P., Jorissen, A., *et al.* 2017, *A&A*, 601, 10
Wedemeyer-Böhm, S., & Rouppe van der Voort, L. 2009, *A&A* 503, 225
Wittkowski, M.; Hofmann, K.-H.; Höfner, S., *et al.* 2017, *A&A*, 601, 3

Why Galaxies Care About AGB Stars: A Continuing Challenge through Cosmic Time
Proceedings IAU Symposium No. 343, 2019
F. Kerschbaum, M. Groenewegen & H. Olofsson, eds.

doi:10.1017/S1743921318006245

Imaging the dust and the gas around Mira using ALMA and SPHERE/ZIMPOL

Theo Khouri[1], Wouter H. T. Vlemmings[1], Hans Olofsson[1], Christian Ginski[2], Elvire De Beck[1], Matthias Maercker[1] and Sofia Ramstedt[3]

[1]Department of Space, Earth and Environment, Chalmers University of Technology, Onsala Space Observatory, 439 92 Onsala, Sweden
email: theokhouri@gmail.com

[2]Sterrewacht Leiden, P.O. Box 9513, Niels Bohrweg 2, 2300RA Leiden, The Netherlands

[3]Department of Physics and Astronomy, Uppsala University, Box 516, 751 20, Uppsala, Sweden

Abstract. The mass-loss mechanism of asymptotic giant branch stars has long been thought to rely on two processes: stellar pulsations and dust formation. The details of the mass-loss mechanism have remained elusive, however, because of the overall complexity of the dust formation process in the very dynamical pulsation-enhanced atmosphere. Recently, our understanding of AGB stars and the associated mass loss has evolved significantly, thanks both to new instruments which allow sensitive and high-angular-resolution observations and the development of models for the convective AGB envelopes and the dust formation process. ALMA and SPHERE/ZIMPOL on the VLT have been very important instruments in driving this advance in the last few years by providing high-angular resolution images in the sub-mm and visible wavelengths, respectively. I will present observations obtained using these instruments at the same epoch (2.5 weeks apart) of the AGB star Mira that resolve even the stellar disk. The ALMA data reveals the distribution and dynamics of the gas around the star, while the polarised light imaged using SPHERE shows the distribution of the dust grains expected to drive the outflows. Moreover, the observations show a central source surrounded by asymmetric distributions of gas and dust, with complementary structures seen in the two components. We model the observed CO $v = 1, J = 3-2$ line to determine the density, temperature and velocity of gas close to the star. This model is then used to estimate the abundance of AlO. Our results show that only a very small fraction of aluminium ($\lesssim 0.1\%$) is locked in AlO molecules. We also calculate models to fit the observed polarised light based on the gas densities we find. The low level of visible-light polarisation detected using ZIMPOL implies that, at the time of the observations, aluminium atoms are either not efficiently depleted into dust or the aluminium-oxide grains are relatively small ($\lesssim 0.02\ \mu$m).

---

1. Introduction

At the end of their lives, low- and intermediate-mass stars evolve through the asymptotic giant branch (AGB) phase. In the AGB, these stars return a large fraction of their initial mass to the interstellar medium through slow ($\upsilon_\infty \sim 10$ km/s) but massive (10^{-8} M$_\odot$/yr $\lesssim \dot{M} \lesssim 10^{-4}$ M$_\odot$/yr) outflows. The wind-driving mechanism is thought to consist of a two-step process. First, large convective cells and stellar pulsations lift material up to distances where the temperatures are low enough for dust condensation to happen. Then, radiation pressure acting on the newly-formed dust grains leads to an outflow (Höfner & Olofsson 2018). Predicting the mass-loss rate of a given star

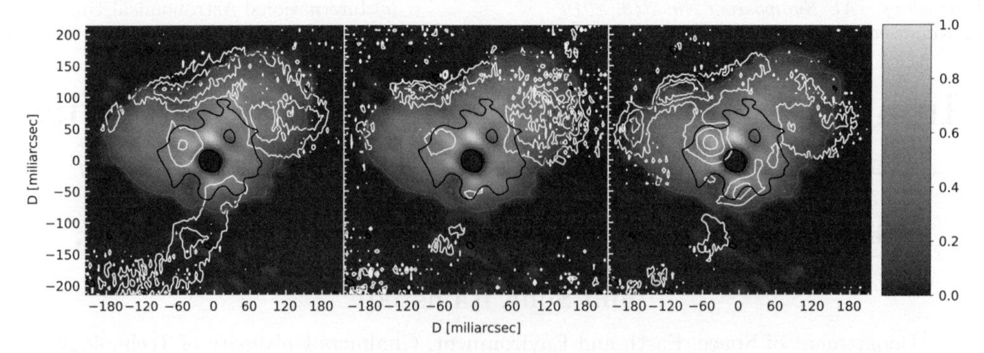

Figure 1. The colour scale shows the normalised emission in the SO $N_J = 8_8 - 7_7$ line imaged using ALMA and the white contours show the polarisation degree observed using SPHERE with three filters (*left*: 0.65 μm, *middle*: 0.75 μm, and *right*: 0.82 μm). The contours show the 1.5, 2.5, and 3.5% levels. The ALMA image is normalised to its peak value. The red and black lines marks the 10%-level contour of the SO $N_J = 8_8 - 7_7$ and CO $v = 1, J = 3 - 2$ emission lines, respectively. The contour of the SO line shows the boundary of the high-gas-density region detected by ALMA. The accumulation of dust along the edge of this high-gas-density region is particularly seen between the north and west directions. Absorption against the star is seen as negative fluxes in the centre of these stellar-continuum-subtracted images.

is not yet possible at present, because the modelling of convection is difficult and the stellar-pulsation mechanism and the details of the dust formation process are still not understood.

For stars richer in oxygen than carbon (O-rich), the outflows are thought to be driven by *scattering* of radiation off dust grains. This is only possible if the dust grains grow to relatively large sizes, ~ 0.3 μm in radius (Höfner 2008). Observations revealed polarised light produced by scattering of radiation off dust grains with sizes within the required range (Norris *et al.* 2012). SPHERE/ZIMPOL observations now allow for studies of the distribution and amount of such large grains around several O-rich AGB stars (Khouri *et al.* 2016; Ohnaka *et al.* 2017). For the closer-by sources, the dust-formation region, and even the stellar disc, is clearly resolved (see Fig. 1). The dust condensation sequence is not well understood but the current picture is that aluminium oxide grains from close to O-rich AGB stars (Gobrecht *et al.* 2016) and that iron-free silicates condensate at slightly larger radii on top of the aluminium-oxide seeds. The aluminium-oxide grains are expected to grow to relative large sizes (~ 0.3 μm) and have large scattering cross-sections at visible wavelengths (Höfner *et al.* 2016). The dust mass of aluminium oxide grains is expected to be too low for driving the outflow, however, because of the low abundance of aluminium (Bladh & Höfner 2012). To study the dust-formation and wind-driving mechanisms, we observed the O-rich AGB stars Mira using SPHERE/ZIMPOL and ALMA. The work discussed here is present in more depth in Khouri *et al.* (submitted).

At 107 parsecs, Mira is the archetypal object of the class of Mira variables. Mira has a companion (thought to be a white dwarf) with an orbital separation of ~ 90 AU and a projected distance of 0.5 arcseconds (e.g., Ireland *et al.* 2007). Mira B is not expected to have a significant gravitational effect on material in the wind-acceleration region (at distance $\lesssim 10$ AU from Mira A, e.g., (Mohamed & Podsiadlowski 2012).

2. Observations

ZIMPOL observations of Mira were acquired using filters NR, cnt748, and cnt820 on 27 November 2017 (ESO program ID 0100.D-0737, PI: Khouri). The ALMA observations were obtained 18 days before the ZIMPOL observations on 9 November 2017 (project 2017.1.00191.S, PI: Khouri). The data were acquired at post-minimum light phase ($\varphi \approx 0.7$) of Mira A. The data reduction followed standard procedures for both instruments (Khouri *et al.*, submitted).

The ZIMPOL data reveal dust grains around Mira A, but also around Mira B and in a dusty trail that connects the two stars. Within an aperture with radius of 80 mas around Mira A, we find a polarisation degree of at most $\sim 2\%$ (see discussion in Khouri *et al.*, submitted). The degree of polarisation measured around Mira A peaks at ≈ 3.2 % in NR, ≈ 3.5 % in cnt748 and ≈ 4.0 % in cnt820, being is relatively lower than what is seen for other close-by O-rich AGB stars, such as R Dor (Khouri *et al.* 2016), W Hya (Ohnaka *et al.* 2017), and R Leo (archival data). The polarization degree around these other AGB stars peaks at $> 7\%$ and is particular high ($> 10\%$) close to minimum-light phase.

In the reduced ALMA data, we obtained a root-mean squared (rms) noise of ~ 1.4 mJy for a spectral resolution of ~ 1.7 km s^{-1}. The continuum images have an rms noise of 0.4 mJy/beam and the continuum flux density of Mira A is measured to be 250.6 ± 0.2 mJy. The ALMA observations reveal more than one hundred spectral lines in the observed spectral windows, 329.25 – 331.1 GHz, 331.1 – 333.0 GHz, 341.35 – 343.25, and 343.2 – 345.1 GHz.

A well-define emission region (see Fig. 1) is seen in low-excitation lines of several species, such as SO, ^{13}CO, SO_2, and to a lesser degree AlO and PO. We interpret this region to be delimited by a steep density decline at its outer edge. Interestingly, the polarization degree measured using ZIMPOL show peaks that follow the edge of this molecular-line emission region, especially in the north and northeast regions (see Fig. 1). The cause for both this well-defined molecular-line emission region and the apparent accumulation of dust at its outer edge is not obvious from our data. We speculate that a recent increase in the mass-loss rate could have led to higher densities close to the star. The mass-loss rate burst could be connected to a recent X-ray outburst observed towards Mira A in 2003 (Karovska *et al.* 2005). As pointed out by the authors, such an X-ray burst could have significant consequences for the mass-loss rate of Mira A. The edge of the molecular-line emission region we see is at a distance of ~ 18 au of the star. Therefore, gas would need to travel at a speed of ~ 6 km/s to cover this distance over the period of 14 years separating the observations reported by us from those by Karovska *et al.* Monitoring the evolution of the molecular-line emission region will help determine whether this scenario is correct.

3. Models

To study the distributions of gas and dust close to Mira A, we calculated radiative-transfer models to fit both the emission in the line CO $v = 1, J = 3 - 2$ and the polarization degree observed in the same region around Mira A. The CO $v = 1, J = 3 - 2$ line is expected to be only excited in the high-density extended atmospheres and close environment of AGB stars and is a good tracer of gas in this region (e.g., Vlemmings *et al.* 2017). The details of the fitting procedure are given in Khouri *et al.* (submitted). Below, we summarise the main findings.

Our best if model to the CO $v = 1, J = 3 - 2$ is shown in Fig. 2. Our results implie a gas mass in the region traced by the vibrationally-excited CO line of $(3.8 \pm 1.3) \times 10^{-4} M_\odot$. A comparison of the prediction from our best models to the observed ^{13}CO line in this inner region suggests that models with gas masses in the lower end of our uncertainty interval are preferred. For these calculations, we assumed a ^{12}C-to-^{13}C ratio of 10 ± 3 (Hinkle *et al.* 2016).

We used this model to constrain the abundance of AlO and we find that only $\sim 0.1\%$ of the aluminium atoms are accounted for by AlO molecules, assuming solar composition

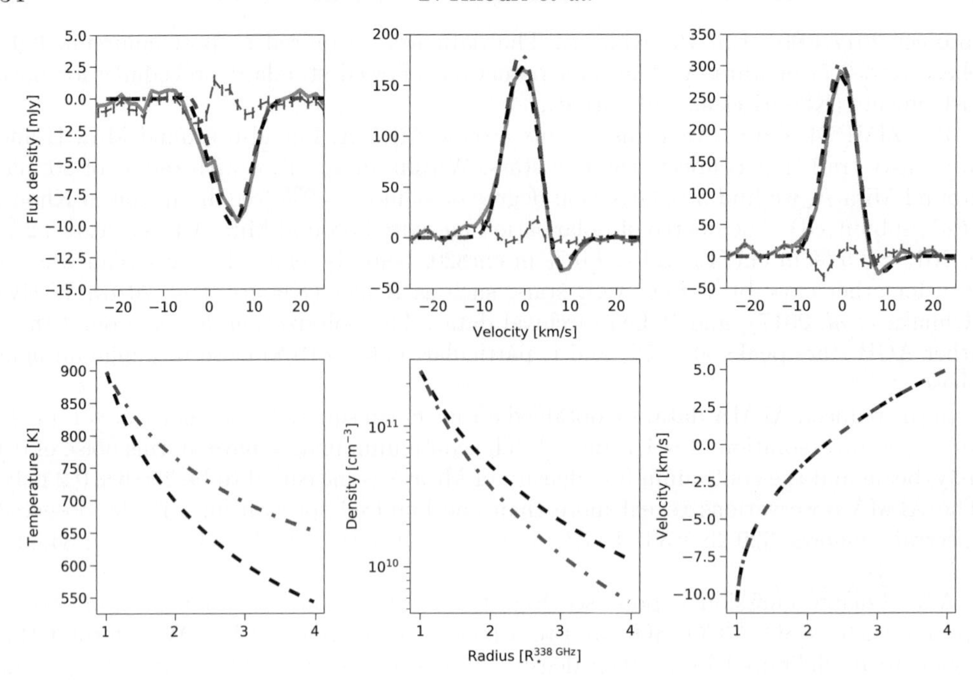

Figure 2. Models for the CO $v=1, J=3-2$ line emission. The black dashed line shows the best-fitting model and the dashed red line shows the preferred model when the ^{13}CO $J=3-2$ line is also considered. *Upper panels:* Best model fits to the observed CO $v=1, J=3-2$ line (solid blue line) extracted using apertures of 20, 50, and 100 mas, from left to right. The thin, dashed blue line shows the residuals with errorbars. *Lower panels:* temperature, density and velocity profiles of the best-fit and preferred models.

(Asplund *et al.* 2009). Hence, there is ample room for aluminium condensation into aluminium oxide grains as expected to happen close to O-rich AGB stars from observations (e.g., Zhao-Geisler *et al.* 2011; Karovicova *et al.* 2013; Khouri *et al.* 2015) and supported by theoretical calculations (Gobrecht *et al.* 2016; Höfner *et al.* 2016). We note that a definitive answer on the amount of aluminium depletion cannot be obtained based on this data because the abundance of other aluminium-bearing molecules was not determined. Nonetheless, studies that have target molecules that could account for a significant fraction of aluminium atoms have not found any species that appears to have significantly higher abundance than AlO (e.g., Kamiński *et al.* 2016; Decin *et al.* 2017).

We used the radiative transfer code MCMax (Min *et al.* 2009) to calculate the polarisation degree from models with aluminium oxide dust grains. The density profile of the dust was derived from the gas density profile in our best-fitting models and scaled assuming different aluminium depletion levels. We calculated the opacities using data from different authors for aluminium oxide grains (Suh 2016; Koike *et al.* 1995; Harman *et al.* 1994; Edlou *et al.* 1993) and a hollow-spheres grain model (Min *et al.* 2003). We considered a single grain size for each model calculation, varying it between 0.01 and 1.0 μm. We find that the models reproduce the low level of visible-light polarisation in the vibrationally-excited CO region only if the grains are very small (sizes $\lesssim 0.02$ μm) or account for a small fraction of aluminium atoms ($\sim 1\%$ for 0.3 μm grains).

References

Asplund, M., Grevesse, N., Sauval, A. J., & Scott, P. 2009, *ARAA*, 47, 481
Bladh, S. & Höfner, S. 2012, *A&A*, 546, A76

Decin, L., Richards, A. M. S., Waters, L. B. F. M., *et al.* 2017, *A&A*, 608, A55
Edlou, S. M., Smajkiewicz, A., & Al-Jumaily, G. A. 1993, *Applied Optics*, 32, 5601
Gobrecht, D., Cherchneff, I., Sarangi, A., Plane, J. M. C., & Bromley, S. T. 2016, *A&A*, 585, A6
Harman, A. K., Ninomiya, S., & Adachi, S. 1994, *Journal of Applied Physics*, 76, 8032
Hinkle, K. H., Lebzelter, T., & Straniero, O. 2016, *ApJ*, 825, 38
Höfner, S. 2008, *A&A*, 491, L1
Höfner, S., Bladh, S., Aringer, B., & Ahuja, R. 2016, *A&A*, 594, A108
Höfner, S. & Olofsson, H. 2018, *AAR*, 26, 1
Ireland, M. J., Monnier, J. D., Tuthill, P. G., *et al.* 2007, *ApJ*, 662, 651
Kamiński, T., Wong, K. T., Schmidt, M. R., *et al.* 2016, *A&A*, 592, A42
Karovicova, I., Wittkowski, M., Ohnaka, K., *et al.* 2013, *A&A*, 560, A75
Karovska, M., Schlegel, E., Hack, W., Raymond, J. C., & Wood, B. E. 2005, *ApJ* (Letters), 623, L137
Khouri, T., Maercker, M., Waters, L. B. F. M., *et al.* 2016, *A&A*, 591, A70
Khouri, T., Waters, L. B. F. M., de Koter, A., *et al.* 2015, *A&A*, 577, A114
Koike, C., Kaito, C., Yamamoto, T., *et al.* 1995, *Icarus*, 114, 203
Min, M., Dullemond, C. P., Dominik, C., de Koter, A., & Hovenier, J. W. 2009, *A&A*, 497, 155
Min, M., Hovenier, J. W., & de Koter, A. 2003, *A&A*, 404, 35
Mohamed, S., & Podsiadlowski, P. 2012, *Balt. Ast.*, 21, 88
Norris, B. R. M., Tuthill, P. G., Ireland, M. J., *et al.* 2012, *Nature*, 484, 220
Ohnaka, K., Weigelt, G., & Hofmann, K.-H. 2017, *A&A*, 597, A20
Suh, K.-W. 2016, *Journal of Korean Astronomical Society*, 49, 127
Vlemmings, W., Khouri, T., O'Gorman, E., *et al.* 2017, *Nat. Ast.*, 1, 848
Zhao-Geisler, R., Quirrenbach, A., Köhler, R., Lopez, B., & Leinert, C. 2011, *A&A*, 530, A120

Discussion

LILJEGREN: Would your results for the grain size change if you considered grains consisting of an aluminium-oxide core with silicate mantle?

KHOURI: I think that would make it worse, actually, since in that case aluminium atoms would account for a relatively smaller mass fraction of the grains. Considering that the scattering properties of silicates are not too different from those of aluminium oxide, I would expect a stronger polarization signal for the same amount of aluminium depletion for core-mantle grains.

SCICLUNA: You can also constrain the grain sizes based on the spectral dependence of the polarized light, right? Why haven't you tried that?

KHOURI: Our observations were affected by the beam-shift effect of ZIMPOL, which makes the polarized light close to the star somewhat uncertain. Since the levels of polarization degree were already relatively low, we choose to consider an upper limit on the polarization degree and calculate the expected level of polarization for different assumed grain sizes and depletion factors.

Why Galaxies Care About AGB Stars:
A Continuing Challenge through Cosmic Time
Proceedings IAU Symposium No. 343, 2019
F. Kerschbaum, M. Groenewegen & H. Olofsson, eds.

doi:10.1017/S1743921318007330

Evolutionary timescales from the AGB to the CSPNe phase

Marcelo M. Miller Bertolami[1,2]

[1]Instituto de Astrofísica de La Plata, CCT-La Plata, CONICET-UNLP,
Paseo del Bosque s/n (B1900FWA), La Plata, Argentina
email: mmiller@fcaglp.unlp.edu.ar

[2]Universidad Nacional de La Plata
Paseo del Bosque s/n (B1900FWA), La Plata, Argentina

Abstract. The transition from the asymptotic giant branch (AGB) to the final white dwarf (WD) stage is arguably the least understood phase in the evolution of single low- and intermediate-mass stars ($0.8 \lesssim M_{\rm ZAMS}/M_{\odot} \lesssim 8...10$). Here we briefly review the progress in the last 50 years of the modeling of stars during the post-AGB phase. We show that although the main features, like the extreme mass dependency of post-AGB timescales were already present in the earliest post-AGB models, the quantitative values of the computed post-AGB timescales changed every time new physics was included in the modeling of post-AGB stars and their progenitors. Then we discuss the predictions and uncertainties of the latest available models regarding the evolutionary timescales of post-AGB stars.

Keywords. stars: AGB and post-AGB, planetary nebulae: general, stars: evolution, stars: mass loss

---

1. Modeling the evolution after the AGB. A historical introduction

The transition from the asymptotic giant branch (AGB) to the final white dwarf (WD) stage is arguably the least understood phase in the evolution of single low- and intermediate-mass stars ($0.8 \lesssim M_{\rm ZAMS}/ \lesssim 8...10$). This transition phase includes the so-called proto-planetary nebulae (PPNe) central stars and some OH/IR objects (also collectively known as *post-AGB stars*, van Winckel 2003†) as well as the hotter central stars of planetary nebula (CSPNe) and other UV-bright stars. In the most simple picture low- and intermediate-mass stars undergo strong and slow stellar winds ($10^{-8} M_{\odot}/{\rm yr} \lesssim \dot{M} \lesssim 10^{-4} M_{\odot}/{\rm yr}$ and $3\ {\rm km/s} \lesssim v_{\rm wind} \lesssim 30$ km/s, see Höfner & Olofsson 2018) at the end of the AGB which lead to the almost complete removal of the H-rich envelope of the AGB star—see Shklovsky (1957), Abell & Goldreich (1966) and Paczyński (1970). After this point, the stars contract at constant luminosity ($L_{\star}$) increasing their effective temperatures ($T_{\rm eff}$) by almost two orders of magnitude, and if there is enough material surrounding the star, a PN is formed in the process.

Since the very first stellar evolution models of CSPNe were computed by Paczyński (1970) it became clear that the evolution from the AGB to the WD phase is extremely mass dependent. In fact, the early models of Paczyński (1970) suggested that the time to "cross" the HR diagram decreased by 4 orders of magnitude just by increasing the mass of the CSPN model by a factor of two (see Fig. 1). The early models by Paczyński (1970,

† Note that throughout this paper we refer to the whole evolutionary stage between the end of the AGB and the beginning of the WD phase as the *post-AGB phase*. This should not be confused with the so called *post-AGB stars* which are defined as stars that have already departed from the AGB but are not yet hot enough to ionize the circumstellar material, i.e. $T_{\rm eff} \lesssim 30\,000$K.

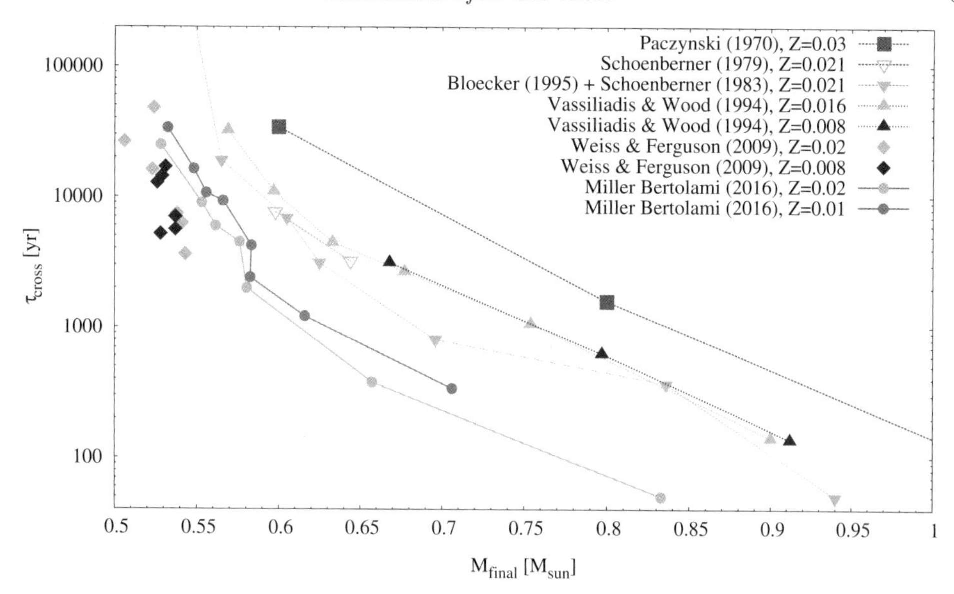

Figure 1. Evolutionary post-AGB time between the point at $\log T_{\rm eff}=4$ to the point of maximum effective temperature in the HR diagram; $\tau_{\rm cross}$.

1971) were constructed by artificially fitting H-envelopes to core structures all obtained from a flash suppressed AGB sequence of $M_{\rm i}=3M_{\odot}$ ($Z=0.03$) and evolved to obtain CO-cores at desired final remnant mass ($M_{\rm f}=0.6, 0.8, 1.2M_{\odot}$). Schönberner (1979, 1981) computed for the fist time the transition from the AGB to the CSPNe phase by assuming a steady wind according to the mass loss prescription by Reimers (1975) and including a detailed computation of the thermally pulsating (TP) AGB phase. This was done for two full sequences with initial masses $M_{\rm i}=1$ and 1.45 $M_{\odot}$ (final masses $M_f=0.598$ and $0.644M_{\odot}$ respectively) and $Z=0.021$. These two sequences already showed post-AGB timescales to be about 4.5 times faster than predicted by the early Paczyński (1970) models (see Fig. 1). Later, Schönberner (1983) computed two more sequences of initial masses $M_i=0.8$ and $1M_{\odot}$ (initial metallicity $Z=0.021$, final masses $M_f=0.546$ and $0.565M_{\odot}$ respectively) by including for the first time a "superwind" phase at the end of the AGB with mass-loss rates of $\dot{M}\gtrsim 10^{-4}M_{\odot}/{\rm yr}$. Although it only covered a small mass range, Schönberner's post-AGB models were the first to include a detailed treatment of the TP-AGB, showing the importance of AGB modeling for the computation of accurate post-AGB stellar models Schönberner (1987). The next grid of models, which covered a wider range of remnant masses ($0.6\leqslant M_f/M_{\odot}\leqslant 0.89$), was computed by Wood & Faulkner (1986). These models were constructed by artificially stripping most of the H-envelope from red giant models computed through many thermal pulses on the AGB but from a single progenitor sequence of $M_i=2M_{\odot}$ ($Z=0.02$). Wood & Faulkner (1986) computed the end of the TP-AGB by assuming two different extreme mass-loss rates in their computations. However, the lack of a realistic initial-final mass relation (IFMR Weidemann 1987) had consequences in the predicted post-AGB evolution and was criticized by Blöcker & Schönberner (1990).

The following generation of post-AGB models came in the 90's when both Vassiliadis & Wood (1994) and Blöcker (1995) published grids of post-AGB models, covering the whole mass range of CSPNe, which included a detailed treatment of the winds during the TP-AGB phase—see Vassiliadis & Wood (1993) and Blöcker (1995). In particular these grids adopted different initial masses to produce different CSPNe, as expected from early determinations of the Initial/Final Mass Relation (see Weidemann 1987 and references therein). These grids of post-AGB stellar evolution models represented a great improvement over the previous Paczyński (1970, 1971) and Wood & Faulkner (1986) models and confirmed the previous result by Schönberner (1979, 1981) that post-AGB timescales were several times shorter than predicted by Paczyński (1970), as can be seen in Fig. 1. It is worth noting, however, that neither Vassiliadis & Wood (1994) nor Blöcker (1995) incorporated the impact of core-overshooting in the upper main sequence, which was already know at the time to be significant, e.g. Schaller *et al.* (1992). Neither did these models make use of the updated radiative opacities computed by the OPAL (Rogers & Iglesias 1992) and OP (Seaton *et al.* 1994) projects that revolutionized stellar astrophysics during the early nineties.

About a decade later another significant improvement in AGB and post-AGB stellar evolution models was made by Kitsikis (2008) (later published in Weiss & Ferguson 2009, from now on KWF). These authors incorporated several features for the first time, both in the computation of the AGB and post-AGB stellar evolution models. First, following Marigo (2002), these authors included for the first time C-rich molecular opacities in the computation of full AGB stellar evolution models. In addition, they also incorporated a separated treatment of C-rich and O-rich dust-driven winds. Most importantly, these authors included both the impact of convective boundary mixing on the main sequence, helium core-burning stage and TP-AGB evolutionary stages, as well as the inclusion radiative opacities from the OPAL project. Probably because of the latter, many convergence problems prevented KWF from computing a large grid of post-AGB stellar evolutionary models. In spite of the lack of a large grid of post-AGB sequences, the models computed by KWF already showed a clear trend, the post-AGB evolution of these models was much faster than those computed by Vassiliadis & Wood (1994) or Blöcker (1995), see Fig. 1. Again, as it happened with Schönberner's post-AGB models more than two decades before, an improvement in the modeling of previous evolutionary stages lead to much shorter post-AGB timescales. Finally, following the approach of KWF, Miller Bertolami (2015, 2016) computed a larger grid of post-AGB stellar evolution models. The main difference between this work and that of KWF is that mixing at convective boundaries from the ZAMS to the TP-AGB were calibrated trying to reproduce several observables on the main sequence and on the TP-AGB and post-AGB evolutionary stages. In particular, the models computed by Miller Bertolami are able to reproduce the width of the main sequence, the C/O ratios of the AGB and post-AGB stars and the He, C and O abundances observed in PG1159 stars. Most importantly, the IFMR of the theoretical models computed by Miller Bertolami (2016) are closer to the semi empirically derived ones than those of KWF (see Fig. 2). In agreement with the findings of KWF the post-AGB models computed by Miller Bertolami (2016) are significantly faster and slightly brighter than earlier models of similar final mass (see Fig. 1).

2. Post-AGB timescales and definitions

The terminology used to define the various stages after the departure from the AGB is sometimes confusing. For example, stellar evolution studies usually refer to the whole stage between the end of the AGB and the beginning of the white dwarf phase as the post-AGB stage (e.g. Vassiliadis & Wood 1994, Blöcker 1995 and Miller Bertolami 2016), while observationally is common to refer as *post-AGB stars* to those stars that have already departed from the AGB but are not yet hot enough to ionize the circumstellar material, i.e. $T_{\rm eff} \lesssim 30\,000$ K (van Winckel 2003; Szczerba *et al.* 2007). Also, from the observational point of view it is usual to split the evolution after the AGB and before the WD phase into the Proto-Planetary Nebulae (PPNe) and the Central Star of Planetary Nebulae (CSPNe) phases. Within this classification the PPNe phase corresponds to the early evolution from the end of the AGB to the beginning of the ionization of the surrounding

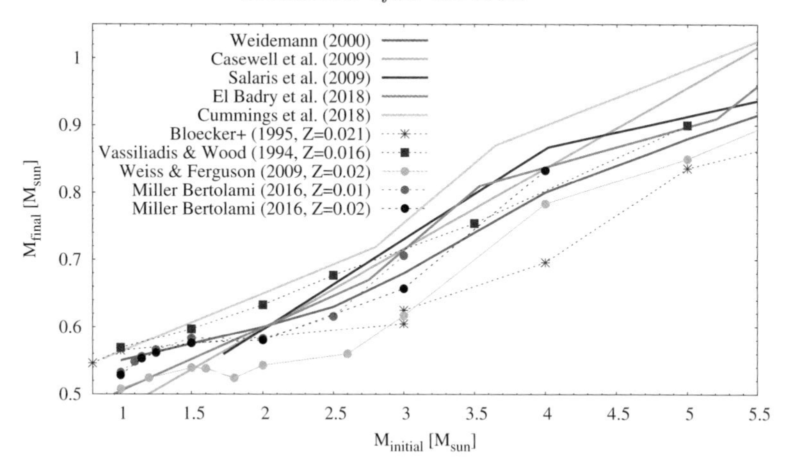

Figure 2. Initial-Final Mass Relation (IFMR) of different post-AGB stellar evolution models as compared with the latest semi empirical determinations Casewell *et al.* (2009); Salaris *et al.* (2009); El-Badry *et al.* (2018); Cummings *et al.* (2018), and the classic semi empirical relation of Weidemann (2000).

nebulae at about $T_{\rm eff} \sim 30\ 000$ K. During the PPNe phase it is expected that many central stars would still be enshrouded in dust and not visible in the optical, see Szczerba *et al.* (2007). The CSPNe phase would then correspond to the phase from the moment the central star attains $T_{\rm eff} \sim 30\ 000$ K to the beginning of the white dwarf cooling track. Yet, from the point of view of the evolution of the central star, this classification is of little use, as it relies on the properties of the surrounding material. Even more, very low mass stars might not evolve fast enough or eject significant amounts of material during the late AGB phase to produce a visible PNe.

In order to be able to quantify the properties of the models during the post-AGB phase precise definitions are required. In particular it is worth noting that the very idea of the end of the AGB is not easy to define from the point of view of stellar evolution, as stars continuously evolve away from the AGB during the late AGB evolution but without any sudden change in the stellar properties From the point of view of the structure of the central star the main change that takes place during the departure from the AGB is the transition from a expanded giant-like configuration into a dwarf-like one. This is caused by the reduction of the H-rich envelope below the critical value required to sustain a giant-like structure (see Fig. 3 and Faulkner 2005 for an extended discussion of this problem). This leads to a continuous increase in the heating rate of the stellar surface from $\dot{T}_{\rm eff} \lesssim 0.1$ K/yr on the AGB to $1{\rm K/yr} \lesssim \dot{T}_{\rm eff} \lesssim 10\ 000$ K/yr once the star attained $T_{\rm eff} \gtrsim 10\ 000$ K, see Fig. 4.

In this context, and in order to discuss the properties of the computed stellar models, different authors choose to divide the transition from the AGB to the WD phase according to different arbitrary definitions. As the relative mass of the envelope is a key feature determining the end of the AGB, Miller Bertolami (2016) choose to define the end of the AGB phase as the moment in which $M_{\rm env}/M_\star = 0.01$ (see Fig. 3). At this moment, models have already moved significantly to the blue ($T_{\rm eff} \sim 3700...5000$ K), which is true at all masses and metallicities. Although this choice defines the end of the AGB in a homogeneous way for all masses and metallicities, and is based on the underlying physical reason behind the departure from the AGB, the choice remains arbitrary.

In order to disentangle the impact of different uncertainties and definitions we define two different timescales: the transition time scale $\tau_{\rm trans}$ corresponding to the early (and

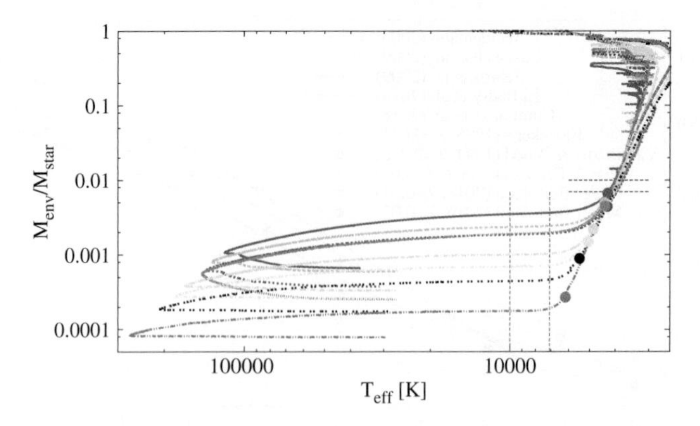

Figure 3. Mass of the H-rich envelope of the H-burning models computed by Miller Bertolami (2016). The vertical dashed lines indicate the zero points at $\log T_{\text{eff}} = 3.85$ ($\log T_{\text{eff}} = 4$) adopted by Miller Bertolami (2016) Vassiliadis & Wood (1994) for the computation of the post-AGB crossing times (τ_{cross}). Horizontal dashed lines indicate two alternative envelope masses adopted by Miller Bertolami (2016) as a definition of the end of the AGB. Color circles indicate the end of the AGB as estimated from the suggestion by Soker (2008).

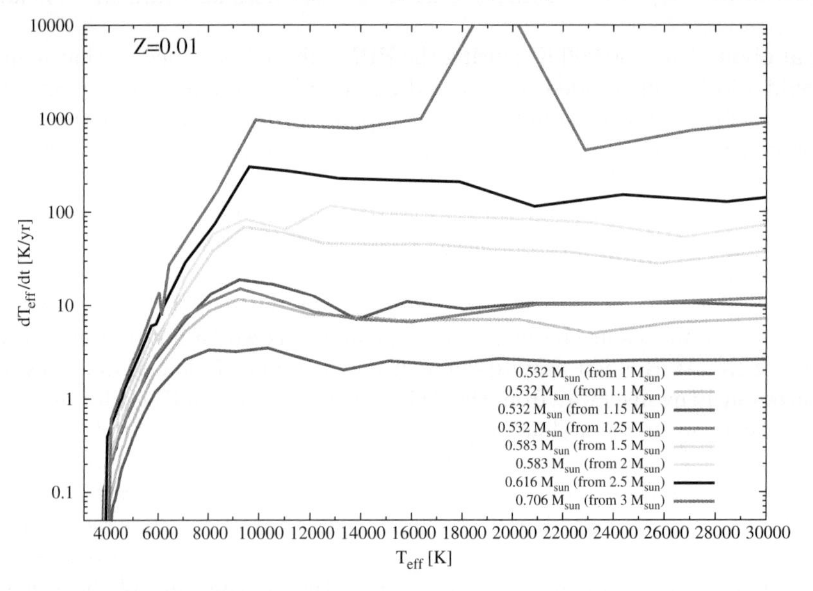

Figure 4. Heating speed of the stellar evolution models during the early post-AGB phase. Note the exponential increase in the heating rate below $T_{\text{eff}} \lesssim 10\,000$ K as the almost constant heating rate at $10\,000\text{K} \lesssim T_{\text{eff}} \lesssim 30\,000$K.

slow) evolution from the end of the AGB ($M_{\text{env}}/M_\star = 0.01$) to the point at $\log T_{\text{eff}} = 3.85$, and the crossing timescale τ_{cross} corresponding to the late (fast) evolution from $\log T_{\text{eff}} = 3.85$ to the point of maximum effective temperature. In what follows we discuss the properties and uncertainties of these two post-AGB timescales.

2.1. *The crossing time:* τ_{cross}

The uncertainties in τ_{cross} (Fig. 1) are mostly related to uncertainties in the previous evolution, and not to uncertainties of the physics during the post-AGB phase itself. In particular, with the exception of very luminous CSPNe one should not expect

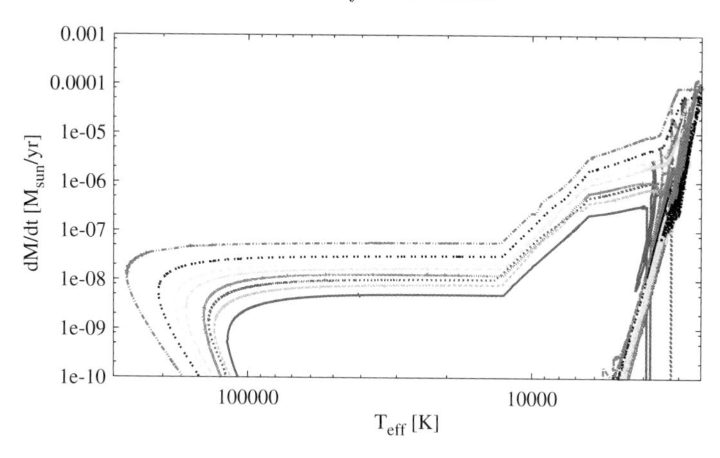

Figure 5. Mass-loss rates of the sequences computed by Miller Bertolami (2016) for $Z = 0.01$.

hot radiative-driven winds to be of any importance for the value of $\tau_{\rm cross}$ in H-burning post-AGB stars. Fig. 5 shows the evolution of the mass-loss rate $\dot{M}^{\rm env}_{winds}$ adopted in the computation of the sequences computed by Miller Bertolami (2016) for $Z = 0.01$. The rate of reduction of the H-rich envelope by winds have to be compared with the rate of H-consumption by nuclear burning which in the case of CNO-burning is of $\dot{M}_{\rm H}/(M_{\odot}{\rm yr}^{-1}) \sim 10^{-11} L_{\rm H}/L_{\odot}$. Consequently, for typical post-AGB stars, in the range $\log L/L_{\odot} = 3...4$, the H-rich envelope is consumed by nuclear burning at $\dot{M}^{\rm burn}_{\rm env} \sim 1.4 \times 10^{-8}...2 \times 10^{-7}$. Consequently, and with the exception of the more massive and luminous model, winds affect the rate of consumption of the H-rich envelope by less than a 20%, see Table 3 of Miller Bertolami (2016), and do not play a significant role in the determination of $\tau_{\rm cross}$. The value of $\tau_{\rm cross}$ is consequently determined by the mass of the H-rich envelope at which the model departs from the AGB and the intensity of the H-burning shell. While these two properties are to some extent affected by the phase of the thermal pulse cycle at which the star departs from the TP-AGB, they are much more affected by the degeneracy level of the CO-core and intershell (see Blöcker 1995) as well as by the chemical composition of the H-rich envelope (see Tuchman *et al.* 1983 and Marigo *et al.* 1999). In turn these properties are mostly affected by the micro physics adopted in the models and the macro physics (winds and convective boundary mixing prescriptions) which affect the efficiency of third dredge up as well as the length of the TP-AGB phase and the initial-final mass relation. It is worth emphasizing that convective boundary mixing the main sequence also affects the final post-AGB timescales, as it has an important impact in the initial final mass relation, see Salaris *et al.* (2009) and also Wagstaff *et al.* in preparation.

Additionally, due to the fast evolution from $T_{\rm eff} = 7\,000$ K to 10 000K (see Fig. 6) different choices of the zero point (like those of Vassiliadis & Wood 1994, Blöcker 1995 and Miller Bertolami 2016) have a negligible impact in the actual value of $\tau_{\rm cross}$. This, together with the previously discussed points, link all important uncertainties and differences in the computations of $\tau_{\rm cross}$ (see Fig. 1) directly to the modeling of previous evolutionary stages.

2.2. *The transition time:* $\tau_{\rm trans}$

At variance with what happens with $\tau_{\rm cross}$ the value of the transition time $\tau_{\rm trans}$ is directly affected by the intensity of stellar winds during this phase. As can be seen

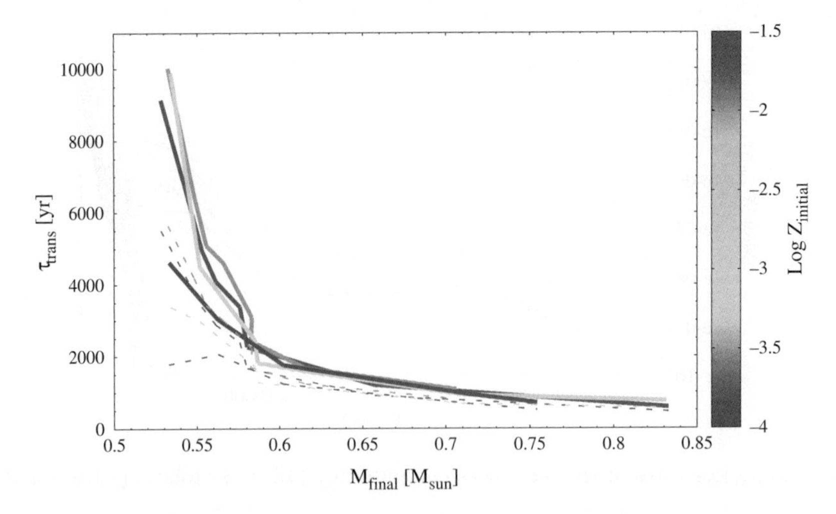

Figure 6. Solid lines: Transition times $\tau_{\rm trans}$ from the end of the AGB defined at $M_{\rm env} = 0.01 M_\star$ to the point at $\log T_{\rm eff} = 3.85$ during the post-AGB evolution for sequences of different final masses and metallicities Miller Bertolami (2016). Dashed lines: Same as the solid lines but with the end of the AGB defined as the point at $M_{\rm env} = 0.007\ M_\star$.

in Fig. 5 mass-loss rates are well above the threshold of $10^{-8}...10^{-7} M_\odot/{\rm yr}$ and the evolution is then dominated by the intensity of stellar winds. To make things worst, winds during this transition phase are completely uncertain. Also, attempts to measure the evolutionary speed of these objects by means of the study of the period drift in PPNe are not still possible (see Hrnivak *et al.*, these proceedings). For example, Miller Bertolami (2016) adopted during this stage mostly the wind prescription for cold giant winds by Schröder & Cuntz (2005) which have only been validated for much cooler stars $T_{\rm eff} \lesssim 4500$ K (see Schröder & Cuntz 2007), so inaccuracies of a factor of a few are not unthinkable. Note that, as $\tau_{\rm trans}$ is basically determined by speed of the reduction of the remaining H-rich envelope by winds, any error in the wind intensity in this transition regime (4000 K $\lesssim T_{\rm eff} \lesssim$ 7000 K) will directly translate into errors in the computed value of $\tau_{\rm trans}$.

In addition to our current lack of knowledge of winds during this early post-AGB phase the arbitrariness in the definition of the end of the AGB (an thus of the beginning of this early stage) directly affects the value of $\tau_{\rm trans}$. Fig. 6 shows that the value of $\tau_{\rm trans}$ would change by a factor of ~ 2 if the end of the AGB would have been defined as $M_{env} = 0.007 M_\star$ (dashed lines) as compared with the $M_{env} = 0.01 M_\star$ adopted by Miller Bertolami (2016). In this connection it is worth recalling the suggestion by Soker (2008) of quantitative definition for the end of the TP-AGB based on the ratio Q of the dynamical and envelope timescales of the star. Fig. 3 shows a simple estimation of the point at which Q reaches its maximum value (from Eq. 6 in Soker 2008 and under the assumption of $\beta = 1$). Fig. 3 suggests that the criterion proposed by Soker (2008) might indeed be able to capture key aspect of the transition from the AGB to the post-AGB, as it defines the end of the AGB close to the point where the fast post-AGB phase starts. It remains to be seen to which extent it agrees with the observational definitions of the post-AGB phase, but it certainly deserves further examination.

In view of the previous discussion, the values of $\tau_{\rm trans}$ are only qualitatively useful. In particular, an interesting result from Fig. 6 is that $\tau_{\rm trans}$ changes by more than an order of magnitude when going from $M_f \sim 0.5 M_\odot$ to $M_f \gtrsim 0.7 M_\odot$. Note, in particular, that for $M_f \gtrsim 0.6 M_\odot$ this stage lasts for $\tau_{\rm trans} \lesssim 2000$ yr for all metallicities.

3. Final comments

During the last 50 years our modeling of post-AGB stars has been slowly but continuously improved as better physics (both micro- and macro-physics) have been included in the modeling both of the post-AGB and previous evolutionary phases. In particular, it seems that with each current improvement post-AGB timescales became shorter. Current models that have been calibrated to reproduce several observables in the evolution of low- and intermediate-mass stars Miller Bertolami (2016) indicate that the time required to cross the HR-diagram from $T_{\rm eff} \sim 7000$ K to $T_{\rm eff} \gtrsim 7000$ K is of only $\tau_{\rm cross} \sim 10000$ yr for remnant stars of $M_f \sim 0.55 M_\odot$ of $\tau_{\rm cross} \lesssim 2000$ yr $M_f \gtrsim 0.58 M_\odot$ and less than a few hundred years for objects with $M_f \gtrsim 0.70 M_\odot$. The fast post-AGB evolution of this models helps to explain the observed existence of single CSPNe of low mass (e.g. Althaus *et al.* 2008, Henry *et al.* 2018 and Miller *et al.* 2018), as well as to understand the properties of CSPNe in the Galactic Bulge (Gesicki *et al.* 2014). In addition the faster evolution of current post-AGB models might be key to understand the mystery of the invariance of the planetary nebulae luminosity cut-off mystery (Gesicki *et al.* 2018) and, may be, the dearth of post-AGB stars in M32 (Brown *et al.* 2008). However, although the current models are able to reproduce several observables of AGB and post-AGB populations (Miller Bertolami 2016), some significant discrepancies are still present. The most important ones are the inability of the current models to reproduce the total lifetime of intermediate luminosity M-stars and C-stars at about the LMC luminosity (see Miller Bertolami 2016) and the systematically lower final masses of current models in the range $M_i \sim 2...3 M_\odot$ when compared with the latest determinations of the initial-final mass relation, see Fig. 2 (Casewell *et al.* 2009, Salaris *et al.* 2009, El-Badry *et al.* 2018, Cummings *et al.* 2018). A lower intensity of third dredge up processes and a lower intensity of the mass loss during the C-rich phase might help to solve both problems (Wagstaff *et al.* in preparation).

Still, the largest uncertainty in our current understanding of the post-AGB evolution in single stars concerns the intensity of winds during the departure from the AGB 4000K $\lesssim T_{\rm eff} \lesssim 7000$ K which strongly affects the evolutionary speed of the models during the transition stage ($\tau_{\rm trans}$).

Finally it should be mentioned that all post-AGB stellar evolutionary sequences are based on the assumption that the final ejection of the envelope occurs through steady winds. This leads to a well defined relationship between the mass of the remnant and the mass of the remaining H-rich envelope. The strong dependency of the critical mass of the envelope at which the stars depart from the AGB (Fig. 3) is key in the determination of the mass dependency of the crossing timescale ($\tau_{\rm cross}$, Fig. 1). If some objects eject their envelopes by means of a dynamical phase due to binary interaction or, for example, the ingestion of a substellar companion then the remnant might depart from the AGB with smaller envelope masses and our of thermal equilibrium (e.g. Hall *et al.* 2013), and evolve through the post-AGB phase at a much faster pace. In particular, this implies that any comparison of post-AGB stellar models like those computed by Vassiliadis & Wood (1994); Blöcker (1995); Miller Bertolami (2016) with CSPNe in close binary systems should be addressed with strong skepticism.

Acknowledgements

M3B thanks the IAU and the organizers for a travel grant and the waiving of the registration fee which allowed him to attend this exciting Symposium. M3B also acknowledges a travel grant from the Facultad de Ciencias Astronómicas y Geofísicas.

References

Abell, G. O., & Goldreich, P. 1966, *PASP*, 78, 232
Althaus, L. G., Córsico, A. H., Kepler, S. O., & Miller Bertolami, M. M. 2008, *A&A*, 478, 175
Blöcker, T., & Schönberner, D. 1990, *A&A*, 240, L11
Blöcker, T. 1995, *A&A*, 297, 727
Blöcker, T. 1995, *A&A*, 299, 755
Brown, T. M., Smith, E., Ferguson, H. C., *et al.* 2008, *ApJ*, 682, 319–335
Casewell, S. L., Dobbie, P. D., Napiwotzki, R., *et al.* 2009, *MNRAS*, 395, 1795
Cummings, J. D., Kalirai, J. S., Tremblay, P.-E., Ramirez-Ruiz, E., & Choi, J. 2018, arXiv:1809.01673
El-Badry, K., Rix, H.-W., & Weisz, D. R. 2018, *ApJ* (Letters), 860, L17
Faulkner, J. 2005, The Scientific Legacy of Fred Hoyle, 149
Gesicki, K., Zijlstra, A. A., Hajduk, M., & Szyszka, C. 2014, *A&A*, 566, A48
Gesicki, K., Zijlstra, A. A., & Miller Bertolami, M. M. 2018, *Nature Astronomy*, 2, 580
Henry, R. B. C., Stephenson, B. G., Miller Bertolami, M. M., Kwitter, K. B., & Balick, B. 2018, *MNRAS*, 473, 241
Hall, P. D., Tout, C. A., Izzard, R. G., & Keller, D. 2013, *MNRAS*, 435, 2048
Höfner, S., & Olofsson, H. 2018, *ARAA*, 26, 1
Iglesias, C. A., & Rogers, F. J. 1996, *ApJ*, 464, 943
Lagadec, E., Verhoelst, T., Mékarnia, D., *et al.* 2011, *MNRAS*, 417, 32
Kitsikis, A. 2008, Ph.D. Thesis, Ludwig-Maximilians-Universität München
Marigo, P., Girardi, L., Weiss, A., & Groenewegen, M. A. T. 1999, *A&A*, 351, 161
Marigo, P. 2002, *A&A*, 387, 507
Mendez, R. H. 2016, arXiv:1610.08625
Miller, T. R., Henry, R. B. C., Balick, B., *et al.* 2018, arXiv:1809.04002
Miller Bertolami, M. M. 2016, *A&A*, 588, A25
Miller Bertolami, M. M. 2015, in: Dufour, P. and Bergeron, P. & Fontaine, G. (eds.), *19th European Workshop on White Dwarfs*, *ASP-CS* 493, p. 83
Paczyński, B. 1970, *AcA*, 20, 47
Paczyński, B. 1971, *AcA*, 21, 417
Reimers, D. 1975, Problems in stellar atmospheres and envelopes., 229
Rogers, F. J., & Iglesias, C. A. 1992, *ApJ* (Supplement Series), 79, 507
Salaris, M., Serenelli, A., Weiss, A., & Miller Bertolami, M. 2009, *ApJ*, 692, 1013
Seaton, M. J., Yan, Y., Mihalas, D., & Pradhan, A. K. 1994, *MNRAS*, 266, 805
Schaller, G., Schaerer, D., Meynet, G., & Maeder, A. 1992, *A&ASS*, 96, 269
Schönberner, D. 1979, *A&A*, 79, 108
Schönberner, D. 1981, *A&A*, 103, 119
Schönberner, D. 1983, *ApJ*, 272, 708
Schönberner, D. 1987, IAU Colloq. 95: Second Conference on Faint Blue Stars, 201
Schönberner, D. 1987, in: Kwok, S. & Pottasch, S. R., *Late Stages of Stellar Evolution, Dordrecht, D. Reidel Publishing Co.*, p. 337
Schönberner, D., Jacob, R., Steffen, M., & Sandin, C. 2007, *A&A*, 473, 467
Schröder, K.-P., & Cuntz, M. 2005, *ApJ* (Letters), 630, L73
Schröder, K.-P., & Cuntz, M. 2007, *A&A*, 465, 593
Shklovsky, I. S. 1957, in: G. H. Herbig (ed.), *Non-stable stars, proceedings from IAU Symposium no. 3, Cambridge University Press,*, p. 83
Soker, N. 2008, *ApJ* (Letters), 674, L49
Steffen, M., Schönberner, D., & Warmuth, A. 2008, *A&A*, 489, 173
Suárez, O., García-Lario, P., Manchado, A., *et al.* 2006, *A&A*, 458, 173
Szczerba, R., Siódmiak, N., Stasińska, G., & Borkowski, J. 2007, *A&A*, 469, 799
Toalá, J. A., & Arthur, S. J. 2014, *MNRAS*, 443, 3486
Tuchman, Y., Glasner, A., & Barkat, Z. 1983, *ApJ*, 268, 356
van Winckel, H. 2003, *ARAA*, 41, 391
Vassiliadis, E., & Wood, P. R. 1993, *ApJ*, 413, 641

Vassiliadis, E., & Wood, P. R. 1994, *ApJ* (Supplement Series), 92, 125
Weidemann, V. 1987, *A&A*, 188, 74
Weidemann, V. 2000, *A&A*, 363, 647
Weiss, A., & Ferguson, J. W. 2009, *A&A*, 508, 1343
Wood, P. R., & Faulkner, D. J. 1986, *ApJ*, 307, 659

Discussion

D'ANTONA: From your very complete presentation, do I understand correctly that, in spite of the difficulties with the determination of the transition time, massive planetary nebulae nuclei cannot be found in the high luminosity crossing phase, like in the old Paczynski models?

MILLER BERTOLAMI: Yes, indeed. According to my models, remnants with masses between 0.7 and 0.8 $M_\odot$ (the largest ones I computed) cross the HR diagram in only ~ 100 to ~ 10 yr and then start to decrease their luminosity towards the white dwarf phase. So they should be very rare. In addition, one might wonder whether they would not be still highly obscured by circumstellar material.

VENTURA: Which are the typical timesteps and mass-loss rates which you use during the transition time?

MILLER BERTOLAMI: During the transition phase I forced the code not to use the extremely small timesteps our algorithm would naturally suggest (which might be as small as 10^{-4} yr). So I usually forced timesteps to be between 0.1 and 1 yr during the transition phase.

New and future observational perspectives

Why Galaxies Care About AGB Stars:
A Continuing Challenge through Cosmic Time
Proceedings IAU Symposium No. 343, 2019
F. Kerschbaum, M. Groenewegen & H. Olofsson, eds.

doi:10.1017/S1743921318007032

Resolved stellar populations: the outlook for JWST and ELT

Eline Tolstoy

Kapteyn Astronomical Institute, University of Groningen,
PO Box 800, 9700 AV Groningen, The Netherlands
email: etolstoy@astro.rug.nl

Abstract. The study of the resolved stellar populations in nearby galaxies and star clusters through the analysis of colour-magnitude diagrams provides the most detailed and quantitative determination of the star formation histories of these systems. The properties of different age populations provide an insight into distinct physical processes taking place during the entire history of the stellar system. The detection of the oldest main sequence turn-offs is currently restricted to stellar systems within the Local Group due to the limitations in spatial resolution and flux sensitivity of available telescopes. Individual stars need to be detected and accurately distinguished from their neighbours. To improve this situation we need to build new telescopes with larger primary mirrors that can deliver a very stable image quality at the diffraction limit. Over the next decade we can look forward to new larger telescope in space: the *James Webb Space Telescope*, currently scheduled to be launched in 2021; and several large telescope projects, the largest of which is the 39m ESO extremely large telescope on Cerro Armazones in Chile, currently scheduled to start operations in 2024.

Keywords. stars: Hertzsprung-Russell diagram

1. Introduction

The detailed study of resolved stellar populations has a long history. One of the first things that Galileo pointed his telescope at in 1609 was the crowded stellar fields that make up the Milky Way. He showed for the first time that the diffuse light seen with the naked eye is made up of individual stars. In 1785 Caroline and William Herschel made the first estimate of the distribution of stars in the entire Milky Way, using the laborious approach of counting stars by eye. The advent of astronomical photography in the middle of the 19th century made it possible to record images and spectra in a repeatable and consistent way and keep them for later analyses with the facilities and comfort available in a laboratory. At the beginning of the 20th century photographic plates advanced our understanding of both the luminosity distribution and variations in stellar properties in and around the Milky Way, thus providing a dramatic revelation of the size of the Universe and our place in it.

Another major development took place in the 1980s: the advent of low-noise large-format CCD cameras which, unlike photographic plates, can have a linear response to incident light, maintained down to extremely low light levels. This improved the accuracy and repeatability of stellar photometry and allowed much deeper images to be made. An additional important development was adaptive optics, the ability to actively stabilise and dramatically improve image quality, by correcting the distortions coming from the passage of light through the Earth's atmosphere. This has lead to increasingly spectacular images (and resolved spectral data cubes) from ground-based telescopes. Of course the

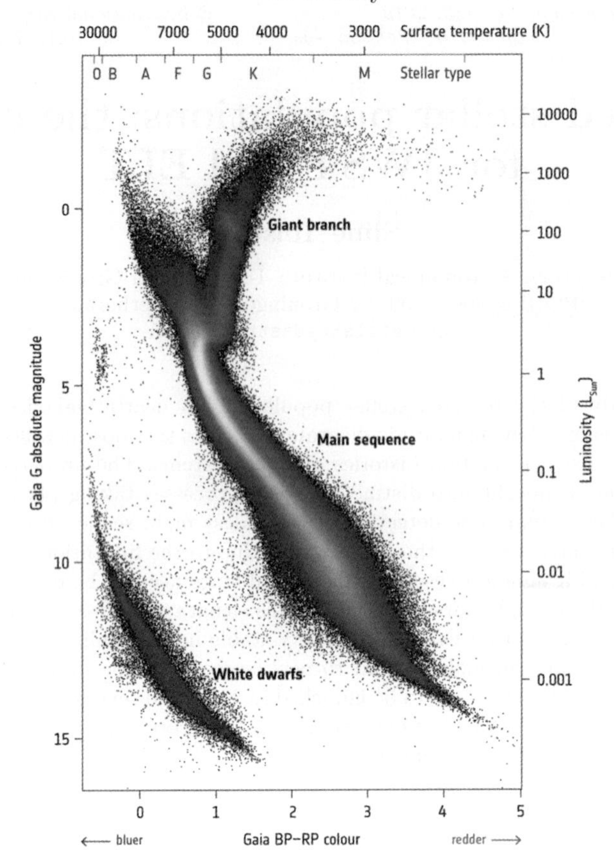

Figure 1. Gaia DR2 Hertzsprung-Russell Diagram of 4 276 690 sources with low extinction. The colour scale represents the square root of the density of stars. Approximate temperature and luminosity equivalents are provided on the top and right axes, to guide the eye. From Gaia Collaboration and Babusiaux *et al.* (2018).

most direct (if not the cheapest) way to remove the effects of the Earth's atmosphere is to move above it, as was shown, by sending a 2.4m diameter optical telescope into space in 1990: the NASA/ESA *Hubble Space Telescope* (HST), equipped with imagers and spectrographs, and for many years it could be refurbished and modernised by NASA/ESA astronauts.

A dramatic recent development is coming from the ESA/Gaia astrometric satellite. It is carrying out a ground-breaking all-sky survey of accurate magnitudes, colours, proper motions and parallaxes of more than 1.7 billion individual stars in and around the Milky Way, extending well out into the stellar halo, and including many nearby dwarf galaxies (Gaia Collaboration and Brown *et al.* 2018). The early results have already started to overturn what we thought we understood about our own Galaxy, and also the small dwarf galaxies orbiting around it. One of the most impressive results coming from the latest data release (DR2), that took place in April 2018, is the most detailed and extensive colour-magnitude diagram (CMD) ever made of the Milky Way (Gaia Collaboration and Babusiaux *et al.* 2018). The accurate distances provided by Gaia parallaxes allow every star to have correct luminosity, leading to the exquisitely detailed CMD shown in Fig. 1.

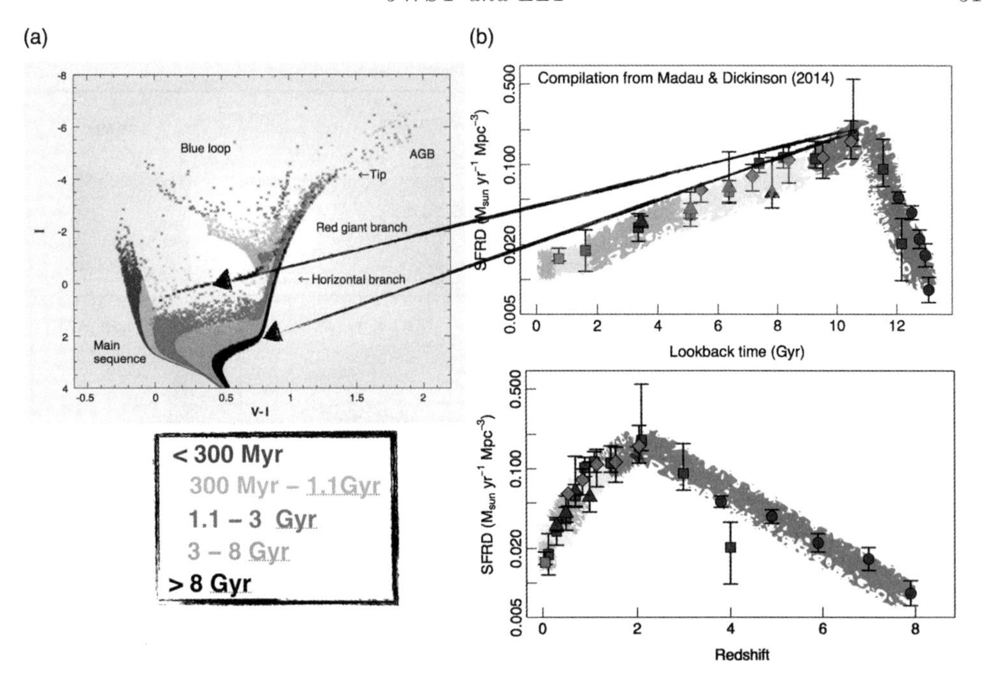

Figure 2. (a) a synthetic CMD for an artificial galaxy with a constant star-formation rate over time and a constant metallicity. The different stellar evolution phases are labelled. The age ranges of these different features are colour coded, as shown below the CMD. The oldest stars are in black and are to be found on the faintest main sequence turnoffs and on the Horizontal Branch, as indicated by black arrows. These stars have formed at look-back times of $\sim$8 Gyr and more (from Tolstoy 2011); (b) the corresponding regions in terms of the evolution of the cosmological star formation rate density (SFRD) are shown, using data taken from Madau & Dickinson (2014). The colour coding is that same as in the CMD. The SFRD is plotted both in terms of redshift and look-back time.

2. Star Formation Histories

Determining the star formation history (SFH) of a stellar system based on CMDs could begin once it was established how the distribution of stars related to the different phases of stellar evolution. The first studies were based on looking at the overall characteristics of the CMD and separating the different populations into two classes: Population I (young stars) and Population II (old stars). This was first applied to two newly discovered dwarf spheroidal galaxies (Sculptor and Fornax) by D. Baade in 1944. This terminology clearly became too simple, as it was realised that stellar populations in galaxies were always more complex than this, as ever deeper and more accurate CMDs were made, and the models of stellar evolution used to interpret them also became more sophisticated (see Fig. 2a).

In the 1990s for the first time model CMDs were created to compare directly to observed CMDs to obtain accurate and well motivated SFHs (e.g. Tosi *et al.* 1991; Greggio *et al.* 1993; Gallart *et al.* 1996; Tolstoy & Saha 1996; Dolphin 1997; Tolstoy *et al.* 1998). Fig. 2a shows an artificial CMD and the colour code shows how different age stellar populations distribute themselves in a CMD of constant star formation rate with time. In principle the interpretation is the straight forward counting of the number of stars in different parts of the diagram corresponding to the different ages. Of course reality isn't

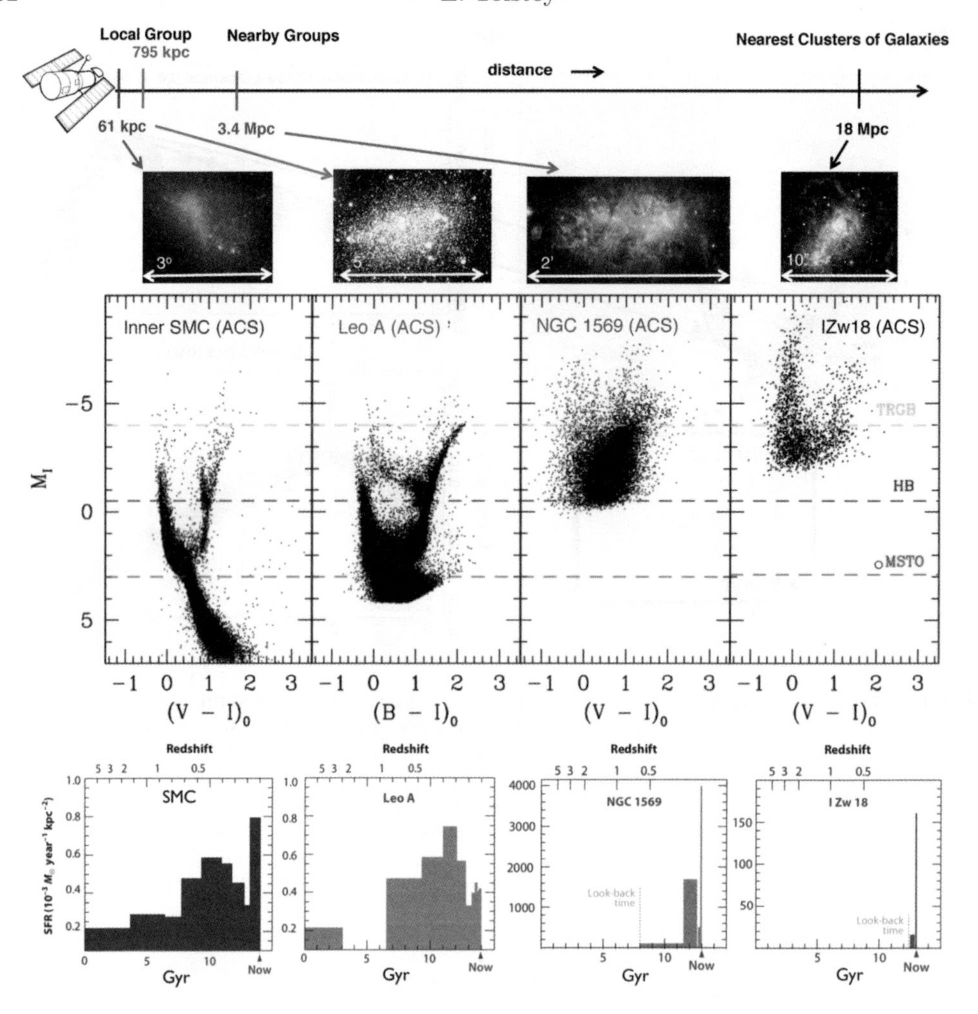

Figure 3. A collection of HST/ACS CMDs and below them the corresponding SFHs for 4 dwarf galaxies at increasing distance from us. The dashed lines show the position of the oldest MSTO, the Horizontal Branch (HB) and the Tip of the Red Giant Branch (TRGB). The photometry and SFH analysis are taken from Cignoni *et al.* 2012 (SMC); Cole *et al.* 2007 (Leo A); Grocholski *et al.* 2012 (NGC 1569) and Aloisi *et al.* 2007 (I Zw 18).

quite this simple, and issues like photometric errors and completeness need to be considered. In addition stellar models have been more reliable in some regions of the CMD than in others. Both the Horizontal Branch (HB) and the Asymptotic Giant Branch (AGB) have proved challenging to model in such away that they can be securely tied to a star formation rate with time. But this is changing rapidly with ever better models.

The most secure way to determine the age of the full range of different populations in a galaxy is to accurately measure the number of stars along the main sequence turnoff (MSTO) region, where the different age populations are clearly well separated by age, as seen in Fig. 2a. However, the oldest turnoffs are very faint, and they get thinner and thinner in luminosity range. It is thus only within the Local Group that it is possible (with often still a large amount of telescope time) to resolve stars going all the way down to the oldest MSTOs, as seen in Fig. 3.

Fig. 3 shows some examples of HST CMDs, all the result of significant amounts of HST/ACS time, and the resulting SFHs coming from them, and how the detail that is possible decreases with increasing distance, as the older MSTOs are not detected. These are among the best CMDs that have come from HST at each of the distances and so this reflects current absolute limits. Thus Fig. 3 shows both the strengths of diffraction limited imaging in the Local Group (for the SMC at $\sim$ 60 kpc and even Leo A at nearly 800 kpc) and the limitations of a small telescope like HST in being able to extend the same detail beyond the Local Group (to NGC 1569 at 3.4 Mpc or I Zw 18 at 18 Mpc).

2.1. *Ancient Main Sequence Turnoffs*

There a number of excellent and detailed studies of large samples of (mostly) dwarf galaxies, determining star formation histories (SFHs) back to the oldest times (e.g. Cole *et al.* 2007; Monelli *et al.* 2010a,b; de Boer *et al.* 2012, 2015; Cignoni *et al.* 2012; Brown *et al.* 2014; Weisz *et al.* 2014; Gallart *et al.* 2015; Skillman *et al.* 2017; Rubele *et al.* 2018; see also Fig. 3). These are primarily with the HST, but for some very nearby galaxies, such as the Magellanic Clouds and dwarf spheroidal galaxies in the halo of the Milky Way, ground-based wide-field imagers with excellent image quality give a more complete overview. These are often complemented by HST zoom-in fields for more precise details of the most ancient populations.

The details with which resolved SFHs can be determined for large numbers of individual galaxies is making it possible to accurately tie the Local Group SFH to the cosmological star formation rate density (SFRD) that is measured in redshift surveys (e.g. Boylan-Kolchin *et al.* 2015) going up to very high redshifts (which correspond to ancient look-back times, see Fig. 2b). This is an important connection, especially for the study of the role of the most numerous low mass dwarf galaxies that redshift surveys will always struggle to detect.

2.2. *The Horizontal Branch & the Asymptotic Giant Branch*

One clear advantage for extending accurate SFHs beyond the Local Group is to be able to look at more luminous features in a CMD. For example the HB to trace ancient populations, and AGB stars to trace intermediate age populations.

More luminous stars have the obvious advantage that they can be detected at greater distances more easily. However they have the disadvantage that their direct relation to SFHs has often been rather rough. This is because both AGB and HB star evolution depend on uncertain and possibly hard to predict mass-loss events before and/or during their evolution. They are late-stages of stellar evolution, and thus are dependent on an understanding of all the stellar evolution processes that proceeded them, each with their own uncertainties, as John Lattanzio reminded us at this meeting.

HB stars offer a uniquely luminous insight into the most ancient stellar populations in a galaxy, and it has recently been shown that they can offer a more accurate look at how star formation rates and metallicity varied at the earliest times (Salaris *et al.* 2013; Savino *et al.* 2015, 2018). It is clear that the HBs in nearby galaxies all look different (see Fig. 4), revealing a variety of early SFHs even in the smallest dwarf galaxies, even more so than had previously been envisaged (Savino *et al.*, *in prep*).

At this conference it has been shown that our understanding of AGB stars is rapidly developing and they are important constituents of any intermediate age stellar population. Their very high luminosities in the infrared mean that they are clearly critical contributors to the total infrared flux of such galaxies.

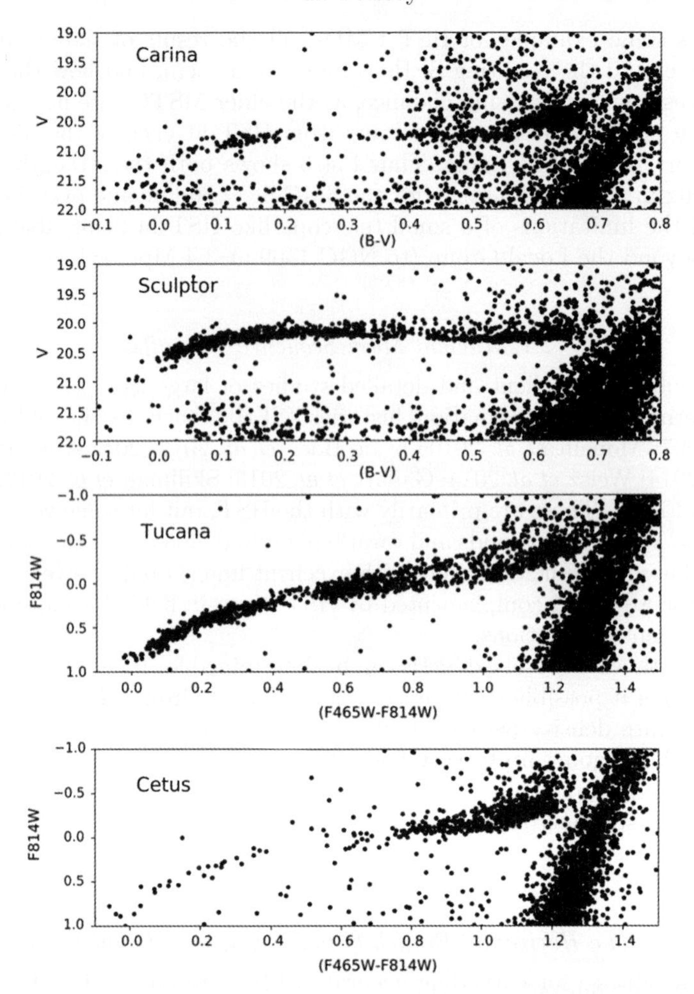

Figure 4. A compilation of the Horizontal Branches of 4 Local Group dwarf spheroidal galaxies. These data come from CTIO/MOSAIC: Carina (Bono *et al.* 2010); Sculptor (de Boer *et al.* 2012); and HST: Tucana (Monelli *et al.* 2010b); Cetus (Monelli *et al.* 2010a). Compilation by A. Savino.

3. Gaia: Proper Motions

Gaia allows uniquely accurate measurements of the positions and motions of resolved stellar populations in real time (Gaia collaboration and Helmi *et al.* 2018). The combination of Gaia's exquisite positions and the stability of HST gives a solid time baseline of many years and means we have the ability to determine both very accurate orbits of dwarf galaxies and globular clusters and also the internal proper motions of stars in these systems (e.g. Massari *et al.* 2018), see Fig. 5. This information is going to continue to increase in accuracy and sophistication over the next series of Gaia data releases.

This detailed information can be combined with metallicity and also SFHs to build up an accurate chemo-dynamical picture of the Milky Way and also some nearby galaxies and globular clusters. It also allows much more detailed CMDs as the proper motions provide a useful method to remove foreground stars, as they will move like the halo or disk of the Milky, and not like member stars in these systems (see Gaia collaboration and Helmi *et al.* 2018). This is clearly a fundamental issue in the study of ultra faint dwarf galaxies (e.g. Simon 2018; Fritz *et al.* 2018; Massari & Helmi 2018).

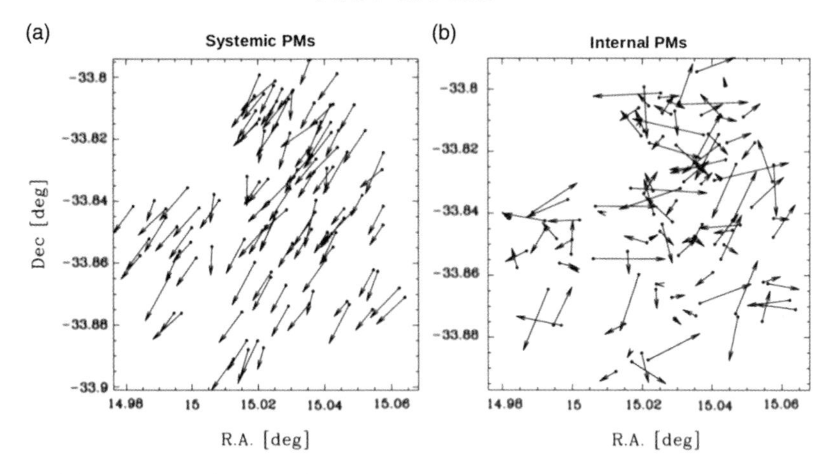

Figure 5. (a) the global (systemic) proper motion of the Sculptor dwarf spheroidal galaxy in an HST field of view, which shows ordered arrows, all of them pointing on average in the direction of the systemic motion; (b) the internal proper motions, after the subtraction of the absolute motion, magnified by a factor of 40. These show the disordered internal motions in the same system. From Massari *et al.* (2018).

4. Future Facilities: JWST & ELT

Making deep and accurate CMDs, tracing ancient populations, will be possible for more distant systems, extending beyond the Local Group, with the advent of future large more sensitive telescopes like the *James Webb Space Telescope* (JWST) in space and MICADO on the ESO Extremely Large Telescope (ELT). These studies benefit tremendously from the combination of extra flux sensitivity (allowing fainter stars to be detected) and spatial resolution coming from a larger primary mirror, allowing these faint stars to be separated and accurately measured in more crowded conditions that naturally arise as we look out to greater distances and into more compact regions, such as Elliptical galaxies and Bulges of Spiral galaxies.

As part of the ELT/MICADO project an image simulator was developed, called SimCADO (Leschinski *et al.* 2016). This simulator takes account of the properties of the source, the atmosphere, the telescope, the instrument and the detector to simulate a realistic MICADO image. This can be used to understand the crowding properties of dense stellar systems, for example, see Fig. 6. At the top a series of simulations are shown, using the dense Galactic globular cluster M4 as a template, which is placed at a range of different distances (from 200 kpc to 2 Mpc). This shows how the crowding increases as the spatial scale becomes increasingly compressed. The extraordinary resolving power of MICADO is critically important to be able to detect individual stars in the heart of Elliptical galaxies (see also Deep *et al.* 2011). This is shown in the lower part of Fig. 6, where the synergy between JWST/NIRCAM and ELT/MICADO is made clear. NIRCAM will have a larger field of view, but also larger pixels than MICADO. This is because JWST is a much smaller telescope than the ELT, and so the diffraction limit will result in a poorer spatial resolution. MICADO has a 6 times better spatial resolution than JWST, which makes a larger field of view very expensive in pixels of the size that properly sample the diffraction limit. The number of stars in the MICADO field is likely to be comparable if not much greater than in the JWST field for a galaxy at the distance of the Virgo cluster.

The superior angular resolution of MICADO images makes crowded field photometry and astrometric applications especially attractive. Exciting opportunities include mapping individual stellar orbits in external stellar systems and stellar orbits and flaring

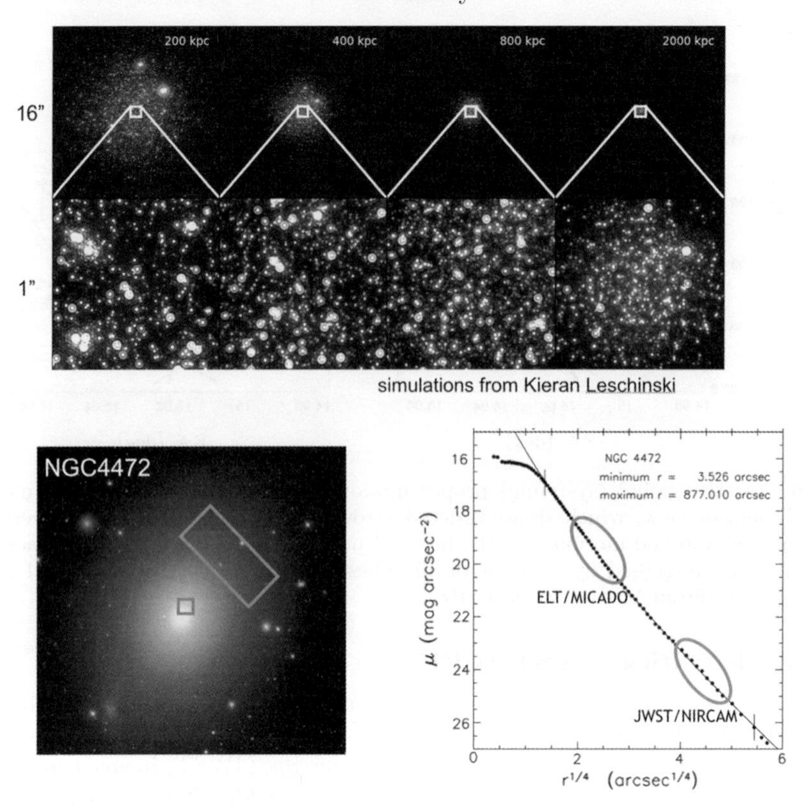

Figure 6. On top are the simulations made by SimCADO of what would be expected from an M4-like globular cluster observed at a range of different distances within the central CCD chip of MICADO, and the zoom-in of the central 1 arcsec. These simulations were carried out by Kieran Leschinski. Below is an image of NGC 4472 with the rough footprint of JWST-NIRCAM (in blue) and ELT-MICADO (in red), and next to this image we show the stellar surface brightness (or stellar density) range that these positions correspond to, which are close to the maximum that these instruments can hope to achieve.

gas fainter and closer to the event horizon of the central black hole in the Galaxy than ever before with direct imaging. Astrometry will be feasible within the central regions of globular clusters over a range of distances and can be extended to monitor black holes in other galaxies. Proper motions of numerous individual stars in resolved stellar systems will allow a more active view of how stellar systems move and change. This will, amongst many other things, enable us to trace the presence (or absence) of black holes in a range of environments, as well as making accurate mass models by combining these results with radial velocity measurements to obtain a 3D view of stellar motions. We will be able to map the dark matter distribution in a variety of environments and for a range of spatial and temporal scales, and set constraints on the physical nature of dark matter particles.

One of the major issues for most people working on interpreting the whole CMD in terms of a SFH is that this is best done in the optical, which better matches the peak of the spectral energy distribution of most stars. However both JWST and ELT will work best in the infra-red, and may not be able to produce anything comparable in the optical. Thus we have to get used to looking infra-red CMDs and adapting our techniques to this. Certain populations stand out very well in the infra-red, and AGB stars are a prime example of this. However main sequence stars and even red giant branch stars typically fall into a smaller region of an infrared CMD, compared to an optical CMD. This requires more accurate photometry to extract the same information as in an optical CMD.

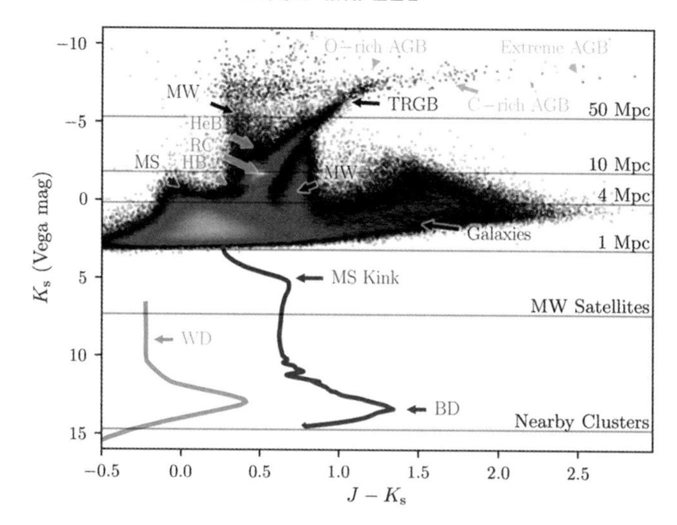

Figure 7. The underlying black open circles are the photometry from the VISTA Magellanic Clouds (VMC) infrared survey of a field in the Large Magellanic Cloud (from Cioni *et al.* 2011). This CMD includes foreground Milky Way stars and also faint background galaxies misidentified as stars. Over plotted are the expected sensitivity limits for similar observations with JWST at a range of different distances. This figure was made by D. Weisz.

In Fig. 7 we show an example of a ground based infrared CMD of the Large Magellanic Cloud, including foreground stellar contamination from the Milky Way and also faint background galaxies that are misidentified as stars. The horizontal lines compare the depths that can be reached with JWST for similar resolved stellar systems at different distances. It is clear that JWST will be a very powerful instrument pushing our understanding of resolved stellar populations far beyond the Local Group, but this will mostly be on the basis of the brighter components of the CMD. As stated above, this is also only possible where the images are not overly crowded. ELT/MICADO should achieve a comparable sensitivity in the best conditions, but with the ability to go into higher surface brightness regions, like the central regions of Elliptical galaxies (see Fig. 6).

So in summary ELT/MICADO and NIRCAM will work over the same infrared wavelength range, but MICADO will be able to resolve structures that NIRCAM cannot. In crowded fields, the MICADO resolution gives an effective sensitivity gain of $\sim$3 magnitudes with respect to NIRCAM, allowing MICADO to probe regions where JWST cannot reach. ELT/MICADO will also achieve astrometric measurements $\sim$6 times faster, or for objects $\sim$6 times more distant, than JWST/NIRCAM (Davies *et al.* 2018). However, as shown by Fig. 6, there will be great complementary between the two facilities and together they will provide us with, among many other things, the first detailed picture of the resolved stellar populations in an Elliptical galaxy.

References

Aloisi, A., *et al.* 2007 *ApJ* (Letters), 667, L151
Gaia Collaboration and Babusiaux, C., *et al.* 2018, *A&A*, 616, A10
Boylan-Kolchin, M., *et al.* 2015, *MNRAS*, 453, 1503
Brown, T.M. *et al.* 2014, *ApJ*, 796, 91
Gaia Collaboration and Brown A.G.A *et al.* 2018, *A&A*, 616, A1
Bono G. *et al.* 2010, *PASP*, 122, 651
Cignoni, M. *et al.* 2012, *ApJ*, 754, 130
Cioni, M.R. *et al.* 2011, *A&A*, 527, A116
Cole, A.A., *et al.* 2007, *ApJ* (Letters), 659, L17

Davies, R., *et al.* 2018, *SPIE*, 10702, 1
Dolphin, A. 1997, *NewA*, 2, 397
de Boer, T.J.L., *et al.* 2012, *A&A*, 539, A103
de Boer, T.J.L, Belokurov, V. & Koposov, S. 2015, *MNRAS*, 451, 3489
Deep, A., *et al.* 2011, *A&A*, 531, A151
Fritz, T.K., *et al.* 2018, *A&A*, in press [arXiv:1805.00908]
Gallart, C., Aparicio, A., Bertelli, G., & Chiosi, C. 1996, *AJ*, 112, 1950
Gallart, C. *et al.* 2015, *ApJ* (Letters), 811, L18
Greggio, L., Marconi, G., Tosi, M., & Focardi, P. 1993, *AJ*, 105, 894
Grocholski, A.J. *et al.* 2012 *AJ*, 143, 117
Gaia Collaboration and Helmi, A., *et al.* 2018, *A&A*, 616, A12
Leschinski, K. *et al.* 2016, *SPIE*, 9911, 24
Madau, P. & Dickinson, M. 2014, *ARAA*, 52, 415
Massari, D. *et al.* 2018, *Nat. Astron.*, 2, 156
Massari, D. & Helmi, A. 2018, *A&A*, in press [arXiv:1805.01839]
Monelli, M., *et al.* 2010a, *ApJ*, 720, 1225
Monelli, M., *et al.* 2010b, *ApJ*, 722, 1864
Rubele, S., *et al.* 2018, *MNRAS*, 478, 5017
Salaris, M., *et al.* 2013, *A&A*, 559, A57
Savino, A., Salaris, M. & Tolstoy, E. 2015, *A&A*, 583, A126
Savino, A., *et al.* 2018, *MNRAS*, 480, 1587
Simon, J.D. 2018, *ApJ*, 863, 89
Skillman, E.D. *et al.* 2017, *ApJ*, 837, 102
Tosi, M., Greggio, L., Marconi, G., & Focardi, P. 1991, *AJ*, 102, 951
Tolstoy, E., & Saha, A. 1996, *ApJ*, 462, 672
Tolstoy, E., Gallagher, J. S., & Cole, A. A., *et al.* 1998, *AJ*, 116, 1244
Tolstoy, E. 2011, *Science*, 33, 176
Weisz, D.R. *et al.* 2014, *ApJ*, 789, 147

Discussion

GENNARO: What do you think the impact of loss of UV capabilities will be in the next decade on studies of stellar populations?

TOLSTOY: We will certainly suffer from it, but we should stay positive and try to learn as much as we can from the new IR capabilities.

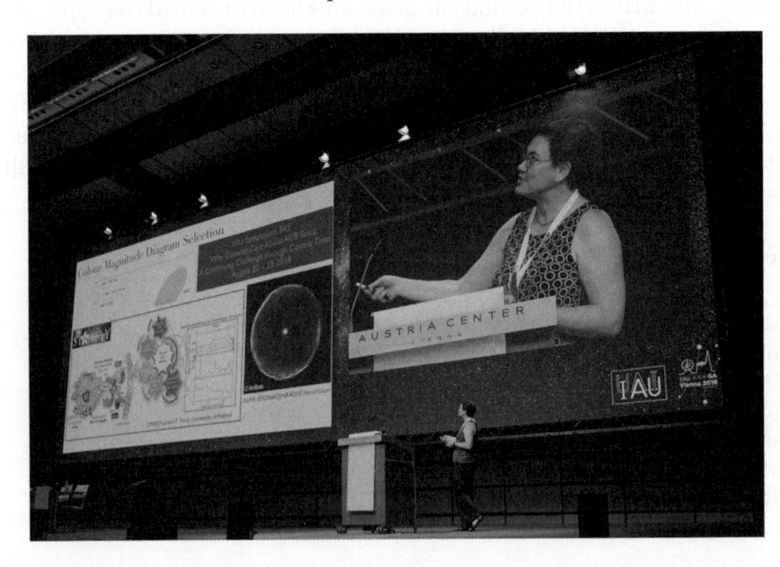

Why Galaxies Care About AGB Stars:
A Continuing Challenge through Cosmic Time
Proceedings IAU Symposium No. 343, 2019
F. Kerschbaum, M. Groenewegen & H. Olofsson, eds.

doi:10.1017/S1743921318005690

LSST: making movies of AGB stars

Željko Ivezić, Krzysztof Suberlak and Owen M. Boberg

Department of Astronomy, University of Washington,
Box 351580, Seattle, WA 98195-1580, USA
email: ivezic@astro.washington.edu

Abstract. LSST (www.lsst.org) will be a large, wide-field ground-based system designed to obtain repeated images covering the sky visible from Cerro Pachón in northern Chile. The telescope will have an 8.4m (6.5m effective) primary mirror, a 9.6 sq.deg. field of view, and a 3.2 Gigapixel camera. In a continuous observing campaign, LSST will cover the entire observable sky every three nights to a depth of $V \sim 25$ per visit (using 30-second exposures and *ugrizy* filter set), with exquisitely accurate astrometry and photometry. Close to a half of the sky will be visited about 800 times during the nominal 10-year survey. The project is in the construction phase with first light expected in 2020 and the beginning of regular survey operations by 2022. We describe how these data will impact AGB star research and discuss how the system could be further optimized by utilizing narrow-band TiO and CN filters.

Keyword. surveys

---

1. Introduction

The progenitors of stars on the asymptotic giant branch (AGB) are red giants; their progeny are planetary nebulae and white dwarfs. AGB stars have a strong impact on the galactic environment; stellar winds blown during this evolutionary phase are an important component of mass return into the interstellar medium and may account for a significant fraction of interstellar dust (for a recent review, see Höfner & Olofsson 2018). These dusty winds reprocess the stellar radiation, shifting the spectral shape towards the IR. Dusty AGB stars are observable in our Galaxy and in its close satellites, most notably the Magellanic Clouds. In addition to its obvious significance for the theory of stellar evolution, the study of AGB winds has important implications for the structure and evolution of galaxies (Girardi & Marigo 2007; Marston *et al.* 2009). In our own Galaxy, its estimated 200,000 AGB stars are good tracers of dominant components, including the bulge (Whitelock & Feast 2000; Jackson, Ivezić & Knapp 2002). AGB stars also have a great potential as distance indicators (Rejkuba 2004).

The last decade has seen fascinating observational progress in optical and infrared imaging surveys. Modern large sky surveys are having a major impact on AGB star research: well-defined statistical samples allow the use of AGB stars as stellar population tracers both in our and other galaxies. In the infrared, the recent Wide-field Infrared Survey Explorer All-sky Survey (WISE; launched in 2010) improved sensitivity by up to three orders of magnitude compared to IRAS, and detected about 560 million objects (Wright *et al.* 2010). In the optical, the "gold standard" SDSS dataset is currently being greatly extended by ongoing ground-based surveys such as Pan-STARRS (Kaiser *et al.* 2010) and the Dark Energy Survey (Flaugher 2008), and most recently by Gaia's Data Release 2.

Optical variability greatly helps with the identification of AGB stars, as well as providing constraints on the physics of AGB phenomenon (via light curve shapes, amplitudes

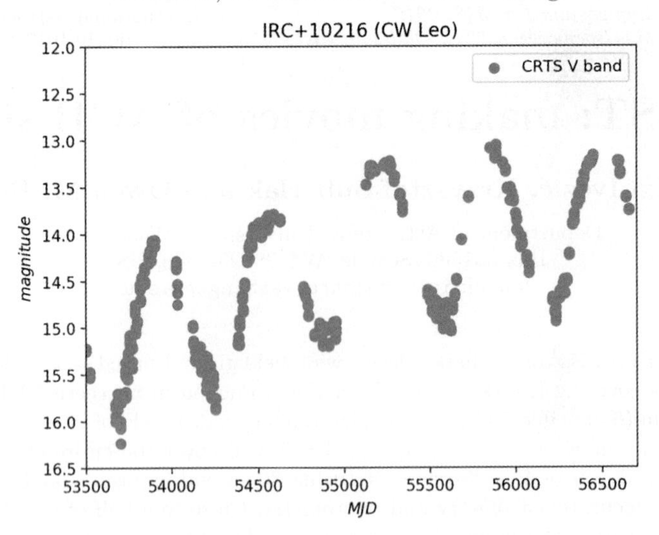

Figure 1. The visual light curve for the carbon-rich AGB star CW Leo (IRC +10 216). It was serendipitously observed by the Catalina Real-time Transient Survey (CRTS, see http://crts.caltech.edu) over 700 times during a period of close to 10 years. LSST will obtain similar light curves in six photometric bandpasses, with comparable signal-to-noise ratios, for close to 100,000 AGB stars.

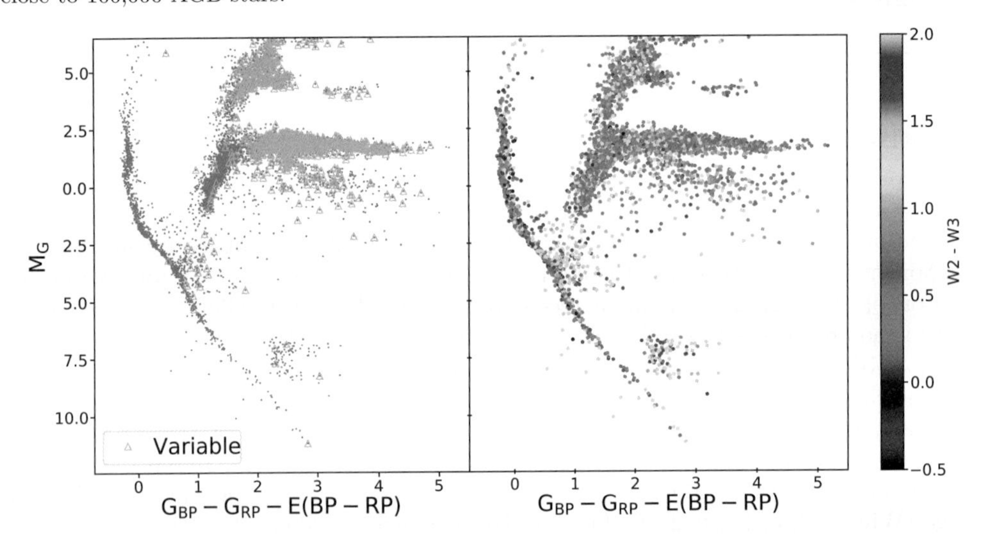

Figure 2. An illustration of the power of combining multiple unbiased surveys using a high-latitude subsample of bright Gaia sources with trigonometric parallaxes. The two panels show the subsample distribution in the Hertzsprung-Russell diagram (absolute Gaia-based magnitude $M_{\rm G}$ vs. reddening-corrected Gaia color). The left panel marks sources where Gaia detected variability, and the right panel is color-coded by the WISE $W2-W3$ color, which indicates dust presence for positive values. Variable sources with $-2.5 < M_{\rm G} < 1$ and $W1-W2 > 0.4$ are likely AGB stars with dusty envelopes.

and periods; see e.g., Wood 2015). With modern sky surveys, light curves for detected AGB stars come for "free". For example, the Catalina Real-time Transient Survey (CRTS, see http://crts.caltech.edu) alone has provided over 700 photometric measurements spanning ∼9 years (see Figure 1) for the famous AGB star CW Leo (IRC10+216).

Particularly clean (low contamination) stellar population samples can be constructed by combining several unbiased surveys. Figure 2 illustrates how a combination of optical

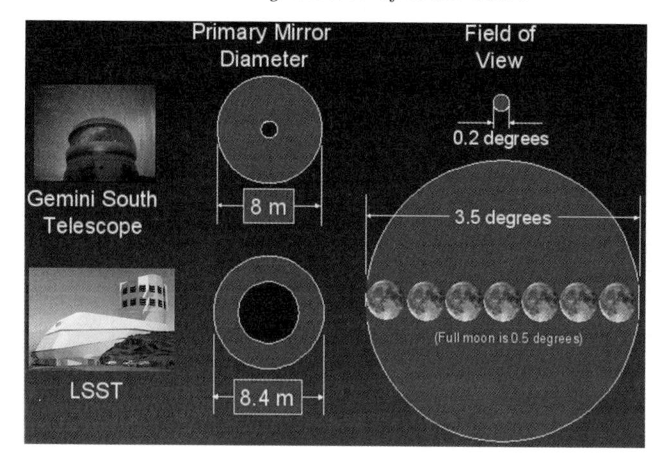

Figure 3. The speed at which a system can survey a given sky area to a given flux limit is determined by the so-called étendue (or grasp): the product of the primary mirror size and the field-of-view. The figure compares the primary mirror size and the field-of-view for LSST and Gemini South telescopes. The system étendue is much larger for LSST. Figure courtesy of Chuck Claver.

variability and red infrared colors efficiently selects dusty AGB stars. The ongoing Zwicky Transient Facility (ZTF; Bellm & Kulkarni 2017) is expected to obtain 300-epoch optical photometry over three quarters of the sky, including the Galactic plane, within next three years. The depth of the ZTF data will be comparable to Gaia's depth ($V \sim G \sim 20.5$). Because dusty AGB stars are heavily extincted at optical wavelengths, and thus can be much fainter, a much deeper optical time-domain survey can open a huge parameter space for studying AGB stars. LSST will be such a survey, extending the depth reached by time-domain ZTF and Gaia surveys by about 4 magnitudes.

2. Brief overview of LSST

The Large Synoptic Survey Telescope (LSST; for an overview see Ivezić *et al.* 2008) is the most ambitious survey currently being constructed or planned in the visible band. The LSST survey power is due to its large étendue (see Figure 3). LSST will extend the faint limit of the SDSS by about 5 magnitudes and will have unique survey capabilities for faint time domain science.

The LSST design is driven by four main science themes: probing dark energy and dark matter, taking an inventory of the Solar System, exploring the transient optical sky, and mapping the Milky Way (for a detailed discussion see the LSST Science Book, LSST Science Collaboration 2009). LSST will be a large, wide-field ground-based system designed to obtain multiple images covering the sky that is visible from Cerro Pachón in Northern Chile. The system design, with an 8.4m (6.5m effective) primary mirror (Gressler *et al.* 2018), a 9.6 deg^2 field of view, and a 3.2 Gigapixel camera (Kahn *et al.* 2010), will enable about 10,000 deg^2 of sky to be covered using pairs of 15-second exposures in two photometric bands every three nights on average, with typical 5σ depth for point sources of $r \sim 24.5$. The system is designed to yield high image quality as well as superb astrometric and photometric accuracy. The LSST camera provides a 3.2 Gigapixel flat focal plane array, tiled by 189 4K$\times$4K CCD science sensors with 10 μm pixels (see Figure 4).

The construction phase of LSST, funded by the U.S. National Science Foundation and Department of Energy, started in 2016 and is progressing according to the planned schedule (for an illustration of the current status, see Figure 5). First light for LSST is

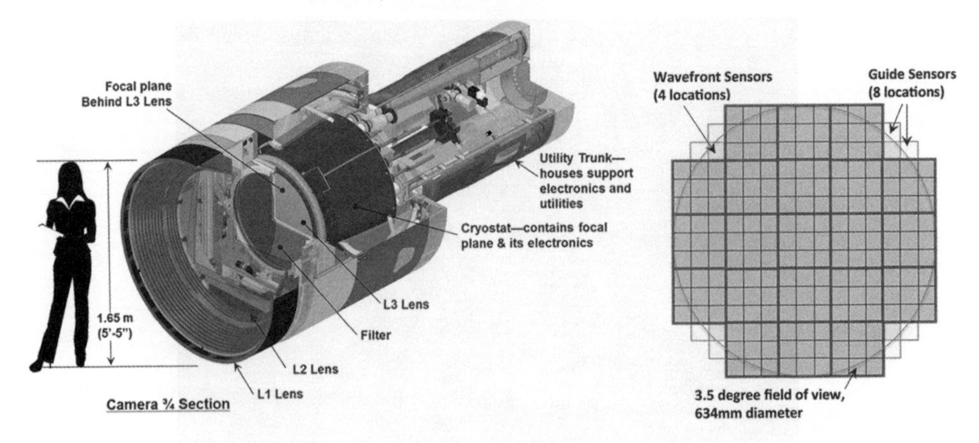

Figure 4. The left panel shows a cutaway view of LSST camera. Not shown are the shutter, which is positioned between the filter and lens L3, and the filter exchange system. The right panel shows the LSST Camera focal plane array. Each cyan square represents one 4K× 4K pixel sensor. Nine sensors are assembled into a raft; the 21 rafts are outlined in red. There are 189 science sensors, for a total of 3.2 Gigapixels. Also shown are the four corner rafts, where the guide sensors and wavefront sensors are located. Adapted from Ivezić *et al.* (2008).

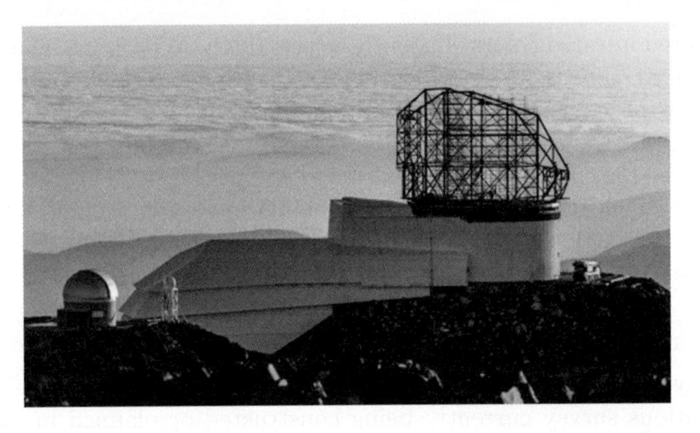

Figure 5. The figure shows a photograph of the LSST summit at sunset, from the direction of Gemini South, at the time of this Symposium (August 2018). First light for LSST is expected in 2020 with a 144 Mpix engineering camera, and with the full 3.2 Gpix camera in 2021. The small dome on the left is an auxiliary telescope that will be used for photometric calibration. For more photographs, see https://www.lsst.org/gallery/image-gallery. Credit: LSST and Gianluca Lombardi.

expected in early 2020 with a small commissioning camera (144 Mpix), with the full 3.2 Gpix camera integrated by the end of 2020.

2.1. *Baseline LSST surveys*

The survey area will be included within 30,000 deg^2 with $\delta < +34.5°$, and will be imaged multiple times in six bands, *ugrizy*, covering the wavelength range 320–1050 nm. About 90% of the observing time will be devoted to a deep-wide-fast survey mode which will observe an 18,000 deg^2 region over 800 times (summed over all six bands) during the anticipated 10 years of operations, and yield a coadded map to $r \sim 27.5$ ("the main survey"). These data will result in databases containing about 20 billion galaxies and a similar number of stars, and will serve the majority of science programs. The remaining 10% of the observing time will be allocated to special programs such as a Very Deep and

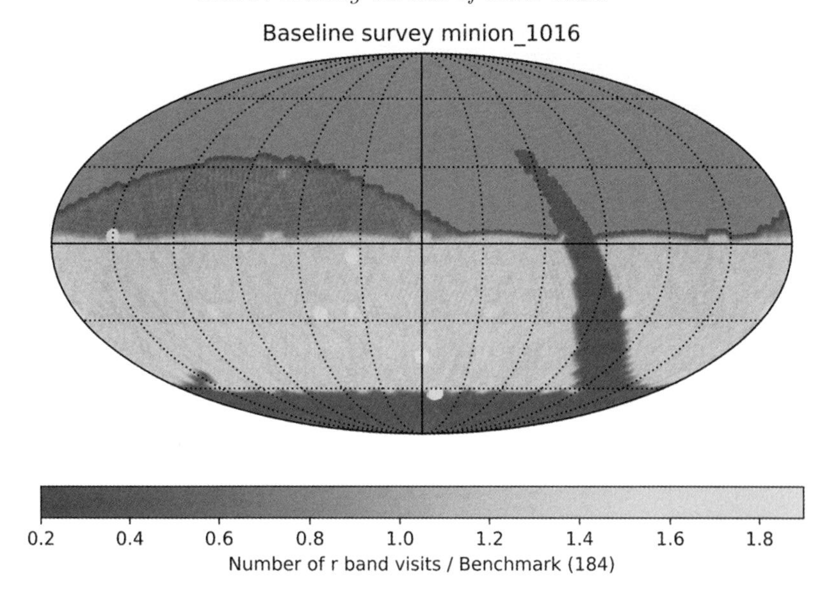

Figure 6. The distribution of the r band visits on the sky for a simulated realization of the baseline cadence (Delgado *et al.* 2014; Jones *et al.* 2014). The sky is shown in the equal-area Mollweide projection in equatorial coordinates (the vernal equinoctial point is in the center, and the right ascension is increasing from right to left). The number of visits for a 10-year survey, normalized to the survey design value of 184, is color-coded according to the legend. The three regions with smaller number of visits than the main survey ("mini-surveys") are the Galactic plane (arc on the right), the region around the South Celestial Pole (bottom), and the so-called "northern Ecliptic region" (upper left; added in order to increase completeness for moving objects). Deep drilling fields, with a much higher number of visits than the main survey, are also visible as small circles (the color scale is saturated; typically, deep drilling fields will be observed more than a thousand times per band). The fields were dithered on sub-field scales and pixels with angular resolution of ~30 arcmin were used to evaluate and display the coverage. Adapted from Ivezić *et al.* (2008).

Fast time-domain survey. For an illustration of the current baseline survey sky coverage, see Figure 6.

3. Anticipated Impact of LSST on AGB star research

During the 10-year survey beginning in 2022, the LSST will cover about half of the sky, with each sky location observed over 800 times in six broad-band filters. These imaging data will be superior to optical survey data currently available (for a comparison of SDSS and LSST-like images, see Figure 7).

The LSST data will be used for finding AGB stars by several complementary methods:

(a) **Optical identifications of IR counterparts.** If, for example, the dust-enshrouded C star IRC +10 216 were 40 kpc away, it would have $i=27$, $z=25$ and $y=23$ (based on SDSS observations), which is brighter than LSST faint limits in these bands. Therefore, even stars with exceedingly thick dust shells and barely detected by IRAS will be detectable in the i, z, and y bands by LSST throughout the Galaxy.

(b) **Search for spatially-resolved envelopes.** As demonstrated by SDSS observations of IRC +10 216, LSST will be able to detect and *resolve* an IRC +10 216-like envelope at a distance of 15 kpc! Stars with resolved envelopes will be especially useful for refining selection criteria based on other methods (such as color and variability).

Figure 7. A comparison of an SDSS image (left, 3.5×3.5 arcmin^2 *gri* composite) showing a relatively random piece of high Galactic latitude sky with a similar *gri* composite image of the same field obtained by the Hyper Suprime Cam survey (HSC, right). The HSC image is about 4-5 mag deeper than the SDSS image and comparable to the anticipated LSST 10-year coadded data). Courtesy of Robert Lupton.

(*c*) **Color selection.** The extremely red optical colors of dusty AGB stars are very distinctive; color-selected LSST samples will be able to trace structure throughout the Local Group and beyond.

(*d*) **Variability.** Thanks to their particularly large light curve amplitudes (several magnitudes), and over 800 observations during 10-year survey, variability will be a powerful detection method for AGB stars (n.b., LSST will detect over one hundred million variable stars).

It is evident that LSST, although driven by different science goals, will be a powerful machine for discovering and characterizing AGB stars. This ability could be further enhanced by utilizing narrow-band filters.

4. Specialized narrow-band filters

According to their photospheric chemical composition, AGB stars can be divided into oxygen-rich (O-rich) and carbon-rich (C-rich) stars; O-rich stars are associated with silicate-rich dust chemistry and C-rich stars with carbonaceous dust grains. The C-to-O stellar count ratio is an important distinguishing characteristic of stellar populations; for example, it is more than two orders of magnitude higher in the Galactic center than in the Large Magellanic Cloud (Nikutta *et al.* 2014). Hence, it is highly desirable for a survey to be able to classify stars into O and C classes.

At infrared wavelengths, the predominant dust type is relatively easily determined thanks to the prominent silicate dust feature at 10 μm. At optical wavelengths, it is harder to separate the two classes. The current LSST baseline design includes six broad-band filters. The system throughput as a function of wavelength for these bandpasses is shown in Figure 8. The ability of LSST to characterize AGB stars (i.e., to separate C from O type stars) could be further enhanced by adding narrow-band filters. For example, the so-called TiO (7780 Å) and CN (8120 Å) filters, see Figure 9, introduced by Wing (1971) have been successfully used by a number of groups (Cook, Aaronson & Norris 1986; Kerschbaum *et al.* 2004; Battinelli & Demers 2005, and references therein) for the identification and characterization of late-type stars in external Local Group galaxies.

The LSST Science Requirements Document (Ivezić & The LSST Science Collaboration 2011) allows for about 10% of the observing time (300 nights) to be allocated to specialized programs. If only 2 nights ($<$0.1% of the total observing time) were allocated to a narrow-band survey, it would be possible to cover about 10,000 sq.deg. of sky in each band. Such a time allocation would match the cost of procuring the filters (of the order

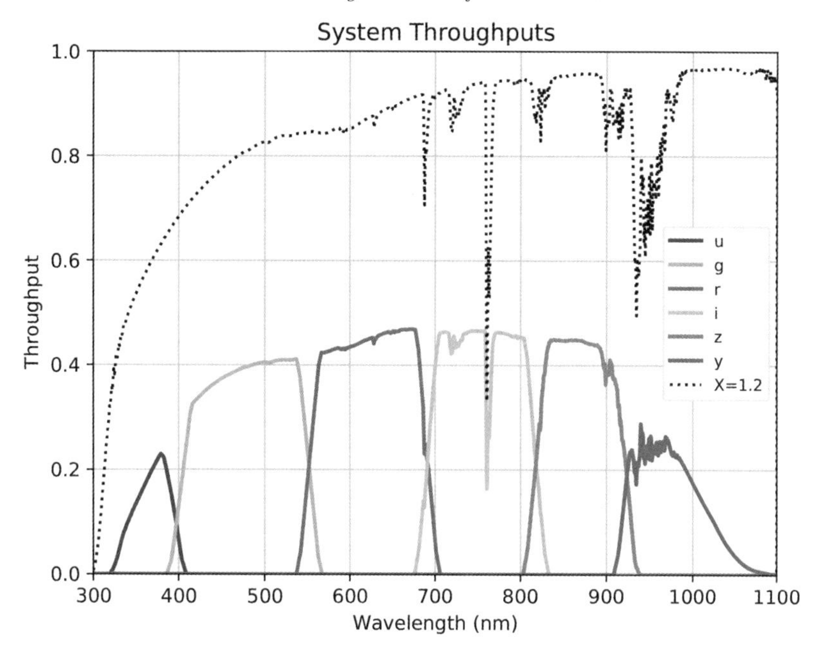

Figure 8. The LSST bandpasses. The vertical axis shows the total throughput. The computation includes the atmospheric transmission (assuming an airmass of 1.2, dotted line), optics, and the detector sensitivity. Adapted from Ivezić *et al.* (2008).

\$1M) to the cost of the LSST system itself (about 400,000 USD per observing night). Given that such data would be useful for extra-galactic astronomy and cosmology, too (C. Stubbs, priv. comm.), it is plausible that more than 2 observing nights could be negotiated for this program.

Assuming 150 Å wide filters, the faint limits would be at about apparent magnitudes 22-22.5. This is about 0.5-1 mag shallower than e.g. a study of And II by Kerschbaum *et al.* (2004), but the surveyed area would be over 1,000,000 times larger! Furthermore, it is noteworthy that the deep and exceedingly accurate broad-band photometry will come for "free" and will include many epochs which can be used to reject foreground Galactic M dwarfs by variability. The same data would enable efficient calibration of the narrow-band survey.

This program may represent an exciting opportunity for the AGB star community. In order to execute such a program, this community may wish to organize a working group which would have three main goals: fundraising for the filter procurement, securing an allocation of observing time from the LSST, and the timely analysis of narrow-band survey data.

5. Opportunities for International Participation in LSST

LSST was designed to be a public project, with full access to all data and data products† open to the entire U.S. and Chileans scientific communities and the public at large. Unlimited immediate access to LSST Data Releases will also be granted to international partners who signed Memoranda of Understanding or Memoranda of Agreement with LSST. For other users of LSST data, there will be a 2-year delay‡. LSST is currently

† The baseline definition of LSST Data Products is available in Jurić *et al.* (2017).

‡ The transient stream, based on image difference analysis, will be available to everyone in the world within 60 seconds from closing the shutter.

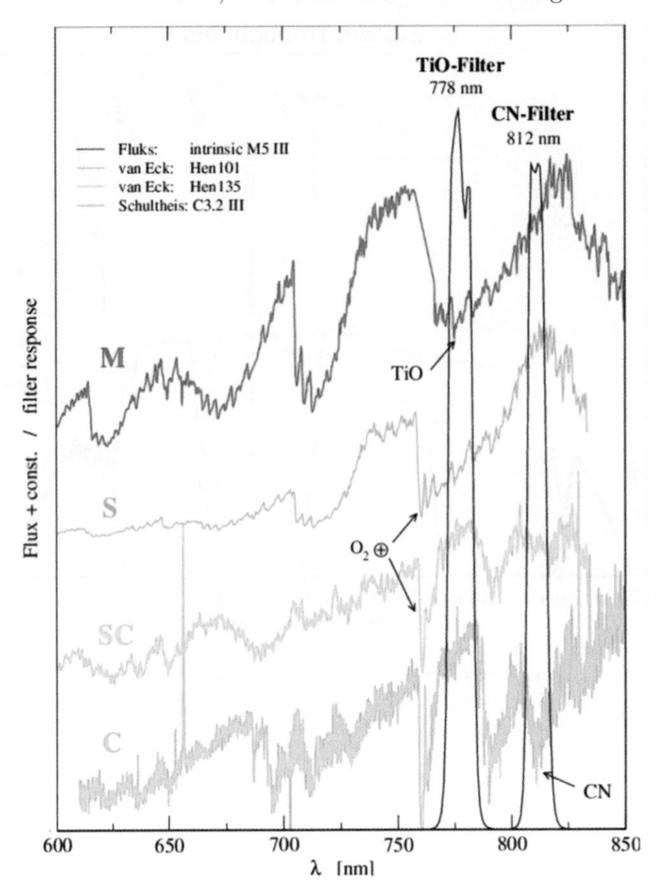

Figure 9. An illustration of the power of TiO (7780 Å) and CN (8120 Å) filters to distinguish stellar spectral features for late-type stars: O-rich (M-type) stars have much redder CN-to-TiO colors than C-type stars. Adapted from Kerschbaum *et al.* (2004).

seeking additional international partners and we encourage the interested colleagues to contact the LSST Corporation. For details about application process for International Affiliates, please see Section 3 in Ivezić, Kahn & Eliason (2014).

6. Conclusions

We are witnessing rapid progress in the availability of large and sensitive sky surveys at all wavelengths. As in many other fields, modern sky surveys are having a major impact on AGB star research: well-defined statistical samples allow the use of AGB stars as stellar population tracers both in our and other galaxies.

LSST, although driven by different science goals, will be a powerful machine for discovering and characterizing AGB stars. As an optical time-domain survey, LSST will be used for finding AGB stars by several complementary methods: as optical identifications of IR counterparts, by searching for spatially-resolved envelopes, by red optical color selection, and by variability.

The ability of LSST to find and classify AGB stars could be further enhanced with narrow-band filters, in particular by utilizing narrow-band TiO and CN filters. In order to make such a narrow-band filter survey a reality, the AGB stars community would have to organize a working group with three main goals in mind: fundraising for the filter procurement, securing an allocation of observing time from the LSST, and the timely analysis of narrow-band survey data. Colleagues interested in this program should contact the first author.

Acknowledgements

We thank Andrew Connolly for careful reading of this manuscript and excellent comments. This material is based upon work supported in part by the National Science Foundation through Cooperative Agreement 1258333 managed by the Association of Universities for Research in Astronomy (AURA), and the Department of Energy under Contract No. DE-AC02-76SF00515 with the SLAC National Accelerator Laboratory. Additional LSST funding comes from private donations, grants to universities, and in-kind support from LSSTC Institutional Members. This work was performed in part at Aspen Center for Physics, which is supported by National Science Foundation grant PHY-1607611.

References

Battinelli, P. & Demers, S. 2005, *A&A*, 442, 159

Bellm, E. & Kulkarni, S. 2017, *Nature Astr.*, 1, 71

Cook, K. H., Aaronson, M. & Norris, J. 1986, *ApJ*, 305, 634

Delgado, F., Saha, A., Chandrasekharan, S., *et al.* 2014, *Proceedings of the SPIE*, Volume 9150, id. 915015

Flaugher, B. 2008, In *A Decade of Dark Energy: Spring Symposium, Proceedings of the conferences held May 5-8, 2008 in Baltimore, Maryland. (USA). Ed. by N. Pirzkal & H. Ferguson.*

Girardi, L. & Marigo, P. 2007, *ASPC*, 378, 7

Gressler, W., Coleman, G., Delgado, F., *et al.* 2018, *Proceedings of the SPIE*, Volume 10700, id. 107002I

Höfner, S. & Olofsson, H. 2018, *A&A*, 26, 1

Ivezić, Ž., Kahn, S.M. & Eliason, P. 2014, *EAS Publications Series*, Volume 67-68, 211 (also arXiv:1502.06555)

Ivezić, Ž., Kahn, S.M., Tyson, J.A., *et al.* 2008, arXiv:0805.2366

Ivezić, Ž. & The LSST Science Collaboration. 2011, LSST Science Requirements Document, LSST Document LPM-17, LSST. https://ls.st/LPM-17

Jackson, T., Ivezić, Ž. & Knapp, G.R. 2002, *MNRAS*, 337, 749

Jones, R.L., Yoachim, P., Chandrasekharan, S., *et al.* 2014, *Proceedings of the SPIE*, Volume 9149, id. 91490B

Jurić, M., Axelrod, T., Becker, A.C., *et al.* 2017, LSST Data Products Definition Document, LSST Document LSE-163, LSST. https://ls.st/LSE-163

Kahn, S.M., Kurita, N., Gilmore, K., *et al.* 2010, *Proceedings of the SPIE*, Volume 7735, id. 77350J

Kaiser, N., Burgett, W., Chambers, K., *et al.* 2010, *Proceedings of the SPIE*, Volume 7733, id. 77330E

Kerschbaum, F., Nowotny, W., Olofsson, H. & Schwarz, H. E. 2004, *A&A*, 427, 613

Marston, C., Strömbäck, G., Thomas, D., Wake, D.A. & Nichol, R.C. 2009, *MNRAS*, 394, 107

Nikutta, R., Hunt-Walker, N., Nenkova, M., Ivezić, Ž & Elitzur, M. 2014, *MNRAS*, 442, 3361

Rejkuba, M. 2004, *A&A*, 413, 903

LSST Science Collaboration 2009, *LSST Science Book*, http://www.lsst.org/lsst/SciBook, arXiv:0912.0201

Whitelock, P.A. & Feast, M.W. 2000, *MNRAS*, 317, 460

Wing, R.F. 1971, *Conference on Late-Type Stars*, ed. G.W. Lockwood & H. M. Dyck, Kitt Peak National Observatory, Contrib. No. 554, p. 145

Wood, P.R. 2015, *MNRAS*, 448, 3829

Wright E. L. *et al.* 2010, *AJ*, 140, 1868

Discussion

QUESTION: What will be the photometric precision of LSST?

IVEZIĆ: Before Gaia, the LSST target for photometric precision was 5 mmag. Now that Gaia is a reality, we may achieve 2-3 mmag photometric precision.

Why Galaxies Care About AGB Stars:
A Continuing Challenge through Cosmic Time
Proceedings IAU Symposium No. 343, 2019
F. Kerschbaum, M. Groenewegen & H. Olofsson, eds.

doi:10.1017/S1743921318005458

Probing stellar evolution with S stars and Gaia

S. Shetye[1,2]†, S. Van Eck[1], A. Jorissen[1], H. Van Winckel[2], L. Siess[1] and S. Goriely[1]

[1]Institut d'Astronomie et d'Astrophysique, Université Libre de Bruxelles, CP 226, Boulevard du Triomphe, B-1050 Bruxelles,Belgium
email: `Shreeya.Shetye@ulb.ac.be`

[2]Instituut voor Sterrenkunde (IvS), KU Leuven, Celestijnenlaan 200D, B-3001 Leuven, Belgium

Abstract. S-type stars are late-type giants enhanced with *s*-process elements originating either from nucleosynthesis during the Asymptotic Giant Branch (AGB) or from a pollution by a binary companion. The former are called intrinsic S stars, and the latter extrinsic S stars. The intrinsic S stars are on the AGB and have undergone third dredge-up events. The atmospheric parameters of S stars are more numerous than those of M-type giants (C/O ratio and *s*-process abundances affect the thermal structure and spectral synthesis), and hence they are more difficult to derive. These atmospheric parameters are also entangled within each other. Nevertheless, high-resolution spectroscopic data of S stars combined with the Gaia Data Release 2 (GDR2) parallaxes and with the MARCS model atmospheres for S-type stars were used to derive effective temperatures, surface gravities, and luminosities. These parameters not only allow to locate the intrinsic and extrinsic S stars in the Hertzsprung-Russell (HR) diagram but also allow the accurate abundance analysis of the *s*-process elements.

Keywords. S stars, AGB stars, s-process nucleosynthesis, HR diagram

1. Introduction

S stars are late-type giants showing ZrO molecular bands along with TiO bands as the most characteristic distinctive spectral features (Merrill 1922). The C/O ratio of S stars ranges from 0.5 to 1 suggesting that they are transition objects between M-type giants (C/O ~ 0.5) and carbon stars (C/O > 1) on the Asymptotic Giant Branch (AGB) (Iben and Renzini 1983). Their spectra show signatures of overabundances in *s*-process elements (Smith and Lambert 1990).

The evolutionary status of S stars as AGB stars was challenged when Tc lines (a *s*-process element with no stable long-lived isotope) were reported as missing in some S stars (Merrill 1952; Smith and Lambert 1986; Jorissen *et al.* 1993). This puzzle of the evolutionary status of S stars was solved when it was perceived that the Tc-poor S stars belong to binary systems (Smith and Lambert 1986; Jorissen *et al.* 1993). S stars can therefore be classified into two different classes: Tc-rich as intrinsic S stars that are genuine thermally-pulsing AGB (TP-AGB) stars and Tc-poor as extrinsic S stars that correspond to the cooler analogues of barium stars and owe their *s*-process element

† This work has made use of data from the European Space Agency (ESA) mission *Gaia* (http://www.cosmos.esa.int/gaia), processed by the *Gaia* Data Processing and Analysis Consortium (DPAC, http://www.cosmos.esa.int/web/gaia/dpac/consortium). Funding for the DPAC has been provided by national institutions, in particular the institutions participating in the *Gaia* Multilateral Agreement.

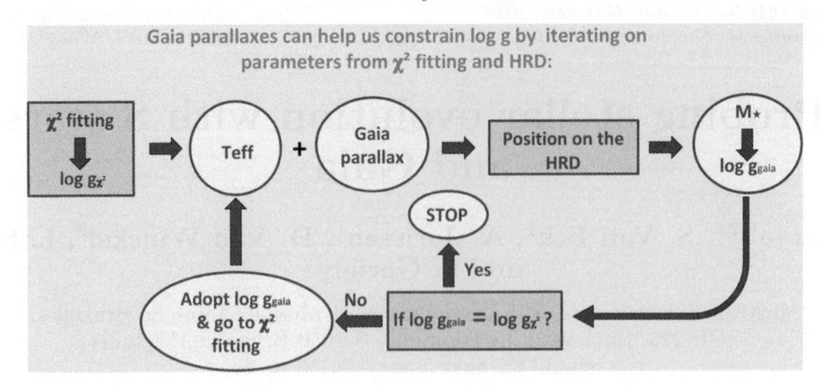

Figure 1. Algorithm adopted to constrain $\log g$ using the locations of the stars in the HR diagram compared to the evolutionary tracks from STAREVOL code.

overabundances to a mass transfer from a former AGB companion which is now a white dwarf.

The thermal structure of the atmospheres of S stars depend on the effective temperature ($T_{\rm eff}$), surface gravity ($\log g$), [Fe/H], C/O as well as [s/Fe] (s-process element abundances). Abundance analysis of S stars thus requires a reliable determination of all these stellar atmosphere parameters.

2. Stellar sample and parameter determination

Our sample consists of S stars from the General Catalog of S stars (Stephenson 1984) with the condition to have a V-band magnitude brighter than 11 and $\delta \geq -30°$, to be observable with HERMES (High Efficiency and Resolution Mercator Echelle Spectrograph, mounted on the 1.2m Mercator Telescope at the Roque de Los Muchachos Observatory, La Palma, Raskin *et al.* 2011). Furthermore, a condition was imposed on the TGAS parallaxes (Gaia Collaboration 2016), considering only those stars with a small error on the parallax ($\sigma_{\bar{\omega}} \leq 0.3\,\bar{\omega}$). With these conditions, the sample amounts to 18 S stars. During the course of our study, the Gaia Data Release 2 (GDR2; Gaia Collaboration 2018) parallaxes were released and these more accurate parallaxes are used in the present study.

The stellar parameters were derived using the MARCS grid of atmospheric models for S stars (Van Eck *et al.* 2017) containing more than 3500 models covering the parameter space in $T_{\rm eff}$, $\log g$, [Fe/H], C/O and [s/Fe] ratios. The comparison between observed and synthetic spectra is then performed by a χ^2-fitting procedure, summing over all spectral pixels in spectral bands approximately 200 Å wide. The model with the lowest χ^2 value is chosen as the best fitting model.

The distance and luminosity of these stars was determined from the GDR2 parallaxes. Comparison between the positions of the stars in the HR diagram constructed from GDR2 parallaxes with the evolutionary tracks from the STAREVOL code (Siess and Arnould 2008) yield the mass, hence the surface gravity of our stars ($\log g_{\rm Gaia}$). Because $\log g_{\rm Gaia}$ and $\log g$ derived from the χ^2 fitting do not always agree, we derive a new surface gravity estimate as explained in Figure 1. This iteration on the stellar parameters ensures that the adopted $\log g$ is consistent with the GDR2 parallaxes.

3. HR diagram of S stars

The final parameters and GDR2 parallaxes along with the STAREVOL evolutionary tracks corresponding to the closest grid metallicity lead to the HR diagram of S stars presented in Figure 2. The intrinsic S stars are cool and luminous objects likely on the TP-AGB in the HR diagram. On the other hand, the extrinsic S stars are hotter and

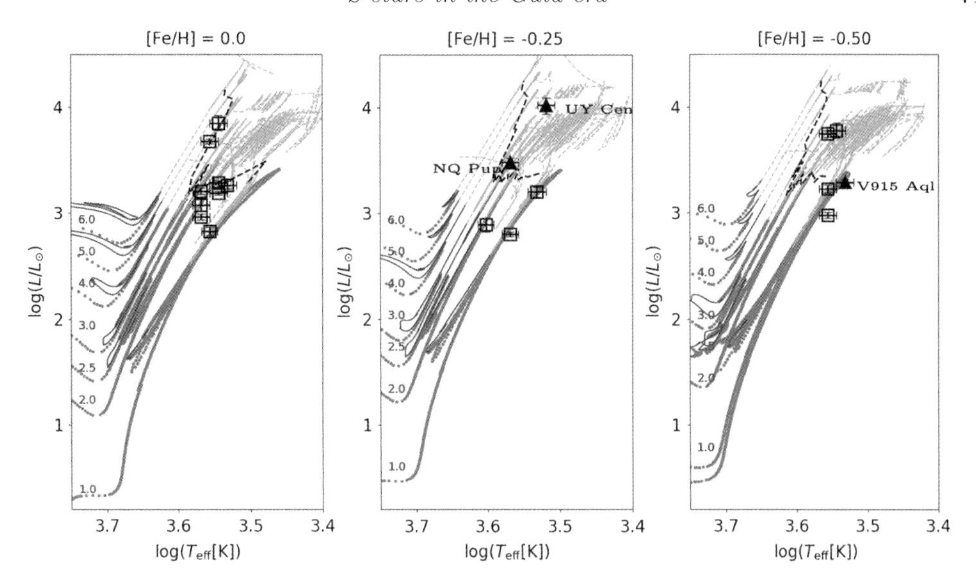

Figure 2. HR diagram of intrinsic (filled triangles) and extrinsic (open squares) S stars along with the STAREVOL evolutionary tracks corresponding to the closest grid metallicity. The red giant branch is represented by the blue dots, the core He-burning phase by the green solid line, whereas the red dashed line corresponds to the AGB tracks. The black dotted line marks the predicted onset of third dredge-up (TDU) from the STAREVOL code.

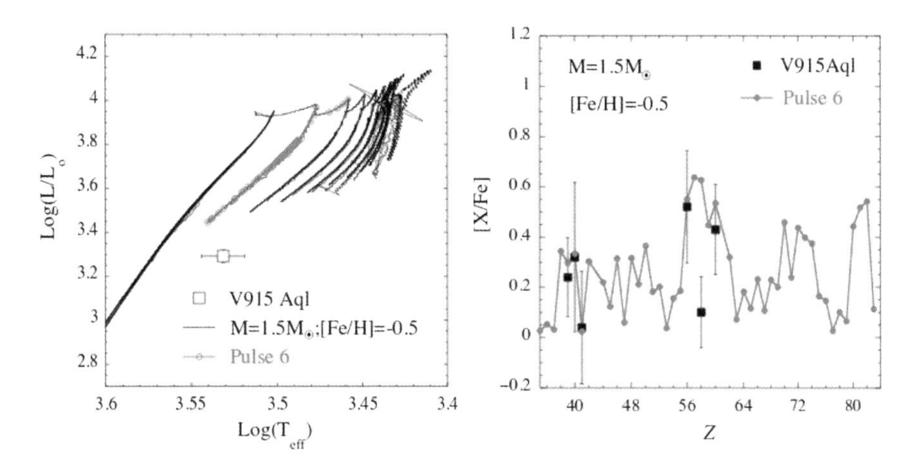

Figure 3. Left panel: Location of the V915 Aql in the HR diagram, compared with STAREVOL tracks of the corresponding metallicity. Right panel: Predicted abundance distribution for this object.

intrinsically fainter on the early AGB or more likely on the red giant branch (RGB). Interestingly, the Tc-poor S stars with $M \leq 2\ M_\odot$ are on the upper-end of the RGB or on the early-AGB while the Tc-poor S stars with $M \geq 2\ M_\odot$ are on the early- AGB. We also find a remarkable agreement between the location of the stars in the HRD and the predicted boundary for the occurence of TDU.

4. A Tc-rich S star of $M \sim 1\ M_\odot$

V915 Aql is an intrinsic S star located on the 1 $M_\odot$ track. It is intriguing because stellar models do not predict the third dredge-up to occur for masses less than 1.3 $M_\odot$ (Karakas and Lugaro 2016). Figure 3 presents the comparison of the observed *s*-process elements abundance with nucleosynthesis predictions from the STAREVOL code (Goriely and Siess 2018). To match V915 Aql surface abundances, a 1.5 $M_\odot$ model had to be used, but such a model does not match so well its position in the HR diagram. Nevertheless, abundance predictions for this model reproduce fairly well the overall pattern of V915 Aql (except for La). The occurence of the third dredge-up for low mass stars (< 1.3 $M_\odot$) was also found in low-luminosity *s*-process rich post-AGB stars (De Smedt *et al.* 2015).

5. Conclusion

Combining the derived parameters with GDR2 parallaxes allows a joint analysis of the location of the stars in the Hertzsprung-Russell diagram and of their surface abundances. The Tc-rich star V915 Aql is challenging as it points at the occurrence of TDU episodes in stars with masses as low as $M \sim 1$ $M_\odot$. Extending the sample to include many more intrinsic S stars with GDR2 parallaxes in order to constrain the luminosity of the first occurrence of the TDU is a work in progress.

References

De Smedt, K. *et al.* 2015, *A&A*, 583, A56
Gaia Collaboration 2016, *A&A*, 595, A2
Gaia Collaboration 2018, *A&A*, 616A, 1G
Goriely, S. and Siess, L. 2018, *A&A*, 609, A29
Iben, Jr., I. and Renzini, A. 1983, *ARA&A*, 21, 271
Jorissen, A. *et al.* 1993, *A&A*, 271, 463
Karakas, A. I. and Lugaro, M. 2016, *ApJ*, 825, 26
Merrill, P. W. 1922, *ApJ*, 56, 457M
Merrill, P. W. 1952, *ApJ*, 116, 21
Raskin, G. *et al.* 2011, *A&A*, 526, A69
Siess, L. and Arnould, M. 2008, *A&A*, 489, 395
Smith, V. V. and Lambert, D. L. 1986, *ApJ*, 311, 843
Smith, V. V. and Lambert, D. L. 1990, *ApJS*, 72, 387
Stephenson, C. B. 1984, *Publications of the Warner & Swasey Observatory*, 3, 1
Van Eck, S. *et al.* 2017, *A&A*, 601, A10

Discussion

UTTENTHALER: You said that intrinsic S-type stars are the first stars on the AGB to show signs of 3DUP. However, there are pure M-type AGB stars that show one sign of 3DUP, namely lines of Tc; they have no ZrO bands. Would these objects be interesting for you to study and do you have plans to do so?

SHETYE: Yes, these objects will be indeed very interesting to study. It will be interesting to compare the positions of these stars with the intrinsic (Tc-rich) S-stars in the HR diagram. Thanks for the suggestion, it will definitely be added to our future plans.

Why Galaxies Care About AGB Stars:
A Continuing Challenge through Cosmic Time
Proceedings IAU Symposium No. 343, 2019
F. Kerschbaum, M. Groenewegen & H. Olofsson, eds.

doi:10.1017/S1743921318005999

AGB stars in *Gaia* DR2

Thomas Lebzelter[1], Nami Mowlavi[2], Paola Marigo[3], Isabelle Lecoeur-Taibi[2], Michele Trabucchi[3], Giada Pastorelli[3], Peter Wood[4] and *Gaia* Collaboration

[1]Institut f. Astrophysik, University of Vienna
Tuerkenschanzstrasse 17, A1180 Vienna, Austria
email: thomas.lebzelter@univie.ac.at

[2]Astronomy Department, University of Geneva, Switzerland
email: Nami.Mowlavi@unige.ch

[3]Dipartimento di Fisica e Astronomia Galileo Galilei, Universit di Padova, Italy

[4]RSAA, Canberra, Australia

Abstract. *Gaia* Data Release 2 (DR2; April 25, 2018) provides astrometric and photometric data for more than a billion stars - among them many AGB stars. As part of DR2 the light curves of several hundreds of thousand variable stars, including many long-period variable (LPV) candidates, are made available. The publication of the light curves and LPV-specific attributes in addition to the standard DR2 products offers a unique opportunity to study AGB stars. In this contribution, we present the first results for AGB stars based on the analysis of the *Gaia* data performed after their release. As an immediate result of the *Gaia* DR2 LPV database we introduce a new photometric index capable of efficiently distinguishing AGB stars of different masses and chemical properties.

Keywords. stars: variability, stars: evolution, stars: AGB and post-AGB

---

1. Introduction

The *Gaia* all-sky survey is expected to add a major contribution to the study of the populations of long-period variables (LPVs) during its five-year nominal mission plus extensions, in particular by covering Galactic field and halo stars. These groups of stars can only be studied (Trabucchi *et al.* 2017) in a way comparable to the major Magellanic Cloud surveys done in the past (Ita *et al.* 2004, Soszinski *et al.* 2009) if the distance is obtained for the stars. Furthermore, the high level of completeness of LPVs expected from the *Gaia* survey will offer the opportunity to study, among other subjects, the frequencies of various groups of LPVs in the extended solar neighbourhood and other parts of the Galaxy. Thus, the *Gaia* database of LPVs will be unique and will provide a major step forward in understanding these variables.

Gaia Data Release 2 (DR2, *Gaia* Collaboration 2018) offers the first opportunity to provide a *Gaia* catalogue of LPV candidates to the scientific community. In this conference paper we present an overview on the first *Gaia* catalogue of LPVs published in DR2. Using this catalogue we developed a new photometric index capable of effectively distinguishing stars of various masses and chemistry on the AGB.

2. The *Gaia* DR2 catalogue of LPVs

This catalogue is described in detail in Mowlavi *et al.* (2018). We have put the priority to ensure an as-low-as-possible level of contamination, without targeting completeness.

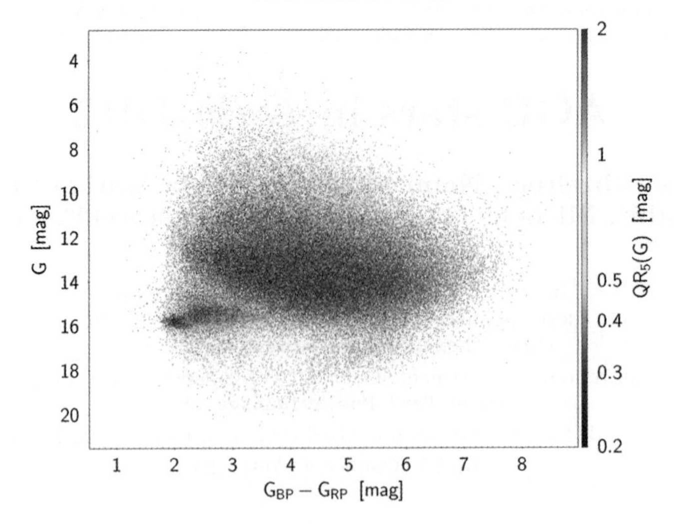

Figure 1. Colour-magnitude diagram of all LPV candidates published in *Gaia* DR2, with the 5–95% quantile range QR_5 of the cleaned G time-series shown in colour according to the colour scale drawn on the right of the figure. The clump observed around $G \simeq 15.6$ mag contains stars from the Magellanic Clouds, while the Sagittarius dwarf galaxy appears around $G \simeq 14.7$ mag.

Knowing that LPVs have periods that can exceed 1000 days, and given the limited 22-month coverage of DR2 combined with the temporal sparsity of *Gaia* measurements due to the spacecraft's scanning law, this first *Gaia* catalogue of LPV candidates is incomplete. We have limited the catalogue to Mira and semi-regular variables (SRVs) with variability amplitudes larger than 0.2 mag in the *Gaia* G-band. Small amplitude red giant variables, detected as a large group in the OGLE database, were excluded at this stage.

The catalogue contains 151 761 candidates. Already at this preliminary stage, the catalogue thus contains about twice the number of LPVs with amplitudes larger than 0.2 mag known before DR2. This shows the great potential of *Gaia* for the detection and characterisation of an even larger set of LPVs in future data releases.

Figure 1 shows the content of the catalogue in a colour-magnitude diagram. The colours extend from $G_{\rm BP} - G_{\rm RP} \simeq 2$ mag to $\sim$8 mag for the reddest LPV candidates. This spread in $G_{\rm BP} - G_{\rm RP}$ largely originates from extinction due to interstellar and/or circumstellar dust. The spread in G magnitudes, on the other hand, is largely due to distance effects. However, among these reddest stars in our sample we also find the large amplitude, Mira-like variables, which constitute about 20% of our catalogue. The LPV candidates from the Magellanic Clouds at $15 \lesssim G$ [mag] $\lesssim 16$ and $1.8 \lesssim (G_{\rm BP} - G_{\rm RP})$ [mag] $\lesssim 3.5$ clearly stand out. The tail of faint outliers visible below the main bulk of the distribution is a result of the magnitude threshold in $G_{\rm BP}$.

The *Gaia* DR2 database includes multi-epoch observations of each object covering a time span of up to 22 months. The resulting light curves in G, $G_{\rm BP}$, and $G_{\rm RP}$ can be extracted from the database as described in Mowlavi *et al.* (2018). In the course of the evaluation of the data, the light curves and derived quantities were compared with ground based observations from various surveys, showing satisfactory agreement.

Finally, the DR2 catalogue of LPVs gives bolometric corrections for all objects included based on their $G_{\rm BP} - G_{\rm RP}$ colour. We note, however, that in DR2 no discrimination was made between O-rich and C-rich stars. Therefore, this difference is not taken into account in the current data release. It has to be mentioned further that the parallaxes have to be seen as preliminary at the moment. Major improvements in these aspects are expected for the forthcoming data release 3.

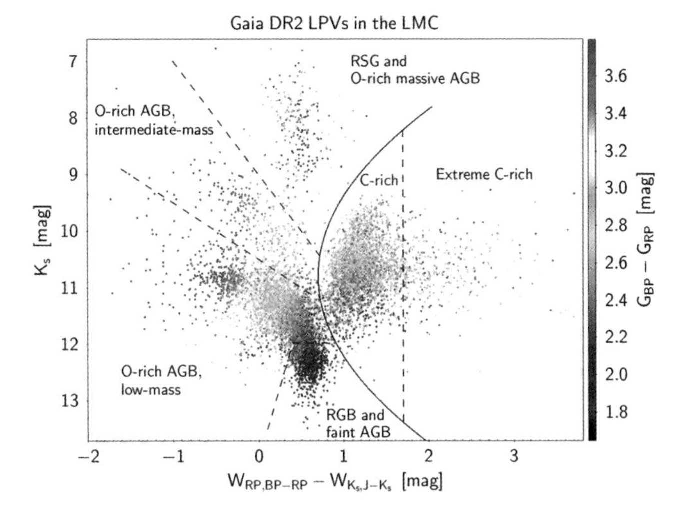

Figure 2. $W_{RP,BP-RP} - W_{K_s,J-K_s}$ versus K_s diagram of *Gaia* DR2 LPVs in the LMC. The markers are coloured with $BP - RP$ according to the colour-scale shown on the right of the figure. The solid line delineates O-rich (left of the line) and C-rich (right of the line) stars, and dashed lines delineate sub-groups as indicated in the figure.

3. A new method to identify subclasses among AGB stars

Large photometric surveys formed the base for fundamental steps forward in our understanding of the AGB evolution. With the *Gaia* catalogue we now have a deep, all-sky survey at hand. Here we report on a first result derived from this dataset, namely the development of a new photometric index useful for the distinction of AGB stars according to their mass and their evolutionary status. This distinction is challenging because the upper giant branch represents a mixture of objects of various masses and evolutionary stages.

As a starting point we selected LPVs belonging to the LMC in terms of sky positions, proper motions, and parallaxes ($\varpi < 0.5$ mas). All data were taken from the *Gaia* catalogue of LPVs described above. Their variability supported their identification as AGB stars. The sample was then cross-matched with the 2MASS catalogue. This allowed the construction of a combination of two Wesenheit functions, namely one using *Gaia* $G_{\rm BP}$ and $G_{\rm RP}$ photometry, and one using 2MASS J and K_s photometry. The final photometric index is of the form

$$W_{RP,BP-RP} - W_{K_s,J-K_s} = G_{\rm RP} - 1.3 \times (G_{\rm BP} - G_{\rm RP}) - K_s + 0.686 \times (J - K_s). \quad (3.1)$$

This index is thus a combination of three colours. It turns out to be very useful because of its different sensitivity to temperature changes for O-rich and C-rich stars. Details are described in Lebzelter *et al.* (2018). In Fig. 2 we plotted this index against K_s. Note that this is not a colour-magnitude diagram in the classical sense since the hottest stars are found in the centre of the x-axis, while cooler stars are found on the left side (if O-rich) or right side (if C-rich) of the diagram.

The various branches in Fig. 2 can be associated with various mass ranges with very little ambiguity. This fact allows the use of this diagram to separate AGB stars according to this parameter and to compare predictions from models with observed populations for studying both stellar evolution and the star formation history. In Fig. 3, the high diagnostic capability of the new diagram is corroborated with the help of state-of-the-art stellar evolution models (`PARSEC` and `COLIBRI` – see Marigo *et al.* 2017 for details).

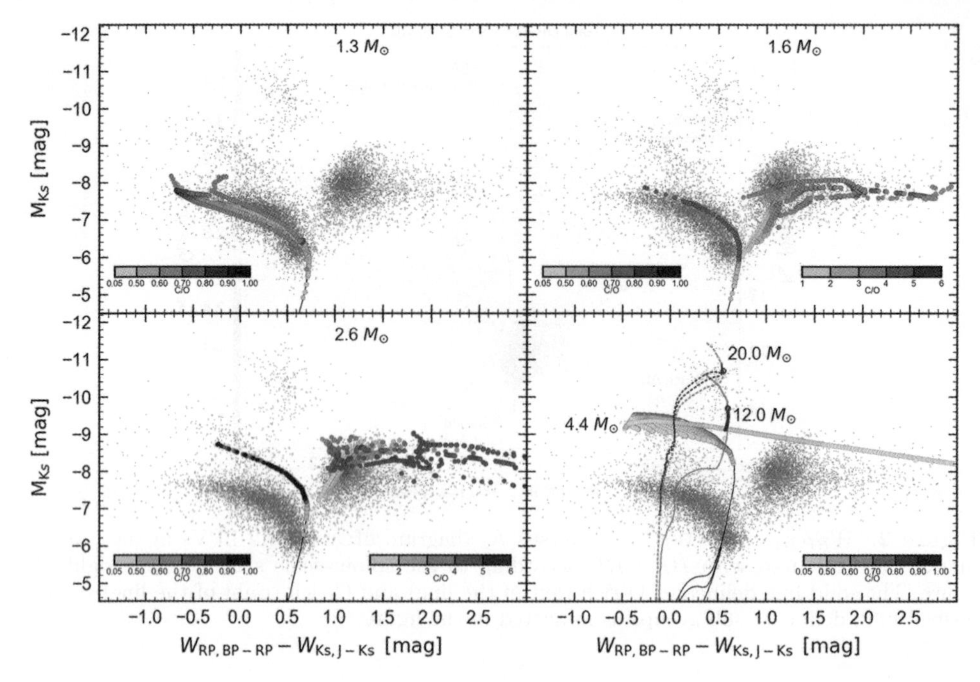

Figure 3. Location of stellar evolutionary tracks computed with the `PARSEC` and `COLIBRI` codes for a few selected values of the initial mass (as indicated). The initial metallicity is $Z = 0.006$ for the tracks with $M_{\rm i} = 1.3$, 1.6, 2.6, and 4.4 $M_{\odot}$, and Z = 0.008 for the massive ones with $M_{\rm i} = 12$ $M_{\odot}$ (solid line) and 20 $M_{\odot}$ (dashed line). The TP-AGB phase is color-coded according to the surface C/O ratio, while the previous evolution is shown with a black line.

4. Conclusions

The first *Gaia* catalogue of LPVs provides the scientific community with a photometric and astrometric database greatly expanding the known number of AGB variables with amplitudes >0.2 mag. While the quality of the results is going to be highly improved in future *Gaia* data releases, this data set already allowed the construction of a new photometric index that is very efficient for identifying AGB stars according to their mass and C or M star status. Further investigation of this index is currently under way.

References

Gaia Collaboration, Brown, A.G.A., van Leeuwen, F., *et al.* 2018, *A&A*, 616, A1
Ita, Y., Tanabe, T., Matsunaga, N., *et al.* 2004, *MNRAS*, 353, 705
Lebzelter, T., Mowlavi, N., Marigo, P., *et al.* 2018, *A&A*, 616, L13
Marigo, P., Girardi, L., Bressan, A., *et al.* 2017, *ApJ*, 835, 77
Mowlavi, N., Lecoeur-Taibi, I., Lebzelter, T., *et al.* 2018, *A&A*, in press
Soszyński, I., Udalski, A., Szymański, M.K., *et al.* 2009, *AcA*, 59, 239

Discussion

MENZIES: Can you indicate how the distribution of points would change in $(W_{\rm RP} - W_{K_{\rm s}})$ vs. $K_{\rm s}$ diagram for a metallicity much less than for the LMC?

LEBZELTER: We did some first tests for the *Gaia* sample of the SMC and Sgr DSph. The structure we see in the $(W_{RP,BP-RP} - W_{K_{\rm s},J-K_{\rm s}})$ vs. $K_{\rm s}$ diagram is the same, but the distribution of stars within the diagram is different.

Nucleosynthesis, mixing, and rotation

Why Galaxies Care About AGB Stars:
A Continuing Challenge through Cosmic Time
Proceedings IAU Symposium No. 343, 2019
F. Kerschbaum, M. Groenewegen & H. Olofsson, eds.

doi:10.1017/S1743921318006166

Nucleosynthesis in stars: The Origin of the Heaviest Elements

Amanda I. Karakas

Monash Centre for Astrophysics, School of Physics & Astronomy,
Monash University, VIC 3800, Australia
email: amanda.karakas@monash.edu

Abstract. The chemical evolution of the Universe is governed by the nucleosynthesis contribution from stars, which in turn is determined primarily by the initial stellar mass. The heaviest elements are primarily produced through neutron capture nucleosynthesis. Two main neutron capture processes identified are the slow and rapid neutron capture processes (s and r processes, respectively). The sites of the r and s-process are discussed, along with recent progress and their associated uncertainties. This review is mostly focused on the s-process which occurs in low and intermediate-mass stars which have masses up to about 8 solar masses ($M_\odot$). We also discuss the intermediate-neutron capture process (or i-process), which may occur in AGB stars, accreting white dwarfs, and massive stars. The contribution of the i-process to the chemical evolution of elements in galaxies is as yet uncertain.

Keywords. nuclear reactions, nucleosynthesis, abundances, stars: AGB and post-AGB, galaxies: abundances

1. Introduction

The story of the origin of the elements is one of the most fascinating in astronomy. Primordial element synthesis during the Big Bang 13.7 billion years ago created hydrogen, helium, and trace amounts of lithium (Li). The rest of the elements came from stars. The quest for the stellar sites that produced the elements is fundamental to modern science because this quest is linked to questions concerning the origins of planetary systems, life and astrobiology, origins of the Universe, and the process of galaxy formation.

Elements from carbon to iron are made by charged-particle nuclear reactions inside stars. By charged-particles reactions we mean those involving protons and α particles (helium atomic nuclei). These are the type of reactions that are taking place in the heart of our Sun, where hydrogen (H) is being fused or burnt into helium (He). Once a star runs out of its main source of fuel, hydrogen, the core contracts until the temperature in the centre is hot enough to fuse helium, the next most abundant element. Helium fusion reactions synthesize carbon and oxygen, and are the last central nuclear source for stars less massive than 8 times the mass of our Sun ($8M_\odot$). After core helium burning the stars enter the asymptotic giant branch (AGB) phase, where they are seen as immense red giants with distended outer envelopes. The outer layers are only tenuously held on and can be lost from the star by outflows of material (winds). Eventually, the winds drive the entire envelope into interstellar space, revealing the hot cores which light up the gas as beautiful planetary nebulae. We refer to reviews of the AGB phase by Karakas & Lattanzio (2014) and Herwig (2005) for details.

The low and intermediate-mass stars that evolve into AGB stars are fairly numerous because the initial mass function peaks at $\approx 1M_\odot$. Of importance for the chemical content of the Galaxy is the fact that the stellar winds contain matter that has been enriched by nuclear reactions deep in the star's interior, and brought to the stellar surface via mixing episodes. AGB stars are therefore crucial contributors to the chemical evolution of elements in galaxies across the Universe (Romano *et al.* 2010; Kobayashi *et al.* 2011). Furthermore, galaxies care about AGB stars because their dusty winds contribute toward the dust budgets of galaxies, and in particular are copious producers of carbon-rich dust even in metal-poor galaxies. Galaxies dominated by intermediate-age stellar populations emit much of their starlight in the infra-red, which is mostly produced by AGB stars.

Spectacular supernova explosions mark the end of stars more massive than about $10M_\odot$. These explosions release vast quantities of energy and α-elements (e.g., oxygen, magnesium, silicon, calcium) as well as iron into the Galaxy. Binary systems that explode as Type Ia supernovae are also responsible for producing substantial metals, mostly in the form of iron-peak elements. Nomoto, Kobayashi & Tominaga (2013) reviewed the explosive nucleosynthesis from supernovae and their contribution toward the Galactic chemical evolution of galaxies.

When considering the sources of nucleosynthesis in galaxies it is important to consider the different lifetimes of contributing sources. Massive stars have short lifetimes and explode almost instantaneously, compared to the lives of galaxies. In contrast, AGB stars span an enormous range in stellar lifetimes. The most massive stars that experience the AGB phase of about $8M_\odot$ have short lifetimes of 30 million years whereas the lowest masses to enter the AGB have lifetimes of 10 billion years, comparable to the age of the Milky Way Galaxy.

In this review I summarise the mechanism for producing elements heavier than iron in nature along with the sites of heavy element nucleosynthesis.

2. Making heavy elements

The origin of elements heavier than iron is not linked to the nuclear reactions that produce energy inside stars. Heavy elements are instead synthesized by reactions that involve *the addition of neutrons* onto Fe-peak elements. The foundations for the origin of heavy elements was laid down in the seminal review papers of Burbidge *et al.* (1957) and independently by Cameron (1957). Forty years later, Wallerstein *et al.* (1997) provided a comprehensive update, based on the latest observations, theoretical models, and nuclear physics data.

It was proposed that two processes can produce the bulk of heavy elements in our solar system (Fig. 1): The *slow* and *rapid* neutron-capture process (or s and r process, respectively). The s-process occurs when the rates of neutron addition are, in general, slower compared to the timescales for the β-decays of radioactive nuclei. This builds up nuclei along the valley of nuclear stability as shown In Fig 2, where the path of the s and r processes are shown in a section of the chart of the nuclides. We can see that the main s-process path goes through the long-lived unstable Tc isotope, ^{99}Tc, which has a half life of $\approx 2 \times 10^5$ years. Tc was first detected in the spectra of long-lived red giant stars in the early 1950's the first observational confirmation that stars can make elements heavier than iron.

During the r-process there are so many neutrons that radioactive nuclei do not have time to decay before capturing another neutron and this produces nuclei far from nuclear stability. Once the neutron flux is gone the unstable nuclei decay until stable nuclei are produced; the result is some of the rarest heaviest elements found in nature including uranium and thorium. Of interest but not part of this story is the origin of the rare proton-rich isotopes such as ^{94}Mo which are only destroyed by neutron-capture reactions. These isotopes are likely synthesized by combinations of proton captures, β-decays and/or spallation reactions occurring in high energy, explosive environments (e.g., core collapse supernovae).

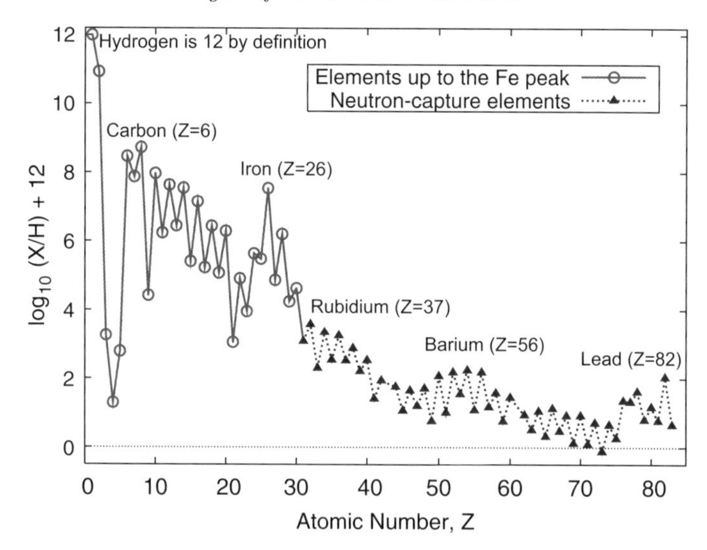

Figure 1. Solar abundance distribution using data from Asplund *et al.* (2009). The main features of the abundance distribution include the hydrogen peak ($Z = 1$) followed by helium ($Z = 2$). The gorge separating helium from carbon, the continuous decrease from carbon to scandium ($Z = 21$), followed by the iron peak, and the gentle downwards slope towards the platinum ($Z = 78$) and lead peaks, ending with the heaviest element found in nature, e.g., uranium ($Z = 92$).

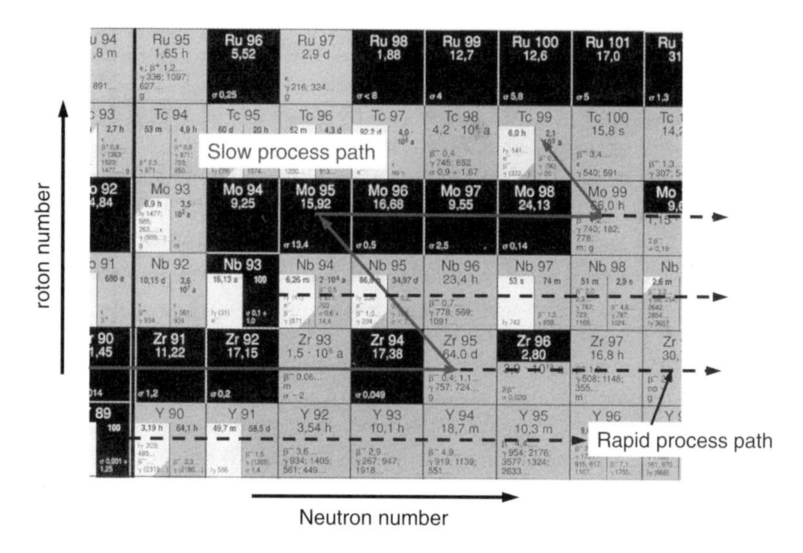

Figure 2. A section of the chart of the nuclides. Neutrons increase on the x-axis and protons on the y-axis. Here we can see isotopes around the Sr to Tc region, where stable isotopes are shown in black, with the solar system fraction provided (e.g., 17.38% for ^{94}Zr). Gray squares show unstable radioactive isotopes, where the half-life is given (e.g., 1.5×10^6 years for ^{93}Zr). The path of the s-process is highlighted by the solid red line, with the r-process path by the dashed blue line.

3. Sites of heavy element nucleosynthesis

The astrophysical site of the s-process in nature is well constrained by observational evidence and theoretical models to occur inside AGB stars. Massive stars may also produce heavy elements via the s-process in hydrostatic evolutionary phases leading up to core collapse. The origin of the r-process was for many years a complete mystery. Owing to recent discoveries coming from the gravitational wave community we have now confirmed merging neutron stars as a site of the r-process.

Here we summarize the sites of heavy element nucleosynthesis in more detail, starting with the r-process.

3.1. *The site(s) of the r-process*

The rapid neutron capture process releases a huge number of neutrons over a few seconds where typical densities are on the order of $N_{\mathrm{n}} > 10^{20}$ n cm^{-3}. Such conditions suggest an explosive site and for a long time core collapse supernovae were the favoured mechanism for producing the r-process. Calculations have so far failed to produce the necessary neutron rich environments (Thielemann *et al.* 2018). Other rarer sites were also proposed including electron-capture supernovae (Wanajo *et al.* 2011), merging neutron stars (Lattimer & Schramm 1976), and magneto-rotationally induced supernova (Winteler *et al.* 2012). Up until 2016 it was unclear what site(s) were responsible (Sneden *et al.* 2008).

In 2016 measurements of Ba and Eu in the ultra-faint dwarf galaxy Reticulum II by Ji *et al.* (2016) provided some hints. The high abundances of these two elements, higher than in other dwarf galaxies, suggested that a single rare r-process event took place that produced the heavy elements. The study ruled out core collapse supernovae; too many of them were needed, which would have blown the already fragile galaxy apart. A complementary study by Wallner *et al.* (2015) of the heavy ^{244}Pu isotope in ocean floor sediments also ruled out core collapse supernovae. Again, a rare source such as merging neutron stars was suggested in order to account for the low abundance of the ^{244}Pu detected.

In 2017 the discovery of gravitational waves by the source GW170817 settled at least part of the mystery. The source of the gravitational waves was determined to originate from a pair of merging neutron stars (Abbot *et al.* 2017). The electro-magnetic counterpart was also discovered to be a red kilonova where the spectral energy distribution was best fit by the decay of radioactive isotopes produced by the r-process (Kilpatrick *et al.* 2017). It has been estimated that $\sim 0.05 M_{\odot}$ of pure r-process elements were expelled (Drout *et al.* 2017).

It is still not clear if merging neutron stars is the *only* site of the r-process in nature. The element Eu is primarily made by the r-process ($\sim 98\%$ in the solar system is attributed to the rapid process). Spectroscopic observations of Eu in old, metal-poor halo stars show high levels of Eu and significant scatter (Sneden *et al.* 2008). Can neutron star mergers occur early enough and frequently enough in the Galaxy to account for the abundances of Eu observed in the halo and disk?

The probability of neutron star mergers as a function of time in galaxies is not well constrained – we only have one confirmed event. Recent studies suggest that the rate of neutron-star mergers determined from theory and observations cannot account for the chemical evolution of Eu in Galactic disk stars (e.g., Hotokezaka *et al.* 2018). This suggests that at least one other, rapid source, of r-process elements may be needed to enrich the Galaxy. Future gravitational wave discoveries from merging neutron stars may help solve this problem, and provide some indication of how normal or unusual GW170817 is.

One issue is that we cannot directly observe the elemental abundances produced in the kilonova. For this reason the best way to constrain models of the r-process is to remove the s-process contribution from the solar system abundances. This technique allows us to obtain an "observational" r-process distribution. This is still the most accurate observational test for comparison to the r-process models, which are extremely uncertain both in the stellar and the nuclear physics.

3.2. *The s-process*

In low and intermediate-mass stars, the s-process can occur during hydrostatic burning of helium in the core or shell, where unstable helium shell burning is characteristic of the AGB phase of evolution. In massive stars the s-process occurs also during hydrostatic burning in the carbon and helium burning shells.

During the s-process neutron densities are typically on the order of $N_{\rm n} \lesssim 10^{13}$ n cm^{-3}, depending on the source of neutrons. In massive stars neutrons are released by the ^{22}Ne(α,n) ^{25}Mg reaction, which operates at temperatures over about 300×10^6 K. The efficiency of the ^{22}Ne neutron source relies on the concentration of ^{22}Ne, which is produced by α-captures onto ^{14}N during He-burning. Normally this is a secondary process and dependent on the initial abundance of CNO nuclei that the star was born with.

However in rapidly rotating massive stars, primary ^{14}N can be produced which in turn leads to an enhanced concentration of ^{22}Ne in the He-shell. We refer to the review by Maeder & Meynet (2012) for details. Rotation means that even low-metallicity massive stars can produce significant s-process elements. Nucleosynthesis calculations include those by Pignatari *et al.* (2008), Frischknecht *et al.* (2016), Choplin *et al.* (2018), and Limongi & Chieffi (2018). The contribution of massive stars toward the s-process is important for elements between Zn and Sr.

3.3. *The s-process: AGB stars*

Theoretical and observational studies of the s-process have a long history. We do not attempt to review the full history here but refer to Busso *et al.* (1999), Herwig (2005), Sneden *et al.* (2008), Käppeler *et al.* (2011), and Karakas & Lattanzio (2014) for references. Below we summarize the operation of the s-process in AGB stars. In Section 4 we review the latest AGB s-process yields predictions.

The first neutron source postulated to occur in AGB stars was also the ^{22}Ne(α,n)^{25}Mg reaction. However because this reaction occurs at temperatures over 300×10^6 K it is only efficient in intermediate-mass AGB stars over about $4M_\odot$. It may also produce a brief burst of neutrons in lower mass AGB stars during their final few thermal pulses. In low-mass AGB stars, neutrons are produced predominantly by the ^{13}C(α,n)^{16}O reaction (e.g., Abia *et al.* 2001). The difficulty with the ^{13}C neutron source is that ^{13}C is absent in the He-shells of AGB stars, unlike ^{22}Ne it is not produced by He-burning reactions. For this reason it has been hypothesised that some mixing of protons from the surrounding envelope occurs into the top layers of the He-shell. These protons are captured by the abundant ^{12}C to form ^{13}C via ^{12}C(p,γ)^{13}N(β^+)^{13}C. Straniero *et al.* (1995) found that the ^{13}C nuclei burn radiatively between pulses, allowing for the release of neutrons and neutron-capture reactions relatively slowly. In AGB stars, typical timescales for neutron captures are on the order of 10^4 years with neutron densities $N_{\rm n} \lesssim 10^8$ n cm^{-3}. AGB stars are particulary important for the production of elements between Sr and Pb.

During the interpulse the neutrons are captured by iron-peak nuclei and converted into heavy elements. The next thermal pulse engulfs the pocket of heavy elements and subsequent third dredge-up mixes some of these heavy elements to the surface (along with C, F etc) where they can be observed. In low-mass AGB stars elements between Sr and Pb can be made in large quantities, where observed [s/Fe] ~ 1 in AGB stars. The exact distribution of heavy elements is dependent on the thermodynamic quantities in the stellar models and depends on the initial mass and metallicity. In Fig 3 we show the final surface composition of low-mass AGB models of [Fe/H] $= -0.7$. In particular the

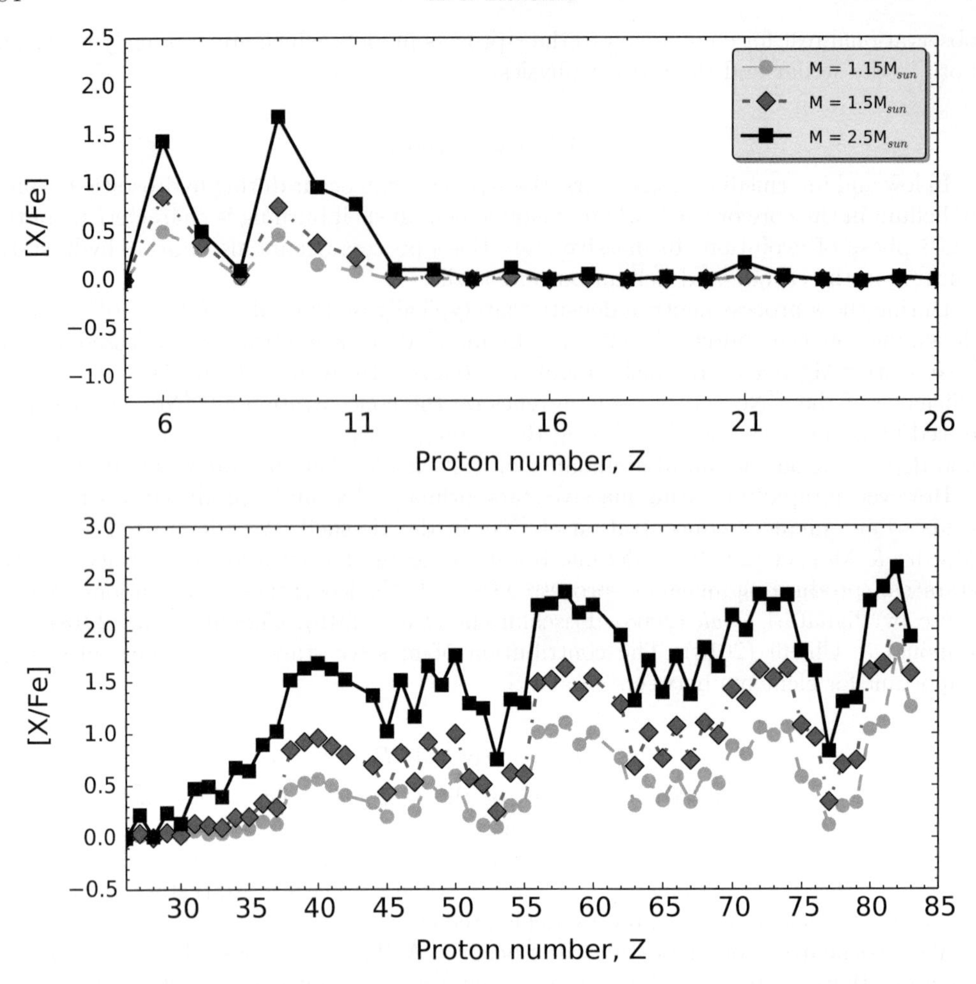

Figure 3. The final surface composition for a selection of low-mass AGB models of [Fe/H] = −0.7. The top panel shows elements lighter than iron while the bottom panel shows elements heavier than iron. Using model predictions from Karakas *et al.* (2018).

bottom panel shows the results for *s*-process elements where large enhancements in Sr ($Z = 38$), Ba ($Z = 56$) and Pb ($Z = 82$) are noticeable. The ratio of Sr/Pb and Ba/Pb in particular depends on the initial metallicity. This is because the amount of ^{13}C made in the He-intershell is primary which means that the ratio of neutrons/Fe increases with decreasing metallicity. This leads to a build up of Pb in low-metallicity AGB models as noticed first by Gallino *et al.* (1998) and later confirmed by observations by van Eck *et al.* (2001).

4. Yields from AGB stars

Stellar yields are an essential ingredient of chemical evolution models. While there are many studies of the *s*-process in AGB stars, it has only been in the last 10 years that tabulated stellar yields including *s*-process elements have been available. For this reason we limit our discussion here to these predictions, noting that other yield sets exist that focus on elements lighter than Fe (e.g., Ventura *et al.* 2013). We also do not discuss the many studies of AGB nucleosynthesis where only surface abundance predictions are given. Bisterzo *et al.* (2010) for example only published surface abundance predictions

Table 1. s-process yields: We only show predictions that include s-process elements as well as yield tables, not just surface abundance predictions.

Reference	Mass Range (in $M_\odot$)	Metallicity Range (in mass fraction, Z)	Downloadable tables?
FRUITY database[a]	1.3–6.0	2×10^{-5} to 0.02	Yes
Monash models[b]	1.0–8.0	1×10^{-4} to 0.03	Yes
NuGrid/MESA[c]	1.5–5.0	0.01, 0.02	Yes

(a) Website: http://fruity.oa-abruzzo.inaf.it/
(b) Data tables available for download from associated papers.
(c) Website: http://www.astro.keele.ac.uk/nugrid/data-and-software/yields/

although the yields from their updated models have been subsequently used in chemical evolution studies (e.g., Bisterzo *et al.* 2017). For a more detailed set of references we refer to Karakas & Lattanzio (2014) noting that this is already 4 years out of date!

In Table 1 we provide a brief summary of the AGB yields available that include s-process elements. The three main groups are the FRUITY models, the Monash/Stromlo calculations and the NuGrid/MESA models. The FRUITY models are described in a series of papers starting with Cristallo *et al.* (2009) and detailed in Cristallo *et al.* (2015). We refer to references given on the FRUITY website (see Table 1) for details of the models. These are currently the only AGB yields of the s-process that include stellar rotation.

The yield tables from the Monash/Stromlo models are available for download from the papers associated with these studies. The main papers are Fishlock *et al.* (2014), Karakas & Lugaro (2016) and Karakas *et al.* (2018). These calculations include the full range of AGB masses up to the CO-core limit of $8M_\odot$ for solar metallicity. Fishlock *et al.* (2014) and Shingles *et al.* (2015) also include a few models of heavy-element yields from super-AGB stars; and the latter paper also includes the effect of helium enrichment on stellar yields.

NuGrid/MESA yields are described in Pignatari *et al.* (2016) although there is an extended set published by Ritter *et al.* (2018). The novelty of the NuGrid/MESA yields is that they also include yields of massive stars calculated using the same codes, which means the same initial abundances and reaction rates. The second set covers a larger range of masses from $1 - 25M_\odot$ and larger range of metallicity, from $Z = 0.02$ to 0.0001, with α-enhancement for lower metallicities. We note that on the NuGrid website given in Table 1 only the first set from Pignatari *et al.* (2016) is currently available.

There are currently considerable gaps in the yields available. These are the most significant at low metallicities, where there are no tabulated yields below $Z \leqslant \text{few} \times 10^{-5}$. Very low metallicity models may experience proton-ingestion episodes at various phases of stellar evolution. Current studies are limited to Cruz *et al.* (2013) and Campbell *et al.* (2010). Furthermore, there are few yields of super-AGB stars that include full s-process calculations, beyond one or two masses discussed already (e.g., the $7M_\odot$, $Z = 0.001$ model in Fishlock *et al.* 2014). The AGB yields also do not consider the effect of a binary companion.

5. Beyond the standard model of nucleosynthesis

We have made considerable progress in understanding the s-process in spite of considerable modelling uncertainties. This is mostly owing to excellent nuclear physics data. The major uncertainty for the s-process is the mechanism for the formation of a ^{13}C-rich region in the He-intershell of AGB stars (Buntain *et al.* 2017). Convective overshoot may be responsible (Herwig *et al.* 1997; Cristallo *et al.* 2009), perhaps helped by magnetic fields (Trippella *et al.* 2016). How stellar rotation affects the s-process is still an open question. Models have found that rotation may inhibit the s-process completely (Herwig *et al.* 2003) or simply modulate the abundances (Piersanti *et al.* 2013).

There are some observations of post-AGB stars that do not appear to fit within the classical s-process scenario in AGB stars, even considering uncertainties associated with modelling and nuclear physics. These include Sakurai's Object, which appears to be best fit by a proton-ingestion episode following a late thermal pulse (Herwig *et al.* 2011). Sakurai's Object was for a long time seen as an anomaly (and it still may be!) and other post-AGB stars were seen as exquisite tracers of nucleosynthesis during the AGB phase (Van Winckel 2003).

The Magellanic Cloud post-AGB stars are some of the most s-process enriched objects known and overall seem to be fit well by s-process AGB nucleosynthesis except for the element Pb (De Smedt *et al.* 2012). There are suggestions that the abundances observed in these post-AGB stars are not made by a typical s-process but would be better fit by an *intermediate* neutron capture process (e.g., Lugaro *et al.* 2015), operating at higher neutron densities and resulting from a proton ingestion episode.

That there may be neutron captures occurring at intermediate neutron densities is not a new idea (Cowan & Rose 1977). Carbon enhanced metal-poor stars with an s and r process enrichment may be better fit by an i-process (Dardelet *et al.* 2015; Hampel *et al.* 2016). Furthermore, accreting white dwarfs may also produce i-process elements, which may be important for chemical evolution (Denissenkov *et al.* 2017). The contribution of the i-process to the Galactic inventory is as yet unknown. Côté *et al.* (2018) suggest the i-process may be important for the elements Sr, Y and Z. Note that Galactic chemical evolution calculations by Prantzos *et al.* (2018) using the yields from the FRUITY database and Limongi & Chieffi (2018) find no need for an extra contribution for the elements Sr, Y and Zr.

6. Summary

In this review we have summarised the sites of heavy element nucleosynthesis in the Galaxy. We have discussed the latest results for the rapid and slow neutron capture processes. While we have a site for the r-process we still require accurate yields of s-process nucleosynthesis in order to better constrain uncertain r-process models. While yields of AGB stars are available, there are still considerable gaps in the parameter space, particularly at low metallicities and at higher masses (e.g., for super-AGB stars). There is evidence for an intermediate-neutron capture process although the site is not well constrained at present. For this reason the contribution of the i-process to the Galactic inventory of heavy elements is currently uncertain.

There is currently an explosion of new stellar abundance data from various large-scale spectroscopic surveys (e.g., the Galah survey, the GAIA-ESO survey, LAMOST, APOGEE etc.). These data will help answer big questions related to the formation and evolution of galaxies but will also provide new data to help constrain stellar physics problems, such as those related to the origin of the elements and chemical evolution in galaxies.

References

Abbott, B. P., *et al.* 2017, *Physical Review Letters*, 119, 161101
Abia, C., *et al.* 2001, *ApJ*, 559, 1117
Bisterzo, S. *et al.* 2010, *MNRAS*, 404, 1529
Bisterzo, S., Travaglio, C., Wiescher, M., Käppeler, F., & Gallino, R. 2017, *ApJ*, 835, 97
Buntain, J. *et al.* 2017, *MNRAS*, 471, 824
Burbidge, E.M., Burbidge, G.R., Fowler, W.A., & Hoyle, F. 1957, *Rev. of Mod. Phys.*, 29, 547
Busso, M., Gallino, R., & Wasserburg, G.J. 1999, *ARAA*, 37, 239

Cameron, A.G.W. 1957, *AJ*, 62, 9
Campbell, S.W., Lugaro, M., & Karakas, A.I., 2010, *A&A*, 522, L6
Choplin, A., *et al.* 2018, *A&A*, in press
Côté, B. *et al.* 2018, *ApJ*, 854, 105
Cowan, J.J., & Rose, W.K. 1977, *ApJ*, 212, 149
Cristallo, S., *et al.* 2009, *ApJ*, 696, 797
Cristallo, S., *et al.* 2015, *ApJ* (Supplement Series), 219, 40
Cruz, M.A., Serenelli, A., & Weiss, A. 2013, *A&A*, 559, A4
Dardelet, L., *et al.* 2015, *Proceedings of Science*, 204, 145
Denissenkov, P.A. *et al.* 2017, *ApJ* (Letters), 834, L10
De Smedt, K. *et al.* 2012, *A&A*, 541, A67
Drout, M. R., *et al.* 2017, *Science*, 358, 1570
Fishlock, C.K., Karakas, A.I., Lugaro, M., & Yong, D. 2014, *ApJ*, 797, 44
Frischknecht, U. *et al.* 2016, *MNRAS*, 456, 1803
Gallino, R. *et al.* 1998, *ApJ*, 497, 388
Hampel, M., Stancliffe, R.J., Lugaro, M., & Meyer, B.S. 2016, *ApJ*, 831, 171
Herwig, F. 2005, *ARAA*, 43, 435
Herwig, F., Bloecker, T., Schoenberner, D., & El Eid, M. 1997, *A&A*, 324, L81
Herwig, F., Langer, N., & Lugaro, M. 2003, *ApJ*, 593, 1056
Herwig, F., *et al.* 2011, *ApJ*, 727, 89
Hotokezaka, K. *et al.* 2018, `arXiv:1801.01141`
Ji, A., Frebel, A., Chiti, A., & Simon, J.D. 2016, *Nature*, 531, 610
Käppeler, F., Gallino, R., Bisterzo, S., & Aoki, W. 2011, *Reviews of Modern Physics*, 83, 157
Karakas, A.I., & Lattanzio, J.C. 2014, *PASA*, 31, e030
Karakas, A.I., & Lugaro, M. 2016, *ApJ*, 825, 26
Karakas, A.I., *et al.* 2018, *MNRAS*, 477, 421
Kilpatrick, C. D., *et al.* 2017, *Science*, 358, 1583
Kobayashi, C., Karakas, A.I., & Umeda, H. 2011, *MNRAS*, 414, 3250
Lattimer, J.M., & Schramm, D.N. 1976, *ApJ*, 210, 549
Limongi, M. & Chieffi, A. 2018, *ApJ* (Supplement Series), 237, 13
Lugaro, M., Karakas, A.I., Stancliffe, R.J., & Rijs, C. 2012, *ApJ*, 747, 2
Lugaro, M., *et al.* 2015, *A&A*, 583, A77
Maeder, A, & Meynet, G. 2012, *Reviews of Modern Physics*, 84, 25
Nomoto, K., Kobayashi, C., & Tominaga, N. 2013, *ARAA*, 51, 457
Piersanti, L., Cristallo, S., & Straniero, O. 2013, *ApJ*, 774, 98
Pignatari, M., *et al.* 2008, *ApJ* (Letters), 687, L95
Pignatari, M., *et al.* 2016, *ApJ* (Supplement Series), 225, 24
Prantzos, N., Abia, C., Limongi, M., Chieffi, A., & Cristallo, S. 2018, *MNRAS*, 476, 3432
Ritter, C., *et al.* 2018, *MNRAS*, 480, 538
Romano, D., Karakas, A.I., Tosi, M. & Matteucci, F. 2010, *A&A*, 522, A32
Shingles, L.J., *et al.* 2015, *MNRAS*, 452, 2804
Sneden, C., Cowan, J.J., & Gallino, R. 2008, *ARAA*, 46, 241
Straniero, O., *et al.* 1995, *ApJ* (Letters), 440, L85
Thielemann, F.-K., *et al.* 2018, *Space Science Reviews*, 214, 62
Trippella, O., Busso, M., Palmerini, S., Maiorca, E., & Nucci, M.C. 2016, *ApJ*, 818, 125
Van Eck, S., Goriely, S., Jorissen, A., & Plez, B. 2001, *Nature*, 412, 793
Van Winckel, H. 2003, *ARAA*, 41, 391
Ventura, P., Di Criscienzo, M, Carini, R., & D'Antona, F. 2013, *MNRAS*, 431, 3642
Wallerstein, G., *et al.* 1997, *Reviews of Modern Physics*, 69, 995
Wallner, A. *et al.* 2015, *Nature Communications*, 6, 5956
Wanajo, S., Janaka, H.-T., & Müller, B. 2011, *ApJ* (Letters), 726, L15
Winteler, C. *et al.* 2012, *ApJ* (Letters), 750, L22

Discussion

QUESTION: It's worth noting that there is currently still no spectroscopic identification of individual elements in the kilonova associated with GW170817.

KARAKAS: I agree. This is an important point.

WIJERS: *Comment:* Lathmer & Schramm perhaps deserve some credit for having predicted mergers as the site of the r-process. *Question:* Do the low- or high-metallicity dominate the total s-process production?

KARAKAS: The low-metallicity stars dominate, for various stellar physics reasons, a.o. because they mix more burnt material into their envelopes.

ZINNECKER: Can you comment on the binary nature of AGB stars and their influence on nucleosynthesis? I recall that SNIa, the nucleosynthetic source of iron, requires binary star evolution. Without binaries, no iron!

KARAKAS: The binary nature of AGB stars is unclear. If a star reaches the AGB with a companion, the orbit will be wide. Hence, mass transfer may occur or even a truncation of the AGB phase. Binaries are likely important for ending the AGB and shaping planetary nebulae. There is still much work to be done on this topic.

HRIVNAK: Regarding i-process and post-AGB stars, are you saying that this formation occurs in the post-AGB phase or that in this phase, with envelope removed, we see these results?

KARAKAS: The i-process occurs when protons are ingested into the convectively burning He-shell. This is more likely to occur in post-AGB stars, which have small envelopes, than in the AGB phase (for metallicities above [Fe/H] ~ -2 dex or so).

D'ANTONA: Again about the i-process. If it occurs only in post-AGB or at the He-core flash, it is important only to explain a few observations, but if it occurs in accreting white dwarfs, it may make an important contribution to abundances evolution. Do we have any observational evidence from observations, say of symbiotic novae or nova ejecta?

KARAKAS: There is no observational evidence for the i-process in novae or symbiotic stars, but observers haven't really looked for this signature.

Why Galaxies Care About AGB Stars: A Continuing Challenge through Cosmic Time
Proceedings IAU Symposium No. 343, 2019
F. Kerschbaum, M. Groenewegen & H. Olofsson, eds.

doi:10.1017/S1743921318005100

The composition of Barium stars and the s-process in AGB stars

Borbála Cseh[1], Maria Lugaro[1,2], Valentina D'Orazi[3], Denise B. de Castro[4], Claudio B. Pereira[5], Amanda I. Karakas[2], László Molnár[1,6], Emese Plachy[1,6], Róbert Szabó[1,6], Marco Pignatari[7,1,8,9] and Sergio Cristallo[10,11]

[1]Konkoly Observatory, Research Centre for Astronomy and Earth Sciences, Hungarian Academy of Sciences, H-1121 Budapest, Konkoly Thege M. út 15-17, Hungary
email: `cseh.borbala@csfk.mta.hu`

[2]Monash Centre for Astrophysics, School of Physics and Astronomy, Monash University, VIC 3800, Australia

[3]INAF Osservatorio Astronomico di Padova, Vicolo dellOsservatorio 5, 35122 Padova, Italy

[4]Department for Astrophysics, Nicolaus Copernicus Astronomical Centre of the Polish Academy of Sciences, 00-716 Warsaw, Poland

[5]Observatorio Nacional, Rua General José Cristino, 77 Sao Cristovao, Rio de Janeiro, Brazil

[6]MTA CSFK Lendület Near-Field Cosmology Research Group

[7]E. A. Milne Centre for Astrophysics, Department of Physics & Mathematics, University of Hull, HU6 7RX, United Kingdom

[8]The NuGrid Collaboration, http://www.nugridstars.org

[9]Joint Institute for Nuclear Astrophysics - Center for the Evolution of the Elements, USA

[10]INAF, Osservatorio Astronomico d'Abruzzo, I-64100 Teramo, Italy

[11]INFN-Sezione di Perugia, I-06123 Perugia, Italy

Abstract. Using abundances from the available largest, homogeneous sample of high resolution Barium (Ba) star spectra we calculated the ratios of different hs-like to ls-like elemental ratios and compared to different AGB nucleosynthesis models. The Ba star data show an incontestable increase of the hs-type/ls-type element ratio (for example, [Ce/Y]) with decreasing metallicity. This trend in the Ba star observations is predicted by low mass, non-rotating AGB models where ^{13}C is the main neutron source and is in agreement with Kepler asteroseismology observations.

Keywords. stars: abundances, nuclear reactions, nucleosynthesis, abundances, stars: AGB and post-AGB

---

1. Introduction

Barium stars are chemically peculiar giants and dwarfs with spectral classes from G to K. These stars are in binary systems where the primary star (a former AGB, now a white dwarf) polluted the companion with *s*-process enhanced material. Since these elements were synthesised during the TP-AGB evolutionary phase, Ba stars can be used to probe AGB *s*-process nucleosynthesis. The temperatures of Ba stars (from over 4000K up to 6500K) are higher than those of late AGB stars ($\simeq$ 3000–4000K), meaning that their spectra is easier to model because of the absence of molecular features in their spectra. Another advantage of using Ba stars instead of AGBs is that many dynamical processes can be present in the atmospheres of AGB stars (such as pulsations, mass loss, and also dust formation) which makes the modelling more complicated (Abia *et al.* 2002, Pérez-Mesa *et al.* 2017).

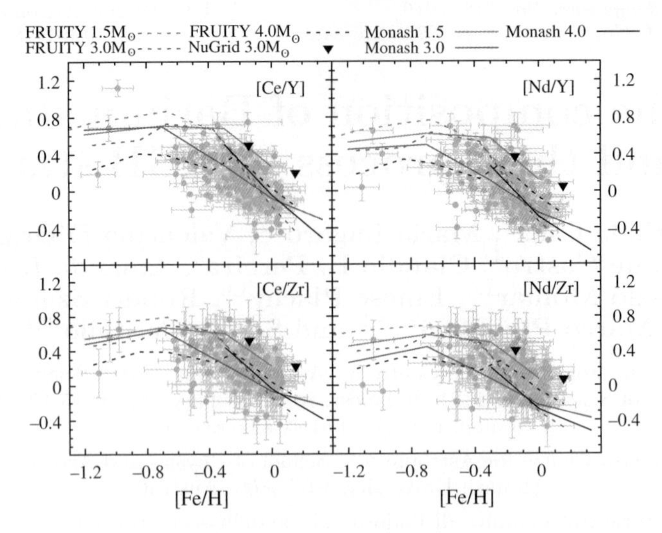

Figure 1. Final surface abundances of different AGB nucleosynthesis models compared to the available hs-like to ls-like elemental ratios for the Ba star sample of de Castro *et al.* (2016). Monash label indicates the models from Karakas & Lugaro (2016) and Fishlock *et al.* (2014), while FRUITY stands for models from Cristallo *et al.* (2015). We show also 3 $M_{\odot}$ type He07 models from Battino *et al.* (2016) for comparison. The dots without error bars represent stars where at least one of the elements have less than three lines. All the models shown here are in the mass range of Ba stars and have [s/Fe] $\geqslant$ 0.25 dex.

The efficiency of the *s*-process has been measured using the abundances of the ls (light-*s*) and hs (heavy-*s*) indexes, taken as the average of different elements belonging to the first (Sr, Y, Zr) and second (Ba, La, Ce, Nd, Sm) *s*-process peaks, respectively. However, the availability and accuracy of the abundances of these elements vary in different studies. Using the largest set of homogeneous observational data of Ba stars published by de Castro *et al.* (2016) we are moving forward from the use of ls and hs to directly use the available elements.

2. Data sample

The largest, self-consistent determination of *s*-process element abundances in Ba giants with masses from 1 to 6 $M_{\odot}$ was published by de Castro *et al.* (2016). In this study a star was considered as Ba star if the limit of [s/Fe] $\geqslant$ 0.25 was reached. This values was calculated as the arithmetic mean of [La/Fe], [Ce/Fe], [Nd/Fe], [Y/Fe] and [Zr/Fe]. Using this homogeneous sample (169 Ba stars in total) we calculated proper errors for the first time for different hs-like to ls-like elemental ratios for each individual star (Cseh *et al.* submitted) instead of using an average value and typical error for the [hs/ls] ratio.

3. Results and conclusions

All of the four calculated combination of hs-to-ls elemental ratios ([Ce/Y], [Ce/Zr], [Nd/Y], [Nd/Zr]) and the calculated error bars based on the sample of de Castro *et al.* (2016) were compared to different AGB nucleosynthesis models. A spread of about a factor of 3 is apparent in the data, indicating that other effects may be present, such as variations in the initial mass of the AGB, magnetic field or rotation.

The four available elemental ratios show a clear increasing trend with decreasing metallicity and are in agreement with the model predictions. All of the models shown in Fig. 1 are the final surface abundances and have [s/Fe] $\geqslant$ 0.25, the same limit as was used for the Ba stars. This value is reached after the first few thermal pulses in the most cases,

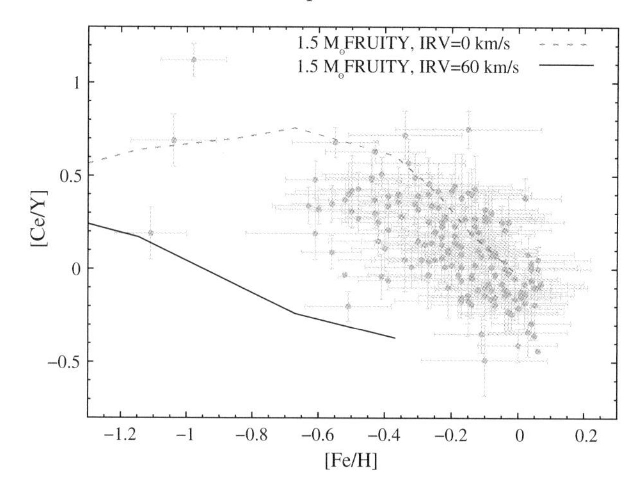

Figure 2. Comparison for [Ce/Y] between the Ba stars and the 1.5 $M_\odot$ rotating and non-rotating (IRV = initial rotational velocity) FRUITY models (Cristallo *et al.* 2015, Piersanti, Cristallo & Staniero 2013). The dots without error bars represent stars for which there are less than 3 lines for one of the elements. All the models shown here have [s/Fe] $\geqslant$ 0.25 dex.

indicating that a Ba star could formed even if the mass transfer occured before the AGB reached the last thermal pulse.

The 1.5 $M_\odot$ rotating models with initial rotational velocity of 60 km/s (Piersanti, Cristallo & Staniero (2013)), producing enough *s*-process elements ([s/Fe] $\geqslant$ 0.25), are shown with their non-rotating counterparts in Fig. 2. The production of [Ce/Y] in the rotating models is much lower than those observed in Ba stars, indicating that strong mixing within the ^{13}C pocket of their AGB companions should not occur. Kepler asteroseismology observations of the cores of red giant stars and of white dwarfs (the ancestors and the progeny of AGB stars, respectively) show low core rotational velocities (Hermes *et al.* 2017), which is in agreement with the independent results from the Ba star data. This points out the need for the existence of a mechanism for angular momentum transport in giant stars, but without mixing the chemical species.

4. Future work

We will continue the analysis of this set of Ba star data with other elements (C, O, and further heavy elements) and compare with other Ba star observations. The possible effect of dilution after the mass-transfer was also not taken into account here. Furthermore, we are planning to compare the observations of different other *s*-process enhanced stars (C, CEMP-, CH- and post-AGB stars) with the Ba star sample in order to identify the signature of other effects (such as the *i*-process).

Acknowledgements

BCs, ML, LM, EP, and RSz acknowledge the support of the Hungarian Academy of Sciences via the Lendület project LP2014-17. BCs and Vd'O acknowledge support from the ChETEC COST Action (CA16117), supported by COST (European Cooperation in Science and Technology). This research was supported in part by the National Science Foundation under Grant No. PHY-1430152 (JINA Center for the Evolution of the Elements). This project was supported by K-115709 grant of the National Research, Development and Innovation Fund of Hungary, financed under the K_16 funding scheme. This project has also been supported by the Lendület Program of the Hungarian Academy of Sciences, project No. LP2018-7/2018. The research leading to these results have been supported by Research, Development and Innovation Office (NKFIH) grants K-115709, PD-116175, and PD-121203. LM and EP were supported by the János Bolyai Research Scholarship of the Hungarian Academy of Sciences. AIK acknowledges financial support from the Australian Research Council (DP170100521). MP acknowledge the support of STFC through the University of Hull Consolidated Grant ST/R000840/1, and ongoing resource allocations on the University of Hulls High Performance Computing Facility viper.

References

Abia, C., Domínguez, I., Gallino, R., Busso, M., Masera, S., Straniero, O., de Laverny, P., Plez, B., and Isern, J.: 2002, *ApJ* 579, 817

Battino, U., Pignatari, M., Ritter, C., Herwig, F., Denisenkov, P., Den Hartogh, J. W., Trappitsch, R., Hirschi, R., Freytag, B., Thielemann, F., and Paxton, B.: 2016, *ApJ* 827, 30

Cristallo, S., Straniero, O., Piersanti, L., and Gobrecht, D.: 2015, *ApJS* 219, 40

Cseh, B., Lugaro, M., D'Orazi, V., de Castro, D. B., Pereira, C. B., Karakas, A. I., Molnár, L., Plachy, E., Szabó, R., Pignatari, M., Cristallo, S., submitted to *A&A*

de Castro, D. B., Pereira, C. B., Roig, F., Jilinski, E., Drake, N. A., Chavero, C., and Sales Silva, J. V.: 2016, *MNRAS* 459, 4299

Fishlock, C. K., Karakas, A. I., Lugaro, M., and Yong, D.: 2014, *ApJ* 797, 44

Hermes, J. J., Gänsicke, B. T., Kawaler, S. D., Greiss, S., Tremblay, P.-E., Gentile Fusillo, N. P., Raddi, R., Fanale, S. M., Bell, K. J., Dennihy, E., Fuchs, J. T.,Dunlap, B. H., Clemens, J. C., Montgomery, M. H., Winget, D. E., Chote, P., Marsh, T. R., and Redfield, S.: 2017, *ApJS* 232, 23

Karakas, A. I. and Lugaro, M.: 2016, *ApJ* 825, 26

Pérez-Mesa, V., Zamora, O., Garca-Hernndez, D. A., Plez, B., Manchado, A., Karakas, A. I., and Lugaro, M.: 2017, *A&A* 606, A20

Piersanti, L., Cristallo, S., and Straniero, O.: 2013, *ApJ* 774, 98

Why Galaxies Care About AGB Stars:
A Continuing Challenge through Cosmic Time
Proceedings IAU Symposium No. 343, 2019
F. Kerschbaum, M. Groenewegen & H. Olofsson, eds.

doi:10.1017/S1743921318007500

Abundances of C, N, and O in AGB Giants and Model Atmospheres

B. Aringer[1], P. Marigo[1], W. Nowotny[2], L. Girardi[1], M. Mečina[2] and A. Nanni[1]

[1]Dipartimento di Fisica e Astronomia Galileo Galilei, Università di Padova, Vicolo dell'Osservatorio 3, I-35122 Padova, Italy
email: bernhard.aringer@unipd.it

[2]Department of Astrophysics, Univ. of Vienna, Türkenschanzstraße 17, A-1180 Wien, Austria

Abstract. Based on hydrostatic models we discuss the effects of molecular opacities and abundance changes concerning C, N or O on the atmospheric structures, spectra and photometric properties of C/M AGB giants.

Keywords. stars: atmospheres, stars: abundances, stars: carbon, stars: AGB and post-AGB

1. Model Atmospheres and Opacities

In this discussion we focus on the warmer AGB stars with effective temperatures above about 2800 to 2900 K showing weak or no pulsation. Their observable properties may be simulated by classical hydrostatic models (e.g. Aringer *et al.* 2009, Aringer *et al.* 2016), while cooler objects with stronger pulsation are dominated by shock waves, dust formation and mass loss plus deviations from spherical symmetry as well as from chemical and local thermodynamic equilibrium. Due to the uncertainties related to the combination of all these complicated phenomena, which are discussed in several contributions presented in this symposium (e.g. Bladh *et al.*, Eriksson *et al.*, Gobrecht *et al.*, Freytag *et al.*, this volume), it remains very difficult or impossible to study abundance effects in the AGB giants with low temperatures and large amplitudes.

In the hydrostatic case the model structures and spectra are mainly determined by a huge number (10^8 to 10^9) of molecular and atomic transitions absorbing a large fraction of the stellar radiation. In Fig. 1 we demonstrate the situation for a typical cool M giant, where TiO and water are the dominant species. Due to the strong influence of the opacities on the atmospheres it is essential to treat the corresponding input data and chemical abundances for the calculation of models and observable properties in a consistent way. Thus, one needs line lists, which are complete enough to derive realistic temperature-pressure structures and accurate enough to get good medium and high resolution spectra. Unfortunately, such data are not always available (especially for carbon stars).

An example for the problems with the molecular opacities in carbon giants can be seen in Fig. 2, where we present C_2H_2 spectra based on various available sources. In addition to the HITRAN 2008 list, which contains only a limited number of measured lines, the plot includes the old SCAN data (Jørgensen *et al.* 1989) still used in our standard C star models and the more recent ASD1000 data (Lyulin & Perevalov 2017). The latter are expected to be more accurate, but above 1000 K they may not be complete. We find a large difference between SCAN and ASD1000, which has also a huge impact on the atmospheric structures. This becomes obvious, when we compare the result of a

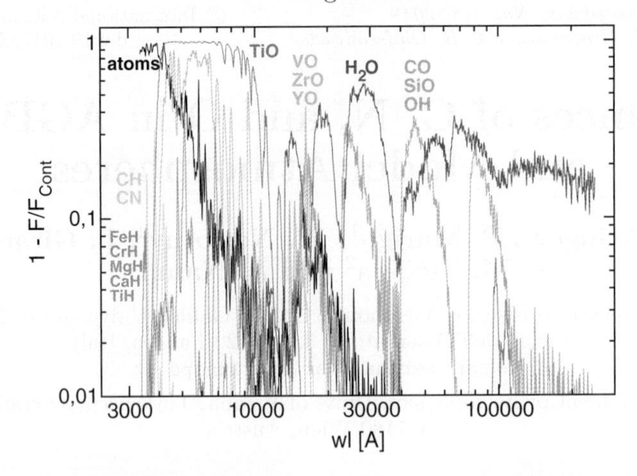

Figure 1. Low resolution $R = 200$ spectra for various species computed from a COMARCS M star model with $T_{\rm eff} = 2800$ K, $\log(g\ [\rm cm/s^2]) = 0.0$, solar mass and abundances. The data are subtracted from a calculation without line opacities.

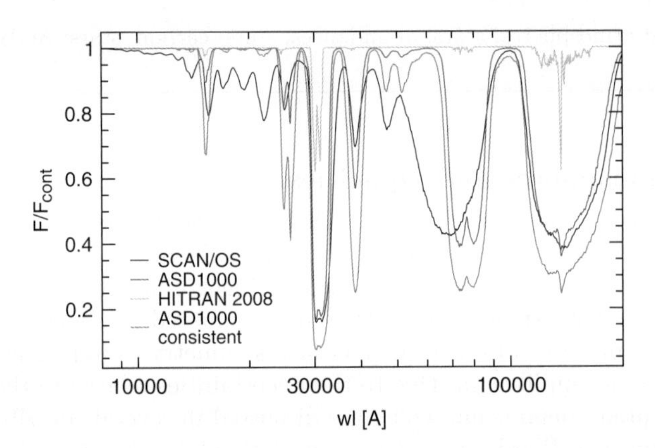

Figure 2. Low resolution R = 200 C_2H_2 spectra computed from a COMARCS carbon star model with $T_{\rm eff} = 2800$ K, $\log(g\ [\rm cm/s^2]) = 0.0$, C/O = 1.4, solar mass and metallicity using different opacity data (see text). The spectra are normalised relative to a calculation without lines.

consistent calculation with ASD1000 to one, where the new list was only taken for the spectral synthesis. Thus, we must conclude that there is still a considerable uncertainty affecting the models of cool C giants.

2. Models with an Enhancement of C, N & O

The study presented here is based on the COMARCS grid of hydrostatic spherical model atmospheres for K, M, S and C stars (Aringer *et al.* 2016). It covers effective temperatures between 2500/2600 and 4500/5000 K, $\log(g\ [\rm cm/s^2])$ values from $-1/0$ to 5 and metallicities in the range $-2 \leqslant [\rm Z/H] \leqslant +1$. C/O ratios between 0.3/0.55 and 2/29 have been considered. In addition, sequences of C and M giant models with deviating oxygen and nitrogen abundances are available in order to derive correction terms, which can be applied to the standard grid, when colours or spectra in synthetic stellar populations are predicted. All spectra and photometric data may be found at http://starkey.astro.unipd.it/atm.

An example for the photometric results can be seen in Fig. 3, where we plot $(J - K)$ as a function of the carbon excess [C−O] for C giant models having 2800 and 3300 K. If

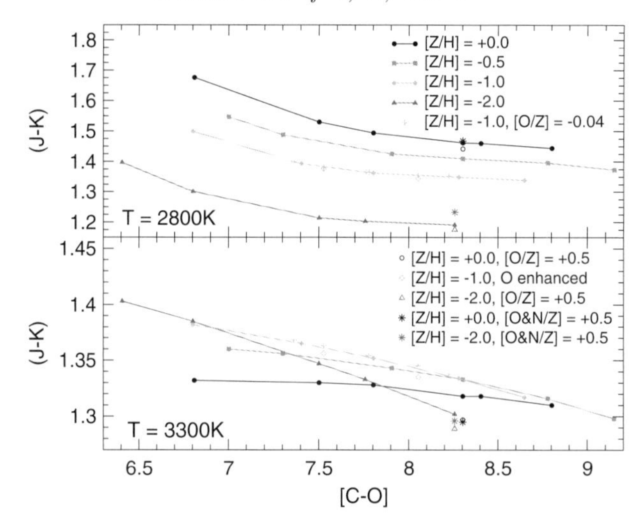

Figure 3. The $(J-K)$ colour as a function of [C$-$O] for COMARCS carbon star models with $\log(g\ [\mathrm{cm/s^2}]) = 0.0$ and one solar mass having $T_{\mathrm{eff}} = 2800$ K and 3300 K. The colours in the plot correspond to different metallicities [Z/H]. Filled symbols connected by lines represent the sequences with standard composition where [O/Z] and [N/Z] remain 0.0. Crosses and open symbols mark results for a deviating oxygen abundance. Asterisks stand for an enrichment of O plus N.

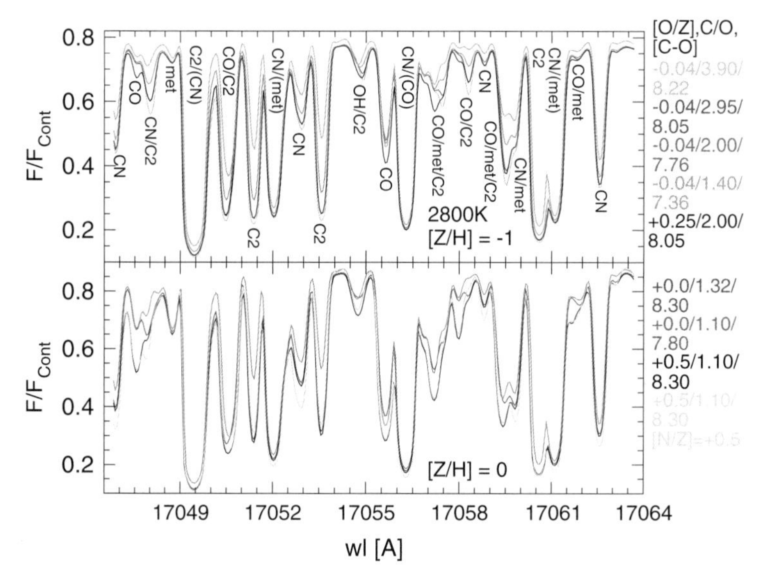

Figure 4. High resolution R = 300 000 spectra based on COMARCS carbon star models with $\log(g\ [\mathrm{cm/s^2}]) = 0.0$ and one solar mass having $T_{\mathrm{eff}} = 2800$ K and [Z/H] = 0 or -1. The data are normalised relative to a calculation without line opacities. The included combinations of [C$-$O], [O/Z] and [N/Z] are shown with different colours. The absorption features have been marked with the species creating them ("met" = metals). If a component of a blend is much weaker than the rest, it was put in brackets.

the amount of oxygen changes, predictions concerning spectra and colours derived from the stellar parameters, metallicity and [C$-$O] are much better than those involving C/O, since [C$-$O] determines the abundances of many important molecules.

Quantities like [C$-$O], [O/Z] or [N/Z] cannot be determined from low resolution spectra and photometric data alone. As we show in Fig. 4, this is in principle possible using high resolution spectra. For C giants [C$-$O] has again a strong effect on many of the lines, which exceeds for the models presented in the plot changes caused by a moderate variation of $T_{\rm eff}$ (100 K) or $\log(g)$ (0.5 dex) and the influence of a thin dust shell. Only the CO transitions depend mainly on the oxygen abundance.

Acknowledgements

This work was supported by the ERC Consolidator Grant funding scheme (project STARKEY, G.A. n. 615604).

References

Aringer, B., Girardi, L., Nowotny, W., Marigo, P., & Lederer, M.T. 2009, *A&A*, 503, 913
Aringer, B., Girardi, L., Nowotny, W., Marigo, P., & Bressan, A. 2016, *MNRAS*, 457, 3611
Jørgensen, U.G., Almlof, J., & Siegbahn, P.E.M. 1989, *ApJ*, 343, 554
Lyulin, O.M., & Perevalov, V.I. 2017, *JQSRT*, 201, 94

Discussion

Andersen: Could you clarify, if you really meant that it is basically impossible to get abundances for AGB stars, or did you just mean "the naughty" ones?

Aringer: We can get abundances for the more hydrostatic stars, if we have good opacities. For the cool pulsating (naughty) objects it will take much longer, since shocks, dust, mass loss, non-LTE, deviations from spherical symmetry and chemical equilibrium come together. It is hard to include everything at the same time in one model!

Pulsation, dynamical atmospheres, and dust formation

Why Galaxies Care About AGB Stars:
A Continuing Challenge through Cosmic Time
Proceedings IAU Symposium No. 343, 2019
F. Kerschbaum, M. Groenewegen & H. Olofsson, eds.

doi:10.1017/S1743921318005926

The evolution of DARWIN: current status of wind models for AGB stars

Sara Bladh

Division of Astronomy and Space Physics, Department of Physics and Astronomy, Uppsala University, Box 516, 751 20 Uppsala, Sweden
email: sara.bladh@physics.uu.se

Abstract. The slow, dense winds observed in evolved asymptotic giant branch (AGB) stars are usually attributed to a combination of dust formation in the dynamical inner atmosphere of these stars and momentum transfer from stellar photons interacting with the newly formed dust particles. Wind models calculated with the DARWIN code, using this mass-loss scenario, have successfully produced outflows with dynamical and photometric properties compatible with observations, for both C-type and M-type AGB stars. Presented here is an overview of the DARWIN models currently available and what output these models produce, as well as future plans.

Keywords. stars: AGB and post-AGB, stars: atmospheres, stars: carbon, stars: mass loss, stars: winds, outflows, shock waves, stellar dynamics

---

1. Introduction

Stellar winds in evolved AGB stars are generally considered to be pulsation-enhanced dust-driven outflows (PEDDRO, see Höfner and Olofsson 2018, for a more detailed explanation). This mass-loss scenario is built on a two-stage process, where in the first stage, the contracting and expanding photosphere of the star triggers shock waves that propagate through the steep density gradient of the atmosphere. The shock waves transfer kinetic energy, originating in the pulsations, into the atmosphere, leading to almost ballistic trajectories of the gas. The result is an extended atmosphere with layers of enhanced density at higher altitudes, and consequently, cooler temperatures: an environment favourable to dust formation. In the second stage of this wind scenario, momentum is transferred to the newly formed dust particles by absorption and scattering of the numerous stellar photons reaching the dust formation zone. Friction between the accelerated dust particles and the surrounding gas then triggers a general outflow. A visualisation of the two-stage process of pulsation-enhanced dust-driven outflows is given in Fig. 1.

2. The DARWIN code

The atmospheres and winds of AGB stars are modeled using the 1D radiation-hydrodynamic code DARWIN (Dynamic Atmosphere and Radiation-driven Wind models based on Implicit Numerics). The DARWIN code produces time-dependent radial structures of the atmospheres and winds of AGB stars by simultaneously solving the hydrodynamic equations (conservation of mass, momentum, and energy), frequency-dependent radiation transfer and time-dependent grain growth. These wind models are spherically symmetric, with an inner boundary situated just below the photosphere and an outer boundary at about 25 stellar radii in models that develop a wind. The initial structure is a hydrostatic

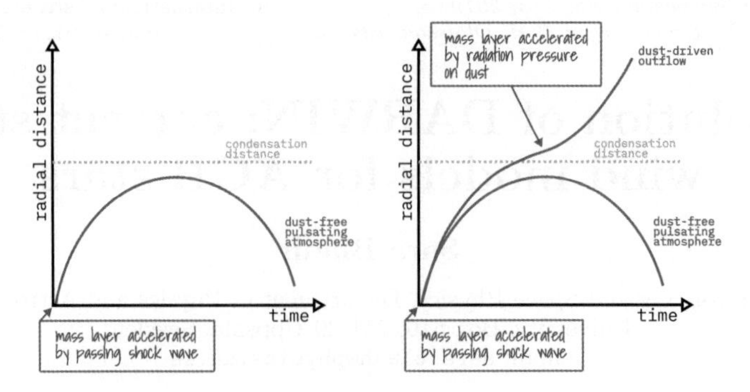

Figure 1. Visualisation of the two-stage process of the pulsation-enhanced dust-driven outflow (PEDDRO) scenario. Image by courtesy of S. Liljegren.

model atmosphere, characterised by the fundamental stellar parameters (current stellar mass $M_\star$, stellar luminosity $L_\star$, effective temperature $T_\star$) and chemical composition. The variability of the star is simulated by a sinusoidal variation (described by the pulsation period P, velocity amplitude u_p, and a scaling factor f_L) of radius and luminosity at the inner boundary. The amplitude of this variation is gradually ramped up during the first pulsation cycles, thereby turning the initial hydrostatic atmosphere into a dynamical atmosphere. Figure 2 shows atmospheric structures of a DARWIN model at maximum and minimum luminosity (at $\phi = 0.0$ and 0.5, respectively). The different panels show gas density (upper panels), dust density (middle panels) and velocity (lower panels). Note the gas moving inwards and outwards in the inner atmosphere, before the material is accelerated outwards by radiation pressure on dust.

A detailed description of the DARWIN code can be found in Höfner *et al.* (2016). The output from these wind models compares well with observed wind properties and photometry (Nowotny *et al.* 2010, 2011; Eriksson *et al.* 2014; Bladh *et al.* 2013, 2015), illustrating their capability to reproduce the overall momentum transfer and spectral energy distribution of AGB stars.

3. The available DARWIN models

The DARWIN code exists in two versions; one for M-type AGB stars, where the wind is driven by photon scattering on Fe-free silicate grains of sizes comparable to the wavelength of the flux maximum, and one for C-type AGB stars, where the wind is driven by photon absorption on amorphous carbon grains. In both versions of the code the dust grains are assumed to be spherical and the optical properties are calculated using Mie theory.

The DARWIN models for C-type AGB stars include a description for the nucleation of amorphous carbon grains, based on classical nucleation theory. The carbon abundance is treated as a free parameter, as carbon may be dredged-up during the thermal pulses when AGB stars evolve. The relevant quantity for wind-driving in carbon stars is the carbon excess, C-O, since it is the amount of free carbon in the atmosphere, i.e. the carbon that is not bound in CO molecules, that is important for dust formation. There exist two generations of DARWIN models for C-type AGB: the first generation of models assumed that dust particles are small compared to the wavelengths at which the stars emit most of their stellar flux and the dust opacity is calculated in the small particle limit (SPL). In the second generation of models this assumption is relaxed and size-dependent dust opacities (SDO) are used instead. The first large model grid for C-type AGB stars was published by Mattsson *et al.* (2010), providing mass-loss rate, wind velocities, and dust

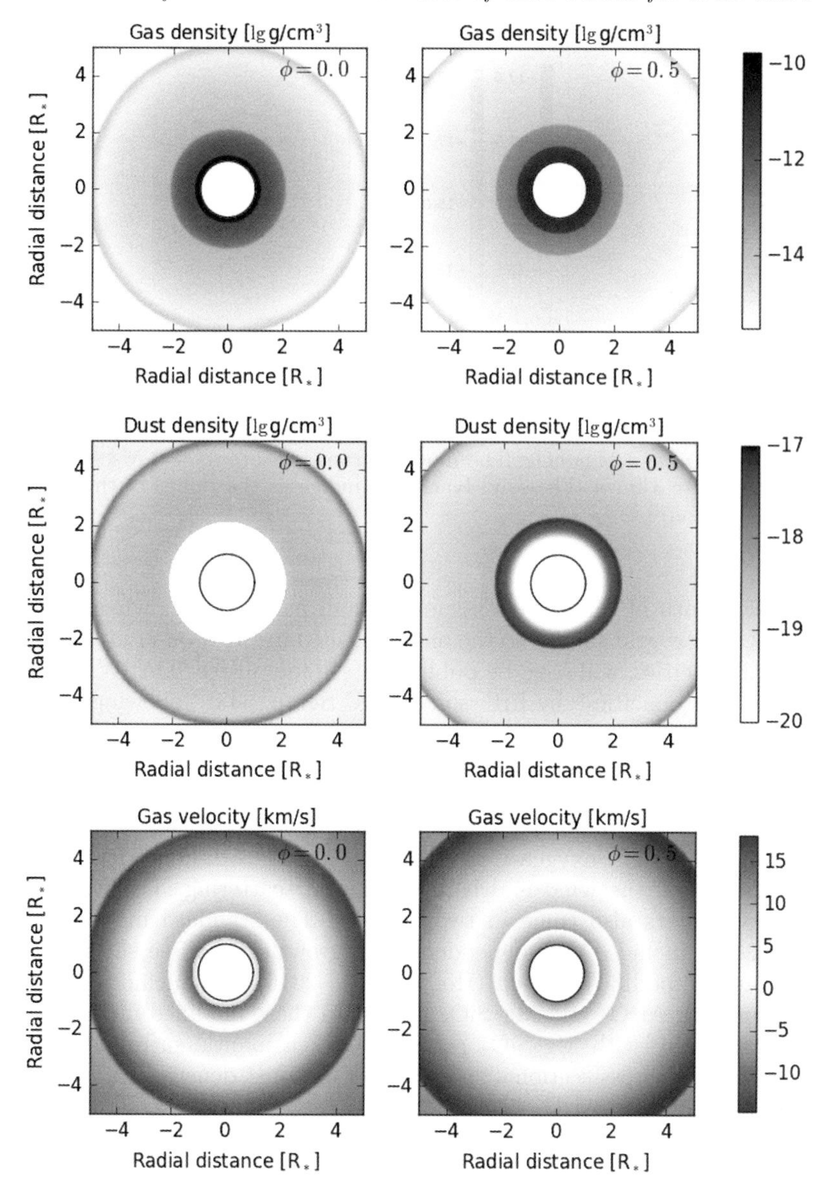

Figure 2. Cross-sectional snapshots of a DARWIN model for a C-type AGB star, with $M_* = 1\,\mathrm{M}_\odot$, $\log L_* = 4.00\,\mathrm{L}_\odot$, $T_* = 2600\,\mathrm{K}$, $u_\mathrm{p} = 2\,\mathrm{km/s}$, and $[\mathrm{Fe/H}] = -1.0$. The upper, middle and lower panels show gas density, dust density, and velocity of the gas, respectively. The left and right panels show cross-sectional snapshots at maximum ($\phi = 0.0$) and minimum ($\phi = 0.5$) luminosity, respectively. The solid black line indicates the radius of the star, calculated from Stefan-Boltzmann's law.

yields for a wide range of stellar parameters (also see contributed talk "Calibrating TP-AGB stellar models and chemical yields through resolved stellar populations in the SMC" by Pastorelli at IAU Symp. 343). This publication also includes a mass-loss routine that can be used in stellar evolution modelling. A slightly smaller grid of DARWIN models for C-type AGB stars, with updated opacities, was published by Eriksson *et al.* (2014). It provides mass-loss rates, wind velocities, and dust yields, but also photometry and a

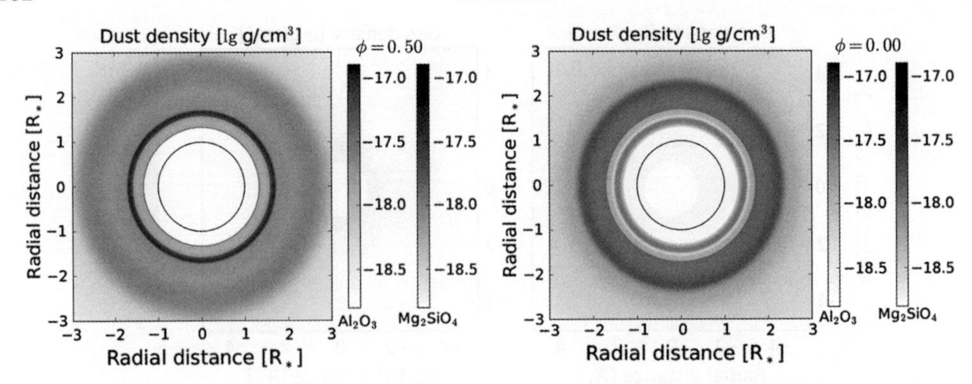

Figure 3. Cross-sectional snapshots at maximum (left panel, $\phi = 0.0$) and minimum (right panel, $\phi = 0.5$) luminosity, showing the dust density of an oxygen-rich DARWIN model with grain growth of both Al_2O_3 and Mg_2SiO_4. The model parameters are $M_* = 1\,M_\odot$, $\log L_* = 3.70\,L_\odot$, $T_* = 2700\,K$, and $u_p = 3\,km/s$. The solid black line indicates the radius of the star, calculated from Stefan-Boltzmann's law.

spectral library. Both of these grids assumed small particle limit when calculating the dust opacities. A new grid similar to the one published in Eriksson *et al.* (2014), but with size-dependent opacities, will soon be published (see the poster "DARWIN C-star model grid with new dust opacities" by Eriksson at IAU Symp. 343), following up on the pilot study by (Mattsson and Höfner 2011)

DARWIN models of M-type AGB stars include pre-existing seed particles that start to grow when the thermodynamical conditions are favourable, since the nucleation processes in oxygen-rich environments is still not fully understood (see e.g. Gail *et al.* 2016; Gobrecht *et al.* 2016). The oxygen-rich models all include size-dependent dust opacities, which is crucial when the wind is driven by photon scattering. The DARWIN models for M-type AGB stars also come in two flavours. The standard model (Höfner 2008; Bladh *et al.* 2015) includes time-dependent dust growth of Fe-free silicates (Mg_2SiO_4), but there is also a modified version (Höfner *et al.* 2016) that includes time-dependent dust growth of both Al_2O_3 and Mg_2SiO_4 grains. In these modified models Al_2O_3 forms the core of the grain, with Mg_2SiO_4 condensing in a mantle surrounding the Al_2O_3-cores. Figure 3 shows cross-sectional snapshots of the dust density of a DARWIN model with grain growth of both Al_2O_3 and Mg_2SiO_4. Note that Al_2O_3 condenses in the close vicinity of the star, with Mg_2SiO_4 growing on top of the Al_2O_3-core a bit further out. The first grid of DARWIN models for M-type AGB stars was published by Bladh *et al.* (2015). This grid only included one solar mass models. A new extensive grid of oxygen-rich DARWIN models, including models with different masses, will soon be published (Bladh *et al.* 2018b).

There is also an ongoing effort to explore wind models in sub-solar metallicity environments. A first exploratory work for carbon-rich DARWIN models was published by Mattsson *et al.* (2008), and a more extensive investigation will be published soon in Bladh *et al.* (2018a), showing that if carbon excess is used as defining parameter, then the mass-loss rates at sub-solar metallicities agree well with those at solar metallicities for the same stellar parameters. From this it can be concluded that as long as AGB stars manage to dredge up sufficient carbon from the interior, they can contribute to dust production even at lower metallicities. Furthermore, this indicates that stellar evolution models can use the C-type DARWIN solar grids for mass-loss rates at sub-solar metallicities if they select models with the same carbon excess. Figure 4 gives an overview of the DARWIN models currently available or soon to be published.

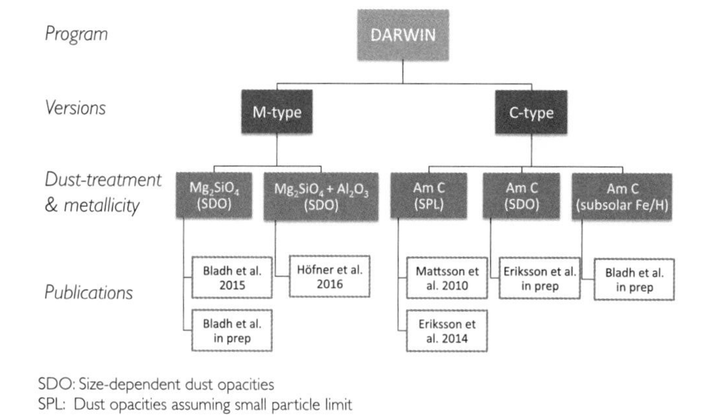

Figure 4. Overview of DARWIN models currently available or soon to be published.

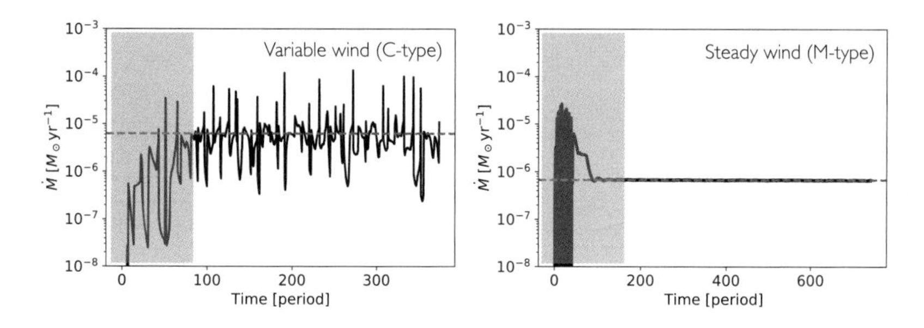

Figure 5. Mass-loss rates at the outermost layers of two DARWIN models as a function of time. The averaged mass-loss rates is marked with a red line. The shaded area indicates the early pulsation periods that are excluded in order to avoid transient effects when calculating the average mass-loss rate.

4. Output from DARWIN models

The DARWIN models consist of long time-series of snapshots of the atmospheric structures. From these time-series we can derive wind properties, i.e., the wind velocity and mass-loss rate at the outermost layers of the model, averaged over typically hundreds of pulsation periods. The early pulsation periods are excluded to avoid transient effects of ramping up the amplitude. Figure 5 shows the wind properties for two such time-series. The left panel is an example of a stellar wind from a DARWIN model where the mass-loss rate over time oscillates quite a lot around the average mass-loss rate (indicated by the red line). The right panel shows a DARWIN model where the mass-loss rate over time instead is extremely steady.

Another important output from the time-series is grain properties. By noting the degree of material condensed into dust in the outer mass layers averaged over time we can calculate the average dust-to-gas ratio for each model. Figure 6 shows the gas-to-dust mass ratios for a collection of oxygen-rich DARWIN models at solar metallicities.

The snapshots of the atmospheric structure can themselves be used to study physical processes in the extended atmosphere, i.e., maser emission or polarisation of light by dust. As an example of this, Aronson *et al.* (2017) investigated the polarisation of light by dust at different phases and wavelengths using DARWIN models for M-type AGB stars. Processes affected by the shocks in the inner atmosphere, like non-equilibrium chemistry, can potentially also be studied using DARWIN models.

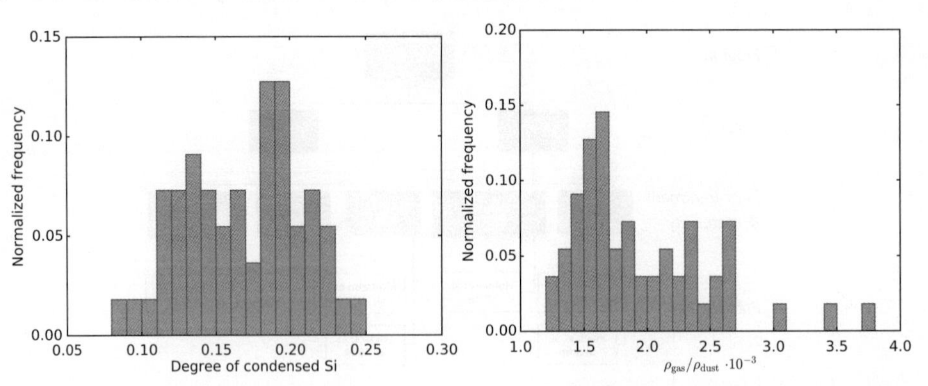

Figure 6. Degree of condensed Si (left panel) and gas-to-dust ratios by mass (right panel) for a subset of the DARWIN models of M-type AGB stars presented in Bladh *et al.* (2015).

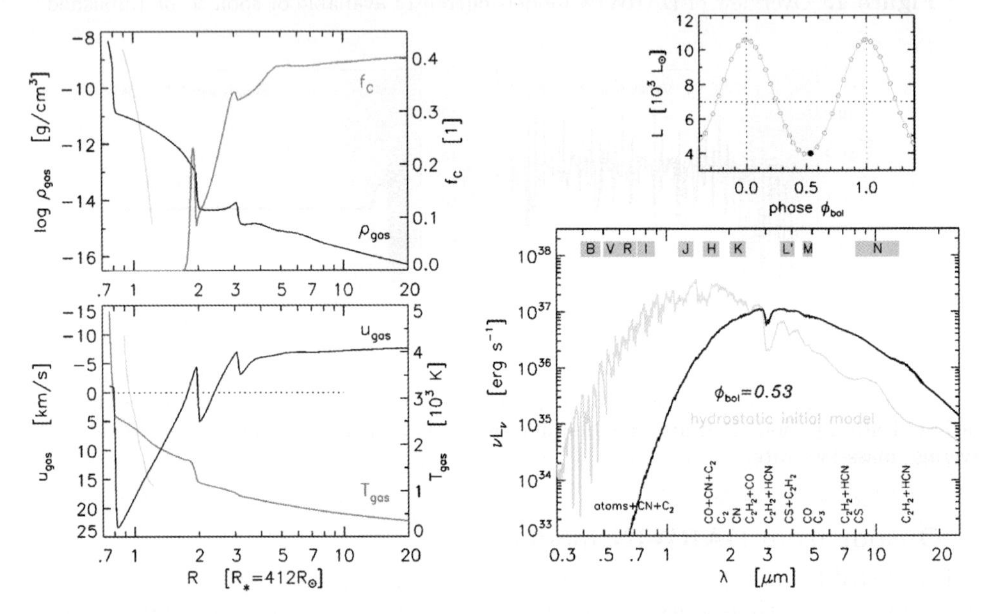

Figure 7. Radial profiles from an atmospheric snapshot of a DARWIN model of C-type (model S from Nowotny *et al.* (2011)), showing gas density and degree of condensed carbon in the top left panel, and gas velocity and temperature in the bottom left panel. The luminosity as a function of bolometric phase is shown in the top right panel, and the spectrum at a bolometric phase of 0.53 is shown in the lower right panel. The corresponding properties for the initial static model is shown in green. Image by courtesy of W. Nowotny.

Detailed *a posteriori* radiative transfer calculations of the atmospheric snapshots, using opacity sampling program COMA (Aringer *et al.* 2016), can produce synthetic spectra, photometry and interferometric visibilities that can be compared with observations directly. Figure 7 shows an example of a snapshot of the atmospheric structure and the corresponding spectrum from a C-type DARWIN model. Comparisons between synthetic and observed photometry have been made for DARWIN models of both C-type and M-type (Nowotny *et al.* 2011; Eriksson *et al.* 2014; Bladh *et al.* 2015). Individual CO rotation-vibration lines, that depend on the shock dynamics, can be synthesised and compared to observations to probe the structure of the inner atmosphere. Such studies have been done for both M-type and C-type AGB stars (Nowotny *et al.* 2011; Liljegren *et al.* 2016, 2017). A first attempt at comparing high-resolution spectra from DARWIN models of M-type and the X-shooter Spectral Library was presented in the contributed

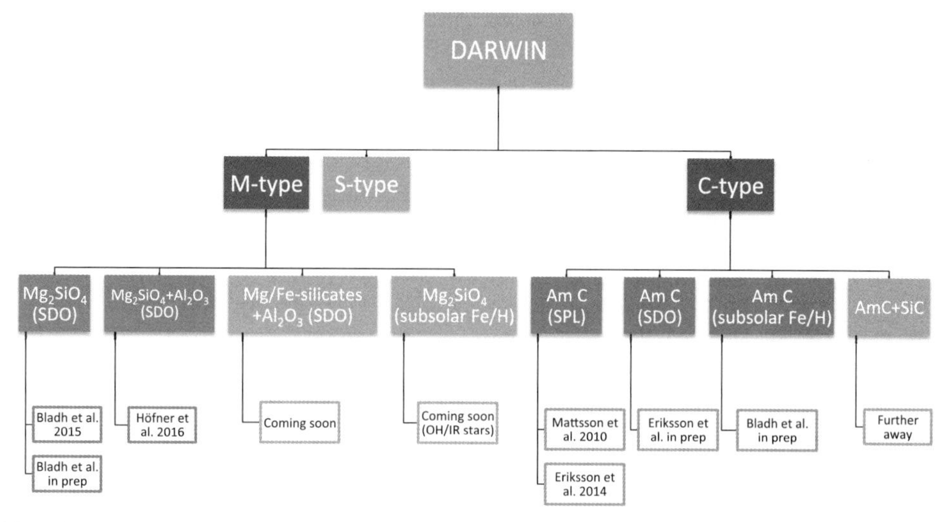

Figure 8. Overview of DARWIN models currently available or to be published soon, with potential future plans marked in orange.

talk "O-rich LPVs in the X-shooter Spectral Library" by Lancon at IAU Symp. 343. Examples of comparisons between observed and synthetic interferometric visibilities can be found in Sacuto *et al.* (2013); Bladh *et al.* (2017) for M-type AGB stars, and in Sacuto *et al.* (2011); Rau *et al.* (2017); Wittkowski *et al.* (2017) for C-type AGB stars.

5. Future plans

There are two kinds of challenges lying ahead: one is to further develop the description in the DARWIN models and the other is to increase the parameter space of the current model grids and producing model grids at different metallicities.

The plan for the immediate future concerning model development is to make a more complete model for M-type AGB stars, by constructing a model version that includes the growth of both alumina and silicates, and allows for a variable Mg/Fe ratio in the silicate grains. With a more complete model for M-type AGB stars, the steps towards a model for S-type AGB stars become more feasible. A bit further down on the to-do list is the plan to include SiC dust in the DARWIN models for C-type AGB stars. There is also an ongoing effort to explore if it is possible to improve the DARWIN models to account for non-symmetrical effects (see, e.g., Liljegren *et al.* 2018, and her talk "Lumpy stars and bumpy winds" at IAU Symp. 343).

Plans exist for expanding the model grids to include more extreme stellar parameters, but also to other metallicities. This is needed for a better understanding of stellar evolution during the AGB phase, as well as dust production in the early Universe. Figure 8 shows the potential future plans marked in orange, in addition to the DARWIN models currently available or soon to be published.

References

B. Aringer, L. Girardi, W. Nowotny, P. Marigo, and A. Bressan. *MNRAS*, 457: 3611–3628, Apr. 2016. doi: 10.1093/mnras/stw222.

E. Aronson, S. Bladh, and S. Höfner. *Astronomy & Astrophysics*, 603:A116, July 2017. doi: 10.1051/0004-6361/201730495.

S. Bladh, S. Höfner, W. Nowotny, B. Aringer, and K. Eriksson. *Astronomy & Astrophysics*, 553: A20, May 2013. doi: 10.1051/0004-6361/201220590.

S. Bladh, S. Höfner, B. Aringer, and K. Eriksson. *Astronomy & Astrophysics*, 575:A105, Mar. 2015. doi: 10.1051/0004-6361/201424917.

S. Bladh, C. Paladini, S. Höfner, and B. Aringer. *Astronomy & Astrophysics*, 607:A27, Oct. 2017. doi: 10.1051/0004-6361/201731090.

S. Bladh, K. Eriksson, B. Aringer, S. Liljegren, and P. Marigo. *In prep.*, 2018a.

S. Bladh, S. Liljegren, S. Höfner, B. Aringer, and P. Marigo. *In prep.*, 2018b.

K. Eriksson, W. Nowotny, S. Höfner, B. Aringer, and A. Wachter. *Astronomy & Astrophysics*, 566:A95, June 2014. doi: 10.1051/0004-6361/201323241.

H.-P. Gail and E. Sedlmayr. *Astronomy & Astrophysics*, 347:594–616, July 1999.

H.-P. Gail, M. Scholz, and A. Pucci. *Astronomy & Astrophysics*, 591:A17, June 2016. doi: 10.1051/0004-6361/201628113.

D. Gobrecht, I. Cherchneff, A. Sarangi, J. M. C. Plane, and S. T. Bromley. *Astronomy & Astrophysics*, 585:A6, Jan. 2016. doi: 10.1051/0004-6361/201425363.

S. Höfner. *Astronomy & Astrophysics*, 491:L1–L4, Nov. 2008. doi: 10.1051/0004-6361:200810641.

S. Höfner and H. Olofsson. *Astronomy & Astrophysics Review*, 26:1, Jan. 2018. doi: 10.1007/s00159-017-0106-5.

S. Höfner, S. Bladh, B. Aringer, and R. Ahuja. *Astronomy & Astrophysics*, 594:A108, Oct. 2016. doi: 10.1051/0004-6361/201628424.

C. Jäger, J. Dorschner, H. Mutschke, T. Posch, and T. Henning. *Astronomy & Astrophysics*, 408:193–204, Sept. 2003. doi: 10.1051/0004-6361:20030916.

S. Liljegren, S. Höfner, W. Nowotny, and K. Eriksson. *Astronomy & Astrophysics*, 589:A130, May 2016. doi: 10.1051/0004-6361/201527885.

S. Liljegren, S. Höfner, K. Eriksson, and W. Nowotny. *Astronomy & Astrophysics*, 606:A6, Sept. 2017. doi: 10.1051/0004-6361/201731137.

S. Liljegren, S. Höfner, B. Freytag, and S. Bladh. *ArXiv e-prints*, Aug. 2018.

L. Mattsson and S. Höfner. *A&A*, 533:A42, Sept. 2011. doi: 10.1051/0004-6361/201015572.

L. Mattsson, R. Wahlin, S. Höfner, and K. Eriksson. *Astronomy & Astrophysics*, 484:L5–L8, June 2008. doi: 10.1051/0004-6361:200809689.

L. Mattsson, R. Wahlin, and S. Höfner. *Astronomy & Astrophysics*, 509:A14, Jan. 2010. doi: 10.1051/0004-6361/200912084.

W. Nowotny, S. Höfner, and B. Aringer. *Astronomy & Astrophysics*, 514:A35, May 2010. doi: 10.1051/0004-6361/200911899.

W. Nowotny, B. Aringer, S. Höfner, and M. T. Lederer. *Astronomy & Astrophysics*, 529:A129, May 2011. doi: 10.1051/0004-6361/201016272.

G. Rau, J. Hron, C. Paladini, B. Aringer, K. Eriksson, P. Marigo, W. Nowotny, and R. Grellmann. *A&A*, 600:A92, Apr. 2017. doi: 10.1051/0004-6361/201629337.

F. Rouleau and P. G. Martin. *ApJ*, 377:526–540, Aug. 1991. doi: 10.1086/170382.

S. Sacuto, B. Aringer, J. Hron, W. Nowotny, C. Paladini, T. Verhoelst, and S. Höfner. *A&A*, 525:A42, Jan. 2011. doi: 10.1051/0004-6361/200913786.

S. Sacuto, S. Ramstedt, S. Höfner, H. Olofsson, S. Bladh, K. Eriksson, B. Aringer, D. Klotz, and M. Maercker. *Astronomy & Astrophysics*, 551:A72, Mar. 2013. doi: 10.1051/0004-6361/201220524.

S. Van Eck, P. Neyskens, A. Jorissen, B. Plez, B. Edvardsson, K. Eriksson, B. Gustafsson, U. G. Jørgensen, and Å. Nordlund. A grid of MARCS model atmospheres for late-type stars. II. S stars and their properties. *A&A*, 601:A10, May 2017. doi: 10.1051/0004-6361/201525886.

M. Wittkowski, K.-H. Hofmann, S. Höfner, J. B. Le Bouquin, W. Nowotny, C. Paladini, J. Young, J.-P. Berger, M. Brunner, I. de Gregorio-Monsalvo, K. Eriksson, J. Hron, E. M. L. Humphreys, M. Lindqvist, M. Maercker, S. Mohamed, H. Olofsson, S. Ramstedt, and G. Weigelt. *A&A*, 601:A3, May 2017. doi: 10.1051/0004-6361/201630214.

Discussion

DECIN: The plots in which you show that the mass-loss rate and wind velocity are independent with respect to metallicity only for carbon-rich stars or for both carbon and oxygen-rich stars?

BLADH: The result that the mass-loss rate is independent of metallicity for a sample of models at solar and sub-solar metallicity, if we compare models with the same input parameters and the same carbon excess, is for carbon stars only. Carbon stars manufacture the constituents needed for the wind-driving dust species internally (namely carbon), irrespectively of the metallicity environment they are in. The wind-driving dust species in M-type AGB stars, however, requires minerals, and a lower abundance of minerals will affect the ability to produce stellar winds at lower metallicities.

SAHAI: How do the wind parameters get affected as one goes towards an S-type composition, i.e. close to C/O=1? Since S-type stars are known to have strong winds, e.g., X Cygni.

BLADH: There seems to be an ongoing debate of what C/O-ratios S-type AGB stars actually have and it may be as low as C/O=0.5 for some early-type AGB stars (Van Eck *et al.* 2017). We have managed to produce winds for models at C/O=0.7, but this was not a systematic investigation. Theoretically, if S-type AGB stars all have C/O-ratio close to 1, then it will not be possible to drive a wind with the current DARWIN models.

QUESTION: You shared pictures of irregular dust grains (amorphous carbon, Mg_2SiO_4) with sharp edges. What approximation or method do you use for calculation of the dust optical properties - e.g. spherical grains, ellipsoidal grains or other?

BLADH: The time-dependent description of grain growth in DARWIN models follows the method of Gail and Sedlmayr (1999). We assume spherical grains, and calculate the dust opacity with Mie theory. Both dust-species are assumed to be amorphous (amC and Mg_2SiO_4). The optical data is taken from Rouleau and Martin (1991); Jäger *et al.* (2003).

Why Galaxies Care About AGB Stars:
A Continuing Challenge through Cosmic Time
Proceedings IAU Symposium No. 343, 2019
F. Kerschbaum, M. Groenewegen & H. Olofsson, eds.

doi:10.1017/S1743921318006038

Molecular dust precursors in envelopes of oxygen-rich AGB stars and red supergiants

Tomasz Kamiński

Harvard-Smithsonian Center for Astrophysics, 60 Garden Street, Cambridge, MA 02138
Submillimeter Array Fellow, email: tkaminsk@cfa.harvard.edu

Abstract. Condensation of circumstellar dust begins with formation of molecular clusters close to the stellar photosphere. These clusters are predicted to act as condensation cores at lower temperatures and allow efficient dust formation farther away from the star. Recent observations of metal oxides, such as AlO, AlOH, TiO, and TiO_2, whose emission can be traced at high angular resolutions with ALMA, have allowed first observational studies of the condensation process in oxygen-rich stars. We are now in the era when depletion of gas-phase species into dust can be observed directly. I review the most recent observations that allow us to identify gas species involved in the formation of inorganic dust of AGB stars and red supergiants. I also discuss challenges we face in interpreting the observations, especially those related to non-equilibrium gas excitation and the high complexity of stellar atmospheres in the dust-formation zone.

Keywords. stars: AGB and post-AGB; circumstellar matter; stars: mass loss; stars: winds, outflows; dust, extinction; ISM: molecules

1. Dust formation and seed particles in O-rich stars

Galaxies certainly care about AGB stars — in the local Universe, galaxies owe huge amounts of dust to these unexhaustive factories of cosmic solids. Inorganic dust, that is, the silicate and alumina dust, is formed primarily in oxygen-rich environments of M-type AGB stars and red supergiants. Despite the crucial role of dust in a broad range of astrophysical phenomena, the formation mechanisms of stardust are poorly understood, mainly because they involve complex problems of dynamic stellar atmospheres and shock-driven chemistry. Observations readily indicate the presence of hot inorganic dust close to stellar photospheres which points to efficient formation of first solids, or *seeds*, out of molecular gas at temperatures of ~1100–1700 K. I describe here observational efforts to identify these gas-phase species and physio-chemical processes involved in the nucleation process close to a pulsating atmosphere. The complexities of the dust nucleation processes in circumstellar envelopes are described in detail in Gail & Sedlmayr (2013).

2. How can we identify the dust precursors?

Theoretical predictions. The quest to identify inorganic seed particles was initiated and is continued by theoretical studies. The strongest requirement for seed particles is that they have to form from gas-phase molecules that are abundant and have high nucleation rates. For instance, Jeong *et al.* (2003) considered Fe, SiO, SiO_2, TiO and TiO_2, and Al_2O_3 as dust precursors but only TiO and TiO_2 appeared to have high enough nucleation rates in the relevant physical conditions [see also Gail & Sedlmayr (1998) and Sharp & Huebner 1990]. Given the low abundance of titanium in material of cosmic composition, efficient seed formation from titanium oxides requires nearly all titanium to be depleted into these first solids, which is a questionable but observationally verifiable assumption.

From the theoretical standpoint, the role of hot silicon oxides in dust formation is debated and was recently reinstated after new laboratory measurements Gail *et al.* (2013). Alumina oxides have long been strong candidates for the seed particles, including in recent studies of Cherchneff (2013), Gobrecht *et al.* (2016), and Höfner *et al.* (2016) (see also Dell'Agli *et al.* 2014). A wide range of other species were considered (e.g. Ferrarotti & Gail 2002, Plane 2013), but most studies strongly favor Al_2O_3 and TiO_2 as the best candidates. Can we verify these theoretical predictions through observations (or experiments)?

Presolar grains. Laboratory measurements of pre-solar grains deliver important clues on the nucleation process. In particular, a handful of presolar grains were found to contain titanium oxides. Many more grains were identified to contain alumina solids. There are cases where alumina cores are surrounded by silicate mantles (Nittler *et al.* 2008), in accord with some of the theoretical expectations. However, the studies of presolar grains have their biases and limitations, for instance, related to the location of the Solar System and analysis methods that give us access to the largest grains only. The field is advancing with an increasing number of analyzed grains of different types (Takigawa *et al.* 2018, Leitner *et al.* 2018).

Mid-infared spectroscopy. One way to study the nucleation process is to directly observe the first hot solids in the closest M-type stars. To distinguish the hot newly-formed dust from the bulk of warm dust produced at larger radii from the star, such studies require very high angular resolutions available only to interferometers. Different chemical types and crystalline/amorphous forms of dust can be investigated through broad spectral features, located mainly in the mid-infrared (MIR). One such study was conducted by Karovicova *et al.* (2013) using the MIDI instrument on the Very Large Telescope Interferometer (VLTI). For the M-type mira GX Mon, they found that alumina dust particles must be present at radii as close as $2.1\,R_\star$ from the star and silicates are only present at distances greater than $5.5\,R_\star$ (where $R_\star$ denotes the radius of the star). This stratification is consistent with what we would expect if alumina dust provides the condensation cores for silicate dust formation. Unfortunately, such studies are still very rare (Norris *et al.* 2012, Khouri *et al.* 2015) and MIR spectroscopy does not always lead to unique identification of the carriers (Decin *et al.* 2017).

Gas depletion. In this contribution, we explore yet another observational possibility to study nucleation, that is, through signatures of depletion of the gas-phase precursors of the first solids that condense. The idea is simple: we should look for simple molecules whose abundance drops down in the dust-formation zone owing to their nucleation and incorporation into solids. This, in principle, should allow us to figure out what kind of compounds are involved in the process and how effective the nucleation is. Let us consider an example that we schematically illustrate in Fig. 1. Assuming the fist solids are formed from titanium oxides, dust formation must start with the simplest molecule, TiO. As a plethora of studies that go back to the earliest days of optical stellar spectroscopy show, titanium monoxide exists already in the cool photospheres of M-type stars (say, at 2000–3000 K). Once it is transported to the circumstellar material (whatever the transport process may be) it reacts with water, very abundant in these O-rich environments, forming TiO_2. The parcel of gas is moved away from the photosphere and, as it becomes cooler, more and more of TiO is converted into gaseous TiO_2. In the LTE calculations of Gail & Sedlmayr (1998), all TiO is oxidized to TiO_2 at a temperature of about 1000 K or distances equivalent to $3\,R_\star$. At even lower temperatures, gaseous TiO_2 (and TiO) may react with H_2 to form first simple solids which may stick to each other and form clusters. These clusters then become the condensation cores for silicate material at even larger distances from the star where the bulk of dust is formed. If we were to observe such an idealized envelope in spectral lines, we would see a hot (say >1400 K) layer of gaseous TiO surrounding the stellar photosphere. It would be embedded in a more extended layer

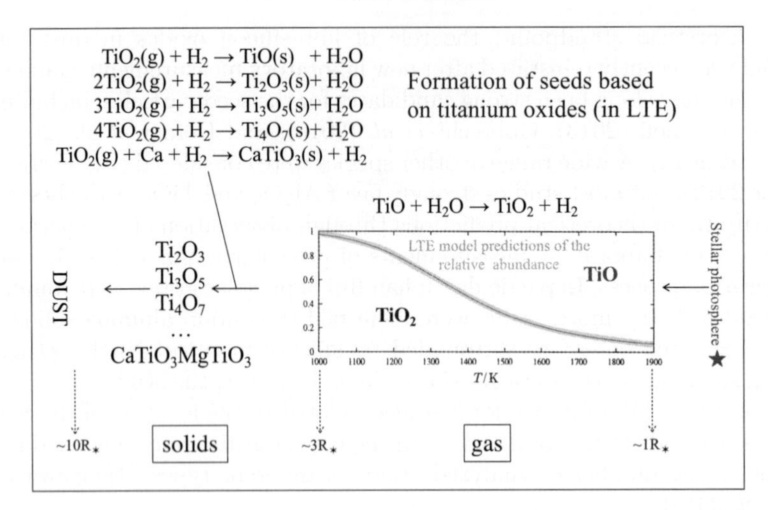

Figure 1. Illustration of seed and dust formation starting from titanium oxides. The sketch shows the stellar photosphere on the right side and the schematic representation of the different chemical processes taking place at different distances from the star. The distances increase to the left and are not in scale. See text for details. Figure is based on Gail & Sedlmayr (1998).

of TiO_2 gas characterized by lower temperatures, of ~1000 K. Spectroscopic signatures of both oxides should disappear at distances larger than $3\,R_\star$, if both species are effectively depleted to form dust. Since the stellar atmospheres of our interest are very much influenced by (pulsation) shocks (e.g. Liljegren *et al.* 2016), the real dust-formation zones are infinitely more complex, especially when it comes to chemical processes involved (Cherchneff 2006, Cherchneff 2012, Cherchneff 2013, Gobrecht *et al.* 2016). However, the nucleation scenario sketched here serves as an illustration of the idea how observations of gas phase molecules can help us understand the dust formation in stars.

About a decade ago, we did not have good direct access to tracers of any of the potential dust precursors in circumstellar gaseous media. A few stars had been known to exhibit circumstellar *emission* in optical electronic bands of TiO and AlO but these bands alone were difficult to interpret and observations were scarce (Kamiński *et al.* 2013a). Seeking the understanding of circumstellar nucleation gave a strong incentive to look for molecular dust precursors especially at submillimeter (submm) wavelengths because (I) pure rotational spectra of simple molecules lead to unambiguous identification of the carriers and (II) with the advent of interferometric arrays with unprecedented angular resolution *and* sensitivity, especially ALMA, we are able to spatially resolve the regions where first solids form. Fortunately, such molecular submillimeter tracers were found in the last few years. The first such observations were acquired towards the iconic red supergiant VY CMa.

3. Titanium oxides in the red supergiant VY CMa and AGB stars

VY CMa is the prime source for first-detection experiments because with an extreme mass-loss rate of $10^{-4}\,M_\odot\,yr^{-1}$ it accumulated a massive (~0.1–1.0 $M_\odot$) envelope rich in molecules and dust. The circumstellar cloud of VY CMa has a characteristically complex substructure seen at optical to millimeter wavelengths (Humphreys *et al.* 2007), which is most likely related to huge convective cells on the surface of this supergiant. The envelope was observed in an interferometric line survey obtained with the Submillimeter Array (SMA; Kamiński *et al.* 2013c). For each channel in the 279.1–355.1 GHz range, a high-sensitivity map at a resolution of about 1 arcsec, corresponding to 1200 AU, was produced, sufficient to resolve emission from most molecules in the envelope. Among the

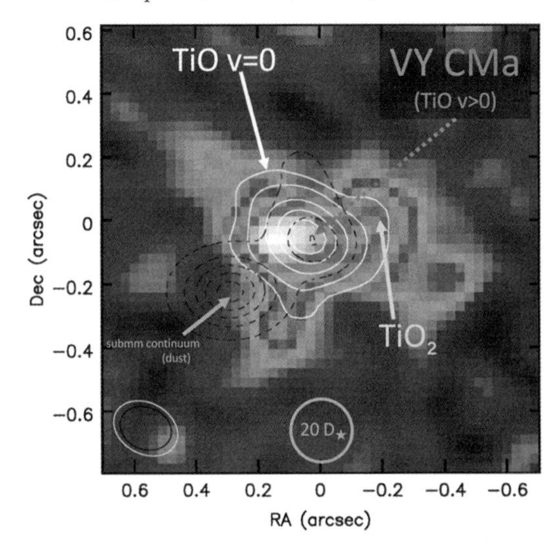

Figure 2. Emission of titanium oxides around the red supergiant VY CMa as seen by ALMA in 2015 (Kamiński *et al.*, in prep.). The position of the star is marked with a dashed arrow and it is the same position where emission of TiO from excited vibrational states was found by Gottlieb *et al.* (in prep.). Solid contours show the emission of TiO and dashed contours show the submillimeter continuum emission of dust. The background image represents TiO_2 emission.

nineteen identified molecules were TiO and TiO_2. This was the first detection of pure rotational emission of TiO in an astronomical object at submillimeter wavelengths and of TiO_2 at any wavelength (Kamiński *et al.* 2013b). The detection was later confirmed with the Plateau de Bure Interferometer (PdBI).

Observing both of these oxides in rotational lines was very encouraging, giving us hope that we can study dust nucleation in stars using submillimeter interferometers. The first observations of VY CMa showed already some interesting features. The emission of TiO_2 was resolved and could be traced out to a radius of about 50 $R_\star$. The excitation temperature of the TiO_2 gas was low, 225 K only. By combining spectra from SMA, PdBI, and UVES at the Very Large Telescope, Kamiński *et al.* derived the extent of the TiO emission to be 30 $R_\star$ and an excitation temperature of about 600 K. These observational characteristics were in stark contrast to the simple prediction outlined here in Sect. 2: both titanium oxides are abundant in the outer, cool parts of the envelope where the dust is already fully formed. This readily indicated that, at least in the violent environment of a supergiant such as VY CMa, titanium oxides do not play a major role in dust nucleation (alternatively, shock sputtering may bring back titanium oxides to the gas phase in the outer envelope; see Kamiński *et al.* 2013b).

Eventually, VY CMa was observed in lines of TiO and TiO_2 with ALMA (De Beck *et al.* 2015; Kamiński *et al.*, in prep). As shown in Fig. 2, the ALMA maps expose both emission regions in great detail. In particular, they show the transition from monoxide- to dioxide-dominated regions. Additionally, it was recently recognized that the SMA survey spectra contain emission features of TiO from vibrationally excites states ($v > 0$; Gottlieb *et al.*, in prep.) giving us an unprecedented view on the gas-phase titanium chemistry throughout the envelope. All these recent observations reinforce the conclusion that titanium oxides are unlikely dust precursors in VY CMa.

4. Titanium oxides in AGB stars

Soon after the first detection of TiO and TiO_2 in VY CMa, both oxides were detected in three M-type AGB stars: *o* Ceti (Kamiński *et al.* 2017), R Dor and IK Tau (Decin *et al.* 2018b). Mira was observed in the greatest detail with multiple instruments, including

APEX, *Herschel*, and ALMA. Emission of both oxides was resolved by ALMA – see Fig. 3. The emission of TiO is rather compact, homogeneous, and well centered on the position of the star. We can trace its emission in a region with an e-folding radius of about $4\,R_\star$ and a typical excitation temperature of 470 K. The emission of TiO_2 is more patchy (or clumpy), not centered on the stellar position, and slightly more extended than TiO with a Gaussian radius of $5.5\,R_\star$; the excitation temperature of TiO_2 is lower, 174 K. Optical spectra additionally reveal the presence of atomic Ti within a few $R_\star$. It actually appears that we trace nearly all of the elemental titanium available within these few $R_\star$ — the derived molecular abundances are very close to the elemental abundance of Ti. With such high abundances of gaseous titanium oxides, it is very unlikely that they provide the first seeds for dust formation in Mira. The theoretical studies emphasize that nucleation starting from titanium oxides requires very efficient condensation of these Ti species, which is clearly not the case in Mira, even if we consider all uncertainties involved in deriving the molecular abundances.

That titanium oxides do not play the major role in dust nucleation was concluded before for VY CMa. Decin *et al.* (2018b) arrived at a similar conclusion after analyzing both oxides in R Dor and IK Tau in ALMA maps. Generalizing, it is therefore very likely that in most evolved O-rich stars the formation of solids does not start with titanium oxides but with some other compounds. This result is surprising because it is widely believed that titanium is almost completely depleted from the gas phase in the interstellar medium and there are certainly presolar AGB grains that contain titanium oxides. Perhaps titanium and titanium oxides are incorporated into grains in the region where silicates form (Kamiński *et al.* 2017).

5. Alumina dust precursors in VY CMa?

Alumina seeds, if present in circumstellar envelopes, were proposed to be composed mainly of Al_2O_3 clusters (Gobrecht *et al.* 2016). It is however unclear what the primitive gas precursors of these solids are (Álvarez-Barcia & Flores 2016); most studies *assume* it is the simplest oxide, i.e. AlO, and observational efforts focused mainly on this molecule, too. The first millimeter-wave observation of rotational transitions of AlO was obtained over a decade ago by Tenenbaum & Ziurys (2009) and the detection experiment was again conducted towards the red supergiant VY CMa. Other Al-bearing molecules were also observed in its complex envelope, including AlOH (Tenenbaum & Ziurys 2010) and AlCl (Kamiński *et al.* 2013c). Although AlO is easiest to observe owing to a combination of abundance and partition-function effects, it is actually AlOH that is the most abundant carrier of Al in VY CMa. These early studies found that the gas-phase molecular carriers of Al spread to large distances from the star, i.e. out to tens of $R_\star$. The bulk of the gas is therefore not directly related to the dust formation close to the star.

Since these pioneering observations, we improved the observational basis of Al-bearing species in VY CMa. The $B - X$ electronic bands of AlO in the optical was analyzed in detail in Kamiński *et al.* (2013a). A range of pure rotational transitions of AlO were observed in wide band surveys with *Herschel* and the SMA (Alcolea *et al.* 2013; Kamiński *et al.* 2013c); most recently, Gottlieb *et al.* (in prep.) were able to recognize pure rotational lines from excited vibrational states of AlO. Very detailed maps of the AlO and AlOH emission regions were obtained with ALMA at ~0.1 arcsec resolution (Kamiński *et al.*, in prep.) showing very intricate distributions of the molecular line emission with respect to each other and with respect to the location of dust clumps and the stellar photosphere (see Fig. 4). It appears that the gas-phase chemistry of Al-bearing species in this supergiant is particularly complex and the relation of the rich observational material to the state-of-the-art chemical models of AGB stars (Gobrecht *et al.* 2016) is unclear and perplexing. There is hope that Al chemistry in "better-behaving" AGB stars is going

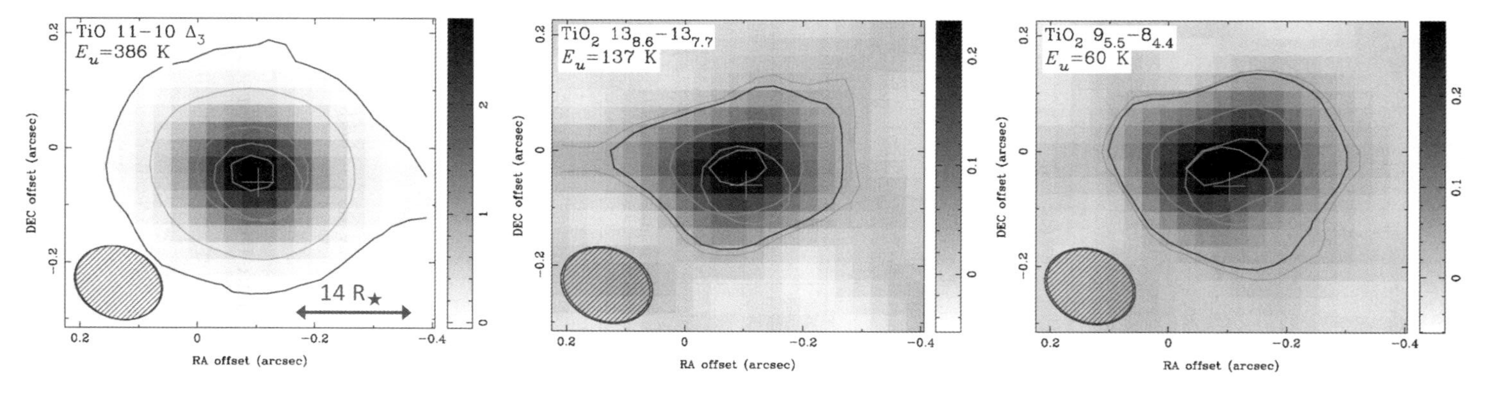

Figure 3. Emission of titanium oxides in Mira as seen by ALMA in 2015 (Kamiński *et al.* 2017). Left panel shows emission of TiO and the two other panels show emission of TiO_2 in the transitions indicated in each panel.The cross marks the position of the star and the associated ellipse represents the extent of the beam-smeared radio-photosphere. Light contours show molecular emission at 10, 50, and 90% of the peak and the dark contours represent the 3×rms noise levels of each map. ALMA beams are also shown (lower left corners).

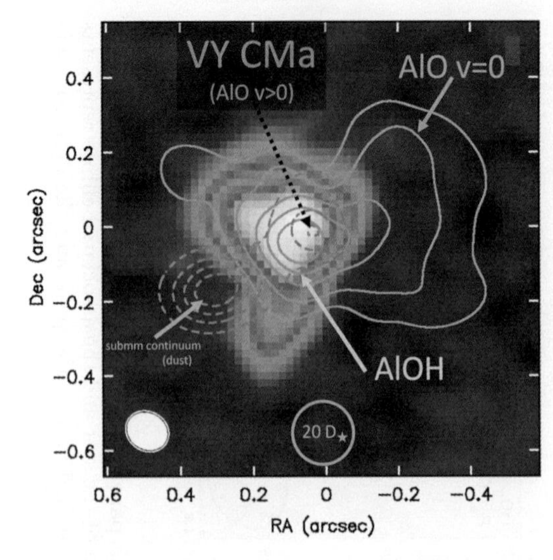

Figure 4. Emission of aluminium oxides around the red supergiant VY CMa as seen by ALMA in 2015 (Kamiński *et al.*, in prep.). The position of the star is marked with a dashed arrow and it is the same position where emission of AlO from excited vibrational states was found by Gottlieb *et al.* (in prep.). Solid contours show the emission of AlO and dashed contours show the submillimeter continuum emission of dust. The background image represents AlOH emission.

Object	silicate index	IR $R_\star$ (mas)	distance (pc)	period (d)	v_{exp} (km/s)	$\dot{M}$ ($M_\odot$/yr)	observed Al-bearing species	main AlO reference
o Ceti	SE8	14	107	333	5	$2 \cdot 10^{-7}$	Al I-II, AlH, AlO, AlOH, AlF?	Kamiński et al. (2016)
R Leo	SE2	20	114	310	7	$4 \cdot 10^{-7}$	AlO, AlOH, AlCl, AlF?	Kamiński et al. (in prep.)
W Hya	SE8	20	78	361	8	$1 \cdot 10^{-7}$	AlO, AlOH	Takigawa et al. (2017)
IK Tau	SE7	20	260	470	18	$50 \cdot 10^{-7}$	AlO, AlOH, AlCl?	Decin et al. (2017)
R Dor	SE7t	57	60	338	6	$2 \cdot 10^{-7}$	AlO, AlOH, AlCl?	Decin et al. (2017)
R Aqr	SE7	10	214	387	15	$3 \cdot 10^{-7}$	AlO	De Beck et al. (2017)
VY CMa	–	6	1200	irregular	45	$1000 \cdot 10^{-7}$	Al I, AlO, AlOH, AlCl	Kamiński et al. (2013b)

Figure 5. A list of evolved oxygen-rich stars observed in Al-bearing species at (sub-)millimeter wavelengths. The main reference reporting the most relevant data is listed in the last column. The stellar and envelope properties listed here are only approximate and were compiled from a variety of sources. The "silicate indexes" are from Sloan & Price (1998).

to be easier to interpret and indeed will allow us to shed more light on the alumina-dust formation.

6. AlO and AlOH as the alumina-dust precursors in AGB stars

Encouraged by the successful detection experiments in VY CMa, there has been several observational studies of gas-phase Al-bearing species in M-type AGB stars (Kamiński *et al.* 2016; De Beck *et al.* 2017; Decin *et al.* 2017; Takigawa *et al.* 2017). The sources with successful detection are listed and briefly characterized in Fig. 5. Most were observed only in AlO but the number and depth of alumina nucleation studies is increasing fast. The list of detected AGB sources covers already a relatively wide range of envelope or stellar properties, including different mass-loss rates, pulsation periods, wind terminal velocities, and types of infrared spectra [as indicated by the silicate-sequence index of Sloan & Price (1998)].

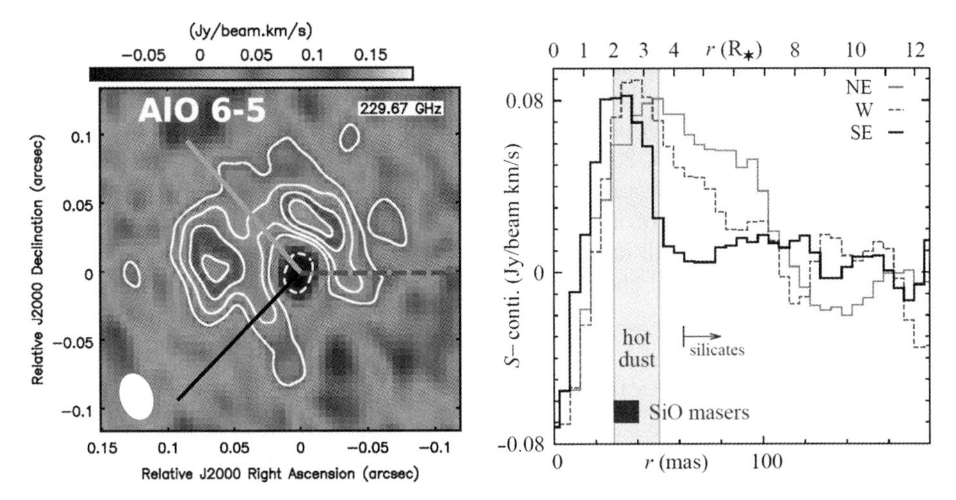

Figure 6. Left: Emission of AlO in Mira as observed with ALMA (background image and contours). The stellar radio photosphere is resolved in these observations and its physical size is illustrated with an ellipse in the center. We observe AlO absorption (the dashed white contour) towards the stellar disk from cool gas located between us and the disk. The emission regions show that the distribution of warm AlO gas is very clumpy. **Right:** Sample spatial profiles of AlO intensity along three cuts shown in the left panel. The drop in intensity at 2.0–3.5 $R_\star$ coincides with the location of hot dust in Mira and may be a signature of AlO depletion into solids. Reproduced from Kamiński *et al.* (2016).

Among these sources, Mira was characterized in most detail and in the largest number of Al-bearing species. In particular, ALMA observations at a resolution of ∼30 mas resolved the radio photosphere and the dust-forming region of Mira. The distribution of AlO emission mapped with ALMA is shown in Fig. 6 (Kamiński *et al.* 2016). The emission is particularly inhomogeneous or patchy. Spatial cuts, also shown in Fig. 6, illustrate that the main emission region extends up to a radius of about 3.5 $R_\star$ beyond which the emission drops significantly. Numerous lines of other species are observed in the same region and do not show such a drop. It was therefore suggested that the observed emission distribution reflects mainly changes in abundance rather than being an excitation effect. Since hot dust (of yet unknown mineralogy) has been directly observed in Mira at radii 2.0–3.5 $R_\star$ in different pulsation phases, it is most tempting to interpret the drop in AlO abundance at ∼3.5 $R_\star$ as a signature of depletion into first solids. However, there is currently no way of knowing whether AlO is converted (in a chain of reactions) to solid Al_2O_3 or to other Al-bearing gas-phase species. In the region where AlO and hot dust are collocated, the abundance of AlO constitutes only 1–10% of the total reservoir of elemental Al. This is fully consistent with effective condensation of alumina solids at 2.0–3.5 $R_\star$ from AlO but, admittedly, other interpretations are possible. Based on observations at lower angular resolutions, Decin *et al.* (2017) derived similarly low abundances (∼1%) of AlO and AlOH in the envelopes of R Dor and IK Tau. Contrary to what was concluded above for seed formation from titanium oxides, it therefore appears that AlO or AlOH may well be the precursor species of the first solids in O-rich AGB stars.

If indeed the clumpy AlO gas in Mira's envelope gives origin to the first solids, the distribution of hot dust should be clumpy too. This is currently investigated in Mira through infrared interferometic imaging with GRAVITY on the VLTI (K. T. Wong, priv. comm.). Moreover, the relative distributions of (polarized) continuum traced with VLT/SPHERE and rotational emission of AlO were also analyzed by Khouri *et al.* (see the contribution in this volume). Objects other than Mira were studied in this context as well. The AGB star W Hya was imaged at high resolution with ALMA and its AlO emission shows similar features as these seen in Mira: it extends only out to $\sim 3\,R_\star$ and shows strong inhomogenities with the azimuthal angle (Takigawa *et al.* 2017). Fortunately, within 150 days of the ALMA observations, W Hya was observed with VLT/SPHERE in scattered light (Ohnaka *et al.* 2016). These optical observations show remarkably similar spatial characteristics of dust and AlO emission. Takigawa *et al.* interpreted this striking resemblance as a signature of nucleation of AlO into solid alumina, but the similarity may simply reflect the overall density variations within the inner envelope. The relation between AlO (or AlOH) and the clumpy nature of the inner envelopes of AGB stars in indeed intriguing.

7. Alternative dust precursors

Gas-phase molecules other than the Ti- and Al-bearing ones were investigated in circumstellar media in order to establish their role in inorganic dust formation, including NaCl (Milam *et al.* 2007; Decin *et al.* 2016), SiO (Wong *et al.* 2016; Takigawa *et al.* 2017), and FeO (a tentative detection of Decin *et al.* 2018a). It appears they are, however, unlikely to be the major providers of seed particles in the majority of O-rich stars.

8. Challenges in interpreting the data and in studying nucleation

Despite the increasing number of observations of the potential gas-phase dust precursors, our understanding of chemistry and nucleation processes taking place close to the stellar photospheres is challenged by the complexity of the circumstellar environments and the physio-chemical processes involved. Taking Al-bearing species as an example, I offer below a brief overview of a few of the most complicating factors that often stop us from drawing firm conclusions from the observations.

Molecular tracers. As noted above, most stars so far have been observed mainly in rotational lines of AlO from within the ground vibrational level. The number of stars with detected lines of AlOH is increasing (see Fig. 5). However, in order to constrain the chemistry leading to nucleation, we should try to trace as many species as possible. For instance, the observed drop in abundance of AlO near $3\,R_\star$ may be interpreted as conversion of AlO into other gas-phase oxides, say to AlO_2, with no associated condensation taking place. Mira remains the source with largest number of species observed – with features of Al I, Al II, AlH, AlO, AlOH, and AlF observed in different wavelength regimes (Kamiński *et al.* 2016, K. T. Wong, priv. comm.). Other observable species include AlS and AlCl but there are gas-phase molecules which we currently are not able to observe, including for instance Al_2, Al_2O, AlO_2, and Al_2O_2. Some of them cannot be observed owing to very weak or lack of rotational transitions; for others, spectroscopic data are incomplete or missing. Because we are interested in envelope regions close to the star, where the prevailing conditions lead to high excitation, the observations should include also electronic, ro-vibrational, and high-J and high-v rotational spectra. The already mentioned identification of rotational lines of AlO and TiO at $v > 0$ levels (Gottlieb *et al.*, in prep.) will certainly add to our understanding of the inner, dust-forming regions of the envelopes.

Variability. Observations of Mira in Al- and Ti-bearing molecules show a very high level of variability. Optical spectra of M-type Mira stars have long been known for their erratic changes in molecular bands of AlO and TiO (Garrison 1997). Decades of optical observations of circumstellar features of AlO and Al I in *o* Ceti (Kamiński *et al.* 2016) do not show any regularity in the intensity changes over the pulsation cycle or on any longer time scales. Multiepoch ALMA observation of the same source in pure rotational transitions also show such an uncorrelated variability. This behavior is most likely caused by the stochastic nature of the pulsation shocks that strongly influence this part of the circumstellar envelope (Cherchneff 2006, Gobrecht *et al.* 2016, Liljegren *et al.* 2016, Liljegren *et al.* 2018). The variability is a major obstacle in determining the physical conditions in the gas since it requires nearly-simultaneous observations in multiple transitions. Such coordinated observations are difficult to arrange, especially when different observing facilities need to be involved. Of course, studying variability in different lines of many different species is also very expensive in terms of telescope time. Similarly, we require coordinated observations of dust tracers, such as optical observation of the (polarized) scattered light and mid-infrared observations of the spectral signatures of solids.

Non-LTE gas excitation. Although optical electronic bands of relevant circumstellar molecules could be observed for many M-type stars (Keenan *et al.* 1969), they are currently of limited use because there are no straightforward ways to derive reliable gas characteristics from the resonantly scattered emission bands (but see Kamiński *et al.* 2013a). Most of the studies thus far have therefore used pure rotational transitions observed at millimeter wavelengths as the diagnostics of gas physical conditions and molecular abundances. The radiative transfer problem for these lines has been solved at different levels of sophistication but all studies ignored the 3-dimensional complexity of the emitting regions (clumping), the influence of variable stellar radiation (radiative pumping), and the influence of shocks which most likely introduce non-LTE effects to molecular excitation. This introduces many uncertainties in the derived column densities. Additionally, without a reliable tracer of the local hydrogen densities, the derived absolute abundances or abundance profiles are highly unreliable. Dust or CO observations are used as a proxy of hydrogen densities but the required conversion factors are also poorly known for individual objects and especially in this very dynamic part of the envelope where dust has not fully formed and active chemistry is taking place. A lot of effort is necessary to build more reliable tools that would translate the observed line intensities into molecular abundances. After forcefully addressing these challenges we will truly be able to trace the depletion of molecules into solids.

Acknowledgements

I would like to thank my collaborators K. M. Menten, I. Cherchneff, N. Patel, and J. M. Winters for providing feedback on an early version of this manuscript.

References

Alcolea, J., Bujarrabal, V., Planesas, P., *et al.* 2013, *A&A*, 559, A93
Álvarez-Barcia, S., & Flores, J. R. 2016, *Physical Chemistry Chemical Physics (Incorporating Faraday Transactions)*, 18, 6103
Cherchneff, I. 2006, *A&A*, 456, 1001
Cherchneff, I. 2012, *A&A*, 545, A12
Cherchneff, I. 2013, *EAS Publications Series*, 175
De Beck, E., Vlemmings, W., Muller, S., *et al.* 2015, *A&A*, 580, A36
De Beck, E., Decin, L., Ramstedt, S., *et al.* 2017, *A&A*, 598, A53.
Decin, L., Richards, A. M. S., Millar, T. J., *et al.* 2016, *A&A*, 592, A76
Decin, L., Richards, A. M. S., Waters, L. B. F. M., *et al.* 2017, *A&A*, 608, A55
Decin, L., Danilovich, T., Gobrecht, D., *et al.* 2018a, *ApJ*, 855, 113
Decin, L., Richards, A. M. S., Danilovich, T., *et al.* 2018b, *A&A*, 615, A28
Dell'Agli, F., García-Hernández, D. A., Rossi, C., *et al.* 2014, *MNRAS*, 441, 1115
Ferrarotti, A. S., & Gail, H.-P. 2002, *A&A*, 382, 256
Gail, H.-P., & Sedlmayr, E. 1998, *Faraday Discussions*, 109, 303
Gail, H.-P., Wetzel, S., Pucci, A., *et al.* 2013, *A&A*, 555, A119

Gail, H.-P., & Sedlmayr, E. 2013, Physics and Chemistry of Circumstellar Dust Shells, Cambridge Astrophysics Series, Cambridge University Press
Garrison, R. F. 1997, *JAAVSO*, 25, 70
Gobrecht, D., Cherchneff, I., Sarangi, A., *et al.* 2016, *A&A*, 585, A6
Humphreys, R. M., Helton, L. A., & Jones, T. J. 2007, *AJ*, 133, 2716
Höfner, S., Bladh, S., Aringer, B., *et al.* 2016, *A&A*, 594, A108
Jeong, K. S., Winters, J. M., Le Bertre, T., *et al.* 2003, *A&A*, 407, 191
Kamiński, T., Schmidt, M. R., & Menten, K. M. 2013a, *A&A*, 549, A6
Kamiński, T., Gottlieb, C. A., Menten, K. M., *et al.* 2013b, *A&A*, 551, A113
Kamiński, T., Gottlieb, C. A., Young, K. H., *et al.* 2013c, *ApJS*, 209, 38
Kamiński, T., Wong, K. T., Schmidt, M. R., *et al.* 2016, *A&A*, 592, A42
Kamiński, T., Müller, H. S. P., Schmidt, M. R., *et al.* 2017, *A&A*, 599, A59
Karovicova, I., Wittkowski, M., Ohnaka, K., *et al.* 2013, *A&A*, 560, A75
Keenan, P. C., Deutsch, A. J., & Garrison, R. F. 1969, *ApJ*, 158, 261
Khouri, T., Waters, L. B. F. M., de Koter, A., *et al.* 2015, *A&A*, 577, A114
Leitner, J., Hoppe, P., Floss, C., *et al.* 2018, *Geochimica et Cosmochimica Acta*, 221, 255
Liljegren, S., Höfner, S., Nowotny, W., *et al.* 2016, *A&A*, 589, A130
Liljegren, S., Höfner, S., Freytag, B., *et al.* 2018, arXiv:1808.05043.
Milam, S. N., Apponi, A. J., Woolf, N. J., *et al.* 2007, *ApJ*, 668, L131
Nittler, L. R., Alexander, C. M. O., Gallino, R., *et al.* 2008, *ApJ*, 682, 1450
Norris, B. R. M., Tuthill, P. G., Ireland, M. J., *et al.* 2012, *Nature*, 484, 220
Ohnaka, K., Weigelt, G., & Hofmann, K.-H. 2016, *A&A*, 589, A91
Plane, J. M. C. 2013, *Phil. Trans. of the Royal Society of London Series A*, 371, 20120335
Sharp, C. M., & Huebner, W. F. 1990, *ApJS*, 72, 417
Sloan, G. C., & Price, S. D. 1998, *ApJS*, 119, 141
Takigawa, A., Kamizuka, T., Tachibana, S., *et al.* 2017, *Science Advances*, 3, eaao2149
Takigawa, A., Stroud, R. M., Nittler, L. R., *et al.* 2018, *ApJ*, 862, L13
Tenenbaum, E. D., & Ziurys, L. M. 2009, *ApJ*, 694, L59
Tenenbaum, E. D., & Ziurys, L. M. 2010, *ApJ*, 712, L93
Wong, K. T., Kamiński, T., Menten, K. M., *et al.* 2016, *A&A*, 590, A12

Why Galaxies Care About AGB Stars: A Continuing Challenge through Cosmic Time
Proceedings IAU Symposium No. 343, 2019
F. Kerschbaum, M. Groenewegen & H. Olofsson, eds.

doi:10.1017/S1743921318006014

On the onset of dust formation in AGB stars

David Gobrecht[1], Stefan T. Bromley[2,3], John M. C. Plane[4], Leen Decin[1] and Sergio Cristallo[5,6]

[1]Institute of Astronomy, KU Leuven, B-3001, Leuven, Belgium
email: david.gobrecht@kuleuven.be

[2]Departament de Ciència de Materials i Química Física & Institut de Química Téorica i Computacional (IQTCUB), Universitat de Barcelona, E-08028 Barcelona, Spain

[3]Institució Catalana de Recerca i Estudis Avançats (ICREA), E-08010 Barcelona, Spain

[4]School of Chemistry, Leeds University, Box 515, GB-75120 Leeds, Great Britain

[5]INAF - Osservatorio Astronomico dAbruzzo, Via mentore maggini, I-64100 Teramo, Italy

[6]INFN - Sezione di Perugia, via A. Pascoli, I-06123, Perugia, Italy

Abstract. A promising candidate to initiate dust formation in oxygen-rich AGB stars is alumina (Al_2O_3) showing an emission feature around $\sim$13μm attributed to Al$-$O stretching and bending modes (Posch *et al.* 1999, Sloan *et al.* 2003). The counterpart to alumina in carbon-rich AGB atmospheres is the highly refractory silicon carbide (SiC) showing a characteristic feature around 11.3μm (Treffers & Cohen 1974). Alumina and SiC grains are thought to represent the first condensates to emerge in AGB stellar atmospheres. We follow a bottom-up approach, starting with the smallest stoichiometric clusters (i.e. Al_4O_6, Si_2C_2), successively building up larger-sized clusters. We present new results of quantum-mechanical structure calculations of $(Al_2O_3)_n$, $n=1-10$ and $(SiC)_n$ clusters with $n=1-16$, including potential energies, rotational constants, and structure-specific vibrational spectra. We demonstrate the energetic viability of homogeneous nucleation scenarios where monomers (Al_2O_3 and SiC) or dimers (Al_4O_6 and Si_2C_2) are successively added. We find significant differences between our quantum theory based results and nanoparticle properties derived from (classical) nucleation theory.

Keywords. dust formation, molecular clusters, nucleation, alumina, silicon carbide, global optimisation, bottom-up, chemical-kinetics

---

1. Introduction

Cosmic dust is crucial for the evolution of galaxies. It impacts the synthesis of complex organic molecules in molecular clouds, the wind-driving of evolved stars and the formation of celestial bodies (e.g. asteroids, planets) in protoplanetary discs (Henning 2010). Asymptotic Giant Branch (AGB) stars represent a major contributor to the global dust budget in galaxies. The chemistry in the inner circumstellar envelopes of AGB stars (i.e. dust formation zone, $R=1-10\,R_*$, where R_* is the stellar radius) is primarily controlled by the carbon-to-oxygen (C/O) ratio. The CO molecule is triple bonded and very stable (dissociation energy of 11.1 eV) (see e.g. Habing & Olofsson 2003). As a consequence the lesser abundant element (C or O) is predominantly locked up in the CO molecule leaving little room to form other molecules than CO. Many molecular abundances can be approximated to a great extent with thermodynamic equilibrium calculations.

However, this simple picture is challenged as carbon-bearing molecules (HCN, CS, CO_2) have been observed in O-rich AGB stars (Lindqvist *et al.* 1988, Decin *et al.* 2010a, Justtanont *et al.* 1998) and oxygen-bearing species (H_2O, SiO) were found in C-rich

AGB atmospheres (Decin *et al.* 2010b, Neufeld *et al.* 2011, Johansson *et al.* 1984). Their presence can only be explained by processes deviating from chemical and thermal equilibrium. CO might be dissociated by shock-induced collisions (Gobrecht *et al.* 2016), or by interstellar photons (Agúndez *et al.* 2010, Van de Sande *et al.* 2018) where free C and O are released to form other unexpected molecular species. In contrast to the above mentioned molecules, the dust condensates are carbonaceous in C/O>1 environments and oxygen-rich in oxygen-dominated (C/O<1) regimes.

In the latter case, metal oxides represent the majority of the solids as there are no stable, pure oxygen compounds larger than ozone (O_3). The main dust component in oxygen-rich AGB stars are silicates. Silicates are composed of a least three elements (Si, O and Mg) and are so called ternary oxides. Iron (Fe) might be incorporated in the Mg-rich silicate grains at a later stage and act as a thermostat for the grain (Woitke 2006). In this case, the dust grains are quaternary oxides consisting of O, Si, Mg and Fe. Measurements on the vapor pressure of Mg-rich silicate grains as well as on pure silicon oxide (SiO) resulted in rather low condensation temperatures (800–1000 K) for pressures that prevail in the circumstellar envelopes (Wetzel *et al.* 2013). Recent ALMA observations have shown that the silicate features around 10 and 18 μm are absent for the low mass-loss rate stars W Hya and R Dor (Decin *et al.* 2017), but they are present in the high mass loss rate object IK Tau. Theses findings indicate that, albeit silicates represent the main component of oxygen-rich dust grains, they are not the first condensates to emerge in the outflows of M-type AGB stars. The first condensate, or *seed* particle, initiating the nucleation process is supposed to have a high condensation temperature, made up from available and abundant elements and molecules, and sufficiently fast growing rates in circumstellar conditions. There are several metal oxides satisfying the above conditions: titania (TiO_2), magnesia (MgO), silicon monoxide (SiO) and alumina (Al_2O_3). In crystalline form, titania exist as anatase and rutile that are both highly refractory. However, the solar titanium abundance is rather low. Silicon monoxide is fairly abundant in the envelopes of AGB stars, though its nucleation is rather unlikely to occur in circumstellar media (Bromley *et al.* 2016). On one hand, a homogeneous nucleation is hampered by energetic barriers (Goumans & Bromley 2012). On the other hand, Si starts to segregate and to form islands inside the most stable clusters for small cluster sizes $(SiO)_n$, $n \geqslant 10$. The nucleation of MgO clusters was investigated by Köhler *et al.* (1997). They found that the nucleation rate of MgO turns out to be too small to form the seed nuclei for silicate dust formation. One reason are the "magic" cluster sizes (2, 4, 6, 9, 12 and 15 MgO units) with enhanced stability that hamper the formation of larger sized clusters. A promising candidate to initiate dust formation in oxygen-rich AGB stars is alumina (stoichiometric formula Al_2O_3) as its nucleation is not constrained by atomic segregation or energy barriers as we will show below. Circumstellar alumina shows a spectral emission feature around ~13 μm attributed to Al–O stretching and bending modes (Posch *et al.* 1999, Sloan *et al.* 2003). This feature has been seen in several oxygen-rich AGB sources (S Ori, R Cnc, GX Mon, W Hya, R Dor) close to the stellar surface at around 2–3 R_* (Karovicova *et al.* 2013, Decin *et al.* 2017). In the case of carbon AGB stars, it is argued that the nucleation of pure or hydrogenated (amorphous) carbon is sufficient to explain the observed mass loss rates (Höfner *et al.* 2003). However, silicon carbide (SiC) may also play an important role in the nucleation of carbon dust owing to its high thermal stability and its presence in meteorites.

2. Global optimisation methods

Top-down chemical methods use a known macroscopic ensemble of atoms (e.g. a crystal structure) and interpolate the unknown properties and characteristics of smaller-sized systems like nanoparticles and molecular clusters. A prime example of a physio-chemical

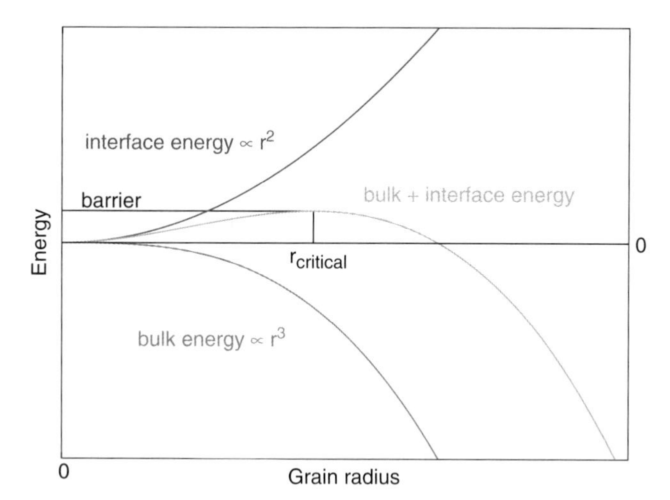

Figure 1. Schematic sketch of the classical nucleation energy versus particle radius (size) in blue. The bulk component is in green, and the interface component is in violet.

top-down approach is Classical Nucleation Theory (CNT). In CNT, many generic properties of a particle with size N can be expressed as a sum of a bulk term and a surface (interface) term. This is also true for the potential and free energies of the particles, i.e. $E_{\rm pot} = E_{\rm surface} + E_{\rm bulk}$. An illustrative example is shown in Fig.1.

The surface (interface) energy can be interpreted as a sort of surface tension. It is repulsive and scales with the particle size (or radius) as r^2. The volume (bulk) term is attractive and scales with the grain radius as r^3. The resulting curve considering both, bulk and interface, has a maximum for $r > 0$. The maximum is located at the so called critical radius $r_{\rm crit}$ and has a value $E_{\rm crit}$ corresponding to an energy barrier. In the CNT regime, $E_{\rm crit}$ represent the energetic bottleneck of the nucleation process. Once this barrier is overcome (i.e. the dust particle has grown to a radius of $r_{\rm crit}$), the subsequent nucleation is energetically favourable. However, it has been shown that the global minima structures of a variety of (sub-)nanosized clusters, including TiO_2 (Lamiel-Garcia *et al.* 2017), SiO (Bromley *et al.* 2016), MgO (Chen *et al.* 2014), and Al_2O_3 (Li & Cheng 2012), as well as SiC (Gobrecht *et al.* 2017) do not exhibit bulk-like geometries. In contrast, potential and free energies, bond lengths and angles, atomic coordination, and formal charges deviate significantly from bulk-like analogues (bulk cuts). We conclude that classical nucleation theory cannot be made to work for small clusters representing the basic building blocks of (circumstellar) dust. We use a bottom-up approach starting with molecules and molecular clusters, and successively build up larger sized dust clusters and dust grains. We are convinced that the formation of condensates are described more realistically using a bottom-up approach, as it mimics the onset of dust formation in expanding circumstellar shells. In order to reduce the enormous computational cost arising in quantum-chemical bottom-up approaches, we generate a number (typically $\sim$100 for each size) of candidate cluster structures among millions of possible structural isomers. Therefore, we apply a couple of semi-classical force-field global optimisation techniques that are presented in the following.

2.1. *Exploration of the Buckingham potential energy landscape*

Alumina. i.e. $(Al_2O_3)_n$, and silicon carbide, i.e. $(SiC)_n$, candidate cluster structures are found with the Monte Carlo - Basin Hopping (MC-BH) global optimization technique

Table 1. The parameter ranges used in this study to compute the inter-atomic Buckingham pair potential. Charges q are given in atomic units, A in eV, B in Å, C in eV Å^{-6}.

q(Al)	q(O)	A(Al-O)	B(Al-O)	C(Al-O)	A(O-O)	B(O-O)	C(O-O)
+3	−2	2409.5	0.2649	0.0	25.410	0.6937	32.32
q(Si)	q(C)	A(Si-C)	B(Si-C)	C(Si-C)	A(C-C)	B(C-C)	C(C-C)
+2	−2	592.34	0.3521	12.897	25.410	0.6937	32.32

(Wales *et al.* 1997) with inter-atomic Buckingham pair potentials. The general form of the inter-atomic Buckingham pair potential reads:

$$U(r_{ij}) = \frac{q_i q_j}{r_{ij}} + A\exp(-\frac{r_{ij}}{B}) - \frac{C}{r_{ij}^6} \tag{2.1}$$

where r_{ij} is the relative distance of two atoms, q_i and q_j the charges of atom i and j, respectively and A, B and C the Buckingham pair parameters. The first term represents the Coulomb law, the second term the short-range, steric repulsion term accounting for the Pauli principle, and the last term describes the van-der-Waals interaction. The steric repulsion term is motivated by the fact that atoms are not dot-like but occupy a certain volume in space.

The Buckingham pair potential parameters for Al-O systems are widely studied and the values are taken from Bush *et al.* (1994). In the case of silicon carbide, parameter sets for the Si-C system are lacking in the literature for several reasons. As an integral part the electrostatic Coulomb potential appears in Eq. 2.1. It describes the repulsion and attraction of charged particles, in this case of Al and O, and Si and C. As lightest Group IV elements in the periodic table, Si and C form strong covalent bonds. However, the Buckingham potential is more suitable for materials with an ionic character,as for example metal oxides. Nonetheless, Watkins *et al.* (2009) have shown the similarity of zincblende ZnO (a cubic crystal type with face-centred lattice points), and β-SiC, despite that the first is generally regarded as an ionic II-VI system and the latter as a covalent IV-IV system. Moreover, they found that the Buckingham parameters for ZnO also describe SiC clusters fairly well. Therefore, we performed MC-BH optimisations with a simplified version of the parameter set for ZnO given by Whitmore *et al.* (2002).

The ZnO forcefield we employ has been shown to be able to stabilize a wide range of different cluster isomers (Al-Sunaidi *et al.* 2008) and bulk polymorphs (Demiroglu *et al.* 2014) which exhibit alternating cation-anion ordering. However, to reduce the probability to miss stable cluster isomers in our searches, we also ran some test calculations for several sizes with a forcefield parameterized for ZnS (Wright & Jackson 1995) which potentially provides an additional source of cluster isomers not easily found with the ZnO forcefield. However, the few structures that we found exclusively with the ZnS parameters had high energies (when converted to SiC clusters) and did not compete with the ZnO cluster analogues. Although the use of force fields is an approximation, their use enables us to perform tractable thorough searches. With our mixed-forcefield approach we hope to have minimized the probability to miss a stable SiC isomer.

2.2. *Tersoff potential simulated annealing*

Albeit the Buckingham pair potential including the Coulomb terms describes the force-field of ionic materials like metal oxides fairly well, it may fail to describe stable cluster configurations that are characterised by covalent rather than ionic bonds.

A simple two-body interaction is thus not sufficient to properly describe the Si-C system. In addition, a three-body potential is needed to describe the covalent character of bond bending and stretching (Stillinger & Weber 1985, Vishishta *et al.* 2007). In order to properly describe internal interactions of the most stable SiC clusters, empirical bond-order potentials are favourable, in particular for small clusters (Erhart & Albe 2005). This class of inter-atomic potentials include the Tersoff- type (Tersoff 1989), the Brenner (Brenner 1990), or ReaxFF (Van Duin *et al.* 2001), which take into account the bonding environment, namely bond lengths, bond angles and the number of bonds. As a consequence of geometry, the bonding angle in a tetrahedrally coordinated system like SiC is Θ = arccos(-1/3) = 109.47°. The general form of a bond-order potential reads:

$$V(r_{ij}) = f_c(r_{ij})\,[V_{rep}(r_{ij}) + b_{ij}V_{att}(r_{ij})] \tag{2.2}$$

where $V_{rep}(r_{ij}) = A_{ij}\exp(-\lambda_{ij}r_{ij})$ is the repulsive part of the bond-order potential and $V_{att}(r_{ij}) = B_{ij}\exp(-\mu_{ij}r_{ij})$ the attractive effective potential. b_{ij} modifies the strength of the bond, depending on the environmental parameter (the bonding angles Θ) as reported in Tersoff 1989. In the Tersoff parametrisation of the inter-atomic Si-C molecular system, which is chosen in our approach, the potential is modified by a taper function f_c. f_c is 1 for inter-atomic distances r_{ij} smaller or equal of typical bonding distances and falls quickly to 0 for distances larger than S and thus restricts the interaction to the first neighbouring atoms within a distance S.

$$f_c(r_{ij}) = \begin{cases} 1, & r_{ij} < R \\ 0.5 + 0.5\cos(\frac{\pi(r_{ij}-R)}{S-R}), & R < r_{ij} < S. \\ 0, & r_{ij} > S \end{cases} \tag{2.3}$$

We use the programme GULP (General Utility Lattice Programme, Gale 1997) which is taylored for the semi-classic parametrisation by Tersoff 1989.

Some SiC cluster structures have been reported in the literature (Pradhan & Ray 2004, Hou & Bin 2008, Duan *et al.* 2013). We tested their stability against (small) distortions in molecular dynamics runs with GULP. Furthermore, we applied the Tersoff potential to these structures. In the majority of the cases, this potential suffices to stabilise the structures. In some cases, however, the Tersoff potential fails to stabilize the clusters, and hand-constructed structures were taken instead for the subsequent computation. In some of these failure cases new, unreported clusters appeared.

2.3. *Quantum-mechanical refinement*

Once pre-optimised, the clusters are refined using quantum-mechanical Density Functional Theory (DFT) calculations to obtain structure-specific infrared spectra (i.e. vibrational frequencies) rotational constants, and zero-point-energies. By comparing the obtained infrared spectra with observational data, the specific isomers present in circumstellar envelopes can be identified. The $(Al_2O_3)_n$ and $(SiC)_n$ cluster structures, so far reported in the literature, rely on various theoretical quantum chemistry methods. They include DFT methods using generalized gradient approximation (GGA), local density approximation (LDA), hybrid functionals (e.g B3LYP, PBE0, M11), respectively, and post-Hartree Fock methods using Møller-Plesset (MP2, MBPT) and coupled-cluster (LCCD, CCSD) techniques.

For DFT methods the computational cost scales with the system size as between the order $\mathcal{O}(N^3)$ and $\mathcal{O}(N^4)$, where N is the number of electrons in the cluster. This means that they can be readily applied to systems containing 10s of atoms. However, many DFT methods can suffer from artificial electron self-interaction that results in overly strong electron delocalisation and too low potential energies. In contrast, Post-HartreeFock methods do not suffer from these effects. However, the computational cost of these latter methods is very high and scales with the system size as $\mathcal{O}(N^5)$-$\mathcal{O}(N^7)$. They are thus prohibitive for systems of more than approximately 10 atoms. Functionals such as B3LYP, M11 or PBE0 attempt to compensate for the above mentioned shortcomings of typical GGA/LDA functionals. The recent extensive benchmark study by Byrd *et al.* (2016) confirms that the M11 functional is able to correctly identify all investigated $(SiC)_n$ ground states. Although B3LYP (Becke 1993) was found to be less accurate than M11 for SiC clusters, we also include data calculated with this widely used functional for comparison. We conclude that, for our purposes and for SiC, the M11 functional method is the best compromise between a reasonable computational cost and the required accuracy.

Owing to its high computational costs, DFT calculations are performed on supercomputers using the well-established and parallelised code *Gaussian 09*. These calculations approximate the wave functions and the energy of a quantum many-body system in a stationary state. *Gaussian 09* optimises cluster structures at a pressure of 0 bar and temperature of 0 K. We account for circumstellar conditions (temperatures and pressures) by applying thermodynamic potential functions (enthalpy, entropy, Gibbs energy) that are evaluated with the help of partition functions. These functions and their derivatives are calculated from the electronic energies, moments of inertia and vibrational frequencies within the rigid-rotor harmonic oscillator approximation (McQuarrie & Simon 1999, Goumans & Bromley 2012).

As a consequence, the relative energy spacings of the individual clusters shift and may cross. This implies that the initial lowest energy isomer may not be the most favourable structure in circumstellar conditions and a different cluster structure is preferred. It is thus necessary to study a range of the energetically lowest-lying structures for each cluster size. The use of partition functions relies on the validity of thermodynamic equilibrium. We note, however, that AGB atmospheres may depart from equilibrium as they are periodically crossed by pulsational shock waves. The resulting Gibbs Free energies thus have limited validity. Nonetheless, they provide a good approximation for the individual cluster stability in circumstellar conditions.

3. Results

3.1. *Alumina clusters*

In Fig. 2, the $(Al_2O_3)_n$, $n = 1-10$, isomers with the lowest potential energies are shown. For $n = 8$, 9, and 10 we found new global minima candidates that are in part reported in Gobrecht *et al.* (2018). The minima isomers exhibit a variety of different structural features, and no attribute, apart from alternating Al-O bonds, is common to all these minima clusters. For this reason, a fit function $f(n) = a + bn^{-\frac{1}{3}}$ to the normalised potential energy oversimplifies the size dependence in energy. Including a second-order term $cn^{-\frac{2}{3}}$ in the fit function describes the size-dependence of the potential energy E(n) better, though not perfectly. Moreover, we show that the nucleation energies by monomer addition, $E_{\rm nuc}(n) = E(n) - E(n-1) - E(1)$, as well as by dimer addition $E_{\rm nuc}(n) = E(n) - E(n-2) - E(2)$, have negative energies and thus nucleation is viable. There are no energetic barriers in homogeneous nucleation scenarios, but the initial steps (dimerisation and dimer coalescense) are expected to be the fastest steps, as they are the most exothermic processes. The vibrational IR spectra is shown in Fig. 3. The most intense vibrations of the clusters are located in a wavelength range between 10 and 12 μm. Around 13 μm, where the alumina dust feature is located, the overall intensity is rather low. Along with other characteristics (bond length, coordination), we conclude that, at the size of the decamers ($n = 10$), the bulk limit for alumina clusters is not yet reached.

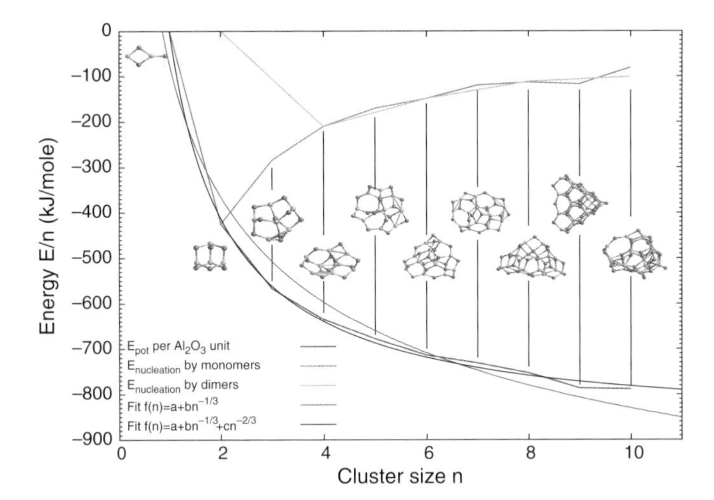

Figure 2. The global minima $(Al_2O_3)_n$, $n=1-10$, candidate isomers and their potential energies (in purple), nucleation energies (by monomers in green, by dimers in blue), and related first (in red) and second (in black) order fits. Oxygen atoms are in red, aluminum atoms in green.

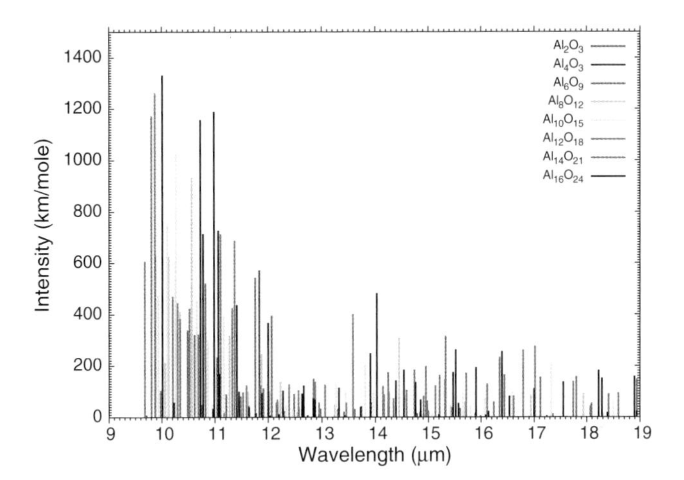

Figure 3. The vibrational IR spectra of the global minima $(Al_2O_3)_n$, $n=1-8$, clusters as a function of wavelength.

3.2. *Silicon carbide clusters*

The silicon carbide clusters have been extensively discussed in Gobrecht *et al.* 2017. The most stable isomers are dominated by two structural families: void cage geometries with strict cation-anion ordering ("bucky"-like) and segregated clusters exhibiting chains and rings of carbon. The binding energy ΔE_b of the lowest-energy Si_nC_n clusters as well as the HOMO (Highest Occupied Molecular Orbital)–LUMO (Lowest Unccupied Molecular Orbital) energy gap ΔE_{MO} are presented in Fig. 4.

The binding energies ΔE_b of the global minima clusters generally increase with cluster size. However, some cluster sizes ($n=9$, 12) appear to be more stable than the next larger cluster of size $n+1=10$, 13. In the HOMO-LUMO energies, no clear trend with

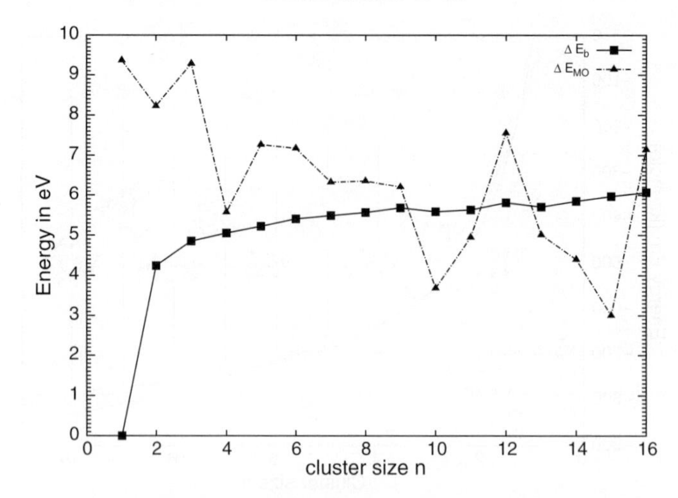

Figure 4. The relative binding energy $\Delta E_{\rm b}$ (filled squares and solid line) of the global minima $\mathrm{Si}_n\mathrm{C}_n$ clusters (normalised to cluster size n) and the HOMO-LUMO energy gap $\Delta E_{\rm MO}$ (triangles and dashed line) of the ground state $\mathrm{Si}_n\mathrm{C}_n$ clusters.

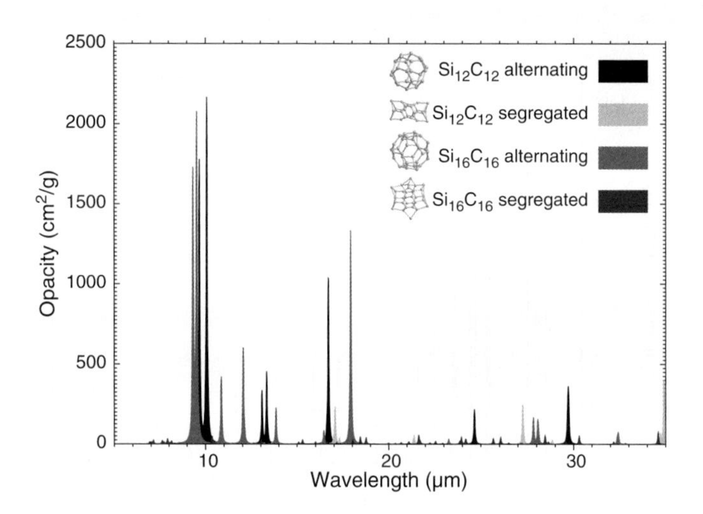

Figure 5. The energetically most favourable $(\mathrm{SiC})_n$ clusters of each structural family (alternating and segregated) and their relative potential energies in kJ/mole. Silicon atoms are in pink, carbon atoms in green.

cluster size n is observed. The "bucky"-like clusters show strong vibrational IR signatures, compared with their segregated counterparts. This fact is illustrated in Fig. 5, where the structure-specific opacity, derived from the IR intensity, is displayed as a function of wavelength.

References

Al-Sunaidi, A. A., Sokol, A. A., Catlow C. R. A., Woodley, S. M. 2008, *JPCC*, 112, 18860

Agúndez, M., Cernicharo, J., Guélin, M., 2010, *ApJ*, 724, 133

Becke, A. D. 1993, *JCP*, 98, 1372

Brenner, D. W. 1990, *Phys. Rev. B*, 42, 9458

Bromley, S. T., Gómez-Martín, J. C., Plane, J. M. C. 2016, *PCCP*, 18, 26913

Bush, T. S., Gale, J. D., Catlow, C. R. A., Battle, P. D. 1994, *JMC*, 4, 831

Byrd, J. N. and Lutz, J. J., Jin, Y. Ranasinghe, D. S., Montgomery, J. A., Perera, A., Duan, X. F., Burggraf, L. W., Sanders, B. A., Bartlett, R.J. 2016, *JCP*, 145, 24312
Chen, M., Felmy, A. R., Dixon, D. A. 2014, *JPCA*, 118, 3136
Decin, L., De Beck, E., Brnken, S., Mller, H. S. P., Menten, K. M., Kim, H., Willacy, K., de Koter, A., Wyrowski, F. 2010, *A&A*, 516, 69
Decin, L., Justtanont, K., De Beck, E., Lombaert, R., de Koter, A., Waters, L. B. F. M., Marston, A. P., Teyssier, D., Schier, F. L., Bujarrabal, V., Alcolea, J., Cernicharo, J., Dominik, C., Melnick, G., Menten, K., Neufeld, D. A., Olofsson, H., Planesas, P., Schmidt, M., Szczerba, R., de Graauw, T., Helmich, F., Roelfsema, P., Dieleman, P., Morris, P., Gallego, J. D., Dez-Gonzlez, M. C., Caux, E. 2010, *A&A*, 521, 4
Decin, L., Richards, A. M. S., Waters, L. B. F. M., Danilovich, T., Gobrecht, D., Khouri, T., Homan, W., Bakker, J. M., Van de Sande, M., Nuth, J. A., De Beck, E. 2017, *A&A*, 608, 55
Demiroglu, I., Tosoni, S. and Illas, F., Bromley, S. T. 2014, *Nanoscale* 6, 1181
Duan, X. F., Burggraf, L. W., Huang, L. 2013, *Molecules*, 18, 8591
Erhart, P., Albe, K. 2005, *Phys. Rev. B*, 71, 35211
Gale, J. D. 1997, *JCS, Faraday Trans.*, 93, 629
Gobrecht, D., Cherchneff, I., Sarangi, A., Plane, J. M. C., Bromley, S. T. 2016, *A&A*, 585, 15
Gobrecht, D., Cristallo, S., Piersanti, L., Bromley, S. T. *ApJ*, 840, 117
Gobrecht, D., Decin, L., Cristallo, S., Bromley, S. T. *CPLett*, 711, 138
Goumans, T. P. M., Bromley, S. T. 2012, *MNRAS*, 420, 3344
Habing, H. J., Olofsson, H. 2003, *Asymptotic giant branch stars, Astronomy and astrophysics library, New York, Berlin: Springer*
Henning, Th. 2010, *Lecture Notes in Physics*, 815
Hoefner, S., Gautschy Loidl, R., Aringer, B., Jørgensen, U.G. 2003, *A&A*, 399, 589
Hou, J., Song, B. 2008, *JCP*, 128,154304
Johansson, L. E. B., Andersson, C., Ellder, J., Friberg, P., Hjalmarson, A., Hoglund, B., Irvine, W. M., Olofsson, H., Rydbeck, G. 1984, *A&A*, 130, 227
Justtanont, K., Feuchtgruber, H., de Jong, T., Cami, J., Waters, L. B. F. M., Yamamura, I., Onaka, T. 1998, *A&A*, 330, 17
Karovicova, I., Wittkowski, M., Ohnaka, K., Boboltz, D. A., Fossat, E., Scholz, M. 2013, *A&A*, 560, 75
Koehler, T. M., Gail, H.-P., Sedlmayr, E. 1997, *A&A*, 320, 553
Lamiel-Garcia, O., Ko, K. C., Lee, J. Y., Bromley, S. T., Illas, F. *JCTC*, 13, 1785
Li, R., Cheng, L. 2012, *CTP*, 996, 125
Lindqvist, M., Nyman, L.-A., Olofsson, H., Winnberg, A. 1988, *A&A*, 205, 15
McQuarrie, D. A., Simon, J. D. 1999, *University Science Books*
Neufeld, D. A., Gonzlez-Alfonso, E., Melnick, G. J., Szczerba, R., Schmidt, M., Decin, L, de Koter, A., Schier, F., Cernicharo, J. 2011, *APJL*, 727, 28
Peverati, R., Truhlar, D. G. 2012, *JPCL* 3, 117
Posch, T., Kerschbaum, F., Mutschke, H., Fabian, D., Dorschner, J., Hron, J. 1999, *A&A*, 352, 609
Pradhan, P., Ray, A. K. 2004, *ArXiv Physics e-prints*, 0408016
Sloan, G. C., Kraemer, K. E., Goebel, J. H., Price, S. D. 2003, *ApJ*, 594, 483
Stillinger, F. H. & Weber, T. A. 1985, 31, 5262
Stroud, R. M., Nittler, L. R., Alexander, C. M. O'D. 2004, *Science* 305, 1455
Tersoff, J. 1989, *Phys. Rev. B*, 39, 5566
Treffers, R., Cohen, M. 1974, *ApJ* 188, 545
Van de Sande, M., Sundqvist, J. O., Millar, T. J., Keller, D., Homan, W., de Koter, A., Decin, L., De Ceuster, F. 2018, *A&A*, 616, 106
Van Duin, A. C. T., Dasgupta, F., Lorant, F. Goddard, W. A. 2001, *JPCA*, 105, 9396
Vashishta, P., Kalia, R. K., Nakano, A., Rino, J. P. 2007, *JAP* 10, 101
Wales, D. J., Doye, J. P. K. 1997, *JPCA*, 101, 5111
Watkins, M. B., Shevlin, S. A., Sokol, A. A., Slater, B., Catlow, C. R. A., Woodley, S. M. *PCCP*, 11, 3186

Wetzel, S., Klevenz, M., Gail, H.-P., Pucci, A., Trieloff, M. *A&A*, 553, 92
Whitmore, L., Sokol, A. A., Catlow, C. R. A. 2002, *Surface science*, 498, 135
Wright, K., Jackson, R. A. 1995, *JMC*, 5, 2037
Woitke, P. 2006, *A&A*, 460, 9

Discussion

DeBeck: Where in the spectrum do you expect the vibrational modes of the alumina clusters?

Gobrecht: The most intense vibration bands are located in a wavelength range between ~ 10–$11\,\mu$m, but I also find less intense IR modes at $\sim 13\,\mu$m (where the observed Al-O stretching modes are found).

Why Galaxies Care About AGB Stars: A Continuing Challenge through Cosmic Time
Proceedings IAU Symposium No. 343, 2019
F. Kerschbaum, M. Groenewegen & H. Olofsson, eds.

doi:10.1017/S1743921318005094

Dynamics, temperature, chemistry, and dust: Ingredients for a self-consistent AGB wind

J. Boulangier[1], D. Gobrecht[1] and L. Decin[1,2]

[1]Institute of Astronomy, KU Leuven, Celestijnenlaan 200D, 3001 Leuven, Belgium
email: jels.boulangier@kuleuven.be

[2]University of Leeds, School of Chemistry, Leeds LS2 9JT, United Kingdom

Abstract. Understanding Asymptotic Giant Branch (AGB) stars is important as they play a vital role in the chemical life cycle of galaxies. AGB stars are in a phase of their life time where they have almost ran out of fuel and are losing vast amounts of material to their surroundings, via stellar winds. As this is an evolutionary phase of low mass stars, almost all stars go through this phase making them one of the main contributors to the chemical enrichment of galaxies. It is therefore important to understand what kind of material is being lost by these stars, and how much and how fast. This work summarises the steps we have taken towards developing a self-consistent AGB wind model. We improve on current models by firstly coupling chemical and hydrodynamical evolution, and secondly by upgrading the nucleation theory framework to investigate the creation of TiO_2, SiO, MgO, and Al_2O_3 clusters.

Keywords. stars: AGB and post-AGB – stars: winds, outflows – hydrodynamics – astrochemistry – methods: numerical

1. Introduction

Detailed understanding of how AGB stars lose their material via stellar winds is hindered by insufficient high resolution observations, simplified models, and a lack of laboratory data. Ample work is being done to improve on all three limitations. This work focuses on the second, improving AGB wind models. The general hypothesis is that the mass loss mechanism of AGB stars is a combination of stellar pulsations and radiative pressure on dust grains. It is slightly more complicated as their interaction encompasses several physical and chemical processes (Fig. 1). Pulsations of the star quickly turn into shock waves, as their velocities exceed the local sound speed, hereby heating up the gas. This influence on temperature affects the chemical composition of the gas, because chemical reactions behave differently at high and low temperatures. In turn, the chemical composition regulates the temperature of the gas, because the efficiency of different heating and cooling mechanisms is determined by the abundance of heating and cooling species. Meanwhile, photons from the star interact with dust grains, transferring their momentum either via absorption or scattering, hereby pushing the grains outwards. The outward moving dust then collides with the nearby gas, dragging it along towards the interstellar medium. Both driving forces, pulsations and radiation on dust grains, have been studied extensively. Yet, the link between them has mainly been ignored. Since the stellar surface does not contain any dust grains, a phase transition from gaseous material to solid grains has to occur. Due to the low density in an AGB wind, this does not happen instantly (as compared to Earth/lab conditions). On a microscopic level, the initiation of a phase transition is called nucleation. This process corresponds to molecules reacting with each other to form larger clusters, either homomolecular (same species) or

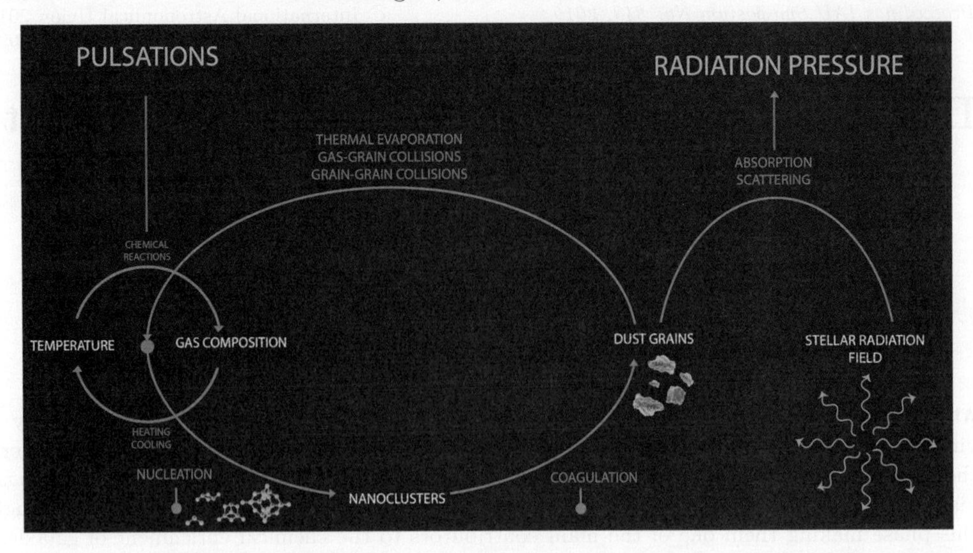

Figure 1. Schematic, yet simplified, overview of the physical and chemical processes occurring in an AGB wind. The two main driving forces are pulsations of the star and radiation pressure on dust grains. Both are self-consistently connected via the mechanisms depicted in the figure.

heteromolecular (different species). From a certain size, physical forces will take over, e.g. van der Waals, and such nanoclusters will coagulate to form macroscopic dust grains. In turn, dust grains can be destroyed by thermal evaporation, gas-grain collisions, and grain-grain collisions, hereby shattering the grains and/or liberating gaseous material. The combination of dynamical and chemical evolution (hydrochemistry), interaction of dust grains with stellar radiation, and the missing link of dust evolution with creation from gas, makes a self-consistent AGB wind model. Such a model will aid in understanding what kind of material is being lost by these stars, and how much and how fast.

2. Hydrochemistry

Driving an AGB wind purely hydrodynamically is unfeasible when using a physically reasonable pulsation behaviour. The gravitational pull of the star is too large for the gas to exceed the local escape velocity. It is however possible to hydrodynamically drive the wind when increasing the pulsational velocity variation at the stellar surface by roughly an order of magnitude. Yet, this is most likely unrealistic. As suggested by the current hypothesis, extra outward force by radiation pressure on dust grain can aid in driving the wind. Because dust formation is highly sensitive to temperature, which also depends on the chemical composition, one has to precisely know the gas composition and its temperature. This can only be achieved by a hydrochemical simulation, since the chemical composition and temperature mutually affect one another, by temperature dependency of chemical reactions and by abundances of heating and cooling species. However, current AGB wind models have ignored this and assume chemical equilibrium (e.g. Bowen 1988, Willson 2000, Woitke 2006, Höfner *et al.* 2016). We improve on these models by introducing non-equilibrium chemical evolution (also done in static AGB wind models, Cherchneff 2012, Marigo *et al.* 2016, Gobrecht *et al.* 2016) and couple this to the dynamics of the wind, called hydrochemistry (Boulangier *et al.* 2018).

We have constructed a chemical reaction network which is applicable to AGB winds, yet still simplified. The bulk of the reactions originates from the UMIST database, extended with a handful from the KIDA database and standalone papers. In total, the network comprises roughly 1700 reactions and 160 species. This has to be reduced when coupled to a hydrodynamical framework, due to computational constraints. After applying a reduction algorithm to identify the most important reactions, we end up with roughly 250 reactions and 70 species, resulting in a speedup factor of 20. The algorithm makes sure that the non-reduced results are reproduced within a certain accuracy, for an AGB wind environment.

We have modelled a 1D AGB wind using the hydrodynamics code MPI-AMRVAC (Keppens *et al.* 2012) and chemistry code KROME (Grassi *et al.* 2014). The former had to be extended to be able to handle conservation of chemical species (Plewa & Müller 1999) and have a variable adiabatic index that depends on the local chemical composition.

3. Nucleation theory

Current AGB wind models assume some dust growth mechanism starting from the artificial presence of dust seed particles (typically molecules consisting of 100 – 1000 of monomers), e.g., Höfner *et al.* 2016. Such seed particles actually have to form by nucleation of gas molecules. Several prescriptions have been used in closed systems in the literature: classical, modified, and kinetic steady state nucleation theory (Helling & Woitke 2006, Gail *et al.* 2013, Köhler *et al.* 1997, Patzer *et al.* 1998, Bromley *et al.* 2016, Goumans *et al.* 2012, Plane 2013). Also outside of AGB modelling, e.g. brown dwarf atmospheres (Lee *et al.* 2018), and supernovae (Nozawa & Kozasa 2013, Lazzati & Heger 2016). We perform a two-step improvement on such models, firstly with non-equilibrium time-dependent nucleation, and secondly coupling this to a large chemical network (Boulangier *et al.* in prep.); its closest resemblance is the "molecular nucleation theory" of Sluder *et al.* (2018)). Our first improvement "non-equilibrium time-dependent nucleation" corresponds to clusters growth reactions by monomer addition and destruction reactions based on the assumption of detailed balance. This latter uses the Gibbs free energy of the clusters, which are calculated from first principles. Density functional theory is used to determine electronic structures, rotational, and vibrational degrees of freedom, all of which are needed to infer the Gibbs free energy of the molecules. This part has also been done in the past (Goumans *et al.* 2012, Lee *et al.* 2018, Köhler *et al.* 1997). However, these papers assume steady state nucleation, meaning in chemical equilibrium, whereas we treat each growth/destruction reaction separately, meaning time-dependent non-equilibrium evolution. Our second improvement step is applying this nucleation theory to a chemical reaction network, starting from an atomic composition rather than the nucleating monomer. This removes the assumption of the monomer being (abundantly) present, and species can compete for chemical resources.

We consider four nucleation candidates that have been proposed in the literature: TiO_2, SiO, MgO, and Al_2O_3. The largest clusters considered consist of roughly 30 atoms. This limit is either due to lack of data/computational constraints on larger clusters or where the nucleation-by-monomer principle breaks down. Gibbs free energies of all clusters are calculated by determining the quantum chemical properties via density functional theory where we started from minimum energy configurations found in the literature. Additionally, we have searched the literature thoroughly for chemical reactions of Ti-, Mg-, Si-, and Al-bearing species to, among other, bridge the gap from atoms to nucleation monomers. We first perform 'pure nucleation' models for each candidate, where such a model consists of starting with a monomer abundance and evolving a system of only nucleation growth and destruction reactions over a certain period of time. After performing a grid of 'pure nucleation' models from 500 to 3000 K and 10^{-6} to 10^{-10} kg m^{-3} for one year time, we find that each candidate has a different temperature cut-off above which the largest clusters do not form. For Al_2O_3 this is around 1800 to 2200 K, for MgO around 1500 to 1700 K, for TiO_2 around 1200 to 1300 K, and for SiO around 500 K. According to these models, Al_2O_3 is the best candidate to form dust because it is desirable to form dust as fast as possible while the gas is cooling down, so as quickly as possible have an extra outwards force by radiation pressure. However, the results differ when performing a grid of models with our second improvement, starting from an atomic composition and nucleation included in a large chemical network. Here, the Al_2O_3 monomer does not form at all, meaning non of its bigger clusters form either. MgO clusters have the same problem, namely, all Mg stays atomic, so no MgO monomers, meaning no larger clusters either. SiO clusters are equally inefficient as in the previous model, and are not of interest due to its low temperature cut-off. However, TiO_2 monomers form easily and so do their larger clusters. The abundance of TiO_2 clusters is the same as in the previous model. Thus, according to these models, TiO_2 is the favourable dust precursor. Yet, one has to be careful with drawing conclusions from these models. It is not because the models cannot produce Al_2O_3 clusters, that they don't exist in AGB winds. It does mean that with the current model we cannot form them. As Al_2O_3 clusters are abundantly present in presolar grains, originating from AGB winds, the model is most likely incomplete. It is most probably missing chemical pathways to form such clusters, others than starting from the Al_2O_3 monomer which is chemically unstable due to its triplet ground state. Our current models predict most Al stays atomic ($\sim 99\%$) and the most abundant Al-bearing molecules are AlO, AlH, AlC, AlOH, $Al(OH)_2$, and $Al(OH)_3$. These molecules are then good starting points for determining/calculating different reactions to form Al_2O_3 clusters, skipping its monomer.

Acknowledgements

The authors acknowledge support from the ERC consolidator grant 646758 AEROSOL.

References

Bowen G. H. 1988, *ApJ*, 329, 299
Boulangier J., Clementel N., *et al.* 2018, *MNRAS*, accepted
Boulangier J., Gobrecht D., *et al.* (in prep.)
Bromley, S. T., Gómez Martín, J. C., & Plane, J. M. C. 2016, *Phys.Chem.Chem.Phys.*, 18, 26913
Cherchneff I. 2012, *A&A*, 545, A12
Gail, H.-P., Wetzel, S., Pucci, A., Tamanai, A. 2013, *A&A*, 555, A119
Gobrecht D., Cherchneff I., Sarangi A., Plane J. M. C., Bromley S. T. 2016, *A&A*, 585, A6
Goumans, T. P. M., Bromley, S. T. 2012, *MNRAS*
Grassi, T. and Bovino, S. el al. 2014, *MNRAS*, 439, 2386
Helling, Ch., Woitke, P. 2006, *A&A*, 455, 325
Höfner, S. and Bladh, S. and Aringer, B. and Ahuja, R. 2016, *A&A*, 594, A108
Keppens R., Meliani Z., *et al.* 2012, *J. Comput. Phys.*, 231, 718
Köhler, T. M., Gail, H.-P., Sedlmayr, E., 1997, *A&A*, 1997, 320, 553
Lazzati, D., & Heger, A. 2016, *ApJ*, 817, 134
Lee, G. K. H., Blecic, J., Helling, Ch. 2018, *A&A*, 614, 126
Marigo P., Ripamonti E., Nanni A., Bressan A., Girardi L. 2016, *MNRAS*, 456, 23
Nozawa, T., Kozasa, T., & Habe, A. 2006, *ApJ*, 648, 435
Patzer, A. B. C., Gauger, A., Sedlmayr, E. 1998, *A&A*, 337, 847
Plane, J. M. C. 2013, *Phil. Trans. of the Royal Society of London Series A*, 371, 20120335
Plewa, T. and Müller, E. 1999, *A&A*, 342, 179
Sluder, A., Milosavljević, M., & Montgomery, M. H. 2018, *MNRAS*, 480, 5580
Willson A. L. 2000, *ARA&A*, 38, 573
Woitke P. 2006, *A&A*, 452, 537

Discussion

QUESTION: Do you update the chemical reaction network during your hydrochemical model?

BOULANGIER: No, we determine a reduced network prior to the hydrochemical network that is valid for the parameter space of the AGB wind model (temperature, density, time).

QUESTION: Do you include ions in your calculations and what role do they play?

BOULANGIER: Yes. They play an important role in cooling the gas at high temperatures. Their importance for the composition is not immediately clear, and we have not investigated this.

QUESTION: How much confidence do you have that the TiO clusters are most relevant, given that the models reveal difficulties producing the same of the more elementary molecules?

BOULANGIER: We do not claim that TiO clusters are the most relevant. Our current models can however only produce TiO clusters efficiently. However, we mainly think that this is due to the lack of formation pathways of the other cluster candidates. For example, we find more Al_2O_3 than TiO clusters in pre-solar grains, yet we cannot produce it with the models suggesting that our model is not comprehensive enough. A possibility might be to form a larger Al_2O_3 cluster not via its monomer but by a different chemical reaction, skipping the unstable and difficult to form monomer.

QUESTION: Do you rule out that formation of alumina dust can occur, at least partly, on Al-bearing seed grains?

BOULANGIER: We do no rule this out because we do not take this into account. It might be that there is an efficient alumina dust formation pathway that is not the nucleation by monomer addition which we are currently only considering. The only way of knowing is adding more and more formation pathways and then see which ones are most efficient.

Why Galaxies Care About AGB Stars:
A Continuing Challenge through Cosmic Time
Proceedings IAU Symposium No. 343, 2019
F. Kerschbaum, M. Groenewegen & H. Olofsson, eds.

doi:10.1017/S1743921318006117

Lumpy stars and bumpy winds

Sofie Liljegren, Susanne Höfner, Bernd Freytag and Sara Bladh

Division of Astronomy and Space Physics, Department of Physics and Astronomy, Uppsala University, Box 516, 751 20 Uppsala, Sweden
email: `sofie.liljegren@physics.uu.se`

Abstract. The wind-driving process of AGB stars is thought to be a two-step process: first matter is levitated by shock waves, and then accelerated outwards by radiation pressure on newly condensed dust grains. When modelling such a wind, spherical symmetry is usually assumed. This is in stark contrast with recent observations, which shows significant non-spherical structures. Giant convection cells cover the surface of the star, and matter is being ejected into the atmosphere where it condenses into lumpy dust clouds. We try to quantify the differences between what is simulated in the 3D star-in-a-box models (CO5BOLD code) and the 1D dynamical atmosphere and wind models (DARWIN code). The impact of having a non-spherical star on the wind properties is also investigated. We find that the inherent non-spherical behaviour of AGB stars might induce a dust-driven weak wind already early on the AGB, and including that the star is anisotropic when simulating the wind leads to large time variations in the density of the outflow. Such variations might be observable as small-scale structures in the circumstellar envelope.

Keywords. Evolved stars, AGB stars, Stellar winds

1. Introduction

From both 3D models and from high-angular observations it has become increasingly obvious that AGB stars are not only highly variable, but they also have a complex non-spherical morphology. Near-IR interferometric imaging of the surface layers shows variable bright spots, and inhomogeneous distributions of molecules (Haubois *et al.* 2015), which indicates the presence of giant convection cells. As mentioned by Kamiński *et al.* (2016), the inhomogeneous distribution of molecules should influence the dust formation around AGB stars. At distances of 2-3 stellar radii ($R_\star$) dust forms intermittently in clumpy clouds, instead of in spherical shells (Khouri *et al.* 2016). Such non-spherical structures in both the surface layers and in the inner atmosphere also arise in the 3D models of AGB stars. When modelling the AGB star dust-driven wind, however, spherical symmetry is usually assumed. While there are no 3D wind models (yet), attempts to evaluate the effects of a non-spherical star on the stellar wind are outlined here. For an in detail description of the results, see Liljegren *et al.* (2018).

2. Anisotropic morphology and stellar wind

The wind-driving mechanism in AGB stars is a two-stage process; the pulsations of the star eject material into the atmosphere, and when the temperature is low enough dust will condense. This dust interacts with the radiation field through either absorption or scattering, and is accelerated outwards, which induces a wind. Here the dynamical atmosphere and wind driving are described with the DARWIN code, which features frequency-dependent (non-grey) radiation hydrodynamics and includes time-dependent non-equilibrium growth of dust (for details see Höfner *et al.* 2016). The inner boundary of

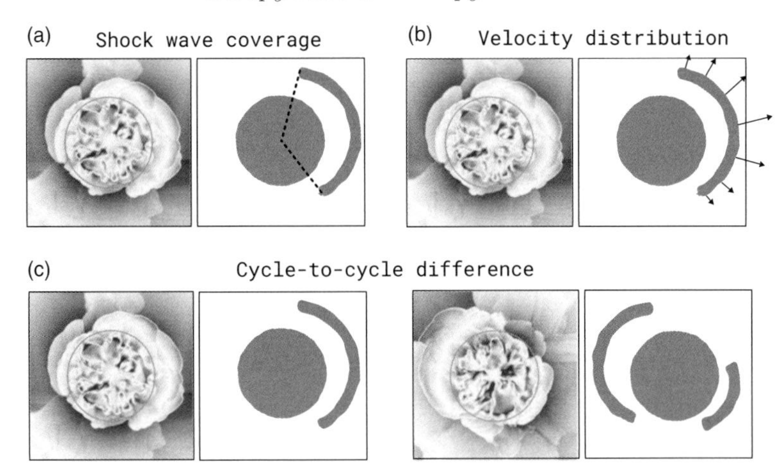

Figure 1. Plot showing three processes present in the 3D models, but not in the 1D models. The left subplot is the radial velocity (blue: outgoing gas, red: in-falling gas), while the right subplot is a schematic picture with the red circle representing the star and the blue arc representing the shockwave of a 3D model.

the DARWIN models is set just below the photosphere at $\sim 0.9\,R_\star$ and the outer boundary is typically located around $\sim 25\,R_\star$, where the wind has reached its final velocity v_∞. It is however unknown to what extent any deviation from spherical symmetry, which is a core assumption in the DARWIN models, will affect the mass-loss rate predictions.

While the extended atmosphere and the wind driving cannot yet be modelled in 3D, there are 3D models of the interior and the inner atmosphere. The CO^{5}BOLD radiation hydrodynamical code has been used in e.g. Freytag *et al.* (2017) to simulate the AGB stars. These are 3D star-in-a-box models where the full star is described. While some parts of the atmosphere are covered as the models reach as far as $\sim 2\,R_\star$, no dust-gas interaction is included in the code.

These two modelling methods describe vastly different parts of an AGB star. However, they have a region above the stellar photosphere, but below the dust condensation distance at $\sim 1-2\,R_\star$, where they overlap. In this region the pulsations and the ballistic motions of the gas dominate the dynamics. In the DARWIN models the behaviour at these distances is very much a consequence of the choice of inner boundary conditions. The atmospheric dynamics of the CO^{5}BOLD models, however, depend on shock waves triggered by pulsations that emerge in the simulations.

This overlap region can be used to compare the gas dynamics in the two model approaches, to try to deduce what impact the 3D effects present in the CO^{5}BOLD models might have on the wind driving. There are three major differences;

(*a*) **Shock wave coverage.** The anisotropy present in the 3D models also means that while the shock waves are of global scale they do not cover the full surface, which is clearly seen in Fig. 1, panel a).

(*b*) **Velocity distribution in the shock.** In contrast to the 1D models, the shock fronts in the 3D models display non-spherical structures. In the DARWIN models the velocity in the shockwave at a certain time and distance is always the same, while this is not the case in the 3D models (Fig. 1, panel b)).

(*c*) **Cycle-to-cycle variations.** In the 3D models there are also large cycle-to-cycle variations in both the direction and size of the shockwave. This can be seen in panel c) of Fig. 1.

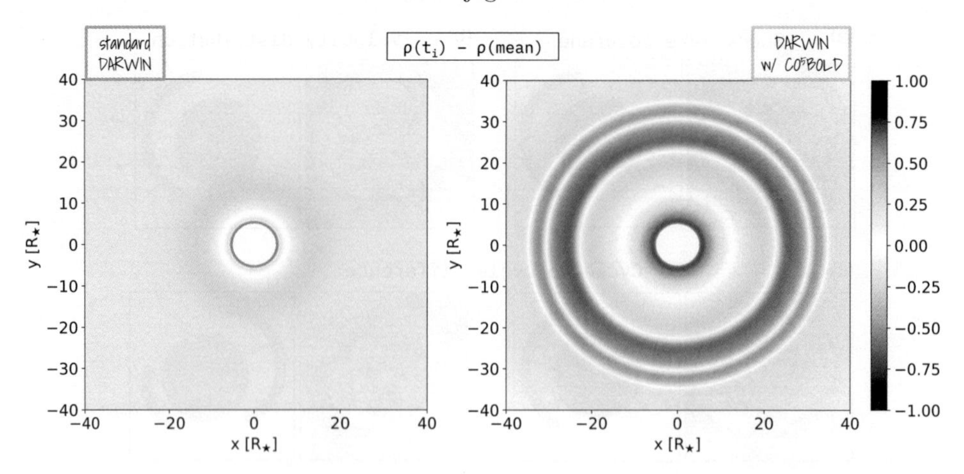

Figure 2. The density fluctuation in the wind of a snapshot of a standard DARWIN model (left), and a DARWIN model with CO^{5}BOLD input (right).

3. Wind models with 3D input

As there are no 3D wind models available (yet) we try to imitate reality by constructing 1D DARWIN models with 3D CO^{5}BOLD model input, to estimate the effect of non-spherical morphology. By using the light curves from the CO^{5}BOLD models combined with derived radial motions, the impact of the anisotropic structures and irregular behaviour on the wind properties can be studied. The inner boundary condition is then also derived from independent modelling of the pulsations, in contrast to the parameterised standard boundary condition.

When comparing the standard DARWIN models with DARWIN models with 3D input, we find that there are no large differences in the averaged wind properties. Rather, the models with boundary conditions derived from the CO^{5}BOLD models have both similar mass-loss rates and wind velocities to the standard case. There is however a large difference when looking at the time evolution of the wind properties. For the standard DARWIN models the mass-loss rates typically vary little with time, indicating a steady wind. However models based on CO^{5}BOLD input can have drastic variations in both wind velocity and mass-loss rates over the simulation period, with the mass-loss rate varying up to an order of magnitude on time scales of 10-20 years. It seems like the anisotropy of the stellar interior and the atmosphere, present in the CO^{5}BOLD-derived model, is imprinted on the outgoing wind. Fig. 2 shows the density fluctuation $(\rho(t_i) - \overline{\rho})$ of a standard DARWIN model and a DARWIN model modified with CO^{5}BOLD input. As seen, there are significant density variation in the DARWIN model with CO^{5}BOLD input, which might be large enough to create observable structures further out in the circumstellar envelope. Wind-wind interaction models with larger distance range than the standard DARWIN models are however needed for any decisive conclusions.

4. Dust-driven wind of less evolved AGB stars

A shock front moving through the atmosphere in a DARWIN model will have the same velocity at a certain time. In the 3D models, however, the anisotropy means that the velocities in the shock front depend on direction. The shock front velocity will then be a distribution, which could have consequences when trying to model the wind for less evolved AGB stars (Fig. 3). While the mean of such a distribution, which is essentially what the DARWIN models try to emulate, might be too low to induce a dust-driven wind, the velocity can locally be larger. It is then possible that some material is levitated to

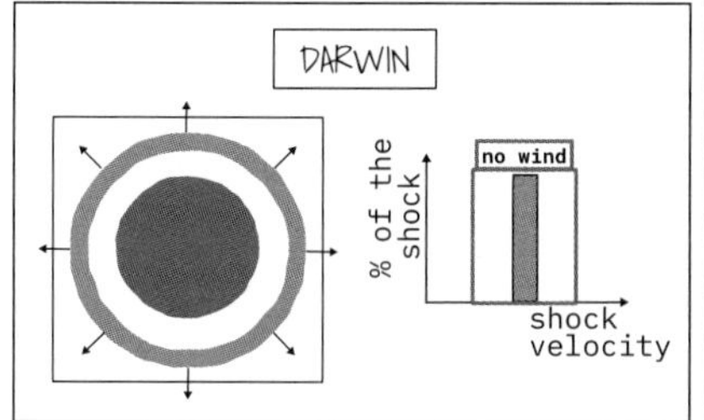

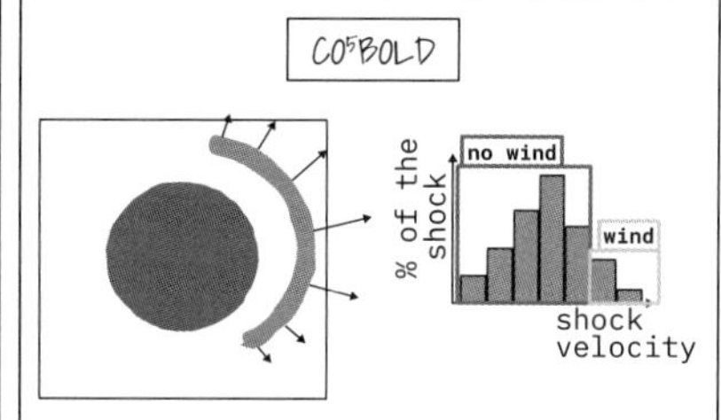

Figure 3. The shock front in the 1D case (left) and in the 3D case (left), with the associated velocity distribution in the shock.

distances where dust can form, creating a weak and maybe intermittent dust-driven wind earlier on the AGB than indicated by the standard DARWIN models. This is, however, still speculative, and both CO^{5}BOLD models and DARWIN models in this parameter space are needed to investigate this further.

References

B. Freytag, S. Liljegren, and S. Höfner. *A&A*, 600:A137, 2017

X. Haubois, M. Wittkowski, G. Perrin, P. Kervella, A. Mérand, E. Thiébaut, S. T. Ridgway, M. Ireland, and M. Scholz. *A&A*, 582:A71, 2015

S. Höfner, S. Bladh, B. Aringer, and R. Ahuja. *A&A*, 594:A108, 2016

T. Kamiński, K. T. Wong, M. R. Schmidt, H. S. P. Müller, C. A. Gottlieb, I. Cherchneff, K. M. Menten, D. Keller, S. Brünken, J. M. Winters, and N. A. Patel. *A&A*, 592:A42, 2016

T. Khouri, M. Maercker, L. B. F. M. Waters, W. H. T. Vlemmings, P. Kervella, A. de Koter, C. Ginski, E. D. Beck, L. Decin, M. Min, C. Dominik, E. O'Gorman, H.-M. Schmid, R. Lombaert, and E. Lagadec. *A&A*, 591:A70, 2016

S. Liljegren, S. Höfner, B. Freytag, and S. Bladh. *ArXiv e-prints*, Aug. 2018

Discussion

LANÇON: The 3D models predict large variations in $\dot{M}$. Do you have enough models to see if there is a correlation between dust extinction along a line of sight to the star, and either phase or the presence/strength of shocks.

LILJEGREN: We have not yet produced synthetic observables for these models, so we don't know. Such investigation are however in future plans.

SAHAI: How do your results compare to those by an older work by V. Icke which also created density variations/$\dot{M}$ variations, which I proposed could be an explanation for the shells in the AGB envelope of the Egg Nebula as seen in HST imaging (Sahai et al. 1998,1999). The time scale for variations there is $\approx$ 100–200 yr.

LILJEGREN: We don't yet know what happens with the density variations further out in the circumstellar envelope, but we are very interested to try to model these structures.

Circumstellar envelopes of AGB stars and their progeny, planetary nebulae

Why Galaxies Care About AGB Stars: A Continuing Challenge through Cosmic Time
Proceedings IAU Symposium No. 343, 2019
F. Kerschbaum, M. Groenewegen & H. Olofsson, eds.

doi:10.1017/S1743921318007007

High angular resolution observations of AGB stars

Eric Lagadec

Université Côte d'Azur, Observatoire de la Côte d'Azur, CNRS, Lagrange, France
email: elagadec@oca.eu

Abstract. Mass loss of AGB stars is a key process for the late stages of evolution of low and intermediate mass stars and the chemical enrichment of galaxies. It is not fully understood yet, as it is the result of a complex combination of pulsation, convection, chemistry, shocks and dust formation.

In this review I present what high angular resolution observations can teach us about this mass-loss process. Instruments such as SPHERE/VLT, Gravity and AMBER at the VLTI, and ALMA give us the possibility to map AGB stars from the optical to millimetre wavelengths with resolutions down to 1 milliarcsec. Moving from the surface of the star outwards, I present how high angular resolution observations can now produce images of the surface of the closest AGB stars and study convective motion at their surfaces, map their extended molecular atmospheres and the seeds for dust. The dust formation zone can also be mapped and its dust content characterized with mid-infrared interferometry, while ALMA can map the gas and its kinematics. I will conclude by showing how high angular resolution can help us study the impact of a companion on mass loss.

Keywords. techniques: high angular resolution, stars: AGB and post-AGB, stars: winds, outflows, stars: mass loss

1. Introduction

During the late stages of their evolution, low and intermediate mass stars reach the Asymptotic Giant Branch (AGB) phase, where they develop a high mass loss, leading to the formation of an envelope made of dust and gas. It is during this mass-losing phase that newly formed dust and gas, enriched with products from stellar nucleosynthesis, are enriching the interstellar medium. This mass loss is not fully understood yet and stellar evolution codes are not able to reproduce it without using ad hoc physics.

Our current understanding is that this mass loss is due to a combination of physical processes with different spatial and time scales (see e.g. Höfner & Olofsson 2018). Pulsation and convection will levitate the gas and extend the atmosphere. This will create shocks where cooler and denser gas will condense and form dust. Radiation pressure (due to absorption or scattering) will then trigger the mass loss, the gas being carried away via friction.

In this review, I will show how new instruments such as optical interferometry, extreme adaptive optics, and ALMA are revolutionising the field. We have entered the milliarcsec resolution era, see Fig. 1, and are now able to map nearby AGB stars and their circumstellar environments, thus bringing key constraints on the physical processes involved in the mass loss.

In the following sections I will present the different techniques and show what they tell us about the physics of the mass loss. I will show how high angular resolution observations

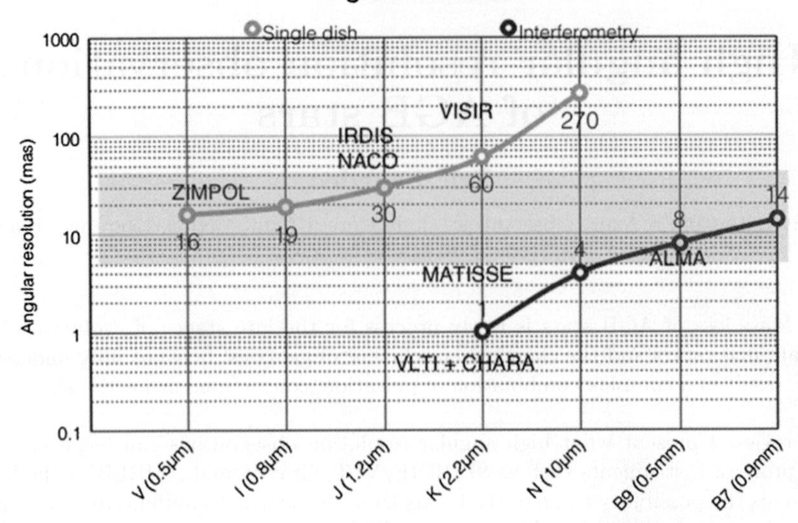

Figure 1. Angular resolution of current high angular resolution instruments operating from the optical to the millimetre wavelengths range, with both single dish and interferometric instruments. Credit: P. Kervella

can help answering key questions about the physics of the mass loss from AGB stars, and how they are bringing new constraints on the models of AGB stars.

With this aim in mind, we will be traveling from the surface of the AGB star outwards.

2. Surface of the star: convection

Theoretical models (see e.g. Freytag *et al.* 2017) predict that the surface of AGB stars is covered by a few large convective cells. Observations with high angular resolution can tell us about the timescales of these cells, their sizes, and the contrast between the cells and their surroundings, giving us constraints on the 3D theoretical models of AGB stars.

This can be achieved with IR/optical interferometry, with instruments such as AMBER/VLTI, PIONIER/VLTI, GRAVITY/VLTI, IOTA, and CHARA. The VLTI can now recombine light from up to 4 telescopes, and CHARA up to 6. This, combined with the advancement of image reconstruction codes (Monnier *et al.* 2014) means that optical/IR interferometry can now produce images with a resolution down to the milliarcsec scale. This means that we can now map the surface of the closest AGB stars.

Using this technique, (Paladini *et al.* 2018) obtained a very spectacular image of the surface of the nearby AGB star π^1 Gruis, Fig. 2. The PIONIER/VLTI observations of this compact, dust free AGB star reveal a complex convective cell pattern at its surface with an average contrast of 12 %. The characteristic size of the convective cells represent ∼27 % of the diameter of the star.

More and more convective patterns are being imaged at the surface of evolved stars with optical/IR interferometry, like for the AGB star R Scl (Wittkowski *et al.* 2017), and the red supergiant (RSG) stars Antares (Ohnaka *et al.* (2017a), Montargès *et al.* 2017), Betelgeuse (Haubois *et al.* 2009) and CETau (Montargès *et al.* 2018).

If the interferometric instrument has a spectroscopic resolution of a few thousands, one can also map the convection across spectral lines. Ohnaka *et al.* (2017a) applied this technique to map convection across the CO line for the RSG star Antares, and thus map the movement of the convective cells via the Doppler effect. The atmosphere appears

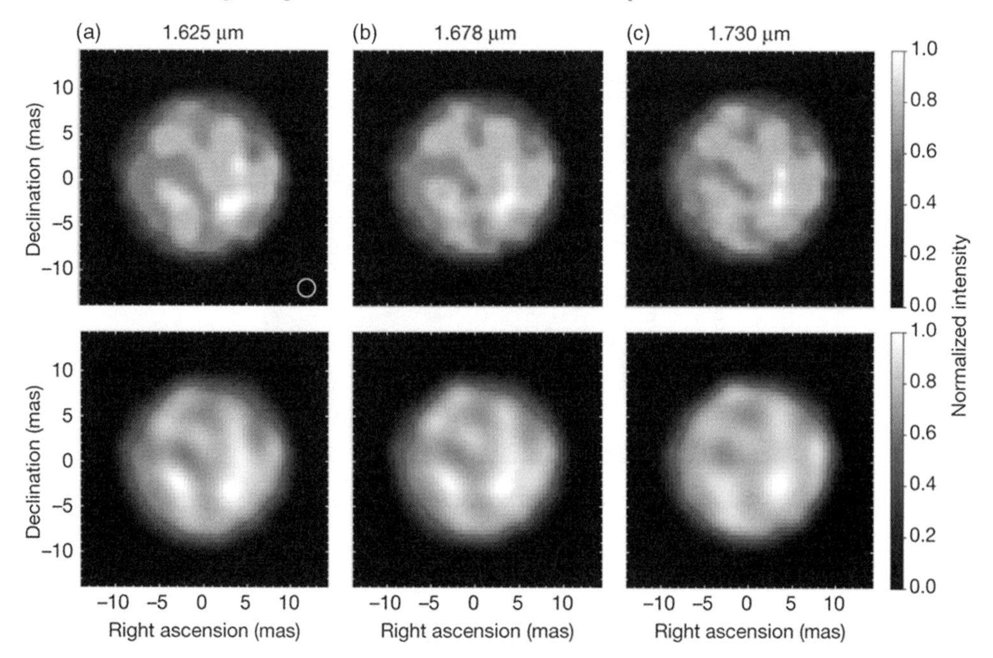

Figure 2. Image reconstruction of the surface of the AGB star π^1 Gruis from Pionier/VLTI observations at three different wavelengths Paladini *et al.* (2018). The two rows are reconstruction using different models. The similar morphology observed in the six panels show that the image reconstruction is robust and reveal a complex surface, covered by a few convective cells.

different across the CO line profile and the velocity field is inhomogeneous. Changes in the convection pattern are seen within a year.

Optical/IR interferometry is thus becoming a great tool to map convection at the surface of AGB stars and to see its evolution with time, bringing key constraints for theoretical models of AGB stars.

3. The gas phase

As mentioned before, the combination of pulsation and convection will expand the atmosphere of the star, where molecules will form. The chemistry in the gas phase will depend on the abundance ratio between carbon and oxygen, as most of the under-abundant species will be locked in the stable CO molecule. We will thus have two kind of AGB stars, oxygen-rich and carbon-rich stars. High angular resolution observations can now help answering the following questions:

- What is the gas distribution above the surface?
- How does it move?
- How does dust form from the gas phase?

Infrared interferometry with high spectral resolution (a few thousands) in the K band can lead to the determination of the size of the CO and H_2O line emitting zone (at 2.0 and 2.29 μm). Ohnaka *et al.* (2016) studied the O-rich AGB star W Hya, revealing an extended molecular atmosphere (MOLSPHERE) with water molecules extending up to 1.5 R_* and CO up to 3 R_*.

Wittkowski *et al.* (2016) performed similar observations of AGB stars (and RSGs) and compared it with state of the art models. Both 1D CODEX (Ireland *et al.* 2011) and 3D CO5BOLD (Freytag & Höfner 2008) models are able to reproduce the extent of the MOLSPHERE for AGB stars, but not for RSGs (some extra mechanisms, such as e.g.

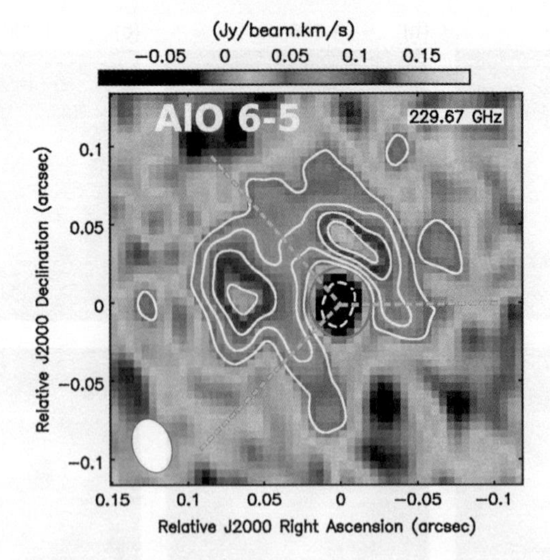

Figure 3. ALMA map of AlO (potential seed for dust formation) around the AGB star *o* Ceti. AlO is located in an incomplete ring at $\sim 2\,R_*$ (Kamiński *et al.* (2016)).

radiation pressure on lines must occur in RSGs and are not taken into account in the models).

So, we can now map extended emission from molecules around AGB stars, but one of the key process to understand is how to condense gas to dust. Dust forms after a chain of chemical reactions that will form larger and larger species.

This is quite well understood for C-rich stars (see e.g. Gautschy-Loidl *et al.* 2004).

For oxygen-rich stars, titanium oxides (TiO, TiO_2) and aluminium oxides have been proposed as the first seeds for dust formation.

Some recent works on oxygen-rich stars with ALMA lead to the detection of aluminium oxides. Kamiński *et al.* (2016) mapped a ring of AlO at $\sim 2\,R_*$ around Mira, Fig. 3. The ring is incomplete, as there is no emission from the south-east of the ring, and the lack of AlO emission at $4\,R_*$ could be a sign of depletion into dust. To confirm that, one would need contemporary monitoring with ALMA (to map AlO depletion) and an infrared interferometer (such as MATISSE/VLTI) to see AlO emission fading where dust is forming. ALMA observations of R Dor and IK Tau revealed that emission of AlO and AlOH, precursors of alumina (Al_2O_3) dust, was extending well beyond the dust condensation radiation, so that the condensation cycle of oxides of aluminium is not fully efficient (Decin *et al.* 2017). However, there are clear observational signs that Al_2O_3 seem to be the first dust to form around AGB stars, and the seeds leading to form dust seem to be aluminium oxides.

4. The dust phase

As mentioned in the introduction, dust is forming behind shocks, where the gas is cold and dense enough to condense into dust.

The questions we want to answer with high angular resolution are:

- Where does dust form?
- What kind of dust forms?
- What is the size of the dust grains?

ALMA observations of W Hya resolved its atmosphere (Vlemmings *et al.* 2017). They detected a hot spot in the southwest, with properties indicating that it is due to shocks, Fig. 4. They also observed the CO(v=1, J=3-2) line, to study gas infall and outflow from

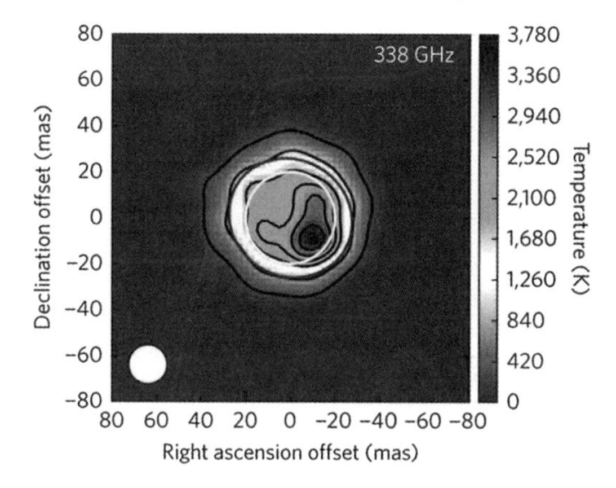

Figure 4. ALMA continuum map at 338 GHz of the surface of the AGB star W Hya (Vlemmings *et al.* 2017). The hotspot in the southwest is unresolved and is certainly due to shocks.

this star. They showed that the total gas mass traced by CO is three orders of magnitude larger than the mass lost by the star during one pulsation period. This means that the CO gas will spend at least 1000 years in the region mapped by the CO observations. As the timescales for shocks and stellar activities are much smaller, all the ejected gas will experience shock heating. Non equilibrium chemistry clearly has to be taken into account in models of dust formation.

Dust forms at less than $3\,R_*$, emits at infrared wavelengths and has spectral features in the mid-infrared. Mid-infrared interferometry is thus the key tool to resolve dust at the dust formation radius and study its properties. Till now, mid-infrared interferometers were using only two telescopes and providing visibilities (i.e., measurements of the size of the emitting zone) and closure phase (i.e., an estimation of the symmetry of the emitting zone) across the N band (between 8 and $13\,\mu$m). Using MIDI/VLTI, Karovicova *et al.* (2013) observed 3 O-rich AGB stars and showed that their observations were consistent with Al_2O_3 grains condensing at $2\,R_*$, while silicates were further away, at 4-5 R_*.

Paladini *et al.* (2017) performed a mid-infrared spectro-interferometric survey of AGB stars and showed that asymmetries were common at the dust formation location of oxygen-rich stars. Mid-infrared interferometry can thus now tell us what kind of dust is forming where, and determine the spatial distribution of the dusty shells.

Direct imaging can also bring constraints on the grain dust properties. The extreme adaptive optics imager SPHERE, installed on the VLT, can perform polarimetric measurements with a resolution down to $\sim$15 milliarcsec in the optical. The instrument is designed to directly image exoplanets and has high resolution and contrast capabilities. One way to enhance the contrast between a planet and its host star is to perform polarimetric observations. The light from the star is not polarised, while the light from the planet is scattered light and thus polarised. This applies to dust close to stars and can thus be used for AGB stars.

Using SPHERE polarimetric observations, Ohnaka *et al.* (2016) resolved clumpy dust clouds at about $2\,R_*$ around the AGB star W Hya. A second epoch of observations enabled them to study the variations of dust properties. Their observations lead to a grains size determination and indicate that small grains ($0.1\,\mu$m) are observed at minimum light, while large grains ($0.5\,\mu$m) are predominant just before the maximum.

It also revealed variation within 9 months, with the appearance of new clouds while some disappeared (Ohnaka *et al.* 2017b). The size of the dust clumps and the timescale of

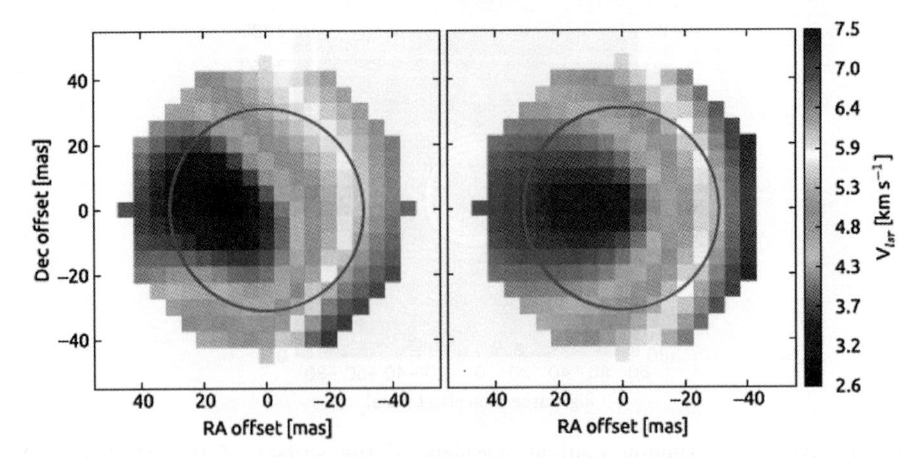

Figure 5. ALMA velocity map (left: observations; right: model) of R Dor showing clear signs of rotation (Vlemmings *et al.* 2018).

their appearance seem to be in agreement with the results of convection, indicating that convection must play a role in the dust formation process (by extending the atmosphere).

Stewart *et al.* (2016) monitored the closest carbon-rich AGB star (IRC +10216) with various high angular resolution instruments (Keck, VLT, and occultation by the rings of Saturn observed by Cassini). They also confirm that dust formation is non isotropic and variable within this period, with clumps fading and appearing.

It thus appears clear that the structure of the extended atmospheres of AGB stars have density inhomogeneities or are clumpy and shocks are present. The chemistry models presented by Van de Sande *et al.* (2018) take these effects into account. This affects the chemistry via the density structure, but also the UV radiation field, which can penetrate deeper in some locations. Thus, species not expected to be present at local thermodynamical equilibrium are formed, such as HCN in O-rich stars and water in carbon stars.

5. Impact of a companion

It is now becoming clear that a majority of planetary nebulae are bipolar, and that the observed bipolarity is due to an extra angular momentum provided by a companion (Boffin *et al.* 2012). This extra angular momentum can lead to the formation of equatorial overdensities such as disks or tori and/or jets, that can be focused by magnetic fields (Balick & Frank 2002).

Such companions should be observable around AGB stars, i.e., in the phase before the planetary nebula phase. Recent high angular observations of AGB stars with ALMA have revealed the presence of spiral structures in their envelopes (see e.g. Maercker *et al.* 2012, Ramstedt *et al.* 2014). This can be explained by the wind roche lobe overflow model proposed by Mohamed & Podsiadlowski (2012). But the binary forming these spirals have a large separation, so that these systems most likely will not form bipolar nebulae due to a lack of angular momentum.

Observations of R Dor both with SPHERE and ALMA are interesting for understanding the potential effect of a nearby companion. SPHERE observations of R Dor resolved the stellar surface, and showed that its morphology was varying with time scales of months, certainly due to a variation of opacity due to TiO molecules (Khouri *et al.* 2016). No sign of interaction with a companion are seen in these data. However, two independent ALMA observations reveal a clear rotation pattern for R Dor, Fig. 5. It was previously, using lower angular resolution data, attributed to a disk in rotation

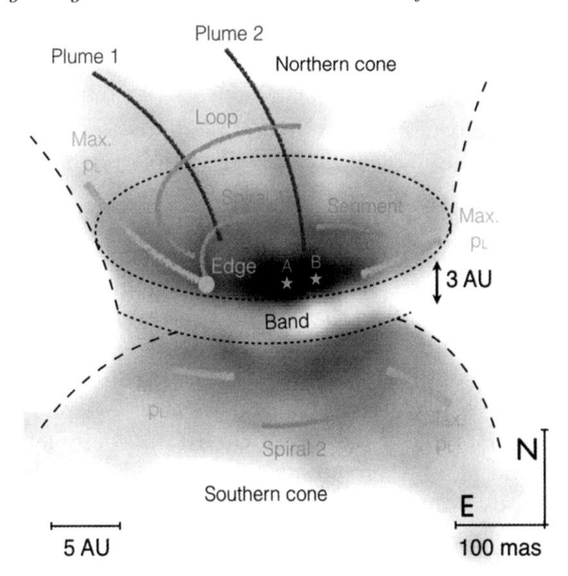

Figure 6. SPHERE optical image of the AGB star L_2 Pup, with the presence of an equatorial disc revealed in scattered light (Kervella *et al.* 2015). A companion is also discovered, and plumes of material are being ejected in a plane perpendicular to the disc.

(Homan *et al.* 2018). Higher angular resolution observations indicate that this is instead due to rotation of the stellar surface (Vlemmings *et al.* 2018). The angular momentum needed to explain the observed rotation requires the presence of a yet unseen companion. This star might form a bipolar nebula, but the best AGB candidate to form a bipolar PN is certainly L_2 Puppis.

L_2 Puppis is a nearby AGB star (64 pc), and it was observed by SPHERE (Kervella *et al.* 2015) in the optical, both in classical and polarimetric imaging. An edge-on disc is clearly seen in scattered light (thus appearing as a lane with no emission) on the SPHERE image, Fig.6, where a companion is also seen at 2 AU. The polarimetric map reveal that inner rim of the disk is located at 6 AU and that plumes of material are being ejected in a plane perpendicular to the disc.

ALMA observations of L_2 Puppis, with an angular resolution similar to those of the SPHERE observations (~ 15 mas), were performed by Kervella *et al.* (2016) to study the gas distribution and dynamics around this AGB star. They confirmed the presence of the disc and that it is in keplerian rotation. This led them to an estimate of $\sim 0.7\,M_\odot$ for the central star of the system and $\sim 12\,M_{\rm Jup}$ for the companion. The companion appears thus to be a planet and not a star. From evolutionary models, they estimated that the initial mass of the central star was very close to the mass of the sun, and that the ALMA observations might be giving us an idea of what will happen to the sun in 5-6 billion years. Let's take bets and wait and see?

6. Conclusion and perspectives

Thanks to new instrumentation such as the extreme adaptive optic system SPHERE on VLT, optical/infrared interferometry, and ALMA, high angular resolution observations are helping us understand the complex mass-loss process of AGB stars. It is now possible to map convection at the surface of stars, shocks above this surface due to convection and pulsation, and dust formation behind these shocks. The atmospheres of AGB stars are not spherical and isotropic where dust is forming, as dust appears to form in clumps above convective cells or shocks. Companions, or hints of companions, of AGB stars are being found, and it is now clear that a non negligible fraction of AGB stars will interact with a companion. Even if not all these interactions will lead to mass transfer, the density structure of the circumstellar envelope will be affected. Impressive developments on theoretical models have been, and are being, made to use this information and take into account the effect of non-spherical circumstellar envelopes, clumpiness, and non-LTE effects.

From an observational point of view, two kinds of observing programs need to be performed, either to better understand the physics of the mass loss from AGB stars or determine the quantitative impact of AGB stars to the enrichment of the Galaxy. Both programs should be very ambitious as they require observations from multiple telescopes using different techniques and operating at various wavelengths.

To understand the physics of the mass loss, time series observation of a representative sample of AGB stars (O-rich, C-rich, with and without companions, and at different evolutionary stages). Combining ALMA, optical/IR interferometry would then allow monitoring of these stars along their pulsation cycle and map the convection, shocks, gas depletion into dust, dust formation, dust properties, and wind acceleration.

Quantitavely determining the impact of AGB stars on the chemical enrichment of nearby galaxies can be achieved as we are outside of these galaxies and can map them fully rather rapidly and we have a good estimate of the distance of stars in the galaxies, which can be assumed to be the same as the galaxies themselves. A similar method has been applied e.g. to determine the dust input from AGB stars to the interstellar medium in the Large Magellanic Cloud (Srinivasan *et al.* 2009, Matsuura *et al.* 2009). In the Galaxy, it is difficult to get the distance to all the AGB stars, and due to extinction in the Galactic plane, some are too obscured to be observed. But thanks to Hipparcos and Gaia, distances to the closest AGB stars is now known (within 200-300 pc). Observations of all the stars in a volume limited and complete sample of nearby stars would help us getting statistics about the properties of these stars. Observations at millimetre wavelengths will tell us about the gas properties, in the infrared about the dust, and high angular resolution can teach us about the density distribution. It is thus important to obtain observations of AGB star in such a sample and combine this with state of the art radiative transfer model to quantify the impact of AGB stars on the chemical enrichment of the Galaxy.

Finally, dust in AGB circumstellar envelopes emit in the mid-infrared and have spectral signal in the N band (between 8 and 13 μm). The best way to study the dust spatial distribution is thus to obtain high angular resolution observations of AGB stars in the mid-infrared. The new generation VLTI instrument MATISSE (Lopez *et al.* 2014) will offer a unique opportunity. It will be offered to the community for observations starting in 2019. It will enable us to obtain images of the different dust species around AGB stars from the dust formation radius outwards.

References

Balick, B., & Frank, A. 2002, ARA&A, 40, 439
Boffin, H. M. J., Miszalski, B., Rauch, T., *et al.* 2012, Science, 338, 773
Decin, L., Richards, A. M. S., Waters, L. B. F. M., *et al.* 2017, A&A, 608, A55
Freytag, B., & Höfner, S. 2008, A&A, 483, 571
Freytag, B., Liljegren, S., & Höfner, S. 2017, A&A, 600, A137
Gautschy-Loidl, R., Höfner, S., Jørgensen, U. G., & Hron, J. 2004, A&A, 422, 289
Haubois, X., Perrin, G., Lacour, S., *et al.* 2009, A&A, 508, 923
Homan, W., Danilovich, T., Decin, L., *et al.* 2018, A&A, 614, A113
Höfner, S., & Olofsson, H. 2018, ARA&A, 26, 1
Ireland, M. J., Scholz, M., & Wood, P. R. 2011, MNRAS, 418, 114
Kamiński, T., Wong, K. T., Schmidt, M. R., *et al.* 2016, A&A, 592, A42

Karovicova, I., Wittkowski, M., Ohnaka, K., *et al.* 2013, A&A, 560, A75
Kervella, P., Homan, W., Richards, A. M. S., *et al.* 2016, A&A, 596, A92
Kervella, P., Montargès, M., Lagadec, E., *et al.* 2015, A&AL, 578, A77
Khouri, T., Maercker, M., Waters, L. B. F. M., *et al.* 2016, A&A, 591, A70
Kim, H., Trejo, A., Liu, S.-Y., et al. 2017, Nature Astronomy, 1, 0060
Lopez, B., Lagarde, S., Jaffe, W., *et al.* 2014, The Messenger, 157, 5
Lykou, F., Zijlstra, A. A., Kluska, J., *et al.* 2018, MNRAS, 480, 1006
Maercker, M., Mohamed, S., Vlemmings, W. H. T., *et al.* 2012, Nature, 490, 232
Matsuura, M., Barlow, M. J., Zijlstra, A. A., *et al.* 2009, MNRAS, 396, 918
Mohamed, S., & Podsiadlowski, P. 2012, Baltic Astronomy, 21, 88
Monnier, J. D., Berger, J.-P., Le Bouquin, J.-B., *et al.* 2014, SPIE, 9146, 91461Q
Montargès, M., Chiavassa, A., Kervella, P., *et al.* 2017, A&A, 605, A108
Montargès, M., Norris, R., Chiavassa, A., *et al.* 2018, A&A, 614, A12
Ohnaka, K., Weigelt, G., & Hofmann, K.-H. 2016, A&A, 589, A91
Ohnaka, K., Weigelt, G., & Hofmann, K.-H. 2017, Nature, 548, 310
Ohnaka, K., Weigelt, G., & Hofmann, K.-H. 2017, A&A, 597, A20
Paladini, C., Baron, F., Jorissen, A., *et al.* 2018, Nature, 553, 310
Paladini, C., Klotz, D., Sacuto, S., *et al.* 2017, A&A, 600, A136
Ramstedt, S., Mohamed, S., Vlemmings, W. H. T., *et al.* 2014, A&A, 570, L14
Srinivasan, S., Meixner, M., Leitherer, C., *et al.* 2009, AJ, 137, 4810
Stewart, P. N., Tuthill, P. G., Monnier, J. D., *et al.* 2016, MNRAS, 455, 3102
Van de Sande, M., Sundqvist, J. O., Millar, T. J., *et al.* 2018, A&A, 616, A106
Vlemmings, W. H. T., Khouri, T., Beck, E. D., *et al.* 2018, A&A, 613, L4
Vlemmings, W., Khouri, T., O'Gorman, E., *et al.* 2017, Nature Astronomy, 1, 848
Wittkowski, M., Chiavassa, A., Freytag, B., *et al.* 2016, A&A, 587, A12
Wittkowski, M., Hofmann, K.-H., Höfner, S., *et al.* 2017, A&A, 601, A3

Why Galaxies Care About AGB Stars:
A Continuing Challenge through Cosmic Time
Proceedings IAU Symposium No. 343, 2019
F. Kerschbaum, M. Groenewegen & H. Olofsson, eds.

doi:10.1017/S174392131800724X

The mass-loss characteristics of AGB stars An observational view

Sofia Ramstedt

Division of Astronomy and Space Physics, Department of Physics and Astronomy, Uppsala University
email: sofia.ramstedt@physics.uu.se

Abstract. The massive outflows of gas and dust which characterize giant stars on the Asymptotic Giant Branch (AGB), build cool circumstellar envelopes readily observed at infrared (IR) and sub-millimeter wavelengths. The observations will give the amount of matter lost by the star, the wind velocity (in the case of spectral line observations), and, when the spatial resolution is sufficient, the wind evolution over time. To gain detailed insight into the mass-loss process, we study the nearby (closer than 1 kpc) stars. Through these investigations we aim to determine the best constrained wind properties available. By combining this with theoretical results, mass-loss estimates for more distant sources can also be significantly improved. ALMA has opened up new opportunities to study the winds of AGB stars. The DEATHSTAR project (www.astro.uu.se/deathstar) has mapped the circumstellar CO emission from so far ~50 nearby M- and C-type AGB stars. The data will initially be used to give a definitive mass-loss prescription for the sample sources, but the large-bandwidth observations opens for many different legacy projects. The current status and results are presented.

Keywords. AGB and post-AGB, circumstellar matter, mass-loss

1. Introduction

Molecular line emission in the sub-millimeter and millimeter regime is considered to be the most reliable indicator of the physical properties of circumstellar envelopes (CSEs) of AGB stars. Even so, the uncertainties of state-of-the-art mass-loss-rate estimates will reach as high as a factor of three (Ramstedt *et al.* 2008). As emphasized by recent high-spatial resolution observations (e.g., Guélin *et al.* 2018), an important factor is the impact of 3D effects such as asymmetries and clumpy structures in the outflows. Additional uncertainties are related to the thermodynamics of the CSEs and data calibration errors. Furthermore, this method has a limited reach due to the still restricted spatial resolution and signal-to-noise ratio that can be achieved in this long-wavelength regime, and therefore, methods used for more distant sources also have to be further developed. For sources beyond more than 1 kpc, the mass-loss-rate estimates are usually based on IR observations that probe the circumstellar dust emission, and an assumed gas-to-dust mass ratio. Recent theoretical results (e.g., Eriksson *et al.* 2014; Bladh *et al.* 2015; *this volume*; Nanni *et al.* 2018) show that the usually assumed gas-to-dust ratio of 200, and thereby the estimated mass-loss rates, could be significantly underestimated.

2. Methods to estimate mass-loss rates from single-pointing observations

2.1. *Observations of circumstellar envelopes*

The main constituent of AGB CSEs is cool gas, mainly in molecular form, with a gas-to-dust-mass ratio usually assumed to be of the order of 200. The most abundant

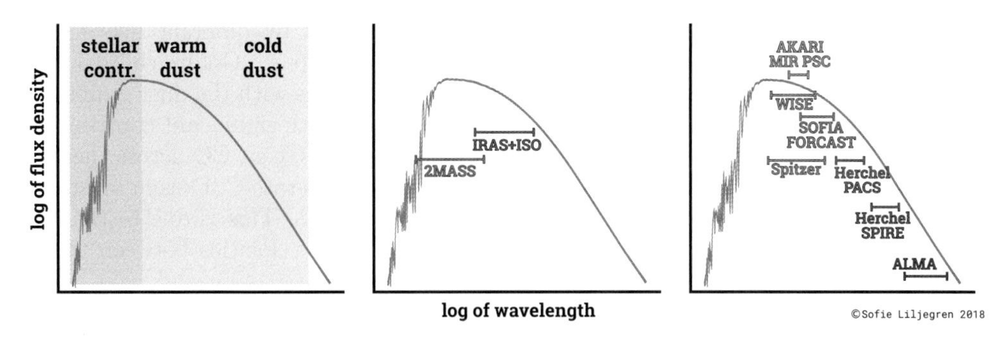

Figure 1. Left: Schematic overview of a typical SED (in log-scale) of a (carbon-rich) AGB star. Middle: Wavelength coverage of publicly available IR data before the turn of the century. Right: Wavelength coverage of recently observed and publicly available IR data. See text for references. Image credit: Sofie Liljegren.

molecule is H_2, but since it is not readily observable, CO rotational line emission is considered to be the most reliable probe for the physical conditions in CSEs. Single-pointing observations of low-J (<5) transitions have been collected for the AGB stars within a distance of about 1 kpc over the last several decades (see e.g., Teyssier *et al.* 2006; Ramstedt *et al.* 2009; De Beck *et al.* 2010, for recent examples). The Herschel/HIFI instrument also made higher-J transition line emission available for a relatively large fraction of the nearby sources (e.g., Justtanont *et al.* 2012; Danilovich *et al.* 2015b). The mass-loss-rate estimates based on these types of observations are representative of the average mass-loss rate over the time of the creation of the emitting region and are not necessarily a good measure of the current mass-loss rate from a star. With the Atacama Large Millimeter/submillimeter Array (ALMA) it has also for the first time been possible to resolve and detect CO line emission toward a few individual AGB stars in the Large Magellanic Cloud (LMC; Groenewegen *et al.* 2016).

The spectral energy distribution (SED) of a typical AGB star will show emission from a few tenths of a μm to some thousand μm and peak at around 1 μm. The dust emission will contribute significantly beyond $\sim$5 μm. Observations in this wavelength range require space-based facilities and several instruments have added to the available data base of IR observations of AGB stars in recent years (Fig. 1): AKARI (e.g., Ishihara *et al.* 2011), Spitzer (e.g., Meixner *et al.* 2006; Matsuura *et al.* 2013), WISE (e.g., Lian *et al.* 2014), Herschel PACS and SPIRE (Groenewegen *et al.* 2011), and SOFIA (e.g. Hankins *et al.* 2018). A comprehensive overview of the far-IR emission from nearby AGB stars is given in Cox *et al.* (2012) with images at 70 and 160 μm from Herschel/PACS. The authors focus on large-scale structure and wind-ISM interaction and classify the sources in four distinct categories (fermata, eyes, irregular, rings). Bow shocks are detected for 40% of the sources.

2.2. *Radiative transfer models of circumstellar envelopes*

Most estimates of mass-loss rates and other physical and chemical parameters are based on a standard description of the CSE: "The standard model" (Höfner & Olofsson 2018). In this model, the CSE is assumed to be homogeneous, spherical, and created by a constant stellar mass-loss rate. It is expanding at a constant velocity and in some calculations a thin inner acceleration region is included (e.g., Danilovich *et al.* 2015b). Micro-turbulent (of the order of 1.0 km s^{-1}) and local thermal contributions are added to the line broadening. The mass-loss rate is given by the conservation of mass: $\dot{M} = 4\pi r^2 v(r)\rho(r)$, where $v(r)$ is just the terminal expansion velocity, v_∞, if a constant expansion velocity is assumed.

The circumstellar temperature distribution is determined by different heating and cooling processes according to the energy-balance equation (e.g., Goldreich & Scoville 1976). The heating of the gas is predominantly due to collisions with the dust grains. The cooling is mainly due to the adiabatic expansion of the gas with significant contributions from line cooling from the most abundant radiating molecules, e.g., CO across the CSE, and H_2O and HCN in the inner CSE. The gas in low-mass-loss-rate CSEs is in most cases found to be warmer than the gas in high-mass-loss-rate CSEs. This is due to the more efficient line cooling, and less efficient heating at lower drift velocities between the gas and dust, in the high-mass-loss-rate CSEs.

The CO abundance distribution is commonly calculated using the model presented in Mamon *et al.* (1988). There is however growing evidence that this approach needs to be revised (Li *et al.* 2014; 2016; Groenewegen 2017; Saberi et al. *this volume*). Under typical CSE conditions it is sufficient to include the ground and first vibrational states and ∼40 rotational levels in the description of the CO molecule as found in the LAMBDA† data base (Schöier *et al.* 2005). The radiation from the central source is usually included in the form of a blackbody with the stellar effective temperature. In addition, thermal emission from the dust grains distributed across the CSE, as well as emission from the cosmic microwave background (mainly affecting the outer CSE) is included (see Höfner & Olofsson 2018, and references therein).

Observations of the broad continuum observations covering the SED can be modelled to estimate the dust optical depth. A dust-mass-loss-rate is then calculated through assumptions about the dust optical properties and the dust expansion velocity. The dust temperature distribution is determined by the balance between absorption and emission of radiation from the dust grains. The estimated mass-loss rate is more reliable, and averaged over a longer timescale, if the model is constrained using data covering the full breadth of the SED. An even higher accuracy is achieved if the continuum observations are complemented with some line measure to constrain the expansion velocity (e.g., OH maser emission lines), since the dust kinematics cannot be directly measured. Emission from the central source is usually included in the form of output from a hydrostatic stellar atmosphere model (e.g., MARCS; Gustafsson *et al.* 2008, or COMARCS; Aringer *et al.* 2009) when modelling the SED. For sources beyond ∼1 kpc, IR observations are essentially the only available probe that can be used to estimate mass-loss rates from AGB stars. Recent examples from the Local group are given in e.g., Gullieuszik *et al.* (2012) and Groenewegen & Sloan (2018).

3. Current results from CO line observations

AGB stars are divided into three main spectral types depending on the relative strength of molecular bands indicative of the C/O-ratio in their atmospheres: M-type stars ($C/O < 1$), S-type stars ($C/O \lesssim 1$), and C-type or carbon stars ($C/O > 1$). Mass-loss rates of nearby AGB stars from radiative transfer modelling of CO line observations are given in Schöier & Olofsson (2001; carbon stars), González-Delgado *et al.* (2003; M-type stars), and in Ramstedt *et al.* (2006; 2009; S-type and the summary of the full nearby sample). The three samples are essentially complete out to 500 pc. Mass-loss rates versus the terminal velocities (determined from the width of the CO lines taking other line broadening mechanisms into account) are shown in Fig. 2 together with the recent estimates for the LMC ([Fe/H]=-0.37 dex; open symbols) sources observed with ALMA (see above; Groenewegen *et al.* 2016). It is not straight-forward to compare wind properties at different metallicities, as it is not appropriate to directly compare CO-line-emission and

† `http://home.strw.leidenuniv.nl/~moldata/`

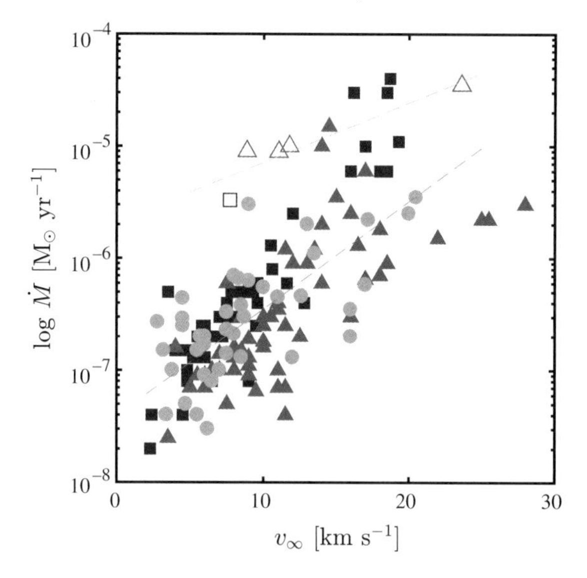

Figure 2. Mass-loss rates and wind velocities from CO line observations for carbon stars (red triangles), M-type stars (blue squares), and S-type stars (green dots). Filled symbols represent nearby, Galactic sources (Ramstedt *et al.* 2009, and references therein), and open symbols mark the recently observed LMC sources (Groenewegen *et al.* 2016). Basic linear fits to the LMC (dashed cyan) and nearby, Galactic (dashed magenta) samples are shown to guide the eye.

dust-continuum estimates, unless explicit care has been taken to calculate them consistently and therefore the comparison is made only for the mass-loss rates estimated using CO observations (for the LMC sources, using the relation given in Ramstedt *et al.* 2008). As seen in Fig. 2, the nearby Galactic sources of all three spectral/chemical types cover the same ranges in, and follow a similar correlation between, the wind properties. This suggests that the wind is driven by the same processes regardless of chemistry. However, as also seen in Fig. 2, it seems that the correlation between the wind parameters could be different at lower metallicity, as also suggested by Groenewegen *et al.* 2016, but this has to be confirmed by further CO observations of resolved LMC sources (or updated dust mass-loss-rate models for the nearby sources).

4. Overview from larger samples

It is clear that reliable mass-loss-rate estimates from larger samples of stars and from lower-metallicity sources are necessary to be able to determine the impact of the AGB stars on the galaxies they live in. The relation between the mass-loss rate of a star and it's basic stellar parameters in some sense hold the key to stellar evolution modelling and the derivation of the stellar yields from a theoretical perspective. There has therefore been attempts to derive empirical correlations between the mass-loss rates and stellar parameters (e.g., van Loon *et al.* 2005; Cummings *et al.* 2016) using different types of mass-loss-rate estimates. However, the outcome is inconclusive. The functional dependence of the mass-loss rate on the stellar luminosity (L) for instance ranges from almost linear for a sample of oxygen-rich giants in the LMC (from SED modelling; van Loon *et al.* 2005), to as steep as L^5 for a sample of nearby stars (from CO line modelling; Danilovich *et al.* 2015b). Another example is given by Riebel *et al.* (2012). Photometry observations (optical to mid-IR) of 30000 LMC stars are fitted using a large grid of dust radiative transfer models with varying stellar parameters and dust properties (GRAMS;

e.g., Sargent *et al.* 2011). In this work, no clear trend between the dust mass-loss rates and stellar luminosities is found. The discrepancies between the different results and methods show that there is still a lot of work that needs to be done to evaluate and improve the reliability of mass-loss-rate estimates, in particular for more distant sources. Some of the problems are likely due to that wind formation, and its evolution over time, gives rise to significant 3D effects.

5. Gas-to-dust mass ratios

Models of the dust continuum emission will give the dust density distribution and, with an assumption about the dust expansion velocity, the dust mass-loss rate (see continuity equation given above). To calculate the total mass-loss rate, the gas-to-dust mass ratio needs to be assumed. The standard value used in e.g. the widely spread public code DUSTY (Ivezic *et al.* 1999) is 200 (Groenewegen 2006; Groenewegen *et al.* 2007; 2009; Gullieuszik *et al.* 2012; Boyer *et al.* 2012; Srinivasan *et al.* 2006). Radiation-hydrodynamic wind models, including grain formation and growth where wind properties are calculated from first principles (Eriksson *et al.* 2014; Bladh *et al.* 2015; *this volume*), give a significantly higher value (of the order of 1000 and above). The same is found from recent estimates of the dust production rate at different metallicities based on stellar evolution models (e.g., Nanni *et al.* 2018). These model results need to be constrained by observations. Comparisons between consistent calculations of gas and dust mass-loss rates from observations give similar high values (Ramstedt *et al.* 2008), but they only exist for a small number of sources. A larger sample of observations would give stronger constraints to the theoretical results, however, this could mean that a large fraction of the mass-loss rates derived for more distant sources are significantly underestimated (by as much as a factor of 5-10). If so, this can have a large impact on stellar yield calculations and on investigations of the dust production across the Universe.

6. High-spatial resolution CO line observations

With earlier generations of submillimeter interferometers, it was possible to resolve the large-scale radial structure of the CSEs in some of the most nearby AGB stars (e.g., Castro-Carrizo *et al.* 2010). In recent years, the exceptional sensitivity and resolving power of ALMA has revolutionized the field by mapping CSEs of nearby AGB stars in remarkable detail (e.g., Maercker *et al.* 2012; Kim *et al.* 2017; Ramstedt *et al.* 2017; Bujarrabal *et al.* 2018), and by enabling studies of dynamical processes even on stellar scales (e.g., Vlemmings *et al.* 2018). The detailed images have opened up the possibility to directly track one of the most important aspects missing from our understanding of late stellar evolution; the evolution of the mass-loss rate and wind kinematics over time. New discoveries on the shaping of CSEs by a binary companion have been made, relevant in particular when studying the formation of planetary nebulae (Jones & Boffin 2017) and supernovae Type Ia (Maoz *et al.* 2014). Of course, the radiative transfer models needed to reproduce the three-dimensional images from interferometric observations require a higher level of complexity than the 1D models discussed above. The analysis of ALMA images has therefore in many cases consisted in comparing structures in the density distribution derived from hydrodynamic models directly with the images, however, some recent attempts at full 3D radiative transfer modelling also exist.

6.1. *Two recent examples: W Aql and π^1 Gru*

W Aql and π^1 Gru are two well-studied S-type AGB stars that are part of a small sample of binary AGB stars observed with ALMA. The sample (also including R Aqr and

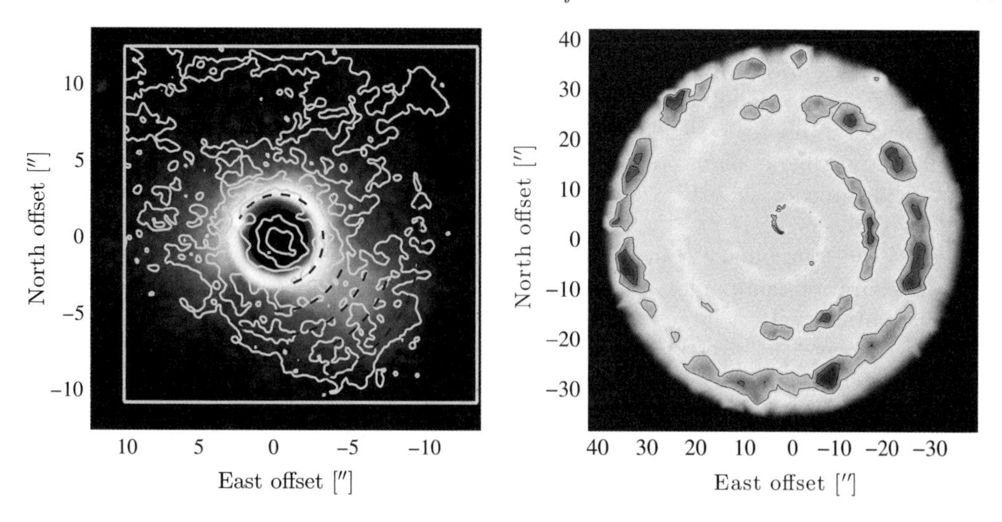

Figure 3. Left: The polarized scattered light emission from W Aql tracing the dust distribution overlayed with contours of the central-channel CO($J=3-2$) emission from ALMA. The stars are covered by a coronograph and blue dashed lines are drawn to emphasize the arcs seen in the gas distribution. Right: Predicted CO($J=3-2$) emission from a 3D radiative transfer calculation of a hydrodynamical wind-RLOF model of the system assuming an orbit eccentricity of e=0.2. See text for details and references.

Mira; Ramstedt *et al.* 2014; 2018) was selected so that the sources are observable from ALMA, have binary companions, and that some constraints on the binary orbit exists. The aim was to provide observational constraints for hydrodynamical models studying the interaction between the wind and a binary companion (wind-Roche-Lobe-OverFlow (wind-RLOF) models; Mohamed & Podsiadlowski 2012) by covering a range in binary separation and in wind velocity. With the data and analysis at hand, it becomes apparent that reproducing the CSEs of the sources is a too ambitious goal since the molecular gas distribution is affected by several effects that are not always accounted for in the models, i.e., wind from the companion (Mira; Ramstedt *et al.* 2014), radiation field of the companion (Mira, R Aqr; Ramstedt *et al.* 2018; Bujarrabal *et al.* 2018), additional companions (W Aql, π^1 Gru; Doan *et al.* 2017), and possible wind-variations over time (Mira, W Aql; Ramstedt *et al.* 2017).

Observations of the circumstellar dust and gas distributions around W Aql are shown in Fig. 3 (left). The binary pair was resolved by HST observations with a binary separation of 0.46" and the companion with a projected position southwest of the AGB star (Ramstedt *et al.* 2011; Danilovich *et al.* 2015a). The circumstellar dust distribution from observations of polarized scattered light is asymmetric with more material on the southwest side. ALMA observations of the CO($J=3-2$) emission (Ramstedt *et al.* 2017) show the same southwest enhancement, but essentially the gas component is dominated by smooth, extended emission as would be expected from "the standard model". In addition, the velocity-resolved ALMA observations revealed a pattern of higher-density arcs across the southwest density enhancement. Figure 3 (right) show the predicted CO($J=3-2$) emission from a 3D radiative transfer model (LIME; Brinch & Hogerheijde 2010) using the density and temperature distribution from a 3D SPH wind-RLOF model with the known parameters of the W Aql-system and an orbit eccentricity of $e=0.2$. The distribution seen in the observations and the emission pattern predicted from the models are similar with enhanced arcs found on the west and southwest side of the binary pair. However, as seen in Fig. 3 the scales are different and the arc-separation is not the same in the two images. Closer investigation of the ALMA images showed an even weaker additional arc-pattern with the same separation as predicted from the known companion. In conclusion, the system shows a circumstellar distribution very similar to that predicted from the model, however, in addition to that, smaller separation arcs are found with a creation timescale (~200 yrs) which is inconsistent with any known process of the system. A possible explanations for the smaller-separation arcs could be 3D pulsation effects (Liljegren *et al.* 2017, *this volume*) or an additional, closer, previously unknown companion.

Out of the four binary sources observed as a sample with ALMA, π^1 Gru has the largest-separation companion (2.6" $\approx$ 400 AU at a distance of 395 pc). The circumstellar distribution, with a large, equatorial, slowly expanding ($v_\infty \approx 15\,\mathrm{km\,s^{-1}}$) torus and a higher-velocity ($v_\infty \approx 60\,\mathrm{km\,s^{-1}}$) bi-polar outflow, already early lead to the suggestion that a closer, undetected companion must be present (Sahai 1992). The combined CO line interferometry data of the system (CO($J = 2-1$) from SMA; Chiu *et al.* 2006, CO($J = 3-2$) from ALMA-ACA) is analysed in Doan *et al.* (2017). The new, high-resolution ALMA observations show arcs or a spiral density pattern across the torus very likely caused by the closer companion. The high-resolution data is analysed in detail in an upcoming paper by Doan et al..

6.2. *Outlook: The DEATHSTAR project*

The DEATHSTAR† (DEtermining Accurate mass-loss rates for THermally pulsing AGB STARs) project on ALMA is currently gathering observations of the CO $J = 2-1$ and $3-2$ emission from 67 nearby (closer than 1 kpc) M-, C- and S-type AGB stars using the Atacama Compact Array (ACA). The data will provide the necessary observational constraints needed for the radiative transfer models used to determine the physical parameters (mass-loss rate and temperature distribution) of the CSEs to not be dependent on the outdated photodissociation model by Mamon *et al.* (1988). Instead, the size of the emitting regions are measured directly.

So far, about 50 sources have been observed and the data have been delivered and reduced. For most of the sources, the interferometer recovers the main part of the flux measured in previous single-dish observations. Using a fitting tool developed at the Nordic ALMA Regional Center at Onsala Space Observatory (UVMULTIFIT; Martí-Vidal *et al.* 2014), initial tests to investigate the emission distributions have been performed. Although the majority of the sources are well-fitted by a Gaussian emission distribution, a significant fraction (~20%) cannot be fitted, meaning that they exhibit a more complex circumstellar morphology. Furthermore, for some sources that show Gaussian emission distributions in all channels, the peak moves significantly between the channels. Some sources also display more complicated line profiles, indicative of a more complex velocity distribution than assumed in "the standard model", and now revealed by the extremely high sensitivity achieved compared to previous single-dish observations (Fig. 4). The observations will be analysed using a more detailed setup comparing the predicted emission from radiative transfer models with the measurements in upcoming publications giving the most accurate mass-loss-rate estimates available for these nearby sources. Sources that cannot be fitted using "the standard model" will be analysed using 3D radiative transfer along the lines described above. In addition to the CO line emission, line emission from an additional ~20 molecular species have been detected across the two Bands (6 and 7) observed. These data will be made available to the community on the DEATHSTAR webpages (see footnote) and can be used for legacy projects.

† `www.astro.uu.se/deathstar/`

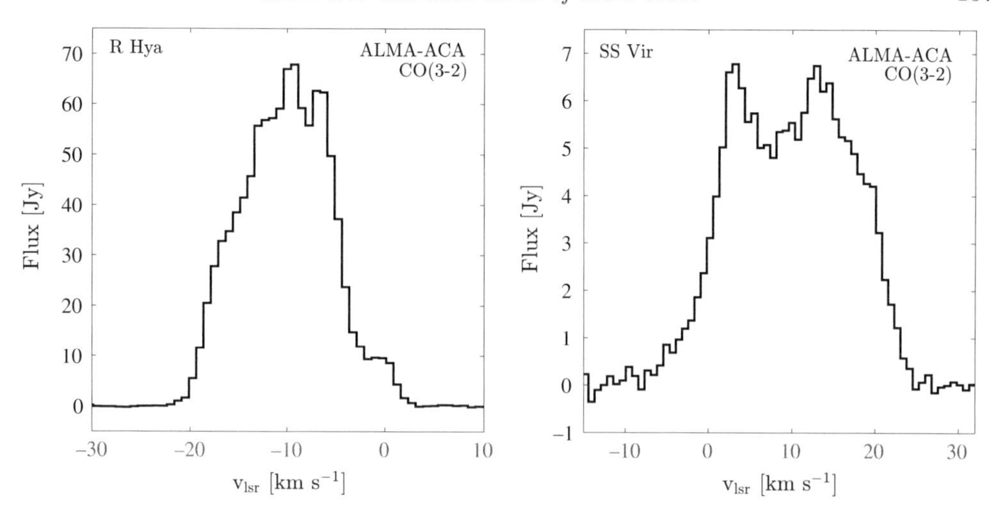

Figure 4. ALMA-ACA CO($J = 3-2$) line profiles for two DEATHSTAR sources (one M-type, left; one C-type, right) showing indications of a more complex velocity structure than described in "the standard model". The names are given in the upper left corner of each panel. The emission distributions of both sources can be fitted with a Gaussian in each channel.

References

Aringer *et al.* 2009, *A&A*, 503, 913
Bladh *et al.* 2015, *A&A*, 575, 105
Boyer *et al.* 2012, *ApJ*, 748, 40
Brinch & Hogerheijde 2010, *A&A*, 523, 25
Bujarrabal *et al.* 2018, *A&A*, 616, L3
Castro-Carrizo *et al.* 2010, *A&A*, 523, 59
Chiu *et al.* 2006, *ApJ*, 645, 605
Cummings *et al.* 2016, *ApJ*, 818, 84
Danilovich *et al.* 2015a, *A&A*, 574, 23
Danilovich *et al.* 2015b, *A&A*, 581, 60
De Beck *et al.* 2010, *A&A*, 523, 18
Doan *et al.* 2017, *A&A*, 605, 28
Eriksson *et al.* 2014, *A&A*, 566, 95
Goldreich & Scoville 1976, *ApJ*, 205, 384
González-Delgado *et al.* 2003, *A&A*, 411, 123
Groenewegen 2006, *A&A*, 448, 181
Groenewegen *et al.* 2007, *MNRAS*, 376, 313
Groenewegen *et al.* 2009, *A&A*, 506, 1277
Groenewegen *et al.* 2011, *A&A*, 526, 162
Groenewegen *et al.* 2016, *A&A*, 596, 50
Groenewegen 2017, *A&A*, 606, 67
Groenewegen & Sloan 2018, *A&A*, 609, 114
Guélin *et al.* 2018, *A&A*, 610, 4
Gullieuszik *et al.* 2012, *A&A*, 609, 114
Gustafsson *et al.* 2008, *A&A*, 486, 951
Hankins *et al.* 2018, *ApJ*, 852, 27
Höfner & Olofsson 2018, *A&ARv*, 26, 1
Ishihara *et al.* 2011, *A&A*, 534, 79
Ivezic *et al.* 1999, *astro-ph:* /9910475
Jones & Boffin 2017, *Nat. Astronomy*, 1, id. 0117
Justtanont *et al.* 2012, *A&A*, 537, 144

Kim *et al.* 2017, *Nat. Astronomy*, 1, id. 0060
Li *et al.* 2014, *A&A*, 568, 111
Li *et al.* 2016, *A&A*, 588, 4
Lian *et al.* 2014, *A&A*, 564, 84
Liljegren *et al.* 2017, *A&A*, 606, 6
Maercker *et al.* 2012, *Nature*, 490, 232
Maoz *et al.* 2014, *ARA&A*, 52, 107
Mamon *et al.* 1988, *ApJ*, 328, 797
Martí-Vidal *et al.* 2014, *A&A*, 563, 136
Matsuura *et al.* 2013, *MNRAS*, 429, 2527
Meixner *et al.* 2006, *AJ*, 132, 2268
Mohamed & Podsiadlowski 2012, *BaltA*, 21, 88
Nanni *et al.* 2018, *MNRAS*, 473, 5492
Ramstedt *et al.* 2006, *A&A*, 454, L103
Ramstedt *et al.* 2008, *A&A*, 487, 645
Ramstedt *et al.* 2009, *A&A*, 499, 515
Ramstedt *et al.* 2011, *A&A*, 531, 148
Ramstedt *et al.* 2014, *A&A*, 570, L14
Ramstedt *et al.* 2017, *A&A*, 605, 126
Ramstedt *et al.* 2018, *A&A*, 616, 61
Riebel *et al.* 2012, *ApJ*, 753, 71
Sahai 1992, *A&A*, 253, L33
Sargent *et al.* 2011, *ApJ*, 728, 93
Schöier & Olofsson 2001, *A&A*, 368, 969
Schöier *et al.* 2005, *A&A*, 432, 369
Srinivasan *et al.* 2006, *AAS*, 38, 1121
Teyssier *et al.* 2006, *A&A*, 450, 167
van Loon *et al.* 2005, *A&A*, 438, 273
Vlemmings *et al.* 2018, *A&A*, 613, L4

Discussion

WHITELOCK: Fascinating that you choose R For as the example showing uniform mass loss. Near IR-photometry (Feast *et al.* 1984, MNRAS 211, 331) shows very non-uniform dust mass-loss rate.

RAMSTEDT: Indeed it is not the case for a lot of the sources where we know of asymmetries from probes of different (smaller) scales that we see any signs of this on the large (100s of AU) scales probed by the CO line profiles.

SAHAI: In heating-cooling models of CO emission from CSEs (e.g. Sahai 1990), CO/H_2 and dust properties are coupled-in, so with good enough data, one can constrain these as well. ALMA observations (e.g. with DEATHSTAR) is now hopefully going to provide such data!

RAMSTEDT: Yes, indeed, increasing the number of lines, also from other molecules than CO, allows for a better constrained model of the CSE.

Why Galaxies Care About AGB Stars:
A Continuing Challenge through Cosmic Time
Proceedings IAU Symposium No. 343, 2019
F. Kerschbaum, M. Groenewegen & H. Olofsson, eds.

doi:10.1017/S1743921318005355

Circumstellar dust, IR spectroscopy, and mineralogy

Kyung-Won Suh

Dept. of Astronomy and Space Science, Chungbuk National University,
Cheongju, Chungbuk 28644, Republic of Korea
email: kwsuh@chungbuk.ac.kr

Abstract. We review the mineralogy of circumstellar dust grains around AGB stars as investigated through infrared spectroscopic studies. The expanding envelopes of AGB stars are chemically fresh because of the strong binding force of CO molecules. O-rich dust grains (silicates and oxides) form in O-rich envelopes and C-rich dust grains (amorphous carbon and SiC) form in C-rich envelopes. Amorphous silicate grains can be crystallized by annealing processes in various environments of AGB stars. We also discuss dust mineralogy for objects that have undergone chemical transition processes.

Keywords. circumstellar matter, stars: AGB and post-AGB, dust-extinction, infrared: stars, radiative transfer

1. Introduction

Asymptotic giant branch (AGB) stars are in the last evolutionary phases of low mass stars (M $\leq 10\ M_{\odot}$). They are long period pulsating variables with extended outer layers that provide good conditions for dust formation. The radiation pressure on newly formed dust grains may drive dusty stellar winds with mass-loss rates of $10^{-8} - 10^{-4} M_{\odot}\ \mathrm{yr}^{-1}$ (e.g., Suh 1999).

Since Wolf & Ney (1968) discovered and identified the 10 μm silicate feature from many M-type giant stars, the mineralogy of circumstellar dust around AGB stars has been one of the key problems in astronomy. Various species of O-rich dust grains (silicates, oxides, and water ice) in O-rich envelopes and C-rich dust grains (amorphous carbon and SiC) in C-rich envelopes are studied. We review the mineralogy of circumstellar dust grains around AGB stars, which is investigated by infrared observations, laboratory measurements, and theoretical models.

2. Infrared observations of circumstellar dust

For a large sample of AGB stars, infrared observational data are available from the Infrared Astronomical Satellite (IRAS), Infrared Space Observatory (ISO), Midcourse Space Experiment (MSX), Two-Micron All-Sky Survey (2MASS), AKARI space telescope, and Wide-field Infrared Survey Explorer (WISE).

2.1. *IR spectroscopy*

The IRAS Low Resolution Spectrograph (LRS; λ = 8-22 μm) data are useful to identify important features of O-rich and C-rich dust grains in AGB stars (e.g., Kwok, Volk, & Bidelman 1997). The LRS data were used to identify many O-rich and C-rich AGB stars in our Galaxy.

Table 1. Dust species in AGB stars

Acronym	Size [μm]	Description	Density [g cm^{-3}]	Reference[a]
SILw	0.1	amorphous warm silicate	3.3	Suh (1999)
Alu	0.1	amorphous alumina	3.2	Begemann *et al.* (1997)
FMO	0.1	$Fe_xMg_{1-x}O$ (x=0.4-1.0)	3.59−5.7	Henning *et al.* (1995)
SILc	0.1	amorphous cold silicate	3.3	Suh (1999)
ICE	0.1	crystalline water ice	0.92	Bertie *et al.* (1969)
SWC	0.1−0.2	SILc core, ICE mantle	1.22	core and mantle
FK	–	crystalline forsterite	3.27	Koike *et al.* (2003)
EK	–	crystalline enstatite	3.27	Chihara *et al.* (2002)
SFE10	–	SILc+FK(5%)+EK(5%)	3.3	a simple mixture
AMC	0.1	amorphous carbon	2.0	Suh (2000)
SiC	0.1	α-SiC	3.26	Pégourié (1988)

[a]Except for FK and EK for which mass absorption coefficients are given, all references provide optical constants.

The ISO Short Wavelength Spectrometer (SWS; λ = 2.4−45.4 μm) and the Long Wavelength Spectrometer (LWS; λ = 43−197 μm) provided high resolution spectra for many AGB stars in our Galaxy (Sylvester *et al.* 1999). The ISO spectral observations identified many fine spectral features of crystalline dust.

The Spitzer Infrared Spectrograph (IRS; λ = 5.2−38 μm) took high resolution spectra for many AGB stars in the Large and Small Magellanic Clouds (LMC and SMC). The IRS observed nearly 800 point sources in the LMC, taking over 1000 spectra (Jones *et al.* 2017).

2.2. *IR two-colour diagrams*

A large number of AGB stars have infrared photometric fluxes from the IRAS, AKARI, MSX, 2MASS, and WISE observations. Although less useful than a full spectral energy distribution (SED), the large number of observations in various wavelength bands can be used to form two-colour diagrams (2CDs) that can be compared to theoretical models. IR 2CDs are useful to distinguish statistically among AGB stars, post-AGB stars, and PNe (e.g., Suh 2015).

3. Mineralogy of circumstellar dust

Mineralogy of dust is revealed by the optical properties. Dust opacity is determined from the optical constants, shape, and size of a dust grain. By comparing the observational data with theoretical radiative transfer models for circumstellar dust, we may identify the mineralogy. The chemical and physical properties (chemical composition, solid structure, shape, and size) of circumstellar dust can provide crucial information on the astrophysical environment. Table 1 lists major dust species for AGB stars.

Due to amorphous silicate dust, low mass-loss rate O-rich AGB (LMOA) stars with thin dust envelopes show the 10 μm and 18 μm emission features. High mass-loss rate O-rich AGB (HMOA) stars with thick dust envelopes show absorbing features at the same wavelengths.

3.1. *Dust in LMOA stars*

Amorphous alumina (Al_2O_3) grains produce a single peak at 11.8 μm (Begemann *et al.* 1997) and influence the shape of the SED at $\lambda \sim$ 10 μm. Comparing the theoretical models with the observations on various IR 2CDs for a large sample of AGB stars in our Galaxy, Suh (2016) found that amorphous alumina dust (about 10-40 %) mixed with amorphous silicates better model the observed data for LMOA stars. From modelling

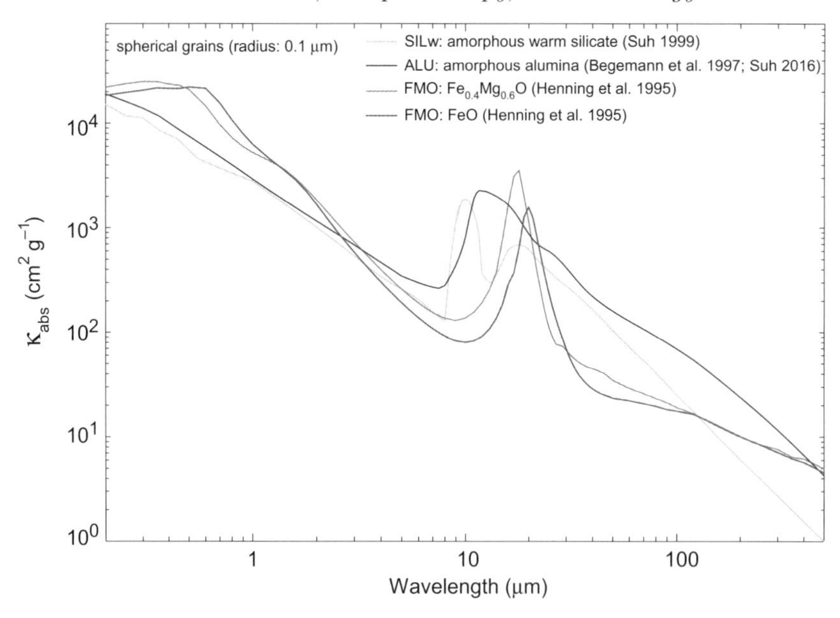

Figure 1. Dust opacity functions for low mass-loss rate O-rich AGB stars. See Table 1.

the Spitzer IRS spectra for O-rich AGB stars in LMC, Jones *et al.* (2014) showed that a mixture of amorphous silicates, amorphous alumina, and metallic iron, provides a good fit to the observed spectra.

Henning *et al.* (1995) presented the optical constants of $Fe_xMg_{1-x}O$ (x=0.4-1.0). Using ISO SWS data, Posch *et al.* (2002) detected and tentatively identified a broad dust emission feature peaking at 19.5 μm, which is especially prominent in LMOA stars, to be $Fe_{0.9}Mg_{0.1}O$.

Figure 1 shows the opacity functions of major dust species for LMOA stars. For the dust species for which the optical constants are available, the opacity functions are calculated for spherical dust grains using Mie theory (Bohren & Huffman 1983).

3.2. *Dust in HMOA stars*

The high resolution ISO spectroscopic observations detected prominent emission features of crystalline silicates in the spectra of HMOA stars but not from the spectra of LMOA stars (Sylvester *et al.* 1999). Fabian *et al.* (2000) investigated the thermal evolution of amorphous silicates and found that annealing at a temperature of 1000 K transformed amorphous silicate grains to crystalline ones on relatively short time scales. For HMOA stars, for which the dust formation temperature is about 1000 K (Suh 2004), the inner region of the outflowing envelope is warm (about 900−1000 K) during an extended period of time (several hundred days) for the annealing process to work (e.g., Suh 2004).

Water-ice features are found in SEDs of some HMOA stars (Justtanont *et al.* 2006). Suh & Kwon (2013) found that dust shell models with about 10% of crystalline silicates (Chihara *et al.* 2002; Koike *et al.* 2003), crystalline water ice (Bertie *et al.* 1969), and amorphous silicate can reproduce the observed SEDs for many HMOA stars (see Figure 2).

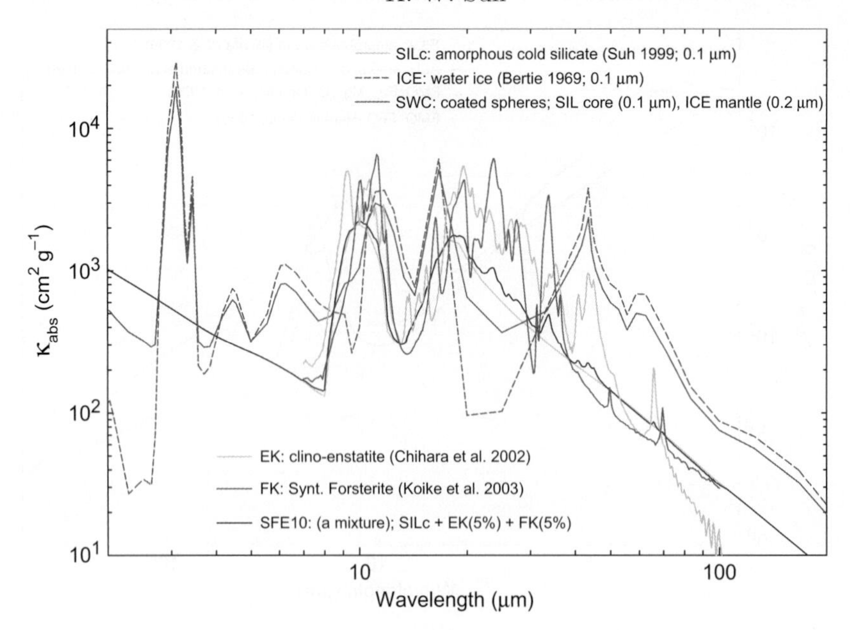

Figure 2. Dust opacity functions for high mass-loss rate O-rich AGB stars. See Table 1.

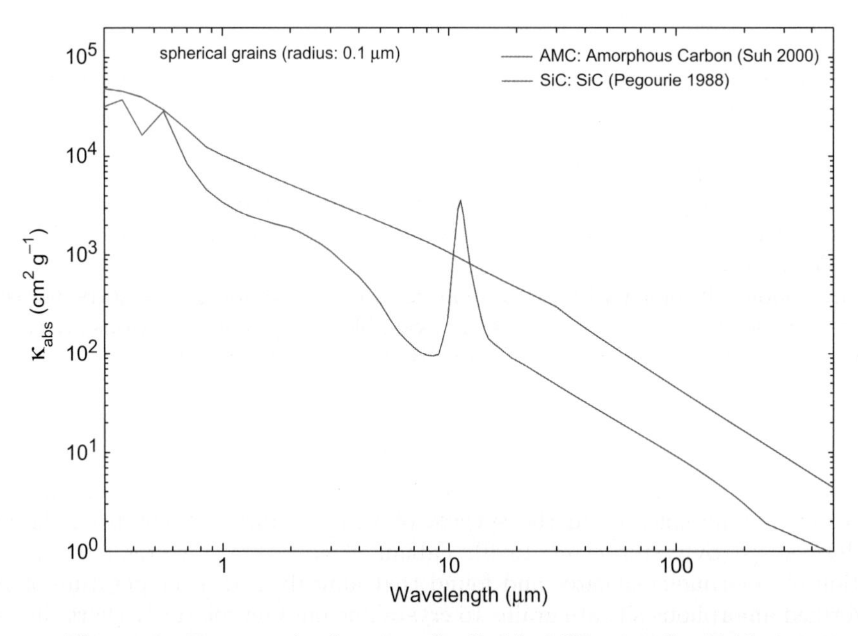

Figure 3. Dust opacity functions for C-rich AGB stars. See Table 1.

3.3. *Dust in C-rich AGB stars*

The main components of dust in the envelopes around carbon stars are believed to be featureless amorphous carbon (AMC) grains and SiC grains producing the 11.3 μm emission feature (Pégourié 1988; Suh 2000; see Figure 3). The carbon stars with SiC grains belong to the IRAS LRS class C.

Unlike AGB stars, C-rich post-AGB stars typically show polycyclic aromatic hydrocarbon (PAH) dust features. This could be due to UV radiation from the hot central stars.

3.4. *Dust in silicate carbon stars*

Silicate carbon stars show the characteristics of a carbon star and circumstellar silicate dust features. A possible scenario for the origin of most silicate carbon stars would be that O-rich material was stored in a circumstellar (or circumbinary) disc and remains after the chemical transition from O to C. Up to now, about 29 silicate carbon stars have been identified in our Galaxy (e.g., Kwon & Suh 2014). Kraemer *et al.* (2017) identified a new silicate carbon star in the SMC using the Spitzer IRS data.

References

Begemann, B., Dorschner, J., Henning, T., *et al.* 1997, *ApJ*, 476, 199
Bertie, J. E., Labbé, H. J., & Whalley, E. 1969, *J. Chemical Phys.*, 50, 4501
Bohren, C. F., & Huffman, D. R. 1983, *Absorption and Scattering of Light by Small Particles* (New York: Wiley)
Chihara, H., Koike, C., Tsuchiyama, A., Tachibana, S., & Sakamoto, D. 2002, *A&A*, 391, 267
Fabian, D., Jäger, C., Henning, T., *et al.* 2000, *A&A*, 364, 282
Jones, O. C., Woods, P. M., Kemper, F., *et al.* 2017, *MNRAS*, 470, 3250
Justtanont, K., Olofsson, G., Dijkstra, C., & Meyer, A. W. 2006, *A&A*, 450, 1051
Henning, T., Begemann, B., Mutschke, H., & Dorschner, J. 1995, *A&AS*, 112, 143
Koike, C., Chihara, H., Tsuchiyama, A., *et al.* 2003, *A&A*, 399, 1101
Kraemer, K. E., Sloan, G. C., Wood, P. R., *et al.* 2017, *ApJ*, 834, 185
Kwok, S., Volk, K., & Bidelman, W. P. 1997, *ApJS*, 112, 557
Kwon, Y.-J., & Suh, K.-W. 2014, *JKAS*, 47, 123
Pégourié, B. 1988, *A&A*, 194, 335
Suh, K.-W. 1999, *MNRAS*, 304, 389
Suh, K.-W. 2000, *MNRAS*,315, 740
Suh, K.-W. 2004, *ApJ*, 615, 485
Suh, K.-W., & Kwon, Y.-J. 2013, *ApJ*, 762, 113
Suh, K.-W. 2015, *ApJ*, 808, 165
Suh, K.-W. 2016, *JKAS*, 49, 127
Sylvester, R. J., Kemper, F., Barlow, M. J., *et al.* 1999, *A&A*, 352, 587
Posch Th., Kerschbaum F., Mutschke, H., *et al.* 2002, *A&A*, 393, L7
Woolf, N. J., & Ney, E. P. 1969, *ApJ.* 155, 181

Discussion

L. JONES: The FeO identification at 20 μm is hard - just one resonance. This feature appears with several others - at 11, 13, and 28 μm. Ben Sargent has a paper coming out identifying all of these with the same alumina related carrier.

SUH: The two-colour diagram showed the effect of just the 20 μm feature. Of course, we can consider other features also.

O. JONES: How confident are you that FeO is present in O-rich AGB stars, given that the 20 μm amorphous silicate feature is so varied and their wavelengths overlap?

SUH: We need to consider various oxides (alumina and $Fe_xMg_{1-x}O$: x=0.4-1.0) as well as silicates to fit the observations in the wavelength range for low-mass loss rate O-rich AGB stars. Among all the oxides, this work showed that the FeO dust is especially useful to fit the WISE W4 data.

Why Galaxies Care About AGB Stars: A Continuing Challenge through Cosmic Time
Proceedings IAU Symposium No. 343, 2019
F. Kerschbaum, M. Groenewegen & H. Olofsson, eds.

doi:10.1017/S1743921319001819

Planetary Nebulae, Morphology and Binarity, and the relevance to AGB Stars

Raghvendra Sahai

Jet Propulsion Laboratory MS 183-900, California Institue of Technology,
Pasadena, CA 91109, USA
email: raghvendra.sahai@jpl.nasa.gov

Abstract. The dramatic transformation of the spherical outflows of AGB stars into the extreme aspherical geometries seen during the planetary nebula (PN) phase is widely believed to be linked to binarity and is likely driven by the associated production of fast jets and central disks/torii. We first briefly summarize results from the imaging surveys of large samples of young PNe and pre-PNe with HST which show that almost all objects have bipolar, multipolar and elliptical morphologies, with widespread presence of point-symmetric structure. We describe a relatively new technique of using UV photometic observations of large AGB star samples to search for binarity and associated accretion activity, and follow-up studies using UV spectroscopy and X-ray observations. We present results from studies of individual objects in transition to the PN phase, highlighting observational techniques being used to determine jet properties that can constrain the accretion modes that power these jets.

Keywords. stars: AGB and post-AGB, instrumentation: High angular resolution, ISM: planetary nebulae:general

---

1. Introduction

Most stars in the Universe that evolve in a Hubble time (i.e., those with main-sequence masses 1-8$M_\odot$) die extraordinary deaths, ejecting half or more of their total mass in the form of nucleosynthetically-enriched material into the interstellar medium (ISM) – a process which dramatically alters the course of stellar evolution, and plays a key role in the chemical evolution of galaxies. The death throes of these stars enrich the ISM in biogenically important elements like C and N, and sow the seeds for the birth of new stars and solar systems, yet this stage remains very poorly understood. Young planetary nebulae (YPNe) represent the bright end-stages of these stars and provide valuable diagnostics on their demise, however, PN formation, and the phase of extreme mass-loss at end of the asymptotic red giant (AGB) phase that leads to it, is very poorly understood. Most YPNe have bipolar or multipolar shapes and very few are round, and more than half of all PNe show collimated lobes and dense, dusty, equatorial waists, which are important morphological features of this class of objects (e.g., Sahai & Trauger 1998 [ST98], Sahai *et al.* 2011a [SMV11]). Binarity is widely believed to be the (likely) underlying cause for the dramatic AGB-to-PN transition, e.g, directly via common-envelope ejection (Ivanova *et al.* 2013) and/or accretion-disk and torus formation, or indirectly via increased rotation and the generation of strong magnetic fields (e.g., Soker 2015, De Marco 2009, Chamandy *et al.* 2018, Jones 2018).

Three morphologically-unbiased HST imaging surveys have observationally bracketed the evolutionary phases that span the spherical to aspherical transition. These surveys imaged (1) YPNe (i.e., compact, with size $<$ 5–10 arcseconds, and the flux ratio [OIII]/H$\alpha \lesssim 1$) (ST98, SMV11), (2) Pre-Planetary Nebulae or PPNe (IRAS 25-to-12μm

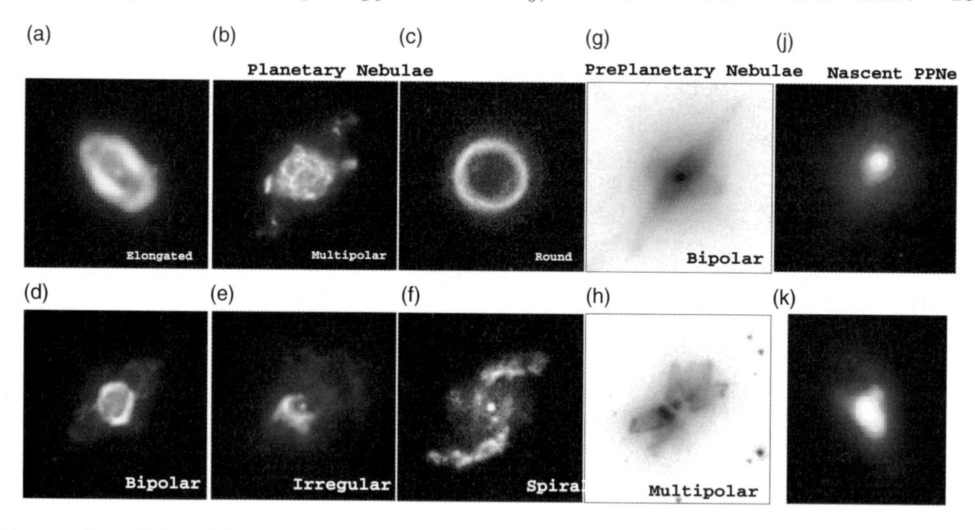

Figure 1. YPNe, PPNe and nascent-PPNe images from HST – (*a–f*) Hα (or [NII]) images of PNe belonging to 6 primary classes in the Sahai, Morris & Villar (2011) classification system, which consists of 7 primary classes based on the overall nebular shape, and several categories of secondary characteristics related to specific properties of the lobes, waist, and haloes, and the presence of point-symmetry; (*g,h*) 0.6 μm images of PPNe belonging to 2 primary classes; and (*j,k*) 0.6 μm images of nascent PPNe. (a) PNG 014.3 − 05.5 ($4.''0 \times 4.''0$), (b) PNG 027.6 − 09.6 ($11.''4 \times 11.''4$, (c) PNG 357.2 + 02.0 ($2.''67 \times 2.''67$), (d) PNG 356.5 − 03.9 ($6.''84 \times 6.''84$, (e) PNG 068.3 − 02.7 ($3.''8 \times 3.''8$), (f) PNG 032.1 + 07.0 ($4.''67 \times 4.''67$), (g) IRAS 04296+3429 ($3.''7 \times 3.''7$), (h) IRAS 19024+0044 ($6.''3 \times 6.''3$), (j) IRAS 23320+4316 ($0.''48 \times 0.''48$), & (k) IRAS 15082-4808 ($0.''69 \times 0.''8$).

flux ratio $F_{25}/F_{12} > 1.4$, to select for objects lacking warm dust, as an indicator that dense AGB mass loss has stopped recently) (Sahai *et al.* 2007), and (3) Nascent PPNe ($1 < F_{25}/F_{12} < 1.4$, to select for objects lacking hot dust, as an indicator that dense AGB mass loss is on the wane) (Sahai 2009). The PPN survey led to a systematic characterization scheme for the observed morphologies with 4 primary classes (B: bipolar, M: multipolar, E: elongated, and I: irregular) (e.g., Fig. 1(g,h)), and a number of secondary descriptors, relating to, e.g., the presence of point-symmetry, ansae, halos, etc. But the survey did not find a single round object.

The scheme was adapted to the morphological classification of YPNe, by adding 3 new primary classes – L (collimated lobe pair, but not pinched-in at the waist), R (round), and S (spiral arm), and additional secondary descriptors (SMV11, and Fig. 1(a-f)). Many of these secondary descriptors are related to features arising from evolutionary effects on the central region, due to (i) the ionizing flux and (ii) hydrodynamic action of the fast radiative wind from the PNe central star on this region. In contrast, the morphologies seen during the nascent PPN phase, based on our HST imaging survey, are rather different, with compact ($\lesssim few \times 0.1''$), one-sided collimated structures being predominant, suggesting that the collimated-outflow phase has just begun (Fig. 1(j,k), Sahai 2009).

The YPN survey led to the hypothesis that PN formation is preceded by the generation of collimated, high-speed outflows that sculpt the progenitor AGB envelope from the inside out, during the very late-AGB or nascent PPN phase (ST98). This hypothesis is supported by molecular line observations of PPNe that directly show the presence of collimated, high-speed outflows morphology (e.g., Bujurrabal *et al.* 2001, Sahai *et al.* 2006, Sánchez Contreras *et al.* 2006, Imai 2007, Sahai *et al.* 2017a; Sahai *et al.* 2017b). Binarity is widely believed to be the most probable cause for producing such outflows, which must start operating during the nascent PPN or late-AGB phase. But observational

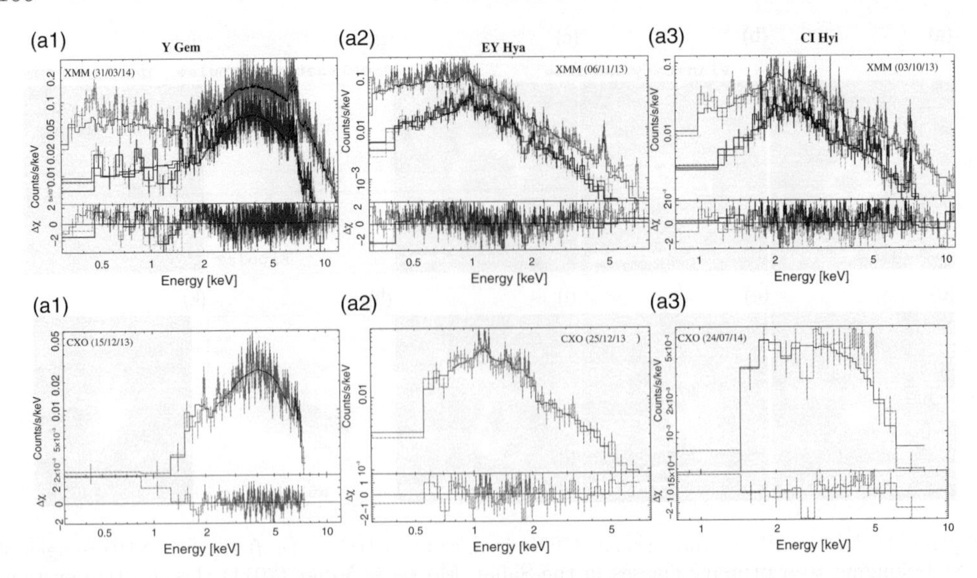

Figure 2. X-ray spectra for 3 fuvAGB stars, taken with (a) XMM/EPIC (pn: *red*, MOS1: *green*, MOS2: *blue*), and (b) CXO ACIS-S (red). APEC model fits are shown in *black*. APEC modeling of the CXO and XMM spectra reveals long-term variations in the emission properties (*adapted from Setal15*).

evidence for binarity and associated accretion-activity has generally been hard to come by. Standard methods for detecting binarity such as periodic radial-velocity or photospheric variations due to a companion cannot be used for AGB stars that are thousands of time more luminous than their expected companions and exhibit strong instrinsic atmospheric pulsations.

2. Binarity and Accretion Activity in AGB Stars

We developed an innovative technique of using UV observations to search for binarity and associated accretion activity – since most AGB stars are relatively cool ($T_{\rm eff} \lesssim 3000$K) objects (spectral types ~M6 or later), whereas any (likely) stellar companions (i.e., white-dwarfs [WD] or main-sequence [MS]) and/or accretion disks around them are likely significantly hotter ($T_{\rm eff} \gtrsim 6000$K), favorable secondary-to-primary flux contrast ratios ($\gtrsim$10) are reached in the GALEX FUV (1344-1786Å) and NUV (1771-2831Å) bands for a source with $T_{\rm eff} \gtrsim$6000 K and luminosity, $L \gtrsim 1 L_\odot$ (companion and/or disk). The feasibility of this technique was demonstrated in a number of recent studies (Sahai *et al.* 2008 [Setal08]; Sahai *et al.* 2011b; Sahai *et al.* 2016a). Setal08 detected emission from 9/21 objects in the FUV; since such objects (hereafter fuvAGB stars) also show significant UV variability, Setal08 concluded that the UV source is unlikely to be solely a companion's photosphere, and is likely dominated by emission related to variable accretion activity. It should be noted that fuvAGB stars in general (based on their optical spectra), do not belong to the well-studied class of symbiotic stars (red giant stars with WD companions) and have never been classified as such. So if the compact companions in fuvAGB star systems are WDs, then these must be quite cool ($\lesssim$ 20 000 K) (Sahai *et al.* 2015 [Setal15]).

Our small X-rays surveys using XMM-Newton and Chandra support this hypothesis, finding X-ray emission in about 50% of the fuvAGB stars. The X-ray emission is characterised by relatively high luminosities $L_{\rm x} \sim (0.002 - 0.11)\, L_\odot$, and very high plasma temperatures $T_{\rm x} \sim (35 - 160) \times 10^6$ K (Setal15), Fig. 2. Amongst the fuvAGB stars, objects with large FUV/NUV ratios, $R_{\rm fuv/nuv} > 0.17$, have a much higher probability

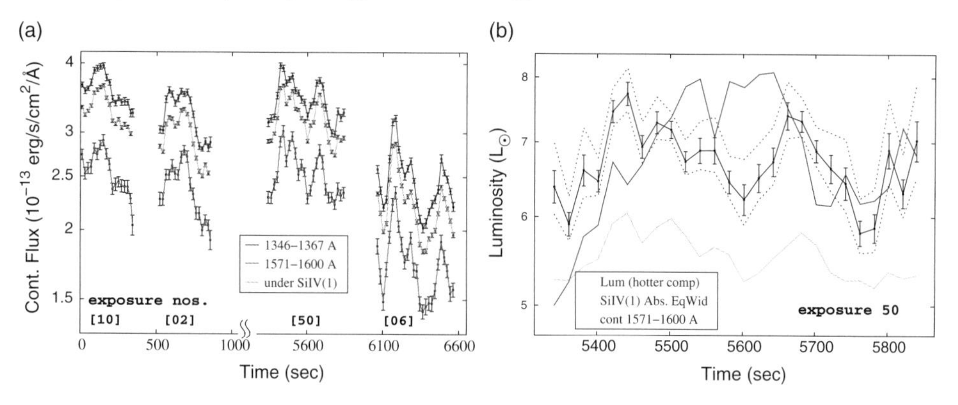

Figure 3. Short-term time variations in the (a) FUV continuum: line-free continuum in the bands 1346 – 1367 Å (blue) and 1571 – 1600 Å (red), and continuum underlying the Si IV(1) line (green), and (b) luminosity (black) of the high-temperature accretion component, derived using a two-blackbody fit to the continuum observed in the subexposures within exposure 50: the dashed curves show upper and lower bounds on the luminosity due to the estimated uncertainties in the cooler blackbody's luminosity and temperature. The 1571 – 1600 Å continuum, scaled up by a factor 1.1×10^7 (cyan), and the square-root of the Si IV(1) absorption line equivalent width, scaled up by a factor 5.2 (blue), are shown for comparison (*adapted from Setal18*).

of being detected in X-ray emission (Sahai *et al.* 2016a), and are almost certainly binaries with accretion activity powering the high-energy emission. A recent STIS spectroscopic study of the prototype high FUV/NUV ratio star, Y Gem, shows the presence of flickering and high-velocity infall and outflows, and thus directly supports the binary/accretion hypothesis (Sahai *et al.* 2018a [Setal18]). STIS observations of Y Gem show several UV lines from species such as Si IV and C IV, with broad (FWHM $\sim 300 - 700$ km s^{-1}) emission and absorption features that are red- and blue- shifted by velocities of $\sim$500 km s^{-1} from the systemic velocity. Time-tag analysis reveals strong flickering in the UV continuum on time-scales of $\lesssim 20$ s (Fig. 3a), characteristic of an active accretion disk. A two blackbody model that fits the G140L and G230L spectra requires two components, a hotter one with luminosity $L(h) = 6.8\,L_\odot$ and temperature $T_{\rm eff}(h) \sim 36600$ K, and a cooler blackbody with $L(c) \sim 6.3\,L_\odot$, $T_{\rm eff}(c) \sim 9940$ K; we find significant variations in $L(h)$ (Fig. 3b), but not in $T_{\rm eff}(h)$. Neither of these two blackbodies fit the properties of a viable stellar companion (WD or MS) to the primary, and it is likely that both the hot and cool UV components arise in the accretion disk.

The proposed model for these (and previous) observations is that material from the primary star is gravitationally captured by a companion, producing a hot accretion disk. The latter powers a fast outflow that produces blue-shifted features due to absorption of UV continuum emitted by the disk, whereas the red-shifted emission features arise in heated infalling material from the primary. The outflow velocities support a previous inference by Setal15 that Y Gem's companion is a low-mass MS star.

But for objects with little or no FUV emission, i.e., with $R_{\rm fuv/nuv} \lesssim 0.1$ (nuvAGB stars), which dominate the population of UV-emitting AGB stars, the UV emission *may have a different source.* From an analysis of the NUV emission in 179 AGB stars, Montez *et al.* (2017) argued that the origin of the GALEX-detected UV emission is intrinsic to the AGB star (chromospheric & photospheric emission), and is unrelated to binarity. However, Ortiz & Guerrero (2016), from a study of a volume-limited sample ($<$0.5 kpc) of 58 AGB stars, concluded that the detection of NUV emission with a very large observed-to-predicted ratio, $Q_{\rm NUV} > 20$, is evidence for binarity in these objects.

Thus a more comprehensive (and easily testable) hypothesis for UV emission from AGB stars is that objects with a close companion produce high FUV/NUV ratios at least some of the time (since accretion can be variable) and/or a very large NUV excess, whereas single AGB stars (or those with large binary separations) should always show low FUV/NUV ratios. UV spectroscopy provides an unambiguous and independent test of this hypothesis because accretion-related activity is expected to produce UV lines with large widths and large Doppler shifts (as shown by the study of Y Gem). In contrast, chromospheric emission typically arising from gas at temperatures of $\lesssim 10^4$ K (e.g., Luttermoser *et al.* 1994), should be characterised by strong low-excitation lines in the NUV (such as the MgII λ2800 doublet), with the high-excitation UV lines being much weaker or absent. In addition, chromospheric line profiles from the AGB star will be relatively narrow ($\lesssim$10 km s^{-1}) and close to the systemic velocity.

2.1. *Additional Probes of Accretion Activity*

Multi-wavelength radio observations can distinguish between thermal and non-thermal emission (Sahai 2018a): since binarity is also expected to generate strong magnetic fields, ionized accretion-related flows can produce non-thermal, variable, emission, as we have found for Y Gem. The VLA lacks the sensitivity to make such observations except for a few ($\lesssim$ 5) of the brightest objects, but the ngVLA should be able to carry out a survey for a statistical sample of AGB stars (i.e., several 100) with UV emission with a modest time expenditure ($\sim$ 1 hr per object to cover 4 bands in the 8–90 GHz range with a 5 σ sensitivity of about 3.5 μJy) (Sahai 2018a).

We may be able to detect optical flickering with the TESS mission, with its extreme photometric accuracy – 1σ noise sensitivity of 690 (340) ppm in 2 min exposures for our faintest (median brightness) object – thus sensitive to fractional variations that are a factor 200 or larger below that in the FUV (peak-to-peak of 40%). We note that optical flickering has been detected by Snaid *et al.* (2018) in Y Gem (0.06 mag peak-to-peak in the u' band on typical timescales of $\sim$10 min).

3. Binarity and late-AGB & early post-AGB Evolution

Binary interactions can, directly or indirectly, produce dense waists and collimated fast outflows or jets (e.g., De Marco 2009, Kastner *et al.* 2012). Hydrodynamic sculpting by these jets of the superwind ejecta can then produce the variety of PN shapes observed (ST98). In order to assess this scenario for PN shaping, we need to understand the origin and properties of these jets (e.g., the associated mass, speeds, opening angles). We need to know whether the jets are episodic and if the jet axis precesses or wobbles, and if so, on what time-scales, and the role of magnetic fields in launching, accelerating and collimating the high-speed outflows. Unlike AGB outflows, for the small sample of PPNe outflows (with typical speeds of $\sim$ *few* $\times$ 100 km s^{-1}) studied so far, the associated momenta, as inferred from an analysis of (mostly single-dish) CO and ^{13}CO line profiles in PPNe, are far too high for these to be driven radiatively (e.g., Bujurrabal *et al.* 2001). We also need to understand the origin and properties of equatorially-dense structures, i.e., the waists, and whether these are bound or expanding.

Quantitative models for the binary interaction that can explain the jet and waist formation are lacking, partly because simulations have long struggled to describe binary interactions, even in the simplest cases. However, Blackman & Lucchini (2014: BL14) show how the properties of the high-velocity flows in PPNe can be used to constrain jet-engine paradigms using an analytical modelling approach, and provide a classification of the interactions and their observational signatures, thus distinguishing the type of engines that eject mass via jet-like collimated outflows.

In order to address these questions, it is important to focus on the late-AGB, nascent-PPN and PPN phases. Some notable recent studies of AGB stars are that of L$_2$ Pup, revealing a dusty disk with Keplerian rotation, a low-mass candidate companion (giant planet or low-mass brown dwarf) using polarimetric imaging with the SPHERE+ZIMPOL instrument of the VLT and ALMA (Kervella *et al.* 2015, 2016) and π^1 Gru, revealing a flared disk using ALMA (Nhung *et al.* 2016). In this paper, I summarize results on three key objects that cover the nascent PPN, YPN and PPN phases.

3.1. *The Carbon Star V Hya: A Nascent PPN*

The carbon star, V Hya, is a key object in understanding the early transition of AGB stars into aspherical PNe, as it shows the presence of high-speed, collimated outflows and dense equatorial structures. This star is ejecting massive high-speed compact clumps (hereafter bullets) periodically, leading to a model in which the bullet ejection is associated with the periastron passage of a close binary companion in an eccentric orbit around V Hya with an orbital period of ~8.5 yr (Sahai *et al.* 2016b, hereafter Setal16). The detailed physical properties of this ejection suggests that the companion approaches the primary very close to the latter's stellar envelope at every periastron passage, suggesting that V Hya is a good candidate where the binary interaction will result in a CE configuration. A hot, central disk-like structure of diameter $0\farcs6$ (240 AU at V Hya's distance of 400 pc) expanding at a speed of $10-15\,\mathrm{km\ s^{-1}}$ was found earlier by Sahai *et al.* (2003), using HST STIS observations. High resolution (~ 100 milliarcsecond) imaging of the central region of V Hya using the coronagraphic mode of the Gemini Planet Imager (GPI) in the 1 μm band shows a larger (size ~ 250 AU), central dusty disk (see Sahai *et al.* 2018b, this volume). The bullet ejection activity has likely been ongoing for several hundered years, and has carved out a bipolar cavity; detailed modelling of the most recently ejected bullets suggests that these are being shepherded towards the overall symmetry axis, and entrain ambient circumstellar material, as a result of glancing encounters with the walls of this cavity (Scibelli *et al.* 2018). Recent hydrodynamical simulations of the evolution of these bullets show that they carry embedded toroidal magentic fields within them that keep them confined laterally (Huang & Sahai 2018), as was found for the bullet-like clumps in the jet of the young PN, He 2-90 (Lee & Sahai 2004).

3.2. *IRAS 16342-3814: A young PPN*

IRAS 16342-3814 (hereafter IRAS 16342) belongs to a class of young PPNe with unusually fast radial H_2O outflows with $V_{\rm exp} \gtrsim 50\,\mathrm{km\ s^{-1}}$ ("water-fountain" PPNe) showing that jet activity is extremely recent ($\lesssim$ 100 yr), and indicating that these objects have become PPNe fairly recently. As the best studied and nearest (~2 kpc) example of this class, its morphology has been well-resolved with optical (HST) and near-infrared (Keck Adaptive Optics) imaging (Sahai *et al.* 1999, Sahai *et al.* 2005). Radio interferometry (VLA, VLBA) shows water masers spread over a range of radial velocities encompassing $270\,\mathrm{km\ s^{-1}}$ (Sahai *et al.* 1999, Claussen *et al.* 2009).

From an ALMA study in which emission from ^{12}CO J=3–2 and other molecular lines was mapped with ~ $0\farcs35$ resolution, Sahai *et al.* (2017a, hereafter Setal17) inferred the presence of two very high-speed, knotty, jet-like molecular outflows, and a central expanding torus. The outflows include the Extreme High Velocity Outflow (EHVO) and the High Velocity Outflow (HVO) with (deprojected) expansion speeds of $360-540$ and $250\,\mathrm{km\ s^{-1}}$ and ages of $130-305$ yr and $\lesssim 110$ yr; their axes are not colinear. The spiral structure seen in a position-velocity (PV) plot of the HVO most likely indicates emission from a precessing high-velocity bipolar outflow, as inferred previously from near-IR imaging (Sahai *et al.* 2005), that entrains material in the near and far bipolar lobe walls. The progenitor AGB star of IRAS 16342 underwent a sudden, very large increase (by a factor >500) in its mass-loss rate ~455 yr ago, with an average value over this period of $> 3.5 \times 10^{-4}\,M_\odot\ \mathrm{yr}^{-1}$. The measured expansion ages of the outflows and torus imply that the torus (age∼160 yr) and the younger high-velocity outflow (age∼110 yr) were formed soon after the sharp increase in the AGB mass-loss rate.

3.3. *The Boomerang Nebula: A post-RGB PPN*

The Boomerang Nebula, is an "extreme" bipolar pre-planetary Nebula (PPN), and also the coldest known object in the Universe, with a massive high-speed outflow that has cooled significantly below the cosmic background temperature (Sahai & Nyman 1997). ALMA observations have confirmed this finding, revealed unseen distant regions of this ultra-cold outflow (UCO) out to $\gtrsim 120\,000$ AU, and found that the expansion velocity in the UCO is not constant, but increases with radius (Sahai *et al.* 2013, Sahai *et al.* 2017b). The very large mass-loss rate ($\sim 0.001\,M_\odot\ \mathrm{yr}^{-1}$) characterising the UCO and the central star's very low-luminosity ($300\,L_\odot$) are unprecedented, making it a key object for testing theoretical models for mass-loss during post-main sequence evolution and for producing the dazzling variety of bipolar and multipolar morphologies seen in PNe (e.g., SMV11). The mass-loss rate ($\dot{M}$) increases with radius in the UCO, similar to its expansion velocity (V) – taking $V \propto r$, Sahai *et al.* (2017b) find $\dot{M} \propto r^{0.9-2.2}$. The mass in the UCO is $\gtrsim 3.3\,M_\odot$, and the Boomerang's MS progenitor mass is $\gtrsim 4\,M_\odot$. The UCO's kinetic energy is very high, $KE_{\mathrm{UCO}} > 4.8 \times 10^{47}$ erg, and the most likely source of this energy is the gravitational energy released via binary interaction in a common envelope event (CEE). Sahai *et al.* (2017b) concluded that the Boomerang's primary was an RGB or early-AGB star when it entered the CE phase; the companion finally merged into the primary's core, ejecting the primary's envelope that now forms the UCO. Such strong binary interactions on the RGB may in fact be a common evolutionary channel, given that Kamath *et al.* (2015, 2016) have recently identified a large sample of post-RGB objects in the Large and Small Magellanic Clouds. Many PPNe in our Galaxy, for which distances are not known, may really be post-RGB objects!

4. Accretion Modes

The objects discussed above show evidence for episodic, collimated jet-like outflows, but with significant differences. The accuracy of BL14's analysis (see § 3) depends on how well we can determine the physical properties of the fast outflows in PPNe (especially the jet momentum, $M_{\mathrm{j}}\,V_{\mathrm{j}}$, and the accretion time-scale, t_{acc}), and the history of the progenitor AGB's mass-loss in the last $\sim 100 - 1000$ yr. Using the limited outflow data known to date at "face-value", BL14 determine the minimum required mass-accretion rates ($\dot{M}_{\mathrm{a}} \propto M_{\mathrm{j}}\,V_{\mathrm{j}}/t_{\mathrm{acc}}$) and rule out some accretion modes such as Bondi-Hoyle-Lyttleton (BHL) wind-accretion and wind Roche lobe overflow (wRLOF). In BL14's approach, the intrinsic jet momentum is estimated from the observed fast outflow's momentum, assuming that the interaction between the intrinsic jet outflow and the ambient circumstellar envelope is momentum-conserving.

In Y Gem, Setal18 conclude, from their UV spectroscopic study, that material from the primary star is gravitationally captured by a low-mass MS companion producing a hot accretion disk around the latter. The disk powers a fast outflow that produces blue-shifted features due to absorption of UV continuum emitted by the disk, whereas the red-shifted emission features arise in heated infalling material from the primary. The accretion luminosity implies a mass-accretion rate $> 5 \times 10^{-7}\,M_\odot\ \mathrm{yr}^{-1}$; Setal18 conclude that Roche lobe overflow is the most likely binary accretion mode since Y Gem does not show the presence of a wind.

In V Hya, the collimated ejection of material is in the form of bullets, and the ejection axis flip-flops around an average orientation, in a regular manner. These data support a model in which the bullets are a result of collimated ejection from an accretion disk (produced by gravitational capture of material from the primary) that is warped and precessing, and/or that has a magnetic field that is misaligned with that of the companion or the primary star (Setal16). The average momentum rate of the bullet ejections is ($\sim 8.2 \times 10^{25}$ g cm s^{-2}), implying a minimum required accretion rate of $\dot{M}_{\rm a} \sim 3.3 \times 10^{-8}\, M_\odot$ yr^{-1}. Since the secondary must get very close to the primary's stellar envelope at periastron passage (Setal16), the binary separation is comparable to the stellar radius (at 400 pc, $R_* \sim 2$ AU: Knapp *et al.* 1997), i.e., $a_{\rm or} \sim 2$ AU, so the Roche-lobe overflow mode (RLOF) is the appropriate accretion mode, which can easily supply the required accretion rate.

In IRAS 16342, the relatively high momentum rate for the dominant jet-outflow ($> 5 \times 10^{28}$ g cm s^{-2}) implies a high minimum accretion rate, $\dot{M}_{\rm a} = 1.9 \times 10^{-5}\, M_\odot$ yr^{-1}. Comparing this rate with the expected mass-accretion rates derived for different accretion models shown in BL14's Fig. 1, Setal17 concluded that standard Bondi-Hoyle-Lyttleton (BHL) wind-accretion and wind-RLOF models with WD or MS companions were unlikely; enhanced RLOF from the primary or accretion modes operating within common envelope evolution were needed. But this conclusion was revised by Sahai (2018b) because the BHL rate shown in BL14's Fig. 1b is derived using the primary's AGB mass-loss properties $\dot{M}_{\rm w} = 10^{-5}\, M_\odot$ yr^{-1} (and $V_{\rm w} = 10$ km s^{-1}, together with assumed orbital separation $a_{\rm or} = 10$ AU, companion mass $M_{\rm c} = 0.6\, M_\odot$, primary mass $M_{\rm p} = 1.0\, M_\odot$) and not the much higher value of $\dot{M} \sim 1.3 \times 10^{-4}\, M_\odot$ yr^{-1} (comparable to the mass-loss rate in IRAS 16342), mentioned in BL14's §3.1 and referenced in the figure caption (E. Blackman, *priv. comm.*). Thus, the expected BHL accretion rate could be quite high in IRAS 16342. However, Sahai (2018b) show that the BHL accretion rate, which is valid when the orbital separation is much larger than the Roche-lobe radius, is $\dot{M}_{\rm BHL} \lesssim (0.3 - 1) \times 10^{-5}\, M_\odot$ yr^{-1}, and therefore the HVO in IRAS 16342 is not driven by accretion via BHL, but requires wind-RLOF or modes that provide higher accretion rates.

The Boomerang's central region has an overall bipolar structure, but in detail this structure is comprised of multiple, highly collimated lobes on each side of the central disk, both in scattered light and in ^{13}CO J=3–2 emission. The velocity of the molecular material in the dense walls of the collimated lobes is not particularly high, and we expect it is likely to be substantially lower than the velocity of the unshocked jet-outflow that has carved out these lobes, since the jet outflow has interacted with a very massive envelope (the UCO). Optical spectroscopy, indicating that the pristine fast outflow may have a speed of about 100 km s^{-1} (Neckel *et al.* 1987), support this expectation. Assuming momentum-conservation to derive the fast outflow momentum (as in BL14) is likely to provide a severe underestimate of the intrinsic jet momentum – numerical simulations are needed. However, in this object, the UCO's extreme properties directly imply CE evolution.

5. Summary

(1) Binarity and associated accretion-activity on the AGB is best studied using observations at UV and X-Ray wavelengths, as demonstrated by our pilot studies. A focussed UV spectroscopic study of the prime example of this phenomenon, Y Gem, shows high speed outflow and infall, and the presence of a hot accretion disk. But more extensive surveys are now needed in order to determine whether binarity and accretion-activity are ubiquitous and to understand these phenomena in detail. An HST UV spectroscopic survey of AGB stars with high and low FUV/NUV ratios is needed to study the relative

contributions to the UV due to accretion-activity, which would result from binarity, as opposed to chromspheric/coronal emission from a single star.
(2) The effects of binary interactions on post-AGB evolution is best studied using high-angular-resolution imaging & spectroscopic studies. An ALMA survey of PPNe covering the variety of morphologies seen in these objects is now needed in order to determine jet momenta and AGB mass-loss rates just prior to the post-AGB phase accurately, as well as to determine the nature of central disks/torii (when these are present). The nature and origin of these central disks/torii remains a mystery. Focussed studies of three objects in this transition phase have provided new details, leading to new insights:
(a) Our HST studies of V Hya implies the presence of a companion in an eccentric orbit, with a strong interaction at periastron passage leading to a stream of episodic, high-speed bullet-like ejections. Resonant precession of the disk/jet-axis as inferred for V Hya may explain the presence of multipolar morphologies, whereas non-resonant precession of disk/jet-axis can explain bipolar nebulae with point-symmetric knots. The bullets likely have an embedded toroidal magnetic field, that keeps them confined laterally.
(b) Our ALMA study of the young PPN IRAS 16342 shows two highly collimated, episodic, high-speed outflows which have different orientations, an expanding torus, and a large, sudden increase in AGB mass-loss rate over the past 500 years. Although such a high mass-loss rates potentially allows BHL accretion to power the dominant high-velocity outflow in this object, the wind-RLOF mode or modes that provide higher accretion rates are required for IRAS 16342. Our IRAS 16342 study underscores the importance of measuring AGB mass-loss rates immediately prior to the post-AGB phase.
(c) Our ALMA studies of the PPN, the Boomerang Nebula (also the coldest object in the Universe) show that it is a post-RGB object, resulting from a strong binary interaction on the RGB that ejected its envelope. It is thus possible that many low-luminosity PPNe may be post-RGB rather than post-AGB objects.

Acknowledgements

I thank Eric Blackman for a helpful discussion related to binary acretion modes. I thank my collaborators in the studies that have contributed to this paper, including C. Sánchez Contreras, J. Sanz-Forcada, W. Vlemmings, M. Morris, M. Claussen, J. Sanz-Forcada, L-Å. Nyman, and S. Scibelli. Special thanks to Orsola de Marco and Noam Soker for their quick responses to my questions related to binary evolution. The author's research described here was carried out at the Jet Propulsion Laboratory, California Institute of Technology, under a contract with NASA, and funded in part by NASA via ADAP awards, and multiple HST GO awards (GO 12227.001, 12664.001, 12519.001, 14713.001) from the Space Telescope Science Institute.

References

Blackman, E. G., & Lucchini, S. 2014, *MNRAS*, 440, L16
Bujarrabal, V., Castro-Carrizo, A., Alcolea, J., & Sánchez Contreras, C. 2001, *A&A*, 377, 868
Chamandy, L., Frank, A., Blackman, E. G., *et al.* 2018, *MNRAS*, 480, 1898
Lee, C.-F., & Sahai, R. 2004, *ApJ*, 606, 483
Claussen, M.J., Sahai, R., & Morris, M., 2009, ApJ, 691, 219
De Marco, O. 2009, *PASP*, 121, 316
Huang, P.-S. & Sahai, R. 2018, *in prep*
Imai, H. 2007, *Astrophysical Masers and their Environments, IAU Symposium 242*, 242, 279
Ivanova, N., Justham, S., Chen, X., *et al.* 2013, *A&ARv*, 21, 59
Jones, D. 2018, arXiv:1806.08244
Kamath D., Wood P. R. & Van Winckel H. 2015, *MNRAS*, 454, 1468
Kamath D., Wood P. R., Van Winckel H. & Nie J. D. 2016, *A&A*, 586, L5

Kastner, J. H., Montez, R., Jr., Balick, B., *et al.* 2012, *AJ*, 144, 58
Kervella, P., Montargès, M., Lagadec, E., *et al.* 2015, *A & A*, 578, A77
Kervella, P., Homan, W., Richards, A. M. S., *et al.* 2016, *A & A*, 596, A92
Knapp, G. R., Jorissen, A., & Young, K. 1997, *A & A*, 326, 318
Luttermoser, D. G., Johnson, H. R., & Eaton, J. 1994, *ApJ*, 422, 351
Montez, R., Jr., Ramstedt, S., Kastner, J. H., Vlemmings, W., & Sanchez, E. 2017, *ApJ*, 841, 33
Neckel, T., Staude, H. J., Sarcander, M., & Birkle, K. 1987, *A & A*, 175, 231
Nhung, P. T., Hoai, D. T., Diep, P. N., *et al.* 2016, *Research in Astronomy and Astrophysics*, 16, 111
Ortiz, R. & Guerrero, M. 2016, *MNRAS*, 3036, 3046
Sahai, R. 2009, *Asymmetric Planetary Nebulae IV proc., I.A.C. electronic publication, eds. R.L.M. Corradi, A. Manchado & N. Soker*, http://www.iac.es/proyecto/apn4/pages/proceedings.php
Sahai. R., Findeisen, K., Gil de Paz, A., Sáanchez Contreras, C. 2008, ApJ, 689, 1274 [Setal08]
Sahai, R., Le Mignant, D., Sánchez Contreras, C., Campbell, R. D., Chaffee, F. H. 2005, ApJ, 622, L53
Sahai, R., Morris, M. R., & Villar, G. G. 2011a, *A&A*, 141, 134
Sahai, R., Morris, M., Knapp, G.R., Young, K., Barnbaum, C. 2003, *Nature*, 426, 261
Sahai, R., Morris, M., Sánchez Contreras, C., & Claussen, M. 2007, *AJ*, 134, 2200
Sahai, R., Neill, J. D., Gil de Paz, A., & Sánchez Contreras, C. 2011b, *ApJ Let*, 740, L39
Sahai, R. & Nyman, L-Å. 1997, ApJ, 487, L155
Sahai, R., Sanz-Forcada, J., & Sánchez Contreras, C. 2016a, *Journal of Physics Conf. Ser.*, 728, 042003
Sahai, R., Scibelli, S., & Morris, M. 2016b, *ApJ*, 827, 92
Sahai, R., Sugerman, Ben, E.K., Hinkle, K. 2009, *ApJ*, 699, 1015
Sahai, R., Sanz-Forcada, J., Sánchez Contreras, C., & Stute, M. 2015, ApJ, 810, 77 [Setal15]
Sahai, R., Te Lintel Hekkert, P., Morris, M., Zijlstra, A., Likkel, L. 1999, ApJ, 514, L115
Sahai, R. & Trauger, J.T. 1998, *AJ*, 116, 1357
Sahai, R., Vlemmings, W. H. T., Gledhill, T., *et al.* 2017a, *ApJ Let*, 835, L13 [Setal17]
Sahai R., Vlemmings W. H. T., Huggins P. J., *et al.* 2013, *ApJ*, 777, 92
Sahai, R., Vlemmings, W. H. T., & Nyman, L.-Å. 2017b, *ApJ*, 841, 110
Sahai, R., Young, K., Patel, N. A., Sánchez Contreras, C., & Morris, M. 2006, *ApJ*, 653, 1241
Sahai, R., Sánchez Contreras, C., Mangan, A. S., *et al.* 2018a, *ApJ*, 860, 105 [Setal18]
Sahai, R., Rajagopal, J., Hinkle, K., Joyce, R. & Morris, M. 2018b, this volume
Sahai, R. 2018a, arXiv:1810.06685
Sahai, R. 2018b, Galaxies, 6, 102
Sánchez Contreras, C., Bujarrabal, V., Castro-Carrizo, A., Alcolea, J., & Sargent, A. 2006, *ApJ*, 643, 945
Scibelli, S., Sahai, R., & Morris, M. 2018, *ApJ*, submitted
Snaid, S., Zijlstra, A. A., McDonald, I., *et al.* 2018, *MNRAS*, 477, 4200
Soker, N. 2015, *ApJ*, 800, 114

Why Galaxies Care About AGB Stars:
A Continuing Challenge through Cosmic Time
Proceedings IAU Symposium No. 343, 2019
F. Kerschbaum, M. Groenewegen & H. Olofsson, eds.

doi:10.1017/S1743921318005938

Planetary nebulae in the (extra)-galactic context: Probing chemical evolution in star-forming galaxies

Letizia Stanghellini

National Optical Astronomy Observatory
950 N. Cherry Avenue
Tucson, Arizona 85719 (USA)
email: lstanghellini@noao.edu

Abstract. The populations of planetary nebulae (PNe) probe metallicity and chemical content (and its evolution) of the parent galaxy, giving clues to galaxy formation and evolution. This sub-field of extra-galactic PN research has been particularly active in the recent years. Comparison of data and models yielded estimates of global cosmic enrichment and provided constraints to galaxy formation history. In external spiral galaxies, the chemical contents of PNe and H II regions can be compared to disclose possible evolution of the radial metallicity gradient, which is, in turn, a powerful constraint to galactic chemical evolutionary models. In the Milky Way, recent PN progenitor dating and new chemical abundances offer an updated look into our own Galaxy. Collectively, Galactic and extra-galactic radial metallicity gradients from emission-line probes (PNe and H II regions) can be compared to have a cosmological outlook on galactic evolution.

Keyword. Planetary nebulae

---

1. Introduction

The field of extragalactic planetary nebulae (PNe), or, more broadly, the field of PNe studied as populations rather than individually, has been active since the late 70s, when it became clear that the bright [O III] line at 5007 Å could be observed in PNe of external galaxies, and the first extra-galactic PN catalogues became available (Sanduleak *et al.* 1978). Nonetheless, there have not been many conferences, or journal reviews, dedicated specifically to this field. The most recent meeting on extra-galactic PNe took place in 2015 in Honolulu, within the XXIX IAU General Assembly (Stanghellini *et al.* 2016). We have to look back more than a decade to encounter the previous extra-galactic PN meeting, that took place at ESO in 2004 (Stanghellini *et al.* 2006), and was the first full science meeting dedicated to extra-galactic PNe. Since 1975 the field has produced more than 450 scientific papers (source: ADS), indicating five major sub-fields: (i) surveys and discoveries of extra-galactic PNe; (ii) the study of Magellanic Cloud PNe; (iii) studies that use PNe as dynamical probes; (iv) PNe as chemical evolution and stellar evolution probes; (iv) the PN luminosity function, and the determination of the extra-galactic distance scale. All these sub-fields are currently active, but with peaks of activity that depend on the scientific capabilities available at different epochs. Furthermore, the sub-field of chemical abundances in extra-galactic PNe is reaching the observability limit with the current technology, especially concerning direct abundances. In fact, most of the nearby galaxies where direct abundances of PNe could be measured have already been studied, and a renaissance of this field is to be expected with the advent of extremely large telescopes (ELTs) spectrographs. This is a review of the field of PN populations, both in the Milky Way and in external galaxies, and in particular the studies of chemical abundances as constraints to the evolution of the parental galaxy. Section 2 introduces the aim and goals of this type of research, Sect. 3 covers the characterization of the ages of the emission-line probes, Sect. 4 gives some examples of how PN and H II region abundances can advance our knowledge of galaxy evolution, and Sect. 5 gives an outlook on future advances of this field, especially for what the ELTs will bring to the field.

2. Constraining chemical evolution in star-forming galaxies with emission-line probes

Radial metallicity gradients of populations probing different epochs in the galactic evolution are crucial to constrain chemical evolutionary models of galaxies. PNe have been used to probe the chemistry of the old stellar populations, both in the Milky Way and in nearby galaxies, while H II regions probe the young populations. Direct abundances (i.e., through plasma diagnostics) of individual PNe are currently measurable out to distances of $\sim$3.5 Mpc. By comparing the metallicity (typically through oxygen abundance) of PN and H II region, one can also trace the global galactic chemical enrichment.

In Fig. 1 we show the radial metallicity gradient comparison for PN and H II regions in M33, which is the best-studied case of a star-forming galaxy. PNe and H II regions have been chosen to spatially sample the whole galactic disk, and their abundances have been measured with the direct method. The resulting gradients are similar for the old and young populations if we used binned abundances (H II regions have slightly steeper gradients than PNe if abundances are left unbinned). A similar analysis can be applied to all local galaxies where we can observe PN and H II regions in a reasonable observing time. A recent review by Bresolin (2017) indicates that, in the current scenario and for galaxies where oxygen gradients have been measured directly, the radial oxygen gradient is often flatter for PN populations than for H II regions or young stars, indicating a steepening of the gradient with time since galaxy formation.

There are challenges associated with this type of studies. First, in order to determine direct PN and H II region abundances one needs to measure the [O III] auroral λ4363 line, which can be done not much farther away than in M81 with 8m-class telescopes. Also, oxygen abundance could vary during AGB evolution; Ar and Cl are safer probes of radial metallicity for the AGB stars, but they are more difficult to measure in PNe (Delgado-Inglada *et al.* 2015). Finally, H II regions probe the ISM chemistry after the galaxy has gone through its whole evolution, and PNe probe the chemistry of the ISM at the time of the progenitor's formation, but it is difficult to date PN progenitors precisely, as illustrated in the next section.

3. Characterizing and dating the progenitors of emission-line probes

Emission-line targets probe both the young and the old stellar populations. H II regions are related to massive stars, whose look-back time is $t_\star = 0$. On the other hand, PN progenitors can be associated with a range of look-back times into the galaxy history, depending on their mass. PN progenitor mass larger than 2-4 $M_\odot$ were formed only 1 Gyr ago (depending on metallicity), thus they mark a time close to that of the H II regions, while PNe whose progenitor mass is smaller than 1.2 $M_\odot$ probe the galaxy 5 Gyr ago or earlier. We will examine a couple of current ways to date PN progenitors, an essential step toward the study of galaxy evolution through emission-line probes.

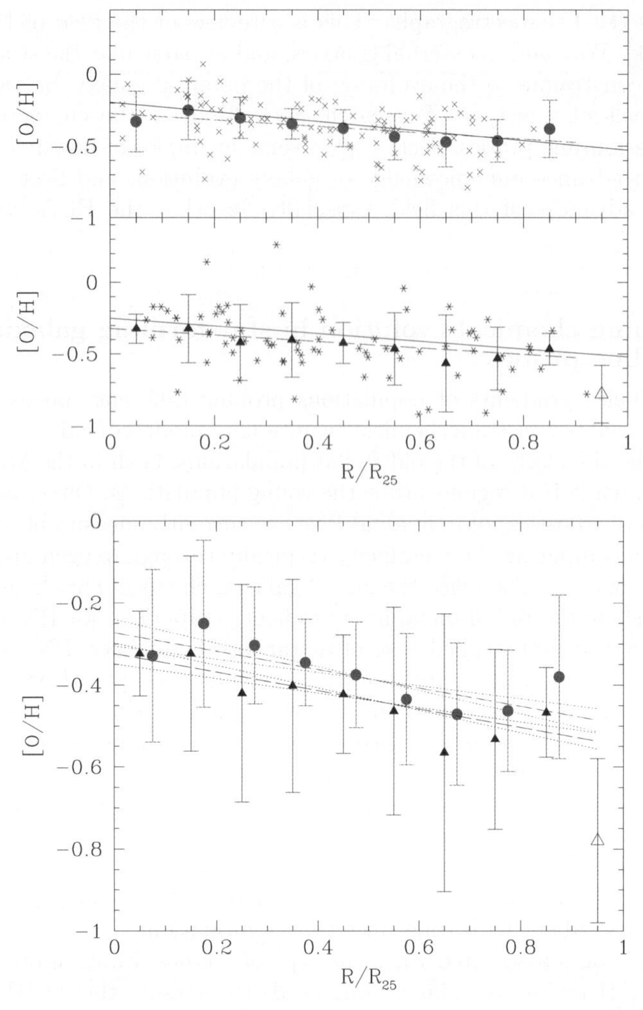

Figure 1. Radial oxygen gradients in M33. Upper panels: Individual and binned results for H II regions (top) and PNe (bottom). The continuous black lines are the gradient of Magrini *et al.* (2010) for H II regions (their whole sample) and the non-Type I PN sample of Magrini *et al.* (2009). Lower panel: weighted linear fits of PN and H II region binned metallicities [adapted from Magrini *et al.* 2016].

3.1. *Dating PN progenitors through the yields of stellar evolution*

In the galaxy and a few nearby galaxies, we can study CNO abundances of individual PNe and compare them with the final yields of stellar evolution models to determine their progenitor ages/masses, given the population metallicity. Let us compare the yields from the late thermal pulses of AGB evolution, such as those in Fig. 1 in Ventura *et al.* (2017), where the calculation is available for a broad selection of initial stellar masses and metallicities. We can select PNe whose observed abundances in the N/H vs. O/H plane correspond to yields of parent stars with $t_{\star} < 1$ Gyr, and those with $t_{\star} > 7.5$ Gyr for a broad range of metallicities. In this way we populate two distinct PN groups, one with young progenitors (YPPNe, $t_{\star} < 1$ Gyr) and another with old progenitors (OPPNe, $t_{\star} > 7.5$ Gyr). More details on this classification are given in Stanghellini & Haywood (2018). The age separation is essential for the study of the evolution of the Milky Way

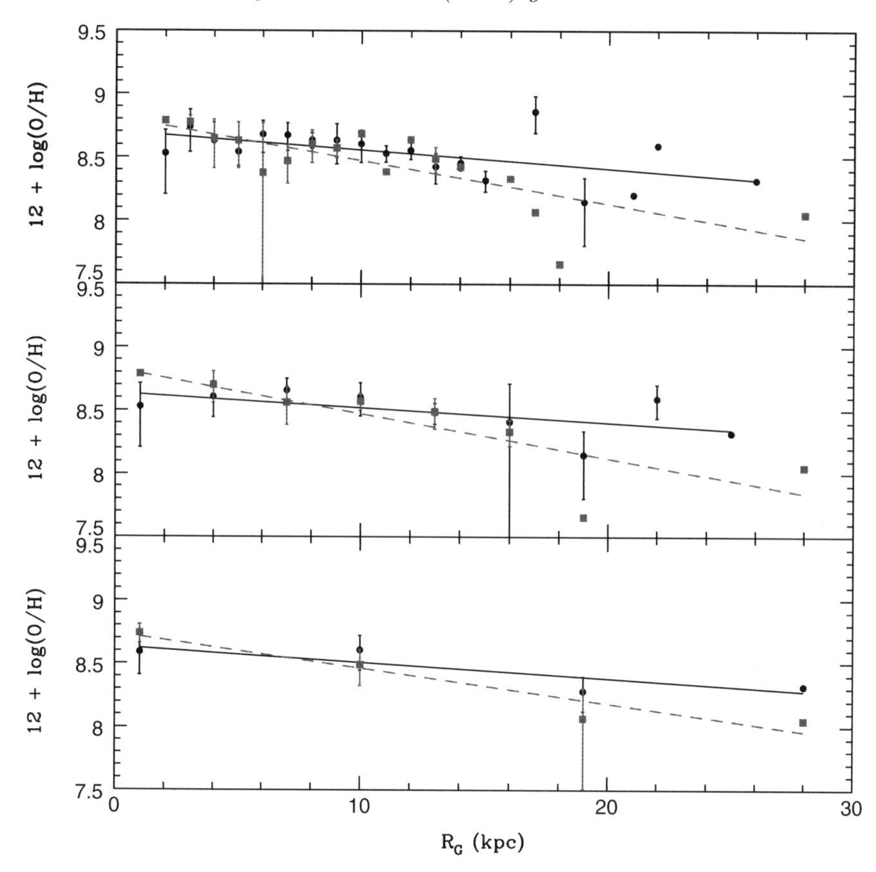

Figure 2. Binned gradients for OPPNe (filled circles and solid lines) and YPPNe (filled squares and dashed lines). Oxygen abundances have been averaged over bins of 1 (top panel), 3 (middle panel), and 9 (bottom panel) kpc. Error bars represent abundance ranges within each bin [adapted from Stanghellini & Haywood (2018)].

and can be applied to the radial metallicity gradients in the Galaxy and in nearby galaxies as well.

3.2. *The Galactic case: evolution of the radial oxygen gradient*

The age discrimination discussed in the previous section has been directly applied to a large number of Galactic PNe. Stanghellini & Haywood (2018) studied two populations of Galactic disk PNe within these progenitor age ranges and found that the radial oxygen gradient measured with YPPNe is steeper ($\Delta\log(O/H)/\Delta R_{\rm G} \sim -0.027$ dex kpc^{-1}) than the one measured with OPPNe ($\Delta\log(O/H)/\Delta R_{\rm G} \sim -0.018$ dex kpc^{-1}), meaning that the gradient has steepened since the formation of the Galaxy. In Fig. 2 we show the gradient obtained with binned data, where the difference in gradient slope between the two populations is independent on the binning and larger than the uncertainties, thus the result is robust.

3.3. *Use target dynamics to date PN progenitors*

While it is possible to date progenitors of Galactic disk PNe through their CNO abundances, this is not always possible with extragalactic PN populations farther away than the Magellanic Cloud. In external galaxies, target dynamics has proven to be the best

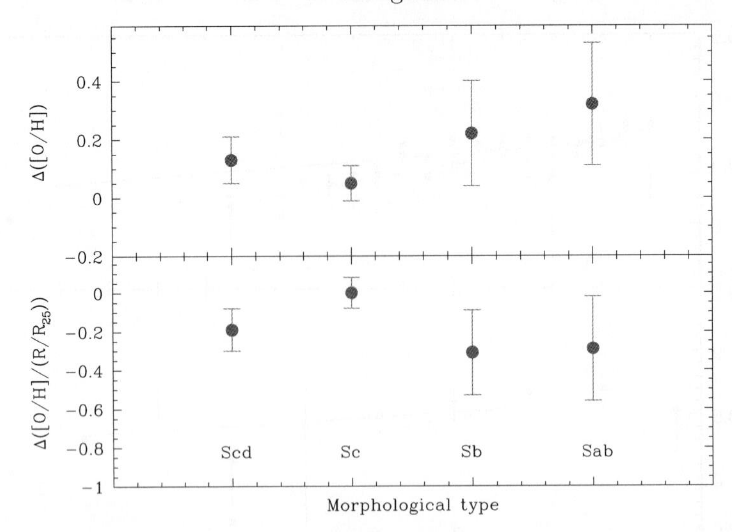

Figure 3. Global enrichment (upper panel) and variation of the slope of the radial oxygen gradient (lower panel) as a function of the morphological type. Galaxies examined are (left to right) NGC300 (Scd), M33 (Sc), M81 (Sb), and M31 (Sab) [from Magrini *et al.* (2016)].

method to determine whether the progenitors of a PN population is old or more recent. For example, Aniyan *et al.* (2018) show the separation of hot (old progenitors) and cold (young progenitors) components of the PN population in NGC628, a face-on nearby spiral galaxy, 9 Mpc away. At this distance, it would be hard to separate OPPNe and YPPNe based on their abundances. PNe in this galaxy have been classified as old or young progenitors based on their dispersions from the PN velocity histogram. Following this classification, one can study for instance the oxygen distribution of the two populations and track their evolution. Given that the PN populations are faint at this distance, it is expected that stacked spectra will be used to determine the metallicity.

4. Galactic and extra-galactic emission-line probes: constraining the chemical evolution of galaxies

The analysis of global enrichment and the radial oxygen gradients from emission-line probes in star-forming galaxies has been explored in several galaxies where the abundances have been measured directly, i.e., through plasma diagnostics. Magrini *et al.* (2016) (see Fig. 3 below) show, in the upper panel, the global oxygen enrichment respectively in (left to right) NGC300, M33, M81, and M31, measured from the relative oxygen abundances in H II regions and PNe. The result is that all studied star-forming galaxies have experienced global oxygen enrichment. The lower panel of Fig. 3 shows the differences between the H II region and PN radial oxygen gradients. For most galaxies the difference is negative (i.e., the radial oxygen gradients from H II regions is steeper than that from PNe), meaning that the gradient is steepening with time since galaxy formation. In the case of M33 the difference is negligible (see also Fig. 1). It is worth noting that the figure includes only the galaxies with plasma diagnostic abundances determined for adequate samples of H II regions and PNe, thus the trend seems universal where tested.

Another way to study these trends is to place the radial oxygen gradients on a time (or redshift) scale. In Fig. 4, the H II region radial oxygen gradients (i.e., the slopes) are placed at the redshift of their parent galaxy. On the other hand, PN populations in the Milky Way and in $z = 0$ galaxies have been placed at the equivalent redshift derived from the look-back time of the parent stellar population of the PNe. In the figure, lines are representative of inside-out chemical galaxy evolution models in a cosmological context:

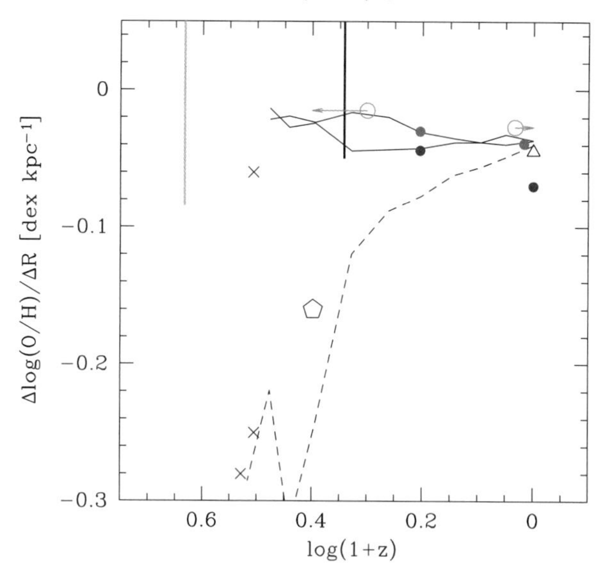

Figure 4. Radial oxygen gradient from star-forming galaxies vs. redshift. Data points: triangle is gradient slope from Galactic H II regions (Balser *et al.* 2011). Red circles: M33 PNe and H II region gradients (Magrini *et al.* 2009, Magrini *et al.* 2010). Blue circles: gradients from PNe and H II regions in M81 (Stanghellini *et al.* 2010, Stanghellini *et al.* 2014). Pentagon: Yuan *et al.* (2011). Crosses: Jones *et al.* (2013). Vertical lines: ranges of gradient slopes from Cresci *et al.* (2010) and Sánchez *et al.* (2014). Open circles: OPPNe and YPPNe (see text) [adapted from Stanghellini & Haywood (2018)].

solid lines, with enhanced feedback; dashed lines, no feedback (Gibson *et al.* 2013). Note that several of the high-redshift galaxies are lensed, thus the plot assumes that we can directly compare external (i.e., lensed) and nearby galaxies. there are a few notable exceptions to this scenario of course, which involve high-redshift galaxies and need to be addressed in the future. From this plot, we infer that measured oxygen gradients agree with enhanced feedback disk formation, and gradients are steepening with time since galaxy formation.

5. Future outlook

Radial metallicity gradient evolution in star-forming galaxies, based on direct abundances of emission-line probes, are currently available for a few nearby spirals and the Milky Way. They could be determined for a few additional galaxies with current technology but need excellent observing conditions and long exposures. PN dating is possible from the comparison of stellar evolutionary yields with observed PN abundances. In more distant galaxies, dynamical dating is also promising. We can then investigate the evolution of metallicity gradients by comparing gradients of nearby galaxies, and those of redshifted galaxies, but it is challenging to understand which $z > 1$ galaxies are comparable to nearby spirals.

Progress can certainly be made in the era of the ELTs. In multi-object spectroscopy (MOS) mode, and by using direct abundances, we can foresee that an instrument such as the WFOS for TMT can make a breakthrough. The FoV allows good sampling of galactic probes across spiral disks, making this mode ideal for this type of science. For example, spectra of the brightest PNe in M81 could be acquired in a few minutes down to auroral lines for plasma diagnostics, as opposed to the two nights used for MMT observations. Much fainter PNe and H II regions are also within reach with reasonable exposures.

Apart from the local galaxies, progress can be made by reaching out to Virgo galaxies, where the variety of galaxy types can be examined in detail for their chemical content though emission-line probes. Direct abundances in Virgo galaxies can be observed rapidly for any resolved spiral galaxy. Another approach to the problem is to populate the gradient slope-redshift diagram with near-IR IFU mode and observing $z > 1$ galaxies. Direct abundances of H II regions in galaxies with $z \sim 1.6$ may be possible ([O II] λ3727; [O III] λ4363, λ4959, λ5007; Hβ, Hα, [N II] λ6548, λ6584 are seen in YJH bands for these redshifts). Finally, for galaxies with $2 < z < 2.6$, strong-line diagnostics lines can be observed in the JHK bands.

References

Aniyan, S., Freeman, K. C., Arnaboldi, M., *et al.* 2018, MNRAS, 476, 1909

Balser, D. S., Rood, R. T., Bania, T. M., & Anderson, L. D. 2011, ApJ, 738, 27

Bresolin, F. 2017, Planetary Nebulae: Multi-Wavelength Probes of Stellar and Galactic Evolution, 323, 237

Cresci, G., Mannucci, F., Maiolino, R., *et al.* 2010, Nature, 467, 811

Delgado-Inglada, G., Rodríguez, M., Peimbert, M., Stasińska, G., & Morisset, C. 2015, MNRAS, 449, 1797

Gibson, B. K., Pilkington, K., Brook, C. B., Stinson, G. S., & Bailin, J. 2013, A&A, 554, A47

Jones, T., Ellis, R. S., Richard, J., & Jullo, E. 2013, ApJ, 765, 48

Magrini, L., Stanghellini, L., & Villaver, E. 2009, ApJ, 696, 729

Magrini, L., Stanghellini, L., Corbelli, E., Galli, D., & Villaver, E. 2010, A&A, 512, A63

Magrini, L., Coccato, L., Stanghellini, L., Casasola, V., & Galli, D. 2016, A&A, 588, A91

Sánchez, S. F., Rosales-Ortega, F. F., Iglesias-Páramo, J., *et al.* 2014, A&A, 563, A49

Sanduleak, N., MacConnell, D. J., & Philip, A. G. D. 1978, PASP, 90, 621

Stanghellini, L., Walsh, J. R., & Douglas, N. G. 2006, Planetary Nebulae Beyond the Milky Way

Stanghellini, L., Magrini, L., Villaver, E., & Galli, D. 2010, A&A, 521, A3

Stanghellini, L., Magrini, L., Casasola, V., & Villaver, E. 2014, A&A, 567, A88

Stanghellini, L., Peña, M., & Méndez, R. 2016, IAU Focus Meeting, 29, 3

Stanghellini, L., & Haywood, M. 2018, ApJ, 862, 45

Ventura, P., Stanghellini, L., Dell'Agli, F., & García-Hernández, D. A. 2017, MNRAS, 471, 4648

Yuan, T.-T., Kewley, L. J., Swinbank, A. M., Richard, J., & Livermore, R. C. 2011, ApJL, 732, L14

Why Galaxies Care About AGB Stars:
A Continuing Challenge through Cosmic Time
Proceedings IAU Symposium No. 343, 2019
F. Kerschbaum, M. Groenewegen & H. Olofsson, eds.

doi:10.1017/S1743921318005720

Extended Dust Emission from Nearby Evolved stars

Thavisha E. Dharmawardena[1,2], Francisca Kemper[1], Peter Scicluna[1], Jan G. A. Wouterloot[3], Alfonso Trejo[1], Sundar Srinivasan[1], Jan Cami[4,5], Albert Zijlstra[6,7], Jonathan P. Marshall[1] and the NESS collaboration

[1]Academia Sinica Institute of Astronomy and Astrophysics, 11F of AS/NTU Astronomy-Mathematics Building, No.1, Sect. 4, Roosevelt Rd, Taipei 10617, Taiwan, R.O.C.

[2] Graduate Institute of Astronomy, National Central University, 300 Zhongda Road, Zhongli 32001, Taoyuan, Taiwan, R.O.C.
email: `tdharmawardena@asiaa.sinica.edu.tw`

[3]East Asian Observatory, 660 N A'ohoku Place, Hilo, Hawaii 96720, USA

[4]Department of Physics and Astronomy and Centre for Planetary Science and Exploration, The University of Western Ontario, London, ON N6A 3K7, Canada

[5]SETI Institute, 189 Bernardo Ave, Suite 100, Mountain View, CA 94043, USA

[6]Jodrell Bank Centre for Astrophysics, School of Physics and Astronomy, University of Manchester, Oxford Road, Manchester M13 9PL, UK

[7]Laboratory for Space Research, University of Hong Kong, Pokfulam Road, Hong Kong

Abstract. We derive azimuthally-averaged surface-brightness profiles of 16 AGB stars in the far-IR and sub-mm with the aim of studying the resolved historic mass loss in the extended circumstellar envelope. The PSF-subtracted extended component fluxes were found to be $\sim 40\%$ of the total source flux. By fitting SEDs at each radial point we derive the dust temperature, column density and spectral index of emissivity via Bayesian inference. The measured dust-to-gas ratios were somewhat consistent with canonical values however with a large scatter.

Keywords. stars: AGB and post-AGB - stars: circumstellar matter - stars: mass-loss

---

1. Introduction

At present many studies involving evolved-star dust production take into account only the present day mass-loss rates via mid-IR observations. This treatment ignores an older cooler dust component emitted during historic mass loss events. Further, in many cases evolved stars are treated as quasi-stable systems not taking into account the variations of historic mass loss. This cooler dust component and its variations are only observable at longer wavelengths such as in the far-IR and sub-mm. Therefore at present we do not have a clear understanding of the extended circumstellar envelopes of evolved stars formed as a result of the historic mass loss. Evidence for the presence of this historic dust mass component is reported by Cox *et al.* (2012) who presented a large sample of evolved stars in the far-IR using Hershcel/PACS observations at 70 μm and 160 μm. These observations show dust envelopes with extensions out to $\sim 1' - 2'$.

We analyse sixteen evolved stars (5 C-rich, 9 O-rich, and 1 S-type Asymptotic Giant Branch (AGB) stars, and 1 red supergiant (RSG)) with the goal of understanding the historic mass loss component (Dharmawardena *et al.* 2018, and Dharmawardena et al., in prep). The analysis was carried out using new sub-mm observations with the JCMT/SCUBA-2 instrument combined with existing far-IR Herschel/PACS observations

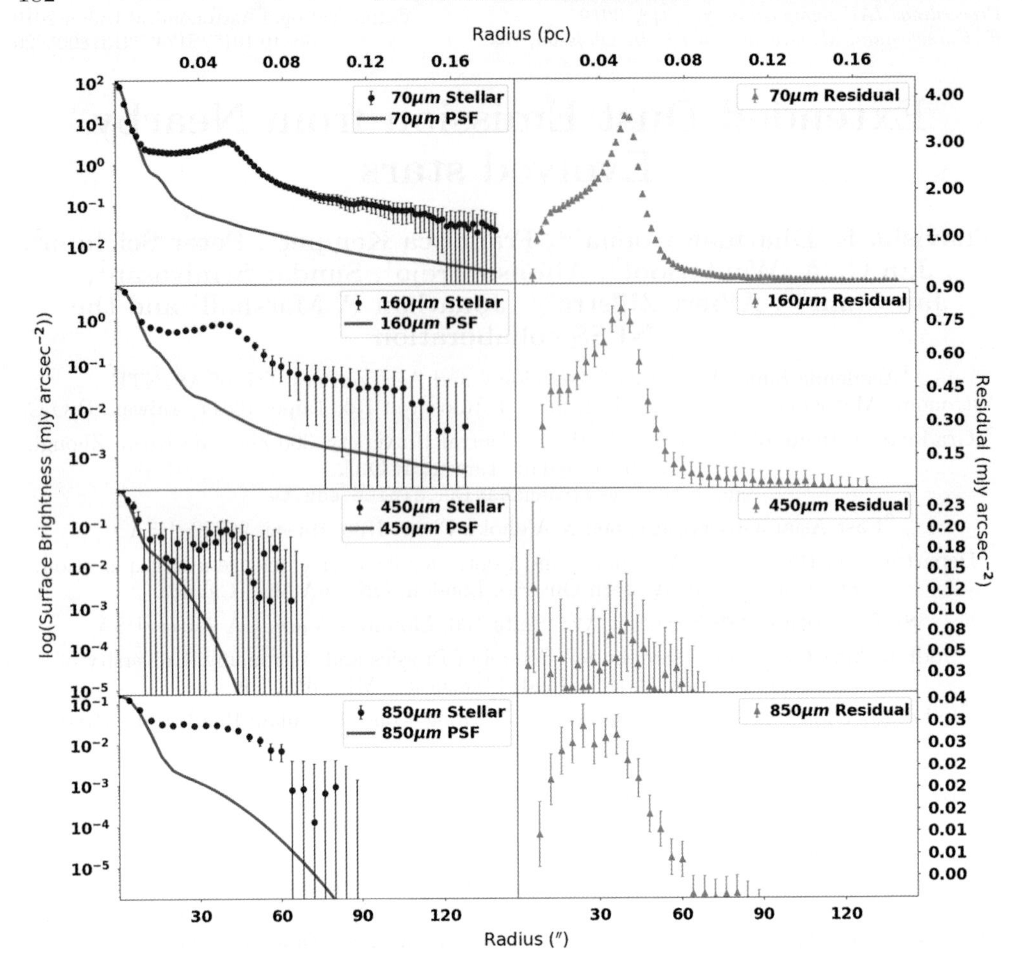

Figure 1. Surface brightness and residual profiles of U Ant. Right hand panels: The blue dashed lines represent the source surface brightness profiles and the grey solid lines represent the PSF profile of the instrument at the given wavelength; Left hand panels: PSF subtracted residual profiles for each wavelength (Dharmawardena et al., in prep.).

from the MESS survey. This analysis was a pilot study for the Nearby Evolved Stars Survey (NESS, https://evolvedstars.space) which aims to statistically analyse the extended dust and gas emissions in a volume limited sample of nearby evolved stars.

2. Azimathally-averaged surface-brightness profiles

By generating azimuthally-averaged surface-brightness profiles we were able to measure the extent of the circumstellar envelope and its surface-brightness properties. The sources on average have radii at 3σ surface brightness limits of $\sim 40'' - 50''$ with a maximum of $80''$ at the SCUBA-2 wavelengths. A significant component of the total flux, an average of $\sim 40\% - 50\%$, was concentrated within the PSF-subtracted extended source component. This shows the significance of the cool historic dust component when analyzing the mass-loss rates and dust productions rates of these evolved stars even in a statistical sense. Additionally, the surface brightness profiles, as well as the PSF-subtracted residual profiles, show circumstellar envelope structure for several sources. For example, as shown in Fig. 1 we see clear evidence for the detached dust shell of U Ant in the form of enhancements in the surface-brightness and residual profiles at all four wavelengths.

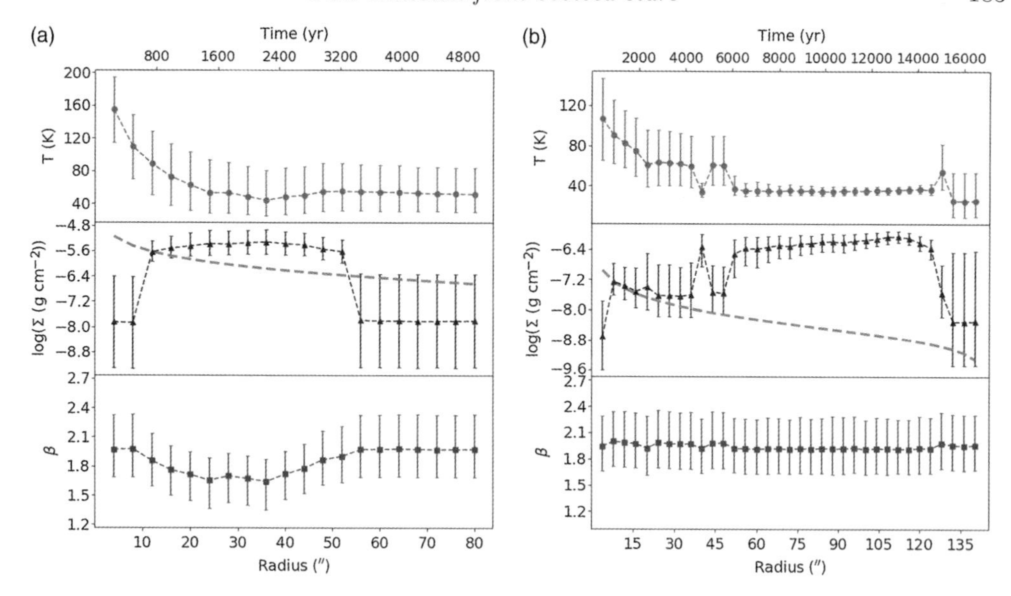

Figure 2. SED fitting results of U Ant (left) and U Hya (right); Top: Temperature (T) radial profiles; Middle: Dust mass column density (Σ) radial profile. The orange dashed line represents the expected dust mass column density for a uniform mass-loss rate; Bottom: The spectral index of dust emissivity (β) profile (Dharmawardena *et al.* 2018, and Dharmawardena *et al.*, in prep)

3. Radial Variation in the Dust Properties

We derive dust mass properties as a function of radius for each source by fitting a single-temperature modified blackbody at each radial point to the four point SED created by combining the residual profiles at all four wavelengths. The SED fitting was carried out using the python MCMC package *emcee*. The derived radially dependent dust temperature (T), spectral index of the dust emissivity (β), and the dust column density (Σ) profiles are used to probe the dust mass-loss history and detect changes in physical properties of dust as a function of radius, thus time.

The derived parameter profiles for U Ant and U Hya are presented in Fig. 2. All three parameter profiles had common features which were present throughout most of the sample. The T profiles initially show a decreasing gradient consistent with a centrally heated optically thin dust mass. This is followed by a near flat gradient consistent with a dust mass heated by the interstellar radiation field (ISRF).

The Σ profiles deviate from the corresponding profile expected for constant mass-loss rate (MLR), hence showing that these sources underwent mass-loss modulations during their lifetimes. The features seen in the profiles, for example the enhancements seen in the middle panels of Fig. 2 of the detached shell sources U Hya and U Ant, correspond to these mass loss modulations. Using these profiles we can therefore derive the look-back time at which the mass loss modulations occur. We find that the thermal pulse which gave rise to the detached shells of U Hya and U Ant occurred $\sim 14000 \pm 1000$ and $\sim 3200 \pm 500$ years ago in look-back time. We observe large scale mass-loss variations up to time scales of $\sim 7400 \pm 1000$ yrs, comparable to thermal pulse time scales reported by Olofsson *et al.* (1990), but much larger than modulations reported by Marengo *et al.* (2001).

Due to the smaller extent of the detections in the SCUBA-2 observations when compared to the PACS observations, the majority of our β profiles were prior dominated following a flat profile similar to what is observed in the case of U Hya (see bottom

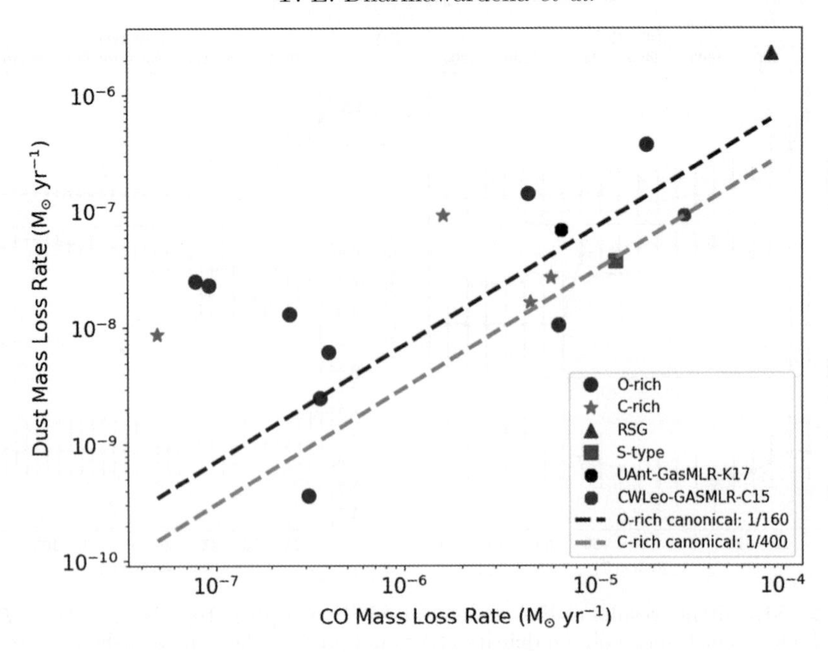

Figure 3. Gas MLR versus dust MLR. Blue circles: O-rich AGB stars; crimson stars: C-rich AGB stars; green square: S-type AGB star; maroon triangle: RSG; orange dashed line: the accepted dust-to-gas ratio of 1/400 for C-rich sources; and blue dashed line: the accepted dust-to-gas ratio of 1/160 for O-rich sources. (Dharmawardena *et al.* 2018)

panel of Fig. 2b). However for several sources with SCUBA-2 detection extents comparable to those of the PACS extents we see the variation of dust properties with time reflected in the variations in β profile. We find that the β limits in the detectable region of IRC+10216 is consistent with the presence of graphitic dust, and in the case of U Ant the dust is composed of amorphous carbon grains (see bottom panel of Fig. 2a).

4. Dust masses and dust-to-gas ratios

We derive the dust masses of the extended component by integrating the Σ profile. We find that these masses are on average ~ 5 times greater than those predicted by GRAMS radiative transfer models (Srinivasan et al., 2012). We attribute this to the fact that GRAMS utilizes only the mid-IR observations which are sensitive only to the warmer central present day mass-loss, not taking into account historic mass loss or its modulation. This shows the importance of taking into account the cooler historic dust-mass component when analyzing mass output by evolved stars into the ISM.

By extrapolating the central point gas mass loss rates reported in De Beck *et al.* (2010) we derive the total gas masses up to the same look-back time period as that of the dust masses we derive. As shown in Fig. 3, the derived dust-to-gas ratios are somewhat consistent with the canonical dust-to-gas ratios presented in Knapp & Morris (1985). However we also see a large scatter in the derived ratios. Again, this is most likely due to the fact that while we take into account the historic mass loss modulations when determining dust masses, it is not possible to do so for the gas masses as we only had central present day gas mass-loss rates. For the case of U Ant and IRC+10216 where we did have a gas mass-loss rate determined via resolved observations which took into account its extended gas emission we find the dust-to-gas ratios changing significantly from the previous measurement and being more consistent with the canonical.

5. Summary

By analysing the resolved extended dust emission of sixteen evolved stars in the far-IR and sub-mm we find an extended cooler dust component resulting from historic mass loss. We find a significant portion of the total source flux is found within this extended component. The variations in total dust masses and dust-to-gas ratios from those derived from present day mass loss rates and mid-IR observations, show the importance of taking this historic cooler dust mass component and the resulting mass loss modulation effects (variations in the circumstellar envelope structure) into account when determining mass output to the ISM by evolved stars.

References

Cox N. L. J., *et al.*, 2012, A&A, 537, A35

De Beck E., Decin L., de Koter A., Justtanont K., Verhoelst T., Kemper F., Menten K. M., 2010, A&A, 523, A18

Dharmawardena T. E., *et al.*, 2018, Monthly Notices of the Royal Astronomical Society, 479, 536

Knapp G. R., Morris M., 1985, ApJ, 292, 640

Marengo M., Ivezić Ž., Knapp G. R., 2001, MNRAS, 324, 1117

Olofsson H., Carlstrom U., Eriksson K., Gustafsson B., Willson L. A., 1990, A&A, 230, L13

Discussion

SLOAN: You compared the dust mass in the mid-IR to the far-IR and got 5 times more dust in the far-IR. Perhaps it would be better to think in terms of rates. If the far-IR region samples 5 times the time, then the rates are the same. How much more time does the far-IR sample?

DHARMAWARDENA: We calculate the mid-IR dust masses by extrapolating to the same look-back time as those of the far-IR and sub-mm. Therefore the dust masses both are measured upto the same time and therefore analogous to measuring the differences in the rates.

MAERCKER: The density profile for U Ant goes in a lot further than would be expected from a detached-CSE.

DHARMAWARDENA: We find the SCUBA-2 850 μm observations peaks in the inner region as a result of the substructure of the outer shells projected to the inner region. The PACS on the other hand peaks at the location of the outer shells. As a result when we combine and compute SEDs as a function of radius we see a density enhancement which is wider than the thin outer shells.

SAHAI: What were the 'canonical' dust-to-gas ratios used for O-rich and C-rich stars, and did you use the standard dust opacities (if so, which ones?)?

DHARMAWARDENA: 1. We use canonical values of 0.003 and 0.007 for C-rich and O-rich stars reported in Knapp and Morris 1985. 2. We use $\kappa^{s}_{\mathrm{eff},160} = 26$ cm^2 g^{-1} for C-rich and 8.8 cm^2 g^{-1} for O-rich (including RSG and S-type) by computing cross sections for spherical grains of $\sim$0.1 μm using ACAR optical constants for amorphous C-grains.

Why Galaxies Care About AGB Stars: A Continuing Challenge through Cosmic Time
Proceedings IAU Symposium No. 343, 2019
F. Kerschbaum, M. Groenewegen & H. Olofsson, eds.

doi:10.1017/S1743921318005045

On the circumstellar envelopes of semi-regular long-period variables

J. J. Díaz-Luis[1], J. Alcolea[1], V. Bujarrabal[1], M. Santander-García[1], M. Gómez-Garrido[2] and J.-F. Desmurs[1]

[1]Observatorio Astronómico Nacional (IGN), Spain
email: jjairo@oan.es

[2]Instituto Geográfico Nacional (IGN), Centro de Desarrollos Tecnológicos, Observatorio de Yebes, Spain

Abstract. The mass loss process along the AGB phase is crucial for the formation of circumstellar envelopes (CSEs), which in the post-AGB phase will evolve into planetary nebulae (PNe). There are still important issues that need to be further explored in this field; in particular, the formation of axially symmetric PNe from spherical CSEs. To address the problem, we have conducted high S/N IRAM 30 m observations of ^{12}CO J=1-0 and J=2-1, and ^{13}CO J=1-0 in a volume-limited unbiased sample of semi-regular variables (SRs). We also conducted Yebes 40 m SiO J=1-0 observations in 1/2 of the sample in order to complement our ^{12}CO observations. We report a moderate correlation between mass loss rate and the ^{12}CO(1-0)-to-^{12}CO(2-1) line intensity ratio, introducing a possible new method for determining mass loss rates of SRs with short analysis time. We also find that for several stars the SiO profiles are very similar to the ^{12}CO profiles, a totally unexpected result unless these are non-standard envelopes.

Keywords. planetary nebulae: general, stars: AGB and post-AGB, millimeter

---

1. Introduction

For low and intermediate mass stars (~1-8 $M_\odot$), the latest stages of their evolution are controlled by the mechanisms of mass loss that occur at these phases. In the AGB (and in general in the red giant or super-giant phases), these mass losses, with rates over 10^{-5} $M_\odot$ yr^{-1}, are crucial for the formation of circumstellar envelopes (CSEs), which in the post-AGB phase will evolve into planetary nebulae (PNe); see e.g. Olofsson (1999). There are still important issues that need to be further explained; in particular, the formation of axially symmetric PNe from spherical CSEs. CSEs around AGB stars are in general quite spherical, and expand isotropically at moderate speeds (10-25 km s^{-1}), at least on large scales (Neri *et al.* 1998; Castro-Carrizo *et al.* 2010). However, PNe display a large variety of shapes with high degree of symmetry; about 4/5 of the cases are far from showing spherically symmetric envelopes (Parker *et al.* 2006). This transformation from spherical CSEs to axially symmetric PNe is still a matter of debate since more than two decades (Balick & Frank 2002), though in the last years there is growing consensus that the presence of binary systems is the most likely explanation; see e.g. Jones & Boffin (2017). This is supported by the large prevalence of multiple systems in solar-type stars in general, and in the AGB population in particular. Some studies reveal the presence of spiral patterns and bipolar outflows at smaller scales, providing observational support for the binary system scenario as the reason for such structures (e.g. Morris *et al.* 2006; Mauron & Huggins 2006; Maercker *et al.* 2012; Kim *et al.* 2015). Moreover, recent studies involve searching for UV excess emission, which is believed to be due to a main-sequence companion, from AGB stars using the GALEX archive (e.g. Sahai *et al.* 2008; Ortiz & Guerrero 2016). Almost 60% of the sample has a main-sequence companion of spectral type earlier than K0. It is believed that more than 50% of PNe harbour binary systems (Douchin *et al.* 2015) and almost 50% of solar-type stars in the solar neighborhood seem to form in binary or multiple systems (Raghavan *et al.* 2010). The main problem is that there are no complete and unbiased samples that resolve the mystery.

During the last years, the results from the increasing number of good interferometric maps of the CSEs of a particular type of evolved stars have attracted our attention. Semi-regular variables (SRs) are red giant stars showing quasi-regular or some irregular variations in the optical range (SRa and SRb variables, respectively), in contrast to the regular pulsators, the Mira-type variables (note that both SRs and Miras constitute the Long-Period Variable stars, LPVs). In principle, the differences in their pulsation mode should not affect the envelope geometry and kinematics. However, a clear axial symmetry has been found in almost all CSEs around SRa and SRb variables that have been well mapped (using mm-wave interferometric observations of CO; see e.g. Castro-Carrizo *et al.* 2010; Hirano *et al.* 2004; Homan *et al.* 2017, 2018), in strong contrast to the CSEs of regular pulsators (Miras). In spite of a strong prevalence of non-spherical CSEs in SRs, we must take into account that the sample is limited and strongly biased. Hence, the question of whether SRs really have CSEs different from those of regular variables remains open.

2. Observations

Following these ideas, we conducted high S/N IRAM 30 m observations of ^{12}CO J=1-0 and J=2-1, and ^{13}CO J=1-0, as well as Yebes 40 m SiO J=1-0 observations, in a volume-limited unbiased sample of well characterized SRs. We selected all the variables in the General Catalog of Variable Stars (GCVS; Samus *et al.* 2012) with declinations above -25°, and with information on the spectral/chemical type and variability period, for which the Hipparcos parallaxes (van Leeuwen 2007) are larger than 2 mas (all sources are closer than 500 pc; note that Gaia parallaxes were not available by the time we started the project). Moreover, we have further refined our sample by selecting only targets with IRAS 60 μm fluxes larger than 4 Jy (and with 12, 25, and 60 μm fluxes of quality 3; IRAS PSC 1988). This resulted in a list of 49 sources. We characterized the main properties of the CSEs, such as expansion velocity, mass loss rate, wind density, etc.

3. Preliminary results

Lines were classified into four groups based on the ^{12}CO line profile. Type 1 profiles are broad and symmetric lines with expansion velocities between 9 and 13 km s^{-1}. Type 2 profiles are narrow and symmetric lines with expansion velocities between 3 and 9 km s^{-1}. Type 3 profiles, however, are very strange line profiles, with very pronounced asymmetries and not compatible with standard optically thin lines already studied, with expansion velocities between 9 and 17 km s^{-1}. Finally, type 4 profiles are those lines which show two different components: one narrow component with low expansion velocity and another broad component with higher expansion velocity (like the addition of type 2+3 profiles).

We derived gas mass loss rates as discussed in Knapp & Morris (1985), by considering optically thick and optically thin envelopes. We also know that the envelope size is limited by the photodissociation of CO by the ambient interstellar radiation field and $T_{\rm mb}$ as a function of the photodissociation radius is different for optically thick and optically thin cases. For optically thick cases, this variation depends a little on $\dot{M}$ and abundance

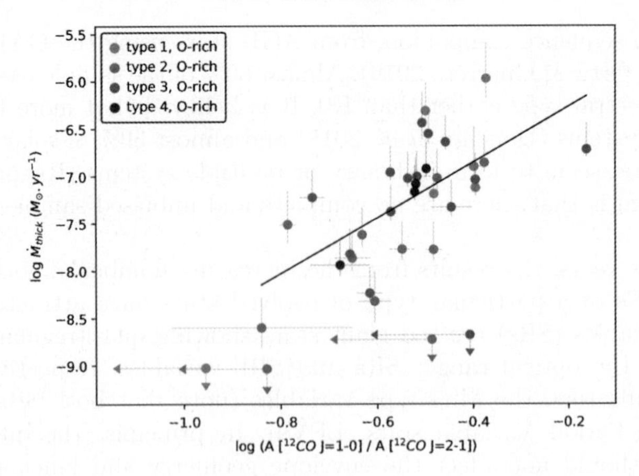

Figure 1. Mass loss rate vs. ^{12}CO(1-0)-to-^{12}CO(2-1) line intensity ratio in oxygen-rich SRs. Note the moderate correlation between the two quantities (r=0.66).

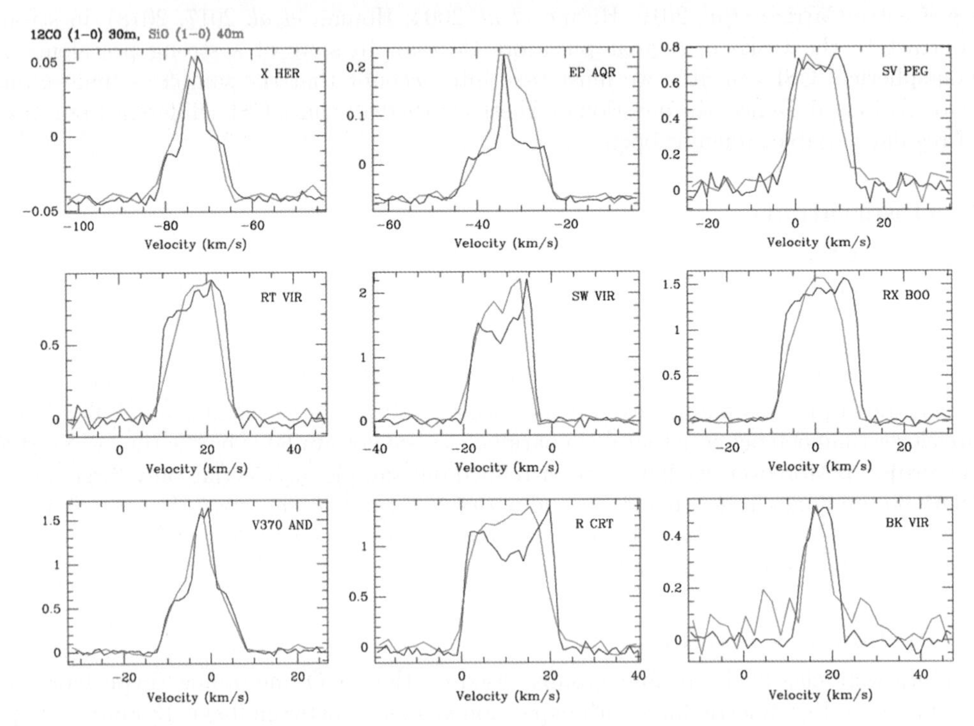

Figure 2. IRAM 30 m ^{12}CO J=1-0 (in black) and Yebes 40 m SiO J=1-0 (in red) spectra from the 9 sources where SiO is detected with high S/N ratio. ^{12}CO units are in T_{mb} scale while SiO spectra have been scaled to fit the box; note the similarities between these profiles in some sources in spite of totally different chemical distributions of the two species.

of CO. The photodissociation radius has been estimated by using the latest results of Groenewegen (2017), which take into account the strength of the interstellar radiation field. Therefore, we calculated consistently mass loss rates and photodissociation radii by numerically solving the system in both cases.

We found a moderate correlation between mass loss rates and ^{12}CO line intensity ratio for O-rich SRs (r=0.66; see Fig. 1). For each star the optically thick estimate of mass loss rate is aproximately two times higher than the optically thin estimate. The correlation

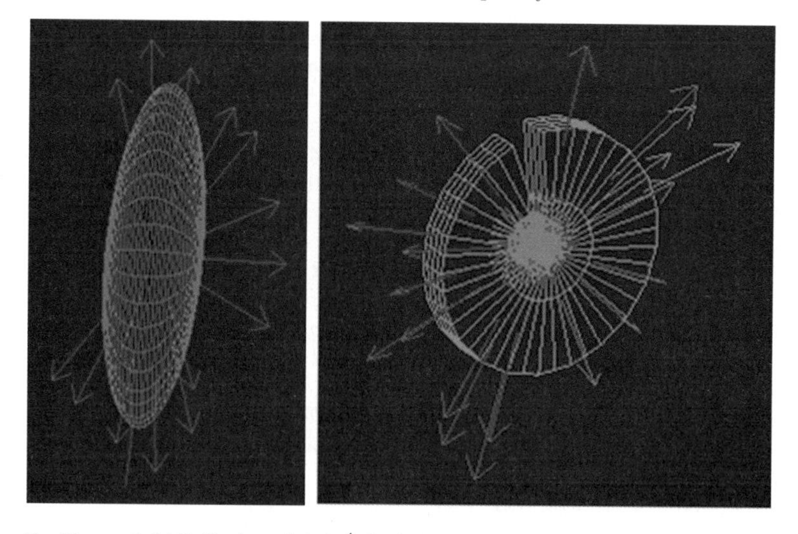

Figure 3. Types 1 (AQ Sgr) and 3 (τ^4 Ser) line profiles may be produced by spheroids and cylinders, respectively.

thus holds between mass loss rates and ^{12}CO line intensity ratio. Therefore, we introduce a new possible method for determining mass loss rates with short analysis time.

The ^{12}CO(1-0)-to-^{12}CO(2-1) intensity ratio is, at the same time, correlated with the wind density ($\dot{M}/v_e$; r=0.63) and the photodissociation radius (R_{CO}; r=0.64) for O-rich SRs. These results are in good agreement with the correlation between $\dot{M}$ derived from CO observations and dust infrared properties reported by Loup *et al.* (1993). Concerning C-rich SRs, the correlation between $\dot{M}$ and the ^{12}CO(1-0)-to-^{12}CO(2-1) intensity ratio, as well as dust infrared properties, is less clear. The observed behavior of $\dot{M}$ is probably explainable by the effects of saturation of T_{mb}(1-0) for large mass loss rates and low kinetic temperature. There could also be possible variations of the gas-to-dust ratio or changes in $\dot{M}$.

We found that for several stars the SiO line profile is very similar to the CO line profile, an result that is not viable unless these are non-standard envelopes (see Fig. 2).

According to SHAPE+SHAPEMOL (Steffen *et al.* 2011; Santander-García *et al.* 2015), types 1 and 2 profiles may be produced by thick and flattened spheroids, respectively. Type 3, however, may be produced by toruses and cylinders (see Fig. 3). Type 4 profiles may be produced by very complex structures. Types 3 and 4 may harbor binary systems ($\sim$41% of the sample).

References

Balick, B., & Frank, A. 2002, *ARA&A*, 40, 439

Castro-Carrizo, A., Quintana-Lacaci, G., Neri, R., *et al.* 2010, *A&A*, 523, A59

Douchin, D., De Marco, O., Frew, D. J., *et al.* 2015, *MNRAS*, 448, 3132

Groenewegen, M. A. T. 2017, *A&A*, 606, A67

Hirano, N., Shinnaga, H., Dinh-V-Trung, *et al.* 2004, *ApJ*, 616, L43

Homan, W., Richards, A., Decin, L., Kervella, P., de Koter, A., McDonald, I., Ohnaka, K. 2017, *A&A*, 601, A5

Homan, W., Richards, A., Decin, L., de Koter, A., Kervella, P. 2018, *A&A*, 616, A34

IRAS, 1988, *The Point Source Catalogue*, version 2.0, NASA RP-1190

Jones, D., & Boffin, H. M. J. 2017, *NatAs*, 1, 0117

Kim, H., Liu, S.-Y., Hirano, N., *et al.* 2015, *ApJ*, 814, 61

Knapp, G. R., & Morris, M. 1985, *ApJ*, 292, 640

Loup, C., Forveille, T., Omont, A., Paul, J. F. 1993, *A&AS*, 99, 291
Maercker, M., Mohamed, S., Vlemmings, W. H. T., *et al.* 2012, *Nature*, 490, 232
Mauron, N., & Huggins, P. J. 2006, *A&A*, 452, 257
Morris, M., Sahai, R., Matthews, K., *et al.* 2006, *in Planetary Nebulae in our Galaxy and Beyond*, ed. M. J. Barlow, & R. H. Mndez, IAU Symp., 234, 469
Neri, R., Kahane, C., Lucas, R., *et al.* 1998, *A&AS*, 130, 1
Olofsson, H. 1999, *in IAU Symposium* 191, 3
Ortiz, R., & Guerrero, M. A. 2016, *MNRAS*, 461, 3036
Parker, Q. A., Acker, A., Frew, D. J., *et al.* 2006, *MNRAS*, 373, 79
Raghavan, D., McAlister, H. A., Henry, T. J., *et al.* 2010, *ApJS*, 190, 1
Sahai, R., Findeisen, K., Gil de Paz, A., *et al.* 2008, *ApJ*, 689, 1274
Samus, N. N., Durlevich, O. V., Kazarovets, E. V., *et al.* 2012, *GCVS*, VizieR On-line Data Catalog: B/gcvs
Santander-García, M., Bujarrabal, V., Koning, N., Steffen, W. 2015, *A&A*, 573, A56
Steffen, W., Koning, N., Wenger, S., Morisset, C., Magnor, M. 2011, *IEEE Transactions on Visualization and Computer Graphics*, 17, 454
van Leeuwen, F. 2007, *A&A*, 474, 653

Why Galaxies Care About AGB Stars:
A Continuing Challenge through Cosmic Time
Proceedings IAU Symposium No. 343, 2019
F. Kerschbaum, M. Groenewegen & H. Olofsson, eds.

doi:10.1017/S1743921318005483

The Impact of UV Radiation on Circumstellar Chemistry

Maryam Saberi[1], Wouter Vlemmings[1], Tom Millar[2] and Elvire De Beck[1]

[1]Dep. of Space, Earth and Environment, Chalmers University of Technology & Onsala Space Observatory, 43992 Onsala, Sweden
email: maryam.saberi@chalmers.se

[2]School of Maths and Physics, Queen's University Belfast, University Road, Belfast BT7 1NN, Northern Ireland

Abstract. UV radiation plays a critical role in the chemistry of circumstellar envelopes (CSEs) around evolved stars on the asymptotic giant branch (AGB). However, the effects of all potential sources of UV radiation have not been included in models. We present preliminary results of adding an internal source of UV to the CSE chemistry and predict large enhancements of atomic and ionic species arising from photo-destruction of parent species. Observations of atomic carbon towards the UV-bright AGB star *o* Cet are consistent with the modelling. In addition, we calculate the precise depth dependence of the CO photodissociation rate in an expanding CSE. We incorporate this within a chemical network active in the outflows of AGB stars, which includes 933 species and 15182 reactions. Our results show that the CO envelope size is about 30% smaller at half abundance than the most commonly used radius reported by Mamon *et al.* (1988).

Keywords. astrochemistry, stars: abundances, stars: AGB and post-AGB, stars: carbon, stars: chromospheres, stars: circumstellar matter, ultraviolet: stars, binaries: general, radiative transfer

---

1. Introduction

In the standard paradigm of a CSE, photodissociation by external UV radiation from the interstellar radiation field has been considered to be the agent that controls molecular photodestruction. However, there is now ample evidence for the presence of extra UV radiation which permeates the inner part of a CSE. Recent GALEX observations revealed 180 asymptotic giant branch (AGB) stars (57% of the observed sample) have detectable far- and/or near-UV emission (Montez *et al.* 2017), indicating the presence of extra UV radiation inside the CSE. The internal UV radiation can originate from stellar chromospheric activity, a hot binary companion, and/or the accretion of material between two stars in a binary system. To constrain the effects of both internal and external UV radiation in the chemical modelling of CSEs, observations of main photodissociation/photoionization products are required. Moreover, comparison of isotopologue ratios of molecular species such as $^{12}CO/^{13}CO$ and $H^{12}CN/H^{13}CN$, which dissociate through different mechanisms, can be used to trace the UV-chemistry (Saberi *et al.* 2017).

2. The isotopic chemistry

To model the chemistry active in the CSE of an AGB star we use an extended version of the rate13-cse code (McElroy *et al.* 2013). In the extended version, we apply the updated version of the UMIST06 chemistry file which includes the ^{13}C and ^{18}O isotopes, all

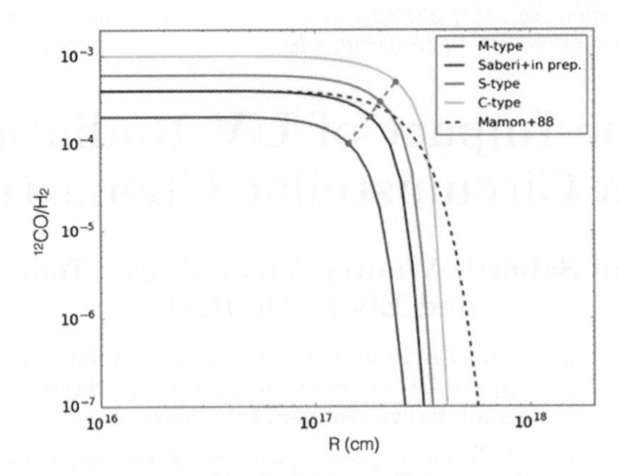

Figure 1. Fractional abundance distributions of CO for models with different initial abundances of f_{CO} which are usually adopted for M-, S-, and C-type AGB stars and the value of the Mamon *et al.* (1988) standard model. The black dashed line shows the Mamon standard model. Comparison of the black solid and black dashed lines shows the difference between our model and that of Mamon *et al.* (1988).

corresponding isotopologues, their chemical reactions, and the properly scaled reaction rate coefficients from Röllig & Ossenkopf (2013). The chemical network includes 933 species and 15182 gas-phase reactions (Saberi *et al.*, in prep.).

3. The CO envelope size

Photodissociation is the dominant process which determines the abundances and distributions of CO and its photodissociation products such as C and C^+ throughout the CSE. The depth dependence of the CO photodissociation mechanism in an expanding spherical CSE was presented by Morris & Jura (1983) through a "1-band approximation". After new laboratory measurement of far-UV absorption and fluorescence cross sections by Letzelter *et al.* (1987), the rate was updated by Mamon *et al.* (1988) considering 34 dissociating bands. In two recent works by Li *et al.* (2014) and Groenewegen (2017), updated CO photodissociation rates in a CSE were presented. However, in both cases they used shielding functions calculated for the ISM by Visser *et al.* (2009). Here, we derive the depth dependence of the photodissociation rate for all dissociating CO lines taking into account the physical properties of a CSE (Saberi *et al.* in prep.). Our derived CO envelope size at half the abundance ($r_{1/2}$) is about 30% smaller compared to that calculated by Mamon *et al.* (1988). Moreover, we find that the CO envelope size depends not only on the mass loss rate and the expansion velocity of the envelope but also on the CO initial fractional abundance f_{CO}. Figure 1 shows how the CO distribution profiles varies for models with different values of f_{CO}. We used the commonly used f_{CO} for M-type, S-type, and C-type AGB stars, and that being used in the standard model of Mamon *et al.* (1988).

4. The internal UV field

The internal UV radiation can arise from a hot binary companion, stellar chromospheric activity, or accretion of matter between two stars in a binary system. This UV field is expected to cause significant photodissociation of CO isotopologues in the inner part of the CSE. This, in turn, has important effects on the chemistry of the gas. The preliminary results of adding an internal source of UV radiation (a blackbody with $T = 18000\,K$) is shown in Fig. 2. As can be seen, there is an enhancement of C and C^+ in the inner CSE. There is also a variation in the fractional abundance distributions of the CO isotopologues, meaning that the circumstellar $^{12}CO/^{13}CO$ ratio can not be a reliable tracer of

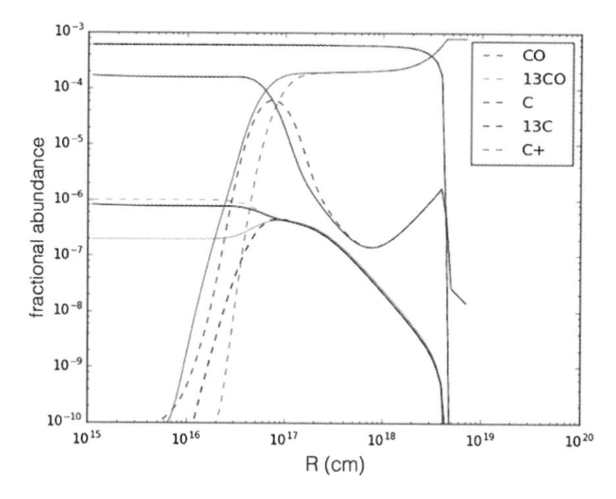

Figure 2. The fractional abundance distributions of CO, C, and C^+ before (dashed lines) and after (solid lines) adding an internal source of UV radiation.

the photospheric $^{12}C/^{13}C$ ratio in the UV-dominated parts of a CSE. A further detailed analysis of the consequences of the presence of the internal UV radiation on the CSE chemistry is in progress (Saberi *et al.* in prep.).

5. An observational tracer of the internal UV radiation

5.1. *The photo-destruction products*

To empirically constrain the effects of internal and external UV radiation sources, observations of the main photo-destruction products, e.g., C and C^+, are required. To date, C has been detected in four post-AGB stars (Knapp *et al.* 2000, and references therein) and three carbon-type AGB stars IRC+10216 (Keene *et al.* 1993; van der Veen *et al.* 1998), R Scl (Olofsson *et al.* 2015), and (tentatively) V Hya (Knapp *et al.* 2000, Saberi *et al.* 2018). Moreover, Saberi *et al.* (2018) reported the detection of C emission from the oxygen-rich AGB star *o* Cet (Fig. 3), and argued that it arises from a compact region near its UV-bright binary companion. To extend our analysis, we have applied for C and C^+ observations with ALMA and SOFIA towards a larger sample of UV-bright AGB stars.

5.2. *The isotopologue ratio*

ALMA observations of R Scl show a discrepancy between the $^{12}CO/^{13}CO$ ratio in the present mass loss (> 60) and the detached shell (~ 19) (Vlemmings *et al.* 2013). The ratio in the inner wind is surprisingly much higher than in the outer region. This may indicate more effective photodissociation by the interstellar radiation field, penetrating deeper into the CSE. More likely, an active binary companion or an active chromosphere of the AGB star would provide an additional source of UV radiation in the inner CSE. The presence of atomic carbon in the inner CSE of R Scl, as reported by Olofsson *et al.* (2015), strengthens the likelihood of the presence of an internal source of UV radiation. Chemical models which consider an internal UV field can explain the variation of the $^{12}CO/^{13}CO$ throughout the CSE. On the other hand, from a detailed excitation analysis of $H^{12}CN$ and $H^{13}CN$, Saberi *et al.* (2017) found a ratio of $H^{12}CN/H^{13}CN \sim 26 \pm 12$ close to the star, in agreement with the photospheric estimate of $^{12}C/^{13}C \sim 19 \pm 6$ reported by Lambert *et al.* (1986). Since CO isotopologues are photodissociated by the UV radiation in well-defined bands and HCN is dissociated in the continuum, we suggested that isotope-selective photodissociation of ^{12}CO and ^{13}CO causes this discrepancy (Saberi *et al.* 2017).

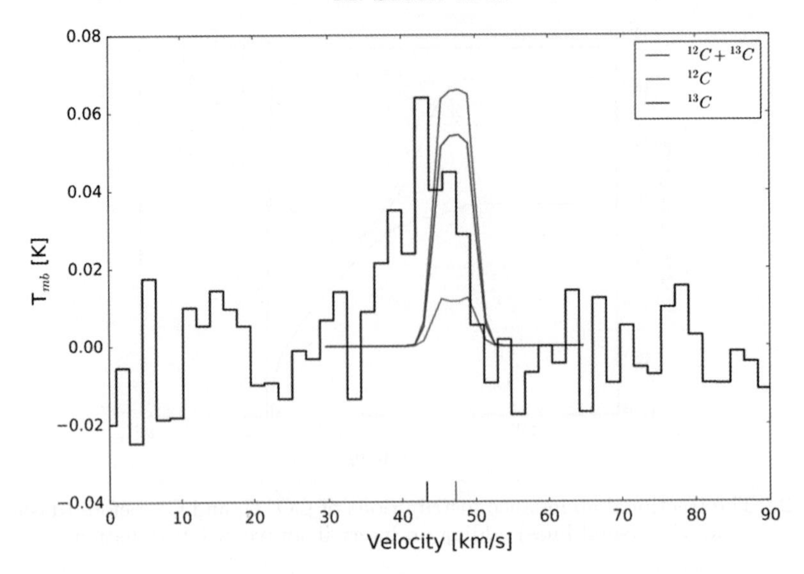

Figure 3. The CI emission towards *o* Cet, 1.8 km/s velocity resolution, in LSR scale (histogram). The stellar, 47.2 km/s, and spectrum peak, 43.4 km/s, velocities are indicated as blue and black vertical markers on the x-axis, respectively. The results of the radiative transfer modelling of the C isotopes are shown in red and green, respectively. The blue profile indicates the total amount of C (from Saberi *et al.* 2018).

This implies that the circumstellar $H^{12}CN/H^{13}CN$ ratio might be a better tracer of the photospheric $^{12}C/^{13}C$ ratio in UV-irradiated CSEs. To confirm this, a more extended analysis towards more UV-bright AGB stars is required.

References

Groenewegen, M. A. T. 2017, *A & A*, 606, 67
Keene, J., Young, K., Phillips, T. G., *et al.* 1993, *ApJ*, 415, 131
Knapp, G. R., Crosas, M., Young, K., Ivezić, Ž 2000, *ApJ*, 534, 324
Lambert, D. L., Gustafsson, B., Eriksson, K., Hinkle, K. H. 1986, *ApJ*, 62, 373
Letzelter, C., Eidelsberg, M., Rostas, F., *et al.* 1987, *Chem. Phys.*, 114, 273
Li, X, Millar, T. J., Walsh, C., *et al.* 2014, *A & A*, 568, 111
McElroy, D., Walsh, C., Markwick, A. J., *et al.* 2013, *A & A*, 550A, 36
Montez, R. Jr., Ramstedt, S., Kastner, J. H., *et al.* 2017, *ApJ*, 841, 33
Morris, M., & Jura, M. 1983, *ApJ*, 267, 179
Olofsson, H., Bergman, P., Lindqvist, M 2015, *A & A*, 582, 102
Röllig, M., Ossenkopf, V. 2013, *A & A*, 550A, 56
Saberi, M., Maercker, M., De Beck, E., *et al.* 2017, *A & A*, 599A, 63
Saberi, M., Vlemmings, W. H. T., De Beck, E., *et al.* 2018, *A & A*, 612, L11
van der Veen, W. E. C. J.; Huggins, P. J.; Matthews, H. E. 1998, *ApJ*, 505, 749
Visser, R., van Dishoeck, E. F., Black, J. H. 2009, *A & A*, 503, 323
Vlemmings, W. H. T., Maercker, M., Lindqvist, M., *et al.* 2018, *A & A*, 556, 1

Discussion

DECIN: How exactly have you calculated the CO photodissociation radius? You presented your new chemical network, but I do not understand your link with the photodissociation radius.

SABERI: I calculate the depth dependence of the CO photodissociation rate for all CO dissociating lines in a spherical envelope which expands with a constant velocity $v_{\rm exp}$. These calculations have been done for the ISM by Visser *et al.* (2009). This is the first time that we calculate this for CSEs. Mamon *et al.* (1988) calculated the CO photodissociation rate using a "band-approximation" considering 34 dissociating bands. Groenewegen (2017) updated the CO photodissociation rate, however, he used the shielding functions that were calculated for the ISM. In our work, we calculate the CO photodissociation rate considering the physical properties of the CSE and explained the environment dependency of the CO photodissociation rate.

DECIN: I do not understand the picture you presented in R Scl where a binary or chromospheric activity can explain the variation in the $^{12}CO/^{13}CO$ fraction in the wind of R Scl. Once ^{13}CO is photo-dissociated (and hence $^{12}CO/^{13}CO$ is 60), it will not recombine further out in the wind due to the low gas density. Hence no explanation is offered for the $^{12}CO/^{13}CO \sim 19$ of the detached shell.

SABERI: Considering penetration of the standard UV field from the ISM, we expect a constant $^{12}CO/^{13}CO$ throughout the CSE which is constant with the photospheric $^{12}C/^{13}C \sim 19$. In case of R Scl, we got a ratio $^{12}CO/^{13}CO \sim 19$ in a detached shell which is consistent with the photospheric $^{12}C/^{13}C$, however the ratio increases $^{12}CO/^{13}CO > 60$ in the inner CSE (present mass loss). To explain the lack of ^{13}CO in the inner part, we speculate that the UV radiation from the binary companion, or stellar activity of R Scl itself, destroys most of the ^{13}CO while the most abundant ^{12}CO gets self-shielded (isotope-selective photodissociation).

Why Galaxies Care About AGB Stars:
A Continuing Challenge through Cosmic Time
Proceedings IAU Symposium No. 343, 2019
F. Kerschbaum, M. Groenewegen & H. Olofsson, eds.

doi:10.1017/S1743921318005914

A Tough Egg to Crack

Jeremy Lim[1] and Dinh-van-Trung[2]

[1]Department of Physics, The University of Hong Kong, Pokfulam Road, Hong Kong
email: jjlim@hku.hk

[2]Institute of Physics, Vietnam Academy of Science and Technology, 10 DaoTan, ThuLe, BaDinh, Hanoi, Vietnam
email: dvtrung@iop.vast.ac.vn

Abstract. The sculpting of the Egg Nebula continues to defy a coherent explanation. Bipolar outflows from the center of the nebula have created bipolar optical lobes that are illuminated by searchlight beams; multiple bipolar outflows orthogonal to the lobes create the appearance of a dark lane; and quasi-circular arcs are imprinted on an approximately spherically-symmetric wind from the progenitorAGB-star. Here, we use archival data from ALMA to study at high angular resolution dust and molecular gas at the center of the nebula. We find that: (i) dust is concentrated in multiple blobs that outline the base of the northern optical lobe; (ii) dense molecular gas forms the wall of a channel swept up and compressed by the outflows that created the bipolar optical lobes; (iii) the expansion and illumination center of the nebula lies at or close to center of the outflow channel. We present a simple working model for the Egg Nebula, and highlight the difficulties that any model face for explaining all the features seen in this nebula.

---

1. Introduction

Figure 1 shows a 1.1 μm/1.6 μm image of the Egg Nebula taken with the HST/WFC3 by Balick *et al.* (2012). The most prominent feature is a pair of optical lobes, which have an expansion age of ~250 yrs (Ueta *et al.* 2006; Balick *et al.* 2012). From HST/WFC2 images, Sahai *et al.* (1998a) had earlier found that the bipolar optical lobes are illuminated at their edges by a pair of searchlight beams emanating from an obscured post-AGB star. They suggested that a precessing bipolar outflow has punched annular holes in a dust cocoon around this star, through which light from the star escapes. This outflow also swept up material in the surrounding envelope to create the bipolar lobes seen in scattered light. At the outer tips of both bipolar optical lobes, Sahai *et al.* (1998b) found line emission at 2.122 μm from molecular hydrogen (H_2). They suggest that the H_2 emission originates from shocked gas at the interface where the bipolar outflow slams into a slower wind ejected when the star was on the AGB. Confirming but also complicating this picture, observations in CO(2-1) by Cox *et al.* (2000) reveal not just one but multiple bipolar outflows along the bipolar optical lobes. Indeed, studies of the spatial expansion of the bipolar optical lobes by Ueta *et al.* (2006) reveal that these lobes actually consist of multiple outflows at distinct inclination angles projected onto each other.

Sahai *et al.* (1998a) found that the bipolar optical lobes of the Egg Nebula are bifurcated by a dark lane that is oriented perpendicular to the symmetry axis of these lobes. This dark lane is commonly attributed to a large equatorial disk of cold dust (Sahai *et al.* 1998b; Weintraub *et al.* 2000). Observations in the continuum at 1.3mm and CO(2-1) by Cox *et al.* (2000), however, fail to detect any emission from either dust or molecular gas in the putative equatorial dust disk. In addition to the 2.122 μm H_2 emission at the outer

Figure 1. A 1.1 μm/1.6 μm image of the Egg Nebula taken with the HST/WFC3 by Balick *et al.* (2012).

tips of the bipolar optical lobes as described above, Sahai *et al.* (1998b) also discovered 2.122 μm H_2 emission on the eastern and western sides of the dark lane. Subsequent observations in CO(2-1) by Cox *et al.* (2000) showed that this H_2 emission coincides with the tips of multiple bipolar molecular outflows oriented roughly orthogonal to the bipolar optical lobes. Thus, just like the 2.122 μm H_2 emission along the bipolar optical lobes, that along the dark lane also originates from shocked gas where (multiple) bipolar outflows slam into the previously ejected AGB-star wind. The central region of the Egg nebula is therefore a source of multiple bipolar outflows along the bipolar optical lobes, as well as multiple bipolar outflows roughly orthogonal to these lobes. How such multiple bipolar outflows having essentially orthogonal orientations arise is one of the great mysteries about the Egg nebula.

In stark contrast to the strongly bipolar morphology described above, Sahai *et al.* (1998a) found a series of quasi-circular concentric arcs imprinted on the optical nebula. These arcs represent approximately equally-spaced density enhancements (by a factor of $\sim$3 over the inter-arc region) spaced in time by $\sim$100 yrs (Sahai *et al.* 1998a; Balick *et al.* 2012). Such arcs are most commonly attributed to the reflex motion of the mass-losing star (in this case, the progenitor AGB star) due to a binary companion, resulting in periodic density enhancements in the wind of the mass-losing star. The morphology and kinematics of the envelope on which the arcs are imprinted has been studied in CO(1-0) by Fong *et al.* (2006), who found the envelope to be roughly spherically symmetric and expanding at a velocity of about 14 km s^{-1}. The bipolar optical lobes are embedded in this envelope, representing features produced by the associated bipolar outflows.

Here, we present observations with ALMA that reveal the position of the source responsible for the bipolar outflows along the bipolar optical lobes. By aligning the ALMA and HST images, we address whether the source of these outflows is spatially coincident with the expansion and illumination center of the nebula. We then advance the simplest working model we can think of for the Egg Nebula: rather than to convince, our purpose is to highlight the inherent difficulties any model face for explaining both the quadrupolar outflows and arcs observed in this proto-planetary nebula. The discrepancies between model predictions and the observed features also plague other bipolar proto-planetary which are not complicated by an additional set of orthogonal bipolar outflows.

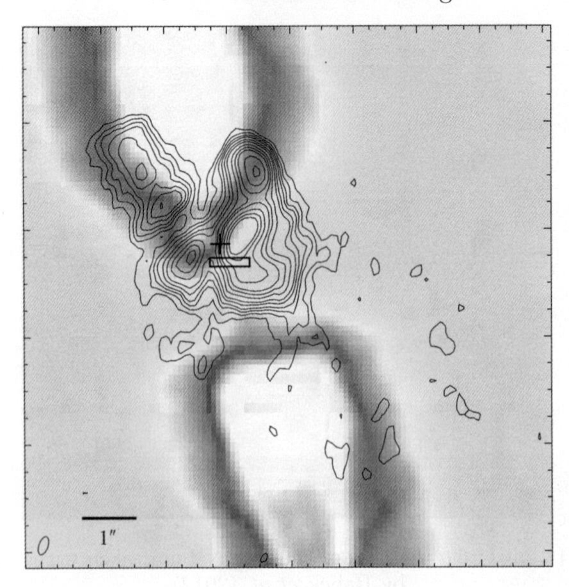

Figure 2. ALMA 1.1 mm dust continuum image (contours) superposed on the HST/WFC3 1.6 μm image (color). The cross marks the expansion center of the nebula determined by Ueta *et al.* (2006), and has arm lengths corresponding to the $\pm 1\sigma$ uncertainty of the registration between the ALMA and HST images. The rectangle indicates the illumination center as derived by Weintraub *et al.* (2000) from the linear polarization of scattered light, and has sides corresponding to their $\pm 1\sigma$ measurement uncertainties.

2. Results

Figure 2 shows an ALMA 1.1 mm dust continuum image (contours) superposed on the HST/WFC3 1.6 μm image (color). To make this overlay, we shifted the ALMA image to the epoch of the 1.6 μm image (2009 October) using the proper motion derived by Ueta *et al.* (2006) of $\alpha = (13.7 \pm 2.0)$ mas yr^{-1} and $\delta = (10.22.0)$ mas yr^{-1}. The expansion center of bipolar lobes as deduced by Ueta *et al.* (2006) is indicated by a cross, which has arm lengths corresponding to the $\pm 1\sigma$ uncertainty of our registration. The illumination center as derived by Weintraub *et al.* (2000) from the linear polarization of scattered light is indicated by a rectangle, which have sides corresponding to their $\pm 1\sigma$ measurement uncertainties. These positions of the expansion and illumination center agree to within about 2σ of the measurement uncertainties. As can be seen, the dust is concentrated in several discrete blobs, none of which coincide with either the expansion or illumination center of the nebula. Instead, the dust blobs seem to outline the base of the northern optical lobe, which is tilted towards us.

Figure 3 shows the same ALMA 1.1 mm dust continuum image (contours) but now superposed on selected ALMA HCN(3-2) channel maps (color). At the systemic velocity (middle panel), the HCN(3-2) emission coincides with the dust blobs. Evidently, both the dust and molecular gas traced in HCN(3-2) comprises material swept up and compressed by a bipolar outflow, which itself – presumably having a much higher velocity than the bipolar molecular outflows – has so-far escaped detection. The expansion center of the nebula – marking the position of the progenitor AGB star – lies precisely at the center of the molecular gas channel carved out by the bipolar outflow. At blueshifted velocities (right panel), the HCN(3-2) emission resembles a ring, as has been seen also in CO(2-1) by Cox *et al.* (2000). The expansion center of the nebula lies at or close to the center of the ring. This ring traces a spherical shell that corresponds to an episode of highly enhanced mass-loss ~200 yrs ago, presumably marking the end of the AGB phase. No ring can be

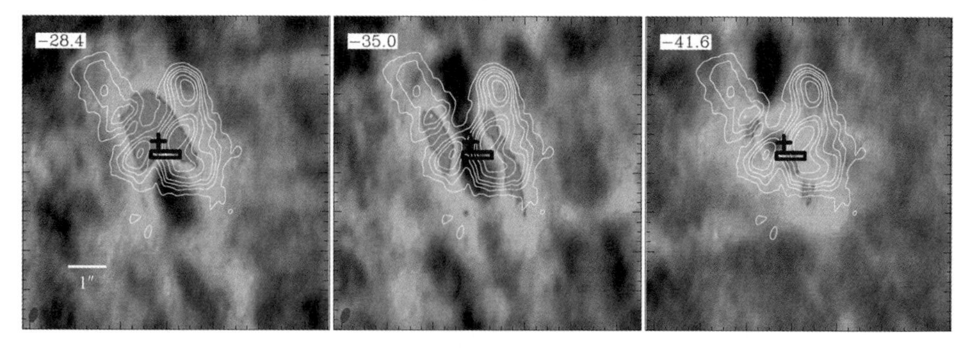

Figure 3. ALMA 1.1 mm dust continuum image (contours) now superposed on selected ALMA HCN(3-2) channel maps (color). The middle panel is at the systemic velocity of the Egg Nebula.

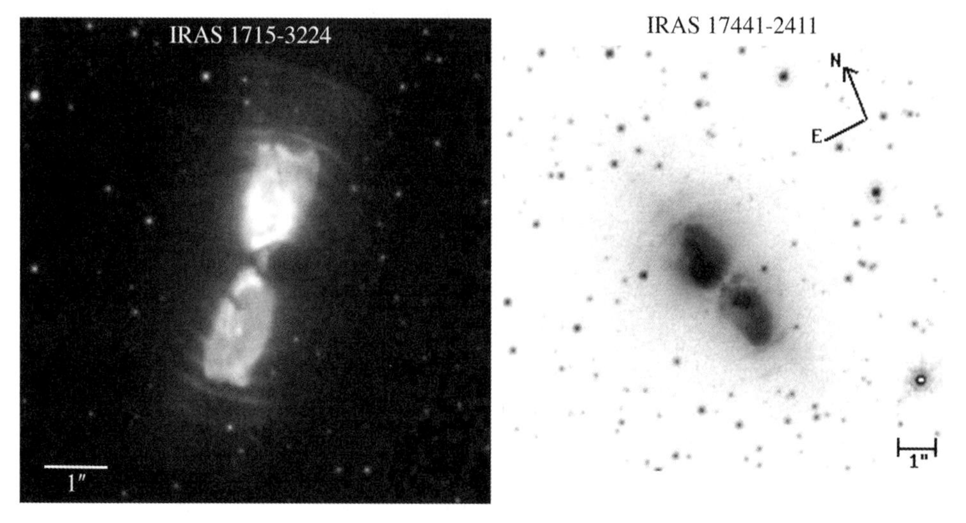

Figure 4. Optical images of two bipolar proto-planetary nebulae that also exhibit circular arcs.

seen in the images at redshifted velocities owing to blending with the swept-up emission in HCN(3-2), but this portion of the ring is clearly revealed in position-velocity diagrams (not shown here).

3. Model Challenges

Figure 3 clearly shows that the progenitor AGB star lies at or close to the source of the bipolar outflow that created the bipolar optical lobes. Sahai *et al.* (1998a) suggested that the annular cavities through which light escapes to create the searchlight beams were carved into the cocoon around the illuminating star by a precessing bipolar outflow from this star. In the picture where the arcs are caused by the reflex motion of the progenitor AGB star by a binary companion, it would be natural to attribute the bipolar outflows along the dark lane to this companion. If the axis of this outflow is perpendicular to the orbital plane, then the orbital plane of this companion would have to be approximately perpendicular to the dark lane. Theoretical models predict that arcs should be interleaved about the orbital axis and hence dark lane - no such interleaved arrangement of arcs, however, is observed along this or any other direction. To be consistent with theoretical predictions, the orbit of the binary companion would have to be in or close to the plane of the sky. If this companion is responsible for the bipolar outflows along the dark lane, then the circumstellar disk around this companion (from which the bipolar outflow of this star is presumably driven) would then have to be perpendicular to the orbital plane, a seemingly unnatural configuration if the material comprising the circumstellar disk of this star was captured from the progenitor AGB star.

A similar problem plagues this model as applied to other bipolar proto-planetary nebulae, but which do not have the complexity of an additional set of bipolar outflows perpendicular to the bipolar optical lobes. In Fig. 4, we show two examples, both of which have clear circular arcs. Once again, for a binary companion to produce these arcs, the orbital plane of the companion would have to be in or close to the plane of the sky, whereas the circumtellar disk from which the outflow originates to produce the bipolar lobes would have to be oriented in a perpendicular direction.

References

Balick, B., Gomez, T., Vinkocíc, D., *et al.* 2012, *ApJ*, 745, 188
Cox, P., Lucas, R., Huggins, P.J., *et al.* 2000, *A&A*, 353, L25
Fong, D., Meixner, M., Sutton, E.C., *et al.* 2006, *ApJ*, 652, 1626
Sahai, S., Trauger, J.T., Watson, A.M., *et al.* 1998, *ApJ*, 493, 301
Sahai, S., Hines, D.C., Kastner, J.H., *et al.* 1998, *ApJ*, 492, L163
Ueta, T., Murakawa, K., & Meixner, M. 2006, *ApJ*, 641, 1113
Weintraub, D.A., Kastner, J.H., & Hines, D.C, *et al.* 2000, *ApJ*, 531, 401

Discussion

D'ANTONA: I wonder whether your presumed binary companion of $0.2\,M_\odot$ with a wide separation would mean that there are different paths to the asymmetric PN, one involving common envelope evolution and the other one avoiding it?

LIM: In my talk, I considered the possibility of a close binary companion driving the bipolar outflows that carved out the bipolar optical lobes, and another more distant companion to drive the bipolar outflows that coincide with the dark lane - one of the many models we considered to come up with a coherent picture for all the features observed in the Egg Nebula. Of course, this model does not exclude common envelope evolution, that is between the AGB star and its nearby binary companion.

Why Galaxies Care About AGB Stars:
A Continuing Challenge through Cosmic Time
Proceedings IAU Symposium No. 343, 2019
F. Kerschbaum, M. Groenewegen & H. Olofsson, eds.

doi:10.1017/S1743921318005446

Newly discovered Planetary Nebulae population in Andromeda (M31): PN Luminosity function and implications for the late stages of stellar evolution

Souradeep Bhattacharya[1]†, Magda Arnaboldi[1], Johanna Hartke[1], Ortwin Gerhard[2], Valentin Comte[1,3], Alan McConnachie[4] and William E. Harris[5]

[1]European Southern Observatory, Karl-Schwarzschild-Str. 2, 85748 Garching, Germany
email: sbhattac@eso.org

[2]Max-Planck-Institut für Extraterrestrische Physik, Giessenbachstrasse, 85748 Garching, Germany

[3]Aix Marseille Université, CNRS, LAM Laboratoire d'Astrophysique de Marseille, 38 rue F. Joliot-Curie, 13388 Marseille, France

[4]NRC Herzberg Institute of Astrophysics, 5071 West Saanich Road, Victoria, BC V9E 2E7, Canada

[5]Department of Physics & Astronomy, McMaster University, Hamilton, ON L8S 4M1, Canada

Abstract. Stars with masses between ~ 0.7 and 8 $M_\odot$ end their lives as Planetary Nebulae (PNe). With the MegaCam at CFHT, we have carried out a survey of the central 16 sq. degrees of Andromeda (M31) reaching the outer disk and halo, using a narrow-band [OIII]5007 and a broad-band *g* filter. This survey extends previous PN samples both in uniform area coverage and depth. We identify ~4000 PNe in M31, of which ~3000 are new discoveries. We detect PNe down to ~6 mag below the bright cut-off of the PN luminosity function (PNLF), ~2 mag deeper than in previous works. We detect a steep rise in the number of PNe at ~4.5 mag fainter than the bright cut-off. It persists as we go radially outwards and is steeper than that seen in the Magellanic clouds. We explore possible reasons for this rise, which give insights into the stellar population of M31.

Keywords. galaxies: individual (M31), planetary nebulae: general

1. Introduction

Having recently left the AGB, PNe are observed to be the glowing shells of gas and dust around stars that are evolving towards the white dwarf stage. The short timescales between the AGB and PN phases imply that the distribution and kinematics of PNe are expected to be drawn from the line-of-sight velocity (LOSV) distribution of their parent population. Due to their relatively strong [OIII] 5007 Å emission and no continuum

† Based on observations obtained with MegaCam, a joint project of CFHT and CEA/DAPNIA, at the Canada-France-Hawaii Telescope (CFHT) which is operated by the National Research Council (NRC) of Canada, the Institut National des Science de l'Univers of the Centre National de la Recherche Scientifique (CNRS) of France, and the University of Hawaii. The observations at the CFHT were performed with care and respect from the summit of Maunakea which is a significant cultural and historic site.

emission, PNe can be easily identified in external galaxies as point sources and studying them as a population provides insight into galactic structure and evolution.

From the integrated [OIII] 5007 Å flux, the m_{5007} magnitude is measured. For an extragalactic PN population, we can then measure its luminosity function (hereafter PNLF) that is observed to have an absolute bright cut-off (Ciardullo *et al.* 1989; C89) mostly invariant with metallicity and age of the parent stellar population, and galaxy type. It is thus used as a secondary distance indicator for determining galactic distances out to ∼20 Mpc. The faint-end of the PNLF can be described by an exponential function (Jacoby 1980) expected from slowly evolving central stars embedded in rapidly expanding, optically thin nebulae (Henize & Westerlund 1963). The PNLF slope has been shown to be correlated with the star formation history of the parent stellar population, with steeper slopes associated with older stellar populations and flatter slopes with younger populations (Ciardullo *et al.* 2004; Longobardi *et al.* 2013).

M31 is the closest giant spiral disk to our Milky Way (MW) at a distance of ∼ 780 kpc. The Pan-Andromeda Archaeological Survey (PAndAS; McConnachie *et al.* 2009) map of the resolved stellar population number counts showed the wealth of substructures, which are residues of past accretion events, present in the M31 halo (see Ferguson & Mackey 2016). Merrett *et al.* (2006; M06) uniformly surveyed the disk and bulge of M31 down to ∼ 3.5 - 4 mag below the bright cut-off. A uniform deep survey of PNe in the halo of M31 and the inner-halo substructures is necessary not only to unambiguously trace the metal-poor halo and inner halo substructures, but also to probe the variation in the PNLF further out from the disk. This will allow us to corroborate the invariant nature of its bright cut-off and observe the effects on its faint end.

2. CFHT MegaCam M31 PNe survey

We survey the inner 16 sq. deg. of M31 (∼20-30 kpc from its center), covering the disk and the inner halo including its substructures. The observations were carried out with the MegaCam wide-field imager (Boulade *et al.* 2003) mounted on the 3.6-m Canada France Hawaii Telescope (CFHT). M31 is observed through a narrow-band [OIII] filter ($\lambda_c =$ 5007 Å, $\Delta\lambda = 102$ Å, on-band) and a broad-band g filter ($\lambda_c = 4750$ Å, $\Delta\lambda = 1540$ Å, off-band). The fields are shown in Fig 1. We used SExtractor (Bertin & Arnouts 1996), a source detection algorithm that detects and measures flux from point-like and extended sources, to detect and carry out photometry of the sources on the images.

Extragalactic PNe can be identified as point-like objects detected in the on-band images but not in the off-band images. To detect PNe, we used the automatic selection procedure developed and validated in Arnaboldi *et al.* (2002, 2003) and later adapted to larger surveys by Longobardi *et al.* (2013) and Hartke *et al.* (2017). We simulate synthetic point-like populations for each field following a PNLF described in C89 onto the on- and off-band images to determine the limiting magnitude of our sample for each field, remove contamination from continuum sources, and correct for incompleteness effects. We identify over 4000 PNe in M31, of which only ∼ 1100 were previously identified by M06. Our survey is uniformly complete down to 5.5 mag below the bright cut-off, reaching 6.4 mag below the bright cut-off in the deepest field. Details of the data reduction and PNe identification are described in Bhattacharya *et al.* (2018; in preparation; Bh+18)

We tested the astrometric quality of our PNe survey, and checked the contamination from HII regions which affected the M06 PNe sample (Sanders *et al.* 2012; Veyette *et al.* 2014) and other sources, by identifying PNe counterparts in the *Hubble Space Telescope* data available from the Panchromatic Hubble Andromeda Treasury (Dalcanton *et al.* 2012). The PHAT survey covers ∼ 1/3 of the star-forming disk of M31, using HSTs imaging cameras (WFC3/IR, WFC3/UVIS, and ACS/WFC cameras) in six bands. We

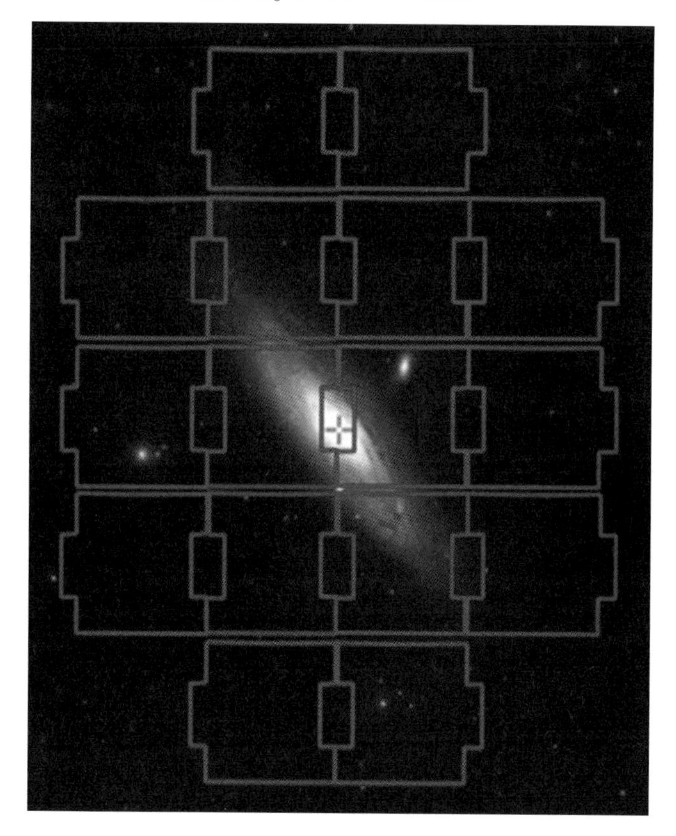

Figure 1. The fields observed with CFHT MegaCam shown in blue. The background image is from SDSS obtained using the Aladin Sky Atlas (Bonnarel *et al.* 2000).

find that of the PNe with counterparts in the PHAT sample, only $\sim 3\%$ may be stellar contamination but none of them were resolved HII regions, owing to the excellent resolution we have with MegaCam (Bh+18 for details).

3. Planetary Nebula Luminosity Function

Fig 2 shows the PNLF for PNe identified in one of the fields in our survey. The PNLF measured from the extended and deep PN catalogue and the extensive catalogue validation are presented in Bh+18. While the bright cut-off and the fitted slope are found to be consistent with those determined by M06 and C89, there is a significant rise in the faint end of the PNLF which remains invariant as we go radially outwards. This indicates that the rise is ubiquitous throughout the surveyed area and not due to crowding or other observational effects. It is steeper than the PNLF observed for the Magellanic clouds (LMC- Reid & Parker 2010; SMC- Jacoby & De Marco 2002). The faint-end rise of the PNLF is also present in the PHAT subsample. This subsample is further investigated for symbiotic stars mimicking PNe in our survey. With HST colours, we can rule out this hypothesis.

The rise in the faint-end of the PNLF appears to be physical and associated with the PN population of M31. It thus reflects the properties of its parent stellar population. The faint-end rise in the PNLF might be due to multiple populations of different ages in M31 that populate different magnitude intervals of the PNLF. Another possible explanation for the rise in the faint-end of the PNLF could be a change in opacity of the nebula of the

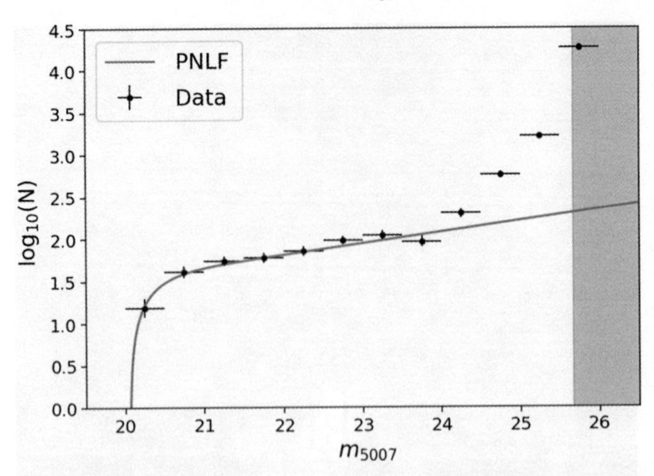

Figure 2. The completeness-corrected PNLF is shown for a single field of our survey, fitted by the C89 analytical LF (in blue). The region beyond the 50% completeness limit is shown in grey.

PNe, previously considered to describe the dip in the PNLF seen in the SMC (Jacoby & De Marco 2002) and also seen in the models by Gisecki *et al.* (2018).

4. Future prospects

We are planning to extend our imaging survey to ~ 50 kpc from the center of M31 to investigate the nature of the PNLF further out in the halo. We are also planning a spectroscopic follow-up of a complete subsample of the M31 PNe with Hectospec at the MMT (Fabricant *et al.* 2005) to obtain their LOSV to kinematically disentangle the M31 spheroid from the substructures and to identify the circumstellar extinction for the brightest PNe, ~ 2 mag below the bright cut-off.

Acknowledgements

SB would like to thank the organizers of the IAU Symposium S343 for the opportunity to present a contributed talk. SB and JH acknowledge support from the IMPRS on Astrophysics at the LMU Munich.

References

Arnaboldi, M., Aguerri, J. A. L., Napolitano, N. R., *et al.* 2002, *AJ*, 123, 760
Arnaboldi, M., Freeman, K. C., Okamura, S., *et al.* 2003, *AJ*, 125, 514
Bertin, E., & Arnouts, S. 1996, *A&AS*, 117, 393
Bonnarel F., Fernique P., Bienaym O., *et al.*, 2000, *A&AS*, 143, 33
Boulade, O., Charlot, X., Abbon, P., *et al.* 2003, *in Instrument Design and Performance for Optical/Infrared Ground-based Telescopes, ed. M. Iye, & A. F. M. Moorwood, Proc. SPIE*, 4841, 72
Ciardullo, R., Jacoby, G. H., Ford, H. C., & Neill, J. D. 1989, *ApJ*, 339, 53 [C89]
Ciardullo, R., Durrell, P. R., Laychak, M. B., *et al.* 2004, *ApJ*, 614, 167
Dalcanton, J. J., Williams, B. F., Lang, D., *et al.* 2012, *ApJS*, 200, 18
Fabricant, D., Fata, R., Roll, J., *et al.* 2005, *PASP*, 117, 1411
Ferguson A. M. N., Mackey A. D., 2016, *Astrophysics and Space Science Library, Vol. 420, Tidal Streams in the Local Group and Beyond. Springer International Publishing, Switzerland*, p. 191
Gesicki, K., Zijlstra, A. A., & Miller Bertolami, M. M. 2018, *Nature Astronomy*, 2, 580
Hartke, J., Arnaboldi, M., Longobardi, A., *et al.* 2017, *A&A*, 603, A104

Henize, K. G., & Westerlund, B. E. 1963, *ApJ*, 137, 747
Jacoby, G.H. 1980, *ApJS*, 42, 1
Jacoby, G. H. & De Marco, O., 2002, *AJ*, 123, 269
Jones, D. & Boffin, H. M. J. 2017, *Nature Astronomy*, 1, 0117
Longobardi, A., Arnaboldi, M., Gerhard, O., *et al.* 2013, *A&A*, 558, A42
McConnachie A. W. *et al.*, 2009, *Nature*, 461, 66
Merrett, H. R., Merrifield, M. R., Douglas, N. G., *et al.* 2006, *MNRAS*, 369, 120 [M06]
Reid, W. A. & Parker, Q. A., 2010, *MNRAS*, 405, 1349
Sanders, N. E., Caldwell, N., McDowell, J. *et al.* 2012, *ApJ*, 758, 133
Veyette, M. J., Williams, B. F., Dalcanton, J. J., *et al.* 2014, *ApJ*, 792, 121

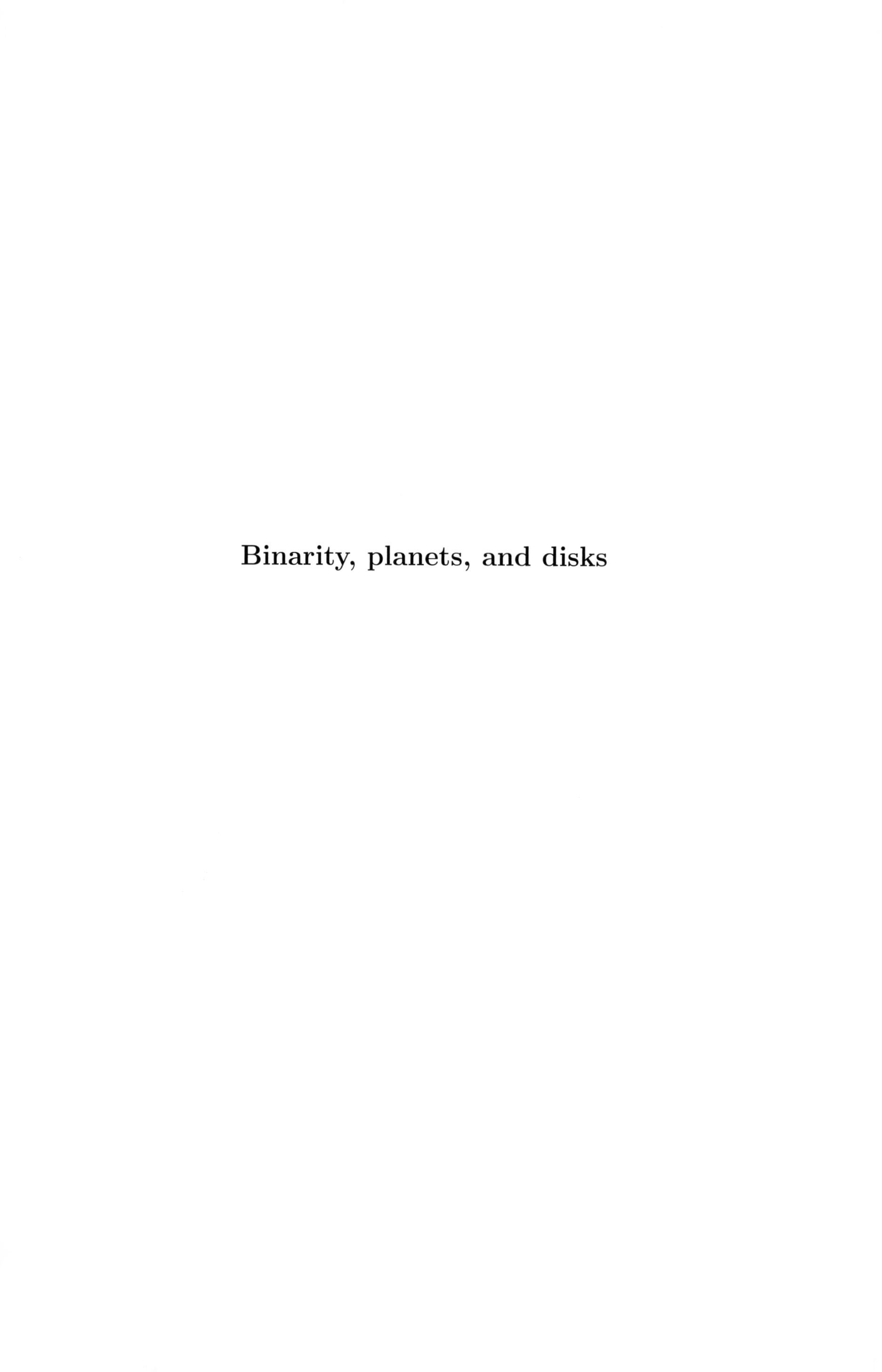

Binarity, planets, and disks

Why Galaxies Care About AGB Stars:
A Continuing Challenge through Cosmic Time
Proceedings IAU Symposium No. 343, 2019
F. Kerschbaum, M. Groenewegen & H. Olofsson, eds.

doi:10.1017/S1743921318006154

Post-RGB and Post-AGB stars as tracers of binary evolution

Devika Kamath[1,2]

[1]Department of Physics and Astronomy, Macquarie University, Sydney, Australia
[2]Astronomy, Astrophysics and Astrophotonics Research Centre, Macquarie University, Sydney, Australia
email: devika.kamath@mq.edu.au

Abstract. Binary interactions can alter the intrinsic properties of stars (such as: pulsation, mass-loss, photospheric chemistry, dust-formation, circumstellar envelope morphology etc.) and can even play a dominant role in determining its ultimate fate. While past studies have shown that binarity can end the AGB life of a star, recent studies have revealed that in specific cases binarity also pre-maturely terminate the RGB evolution. A characteristic feature of evolved binaries is the presence of a Keplerian circumbinary disc of gas and dust which plays a lead role in the evolution of the systems. In this article, I will review our advances in the research landscape of post-RGB and post-AGB binary stars, focussing on their observational properties, spectral energy distribution, photospheric chemistry, the evolution of their stable circumbinary discs, and the evolutionary connection between the enigmatic post-AGB/post-RGB binaries, and other systems whose primary component is a white dwarf.

Keywords. stars: AGB and post-AGB, stars: binaries, stars: evolution, stars: fundamental parameters, stars: chemically peculiar, Galaxy: stellar content, galaxies: Magellanic Clouds.

1. Introduction

It is known that ∼60% of low- and intermediate-mass (LIM) stars evolve in binary systems. It is also established that mass transfer in binary systems result in a range of objects; from energetic systems such as thermonuclear novae, supernovae type Ia, sub-luminous supernovae, gravitational wave sources, to less energetic systems such as sub-dwarf B stars, barium stars, bipolar planetary nebulae, etc. While this brings to light that binarity can alter the ultimate fate of stellar systems, the exact binary interaction mechanisms, and the evolutionary connection between these systems, remain unknown. Therefore, an in-depth study on poorly understood binary interaction processes is essential to constrain stellar (especially binary) evolution as well as the chemical evolution of the Universe. In this article, I will focus on binary stars that have evolved off the giant branches: post-Asymptotic Giant Branch (post-AGB) and post-Red Giant Branch (post-RGB) stars.

2. Post-AGB and Post-RGB binaries

For red giants in binary systems, the binary interaction process is governed by the balance between the Roche-lobe and the stellar radius of the red giant. Binaries with a main sequence separation of less than about two giant radii at the AGB tip will be tidally captured somewhere on the RGB or AGB. Population synthesis models (e.g., Nie *et al.* 2012) predict that a possible outcome of such a binary interaction is stable mass transfer, resulting in the formation of binaries through the following channels: the

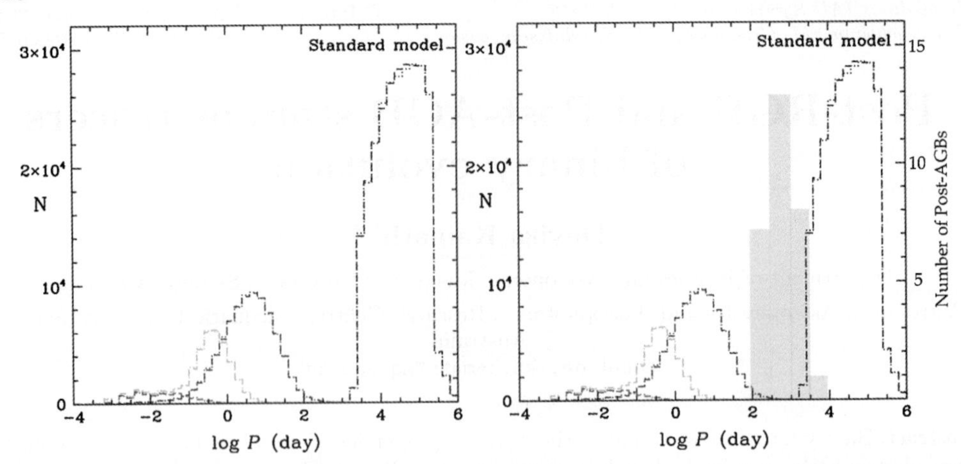

Figure 1. The discrepancy between the observed periods of the Galactic post-AGB and/or post- RGB binaries and the predictions from population synthesis models. Left: The results of the population synthesis models which were normalised to the sequence E (ellipsoidal) variables in the LMC. (The distribution of intermediate period binary post-AGB systems is given in black, the close binary post-AGB stars in blue, the post-RGB and post-EAGB binaries in cyan/grey and the double degenerate secondaries in red. Figure adapted from Nie *et al.* (2012). Right: The histogram shows the distribution of observed orbits for the Galactic systems (e.g., Van Winckel *et al.* 2009; Gorlova *et al.* 2014). All orbits fall in the period range NOT predicted by the models. The observed large eccentricities are also not predicted.

common-envelope (CE) channel and the wind accretion channel. The CE channel results in short-period binaries ($P \approx 1$ day) and the wind accretion channel results in wider systems ($P > 1000$ days), see Fig 1. However, our ongoing large radial velocity monitoring campaign on Galactic post-AGB stars, using the HERMES spectrograph (Raskin *et al.* 2011) mounted on the Flemish 1.2m Mercator telescope, has resulted in the discovery of many evolved binaries with unexpected periods between 100 and 2000 days (see Fig 1, Van Winckel *et al.* 2009; Gorlova *et al.* 2014). These systems are now far from Roche Lobe-filling, but the periods are too short to have previously accommodated a RGB or AGB star. Furthermore, these systems did not suffer dramatic spiral in and therefore the common envelope was either very rapidly expelled or somehow avoided.

The discrepancy between the observations and the predictions are linked to both theoretical and observational limitations. From the theoretical point of view, a wide range of binary interactions are not understood from first principles. Theoretical models have several uncertainties, such as, the increase of the mass-loss prior to contact (e.g., Chen *et al.* 2011; Abate *et al.* 2013), the common-envelope phase (e.g., Izzard *et al.* 2012; Ivanova *et al.* 2013), the efficiency of envelope ejection (e.g., Toonen & Nelemans 2013), the impact of radiation pressure on the shape of the classical Roche potential (e.g., Dermine *et al.* 2009), the impact of eccentricity pumping mechanisms (e.g., Izzard *et al.* 2010; Dermine *et al.* 2013), the assumed mass transfer efficiency and its orbital phase dependency, etc. (referenced here are only the most recent literature on the subject). From the observational point of view, the challenge persists as many parameters involved in predicting the outcome of the binary evolution channels are not well constrained by observations. Additionally, the difficulty in fully interpreting this Galactic sample is that we have no good distances and hence neither luminosities nor core masses (see Sect. 3).

A discrepancy also exists with respect to the observed eccentricities of these systems. Over 70 % of post-AGB binaries (ranging all orbital periods) have significant non-zero eccentricity (Oomen *et al.*, 2018, submitted). This indicates that circularisation has not happened, even though the Roche-lobe radii are smaller than the maximum size of a

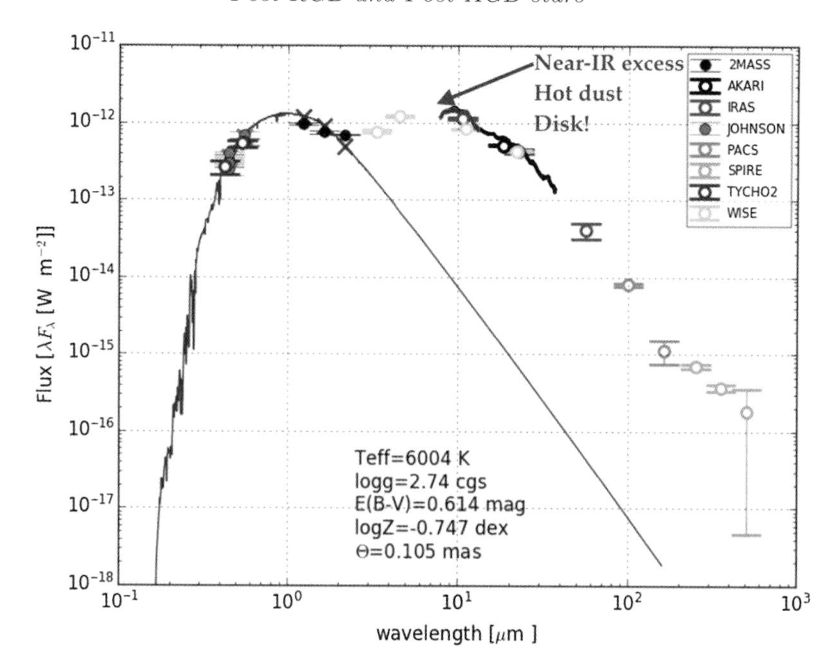

Figure 2. Disc-type SED of a post-AGB binary showing the presence of near-IR excess indicative of a circumbinary disc. The symbols represent the photometric points shown in the legend. The red solid line (in the optical part of te SED) represents the model that best fits the SED. The blue solid line (in the IR part of the SED) represents the IR spectrum of the star from the Spitzer Space Telescope Survey.

typical AGB star and tidal circularisation should have been strong when the objects were on the AGB.

While the binary interaction mechanisms that govern the evolution of post-AGB binaries is unknown, observational studies using photometric, spectroscopic, and interferometric techniques have provided a gateway to investigate these objects.

2.1. *Stable, long-lived circumbinary discs around post-AGB binaries*

Extensive studies of post-AGB stars in our Galaxy and the Magellanic Clouds have shown a group of objects that displayed spectral energy distributions (SEDs) with the onset of a near-IR excess, indicative of hot dust. This indicates that the circumstellar dust must be close to the central star, near sublimation temperature, typically in the form of a disc surrounding the central system (see Fig. 2). Such SEDs are referred to as disc-type SEDs (e.g., Van Winckel 2003; de Ruyter *et al.* 2006; Bujarrabal *et al.* 2013). The evolution of the (binary) star and the formation and evolution of the disc are closely coupled: the primary's AGB phase was abruptly terminated due to mass loss induced by a poorly understood binary interaction process. Part of the ejected material was forced into a circumbinary disc. Stable Keplerian discs are commonly detected towards post-AGB binaries with binary orbital time scales of the order of 100 to 2000 days (see Van Winckel 2017, and references therein). The dust in these discs are processed crystalline and relatively large grains ranging up to 1 μm, typically traced at longer wavelengths. A long life time of the discs is supported by the high degree of dust processing (grain growth, crystallisation Gielen *et al.* 2011; Hillen *et al.* 2016).

While the majority of the post-AGB binaries with circumbinary discs can be recognised by their disc-type SEDs (with an onset of an excess flux at near-IR wavelengths), in

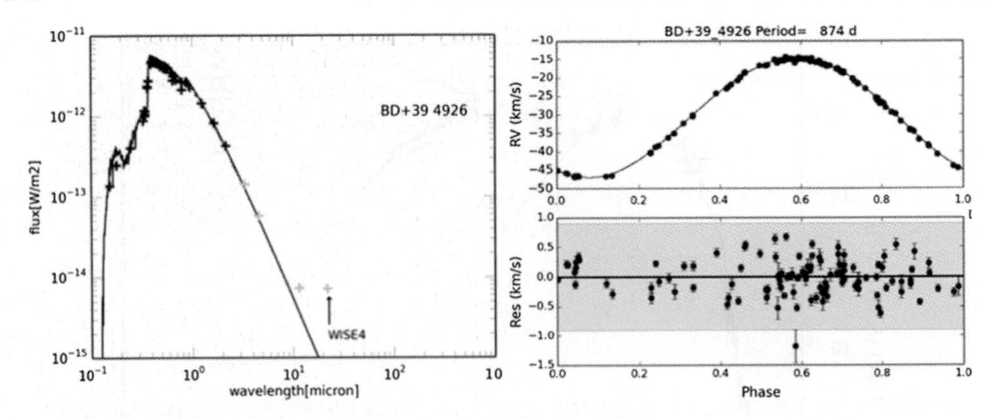

Figure 3. Left: SED of BD+394926, a Galactic post-AGB binary. Right: Radial velocity curve of BD+394926, confirming its binary nature. Figure adapted from Gezer *et al.* (2015).

somes cases, SEDs of post-AGB binaries can have only a mild IR-excess. For example, BD+39492 is a confirmed post-AGB binary with an orbital period of 874^d (see Fig. 3). Thus, we expect this star to have a circumbinary disc. However, its SED does not conform to typical disc-type SED but instead shows only a mild IR-excess, indicative of an evolved disc. This is because, just like the central star, the circumbinary disc also evolves. The evolution of the SED depends more on the disc dissipation rather than the evolution of the star. While the exact evolution of the circumstellar disc is uncertain, it is widely accepted that there is more complex physics involved than just expansion and dissipation of the circumbinary disc.

2.2. *The discovery of dusty post-RGB stars*

Though the post-AGB stars in our Galaxy are observationally well studied, before the Gaia era, the exploitation of these objects has been a challenge owing to the lack of accurate distances (and hence luminosities and initial masses) to these systems (see Sect. 3). On the other hand, since the distances to the Magellanic Clouds are well known, so are the luminosities and initial masses of all the stars within these galaxies. Hence, we initiated a comprehensive survey for post-AGB candidates in the Magellanic Clouds. We performed an extensive low-resolution optical spectral survey with the AAOmega multi-fibre spectrograph mounted on the Anglo Australian telescope, which resulted in a clean and complete census of well-characterised post-AGB objects with spectroscopically determined stellar parameters spanning a wide range in luminosities in the Large Magellanic Cloud, LMC (Kamath *et al.* 2015) and SMC (Kamath *et al.* 2014). The known distances to the Magellanic clouds enabled luminosity estimates for all the objects which led to one of the most important results of this survey: the unexpected discovery of a group of low-luminosity, evolved, low-metallicity, dusty objects in the LMC (119 objects) and SMC (42 objects) (Kamath *et al.* 2016). These objects have mid-IR excesses and stellar parameters similar to those of post-AGB stars (G to A spectral types, low log g values between 0 and 2), and low metallicities (lower than the mean metallicity of young stars in their host galaxy). However, their luminosities are much lower (200 - 2500 $L_\odot$) than that expected for post-AGB stars. We therefore suspect they are 'post-RGB' stars, and these Magellanic Cloud objects are the first examples of such objects.

The $\sim$2:1 ratio of the number of optically visible post-AGBs to post-RGBs in the LMC and SMC proves that not all LIM stars evolve on to the AGB phase before their death. This presents dusty post-RGB stars as new building blocks in the archaeology of galaxies. It also underlines the need to determine the total fraction of these systems and

Table 1. Properties of post-AGB and Post-RGB binaries

Parameters	**Post-AGB binaries**	**Dusty post-RGB binaries**
Initial mass	$\sim 1-8\,M_\odot$	$\sim 1-2\,M_\odot$
Final mass[1]	$\sim 0.48-0.9\,M_\odot$	$\sim 0.28-0.48\,M_\odot$
Observed SED types	mostly disc-type	diverse SEDs[2]: disc-type, double-peaked, mild-IR excess
Orbital periods	100 to 1000 days	yet to be studied
Pulsation periods	$\gtrsim$20 days	$\lesssim$20 days
Metallicity	metal poor in comparison to the young stars in their host galaxy	likely to be metal-rich in comparison to their post-AGB counterparts
Luminosity[1]	$\gtrsim$2500 $L_\odot$	$\lesssim$2500 $L_\odot$
completion of core-He burning	yes	no (for the majority of the objects)
Photospheric chemical depletion	yes	yes
Dust Mass	$10^{-3}\,M_\odot$	yet to be estimated but likely to be greater than $10^{-3}\,M_\odot$

Notes:
[1]Depending upon the metallicity.
[2]This indicates that we do not fully understand the evolution, properties, and circumstellar environment of the dusty post-RGB stars (see Section 2.3)

fully understand their evolution and nucleosynthesis. Table 1 summarizes the properties of post-AGB and Post-RGB binaries.

2.3. *Likely evolution scenario and dust properties of post-RGB stars*

During the RGB phase (where luminosities are $\lesssim 2500\,L_\odot$), single-star mass loss is insufficient to remove the H-rich envelope and produce a dusty post-RGB star (Vassiliadis & Wood 1994). The only way large amounts of mass loss and premature evolution off the RGB can occur is via binary interaction (Han *et al.* 1995). As mentioned in Sect. 2.1, in the Galaxy, the binaries among the post-AGB stars all have a common property: their SEDs display a clear near-infrared excess. It is now well established that this feature in the SED indicates the presence of a stable compact circumbinary disc. We find that many of our new post-RGB candidates, in the LMC/SMC, also show similar disc-type SEDs (see Fig. 3) pointing to binarity in these objects.

For a fraction of the LMC/SMC post-RGB stars, their SEDs are not representative of the expected disc-type SEDs. (see Kamath *et al.* 2016). Since we expect these dusty post-RGB stars to be in binary systems, a possible explanation for these diverse SEDs is the evolution of the disc, as mentioned in Sect. 2.1.

2.4. *The first radial velocity monitoring studies of post-RGBs in the Magellanic Clouds*

While the likely presence of a circumbinary disc, which points to the binary nature of these systems, can be inferred from the SEDs of the LMC/SMC post-RGB stars, a true test of binarity comes the detection of radial velocity changes.

In our recent studies, we have initiated a radial velocity monitoring program carried out with the X-shooter spectrograph on ESO's Very large Telescope (VLT). The targets selected for this study are representative of the entire post-RGB SMC/LMC sample (i.e., objects covering a range of spectral-types and SED-types). The data available so far comes from 4 epochs of observations and covers a period of $\sim$370 days. A detailed radial velocity (RV) analysis of all the data obtained so far revealed significant RV changes in two out of 10 objects, see Fig. 4. While these observations provide a breakthrough result: the first evidence of binarity in two dusty post-RGB stars in the Magellanic Clouds, it is also puzzling since for the remaining objects no significant RV changes were detected. Binary interaction is the only known mechanism for stars to leave the

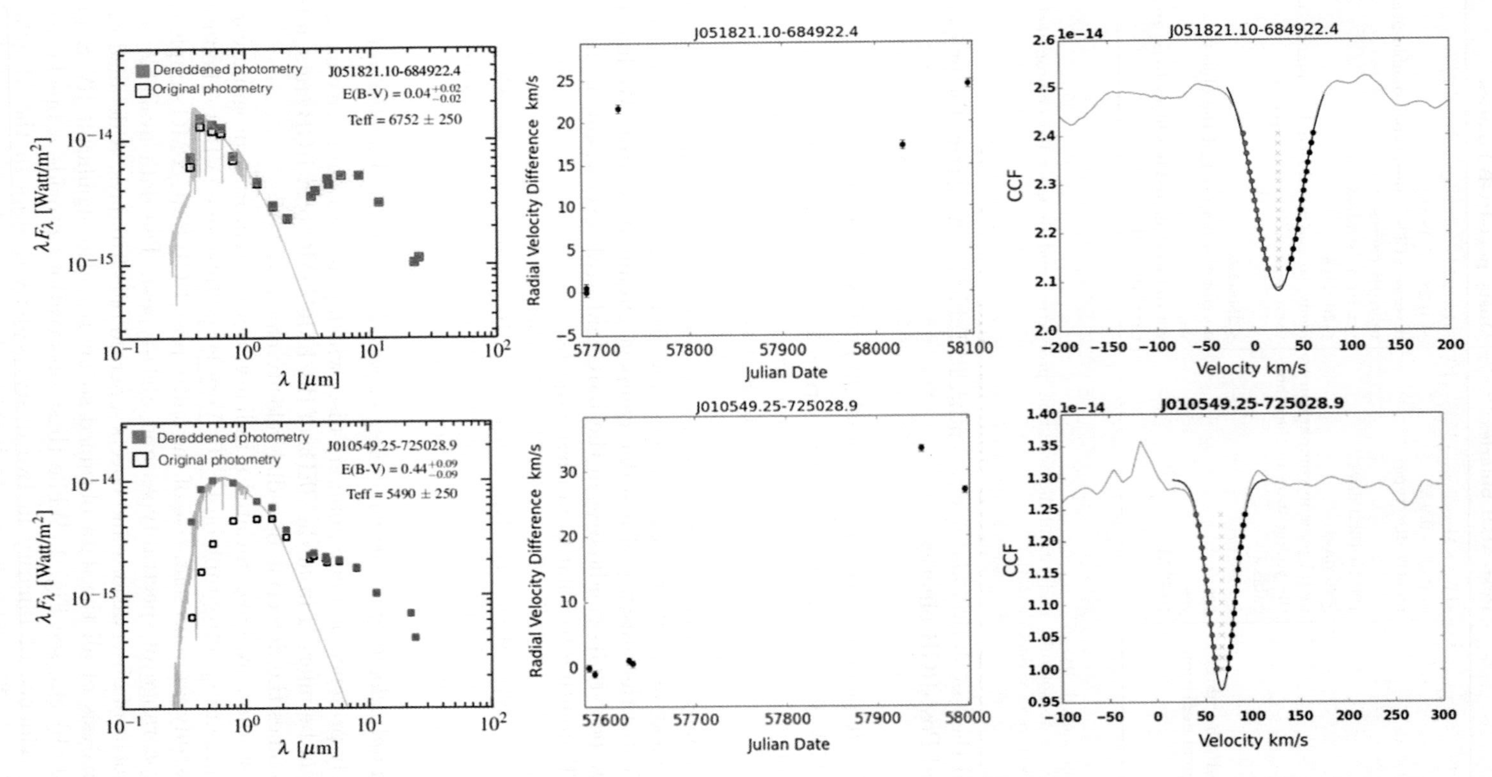

Figure 4. Top-left panel: SED of J051821, an LMC post-RGB source that is representative of the target sample. The SED show a clear near-IR excess, indicative of hot dust in the form of a disc. The grey solid line represents the synthetic spectrum (Munari *et al.* 2005) with the best fit to the observed spectra of the object. Top-center panel: The radial velocity (RV) plot for J051821 (see text for more details). The plot clearly shows the RV changes. Top-right panel: The corresponding cross correlation function (CCF) of the radial velocity measurements. Bottom panels: Same as top panels but for J010549, an SMC post-RGB source.

RGB prematurely. While the proof of binarity is limited to only two of the 10 post-RGBs, our RV monitoring is too sparse to conclude on the nature of these systems. Furthermore, population synthesis models (Nie *et al.* 2012) normalised to the LMC binary red-giants (the likely progeny of these dusty post-RGB objects), predict that our systems are likely to have orbital periods in the range of ~25 to 500 days. Therefore, we are continuing our RV monitoring to obtained an orbital period sampling of $\gtrsim$500 days, which covers the full range of predicted orbital periods for these systems. We have also increased the number of epochs so that we will be able to identify the orbital period for the stars that showed RV variations and look for RV variations in the remaining objects.

As mentioned before, over the last decades, the lack of accurate distances to the objects in our Galaxy has stymied the derivation of luminosities of the Galactic sources. Since luminosity is the only key to disentangling the likely post-RGB stars amongst the Galactic post-AGBs, we are yet to identify the Galactic post-RGB stars. In the following section, we present the progress in this field owing to the recent Gaia DR2 data release (Luri *et al.* 2018).

3. Galactic Post-RGB/Post-AGB binaries and GAIA DR2

The recent release of the Gaia DR2 parallaxes (Luri *et al.* 2018) to many of the Galactic objects, provides a possibility to exploiting these systems. For the single post-AGB stars, the Gaia DR2 parallaxes have provided the opportunity to derive distances and luminosities (Kamath *et al.*, 2018, in-prep). However, care must be taken when using the parallaxes of the binary stars because the parallaxes presented in Gaia DR2 come from astrometric solutions assuming a single star evolutionary nature. Furthermore, for the Galactic post-AGB binaries, the orbital motion amplitude is similar to the parallaxes (1 AU orbit in a period of 1-3 years). Thus for binaries with orbital periods of the order of 1.5 to 2 years (such as post-AGB binaries), the proper motions or parallaxes listed in Gaia DR2 may be quite far from the true values for the system.

4. Nucleosynthesis in binary post-RGB/post-AGB stars

Stellar nucleosynthesis is directly linked and influenced by stellar evolution. Post-AGB binaries commonly show a chemical anomaly in which the photosphere is devoid of refractory elements (e.g., Van Winckel 2003; Venn *et al.* 2014; Gezer *et al.* 2015, and references therein). Although the mechanism to acquire the chemically depleted anomaly is not yet completely understood, the commonly accepted scenario is that, when the circumstellar dust is trapped in a circumbinary disc, the gas, devoid of refractory elements, can be segregated from the dust and be re-accreted onto the photosphere (Waters *et al.* 1992; Van Winckel *et al.* 1998).

Oomen *et al.* (2018, subm.) have shown that post-AGB stars with a high effective temperature (>5500 K) in a wide orbit show depletion, suggesting that re-accretion of material from a circumbinary disc is an ongoing process. It seems, however, that depletion is inefficient for the smaller orbits irrespective of the actual surface temperature.

Kamath & Van Winckel (2018, in prep.) find that photospheric depletion is not only active in luminous post-AGB stars but also in lower luminosity post-RGB stars. This reflects on similar binary properties amongst post-AGB binaries and post-RGB binaries. Figure 5 shows the efficiency of depletion (expressed in terms of the [Zn/Ti] abundance) for both the Galactic and LMC depleted post-AGB/post-RGB sources. The LMC sources (black circles in Fig. 5) have luminosities in the range 2000 $L_\odot$ to 5000 $L_\odot$. This shows that the spread in the efficiency of depletion is independent of luminosity.

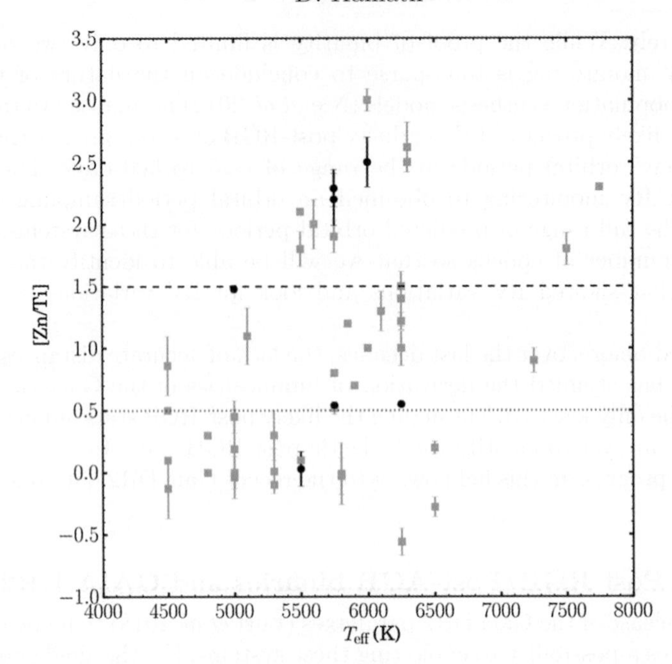

Figure 5. The [Zn/Ti] abundance as a function of $T_{\rm eff}$ for the depleted post-AGB stars in the LMC (black filled circles) and their Galactic analogues (magenta filled squares). The dashed line represent the limit of what we call 'strongly depleted' while the dotted line marks our definition of 'mild depletion'. Figure adapted from Kamath & Van Winckel (2018, in prep.).

Disentangling the likely complex dependence of the efficiency of chemical depletion on a suite of factors such as the effective temperature, luminosity, and the orbital properties of the binary system remains a challenge.

5. The structure and dynamics of the circumstellar environment around post-AGB binaries: the circumbinary disc, jets, and outflows

A review of the state-of-the-art of the studies involving the structure and dynamics of the circumstellar environment around post-AGB binaries is presented in Van Winckel (2017). In this section, I present a few highlights.

IRAS 08544–4431 is a well-studied binary post-AGB star, surrounded by a dusty disc that shows Keplerian rotation. Interferometric studies (Hillen *et al.* 2016; Kluska *et al.* 2018) of IRAS 08544–4431 have imaged the sublimation rim of the circumbinary disc, which has an angular diameter of 15 mas (see attached figure) and is seen at an inclination of 20°. Bujarrabal *et al.* (2018) have identified the rotation of the disc, which has an angular momentum comparable to that of the central binary system. Surprisingly, the system also shows a continuum flux contribution of 4% at 1.65 micron at the location of the secondary component.

Since the photosphere of the secondary is too faint to produce this flux contribution, the exact emitting source remains unknown. Due to the confirmed presence of a high-velocity outflow or jet originating from the secondary component the likely source for the contribution is a circum-companion accretion disc. On-going studies are in place to investigate this.

Ertel *et al.* (2018, in prep.) has resolved the edge-on disc, in the optical, for the post-AGB binary star AR Pup using SPHERE ZIMPOL and IRDIS. This is the second post-AGB disc to be directly imaged after the Red Rectangle.

Our long-term radial velocity monitoring of evolved stars has revealed that the presence of a high-velocity outflow, or jet, is a common feature in post-AGB binaries. Bollen *et al.* (2017) provides a good illustration and geometric models for the jet found in BD+46 442, a post-AGB binary system. Both simulations and model fitting of the observations show that these jets are not strongly collimated, but are rather wide with half-opening angles >40° (Bollen *et al.* 2017). These high-velocity outflows or jets have a great impact on their surrounding environment and are believed to be the key components for the shaping of bipolar and irregular planetary nebulae.

Though our understanding of the geometry and kinematics of the jet has gained progress, the exact source that feeds the accretion disc around the secondary component, from which the jet is launched remains unknown. Furthermore, the jets around post-AGB stars are rather diverse in nature and understanding the nature of jets in different post-AGB stars and obtaining a physical model for the jet, is a part of our on-going study.

While jets around psost-AGB stars are commonly observed, jets around post-RGB systems are yet to be observed and studied, owing to observational limitations.

6. Evolutionary connection between evolved binaries

Standard theories of stellar evolution predict that PNe are the progeny of post-AGB stars. Low-luminosity PNe, such as the Boomerang Nebula, are considered to be the progeny of post-RGB stars (Sahai *et al.* 2017). However, except for a small sample of PNe that house central stars with periods of the order of a few years (Jones *et al.* 2017), spiralled-in systems are commonly observed in central stars of PNe (Miszalski *et al.* 2011). Moreover, post-AGB stars with discs seem to avoid spiral-in. The exact physical mechanism(s) responsible for spiral-in or avoiding a spiral-in remains unknown.

The evolutionary connection between the enigmatic post-AGB and post-RGB binaries, and other systems whose primary component is a white dwarf (WD) is not straight forward. The latter not only include cataclysmic variables and narrow binaries among central stars of PNe (e.g., Miszalski *et al.* 2011) which did suffer a spiral-in phase, but also wide systems such as symbiotic stars, barium stars, CH-stars, and the more extreme CEMP-s stars (Jorissen 2003; Aoki *et al.* 2007). These wide systems show similar observed period distributions as the post-AGB binaries. The former evolution of the current WD in these systems is unknown making them fossils of unidentified binary interaction processes. Additionally, it is generally accepted that Type Ia supernovae are exploding CO WDs. The formation channels are not very well understood but some are thought to be formed by single degenerate explosions which occur when the massive WD accretes material from a companion (see Maoz *et al.* 2014, and references therein).

Piecing together the evolutionary connection between the zoo of evolved binaries is a major obstacle in the domain of binary stellar evolution.

7. Summary

Post-AGB and Post-RGB binary stars are a class of objects that have evolved off the AGB and RGB, respectively, due to binary interactions. These objects have been discovered in our Galaxy and the Magellanic Clouds. They are a significant population of evolved binaries and hence are ideal tracers of the binary evolution. So far, the binary interaction mechanism to form these objects remains uncertain. However, it can be concluded the circumbinary disc plays a key role in the binary interaction process, and the loss of angular momentum is powered by circumbinary disc. The evidence for disc-star interaction is seen in the observed photospheric properties (chemical depletion), orbital properties (intermediate-range periods, high eccentricities), and the structure and kinematics of their circumstellar environment (e.g., jets, outflows). Furthermore, using interferometric and direct imaging techniques, the circumbinary discs are now resolvable and likely to be second generation proto-planetary discs. The evolutionary connection between the post-AGB/post-RGB binaries and their precursors and progeny remains a long-standing astrophysical puzzle.

References

Abate C., Pols O. R., Izzard R. G., Mohamed S. S., de Mink S. E., 2013, *A&A*, 552, A26
Aoki W., Beers T. C., Christlieb N., Norris J. E., Ryan S. G., Tsangarides S., 2007, *ApJ*, 655, 492
Bollen D., Van Winckel H., Kamath D., 2017, *A&A*, 607, A60
Bujarrabal V., Alcolea J., Van Winckel H., Santander-García M., Castro-Carrizo A., 2013, *A&A*, 557, A104
Bujarrabal V., Castro-Carrizo A., Winckel H. V., Alcolea J., Contreras C. S., Santander-García M., Hillen M., 2018, *A&A*, 614, A58
Chen X., Han Z., Tout C. A., 2011, *ApJ*, 735, L31
Dermine T., Jorissen A., Siess L., Frankowski A., 2009, *A&A*, 507, 891
Dermine T., Izzard R. G., Jorissen A., Van Winckel H., 2013, *A&A*, 551, A50
Gezer I., Van Winckel H., Bozkurt Z., De Smedt K., Kamath D., Hillen M., Manick R., 2015, *MNRAS*, 453, 133
Gielen C., *et al.*, 2011, *A&A*, 533, A99
Gorlova N., Van Winckel H., Vos J., Ostensen R. H., Jorissen A., Van Eck S., Ikonnikova N., 2014, ArXiv e-prints
Han Z., Eggleton P. P., Podsiadlowski P., Tout C. A., 1995, *MNRAS*, 277
Hillen M., Kluska J., Le Bouquin J.-B., Van Winckel H., Berger J.-P., Kamath D., Bujarrabal V., 2016, *A&A*, 588
Ivanova N., *et al.*, 2013, *A&ARv*, 21, 59
Izzard R. G., Dermine T., Church R. P., 2010, *A&A*, 523, A10
Izzard R. G., Hall P. D., Tauris T. M., Tout C. A., 2012, in IAU Symposium. pp 95–102
Jones D., Van Winckel H., Aller A., Exter K., De Marco O., 2017, *A&A*, 600, L9
Jorissen A., 2003, in Corradi R. L. M., Mikolajewska J., Mahoney T. J., eds, ASP Conf. Ser. Vol. 303, Symbiotic Stars Probing Stellar Evolution. p. 25
Kamath D., Wood P. R., Van Winckel H., 2014, *MNRAS*, 439, 2211
Kamath D., Wood P. R., Van Winckel H., 2015, *MNRAS*, 454, 1468
Kamath D., Wood P. R., Van Winckel H., Nie J. D., 2016, *A&A*, 586, L5
Kluska, J., Hillen, M., Van Winckel, H., *et al.* 2018, *A&A*, 616, A153
Luri X., *et al.*, 2018, ArXiv e-prints
Maoz D., Mannucci F., Nelemans G., 2014, *ARA&A*, 52, 107
Miszalski B., Napiwotzki R., Cioni M.-R. L., Groenewegen M. A. T., Oliveira J. M., Udalski A., 2011, *A&A*, 531
Munari U., Sordo R., Castelli F., Zwitter T., 2005, *A&A*, 442, 1127
Nie J. D., Wood P. R., Nicholls C. P., 2012, *MNRAS*, 423, 2764
Raskin G., *et al.*, 2011, *A&A*, 526, A69
Sahai R., Vlemmings W. H. T., Nyman L.-Å., 2017, *ApJ*, 841, 110
Toonen S., Nelemans G., 2013, *A&A*, 557, A87
Van Winckel H., 2003, *ARA&A*, 41, 391
Van Winckel H., 2017, in Liu X., Stanghellini L., Karakas A., eds, IAU Symposium Vol. 323, Planetary Nebulae: Multi-Wavelength Probes of Stellar and Galactic Evolution. pp 231–234
Van Winckel H., Waelkens C., Waters L. B. F. M., Molster F. J., Udry S., Bakker E. J., 1998, *A&A*, 336
Van Winckel H., *et al.*, 2009, *A&A*, 505, 1221
Vassiliadis E., Wood P. R., 1994, *ApJS*, 92, 125

Venn K. A., Puzia T. H., Divell M., Côté S., Lambert D. L., Starkenburg E., 2014, *ApJ*, 791, 98
Waters L. B. F. M., Trams N. R., Waelkens C., 1992, *A&A*, 262
de Ruyter S., Van Winckel H., Maas T., Lloyd Evans T., Waters L. B. F. M., Dejonghe H., 2006, *A&A*, 448, 641

Discussion

KAMINSKI: What are the masses of stars and circumstellar discs in the post-RGB systems?

KAMATH: The dusty post-RGB stars have luminosities below the RGB tip. Depending on the metallicity, their luminosities (while on the post-RGB) range between 500 $L_\odot$ to 2500 $L_\odot$. This typically translates into post-RGB masses of 0.2 to 0.45 $M_\odot$. The disc (circumbinary) mass for post-AGB binaries are about 10^{-3} to 10^{-2} $M_\odot$. We expect that post-RGB circumbinary discs will have similar masses.

Why Galaxies Care About AGB Stars: A Continuing Challenge through Cosmic Time
Proceedings IAU Symposium No. 343, 2019
F. Kerschbaum, M. Groenewegen & H. Olofsson, eds.

doi:10.1017/S1743921318007202

AGB stars in binaries and the common envelope interaction

Orsola De Marco[1,2]

[1]Department of Physics and Astronomy, Macquarie University, Sydney, NSW 2109, Australia
email: orsola.demarco@mq.edu.au

[2]Research Centre in Astronomy, Astrophysics and Astrophotonics, Macquarie University, Sydney, NSW 2109, Australia

Abstract. One may argue that, today, proceedings articles are not useful. Results belong in refereed papers and even reviews are best in publications that are long enough to do justice to the topic reviewed. In this short review, reflecting a presentation that was given at the IAU343 symposium, *Why Galaxies Care about AGB Stars*, I have therefore endeavoured to include some practical snippets that, while remaining true to the presentation, also provide a quick look up reference. After a reminder of how few AGB stars actually interact with a companion, and a pictorial summary of the types of interactions that can happen, I list the rapidly growing body of 3D common envelope simulations. Next, I highlight shortfalls and successes of simulations, and then spend some time comparing the two simulations of planetary nebulae from common envelope interactions to date. Finally, I summarise a handful of results pertaining to common envelope interactions between giants and planets.

Keywords. stars: AGB and post-AGB, binaries (including multiple): close, planetary nebulae: general, methods: numerical

1. Introduction

The common envelope (CE) interaction takes place when an expanding star fills its Roche lobe and starts to transfer mass to its companion. For certain configurations, such as a giant transferring mass to a less massive, more compact companion, the mass transfer can be unstable and result in a fast, dynamical-timescale inspiral of the companion through the envelope of the primary (Paczynski 1976, Ivanova *et al.* 2013). The companion and the core of the giant orbit one another within the envelope of the primary and orbital energy and angular momentum are transferred from the orbit to the CE gas via gravitational drag. The inspiral results in a smaller orbital separation and an ejected envelope or in a merger of the companion with the primary's core, where most of the envelope remains bound to the merged core and will settle into a new equilibrium configuration within a thermal timescale.

This interaction can explain the existence of compact binaries where one or both of the components were larger in the past than the present-day orbit. A large number of single and binary star classes may be post-CE objects, such as cataclysmic variables, X-ray binaries, the progenitors of (some) type Ia supernovae, double degenerates including neutron stars and black holes, as well as a number of mergers caught in the act, loosely catalogued under the name of intermediate luminosity red transients (Kasliwal 2012)†.

† This transient class may in fact include quite a heterogeneous group of stars, but several at least have been identified as CE mergers. For more information see De Marco & Izzard (2017).

When a CE interaction takes place during the asymptotic giant branch (AGB) phase of a star, a planetary nebula (PN) may be ejected. We know that at least one in five PN are ejected common envelopes (Miszalski *et al.* 2009). In all cases tested, the axis of symmetry of the nebula coincides with the orbital axis, very likely, in this case, implying causality (Hillwig *et al.* 2016). A nebula may even be ejected in the case of a common envelope on the RGB as is likely the case PN ESO 330-9 (Hillwig *et al.* 2017).

Understanding the CE interaction necessitates hydrodynamic simulations in 3 dimensions. Among the many parameters that we would like to know about, a most essential quantity is the final orbital separation of the emerging binary as a function of stellar and system parameters. On it rests the prediction of the delay times (e.g., Toonen & Nelemans 2013) between binary formation and explosion as type Ia supernova or other types of mergers, including those that emit detectable gravitational waves. The comparison of predicted and observed delay times can inform as to the nature of the stars that underwent the outburst, and hence the past history of the systems, including identifying the progenitors of supernovae Ia. Another consequence of how fast double degenerates merge are the yields of iron (for double white dwarf mergers that give rise to supernovae Ia) and r-process elements manufactured by neutron star-neutron star mergers (Pian *et al.* 2017).

Finally, the common envelope must by necessity interrupt the natural evolution of the giant star along the giant branch. In the case of AGB stars this will prevent a natural termination of the AGB, eliminating or limiting the number of helium shell flashes that characterise the upper AGB. This will mitigate the third dredge-up and the ejection of carbon-rich (as opposed to oxygen-rich) gas into the ISM, as well as the abundance of s-process elements. While this may not have a significant effect on yields in general, because of the scarcity of AGB stars interrupted by a CE interaction, it may mean that post-AGB stars (and their PN) deriving from a CE interaction are not sampling the range of elemental abundances that are more normally ejected into the ISM. Calculating yields using these PN would then by necessity provide a biased value. If these PN were in any way more conspicuous (brighter) and used in preference for yields measurement then, clearly our measured yields would be biased (see discussion in De Marco *et al.* 2009).

In Section 2 we discuss the number of available binaries in a given AGB star population that can undergo a strong interaction. In Section 3 we catalogue the type of interactions that are possible and the type of knowledge that we would like to acquire about these interactions. In Section 4 we talk about hydrodynamic simulation models of the CE interaction. Finally in Sections 5 and 6 we talk about planetary nebulae from CE interactions and planets in common envelopes with giants, respectively. We do not conclude.

2. How many AGB stars are in close binary systems?

Not every peculiarity of AGB and post-AGB stars can be explained by a binary interaction. Yet binary interaction do cause peculiarities in typical stars, sometimes making them more noticeable: more variable, brighter... something that can cause them to draw attention onto themselves, making them selectively more observed and seemingly more common.

According to the recent studies of Raghavan *et al.* (2010), approximately half of all solar-mass stars are in binaries with a period distribution peaked at $\sim$300 years (see cartoon depiction in Fig. 1). Most of these binaries are anyway so wide that they will not interact with their companion, possibly leaving only a quarter of all the main sequence binaries able to interact at all.

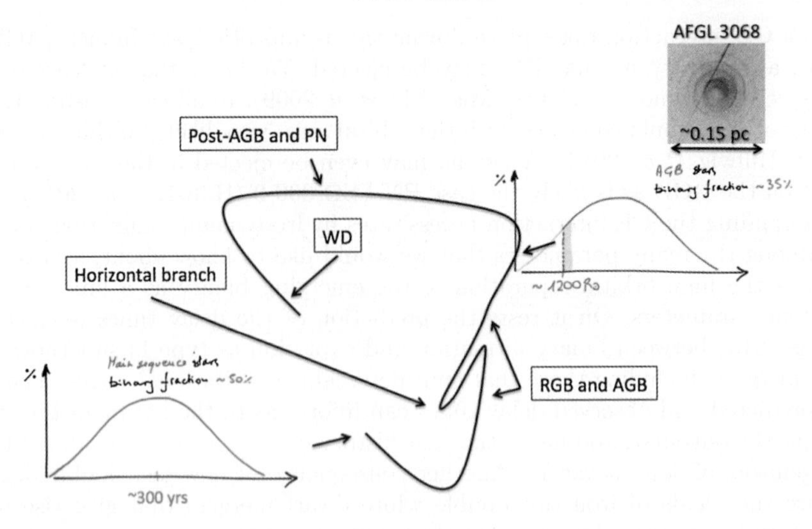

Figure 1. A schematic of the HR diagram, overlaid with a cartoon of the period distribution of sun-mass stars, which are in binaries approximately half the time. On the top right of the diagram we show a cartoon of the approximate period distribution expected of AGB binaries, where short period binaries, i.e., those with a separation smaller than approximately a couple of times the radius of the star, have already entered an interaction, leaving only approximately 35 % of the AGB stars in binaries. However, a much smaller fraction than that are close enough to interact strongly. The image of AFGL 3068 is from Mauron & Huggins (2006).

The closest of these binaries, those with initial orbital separation smaller than a couple of times the maximum RGB radius of the primary will interact during the RGB. For strong interactions like the CE interaction, the binary, if it survives, may never become an AGB star. Its envelope may be too feeble, or in any case the presence of a close companion may prevent expansion. This leaves only the binaries with initial separation between a couple of RGB maximum radii and a couple of AGB maximum radii to interact during the AGB (this is depicted as the pale green stripe in Fig. 1). *Hence, only a few percent of the AGB stars can interact strongly with a companion (Madappatt et al. 2016).* Therefore, of all AGB stars one can expect a small percentage to have a "lurking" companion, ready to interact either when tides decrease the orbital separation, or when the star expands further.

More massive stars have a much higher binary fraction with the most massive stars being in binaries most of the time (e.g., Sana *et al.* 2012). Additionally, the period distribution of more massive stars may have a peak at much shorter period (in virtue of more complex, possibly bimodal distributions; Duchêne & Kraus 2013), increasing the fraction of those stars that interact. This means that, even if the initial mass function dictates that the fraction of massive stars is very small, on balance, the mean mass of interacting AGB binaries may be slightly larger than for single stars.

3. Close binary star interactions on the AGB

A schematic of the strong binary interactions that can take place when an AGB star does have a companion sufficiently close is presented in Fig. 2. We do not include interactions where the companion only accretes wind from the primary star. The figure caption explains the parameters that we would like to know for each type of interaction. Importantly, each interaction can lead to the next one. Tides alter the orbit of a binary such that, particularly in the case of giant-compact star binaries, the two stars may come into close proximity, at which time the giant star can overflow its Roche lobe.

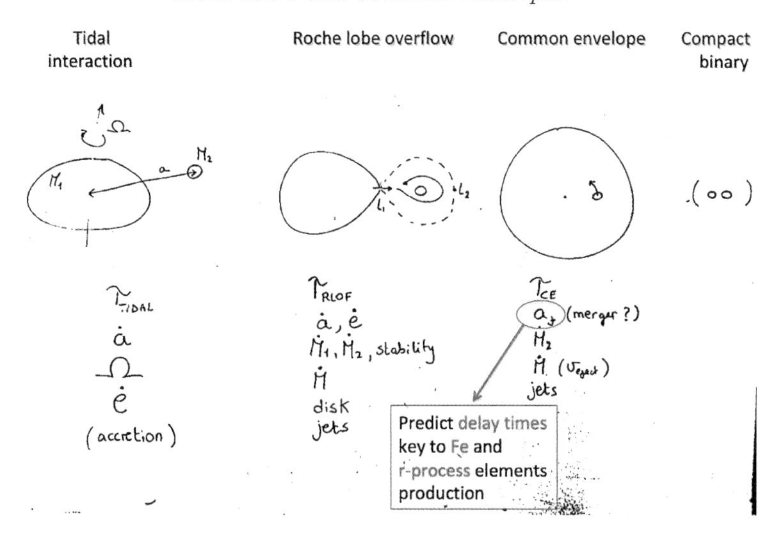

Figure 2. A schematic of the types of close binary interactions that can take place on the AGB and the parameters that they affect. During a tidal interaction (timescale $\tau_{\rm tidal}$) the orbital elements (separation a and eccentricity e) may change as well as the stellar rotation, Ω, and there could be accretion from the wind of the AGB star. During Roche lobe overflow, orbital elements as well as stellar masses, M_1 and M_2, can change, disks and jets may form and the timescale, $\tau_{\rm RLOF}$, and stability of the interaction are all unknown. During the CE interaction, which occurs over a much faster timescale ($\tau_{\rm CE}$) we worry about the final separation of the system, a_f, or if the system merges. We also want to know whether the secondary accretes, how much mass the system loses ($\dot{M}$), the ejection speed, $v_{\rm eject}$, and whether jets are launched.

Mass transfer between a convective giant and a companion is often unstable. If mass is lost (usually via the L2 and L3 Lagrangian points) then the change in angular momentum leads to a shortening of the orbital separation and an increase in mass transfer rate. This is the hallmark of instability, soon leading to the next phase, a CE interaction. The timescales for the tidal, Roche lobe overflow and CE phases are very variable and there is no way that a single simulation of the entire interaction, from the tidal phase to the formation of the compact binary, can be carried out. Attempts at quantifying the Roche lobe overflow phase have been partly successful (MacLeod *et al.* 2018a,b, Reichardt *et al.* 2018) in demonstrating that the behaviour of the RLOF phase leading into the CE phase, are approximately as expected analytically. Yet resolution does play a role in the length of the unstable RLOF phase and the geometry and dynamics of the ejection. Tides in particular are so sensitive to small changes in the envelope shape and on how the energy is dissipated by the convective eddies in the envelope, that we cannot hope to model it in 3D at present. The best modelling tools for this phase are still analytical following prescriptions such as that of Zahn (1966).

4. Simulations: the state of play

Starting with the pioneering paper of Taam *et al.* (1978), which calculated a model of a CE interaction between a 16 $M_\odot$ giant and a neutron star companion in 1D, several papers followed in the same series, calculating a variety of CE models for a range of different stars and with a range of different techniques in 2 and 3D. These early works gave rise to what was for a long time the industry standard for CE simulations, namely the SPH effort of Rasio & Livio (1996) and the Eulerian code calculation of Sandquist *et al.* (1998, 2000). A period of time followed before a new generation of models were carried out, namely by Ricker & Taam (2008, 2012) using Flash (Fryxell *et al.* 2000), a

Table 1. A list of all publications known to the author that include 3D hydrodynamic simulations of the common envelope interaction.

Country/Group	Code[1]	Publications
USA	SPH Nested grid	Terman *et al.* 1994[2], Terman *et al.* 1995 Terman & Taam 1996 Sandquist *et al.* 1998, 2000
USA	SPH	Rasio & Livio 1996
USA	FLASH-AMR	Ricker & Taam 2008, 2012
Australia	Enzo unigrid and AMR, SNSPH, PHANTOM SPH	Passy *et al.* 2012, Kuruwita *et al.* 2016, Staff *et al.* 2016a,c, Iaconi *et al.* 2017, 2018, Galaviz *et al.* 2017, Reichardt *et al.* 2018
Canada	Starsmasher SPH	Nandez *et al.* 2014, 2015; Nandez & Ivanova 2016 Ivanova & Nandez 2016
Germany	AREPO moving mesh	Ohlmann *et al.* 2016a,b, 2017
USA	AstroBEAR AMR	Chamandy *et al.* 2018
USA	Athena++[3] nested grid	MacLeod *et al.* 2018b,a[4]

Notes:
[1]Lagrangian codes implemented with smooth particle hydrodynamics (SPH) techniques and Eulerian codes implemented with a variety of mesh refinement techniques such as adaptive mesh refinement (AMR).
[2]This is paper 3 in a series of 10 papers which tackled simulations of the CE interaction in 1, 2 and 3D. The first paper in the series, by Taam *et al.* (1978), was in 1D.
[3] This is a descendent of the Athena code (Stone *et al.* 2008).
[4] Currently aimed at studying the phase just prior to the in-spiral.

grid code exploiting the adaptive mesh refinement (AMR) techniques, and Passy *et al.* (2012) using the Enzo grid code with no AMR (O'Shea *et al.* (2004), but see also Passy & Bryan (2014)), as well as the smooth particle hydrodynamics (SPH) code SNSPH (Fryer *et al.* 2006). Fortunately, many more efforts followed in quick succession, which are summarised in Table 1.

The main challenge facing simulations is the long run times that often extend into months and the fear that the results are dependent on a resolution that cannot be increased nor properly studied (Iaconi *et al.* 2018). As a consequence of long run times, only a very limited parameter space has been simulated, concentrating on low mass and relatively compact RGB stars (see Iaconi *et al.* (2017) for a list of simulations). Typically, the binary simulation is started with the companion so close to the primary that the primary is already overflowing its Roche lobe. This allows a faster computational time, because the dynamical in-spiral is triggered immediately, but likely leads to differences in the simulations' outcome. Most importantly, MacLeod *et al.* (2018a,b) and Reichardt *et al.* (2018) have recently discussed the effects of calculating the phase of unstable Roche lobe overflow preceding the CE inspiral and how this phase leads to very distinctive ejecta properties and may play a role in the light properties of systems observed (see also discussions in Pejcha *et al.* (2016b,a) and Galaviz *et al.* (2017)).

A persistent issue connected with simulations of the CE interaction is that, unless recombination energy is allowed to contribute†, the common envelope is lifted but not fully unbound. Simulations typically unbind 10% of the entire envelope early in the inspiral, while the remainder is lifted out of the orbital volume, contributing to orbital stabilisation, but if not unbound would fall back onto the binary (see Iaconi *et al.* (2017) for a list of simulations). Even allowing the entire recombination energy budget to do

† For information on the debate of how much energy should be allowed to to contribute rather than being radiated away see (Ivanova 2018; Grichener *et al.* 2018, and references therein)

work, only the envelope of low mass systems becomes unbound ($M \lesssim 2$ M$_\odot$; Nandez & Ivanova 2016), leaving open the question of what other physical mechanism may unbind the envelope in the more massive stars. Solutions or partial solutions have been put forth, such as the action of jets during the dynamical inspiral that can unbind a few times more envelope than if they are not present (Shiber & Soker 2018, and Shiber *et al.* in preparation).

Global magnetic fields may play a role in the envelope dynamics and ejection. Regos & Tout (1995) have discussed that a strong sheer would develop in the CE layers as they are being imparted angular momentum by the inspiralling companion. This sheer would feed an α-Ω dynamo and strengthen a strong, global magnetic field that may play a role in the envelope gas dynamics.

To corroborate or refute this prediction, Tocknell *et al.* (2014) used jets in PN as a way to probe the magnetic field at different stages during the CE interaction. An accretion disk can form around the companion either at the time of Roche lobe overflow or indeed later during the CE phase or shortly after. Some of the observed PN jets can be kinematically dated to have formed just before, while others just after the dynamical ejection. This is taken as an indication that a disk formed around the companion as it accreted mass, and that it was threaded by a magnetic field. If so then physical arguments can be used to connect the measured jet mass loss rates to the magnetic field strength. They deduced local fields of between 1 and 1000 G, depending on whether the jet is launched just before or just after the CE inspiral. In the former case the magnetic field is plausibly that characterising the surface of giants. In the latter case the derived field intensity is in line with predictions of the "wound" and intensified magnetic field by Regos & Tout (1995).

The only 3D magneto-hydrodynamic (MHD) simulation to attempt to study at least one aspect of magnetic fields is that of Ohlmann *et al.* (2016b), where their MHD calculation records field growth locally around the companion. We comment that this magnetic field is not the global magnetic field predicted by Regos & Tout (1995). The Ohlmann *et al.*, (2016b) field is a local field that grows as a result of the magnetorotational instability (Balbus & Hawley 1998) and not one that grows as a result of an α-Ω dynamo – the lack of properly resolved convection in CE simulations would make the *ab initio* simulation of such a field an impossibility. Finally, it has been suggested (Tricco & Price 2012) that the lack of divergence cleaning in the Pakmor & Springel (2013) MHD method utilised by Ohlmann *et al.* (2016b) may lead to *too strong* a field growth. Ultimately, these are very early days for MHD simulations of the CE interactions.

5. Planetary nebulae from common envelopes

When the AGB star suffers a CE interaction that leads to the ejection of the envelope, a PN may form. The CE ejecta will be ploughed up by the fast wind from the heating and luminous post-AGB primary and ionised by its ionising radiation. The shape of the PN will be influenced by the equatorially concentrated CE ejecta.

Two studies, García-Segura *et al.* (2018) and Frank *et al.* (2018), have carried out the experiment of ploughing the CE ejecta with a fast wind supposedly launched by the now hotter, post-AGB central star. While we refer the reader to the work of Reichardt *et al.* (2018) for a detailed comparison, we list here some of the salient similarities and differences between the two simulations (see Table 2).

A difference between the two studies' setups is that García-Segura *et al.* (2018) considered a very compact distribution of CE ejecta (~2 AU; Fig. 3), due to the short duration of the CE simulation of Ricker & Taam (2012; only 57 days of dynamical CE ejection). Ricker & Taam (2012), in addition, started the interaction when the primary was already overflowing its Roche lobe (Table 2), leading to an immediate inspiral and burst-like ejection of the common envelope. On the other hand, Frank *et al.* (2018) started

Table 2. A comparison of the initial setups and outcomes of the two existing simulations of post-common envelope planetary nebula.

Characteristic	García-Segura *et al.* (2018)	Frank *et al.* (2018)
Dimensionality	2D ($\theta = 0 - 90$ deg)	3D
RGB primary's mass	$M_1 = 1.05$ M$_\odot$	$M_1 = 0.88$ M$_\odot$
Companion's mass	$M_2 = 0.6$ M$_\odot$	$M_2 = 0.6$ M$_\odot$
Primary's radius	$R_1 = 32$ R$_\odot$	$R_1 = 90$ R$_\odot$
Initial separation	62 R$_\odot$	218 R$_\odot$
Primary's Roche lobe radius	$R_{RL,1} = 32$ R$_\odot$	$R_{RL,1} = 90$ R$_\odot$
Size of CE ejecta at end CE simulation	~2 AU	~100 AU
Total time of CE simulation	57 days	7000 days
CE simulation time after orbital stabilisation	15 days	2000 days
Total time PN simulation	10 000 years	7000 days
Size of PN at 7 000 days	100–200 au	100 au
Size of the PN at 10 000 years	2–4 pc	–

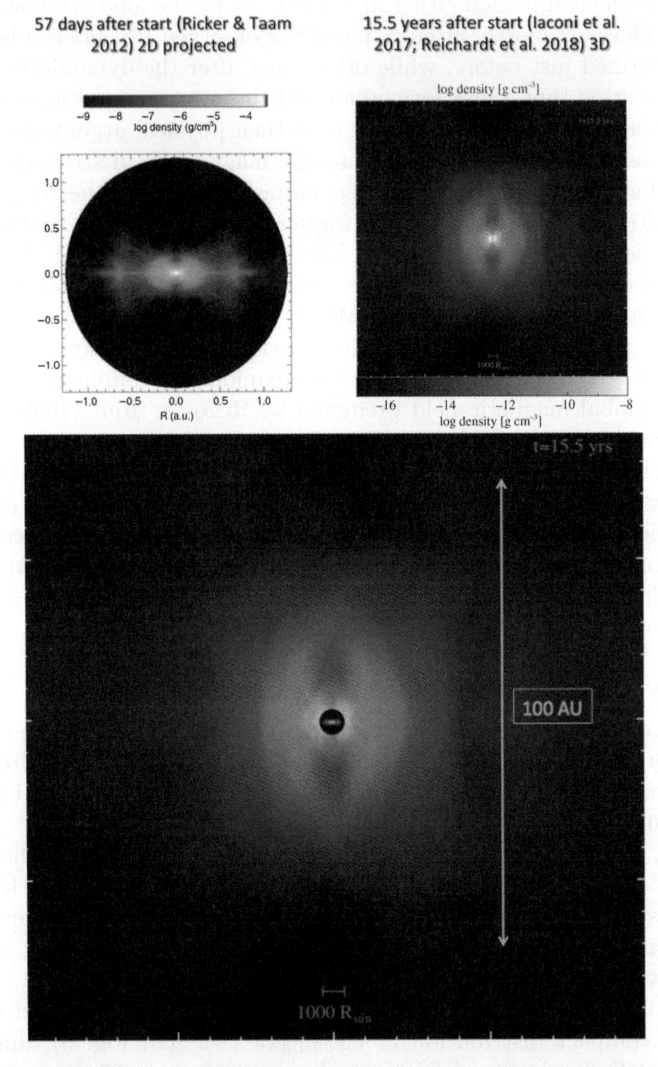

Figure 3. Top left: Figure 1 of García-Segura *et al.* (2018), showing the post-CE gas distribution at the end of the simulation of Ricker & Taam (2012), but projected into 2D and reflected about the horizontal axis. Top right: a slice in the plane perpendicular to the orbital plane at the end of the simulation of Reichardt *et al.* (2018). Lower panel, the top two panels are overlaid and scaled to one another. Both the PN simulated by the two studies extend to about 100 AU after approximately 10 000 days.

with CE ejecta that had more time to expand (15 years) and where the CE simulation was started at the start of Roche lobe overflow. The Roche lobe overflow phase has time to eject a disk from the L2 and L3 Lagrangian points for 7000 days, followed by the burst-like CE ejection, which ploughs into the disk. The evolution of the ejecta is then followed for a further 2000 days, allowing further expansion. This resulted in a very distinctive, overall larger (~100 AU) density distribution (Fig. 3).

Another difference is that García-Segura *et al.* (2018) were able to evolve the PN to the age of 10 000 years thanks to the 2D configuration, while Frank *et al.* (2018) only calculated the initial 10 000 days because of the full 3D treatment.

The time after the ejection of the common envelope when the fast wind starts, as well as the momentum of the fast wind are different for the two studies. These variables matter greatly to the PN that will form and we have not even scratched the surface of how to chose these parameters and how to interface the CE simulation with the PN simulation. We are therefore not going to compare these choices in this review and refer the reader to the two original publications (García-Segura *et al.* 2018; Frank *et al.* 2018).

Both studies observed a very pronounced hydrodynamic collimation thanks to the extremely thin and high density contrast funnel seen in Fig. 3, top left panel and equally present at the core of the top right panel figure. In addition, Frank *et al.* (2018) showed that 3D effects (asymmetry) could be observed if the fast wind momentum is low, though whether the asymmetries remain with further nebular evolution remains a question. Finally, we point out that there is a strong interplay between the disk ejected during the Roche lobe outflow and the mass which is ejected during the faster CE dynamical phase. These two ejections, which both precede the fast wind that later ploughs them, forming the PN, interact with one another generating different degrees of asymmetry and equatorial density concentrations. This interplay may contribute to the variety of PN shapes observed around post-CE binaries.

6. Planets in common envelopes

The interaction between planets and giants has been considered several times in different contexts and with different techniques (e.g., Soker 1998; Carlberg *et al.* 2009; Villaver & Livio 2009; Passy *et al.* 2012, to mention a few). simulated a CE interaction between $3\,M_\odot$ RGB and AGB giants and a $10\,M_J$ companion. The results of those simulations were that the companion inspirals all the way to the core where, presumably, it would get destroyed. It was also determined that the length of the dynamical inspiral is slower than for more massive companions: ~10 years for the RGB star and ~100 years for the AGB case. Yet, it can be showed that during these times the planet is not fully ablated (Passy *et al.* 2012), which presumably would mean that it is tidally disrupted near the core, possibly even forming a disk (Nordhaus & Blackman 2006). The planet inspiral causes a modest expansion of the photosphere but substantial degree of spin-up, in line with certain observations (Carlberg *et al.* 2009).

It is also possible that the presence of the planet in the atmosphere of the giant may trigger secondary effects. In particular the expansion of the stellar envelope may cause a recombination front and some of the relatively high density, low temperature gas farther out may also host dust formation. This may be only of secondary importance because most of the gas would still remain at temperature higher than the condensation temperature. Another effect may have to do with pulsations. AGB stars such as Miras pulsate with periods of a couple of hundred days (Ireland *et al.* 2011), which is similar to the initial orbital period of the planet. Whether an interaction between the pulsations and the energy deposition by the companion during the inspiral is a possibility rests to be determined.

References

Balbus, S. A. & Hawley, J. F. 1998, Reviews of Modern Physics, 70, 1
Carlberg, J. K., Majewski, S. R., & Arras, P. 2009, ApJ, 700, 832
Chamandy, L., Frank, A., Blackman, E. G., *et al.* 2018, MNRAS, 480, 1898
De Marco, O., Farihi, J., & Nordhaus, J. 2009, Journal of Physics Conference Series, 172, 012031
De Marco, O. & Izzard, R. G. 2017, Pub. of the Astronomical Society of Australia, 34, e001
Duchêne, G. & Kraus, A. 2013, ARA&A, 51, 269
Frank, A., Chen, Z., Reichardt, T., *et al.* 2018, ArXiv e-prints arXiv:1807.05925
Fryer, C. L., Rockefeller, G., & Warren, M. S. 2006, ApJ, 643, 292
Fryxell, B., Olson, K., Ricker, P., *et al.* 2000, ApJ Supplement Series, 131, 273
Galaviz, P., De Marco, O., Staff, J. E., & Iaconi, R. 2017, ApJ Supplement Series, 498, 293
García-Segura, G., Ricker, P. M., & Taam, R. E. 2018, ApJ, 860, 19
Grichener, A., Sabach, E., & Soker, N. 2018, MNRAS, 478, 1818
Hillwig, T. C., Frew, D. J., Reindl, N., *et al.* 2017, AJ, 153, 24
Hillwig, T. C., Jones, D., De Marco, O., *et al.* 2016, ApJ, 832, 125
Iaconi, R., De Marco, O., Passy, J.-C., & Staff, J. 2018, MNRAS, 477, 2349
Iaconi, R., Reichardt, T., Staff, J., *et al.* 2017, MNRAS, 464, 4028
Ireland, M. J., Scholz, M., & Wood, P. R. 2011, MNRAS, 418, 114
Ivanova, N. 2018, ApJ, 858, L24
Ivanova, N., Justham, S., Chen, X., *et al.* 2013, The Astronomy and Astrophysics Review, 21, 59
Ivanova, N. & Nandez, J. L. A. 2016, MNRAS, 462, 362
Kasliwal, M. M. 2012, PASA, 29, 482
Kuruwita, R. L., Staff, J., & De Marco, O. 2016, MNRAS, 461, 486
MacLeod, M., Ostriker, E. C., & Stone, J. M. 2018a, ArXiv e-prints arXiv:1808.05950
MacLeod, M., Ostriker, E. C., & Stone, J. M. 2018b, ApJ, 863, 5
Madappatt, N., De Marco, O., & Villaver, E. 2016, MNRAS, 470, 317
Mauron, N. & Huggins, P. J. 2006, A&A, 452, 257
Miszalski, B., Acker, A., Moffat, A. F. J., Parker, Q. A., & Udalski, A. 2009, A&A, 496, 813
Nandez, J. L. A. & Ivanova, N. 2016, MNRAS, 460, 3992
Nandez, J. L. A., Ivanova, N., & Lombardi, J. C. 2015, MNRAS, 450, L39
Nandez, J. L. A., Ivanova, N., & Lombardi, Jr., J. C. 2014, ApJ, 786, 39
Nordhaus, J. & Blackman, E. G. 2006, MNRAS, 370, 2004
Ohlmann, S. T., Röpke, F. K., Pakmor, R., & Springel, V. 2016a, ApJ, 816, L9
Ohlmann, S. T., Röpke, F. K., Pakmor, R., & Springel, V. 2017, A&A, 599, A5
Ohlmann, S. T., Röpke, F. K., Pakmor, R., Springel, V., & Müller, E. 2016b, MNRAS, 462, L121
O'Shea, B. W., Bryan, G., Bordner, J., *et al.* 2004, ArXiv Astrophysics e-prints arXiv:astro-ph/0403044
Paczynski, B. 1976, in IAU Symposium, Vol. 73, Structure and Evolution of Close Binary Systems, ed. P. Eggleton, S. Mitton, & J. Whelan, 75
Pakmor, R. & Springel, V. 2013, MNRAS, 432, 176
Passy, J.-C. & Bryan, G. L. 2014, ApJ Supplement Series, 215, 8
Passy, J.-C., Mac Low, M.-M., & De Marco, O. 2012, ApJ, 759, L30
Pejcha, O., Metzger, B. D., & Tomida, K. 2016a, MNRAS, 461, 2527
Pejcha, O., Metzger, B. D., & Tomida, K. 2016b, MNRAS, 455, 4351
Pian, E., D'Avanzo, P., Benetti, S., *et al.* 2017, Nature, 551, 67
Raghavan, D., McAlister, H. A., Henry, T. J., *et al.* 2010, ApJ Supplement Series, 190, 1
Rasio, F. A. & Livio, M. 1996, ApJ, 471, 366
Regos, E. & Tout, C. A. 1995, MNRAS, 273, 146
Reichardt, T. A., De Marco, O., Iaconi, R., Tout, C. A., & Price, D. J. 2018, ArXiv e-prints arXiv:1809.02297
Ricker, P. M. & Taam, R. E. 2008, ApJ, 672, L41
Ricker, P. M. & Taam, R. E. 2012, ApJ, 746, 74
Sana, H., de Mink, S. E., de Koter, A., *et al.* 2012, Science, 337, 444
Sandquist, E. L., Taam, R. E., & Burkert, A. 2000, ApJ, 533, 984

Sandquist, E. L., Taam, R. E., Chen, X., Bodenheimer, P., & Burkert, A. 1998, ApJ, 500, 909
Shiber, S. & Soker, N. 2018, MNRAS, 477, 2584
Soker, N. 1998, ApJ, 496, 833
Staff, J. E., De Marco, O., Macdonald, D., *et al.* 2016a, MNRAS, 455, 3511
Staff, J. E., De Marco, O., Macdonald, D., *et al.* 2016b, MNRAS, 455, 3511
Staff, J. E., De Marco, O., Wood, P., Galaviz, P., & Passy, J.-C. 2016c, ArXiv e-prints arXiv:1602.03130
Stone, J. M., Gardiner, T. A., Teuben, P., Hawley, J. F., & Simon, J. B. 2008, ApJ Supplement Series, 178, 137
Taam, R. E., Bodenheimer, P., & Ostriker, J. P. 1978, ApJ, 222, 269
Terman, J. L. & Taam, R. E. 1996, ApJ, 458, 692
Terman, J. L., Taam, R. E., & Hernquist, L. 1994, ApJ, 422, 729
Terman, J. L., Taam, R. E., & Hernquist, L. 1995, ApJ, 445, 367
Tocknell, J., De Marco, O., & Wardle, M. 2014, MNRAS, 439, 2014
Toonen, S. & Nelemans, G. 2013, A&A, 557, A87
Tricco, T. S. & Price, D. J. 2012, Journal of Computational Physics, 231, 7214
Villaver, E. & Livio, M. 2009, ApJ, 705, L81
Zahn, J. P. 1966, Annales d'Astrophysique, 29, 313

Discussion

WHITELOCK: As you know symbiotic stars are binaries and some have jets, e.g. R Aqr. They are different from the binaries you are talking about but may be of interest. One of the nebulae you described as a PPN is actually a symbiotic Mira (if I remember correctly).

DE MARCO: The nebula in question, OH231.8+4.2 has a Mira at its centre, not a pAGB star as one might expect. We do not know of a nearby companion though. It is hard to explain. In Staff *et al.* 2016a we tried a CE-merger scenario that launched jets during a CE in-spiral before the merger. The numbers work out approximately, but who knows!

SAHAI: Boomerang is a good example, probably best example of a CE event, where you see *dust formation* (both in ejecta and the disk) formed around the central merged binary. But in this case the wind is not *dust-driven.* Dust has formed in the ejecta, presumably.

DE MARCO: Well, if so, dust in this case may have not contributed to the loss of the envelope. Dust formation on CE transient is observed. Yet it may form only at the equator and not be a way to drive a wind.

KOBAYASHI: You mentioned the impact of a jet on separation. But how much change do you expect on the delay-time of double-degenerate SNIa and NS merger?

DE MARCO: *If* jets really happen and *if* they leave the binary at a much wider orbital separation, then the delay time would become very long or even infinite! It is too early to tell whether the extent to which jets, if they happen, may affect the distribution of delay times.

RICKER: In your CE simulations with jets, have you considered jets at angles other than 90° to the orbital plane? Presumably these would change the shape of the outflow, and if the outflow shapes a later PN phase, the jet angle might contribute to the diversity of PN shapes.

DE MARCO: No, we have not...yet. But before doing so, we would have to justify why the accretion disk around the companion would have an angular momentum vector that is not aligned with the orbital one.

Why Galaxies Care About AGB Stars: A Continuing Challenge through Cosmic Time
Proceedings IAU Symposium No. 343, 2019
F. Kerschbaum, M. Groenewegen & H. Olofsson, eds.

doi:10.1017/S1743921318006403

Orbital properties of binary post-AGB stars

Glenn-Michael Oomen[1,2], Hans Van Winckel[1] and Onno Pols[2]

[1]Instituut voor Sterrenkunde, KU Leuven,
Celestijnenlaan 200D, B-3001, Leuven, Belgium

[2]Department of Astrophysics/IMAPP, Radboud University,
P.O. Box 9010, 6500 GL Nijmegen, the Netherlands
email: glennmichael.oomen@kuleuven.be

Abstract. We present the results of a decade-long radial velocity monitoring campaign of post-AGB binaries. We derived the orbital elements of 33 post-AGB binaries. We find a companion mass distribution centred around 1.09 $M_{\odot}$ with a very large spread. All the post-AGB binaries in our sample are expected to have filled their Roche lobes while at giant dimensions. Current binary evolution models are unable to explain the observed distribution of orbits.

Keywords. stars: AGB and post-AGB, binaries: general

---

1. Introduction

During the end-stages in the evolution of low- and intermediate-mass stars the envelope of the star is removed, either via a stellar wind or via binary interactions. As the mass in the envelope of the star becomes very small ($\sim 0.02\ M_{\odot}$), the radius of the star starts to decrease as a result of mass loss and the star evolves towards higher temperatures. The phase in which the star stops to drive a dust-driven wind but is not yet hot enough to ionise its surroundings to emerge as a planetary nebula is known as the post-asymptotic giant branch (post-AGB) phase (Van Winckel 2003).

The evolution of low- and intermediate-mass stars during the AGB phase is still plagued by many uncertainties. Especially the interaction of evolved stars with a companion is still poorly understood (see reviews by De Marco & Izzard 2017, Jones & Boffin 2017).

There are a number of binary interactions that can impact the evolution of a binary system. Tides play a large role once the radius of the RGB or AGB star increases to the extent that the AGB star fills a significant fraction of its Roche lobe (Zahn 1977). This effect will lead to a synchronisation of the rotational spin of the RGB/AGB star to the orbit and a circularisation of the orbit on a relatively short time scale (Hurley *et al.* 2002). Additionally, the AGB star can interact with its companion via a stellar wind. This can either happen isotropically via the Bondi-Hoyle-Littleton mechanism (Edgar 2004) if the wind velocity is much larger than the orbital velocity, or via a so-called wind-Roche lobe overflow (wind-RLOF, Mohamed & Podsiadlowski 2007). In the latter case, the wind fills the Roche lobe of the AGB star and gets funnelled onto the companion, leading to high accretion rates. The binary can also undergo regular RLOF, but this is usually unstable for AGB stars with deep convective envelopes (Hjellming & Webbink 1987, but see also Pavlovskii & Ivanova 2015). In the unstable case, the binary can enter a common envelope (Izzard *et al.* 2012), leading to spiralling-in of the binary and possibly a merger.

Canonical population synthesis models tend to predict a bimodal period distribution after the interaction (Izzard *et al.* 2010, Nie *et al.* 2012), with on one hand short-period systems that have undergone a common-envelope evolution, and on the other hand wider systems that did not interact much, but where mass loss via a stellar wind led to an increase in the orbital period. However, this is in stark contrast to the observed orbits of binaries that have undergone an interaction involving an evolved star, such as post-AGB binaries (Van Winckel 2003), carbon-enhanced metal-poor (CEMP) stars (Hansen *et al.* 2016, Jorissen *et al.* 2016), Barium stars (Van der Swaelmen *et al.* 2017), and others.

Since the post-AGB phase occurs immediately after the AGB, binary post-AGB stars in intermediate-period orbits (100–3000 days) can be regarded as the immediate products of a binary interaction on the AGB. Consequently, by investigating the orbital properties, chemical properties, and circumstellar environments of these binaries, we aim to learn more about the strong binary interaction on the AGB. In these proceedings, we will focus on the analysis of the orbital properties of post-AGB binaries. For a more in-depth discussion on all the properties of post-AGB binaries, we refer to Oomen *et al.* (2018).

2. Data

In order to investigate the orbital properties of binary post-AGB stars, we have at our disposal the results of a long-term and systematic radial velocity monitoring campaign with the HERMES spectrograph (Raskin *et al.* 2011) mounted on the Mercator telescope in La Palma, Spain (Van Winckel 2015). This high-resolution spectrograph ($R \approx 80000$) is operational since 2009 and has provided us with a large number of radial velocity measurements for many galactic post-AGB stars, selected based on the properties of their spectral energy distribution (SED) (de Ruyter *et al.* 2006, Gezer *et al.* 2015).

Additionally, we have complemented our HERMES data with older data collected from literature taken with several European southern observatory (ESO) telescopes. This allows us to include targets from both the northern and southern hemispheres.

The result from this huge effort is that we managed to fit the orbits for 33 post-AGB binaries. Because we have a large sample of homogeneously derived orbital elements available, we can apply statistical methods to our sample in order to learn more about the general properties of post-AGB binaries.

3. Results

We have derived the distribution of companion masses by making use of the mass functions that are available from the orbital elements. The mass functions are related to the masses of the individual components of the binary according to

$$f(m) = \frac{M_2^3}{(M_1 + M_2)^2} \sin i, \tag{3.1}$$

where M_1 is the mass of the post-AGB star, M_2 is the mass of the companion, and i is the inclination angle.

We can now derive the distribution of companion masses by first assuming a distribution for the post-AGB masses (M_1). This distribution can be estimated since post-AGB stars are stars that have lost their entire envelope, hence we assume that their mass distribution is similar to the observed white-dwarf mass distribution. For that reason, we have used a Gaussian distribution centred around 0.6 $M_\odot$ with standard deviation of 0.05 according to Kilic *et al.* (2018).

Similarly, we can assume a distribution for the inclination angles if we assume that all the orbits are randomly inclined in 3-dimensional space. This is described by

$$i = \arccos(z), \tag{3.2}$$

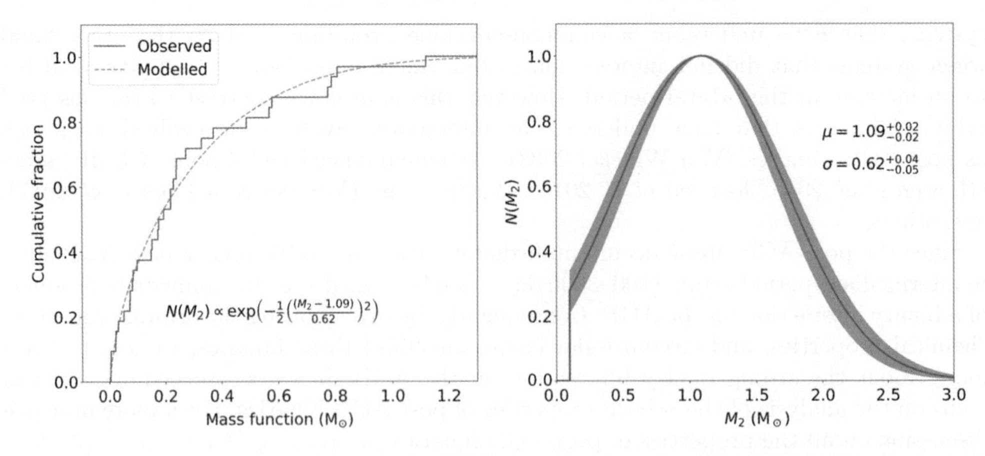

Figure 1. Left: observed vs modelled cumulative mass-function distributions for our sample of post-AGB binaries. Observed mass functions are showed as a blue line, while the simulated distribution is shown as a dashed orange line. Right: distribution of companion masses. The blue curve is described by a Gaussian profile around 1.09 $M_\odot$ with standard deviation of 0.62. The red-shaded region shows the uncertainty. Figures are taken from Oomen *et al.* (2018).

where z is the length of the projection of the unit normal vector along the line-of-sight to the observer, which is uniformly distributed in $[-1, 1]$.

Finally, we can derive the distribution of companion masses, such that the three distributions together result in a mass function distribution according to Eq. 3.1 that best fits the observed cumulative mass function distribution. In this process, we account for observational bias by removing combinations that result in a low velocity semi-amplitude, since we would not have detected those binaries. Moreover, we limit the inclination angle to 75°, since edge-on systems would be obscured by the circumbinary disc, hence would not be observed.

The result of this procedure is that a Gaussian distribution centred around $1.09^{+0.02}_{-0.02}$ $M_\odot$ with standard deviation of $0.62^{+0.04}_{-0.05}$ $M_\odot$ best fits the observed cumulative mass function distribution, which is shown in Fig. 1.

The shape of the companion mass distribution is given on the right panel of Fig. 1. The most remarkable feature of this distribution is the large spread in masses that is observed. The companion masses range from 0.1 $M_\odot$ to about 3 $M_\odot$. This shows the very large diversity in post-AGB binary systems that is observed. Moreover, the large spread in masses leads to a large spread in mass ratios of the binaries, which has strong implications for binary evolution models. Most binary interaction mechanisms, such as the stability of mass transfer, or the orbital evolution of the system in general, depend critically on the mass ratio.

Now that we have derived a distribution of companion masses, we can use the orbital periods of the binaries and Kepler's third law to find a distribution of orbital separations, and finally Roche-lobe radii. If we assume that the separation and masses of the binary did not change much since the end of the interaction on the AGB, then the current distribution of Roche-lobe radii should remain the same.

The distribution of derived Roche-lobe radii is shown in Fig. 2. The Roche lobes range from 0.2 AU to about 1.5 AU, with the majority of Roche lobes being smaller than 1 AU. Since the typical maximum size of an AGB star is 1–2 AU, we expect almost all our post-AGB stars to have filled their Roche lobe at some point in their evolution. This means that all the post-AGB stars in our sample are the result of a strong interaction during the AGB phase.

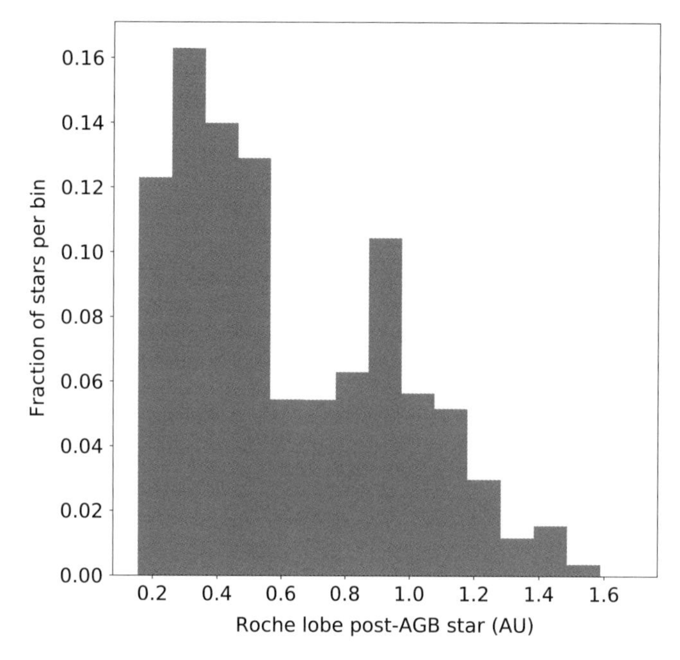

Figure 2. Histogram of the distribution of Roche lobes for our sample of post-AGB binaries. Figure is taken from Oomen *et al.* (2018).

References

De Marco, O., & Izzard, R. G. 2017, *PASA*, 34, 1
de Ruyter, S., van Winckel, H., Maas, T., *et al.* 2006, *A&A*, 448, 641
Edgar, R. 2004, *New Astron. Revs*, 48, 843
Gezer, I., Van Winckel, H., Bozkurt, Z., *et al.* 2015, *MNRAS*, 453, 133
Hjellming, M. S., & Webbink, R. F. 1987, *ApJ*, 318, 794
Hansen, T. T., Andersen, J., Nordström, B., *et al.* 2016, *A&A*, 586, A160
Hurley, J. R., Tout, C. A., & Pols, O. R. 2002, *MNRAS*, 329, 897
Izzard, R. G., Dermine, T., & Church, R. P. 2010, *A&A*, 523, A10
Izzard, R. G., Hall, P. D., Tauris, T. M., & Tout, C. A. 2012, IAU Symposium, 283, 95
Jones, D., & Boffin, H. M. J., 2017, *Nature Astronomy*, 1, 0117
Jorissen, A., Van Eck, S., Van Winckel, H., *et al.* 2016, *A&A*, 586, A158
Kilic, M., Hambly, N. C., Bergeron, P., Genest-Beaulieu, C., & Rowell, N. 2018, *MNRAS*, 479, L113
Mohamed, S., & Podsiadlowski, P. 2007, *ASP-CS*, 372, 397
Nie, J. D., Wood, P. R., & Nicholls, C. P. 2012, *MNRAS*, 423, 2764
Oomen, G.M., Van Winckel, H., Pols, O. R., *et al.* 2018, *A&A*, submitted
Pavlovskii, K., & Ivanova, N. 2015, *MNRAS*, 449, 4415
Raskin, G., van Winckel, H., Hensberge, H., *et al.* 2011, *A&A*, 526, A69
Van der Swaelmen, M., Boffin, H. M. J., Jorissen, A., & Van Eck, S. 2017, *A&A*, 597, A68
Van Winckel, H. 2003, *ARAA*, 41, 391
Van Winckel, H. 2015, *EAS Publications Series*, 71, 121
Zahn, J.-P. 1977, *A&A*, 57, 383

Discussion

SAHAI: You find that many companions have large masses, $\gtrsim 1\ M_\odot$, so that means the primary must have been more massive at the main-sequence; so where is the mass that the primary must have ejected? Why is it not visible?

OOMEN: We expect only a very small fraction of the mass to go to the formation of the circumbinary disc. It is possible that quite some mass gets transferred onto the companion, making it more massive than it initially was. It can also be that the AGB star interacts with the companion at a later stage in the evolution, such that already a large amount of mass is lost via a stellar wind long before the AGB star filled its Roche lobe. However, the short answer is: we do not know.

Why Galaxies Care About AGB Stars:
A Continuing Challenge through Cosmic Time
Proceedings IAU Symposium No. 343, 2019
F. Kerschbaum, M. Groenewegen & H. Olofsson, eds.

doi:10.1017/S174392131800666X

Accretion in common envelope evolution

Luke Chamandy[1], Adam Frank[1], Eric G. Blackman[1], Jonathan Carroll-Nellenback[1], Baowei Liu[1], Yisheng Tu[1], Jason Nordhaus[2,3], Zhuo Chen[1] and Bo Peng[1]

[1]Department of Physics and Astronomy, University of Rochester, Rochester NY 14618, USA

emails: lchamandy@pas.rochester.edu, afrank@pas.rochester.edu, blackman@pas.rochester.edu

[2]National Technical Institute for the Deaf, Rochester Institute of Technology, NY 14623, USA

[3]Center for Computational Relativity and Gravitation, Rochester Institute of Technology, NY 14623, USA

Abstract. Common envelope evolution (CEE) occurs in some binary systems involving asymptotic giant branch (AGB) or red giant branch (RGB) stars, and understanding this process is crucial for understanding the origins of various transient phenomena. CEE has been shown to be highly asymmetrical and global 3D simulations are needed to help understand the dynamics. We perform and analyze hydrodynamic CEE simulations with the adaptive mesh refinement (AMR) code AstroBEAR, and focus on the role of accretion onto the companion star. We bracket the range of accretion rates by comparing a model that removes mass and pressure using a subgrid accretion prescription with one that does not. Provided a pressure-release valve, such as a bipolar jet, is available, super-Eddington accretion could be common. Finally, we summarize new results pertaining to the energy budget, and discuss the overall implications relating to the feasibility of unbinding the envelope in CEE simulations.

Keywords. Binaries: close – accretion, accretion discs – stars: kinematics – hydrodynamics – methods: numerical

---

1. Introduction

When a giant primary overflows its Roche lobe, this can lead to the engulfment of the main sequence (MS) or compact object secondary, resulting in the rapid inspiral of the secondary and dense core of the giant. This process, known as common envelope evolution (CEE), leads to a variety of crucial phenomena in stellar evolution (Paczyński *et al.* 1976, Ivanova *et al.* 2013, Demarco & Izzard 2017; see also O. De Marco, this volume). CEE is needed to explain the bipolar symmetry of many planetary nebulae (PNe) and pre-planetary nebulae (PPNe), and the small separations of their binary central star orbits in several instances (Jones & Boffin 2017 and references therein). Recent simulations (e.g. Ricker & Taam 2012, Ohlmann *et al.* 2016, Iaconi *et al.* 2018) find that only a small fraction $\sim 10\%$ of the envelope of the simulated RGB star becomes unbound during the simulation. However, there are many observations that require the envelope of the giant star to have been ejected during CEE. This has suggested to some that an energy source other than the liberated orbital energy may be required, and one such possibility is the potential energy liberated by accretion of envelope gas onto the secondary.

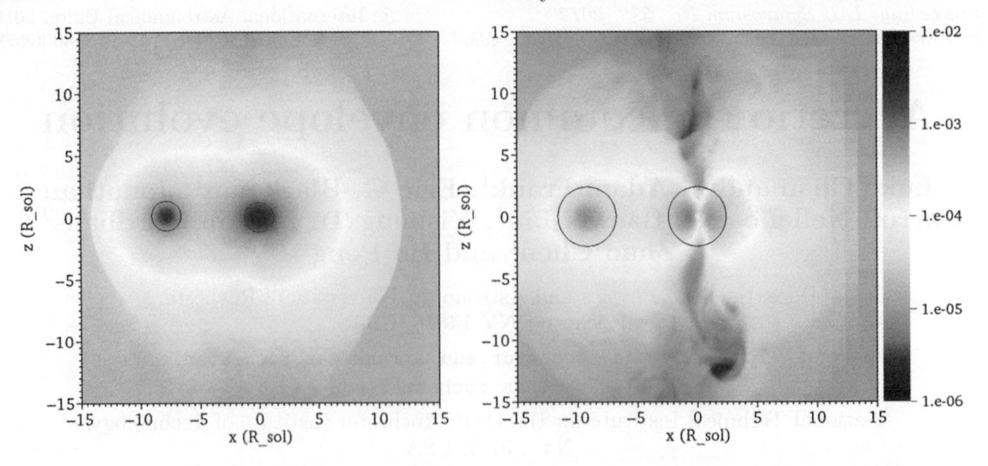

Figure 1. Gas density in $\mathrm{g\,cm^{-3}}$ at $t = 40\,\mathrm{d}$ in a slice through both particles perpendicular to the orbital plane. Model A (no subgrid accretion) is shown in the left-hand panel and Model B (subgrid accretion) is shown in the right-hand panel. The secondary is at the center with the primary particle to its left. Spline softening spheres are shown with green circles.

2. Method and results

In Chamandy *et al.* (2018a) we used the multi-physics AMR code AstroBEAR (Carroll-Nellenback *et al.* 2013) to carry out global simulations of CEE. Our simulation evolves a binary system consisting of a $2\,\mathrm{M_\odot}$ RGB primary with a $0.4\,\mathrm{M_\odot}$ core along with a $1\,\mathrm{M_\odot}$ secondary, initialized in a circular orbit with separation a slightly larger than the RG radius of $48\,\mathrm{R_\odot}$. Our setup and initial conditions are similar to those of Ohlmann *et al.* (2017) and Ohlmann *et al.* (2016). Core and secondary are modeled as gravitation-only point particles. One of our high-resolution runs (Model B) uses a subgrid model for accretion onto the secondary, moving mass from the grid to the particle and removing energy and pressure from the grid (Krumholz *et al.* 2004), while the other run (Model A) does not. We find that while the global morphology and evolution is very similar in the two runs, the rate of mass flow toward the secondary stagnates in the run without subgrid accretion, whereas the accretion rate reaches highly super-Eddington values in the run with subgrid accretion. This demonstrates how very different results for accretion during CEE can be obtained depending on whether or not an inner loss valve is present.

In Fig. 1 we show the difference in morphology that arises near the secondary between the run without (Model A, left) and with (Model B, right) subgrid accretion (see figure caption for details). In Model B, the flow around the secondary has developed a toroidal morphology, while for Model A there is only a hint of a such a torus. Next, Fig. 2 shows the accumulation of mass around the secondary in Model A (left panel) as well as the accretion rate onto the secondary in Model B (right panel). It can be seen that without a mechanism to release the central pressure, the concentration of mass around the secondary reaches a quasi-steady state. We also note that when the softening length is halved suddenly (green vertical line), the concentration of material around the secondary becomes more dense, implying that the simulation is not converged with respect to the softening length, even though the latter is kept below $1/5$ of the inter-particle separation a. By contrast, when a pressure-release valve is implemented in the form of the subgrid accretion model, mass accretes at a rate of 0.2–$2\,\mathrm{M_\odot\ yr^{-1}}$, which is about 2–3 (4–5) orders of magnitude larger than Eddington for a MS (white dwarf (WD)) secondary. Finally, Fig. 3 shows the velocity in the orbital plane in the frame rotating about the secondary with the orbital angular velocity of the particles, normalized by the corresponding

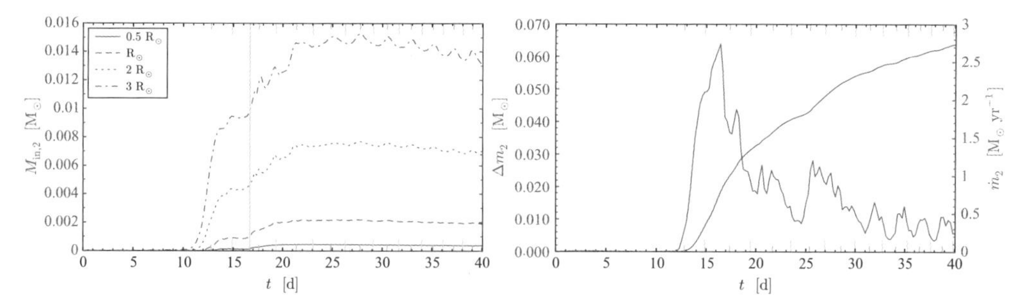

Figure 2. Left: Gas mass contained within spheres of various sizes centered on the secondary for Model A. (The vertical green line shows when the softening length was halved.) Long light blue (short orange) tick marks show the times of apastron (periastron) passage. Right: The accreted mass for Model B (blue, left-hand axis) and the accretion rate (red, right-hand axis).

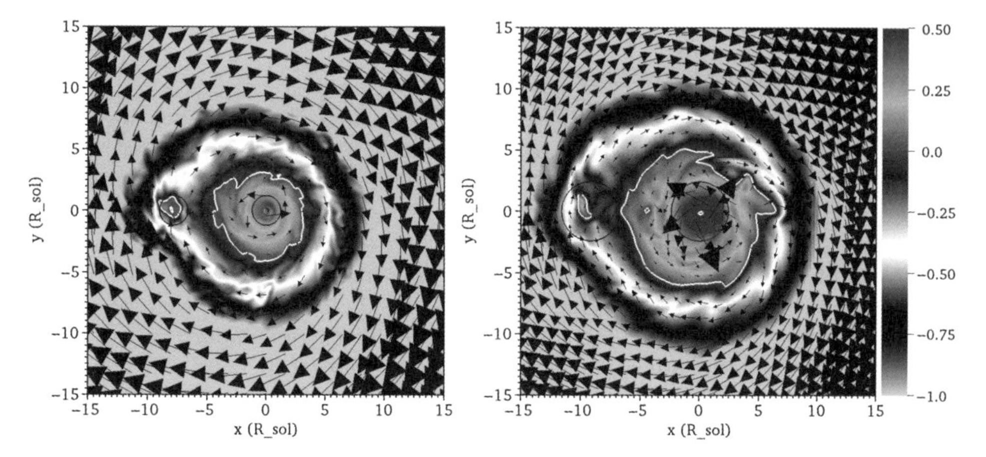

Figure 3. Slice through the orbital plane at $t = 40$ d, where color represents the tangential (with respect to the secondary) velocity component in the frame of reference rotating about the secondary with the instantaneous orbital angular velocity of the particles, normalized by the local Keplerian circular speed around the secondary. Zero tangential velocity is shown using a white contour. Velocity vectors in this frame projected onto the orbital plane are also shown.

local Keplerian value. For both models the gas orbits the secondary but is mainly pressure supported. However, assuming angular momentum to be conserved deeper into the unresolved region (within the softening sphere of the secondary) the flow would become rotationally supported at a radius of ~ 0.05–$0.15\,\mathrm{R}_\odot$. This implies that a thin disc has room to develop around a WD but not a MS secondary. Such discs are likely to be associated with jets that can also act as pressure-release valves *if* they can efficiently transport accretion energy outward so as not to impede the accretion flow (Moreno Méndez *et al.* 2017; Soker 2017). This possibility needs to be explored in global CE simulations. If the secondary was a neutron star, then neutrino transport could remove pressure, allowing for super-Eddington accretion (Armitage & Livio 2000).

3. Energy budget in common envelope evolution

In a separate work, Chamandy *et al.* (2018b), we analyze the transfer of energy between different forms in the simulation of Model A and interpret our results using the so-called energy formalism. We find, in general agreement with previous results, that only about 10–20% of the envelope is unbound during the simulation (with 'unbound' gas defined as that with positive energy density) and that all of the unbinding occurs early on, roughly before the first periastron passage. Counterintuitively, the total energy of the gas remains approximately constant during this time. This can be explained by noting that the plunge-in of the secondary toward the center of the RG causes the kinetic energy of the outer layers to rise, while at the same time resulting in the inner layers being more tightly bound. For $0.1 < \alpha_{\rm CE} < 1$ (see Ivanova *et al.* 2013 for a discussion of this parameter), we find that the envelope is not expected to become completely unbound until the inter-particle separation has reduced to $0.3 < a/{\rm R}_\odot < 3$. Most if not all of this range is currently inaccessible to simulations due to finite resolution and softening length, so it is not really surprising that simulations fail to unbind the envelope, and tend to result in particles with a final separation of order a few softening lengths. Fittingly, considering the topic of this conference, this suggests that binaries involving AGB stars, which are more extended and loosely bound compared with RGB stars, may be more promising targets for studies that hope to simulate the parameter regime for which the end result is an unbound envelope, as opposed to a merger.

4. Summary and conclusions

Observations of bipolar PNe and PPNe imply that many (if not all) such systems have passed through a common envelope phase, resulting in a close binary orbit with typical final separation $a_{\rm f} < 5\,{\rm R}_\odot$ (Iaconi *et al.* 2017). That simulations do not lead to unbound envelopes (or obvious mergers) suggests to us four possibilities: (i) they are not evolved for long enough, (ii) the final states are not fully resolved leading to artificial quasi-stabilization of the orbit, (iii) the parameter regime simulated (almost always involving a RGB rather than AGB star) is more likely to result in a merger than an envelope ejection, and (iv) physics involving an extra source of energy important for envelope unbinding is missing. Our preliminary results suggest that (i), (ii) and (iii) may be part of the explanation. In addition, we have shown that if (iv) turns out to be part of the answer, the potential energy released by accretion of matter onto the companion is a promising candidate. Further simulations are needed to determine whether the jet that could result would act as an efficient pressure valve enabling super-Eddington accretion, or be quenched by the overlying envelope, for a variety of plausible jet turn-on times.

References

Armitage, P. J., & Livio, M. 2000, *ApJ*, 532, 540
Carroll-Nellenback, J. J., Shroyer, B., Frank, A., & Ding, C. 2013, *J. of Comp. Phys.*, 236, 461
Chamandy, L., Frank, A., Blackman, E., *et al.* 2018, *MNRAS* 480, 1898
Chamandy, L., Tu, Y., Blackman, E., Carroll-Nellenback, J., Liu, B., Nordhaus, J., & Frank, A. 2018, In preparation
De Marco, O. & Izzard, R. G. 2017, *PASA*, 34, 1
Iaconi, R., Reichardt, T., Staff, J., *et al.* 2017, *MNRAS*, 464, 4028
Iaconi, R., De Marco, O., Passy, J.-C. & Staff, J. 2018, *MNRAS*, 477, 2349
Ivanova, N., Justham, S., Chen, X., *et al.* 2017, *ARAA*, 21, 59
Jones, D., & Boffin, H. M. J. 2017, *Nat. Ast.*, 1, 117
Krumholz, M. R., McKee, C. F., & Klein, R. I. 2004, *ApJ*, 611, 399
Moreno Méndez, E., López-Cámara, D., & De Colle, F. 2017, *MNRAS*, 470, 2929
Ohlmann, S. T., Röpke, F. K., Pakmor, R., & Springel, V. 2016, *ApJ* (Letters), 816, L9
Ohlmann, S. T., Röpke, F. K., Pakmor, R., & Springel, V. 2017, *A&A*, 599, A5
Paczyński, B. 1976, *IAUS*, 73, 75
Ricker, P., & Taam, R. 2012, *ApJ*, 746, 74
Soker, N. 2017, *MNRAS*, 471, 4839

Why Galaxies Care About AGB Stars: A Continuing Challenge through Cosmic Time
Proceedings IAU Symposium No. 343, 2019
F. Kerschbaum, M. Groenewegen & H. Olofsson, eds.

doi:10.1017/S1743921318005495

The missing mass conundrum of post-common-envelope planetary nebulae

Miguel Santander-García[1], David Jones[2,3], Javier Alcolea[1], Roger Wesson[4,5] and Valentín Bujarrabal[1]

[1]Observatorio Astronómico Nacional, Alfonso XII, 3, 28014, Madrid, Spain
email: m.santander@oan.es

[2]Instituto de Astrofísica de Canarias, E-38205 La Laguna, Tenerife, Spain

[3]Departamento de Astrofísica, Universidad de La Laguna, E-38206 La Laguna, Tenerife, Spain

[4]Department of Physics and Astronomy, University College London, Gower St, London, UK

[5]European Southern Observatory, Alonso de Córdova 3107, Casilla 19001, Santiago, Chile

Abstract. Most planetary nebulae (PNe) show beautiful, axisymmetric morphologies despite their progenitor stars being essentially spherical. Angular momentum provided by a close binary companion is widely invoked as the main agent that would help eject an axisymmetric nebula, after a brief phase of engulfment of the secondary within the envelope of the Asymptotic Giant Branch (AGB) star, known as a common envelope (CE). The evolution on the AGB would thus be interrupted abruptly, its (still quite) massive envelope fully ejected to form the PN, which should be more massive than a PN coming from the same star were it single. We test this hypothesis by deriving the ionised+molecular masses of a pilot sample of post-CE PNe and comparing them to a regular PNe sample. We find the mass of post-CE PNe to be actually lower, on average, than their regular counterparts, raising some doubts on our understanding of these intriguing objects.

Keywords. (ISM:) planetary nebulae: general, (stars:) binaries: close, ISM: jets and outflows

1. Introduction

Most planetary nebulae (PNe) show beautiful, aspherical morphologies with high degrees of symmetry, despite their progenitor stars being essentially spherical. The mechanism behind their shaping, however, is still poorly understood (e.g. Balick & Frank 2002). Angular momentum provided by a close binary companion has been widely invoked as the main shaping agent that would eject an axisymmetric nebula (Jones & Boffin 2017).

The mechanism in close binary systems is thought to be as follows: a star undergoing the Asymptotic Giant Branch (AGB) stage engulfs a companion via Roche-lobe overflow as it expands during the AGB phase. The system then undergoes a very brief (~1 year) common-envelope (CE) stage, where the evolution of the AGB star is abruptly interrupted. Spiraling-in of the secondary and drag forces would then lead to the ejection and shaping of this CE into a bipolar PN whose equator would be coincident with the orbital plane of the binary star, as happens to occur in every single case analysed so far (Hillwig *et al.* 2016).

On theoretical grounds, however, the physics of the CE "friction" and ejection processes remain very elusive. Simulations show most of the gas to be ejected along the equatorial plane, but are unable to gravitationally unbind the whole envelope of the AGB (e.g. Huarte-Espinosa *et al.* 2012, García-Segura, Ricker & Taam 2018). An exception would imply tapping energy from atomic recombination in the envelope (e.g. Ohlmann *et al.* 2016), but then the achieved expansion velocities would likely be too large.

This draws a somewhat uncomfortable big picture: we simply do not understand the physics lying behind the death of a significant fraction of stars in the Universe.

Single star vs. CE evolution: the total nebular mass. It can be argued that CE evolution implies significant differences in the mass-loss history of the primary star.

Let us consider a single AGB star on its way to produce a PN. Most of its envelope's mass is slowly lost along the AGB evolution, and gets too diluted in the Interstellar Medium (ISM) to be detected. In contrast, the mass lost by the star during the superwind phase (last ~500-3000 years), which amounts to ~0.1-0.6 $M_{\odot}$ for a 1.5 $M_{\odot}$ star (see review by Höfner & Olofsson 2018), will form the nebula visible during the PN stage.

On the other hand, let us consider the same AGB star, but now as part of a binary system close enough to engulf its companion and undergo a CE stage. AGB engulfment will thus occur during the last few (~1-20) million years of the AGB stage (e.g. Fig. 1 in MacLeod, Guillochon & Ramirez-Ruiz 2012), effectively interrupting the evolution of the star. All the mass the star did not lose into the ISM during these last million years will be present in the CE, and therefore will *also* be part of the PN as it is suddenly ejected.

In other words, despite the large uncertainties in the mass-loss history along the AGB, *PNe arising from CE events should, on average, be more massive than their single star counterparts.*

This additional mass should be detectable, as it will be close to the central stars during the lifetime of the PN, as opposed to the single star case, where it will be long gone, diluted into the ISM. Testing this hypothesis would lead to a better understanding of the ejection process. Nevertheless, complete mass determinations of post-CE PNe are virtually nonexistent. Here we present the results of a pilot survey of this kind.

2. Sample and Observations

Our pilot sample is composed of 10 post-CE PNe, which amount roughly to 1/6$^{\rm th}$ of the total currently known. It covers a broad range of kinematical ages, central star effective temperatures and luminosities, orbital periods and morphologies. These objects are PM 1-23, Abell 41, Hen 2-428, ETHOS 1, NGC 6778, Abell 63, the Necklace, V458 Vul, Ou 5, and NGC 2346. They lacked any attempt at detecting their molecular content by means of radioastronomical observations, except for NGC 6778 (not detected by Huggins & Healy (1989)), and NGC 2346, already known to host a massive molecular envelope (e.g. Bachiller *et al.* 1989). We therefore carried out spectral observations of the sample (except NGC 2346), in search for ^{12}CO and ^{13}CO J=1-0 and J=2-1 emission, using EMIR on the IRAM 30 m radio telescope. The angular size of the objects of the sample is generally well suited to the telescope Half Power Beam Width at the observed frequencies.

We complemented the mm-range data with archival Hα images and optical spectra of the sample, from various telescopes and instruments, to derive their ionised masses.

3. Results

Molecular content. No object was detected in ^{12}CO or ^{13}CO down to an *rms* sensitivity limit in the range 6-25 mK at 230 GHz, except for NGC 6778. This PN shows a simple, broad ^{12}CO J=1-0 emission profile, as well as double-peaked emission profiles in ^{12}CO and ^{13}CO J=2-1, whose kinematics correspond to the broken, equatorial ring investigated by Guerrero & Miranda (2012). The peak intensity relations lead us to conclude that the ^{12}CO J=1-0 line is optically thin, and the excitation temperatures relatively low. Further analysis of these profiles and the excitation conditions in this nebula will be presented in Santander-García (in prep.).

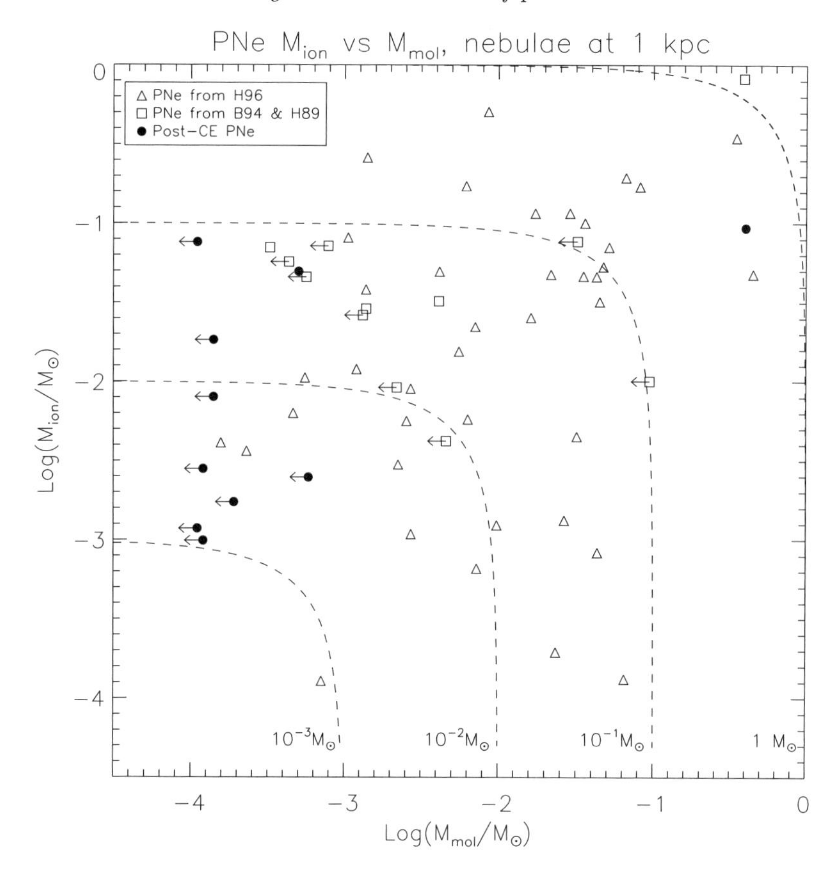

Figure 1. Logarithmic ionised mass vs. logarithmic molecular mass at 1 kpc of our post-CE PNe sample (filled circles), PNe from Huggins *et al.* (1996) (triangles), and a combined sample from Boffi *et al.* (1994) and Huggins *et al.* (1989) (squares). Dashed lines indicate equal total (ionised+molecular) mass; individual nebulae run along these lines as their gas content is progressively ionised.

The ^{12}CO J=1-0 profile of NGC 6778 allows us to derive a molecular mass of $5{\times}10^{-4}\,M_{\odot}$ (at 1 kpc) for this PNe by assuming a representative value of the ^{12}CO abundance of $3{\times}10^{-4}$. On the other hand, the sensitivities achieved in the rest of the observations allow us to derive conservative (3σ) upper limits for the molecular masses of the other objects in the sample.

Ionised content. The ionised mass of NGC 6778, NGC 2346, Abell 41, ETHOS 1, Hen 2-428 and PM 1-23 were derived from their Hβ fluxes and apparent sizes extracted from archival data. Assumptions about the electronic temperatures were made where necessary, in order to produce conservative estimates of the ionised masses of these nebulae (i.e., largest electron temeprature, $T_{\rm e}$, wherever more than one was available). Ionised masses of the Necklace, Abell 63, Ou 5, and V458 Vul were obtained from Corradi *et al.* (2011), Corradi *et al.* (2015), Corradi *et al.* (2015), and Wesson *et al.* (2008), respectively.

Total mass comparison at 1 kpc. Masses found in this work scale with the distance to the nebulae squared. Distances to PNe, however, are still poorly known. Hence, in order to do a proper comparison with PNe not undergoing CE, we must first remove this large dependance by examining the mass every PNe would have at the same distance. Figure 1 shows the ionised and molecular masses of our sample of post-CE PNe at 1 kpc, together with the ionised and molecular masses of a large sample of 44 PNe selected by Huggins *et al.* (1996) in an attempt to approach a volume-limited sample, and another sample of 27 PNe in the Galactic disk, whose ionised/molecular masses were determined by Boffi & Stanghellini (1994) and Huggins & Healy (1989), respectively.

Strikingly, except for NGC 2346, the total masses of the post-CE sample seem similar, if not lower, than those of regular PNe. The median mass at 1 kpc of the combined comparison samples is 0.021 $M_{\odot}$, whereas for the post-CE sample it is $\leqslant$0.0081 $M_{\odot}$.

4. Conclusions

This preliminary work provides a first indication that, contrary to expectations, post-CE PNe seem to be slightly less massive, on average, than their single star counterparts. This discrepancy could however be removed if the molecular gas of these nebulae were too cold (or hot) to be detected, or the ionised gas too hot to emit Hα, but these possibilities seem rather unlikely. Some of the mass could also be in atomic, neutral form, which has not been investigated in this work, and will be part of a future study.

On the other hand, should these results be confirmed by further observations and careful analysis of the possible biases involved, they would present us with the following interesting (and so far speculative) implications. The problem of models unable to unbind such a large mass would be less severe. A fraction of the mass could fall back forming a circumbinary disk (as in Reichardt *et al.* 2018). If any of this material reaches the central stars, it could then be reprocessed perhaps offering an explanation for the correlation between large abundance discrepancy factors and post-CE central stars in PNe (Wesson *et al.* 2018). We can thus wonder whether the CE itself could be not a unique, only-once process, but an episodic, recurrent one. Grazing Envelope Evolution proposed by Soker (2015) and Shiber, Kashi, & Soker (2017) could help explain such a phenomenon.

References

Bachiller, R., Planesas, P., Martín-Pintado, J., *et al.* 1989, *A&A*, 210, 366
Balick, B. & Frank, A. 2002, *ARA&A*, 40, 439
Boffi, F. R., & Stanghellini, L. 1994, *A&A*, 284, 248
Corradi, R. L. M., Sabin, L., Miszalski, B., *et al.* 2011, *MNRAS*, 410, 1349
Corradi, R. L. M., García-Rojas, J., Jones, D., Rodríguez-Gil, P. 2015, *ApJ*, 803, 99
García-Segura, G.; Ricker, P. M.; Taam, R. E. 2018, *ApJ*, 860, 19
Guerrero, M. A., & Miranda, L. F. 2012, *A&A*, 539, 47
Huarte-Espinosa, M., Frank, A., Balick, B., *et al.* 2012, *MNRAS*, 424, 2055
Huggins, P. J., & Healy, A. P. 1989, *ApJ*, 346, 201
Huggins, P. J., Bachiller, R., Cox, P., *et al.* 1996, *A&A*, 315, 284
Hillwig, T. C., Jones, D., de Marco, O., *et al.* 2016, ApJ, 832, 125
Höfner, S.; Olofsson, H. 2018, *A&AR*, 26, 1
Jones, D., & Boffin, H. M. J. 2017, *Nature Astronomy*, 1, 117
MacLeod, M., Guillochon, J., Ramirez-Ruiz, E. 2012, *ApJ*, 757, 134
Ohlmann, S. T.; Röpke, F. K.; Pakmor, R.; Springel, V.; Müller, E. 2016, *MNRAS*, 462, L121
Reichardt, T. A., De Marco, O., Iaconi, R. 2018, MNRAS, submitted (arXiv:1809.02297)
Soker, N. 2015, *ApJ*, 800, 114
Shiber, Sagiv; Kashi, Amit; Soker, Noam 2017, *MNRAS*, 465, L54
Wesson, R., Barlow, M. J., Corradi, R. L. M., *et al.* 2008, *ApJ*, 688, L21
Wesson, R., Jones, D., García-Rojas, J., *et al.* 2018, *MNRAS*, 480, 4589

Discussion

De Marco: A CE interaction must eject the envelope or else the CE cannot be over. So low CE unbinding (ejection) cannot explain the observed low molecular mass. Could instead a scenario be envisaged that CE interactions only happen to AGB stars that already have a low envelope mass?

Santander-Garcia: That's a very interesting idea which could help explain the observations, provided a suitable physical mechanism can be found for the most massive ones to avoid CE interaction (or ejection). It would have some caveats though, since we would still need to explain the few massive cases such as NGC 2346.

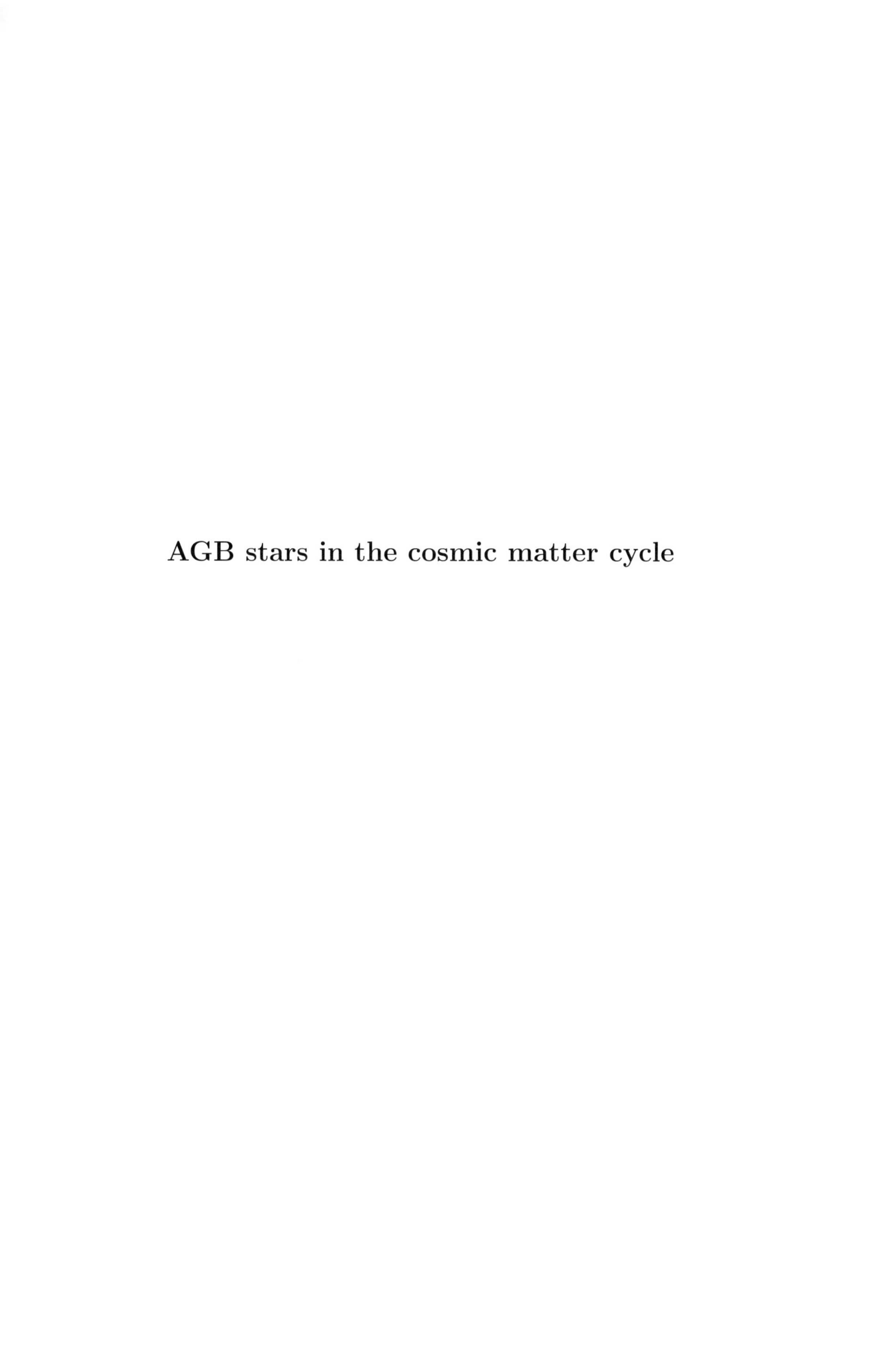

AGB stars in the cosmic matter cycle

Why Galaxies Care About AGB Stars:
A Continuing Challenge through Cosmic Time
Proceedings IAU Symposium No. 343, 2019
F. Kerschbaum, M. Groenewegen & H. Olofsson, eds.

doi:10.1017/S1743921318007184

The role of AGB stars in Galactic and cosmic chemical enrichment

Chiaki Kobayashi, Christopher J. Haynes and Fiorenzo Vincenzo

Centre for Astrophysics Research, University of Hertfordshire,
College Lane, Hatfield, UK
email: c.kobayashi@herts.ac.uk

Abstract. The role of asymptotic giant branch (AGB) stars in chemical enrichment is significant for producing ^{12,13}C, ^{14}N, F, ^{25,26}Mg, ^{17}O and slow neutron-capture process (s-process) elements. The contribution from super-AGB stars is negligible in classical, one-zone chemical evolution models, but the mass ranges can be constrained through the contribution from electron-capture supernovae and possibly hybrid C+O+Ne white dwarfs, if they explode as Type Iax supernovae. In addition to the recent s-process yields of AGB stars, we include various sites for rapid neutron-capture processes (r-processes) in our chemodynamical simulations of a Milky Way type galaxy. We find that neither electron-capture supernovae or neutrino-driven winds are able to adequately produce heavy neutron-capture elements such as Eu in quantities to match observations. Both neutron-star mergers (NSMs) and magneto-rotational supernovae (MRSNe) are able to produce these elements in sufficient quantities. Using the distribution in [Eu/(Fe, α)]–[Fe/H], we predict that NSMs alone are unable to explain the observed Eu abundances, but may be able to together with MRSNe. In order to discuss the role of long-lifetime sources such as NSMs and AGB stars at the early stages of galaxy formation, it is necessary to use a model that can treat inhomogeneous chemical enrichment, such as in our chemodynamical simulations. In our cosmological, chemodynamical simulations, we succeed in reproducing the observed N/O-O/H relations both for global properties of galaxies and for local inter-stellar medium within galaxies, without rotation of stars. We also predict the evolution of CNO abundances of disk galaxies, from which it will be possible to constrain the star formation histories.

Keywords. binaries: general, Galaxy: abundances, galaxies: abundances, hydrodynamics, stars: abundances, stars: AGB and post-AGB, stars: neutron, supernovae: general

1. Introduction

Elemental abundances in the Milky Way Galaxy provide stringent constraints on stellar astrophysics as well as on the formation and evolutionary histories of the Milky Way Galaxy. Elements heavier than helium are synthesized in stars and ejected at their deaths. The next generation of stars forms from gas clouds that include heavy elements from the previous generations. Therefore, stars in the present-day galaxy are fossils that retain the information on the properties of stars in the past. From the elemental abundances of the present-day stars, it is possible to disentangle the star formation history of the galaxy. This approach is called the galactic archaeology. This approach can be applied not only to our Milky Way Galaxy but also to other galaxies (e.g., Kobayashi 2016).

Thanks to the collaboration between nuclear physics and astrophysics, we now have good understanding of the origin of elements in the Universe (e.g., Nomoto *et al.* 2013, hereafter N13). Because of the nature of the triple α reaction, elements heavier with

$A \geqslant 12$ are produced not during the Big Bang, but are instead formed inside stars. Roughly half of the light elements such as C, N and F are produced by low- and intermediate-mass stars ($\sim 0.8 - 8M_\odot$ depending on metallicity) at their asymptotic giant branch (AGB) phase (Kobayashi *et al.* 2011b, hereafter K11). Isotopes such as ^{13}C, ^{17}O, and ^{25,26}Mg are also enhanced by AGB stars, and thus the isotopic ratios can also be used for galactic archaeology (Carlos *et al.* 2018). The α-elements (O, Mg, Si, S, and Ca) are mostly produced in massive stars before being ejected by core-collapse (Type II, Ib, and Ic) supernovae (SNe II, e.g., Kobayashi *et al.* 2006). The production of some elements such as F, K, Sc, and V are increased by neutrino processes in core-collapse supernovae (Kobayashi *et al.* 2011a). Conversely, half of iron-peak elements (Cr, Mn, Fe, Ni, Co, Cu, and Zn) are produced by Type Ia Supernovae (SNe Ia), which are the explosions of C+O white dwarfs (WDs) in binary systems (e.g., Kobayashi & Nomoto 2009). The production of odd-Z elements (Na, Al, P, ... and Cu) depends on the metallicity of the progenitor, as it needs the surplus of neutrons in ^{22}Ne, which is transformed during He-burning from ^{14}N produced in the CNO cycle. The production of minor isotopes (^{13}C, ^{17,18}O, ^{25,26}Mg, ...) also depends on the metallicity.

2. Galactic Chemical Evolution

In Kobayashi, Karakas & Lugaro (2018), we update our galactic chemical evolution (GCE) models in order to match the most recent observations of stars in the solar neighborhood as well as the solar abundance. The basic equations of chemical evolution are described in Kobayashi, Tsujimoto & Nomoto (2000). The code follows the time evolution of elemental and isotopic abundances in a system where the interstellar medium (ISM) is instantaneously well mixed (and thus it is called a one-zone model). No instantaneous recycling approximation is adopted and the chemical enrichment sources with long time-delays are properly included. We adopt the Kroupa initial mass function at $0.01M_\odot \leqslant m \leqslant 50M_\odot$. The parameters for the star formation history of the system are determined to match the observed metallicity distribution function. The nucleosynthesis yields are taken from K11 for supernovae, and from Karakas & Lugaro (2016) for the slow neutron-capture process (s-process).

The fate of stars with initial masses between about $8 - 10M_\odot$ (at $Z = 0.02$) is uncertain and very interesting, although these contributions were not included in K11. The upper limit of AGB stars, $M_{\rm up,C}$, is defined as the minimum mass for carbon ignition, and is estimated to be larger at high metallicity, and also at low metallicity than at $Z \sim 10^{-4}$. Just above $M_{\rm up,C}$, neutrino cooling and contraction leads to off-center ignition of C flame, which moves inward but does not propagate to the center. This may form a hybrid C+O+Ne WD. These WDs can be a progenitor of the sub-classes of SNe Ia, called SNe Iax, which are expected preferably in dwarf galaxies (Kobayashi, Nomoto & Hachisu 2015; Cescutti & Kobayashi 2017). The nucleosynthesis yields of an SN Iax are taken from Fink *et al.* (2014).

Above this mass range, the off-center ignition of C flame moves inward all the way to the center ($\lesssim 9M_\odot$), or stars undergo central carbon ignition ($\gtrsim 9M_\odot$). For both cases, a strongly degenerate O+Ne+Mg core is formed (O+Ne dominant, but Mg is essential for electron capture). If the stellar envelope is lost by winds or binary interaction, an O+Ne+Mg WD may be formed. This upper mass limit is defined as the minimum mass for the Ne ignition, $M_{\rm up,Ne} \sim 9 \pm 1M_\odot$, and is smaller for lower metallicities.

$M_{\rm up,Ne} < M < 10M_\odot$ stars may have cores as massive as $\gtrsim 1.35M_\odot$ and ignite Ne off-centre. If Ne burning is not ignited at the centre (N13), or if off-center Ne burning does not propagate to the center (Jones *et al.* 2014), such a core eventually undergoes an electron-capture-induced collapse. This is the case for the Crab Nebula, and the

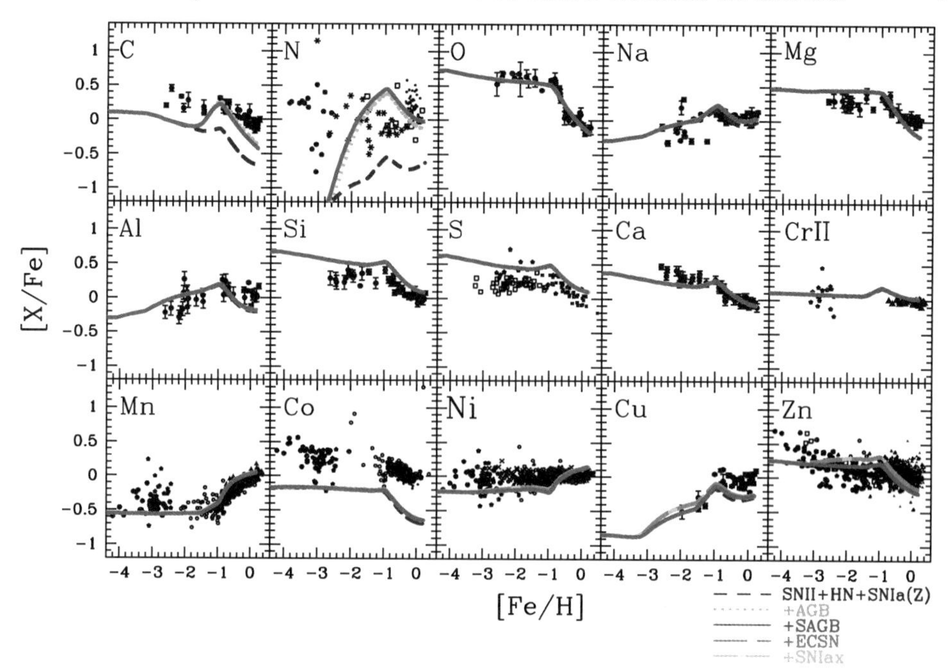

Figure 1. Evolution of elemental abundance ratios [X/Fe] against [Fe/H] in the solar neighborhood for the models with only supernovae (short-dashed lines), with AGB star (dotted lines), with super-AGB stars (solid lines), with ECSNe (long-dashed lines), and with SNe Iax (dot-dashed lines). See Kobayashi, Karakas & Lugaro (2018) for the observational data sources.

electron-capture supernova (ECSN) was one of the candidates of rapid neutron-capture process (r-process). The nucleosynthesis yields of an ECSN are taken from Wanajo *et al.* (2013).

In this paper, nucleosynthesis yields of super AGB stars and the mass ranges of the C+O+Ne WDs, O+Ne+Mg WDs, and ECSNe are taken from Doherty *et al.* (2015). Note that the Ne burning is not followed in Doherty *et al.* (2015); the lower-limit of ECSNe is defined with the temperature $\sim 1.2 \times 10^9$ K, and the upper limit is defined with the core mass $= 1.375 M_\odot$ at the end of carbon burning. These may underestimate the ECSN rate. We also note that these mass ranges are highly affected by convective overshooting, mass-loss, and reaction rates, as well as binary effects, and some of the important physics, such as the URCA process (Jones *et al.* 2014), are also not included. There is no region where the core mass is larger than the Chandrasekhar mass limit in the models compared in Doherty *et al.* (2015); if there were, the stars could explode as so-called Type 1.5 SNe, although no signature of such supernovae has yet been observed (Kobayashi *et al.* 2006).

Figure 1 shows the evolution of elemental abundance ratios [X/Fe] against [Fe/H] in the solar neighbourhood. The contribution to GCE from AGB stars (dotted lines) can be seen mainly for C and N, and slightly for Na. Hence, it would not be easy to explain the observed O-Na anti-correlation with a smooth star formation history as in the solar neighborhood. Although AGB stars produce significant amounts of Mg isotopes (K11), the inclusion of these do not affect the [Mg/Fe]-[Fe/H] relation. The contribution from super-AGB stars (solid lines) are very small; with super AGB stars, C abundances slightly decrease, while N abundances slightly increase. It would be very difficult to put a constraint on super-AGB stars from the evolutionary trends of elemental abundance ratios, but it might be possible see some signatures of super-AGB stars in the scatters

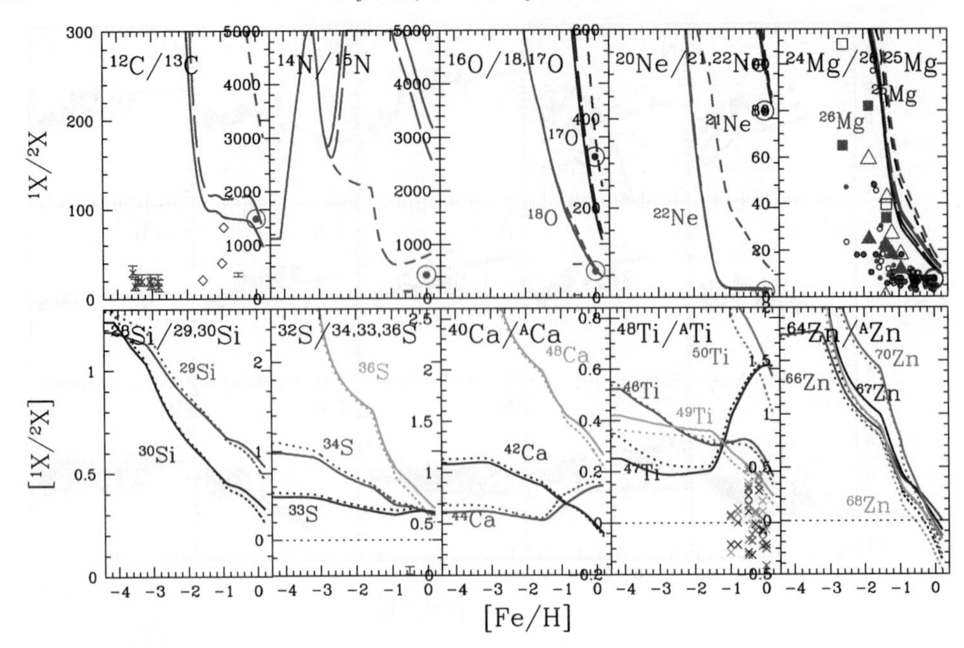

Figure 2. Evolution of isotope ratios against [Fe/H] for the solar neighborhood models: in the upper panels, the models without AGB/super-AGB stars (short-dashed lines), with AGB stars (long-dashed lines), and with super-AGB stars (the fiducial model, solid lines); in the lower panels, the fiducial model (solid lines), and the model without failed SNe (dotted lines). See Kobayashi, Karakas & Lugaro (2018) for the observational data sources.

of elemental abundance ratios. With ECSNe (long-dashed lines), Ni, Cu and Zn may be slightly increased. This is consistent with the high Ni/Fe ratio in the Crab Nebula (N13). Note that the nucleosynthesis yields of ECSNe may be different with neutrino oscillations (Pllumbi *et al.* 2015). No difference is seen with/without SNe Iax (dot-dashed lines) in the solar neighborhood because of the narrow mass range of hybrid WDs. Even with a wider mass range ($\Delta M \sim 1 M_{\odot}$ in Kobayashi, Nomoto & Hachisu 2015), however, the SN Iax contribution is negligible in the solar neighborhood, but can be important at lower metallicities such as in dwarf spheroidal galaxies with stochastic chemical enrichment (Cescutti & Kobayashi 2017).

Figure 2 shows the evolution of isotopic ratios against [Fe/H] for the solar neighborhood models with AGB and super-AGB stars (the fiducial model, solid lines), without super-AGB stars (long-dashed lines), without AGB nor super-AGB stars (short-dashed lines), and without failed SNe (dotted lines). The under-production problem of ^{15}N is known, and may require other sources such as novae (K11). ^{12}C/^{13}C, ^{16}O/^{17}O, ^{20}Ne/^{21}Ne, and ^{24}Mg/^{25,26}Mg ratios become smaller with AGB stars, and even smaller with super-AGB stars, while ^{14}N/^{15}N ratios become larger and larger. With super-AGB stars, ^{12}C/^{13}C is 76.2 at [Fe/H] $=0$ dex, which is lower than the solar ratio of 89.4. ^{17}O is too much produced in AGB and super-AGB stars. Even with super-AGB stars, however, ^{20}Ne/^{21}Ne and ^{24}Mg/^{25,26}Mg are larger than observations. There is no significant difference for ^{16}O/^{18}O and ^{20}Ne/^{22}Ne ratios, and these are determined from SN II/HN yields. Also for the elements heavier than Mg, the isotopic ratios mostly depend on SN II/HN yields, and the small mismatches for the solar ratios still remain.

Figure 3 shows the evolution of neutron-capture elements in the solar neighborhood. As shown for O, there are no differences for the elements up to Ni among these models.

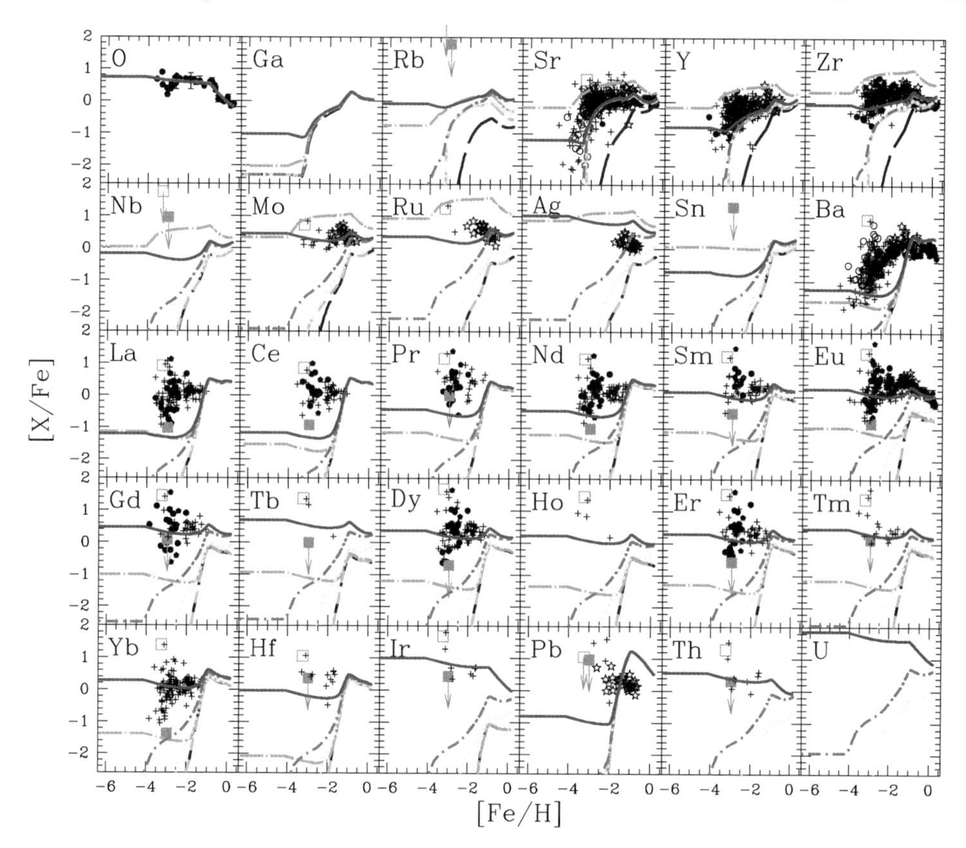

Figure 3. Evolution of the neutron-capture elemental abundances [X/Fe] against [Fe/H] in the solar neighborhood for the models with the s-process from AGB stars only (long-dashed lines) with ECSNe (short-dashed lines), with ECSNe and ν-driven winds (dot-long-dashed lines), with ECSNe and NS-NS mergers (dotted lines), with ECSNe and NS-NS/NS-BH mergers (dot-short-dashed lines), and with ECSNe, NS-NS/NS-BH mergers, and MRSNe (solid lines). See Kobayashi, Karakas & Lugaro (2018) for the observational data sources.

As predicted from nucleosynthesis yields, AGB stars (long-dashed lines) can produce more than enough s-process elements, such as Ba, Hf, and Pb. The contribution appears from [Fe/H] ~ -2 dex for light s-process elements (e.g., Sr, Y, Zr) and only from [Fe/H] ~ -1.5 dex for heavy s-process elements (e.g., Pb). With ECSNe (short-dashed lines), the enhancement is as small as ~ 0.1 dex for [(Cu,Zn)/Fe] (Fig. 1), but a larger enhancement is seen for Rb, Sr, Y, and Zr from [Fe/H] ~ -3 dex, which is enough together with AGB stars. With ν-driven winds (NUW, dot-long-dashed lines), the elements from Sr to Ag are over-produced, which is a crucial problem. The contribution can appear at [Fe/H] $<< -3$ dex, which may explain some of the observation at [Fe/H] $\lesssim -3$ dex, but the ratios between light and heavy neutron-capture elements are not consistent with observations. Therefore, it is better not to include ν-driven winds at all in GCE. With NS-NS mergers (dotted lines), the elements heavier than Mo show a small excess at [Fe/H] ~ -2 dex. The time-delay is shorter for NS-blackhole (BH) mergers (dot-short-dashed lines), and a larger excess is seen at [Fe/H] ~ -4 dex. However, this is still not short enough to explain the observations at [Fe/H] $\lesssim -3$ dex. Finally, with magneto-rotational supernovae (MRSNe, solid lines), it is possible to reproduce the plateau at [Fe/H] $\lesssim -3$ dex for most neutron-capture elements. Although these one-zone models cannot put a strong constraint on the site at [Fe/H] $\lesssim -2.5$ dex, it is better to include both NS-NS/NS-BH mergers and MRSNe in GCE.

3. Chemodynamical Simulations of Milky Way

In a real galaxy, the star formation history is not so simple and in particular the interstellar medium (ISM) is not homogeneous at any time, which is different from one-zone chemical evolution models. The effects of inhomogeneous enrichment can be summarized as follows. i) There is a local variation in star formation and metal flow by the inflow and outflow of the ISM. ii) Heavy elements are distributed via stellar winds and supernovae, and the elemental abundance ratios depend on the mass and metallicity of progenitor stars. iii) There is a mixing of stars due to dynamical effects such as merging and migration. All of these effects are, in principle, included in our chemodynamical simulation, where hydrodynamics, star formation, supernova feedback, and chemical enrichment are solved self-consistently throughout the galaxy formation. Note that the ISM may be mixed before the next star formation by other effects such as diffusion and turbulence, which might be underestimated (see Kobayashi 2014 for more details).

Using a chemo-hydro-dynamical code, Kobayashi & Nakasato (2011) succeeded in reproducing not only the observed trends but also the scatters of elemental abundance ratios from O to Zn. Under the inhomogeneous enrichment, there is only a weak age-metallicity relation of stars. In other words, the most metal-poor stars are not always the oldest stars. Therefore, the contribution of long-lifetime sources such as AGB stars can appear at low metallicities at a later epoch. In order to discuss the role of AGB stars, it is necessary to use a model that can treat inhomogeneous chemical enrichment, such as in our chemodynamical simulations.

In Haynes & Kobayashi (2018, hereafter HK18), we present the distributions of elemental abundance ratios using our chemodynamical simulations of a Milky Way-type galaxy. The code is based on the smoothed particle hydrodynamics (SPH) code GADGET-3, and include all relevant baryon physics (Kobayashi *et al.* 2007). We utilise cosmological zoom-in initial conditions for a Milky Way-type galaxy from the Aquila comparison project (Scannapieco *et al.* 2012) with cosmological parameters as follows: $H_0 = 100h = 73$ km s^{-1} Mpc^{-1}, $\Omega_0 = 0.25$, $\Omega_\Lambda = 0.75$, $\Omega_B = 0.04$. The initial mass of each gas particle is $3.5 \times 10^6 M_\odot$ and the gravitational softening length is $1h^{-1}$ kpc. We choose these conditions because they give a galaxy with morphology, size and merger history reasonably similar to the Milky Way (see the G3-CK model in Scannapieco *et al.* 2012). We also include the following sites of the r-process:

ECSNe: We adopt yields from Wanajo *et al.* (2013) for a $8.8M_\odot$ star and metallicity dependent ($\log Z = -4, -3, -2.4, -2.1, -1.7$) limits of the progenitor mass with the upper (8.4, 8.4, 9.0, 9.65, 9.9 $M_\odot$) and lower (8.2, 8.25, 8.8, 9.5, 9.75 $M_\odot$) bounds (Doherty *et al.* 2015). For metallicities above or below the limits, we assume the rates derived from the upper and lower limits, respectively.

NUW: Ejecta heated by neutrinos from proto-neutron stars (NSs) can provide a site for r-process enrichment. The yields are taken from Wanajo (2013) for stars between 13 and 40 $M_\odot$ with a conversion between the stellar masses (13, 15, 20, 40) $M_\odot$ and the respective NS masses (1.4, 1.6, 1.8, 2.0) $M_\odot$.

NS Mergers: NS mergers provide a site for the r-process in the neutron rich dynamic ejecta created as the binary system merges. We include yield tables for NS-NS mergers (both 1.3 $M_\odot$) from Wanajo *et al.* (2014) and for NS-BH mergers. The NS-NS merger and NS-BH merger rates of simple stellar populations are taken from the binary population synthesis calculations in Mennekens & Vanbeveren (2014) (model 2 for $Z = 0.02$

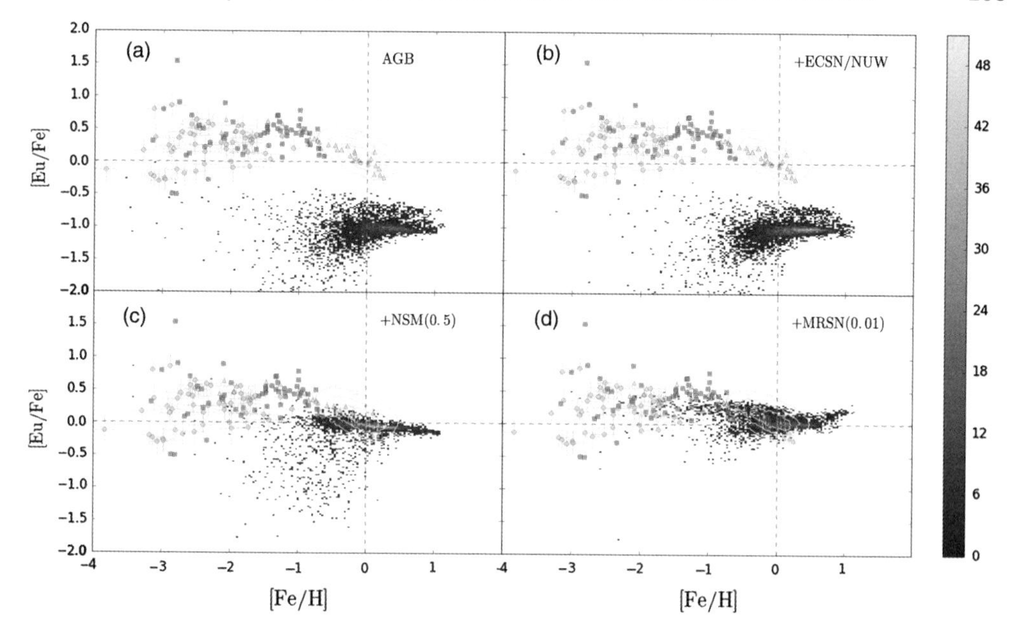

Figure 4. [Eu/Fe] plotted against [Fe/H] for the star particles in the solar neighborhood in our Milky Way simulations at $z = 0$. The panels in order show: supernovae+AGB only, with ECSNe and ν-driven winds, NSMs, and MRSNe. See HK18 for the observational data sources for high-resolution (red squares, green triangles, orange circles) and the HERMES-GALAH survey (cyan contours). The contours show 10, 50 and 100 stars per bin. The red dashed lines denote 0 for both [Eu/Fe] and [Fe/H] which we expect to lie within our data. The color bar shows the linear number of points per bin for our simulation.

and 0.002). We also introduce a free parameter, $f_{\rm NSM}$, representing the fraction of stars in binary systems in our simulations. We choose an initial value of $f_{\rm NSM} = 0.5$, independent of metallicity.

MRSNe: Rapidly rotating massive stars with strong magnetic fields may provide a potential site for r-process nucleosynthesis at the inner boundary of the accretion disc formed around the central collapsed object. We use the yield tables presented in Nishimura *et al.* (2015) for a 25 $M_\odot$ star (B11β1.00 model). This event could be related to HN events, which also require rotation and magnetic fields; recent simulations of supernova explosions have not succeeded in exploding very massive stars ($M \geqslant 25\ M_\odot$). Therefore, we replace a fraction of HN events with MRSNe. The number of stars with suitable conditions for MRSNe is poorly constrained so we introduce a free parameter, $f_{\rm MRSN}$, representing the fraction of stars with the correct conditions. Our initial value is $f_{\rm MRSN} = 0.01$, independent of mass and metallicity, as it gives solar values at [Fe/H]$= 0$ dex and reasonable agreement with observations.

Figure 4 shows the distribution of [Eu/Fe] against [Fe/H]. Panel (a) shows the control simulation with the AGB contributions but no additional r-process sites. The spread of [Eu/Fe] sits well below observational values; [Eu/Fe] is at ~ -1 for [Fe/H] $\gtrsim -1.5$ dex. Panel (b) shows the addition of both NUW and ECSNe to the AGB model. The addition is insufficient to boost [Eu/Fe] to the observed levels and the overall trend remains at [Eu/Fe] ~ -1. This is expected from the input yields: no Eu enrichment from ECSNe and only a small contribution from NUW. Panels (c) and (d) show simulations using NSMs and MRSNe, respectively. Both NSMs and MRSNe increase [Eu/Fe] to ~ 0 at [Fe/H] $= 0$ dex, though with the caveat that both make use of a free parameter that allows the level to be adjusted reasonably freely in our simulations.

At [Fe/H] $\lesssim -1.5$ dex the simulated level and scatter of [Eu/Fe] in the MRSN model matches the observational data much better than the NSM model. This is a result of delay times before NSMs can occur, as NS binaries need to both form and coalesce. MRSNe, on the other hand, are assumed to be a subset of 25–40 $M_\odot$ HNe in our simulations and able to start producing r-process elements almost immediately after the formation of stars. Additionally, the NSM model displays a large amount of scatter below the main trend of at [Fe/H] $\lesssim 0$ dex. We already see a similar scatter at the same [Fe/H] range in the control simulation; the scatter we see in NSMs is likely produced by AGB stars. However, this same pattern of scatter is not present in the MRSN model. In this case it appears that the production of Eu from NSMs is too slow to raise [Eu/Fe] to observed levels at [Fe/H] < -1 dex and instead slowly increases [Eu/Fe] between $-2 \lesssim$ [Fe/H] $\lesssim 0$ dex. However, MRSNe are able to enrich the ISM prior to AGB contributions so no such scatter is seen. Although MRSNe replicate the trend in this region substantially better, it presents a flat trend above [Fe/H] ~ -1 dex where the observational data suggests a downward trend. NSMs have a slight downward trend in this region, though not as steep as the observed data. Both the scatter and average [Eu/Fe] are important for constraining the r-process. SNe Ia produce substantial amounts of Fe which contributes to the scatter in Figure 4, so in order to remove the SN Ia contribution, we also show the [Eu/O] distributions in HK18, and reach the same conclusion as from [Eu/Fe].

4. Cosmological Simulations

On a cosmological scale, chemical enrichment takes place even more dramatically (see a movie by Philip Taylor, https://www.youtube.com/watch?v=jk5bLrVI8Tw). There is gas accretion along the cosmological filaments, where star formation and chemical enrichment are already occurring. This results in strong supernova-driven winds because of the shallow potential in the filaments. As the central galaxy grows through the accretion, a super-massive blackhole also grows (following the so-called M–σ relation), which eventually causes even stronger winds driven by the active galactic nuclei. Metallicity is very spatially inhomogeneous; the center of massive galaxies can reach super-solar metallicity at high redshifts, while the accreted component has only one-hundredth of solar metallicity as it is mainly fed from the intergalactic medium. The wind component has about one-tenth of solar metallicity as it is a mixture of the inflow gas and supernova ejecta. This simulation successfully reproduces various observations of galaxies including the mass-metallicity relation of galaxies (massive galaxies are more metal-rich, Taylor & Kobayashi 2015).

In Vincenzo & Kobayashi (2018b, hereafter VK18b), from another cosmological simulation (with side 10 h^{-1} Mpc) we create a catalogue of 33 stellar systems at redshift $z = 0$, all embedded within dark matter (DM) halos with virial masses in the range $10^{11} \leqslant M_{\rm DM} \leqslant 10^{13} M_\odot$. The mass and spacial resolutions of gas are $6.09 \times 10^6\, h^{-1} M_\odot$ and $0.84\, h^{-1}$ kpc, respectively. Within a galaxy, the metallicity distribution is not uniform either, and the central parts of the galaxies are more metal-rich than the outskirts of the galaxies. This metallicity radial gradient evolves as a function of time. In disk galaxies, the metallicity gradients become steeper at higher redshifts because of inside-out formation of discs (Vincenzo & Kobayashi 2019).

Using our cosmological hydrodynamical simulation, we succeed in reproducing the observed N/O–O/H relation. The observed increasing trend of N/O at high O/H in the individual ISM regions of spatially resolved star-forming disc galaxies (we refer to it as local N/O–O/H relation) can be explained as the consequence of metallicity gradients that have settled in the galaxy ISM, where the innermost regions possess both the highest O/H and the highest N/O ratios (see Fig. 4 of VK18b).

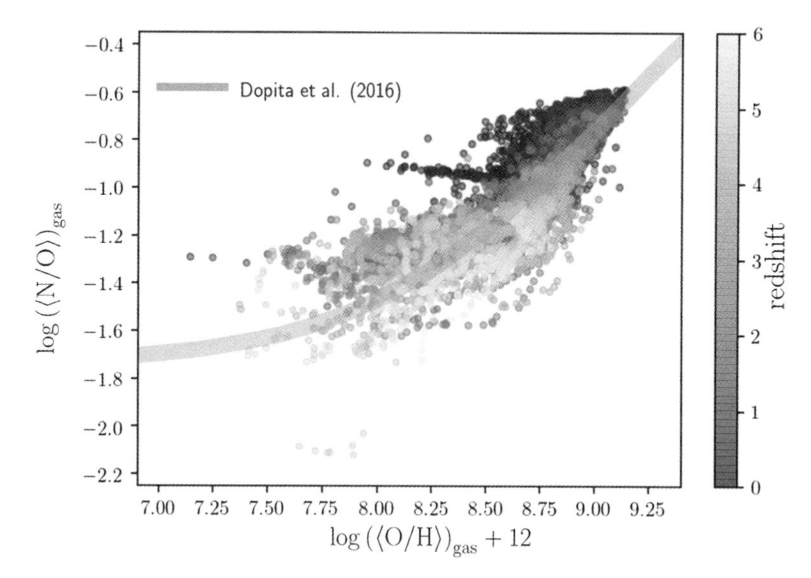

Figure 5. The redshift evolutionary tracks of 33 galaxies from our cosmological simulation in the O/H–N/O diagram. The abundances in the figure correspond to SFR-weighted averages in the gas-phase of the ISM. The color coding represents the redshift. See VK18b for the observational data source (grey bar).

Similar relation is known for average abundances from the whole galaxy ISM. Figure 5 shows the star formation rate (SFR)-weighted averages in the gas-phase of the ISM of 33 galaxies in our catalogue. This global N/O–O/H relation is the consequence of an underlying mass–metallicity relation that galaxies obey as they evolve across the cosmic epochs. In this case, the predicted N/O–O/H relation is an average evolutionary trend that is followed by the chemical evolution tracks of all galaxies at almost any redshift.

Our simulation predicts an almost flat trend of N/O versus O/H at high redshifts, which is caused by an inhomogeneous enrichment of the galaxy ISM with a significant contribution of AGB stars at low metallicity. Our simulation represents an improvement with respect to previous chemical evolution models, both because the N/O–O/H relation was studied with simple one-zone models with instantaneous mixing approximation and because we did not assume any artificial primary N production by massive stars. We note that the N/O plateau would be too high if rapidly rotating massive stars were included.

We then predict how the C, N, and O abundances within the interstellar medium of galaxies evolve as functions of the galaxy star formation history (SFH). In Vincenzo & Kobayashi (2018a, hereafter VK18a), we focus on three star-forming disc galaxies (Galaxy A, B, and C) with different SFHs in our cosmological simulation (left panels of Fig. 6). At the beginning of galaxy formation, CNO are produced by core-collapse supernovae, N is enhanced by intermediate mass AGB stars ($\gtrsim 4 M_\odot$), then C is enhanced by low-mass AGB stars ($\lesssim 4 M_\odot$). In the right panels of Figure 6, we predict that the average N/O and C/O steadily increase as functions of time, while the average C/N decreases, due to the mass and metallicity dependence of the yields of AGB stars; such variations are more marked during more intense star formation episodes. Our predictions on the CNO abundance evolution can be used to study the SFH of disc galaxies with the *James Webb Space Telescope*.

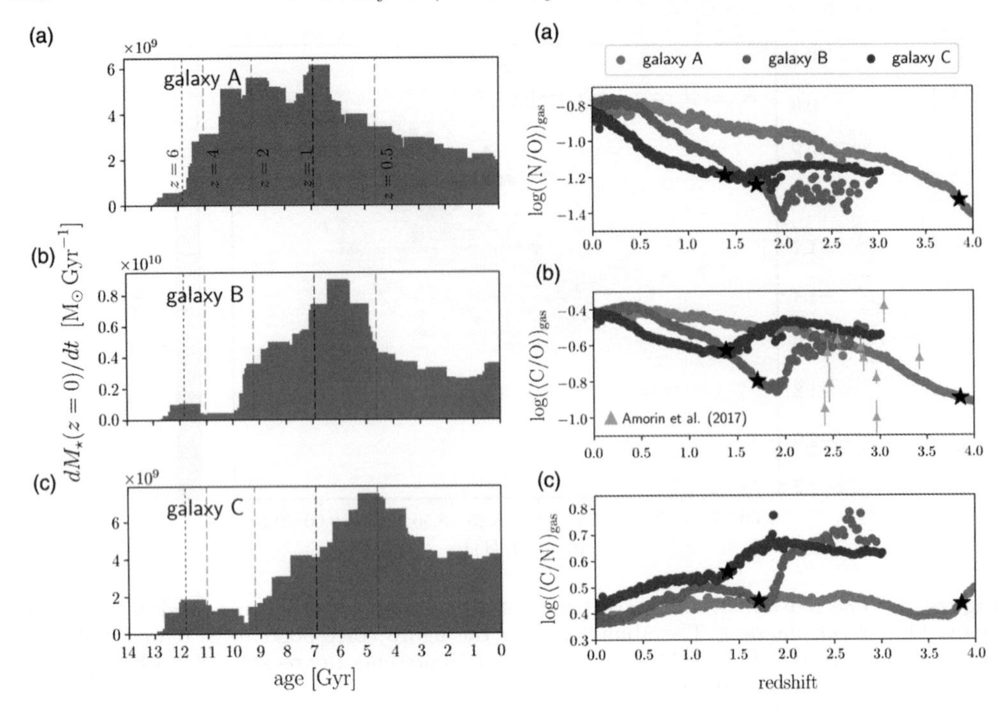

Figure 6. (a) Distribution of the present-day total stellar mass as a function of the star particle age. (b) Average SFR-weighted gas-phase log (N/O), log (C/O) and log (C/N) as functions of redshift. We only show the predictions of our simulation for the redshifts when the galaxy stellar mass $M_* \geqslant 10^8 M_\odot$. The black star symbol on each track marks the redshift when $M_* \simeq 10^9 M_\odot$. See VK18a for the observational data source (grey triangles with the error bars).

References

Carlos, M., Karakas, A. I., Cohen, J. G., Kobayashi, C., & Meléndez, J. 2018, *ApJ*, 856, 161

Cescutti, G., & Kobayashi, C. 2017, *A&A*, 607, 23

Doherty, C. L., Gil-Pons P., Siess L., Lattanzio J. C., & Lau H. H. B., 2015, *MNRAS*, 446, 2599

Fink, M., *et al.* 2014, *MNRAS*, 438, 1762

Haynes, C., & Kobayashi, C. 2018, *MNRAS*, submitted arXiv:1809.10991 (HK18)

Jones, S. Hirschi, R., & Nomoto, K. 2014, *ApJ*, 797, 83

Karakas, A. I., & Lugaro, M., 2016, *ApJ*, 825, 26

Kobayashi, C. 2014, IAU S298, 167

Kobayashi, C. 2016, *Nature*, 540, 205

Kobayashi, C., Izutani, N., Karakas, A. I., *et al.*, 2011a, *ApJ* (Letters), 739, L57

Kobayashi, C., Karakas, I. A., & Lugaro, M. 2018, in preparation

Kobayashi, C., Karakas, I. A., & Umeda, H. 2011b, *MNRAS*, 414, 3231 (K11)

Kobayashi, C., & Nakasato, N. 2011, *ApJ*, 729, 16

Kobayashi, C., & Nomoto, K. 2009, *ApJ*, 707, 1466

Kobayashi, C., Nomoto, K., & Hachisu, I. 2015, *ApJ* (Letters), 804, L24

Kobayashi, C., Springel, V, & White, S. D. M. 2007, *MNRAS*, 376, 1465

Kobayashi, C., Tsujimoto, T., & Nomoto, K. 2000, *ApJ*, 539, 26

Kobayashi, C., Umeda, H., Nomoto, K., Tominaga, N., & Ohkubo, T. 2006, *ApJ*, 653, 1145

Mennekens N., & Vanbeveren D. 2014, *A&A*, 564, A134

Nishimura, N., Takiwaki, T., & Thielemann, F.-K. 2015, *ApJ*, 810, 109

Nomoto, K., Kobayashi, C., & Tominaga, N. 2013, *ARAA*, 51, 457 (N13)

Pllumbi, E., Tamborra, I., Wanajo, S., Janka, H.-T., & Hüdepohl, L. 2015, *ApJ*, 808, 188

Scannapieco, C. *et al.* 2012, *MNRAS*, 423, 1726

Taylor, P., & Kobayashi, C. 2015, *MNRAS*, 448, 1835

Vincenzo, F., & Kobayashi, C. 2018a, *A&A*, 610, L16 (VK18a)
Vincenzo, F., & Kobayashi, C. 2018b, *MNRAS*, 478, 155 (VK18b)
Vincenzo, F., & Kobayashi, C. 2019, IAU FM7, in press
Wanajo, S., 2013, *ApJ* (Letters), 770, L22
Wanajo, S., Janka, H.-T., & Müller, B., 2013, *ApJ* (Letters), 767, L26
Wanajo, S., Sekiguchi, Y., Nishimura, *et al.* 2014, *ApJ* (Letters), 789, L39

Discussion

VENTURA: One of the open points in AGB modelling is the extent of oxygen destruction in massive AGB stars. Is there any possibility of investigating this issue in the context of your modelling?

KOBAYASHI: Most of O is produced from core-collapse SNe, and the signature of oxygen destruction will be washed out in the averaged evolution. However, it may be possible to see some differences in the scatter of N/O ratios.

DE MARCO: Is there a way you can constrain delay-times for SNe (from Fe) or NS-NS mergers (from r-process elements)?

KOBAYASHI: This is done for SNe Ia. For example, binary population synthesis models give too short delay times for double degenerate, which cannot reproduce the sharp decrease of $[\alpha/\mathrm{Fe}]$ from $[\mathrm{Fe/H}] \sim -1$ (Kobayashi & Nomoto 2009). For NSMs, it is not so straightforward as r-process elements can be produced by other sources. The community is working on that and it will be possible in the future.

SAHAI: Is it important to consider and analyze the $^{13}\mathrm{C}/^{12}\mathrm{C}$ enrichment, say as a function of Z, since we know that AGBs definitely can increase $^{13}\mathrm{C}/^{12}\mathrm{C}$ in ISM of galaxies?

KOBAYASHI: Yes, very much. $^{13}\mathrm{C}/^{12}\mathrm{C}$ vary as well as C/N, and it should also be possible to measure with ALMA. There are some measurements at $Z = 0$ already, but the available values do not make sense and maybe are still too uncertain.

Why Galaxies Care About AGB Stars:
A Continuing Challenge through Cosmic Time
Proceedings IAU Symposium No. 343, 2019
F. Kerschbaum, M. Groenewegen & H. Olofsson, eds.

doi:10.1017/S1743921318007561

AGB stars and the cosmic dust cycle

Svitlana Zhukovska

Max Planck Institute for Astrophysics
Karl-Schwarzshild-Str. 1, 85748 Garching, Germany
email: szhukovska@mpa-garching.mpg.de

Abstract. Theoretical and observational studies of dust condensed in outflows of AGB stars have substantially advanced the understanding of dust mixture from individual stars. This detailed information incorporated in models of the lifecycle of interstellar grains provides a flexible tool to study the contribution of AGB stars to the galactic dust budget. The role of these stars in dust production depends on the morphological type and age of galaxy. While AGB stars are sub-dominant dust sources in evolved systems as the Milky Way, the observed relation between the dust-to-gas ratio and metallicity suggests that the dust input in young dwarf galaxies with $7 \lesssim 12 + \log(\mathrm{O/H}) \lesssim 8$ can be dominated by the AGB stars. In application to post-starburst and early-type galaxies, the models for stardust evolution in combination with modern infrared observations give insights in the origin of their high dust content and its implications for their evolutionary scenarios.

Keywords. stars: AGB and post-AGB, dust, extinction, galaxies: evolution

1. Introduction

Cooling envelopes of thermally-pulsing AGB stars are known sites for efficient dust condensation. Presolar grains with isotopic signatures of AGB stars found in meteorites indicate that dust from these stars is transported to the ISM and survives in harsh interstellar environment for 100 – 1000 Myr. Spectroscopic analysis of circumstellar envelopes of oxygen- and carbon-rich AGB stars reveal significant amounts of silicate and carbonaceous grains, respectively, which are thought to be the main dust components in the ISM. AGB stars are therefore considered important sources of interstellar dust, but their actual contribution to the galactic dust budget has remained unclear.

The input of AGB stars to the dust budget depends on (i) their mass and metallicity distributions and (ii) stellar dust yields, i.e. the amount of dust produced by stars during their entire AGB evolution. The former is determined by galactic evolution, namely, by the scenario of galaxy formation and evolution, matter cycle between gas and stars and attendant enrichment of the ISM with heavy elements. These processes are implemented in the models of galactic chemical evolution that follow enrichment history of the ISM and, correspondingly, initial chemical composition of subsequent generations of low- and intermediate-mass stars. The predictive power of these models is significantly enhanced by including a dust evolution model (Dwek 1998). Such models provide a simple, but powerful tool to study populations of AGB stars, their contribution to the interstellar dust budget and evolution of stardust grains in the ISM of galaxies (Gail *et al.* 2009).

Stellar dust yields are the key ingredient of galactic chemical evolution models with dust. The past two decades brought significant progress in the modelling of dust condensation in winds of low- and intermediate-mass stars. Condensation models for winds with carbon and oxygen chemistry were combined with stellar evolution models by Ferrarotti & Gail (2006), who produced the first stellar mass- and metallicity-dependent dust yields. Dust yields are implemented in the models of chemical evolution in a similar way as stellar

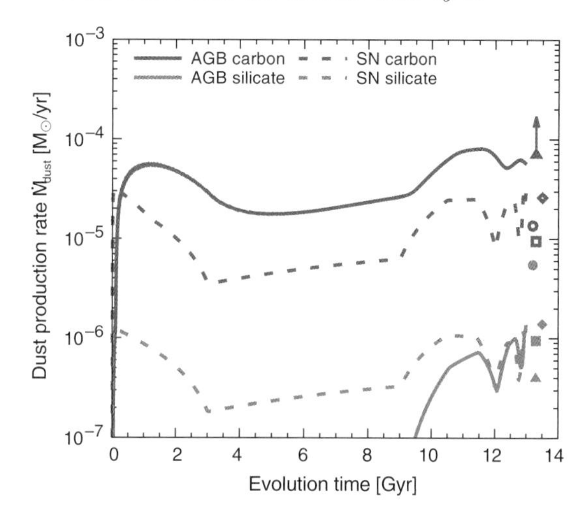

Figure 1. The dust production rate for carbon and silicate dust from AGB stars (red and blue lines, respectively) in the LMC as a function of evolution time. Dust input rate from type II SNe is shown with dashed line for comparison. Red and blue symbols indicate the values for C- and O-rich dust production rates derived from observations (Matsuura *et al.* 2009; Boyer *et al.* 2012; Riebel *et al.* 2012; Srinivasan *et al.* 2009). Adapted from Zhukovska & Henning (2013a).

nucleosynthesis yields, with the difference that dust species can be destroyed in the ISM by SN blast waves and star formation.

The mass of interstellar dust in a galaxy estimated from observations can be compared with the model predictions for grains from AGB stars that survived until the present time. This allows to study the role of AGB stars in the dust and gas recycling (Sect. 2), which in turn can be used to discern galactic evolution scenarios (Sect. 3). A direct comparison of the dust production rates by AGB stars is only possible for a handful of galaxies, for which these rates have been measured observationally for the entire population of dust-forming stars (Sect. 2). In the following we consider applications of dust evolution models for galaxies of various morphological types.

2. Dwarf galaxies

2.1. *Lessons from the Large Magellanic Cloud*

Stardust evolution models for the solar neighborhood find that AGB stars dominate the dust input to the galactic budget only during the early epoch, seeding the ISM with dust grains for the subsequent dust growth by accretion in the ISM (Dwek 1998; Zhukovska *et al.* 2008). This theoretical prediction cannot be directly verified for the Galaxy, since high extinction in the Galactic disk prevents measurements of the dust input rates from all AGB stars. This difficulty is alleviated for the Large and Small Magellanic Clouds (LMC and SMC), the closest dwarf galaxies, whose positions and proximity permitted the first census of their dust-forming stellar population carried out in the *Spitzer* Legacy Program "Surveying the Agents of Galaxy Evolution" (SAGE) (Meixner *et al.* 2006).

SAGE-LMC program resulted in the first observational estimates of the dust production rates in a galaxy (Matsuura *et al.* 2009; Boyer *et al.* 2012; Riebel *et al.* 2012; Srinivasan *et al.* 2009). Zhukovska & Henning (2013a) compare the dust production rate from observations to the theoretically calculated rates based on dust evolution models with AGB stellar populations.

This first comparison allowed to test stellar evolution prescription for the AGB phase used in the calculations of stellar yields and rule out the models with excessive hot bottom burning. Figure 1 shows the time variations of production rates for carbon and silicate

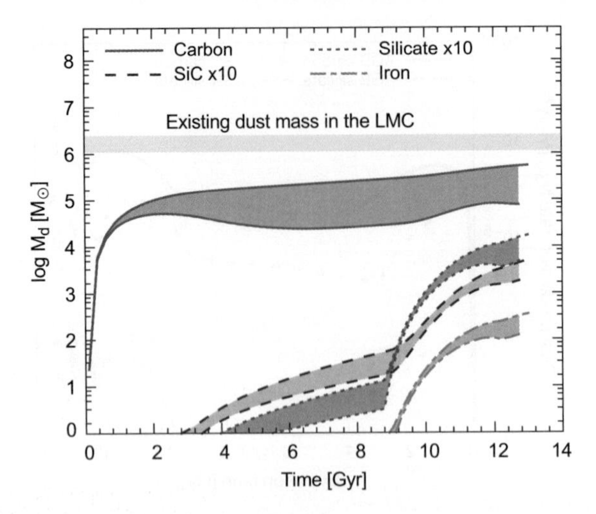

Figure 2. Time evolution of the dust mass ejected by AGB stars over the LMC history for carbonaceous, silicate, SiC, and iron grains. The upper and lower borders of shaded areas show the values with and without destruction, respectively. The grey rectangle indicates the present interstellar dust mass. Adapted from Zhukovska & Henning (2013a).

dust by AGB stars together with the corresponding values from observations plotted for the present time. Dust input from SNe is sub-dominant in the LMC. The results are shown for the dust yields from Ferrarotti & Gail (2006) providing the best match to observations.

Unlike for type II SNe, amount of oxygen-rich dust condensed in winds of AGB stars strongly depends on the stellar metallicity. The rates of O-rich dust production have approached the present-day levels only during the last 3 Gyr owing to the dust supply from stars formed from the gas enriched by the recent burst of star formation.

The total dust mass from AGB stars produced in the LMC during the entire galactic evolution is 8×10^4 $M_\odot$ and 5×10^5 $M_\odot$ with and without destruction in the ISM, respectively (Fig. 2). Even without destruction the accumulated dust mass from AGB stars is lower than the total dust mass in the LMC, $(1.1 - 2.5) \times 10^6$ $M_\odot$. This corroborates the discrepancy between the amount of interstellar dust in the Milky Way and the total dust mass from AGB stars. This is so-called "missing dust problem" is resolved by adding additional dust mass growth in the ISM by accretion of gas-phase species. This discrepancy decreases at subsolar metallicities in the outer Galactic disk and LMC, owing to the longer timescales of dust growth in the ISM (Zhukovska & Henning 2013b).

2.2. *Metal-poor dwarf galaxies*

Models of the lifecycle of cosmic dust predict that the dust content in young galactic systems originates from stellar sources, dust detected in high redshift galactic systems can therefore be injected by evolved stars. Unevolved, metal-poor local dwarf galaxies are considered templates for objects in the young Universe that can provide insights into the origin of their dust content.

Recent far-IR observations by the *Herschel* Space Observatory conducted within the Dwarf Galaxy Survey (DGS) have substantially improved estimates of the dust masses in the most metal-poor galaxies in the local Universe (Rémy-Ruyer *et al.* 2014). This study have for the first time established steepening of the relation between the dust-to-gas ratio (DGR) and the oxygen abundance log(O/H) at metallicities below 1/10 solar metallicity (Fig. 3). Large scatter of the DGR in the metallicity range $7.3 \lesssim 12 + \log(\mathrm{O/H}) \lesssim 8.4$ is another prominent feature of the observed relation. This implies that the fraction of

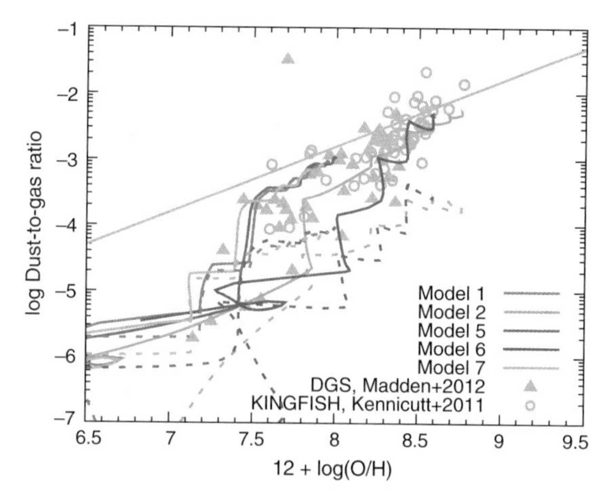

Figure 3. The dust-to-gas ratio as a function of oxygen abundance 12+log(O/H) in the ISM for dwarf galaxies with a wide range of model parameters from Table 1 in Zhukovska (2014) (solid lines). The dust-to-gas ratio for the case of AGB stars as the only dust sources is shown dashed lines. Triangles indicate the dust-to-gas ratios for dwarf galaxies from the Dwarf Galaxy Survey (DGS) and circles show the values of the KINGFISH sample for metal-rich galaxies, for comparison, estimated by Rémy-Ruyer *et al.* (2014). Straight grey line denotes the constant dust-to-metal ratio of 0.4, typical for spiral galaxies.

metals in dust in the DGS sources is generally lower and varies in a wider range than in spiral galaxies, characterised by the dust-to-metal ratio of 0.4 – 0.5.

Because the DGS sample consists of galaxies with large spread in ages and metallicities, dust sources responsible for different trends in the observed DGR–metallicity relation can be studied by comparing it with DGR–metallicity evolutionary tracks for model galaxies (Zhukovska 2014). With a sufficiently large range of galactic parameters and star formation histories, the theoretical tracks with all dust sources included (AGB stars, type II SNe and ISM dust growth) cover the entire range in both DGR and log(O/H) values for dwarf galaxies (Fig. 3). Steps-like behaviour of the DGR is due to the bursty star formation histories, typical for dwarf galaxies. Theoretical tracks with dust production only by AGB stars overlap with the lower part of the observed DGR–metallicity relation, indicating the metallicity range of $7 \lesssim 12 + \log(\mathrm{O/H}) \lesssim 8$ in which AGB stars can dominate the dust production.

AGB stars are the major dust sources in dwarf galaxies during their initial evolution, up to 3 Gyr, before dust growth in the ISM takes over dust production. This is evident from the time variations of mass fraction of dust from AGB stars in the ISM of model galaxies (Fig. 4). The dips in the mass fractions mark bursts of star formation leading to the temporal increases of the SN dust production, while the fraction of AGB stardust increases up to 85% during post-burst evolution. The transition to the ISM-growth-mode of dust formation is indicated by the steep decrease in the mass fraction of AGB dust in all models. The star formation history is more important for this transition than metallicity: the higher is the star formation rate, the younger is the galactic age and the higher is the metallicity when the transition happens. In other words, galaxies with fewer star bursts of low intensity appear leftmost in Fig. 3 and rightmost in Fig. 4.

3. Galaxies with extreme environments

Recent detections of substantial dust amounts in massive galaxies with extreme environments such as post-starburst and early-type galaxies challenge their classical view as

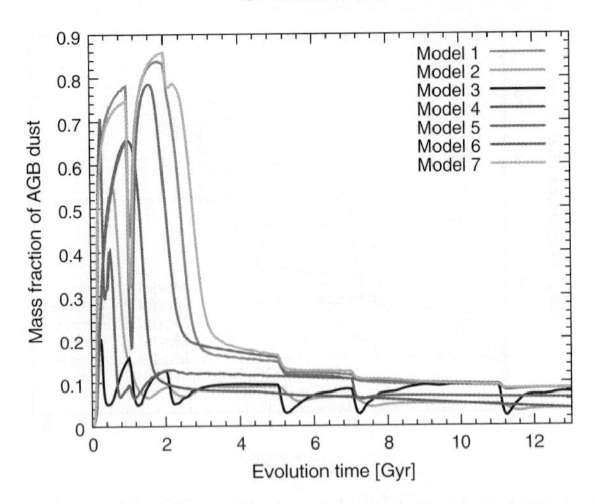

Figure 4. Mass fraction of dust from AGB stars as a function of evolution time (galactic age) for dwarf galaxy models from Zhukovska (2014).

dust- and gas-poor systems and raise questions about the origin and subsequent fate of this dust, which can be addressed with dust evolution models.

The optical spectra of post-starburst galaxies, designated also as E(K)+A galaxies, combine spectral characteristics of a young A-star population and old K-stars suggesting that they experienced a recent burst of star formation that was rapidly quenched several hundred Myr ago. In the traditional picture, post-starburst galaxies are dust- and gas-poor remnants of most violent mergers of gas-rich disks, rapidly transitioning to quiescent spheroids. This picture suggests that their ISM was expelled by the event that nearly truncated their star formation as a result of outflows from starburst or active galactic nuclei activity. The observed gas and dust reservoirs are then explained by replenishment of the ISM by ageing stellar populations. In order to test this scenario, Smercina *et al.* (2018) compare dust and gas masses measured observationally for a sample of 33 post-starburst galaxies with material ejected from a single stellar population (an instantaneous burst of star formation) as a function of its age. Figure 5 shows dust and molecular gas masses per unit burst stellar mass and theoretical curves for matter returned by AGB stars from a single stellar population with metallicity $Z = 0.02$. Since post-starbursts may be in the transition to early-type galaxies filled with hot gas, the models with grain destruction by thermal sputtering in a hot gas are also shown in Fig. 5.

Smercina *et al.* (2018) find that dust amounts in all post-starburst galaxies younger than 400 Myr can not be explained by condensation in stellar winds of AGB stars alone, suggesting that the ISM was not completely expelled by the quenching event and replenished by old stars. This conclusion is corroborated by higher masses of molecular gas in most sources than expected from stellar recycling alone (Fig. 5).

Dust-to-burst stellar mass ratios show a negative trend over three orders of magnitude with galactic age, similar to the models with thermal sputtering of grains in a hot gas on the shortest timescale of 50 Myr. This trend is explained by the higher dust destruction rates than the production rate by stars. For comparison, dust content in young dwarf galaxies increases with time and in spiral galaxies, dust destruction is balanced by formation processes leading to the constant average dust abundances. The dust-to-burst stellar mass ratios for post-starburst ages > 600 Myr are consistent with observations of dusty early-type galaxies ($M_{\rm d}/M_{*,\rm burst}$ of a few 10^{-5}), suggesting an ongoing transition to virialized early-type gas conditions.

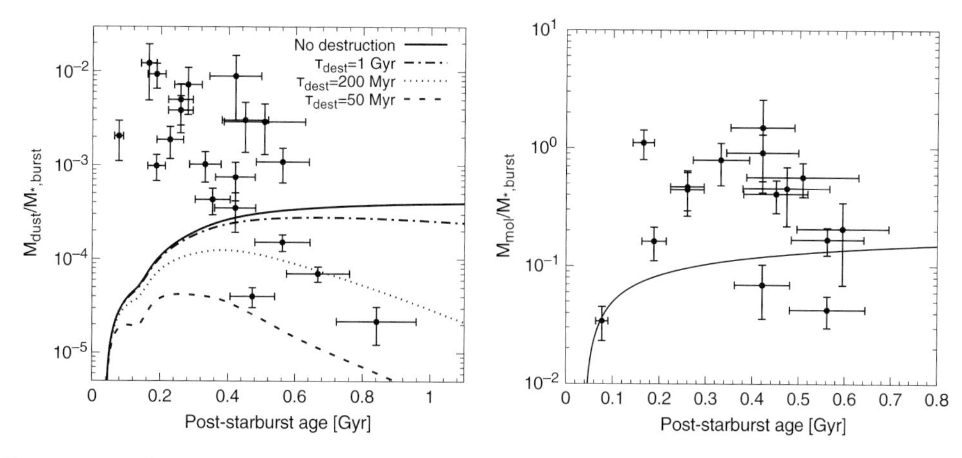

Figure 5. *Left Panel.* Dust mass per unit burst stellar mass for the sample of post-starburst galaxies as a function of time since the burst (error bars). The four curves are the theoretical dust models for a single stellar population with metallicity $Z = 0.02$ without destruction (solid line) and with thermal sputtering on the timescale of 1 Gyr, 200 Myr and 50 Myr (dash-dot, dot, and dash lines, respectively) derived for various representative gas conditions in early-type galaxies. *Right Panel.* Molecular mass per unit burst stellar mass (error bars) also plotted as a function of time since the burst. The solid curve shows total injection of gas by a single stellar population. Adapted from Smercina *et al.* (2018).

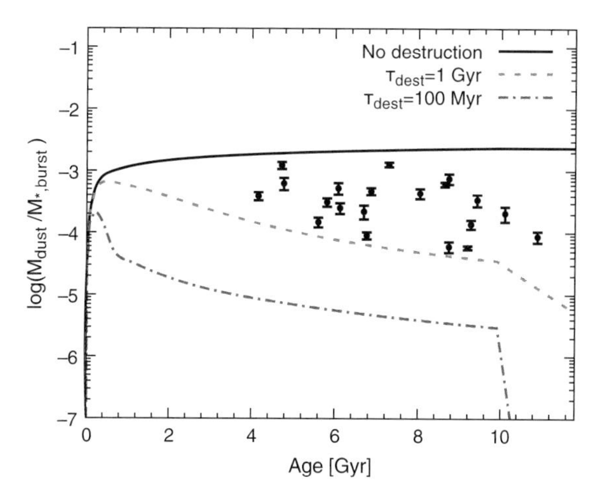

Figure 6. Dust mass per unit stellar mass in the sample of red-elliptical galaxies from Dariush *et al.* (2016) with metallicity $Z > 2.5\ Z_{\odot}$ plotted as a function of the mass-weighted age of their stellar populations (error bars). The three curves are the theoretical dust models for a single stellar population with metallicity $Z = 3\ Z_{\odot}$ without destruction (solid line) and with thermal sputtering on timescale of 1 Gyr and 100 Myr (dash and dash-dot lines, respectively).

Unexpectedly large dust reservoirs were recently detected in sub-millimetre bands in optically red early-type (red-E) galaxies at low-redshifts (Dariush *et al.* 2016). The origin of dust in these systems can be also probed by a simple stellar population model, neglecting the age spread of their old populations. Dust in early-type galaxies is traditionally believed to be formed in outflows of AGB stars. The total dust mass returned by a single stellar population is indeed larger than the dust quantities detected in red-E galaxies. However, dust from AGB stars cannot accumulate over extended time owing to efficient grain destruction by thermal sputtering in a hot gas, ubiquitous for early-type gas conditions (Fig. 6). Even for a low gas density of 10^{-4} cm^{-3} resulting in a long sputtering timescale of 1 Gyr, the models fail to reproduce the observed dust masses. The mismatch between the model predictions and the observed dust masses suggests an external origin (e.g. supplied through mergers or tidal interactions) or dust growth in the cold ISM of these early-type systems.

References

Boyer, M. L., Srinivasan, S., Riebel, D., *et al.* 2012, *ApJ*, 748, 40
Dariush, A., Dib, S., Hony, S., *et al.* 2016, *MNRAS*, 456, 2221
Dwek, E. 1998, *ApJ*, 501, 643
Ferrarotti, A. S., & Gail, H.-P. 2006, *A&A*, 447, 553
Gail, H.-P., Zhukovska, S. V., Hoppe, P., & Trieloff, M. 2009, *ApJ*, 698, 1136
Matsuura, M., Barlow, M. J., Zijlstra, A. A., *et al.* 2009, *MNRAS*, 396, 918
Meixner, M., Gordon, K. D., Indebetouw, R., *et al.* 2006, *ApJ*, 132, 2268
Rémy-Ruyer, A., Madden, S. C., Galliano, F., *et al.* 2014, *A&A*, 563, A31
Riebel, D., Srinivasan, S., Sargent, B., & Meixner, M. 2012, *ApJ*, 753, 71
Smercina, A., Smith, J. D. T., Dale, D. A., *et al.* 2018, *ApJ*, 855, 51
Srinivasan, S., Meixner, M., Leitherer, C., *et al.* 2009, *ApJ*, 137, 4810
Zhukovska, S. 2014, *A&A*, 562, A76
Zhukovska, S., & Henning, T. 2013, *A&A*, 555, A99
Zhukovska, S., & Henning, T. 2013, in The Life Cycle of Dust in the Universe: Observations, Theory, and Laboratory Experiments, ed. A. Andersen *et al.* (Trieste: SISSA), 16
Zhukovska, S., Gail, H.-P., & Trieloff, M. 2008, *A&A*, 479, 453
Zhukovska, S., Petrov, M., & Henning, T. 2015, *ApJ*, 810, 128

Discussion

MAERCKER: In dwarf galaxies, the dust first comes from SNe and AGB stars and then from ISM growth. Would the ISM growth be possible without the first producing stardust? And does the type of ISM dust formed depend on the original stardust?

ZHUKOVSKA: The ISM growth requires seed particles providing initial surface, so stardust is required to initiate the growth. Yes, the type of ISM dust depends on the chemical composition of the original stardust. We assume selective growth, i.e. silicate dust grows on silicate grains and carbon on carbon grains, respectively. The growth rates depend on the abundances of these dust species in the ISM.

Why Galaxies Care About AGB Stars: A Continuing Challenge through Cosmic Time
Proceedings IAU Symposium No. 343, 2019
F. Kerschbaum, M. Groenewegen & H. Olofsson, eds.

doi:10.1017/S1743921318007512

Chemistry and binarity in the early Universe: what is the role of metal-poor AGB stars?

Anke Arentsen[1], **Else Starkenburg**[1], **Matthew D. Shetrone**[2], **Alan W. McConnachie**[3], **Kim A. Venn**[4] **and Éric Depagne**[5]

[1] Leibniz-Institut für Astrophysik Potsdam (AIP), Potsdam, Germany
email: aarentsen@aip.de

[2] McDonald Observatory, The University of Texas at Austin, Austin, USA

[3] NRC Herzberg Institute of Astrophysics, Victoria, Canada

[4] Department of Physics and Astronomy, University of Victoria, Victoria, Canada

[5] Southern African Large Telescope/SAAO, Cape Town, South Africa

Abstract. Carbon-enhanced metal-poor stars are probes of the early universe, that teach us about metal-poor AGB stars and supernovae physics in the very first stars. We find a large fraction of CEMP-no stars with large absolute carbon abundance to be in binary systems. This may be an indication of binary interaction with ultra or extremely metal-poor AGB stars, curiously without enhancement in *s*-process elements.

Keywords. stars: chemically peculiar – binaries: spectroscopic – stars: AGB and post-AGB – Galaxy: halo – galaxies: formation

1. Introduction

It is possible to study the conditions in the early universe using stars in our own neighbourhood. Extremely metal-poor stars have abundance patterns reflecting the chemical composition of the interstellar medium (ISM) out of which they were born and can be used as archeological probes. The most metal-poor stars are often enhanced in carbon and the fraction of these so called carbon-enhanced metal-poor (CEMP) stars increases with decreasing metallicity (e.g. Lee *et al.* 2013). There are two main classes of CEMP stars, the CEMP-s stars and the generally more metal-poor CEMP-no stars, see the left-hand panel of Fig. 1. The former type is enhanced in *s*-process elements and is almost always part of a binary system, while the latter is not enhanced in *s*-process elements and seems to be in binary systems only about 20% of the time (Hansen *et al.* 2016a,b). The enhanced carbon and *s*-process elements in CEMP-s stars can be explained by binary transfer from a former asymptotic giant branch (AGB) star companion. The abundance patterns of CEMP-no stars however are thought represent the ISM out of which the stars have been formed, which has been polluted by e.g. spinstars or faint supernovae with large carbon yields. Yoon *et al.* (2016) have additionally proposed that two groups of CEMP-no stars can be distinguished from each other, the Group II and Group III stars (roughly below and above the dashed line in Fig. 1 respectively).

We have studied more CEMP-no stars with radial velocity monitoring to further constrain the influence of binarity in this type of star, using high-resolution spectra from the CFHT and SALT telescopes.

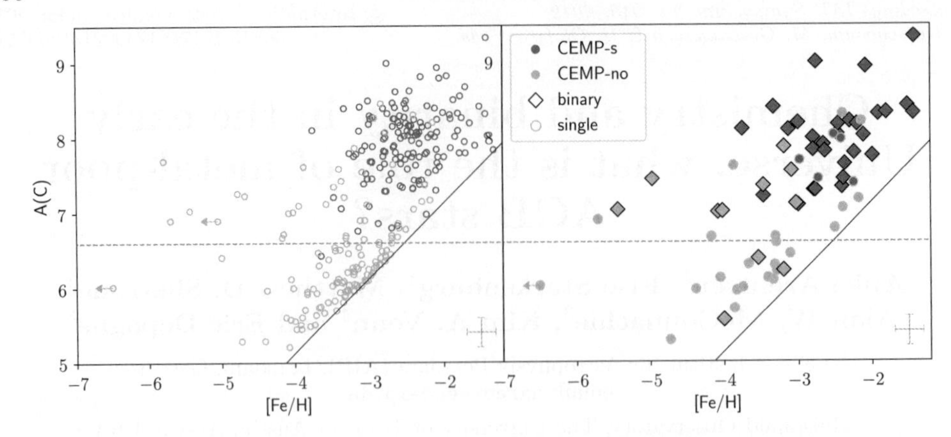

Figure 1. Absolute carbon abundance A(C) vs. metallicity [Fe/H] for CEMP-s and CEMP-no stars, where the former are shown in blue and the latter in orange. In the left-hand panel, we show a compilation of CEMP stars from Yoon *et al.* (2016). In the right-hand panel, we include binary information for 34 stars. Binary stars are indicated with diamonds and single stars with solid dots, where the information in binarity comes from our program combined with the binary flags compiled by Yoon *et al.* (2016) (mostly from Hansen *et al.* 2016a,b). The solid line in both panels indicates the CEMP criterion of [C/Fe] $> +0.7$, and the dashed line at A(C) $= 6.6$ divides the CEMP-no sample with binarity information in two equally sized groups.

2. Results

In our radial velocity monitoring sample we find four new CEMP-no binary systems: HE 2139−5432, SDSS J1422+0031, SDSS J0140+2344 and HE 0107−5240. We add stars from previous radial velocity monitoring programs (Starkenburg *et al.* 2014, Hansen *et al.* 2016b) and three individually studied stars in the literature, which results in a sample of eleven CEMP-no binary systems. From previous studies, there are also 23 known single CEMP-no stars, based on monitoring of their radial velocities. This compilation is inhomogeneous, and a general binary fraction cannot be determined properly. However, if we look at the distribution of the binary stars compared to the single stars in the [Fe/H] – A(C) plane in the right-hand panel of Fig. 1, an interesting pattern starts to emerge. As previously mentioned, almost all CEMP-s stars are in binary systems (however, there are five single CEMP-s stars, which is also interesting to note). Regarding the CEMP-no stars, there seems to be a difference in binary fraction between the higher A(C) and the lower A(C) stars (here separated by a dashed line). For the high A(C) CEMP-no stars, the binary fraction is $47^{+15}_{-14}\,\%$ whereas for the low A(C) stars it is only $18^{+14}_{-9}\,\%$. More data is clearly needed to confirm this hypothesis, but among these low numbers, it is striking that so many of the high A(C) CEMP-no stars are in binary systems, especially below [Fe/H] $= -3$. What could be possible explanations for this phenomenon?

2.1. *Binary interaction*

The CEMP-s stars have such large absolute carbon abundances because they have had an interaction with a binary companion, which previously went through the AGB stage. In the AGB phase of a star, a large amount of carbon is created and finally dredged up to its surface due to thermal pulses in the star. Additionally, the slow neutron capture process creates *s*-process elements which will also be dredged up to the surface of the star. These materials from the surface of an AGB star can be transferred to a binary companion. The orbital periods of the CEMP-no binaries are very similar to those of the CEMP-s stars (typically a few 100 to a few 1000 days). There is a possibility that the high A(C) CEMP-no stars have also experienced mass-transfer from an AGB star, increasing

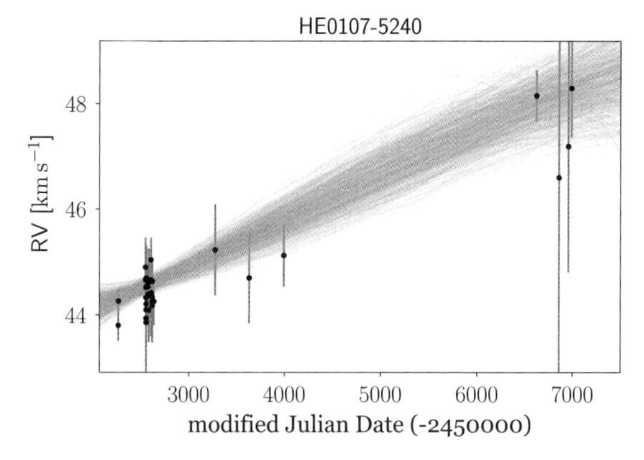

Figure 2. Radial velocities for HE 0107−5240, with possible orbits from *the Joker* (Price-Whelan *et al.* 2017). The four measurements around day 7000 are new from our work (Arentsen *et al.* submitted), which provide evidence that the star is varying in radial velocity over a long timescale.

their carbon abundance to something higher than for the other CEMP-no stars. However, this can only be the case if their companions did not create significant *s*-process elements, since the CEMP-no stars are (by definition) not enhanced in *s*-process elements. Could this indicate that something different is going on inside extremely metal-poor AGB stars?

One hyper metal-poor star whose abundance pattern has previously been explained by interaction with a binary companion is HE 0107−5240 (Suda *et al.* 2004; Lau *et al.* 2007 and most recently Cruz *et al.* 2013). At the time there was no indication that this star was variable in radial velocity so there was no further support for this hypothesis. However, when we add our radial velocity measurements to those from the literature, we find that the star actually varies in radial velocity, see Fig. 2. The orbital period for the star is between 10 000 − 30 000 days (more measurements are needed to constrain it better). This is in the regime where the star could have had an interaction with the wind from an AGB star, which through loss of angular momentum could have widened the orbit to the current state.

If all high A(C) CEMP-no stars would have experienced mass-transfer from a binary companion, this could explain the two different groups of CEMP-no stars (Group II and III). However, some of the high A(C) stars are single. Therefore, even if binary companions do play a role, binarity is likely not the only explanation for the two groups.

2.2. *Other possible explanations for the Group II and III stars*

Based on preliminary comparisons of abundances for some elements, Yoon *et al.* (2016) suggest that the Group II and III stars might have formed from an ISM that has been polluted mainly by faint supernovae or mainly by spinstars for the two groups respectively. A different explanation is that of Chiaki *et al.* (2017) who study the properties of dust in regions of varying carbon enhancement. They find that in the region of the Group III stars carbon dust cooling is most efficient while in the region of the Group II stars it is silicate dust that is more efficient. In between, there is a region where cooling is not efficient at all, producing naturally a gap between the two groups. Another implication of a difference in dust cooling between the two regions could be an effect on the binary fraction of forming stars. If this would be the case (this is truly speculative), that might additionally be able to explain a higher binary fraction among the high A(C) CEMP-no stars. It could also (partly) be a metallicity effect, that at lower metallicity more stars are in binary systems. A last explanation for the two groups is presented by Sharma *et al.*

(2018) who claim that the difference is due to the bursty nature of the progenitors of the Milky Way, where Group II stars formed in the first burst of star formation and the Group III stars formed in a later star formation event when AGB stars have had the time to enrich the ISM in carbon, and in-falling pristine gas has diluted the ISM to a very low overall metallicity.

3. Conclusion

CEMP-no stars are important for constraining physical processes in the early universe, since many of them are among the most metal-poor stars that we know. It is generally assumed that their abundance patterns are directly reflecting physics of the very first stars and their supernova deaths. However, we find that many of the CEMP-no stars with high absolute carbon abundances are in binary systems. Their orbital periods are similar to those of the CEMP-s stars, which are all expected to have been through interaction with a former AGB companion. If the companions of those CEMP-no binary stars are currently white dwarfs, it is very likely that the surfaces of the CEMP-no stars we see today have been polluted by their companion. This should be kept in mind as it complicates the interpretation of their abundance patterns.

References

Arentsen, A., Starkenburg, E., Shetrone, M. D., *et al.* submitted
Chiaki, G., Tominaga, N., Nozawa, T. 2017, *MNRAS*, 472, 115
Cruz, M. A., Serenelli, A., Weiss, A. 2013, *A&A*, 559, 4
Hansen, T. T., Andersen, J., Nordström, B., *et al.* 2016a, *A&A*, 586, A3
Hansen, T. T., Andersen, J., Nordström, B., *et al.* 2016b, *A&A*, 586, A160
Lau, H. H. B., Stancliffe, R. J., Tout, C. A. 2007, *MNRAS*, 378, 563
Lee, Y. S., Beers, T. C., Masseron, T., *et al.* 2013, *AJ*, 146, 132
Sharma, M., Theuns, T., Frenk, C. 2018, arXiv:1805.05342
Starkenburg, E., Shetrone, M.D., McConnachie, A.W., & Venn, K.A. 2014, *MNRAS*, 441, 1217
Suda, T., Aikawa M., Machida, M. N., Fujimoto, M. Y., Iben, I. 2004, *ApJ*, 611, 476
Yoon, J., Beers, T. C., Placco, V. M., *et al.* 2016, *ApJ*, 833, 20
Price-Whelan, A. M., Hogg, D.W., Foreman-Mackey, D., Rix, H.-W. 2017, *ApJ*, 837, 20

Discussion

FEKEL: What is the maximum orbital period of your binary stars?

ARENTSEN: Maximum: 10 000–30 000 days, but typically several 100s–1000s days.

WHITELOCK: May be worth looking at the kinematics, including proper motions, in case they came from disrupted binaries.

ARENTSEN: Thank you, good idea.

LUGARO: What could be the origin of the single CEMP-*s* stars?

ARENTSEN: One idea is that they may be the result of massive spinstars that have produced some *s*-processes in the early universe.

Why Galaxies Care About AGB Stars:
A Continuing Challenge through Cosmic Time
Proceedings IAU Symposium No. 343, 2019
F. Kerschbaum, M. Groenewegen & H. Olofsson, eds.

doi:10.1017/S1743921318006142

Calibrating TP-AGB stellar models and chemical yields through resolved stellar populations in the Small Magellanic Cloud

Giada Pastorelli[1], **Paola Marigo**[1], **Léo Girardi**[1,2] **and the STARKEY project team**

[1] Dipartimento di Fisica e Astronomia Galileo Galilei, Università di Padova, Vicolo dell'Osservatorio 3, 35122 Padova, Italy – email: giada.pastorelli@studenti.unipd.it

[2] Osservatorio Astronomico di Padova – INAF, Vicolo dell'Osservatorio 5, 35122 Padova, Italy

Abstract. Most of the physical processes driving the TP-AGB evolution are not yet fully understood and they need to be modelled with parameterised descriptions. We present the results of the on-going calibration of the TP-AGB phase based on a complete sample of AGB stars in the Small Magellanic Cloud (SAGE-SMC survey). We computed large grids of TP-AGB models with several combinations of third dredge-up and mass-loss prescriptions with the COLIBRI code. The SMC AGB population is modelled with the population synthesis code TRILEGAL according to the space-resolved star formation history derived with the deep photometry from the VISTA survey of the Magellanic Clouds. We put quantitative constraints on the efficiencies of the third dredge-up and mass loss by requiring the models to reproduce the star counts and the luminosity functions of the observed Oxygen-, Carbon-rich and extreme-AGB stars and we investigate the impact of the best-fitting prescriptions on the chemical yields.

Keywords. stars: AGB and post-AGB, stars: mass loss, galaxies: Magellanic Clouds

---

1. Modelling the AGB populations in the SMC

A reliable calibration of TP-AGB models requires: 1) a robust measurement of the Star Formation History (SFH) of the SMC 2) a complete sample of observed AGB stars accurately classified 3) detailed TP-AGB models to be included in the population synthesis code. Rubele *et al.* (2018) derived the SFH, the reddening and the distance of 168 subregions of the SMC, for a total area of 23.57 deg^2, using the deep photometry of the VISTA Survey of the Magellanic Clouds (VMC, Cioni *et al.* 2011). The main advantages of using VMC data are the lower extinction in the near-infrared passbands and the photometry reaching the oldest main-sequence turn-off points, ensuring a robust estimate of the ages. Our calibration is based on the AGB candidate list by Srinivasan *et al.* (2016) (SR16). The AGB population is classified into Carbon-rich (C-rich), Oxygen-rich (O-rich), anomalous (a-AGB) and extreme AGB (X-AGB), on the basis of photometric criteria (Boyer *et al.* 2011) and complemented with the available spectroscopic information (Ruffle *et al.* 2015, Boyer *et al.* 2015). The TP-AGB models are computed with the `COLIBRI` code (Marigo *et al.* 2013) in which we adopt a scheme that considers two regimes of mass loss: (i) "pre-dust mass-loss" ($\dot{M}_{\rm pre-dust}$) and (ii) "dust-driven mass-loss" ($\dot{M}_{\rm dust}$). We test two formalisms for $\dot{M}_{\rm pre-dust}$: the modified Schröder & Cuntz (2005) relation (Rosenfield *et al.* 2016), and the algorithm developed by Cranmer & Saar (2011) (CS11), in which $\dot{M}_{\rm pre-dust}$ is driven by the pressure of Alfén waves in the chromosphere.

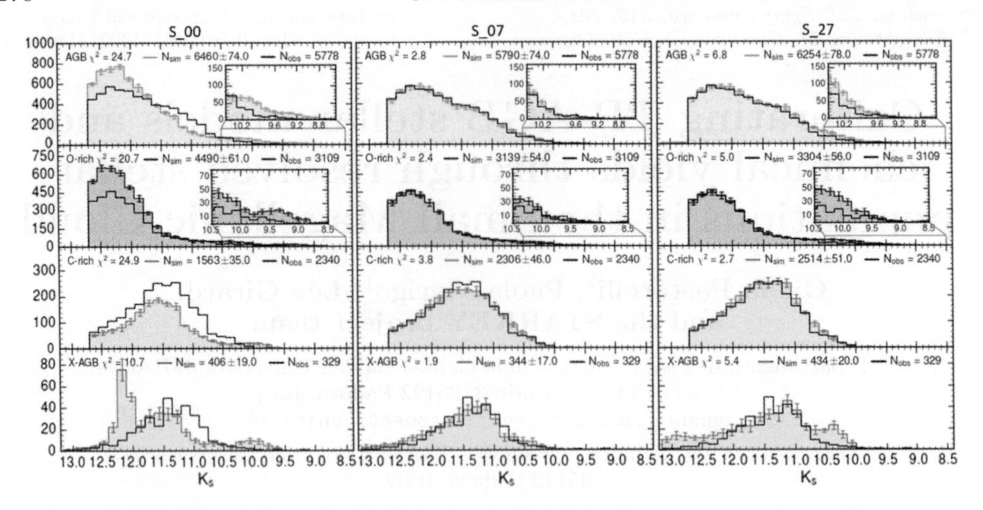

Figure 1. Mean K_s-band LFs of models S_00, S_07 and S_27 with error bars (coloured histograms), as compared to the observations (dark-line histogram) for the entire AGB sample and for the three main classes of AGB stars. The χ^2_{LF} specific to each panel is also reported.

During the dust-driven regime, we investigate a few among the most popular prescriptions of $\dot{M}_{dust}$, i.e. Vassiliadis & Wood (1993) and Blöcker (1995) (BL95). In addition, we use the recent results of dynamical atmosphere models by Mattsson *et al.* (2010) and Eriksson *et al.* (2014) (CDYN) for C-stars. The occurrence of a mixing event is checked with the T_b^{dred} parameter, i.e. the minimum temperature that should be reached at the base of the convective envelope at the stage of the maximum post-flash luminosity. The efficiency of each mixing event is described by the parameter λ, i.e. the fraction of the increment of the core mass during an inter-pulse period that is dredged-up at the next thermal pulse. For any combination of input prescriptions, we compute a complete set of TP-AGB tracks (M_i = [0.5–6] $M_\odot$; Z_i= [0.0005–0.02]), for a total of about 700 evolutionary tracks. Model quantities are converted into the relevant photometry as detailed in Marigo *et al.* (2017). We use the population synthesis code `TRILEGAL` (Girardi *et al.* 2005) to simulate each subregion covered by both VMC and SAGE-SMC surveys. Our primary calibrators are the star counts in the K_s-band Luminosity Functions (LFs) and K_s vs. $J-K_s$ CMD and we require the models to simultaneously reproduce these quantities for the O-, C-, and X-AGB star observed samples.

2. Calibration results

We start by simulating the SMC population using the TP-AGB tracks presented in Marigo *et al.* (2017) (S_00, left panel of Fig. 1). The total number of AGB stars is reproduced to within 10 per cent, in agreement with the previous calibration on low-metallicity dwarf galaxies (Rosenfield *et al.* 2016). However, there are evident discrepancies in the O-rich, C-rich and X-AGB LFs. First, we calculate a series of seven models exploring the effect of different mass-loss prescriptions for both O- and C-rich stars, while keeping the same prescriptions for the 3DU. We adopt the CS11 formalism for $\dot{M}_{pre-dust}$ and for the C-stars in the dust-driven regime we adopt the CDYN results. The sequence of models is computed by increasing the efficiency parameter of the B95 prescription applied to the M-stars, with η_{dust} ranging from 0.02 to 0.06. This results in a relatively modest effect on the total number of O-rich stars, whereas a significant impact shows up on the predicted C-rich LF. The reason is that the bulk of O-rich stars have low-mass progenitors and their lifetimes are mainly affected by the $\dot{M}_{pre-dust}$, whereas the B95 relation mainly

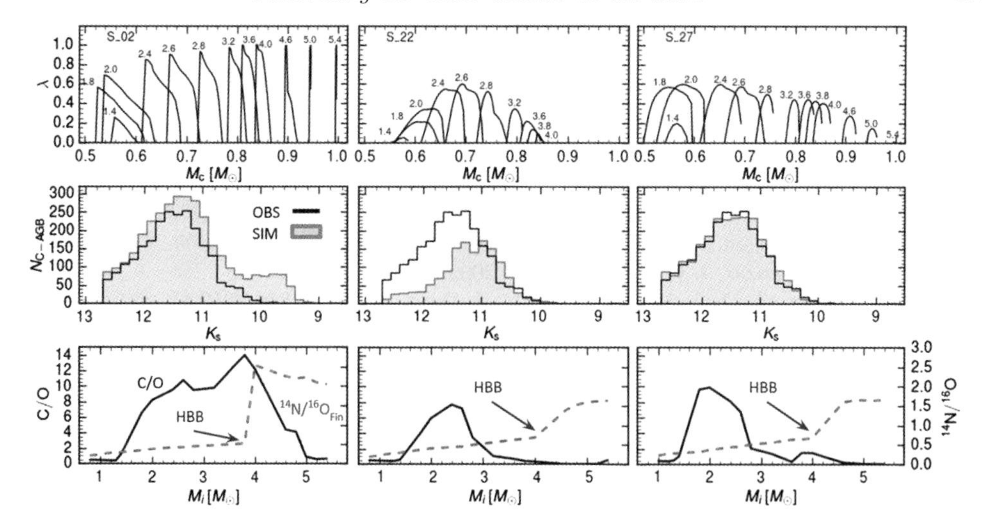

Figure 2. Top rows: efficiency of the 3DU (λ) as a function of the core mass M_c of a few selected evolutionary tracks with Z_i=0.004 and M_i as indicated. Middle rows: observed (black histograms) and simulated C-stars LFs as derived from the corresponding sets of models. The models share the same input prescriptions but for the onset and the efficiency of the 3DU. Bottom rows: final C/O and $^{14}N/^{16}O$ ratio.

affects the evolution of more massive (and brighter) AGB stars due to its high luminosity dependence. The best-fitting model of this first series (S_07, middle panel of Fig. 1) provides reasonably correct TP-AGB lifetimes, for the assumed 3DUP description, but the comparison with semi-empirical initial-to-final mass relations (IFMRs) (El-Badry *et al.* 2018; Cummings *et al.* 2018) reveals that such models tend to underestimate the White Dwarf (WD) masses for $M_i \gtrsim 3M_\odot$. We compute a second series of models with lower mass-loss efficiency ($\eta_{\rm BL95} = 0.01 - 0.03$), focusing on the calibration of 3DU parameters and taking into account also the IFMR constraints. In Fig. 2 we show some of the fundamental steps that lead us to find a reasonable parametrisation of the 3DUP. We find that in order to reproduce the C-star LF (model S_27 in Fig. 2) and the LFs of O-rich, X-AGB, and the AGB sample, in addition to the final WDs masses, we need to keep a lower mass-loss efficiency and a lower efficiency of the 3DU for stars with $M_i \gtrsim 3M_\odot$ (Z_i=0.004). Different 3DU parametrisations have a large impact on the resulting chemical yields as shown in the lower panels of Fig. 2, which we are currently investigating.

Acknowledgements

We acknowledge the support from the ERC *project STARKEY*, G.A. n. 615604.

References

Blöcker, T. 1995, *A&A*, 297, 727

Boyer, M. L., McDonald, I., Srinivasan, S., Zijlstra, A., van Loon, J. T., Olsen, K. A. G., & Sonneborn, G. 2015, *ApJ*, 810, 116

Boyer, M. L., Srinivasan, S., van Loon, J. T., McDonald, I., Meixner, M., Zaritsky, D., Gordon, K. D., Kemper, F., Babler, B., Block, M., Bracker, S., Engelbracht, C. W., Hora, J., Indebetouw, R., Meade, M., Misselt, K. 2011, *AJ*, 142, 103.

Cioni, M.-R. L., Clementini, G., Girardi, L., Guandalini, R., Gullieuszik, M., Miszalski, B., Moretti, M.-I., Ripepi, V., Rubele, S., Bagheri, G., Bekki, K., Cross, N., de Blok, W. J. G., de Grijs, R., Emerson, J. P., Evans, C. J., Gibson, B., Gonzales-Solares, E., Groenewegen, M. A. T., Irwin, M., Ivanov, V. D., Lewis, J., Marconi, M., Marquette, J.-B., Mastropietro,

C., Moore, B., Napiwotzki, R., Naylor, T., Oliveira, J. M., Read, M., Sutorius, E., van Loon, J. T., Wilkinson, M. I., & Wood, P. R. 2011, *A&A*, 527, A116

Cranmer, S. R., & Saar, S. H. 2011, *ApJ*, 741, 54

Cummings, J. D. and Kalirai, J. S. and Tremblay, P.-E. and Ramirez-Ruiz, E. & Choi, J. arXiv:1809.01673

El-Badry, K., Rix, H.-W., & Weisz, D. R. 2018, *ApJ* (Letters), 860, L17.

Eriksson, K., Nowotny, W., Höfner, S., Aringer, B., & Wachter, A. 2014, *A&A*, 566, A95.

Girardi, L., Groenewegen, M. A. T., Hatziminaoglou, E., & da Costa, L. 2005, *A&A*, 436, 895

Karakas, A. I., Lattanzio, J. C., & Pols, O. R. 2002, *PASA*, 19, 515

Marigo, P., Bressan, A., Nanni, A., Girardi, L., & Pumo, M. L. 2013, *MNRAS*, 434, 488

Marigo, P., Girardi, L., Bressan, A., Rosenfield, P., Aringer, B., Chen, Y., Dussin, M., Nanni, A., Pastorelli, G., Rodrigues, T. S., Trabucchi, M., Bladh, S., Dalcanton, J., Groenewegen, M. A. T., Montalbán, J., & Wood, P. R. 2017, *ApJ*, 835, 77

Mattsson, L., Wahlin, R., & Höfner, S. 2010, *A&A*, 509, A14.

Rosenfield, P., Marigo, P., Girardi, L., Dalcanton, J. J., Bressan, A., Gullieuszik, M., Weisz, D., Williams, B. F., Dolphin, A., & Aringer, B. 2014, *ApJ*, 790, 22.

Rosenfield, P., Marigo, P., Girardi, L., Dalcanton, J. J., Bressan, A., Williams, B. F., & Dolphin, A. 2016, *ApJ*, 822, 73.

Rubele, S., Pastorelli, G., Girardi, L., Cioni, M.-R. L., Zaggia, S., Marigo, P., Bekki, K., Bressan, A., Clementini, G., de Grijs, R., Emerson, J., Groenewegen, M. A. T., Ivanov, V. D., Muraveva, T., Nanni, A., Oliveira, J. M., Ripepi, V., Sun, N.-C., & van Loon, J. T. 2018, *MNRAS*, 478, 5017

Ruffle, P. M. E., Kemper, F., Jones, O. C., Sloan, G. C., Kraemer, K. E., Woods, P. M., Boyer, M. L., Srinivasan, S., Antoniou, V., Lagadec, E., Matsuura, M., McDonald, I., Oliveira, J. M., Sargent, B. A., Sewiło, M., Szczerba, R., van Loon, J. T., Volk, K., & Zijlstra, A. A. 2015, *MNRAS*, 451, 3504

Schröder, K.-P. & Cuntz, M. 2005, *ApJ* (Letters), 630, L73

Srinivasan, S., Boyer, M. L., Kemper, F., Meixner, M., Sargent, B. A., & Riebel, D. 2016, *MNRAS*, 457, 2814

Vassiliadis, E., & Wood, P. R. 1993, *ApJ*, 413, 641

Discussion

VENTURA: Which is the fraction of HBB stars within the X-AGB sample of the SMC? Do you believe possible a calibration of Hot Bottom Burning by the study of these bright sources

PASTORELLI: Our two best-fitting models predict a number of HBB stars around ten within the X-AGB sample. A calibration of the HBB using these sources would be possible if such stars could be clearly identified in the observations. In the K_s vs. $J-K_s$ CMD they are not always clearly distinguishable, but we can use other combinations of near- and mid-infrared colours to identify them and and eventually perform a HBB calibration.

SRINIVASAN: Thanks for showing us the excellent agreement between your models and the observed K_s-band LF. Have you compared your prescription for circumstellar reddening with the Spitzer (e.g. 8 μm) LF? Do you find similar agreement?

PASTORELLI: Yes, we find a very good agreement with the Spitzer LFs and also with several combinations of colour-colour diagrams

Resolved and unresolved AGB populations

Why Galaxies Care About AGB Stars:
A Continuing Challenge through Cosmic Time
Proceedings IAU Symposium No. 343, 2019
F. Kerschbaum, M. Groenewegen & H. Olofsson, eds.

doi:10.1017/S1743921318006415

Asymptotic Giant Branch Variables in Nearby Galaxies

Patricia A. Whitelock

South African Astronomical Observatory, P O Box 9, 7935 Observatory, South Africa
Department of Astronomy, University of Cape Town, 7701 Rondebosch, South Africa
email: paw@saao.ac.za

Abstract. Certain types of large amplitude AGB variable are proving to be powerful distance indicators that will rival Cepheids in the *James Webb Space Telescope* era of high precision infrared photometry. These are predominantly found in old populations and have low mass progenitors. At the other end of the AGB mass-scale, large amplitude variables, particularly those undergoing hot bottom burning, are the most luminous representatives of their population. These stars are < 1 Gyr old, are often losing mass copiously and are vital to our understanding of the integrated light of distant galaxies as well as to chemical enrichment. However, the evolution of such very luminous AGB variables is rapid and remains poorly understood. Here I discuss recent infrared observations of both low- and intermediate-mass Mira variables in the Local Group and beyond.

Keywords. stars: AGB and post-AGB, stars: carbon, stars: late-type, stars: mass loss, stars: variables: other, galaxies: dwarf, galaxies: individual (NGC3109, Sgr dIG, NGC4258, M33), (galaxies:) Magellanic Clouds, infrared: stars

---

1. Introduction

Mira variables have the largest amplitudes ($\Delta V > 2$, $\Delta K > 0.4$, $\Delta[4.5] > 0.3$ mag) of any regularly pulsating star, which makes them relatively easy to identify. They pulsate at the fundamental frequency, as do a few semi-regular (SR) variables, although most of the SRs pulsate in one or more overtone modes (e.g., Trabucchi *et al.* 2017). In this presentation I want to address two topics which make Mira variables particularly interesting and important. The first is Miras as distance indicators; there has been a lot of work in this area over the last few years and it becomes increasingly important as we move into the *James Webb Space Telescope* (JWST) era because Miras are such strong infrared sources. At the moment this involves exclusively short period ($P < 400$ days) Miras.

The second topic is the last stage in the AGB evolution of intermediate mass AGB stars. These long period stars, including new Miras that are being discovered in the Local Group and beyond, are not well understood. They can be carbon- or oxygen-rich; many of them have thick circumstellar shells and, depending on their initial metallicity, some are undergoing hot bottom burning. These long period stars are very important to our understanding of mass loss and provide fascinating probes of the late stages of stellar evolution. It is also at this stage of their evolution that the elusive massive, super-AGB, stars are most likely to be found.

In the context of the distance scale it is important to recognise that Miras are really big stars; their angular diameters are approximately twice their parallaxes. Furthermore, they have large convection cells so that their surface features are usually both asymmetric, and variable (e.g., Paladini *et al.* 2017). Thus *Gaia* is not easily going to give us distances.

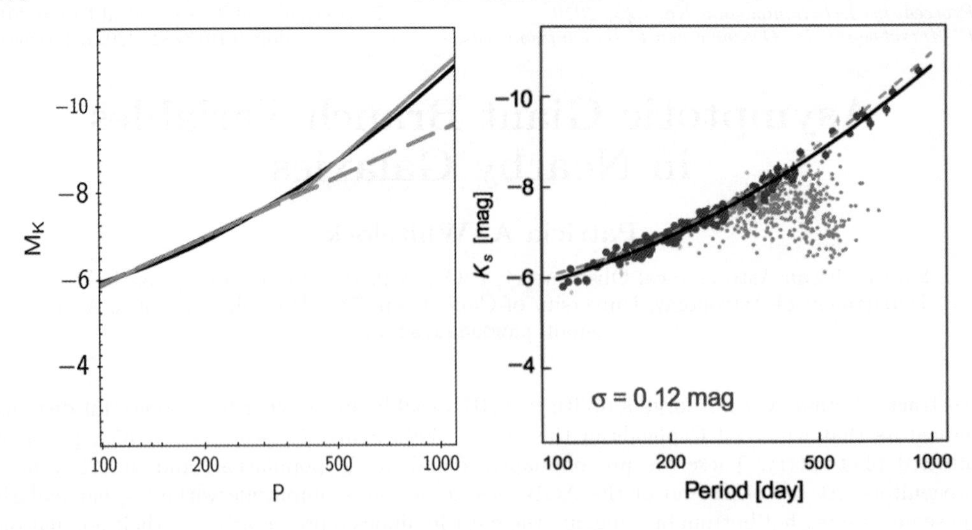

Figure 1. Left: a comparison of the PLRs from Whitelock *et al.* (2008; in magenta), Ita & Matsunaga (2011; in green) and Yuan *et al.* (2017; in black). Right: the Yuan *et al.* (2017) parabolic fit to the LMC O-rich Miras; O-rich and C-rich stars in blue and red, respectively.

It may eventually be possible to disentangle the diameters and the surface features of some very well observed stars, but that sort of analysis will take time and will not be a part of early *Gaia* data releases. van Langevelde (2018) also discussed this point in the context of radio VLBI measures of Miras. VLBI parallaxes are very useful for Miras, but are time consuming to obtain and only applicable to O-rich stars which have OH, H_2O, or SiO Masers.

2. Miras as Distance Indicators

The near-infrared period-luminosity relation (PLR) for the Miras in the LMC has been investigated over many years (Feast *et al.* 1989; Wood *et al.* 1999; Whitelock *et al.* 2008; Ita & Matsunaga 2011; Yuan *et al.* 2017). The details depend on the wavelength, but in general stars with thick dust shells will fall below the PLR because of circumstellar extinction and this is most obvious for C-rich stars and at shorter infrared wavelengths. If the extinction can be corrected these C-stars seem to fall on the same PLR as the O-rich ones.

Figure 1 shows the parabolic K_s-PLR derived by Yuan *et al.* (2017) for O-rich Miras in the LMC, using mean magnitudes, and compares it with the two linear PLRs derived earlier by Ita & Matsunaga (2011), for more or less the same stars although using single observations, and the extrapolated PLR derived for Galactic and LMC mean magnitudes by Whitelock *et al.* (2008). For O-rich stars with $P < 400$ days the three PLRs are essentially identical. For stars with longer periods the extrapolated Whitelock *et al.* relation is different from the other two (noting that none of the curves are well defined for periods much larger then 500 days). The Yuan *et al.* approach nicely illustrates what can be achieved for these large amplitude variables using mean magnitudes.

The details of pulsation in the fundamental mode have not yet been properly modelled (Trabucchi *et al.* 2017), but the linear PLR for short period Miras is presumably a consequence of the core-mass-luminosity relation (Paczynski 1970) very close to the end of the AGB phase in low mass stars. So the pulsation period is a function of the initial mass of these stars, as well as their current mass (e.g. Feast 2009).

Discussions of the bolometric PLR indicated that C-rich stars fall on the same linear PLR as the short period Miras over a large range of periods as seen in, e.g. the dwarf

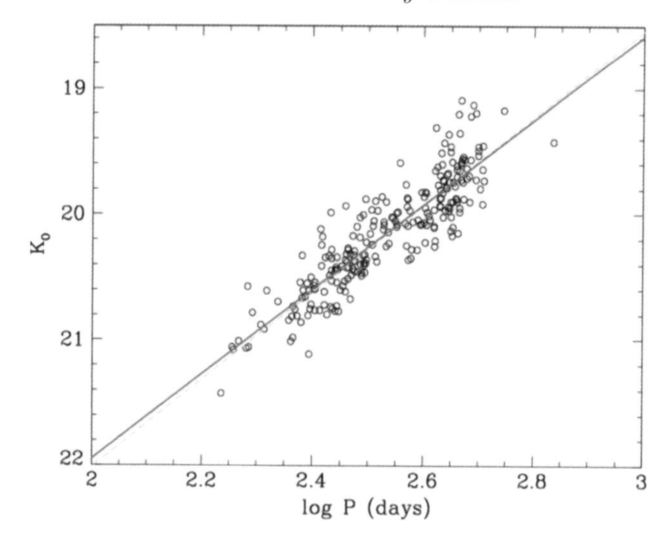

Figure 2. Rejkuba (2004) derived a PLR for Miras with $J - K_s < 1.4$ (presumed O-rich) in the inner halo of NGC 5128. The red line is the best fit PLR while the green line is the LMC PLR (Feast *et al.* 1989).

irregulars NGC 6822 (Whitelock *et al.* 2013) and IC 1613 (Menzies *et al.* 2015). The bolometric magnitudes are calculated by applying a colour dependent bolometric correction to the K magnitude†. The linear PLR can be understood if the C-rich stars obey the same core-mass-luminosity relation as the short period O-rich stars. What is lacking at this stage is a theoretical explanation of why most stars pulsate in the fundamental mode (i.e. become Miras) only at the very end of their AGB lifetimes and hence obey a linear PLR.

Yuan *et al.* (2018) applied the method developed by He *et al.* (2016) to Miras in M33 and used detailed fitting of the I-band light curve to evaluate mean JHK_s magnitudes from a small number of observations. They found that the C-stars lay below the PLR at J and H (as expected due to circumstellar extinction), but very close to it at K_s. The distance they derive is in reasonable agreement with the values found using Cepheids (Gieren *et al.* 2013 and references therein), RR Lyr variables (Sarajedini *et al.* 2006) and eclipsing binaries (Bonanos *et al.* 2006).

2.1. *Mira PLR at long periods*

At periods over about 420 days most LMC O-rich stars are brighter than a linear PLR, fitted to the shorter period stars, would predict. Whitelock *et al.* (2003) suggested that this was due to hot bottom burning (HBB) and subsequent investigations (e.g. Menzies *et al.* 2015) support this interpretation. The additional energy provided by HBB presumably changes the structure of these stars and the characteristics of the PLR, although it is interesting to see that most LMC stars do still fall on a well defined PLR, albeit a different one from non-HBB stars.

There is as yet no detailed understanding of HBB and models differ one to another (e.g., Karakas *et al.* 2018, and references therein). It is not even clear what is the lowest mass at which HBB will occur, although there is consensus that it will depend on metallicity. This is illustrated in the difference between the LMC PLR (e.g., Fig. 1) and that of NGC 5128 which was investigated by Rejkuba (2004) and is shown in Fig. 2. The Miras in NGC 5128

† This does not necessarily provide the best possible measure of bolometric luminosity, but it is simple and can be applied to measurements from different galaxies to derive distances.

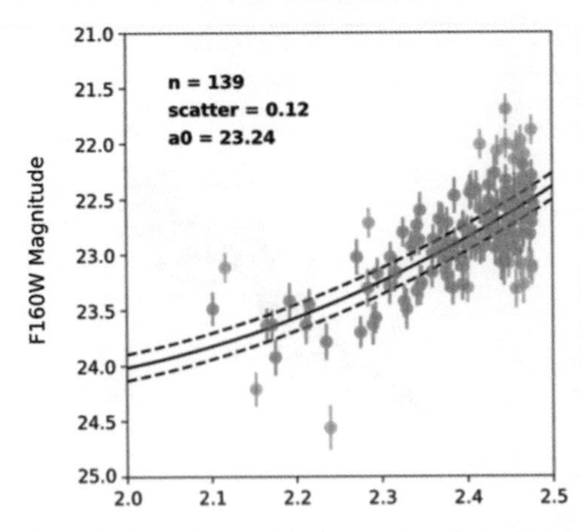

Figure 3. PLR for the best quality sample of Miras in NGC 4258; the red points were those used to derive the final relation, while grey points were removed through 3σ iterative clipping. The solid black curve shows the best fit relation and the dashed curves show the 1σ fit. The functional form of the Yuan *et al.* (2017) fit was used and the zero-point was derived from the fit (Huang *et al.* 2018).

are in the inner halo which has a metallicity of [Fe/H] ~ -0.1, i.e. considerably higher than the LMC and probably has no HBB stars. They obey the same linear PLR as the short period LMC stars.

Blommaert *et al.* (2018) discuss OH/IR stars near the Galactic centre which appear to fall below the bolometric PLR, which they suggest is due to their being more evolved along the AGB than the bulk of Miras. This is an interesting finding that might be expected given the predictions of stellar evolution theory, although, as discussed above, a detailed understanding of the pulsation of stars in the fundamental mode remains elusive. It is important that the existence of Miras with luminosities below the PLR is tested using the same method to measure the bolometric magnitude as was used to establish the calibrating magnitudes for the PLR (ideally using model fits to mean luminosities) and if at all possible in circumstances where interstellar reddening does not add significantly to the uncertainties.

Possibly the most important point to make is that the PLR at $P > 400$ days is going to differ in different galaxies, depending on the metallicity and mass range of that particular population. If the PLR is to be used for distance determination then a linear version should be used and the analysis limited to stars with short periods. Long period Miras are discussed in more detail below. Because they are luminous and generally strong infrared sources they certainly have potential as distance scale probes. However, their behaviour is not yet understood sufficiently well that we can rely on them.

2.2. *Most Distant Miras*

Huang *et al.* (2018) used HST WFC3 observations to find Miras in NGC 4258. This galaxy was selected because it hosts a water megamaser and therefore has a well established distance (Riess *et al.* 2016). Observations of a single field were obtained, through the F125W and F160W filters, at 12 epochs spread over one year. 438 Mira candidates were identified of which 139 fitted the most stringent selection criteria and were used to measure the distance; these all had $P < 300$ days and are illustrated in Fig. 3. Distances were measured relative to the LMC and agreed well with the Cepheids and the megamaser. At 7.5 Mpc these are the most distant Miras to have measured periods and luminosities. This not only shows the potential of Miras to contribute to the distance scale problem, but opens possibilities for studying individual AGB variables to much larger distances with JWST.

3. Miras as Probes of Stellar Evolution

In this section I consider some recent work on Miras with long periods, $P > 400$ days. Most of these have intermediate mass progenitors and the group includes hot bottom burning Miras, and stars that have been identified as 'extreme AGB-stars', because of their thick circumstellar shells and the resultant very red colours. Many of them are C-rich, but some of the most interesting are O-rich and include OH/IR stars with periods over 1000 days.

HBB Miras in NGC 6822, IC 1613 and WLM have been discussed by Whitelock *et al.* (2013), Menzies *et al.* (2015) and Menzies (this volume), respectively. Of particular interest are the very bright long period Miras discovered in the metal deficient Local Group galaxies, Sgr dIG and NGC 3109 which are discussed below. It is also worth noting that there are very long period Miras in the SMC and LMC that were once thought to be supergiants, but which are almost certainly Miras; HV 11417 with $P = 1092$, is one such example. Whether they are undergoing HBB is not clear.

3.1. *Miras in Sgr dIG*

Sgr dIG is a relatively low mass, Local Group, dwarf irregular that was surveyed at JHK_s for variable stars, using the 1.4m InfraRed Survey Facility (IRSF) by Whitelock *et al.* (2018). Three AGB variables were identified, two C-stars with periods of 504 and 670 days and one surprisingly blue ($(J - K_s)_0 \sim 1.3$) O-rich Mira with $P = 950$ days. At a distance of $(m - M)_0 = 25.2$ the IRSF is sensitive only to the brightest variables, so we can be certain that there are many more fainter Miras in this galaxy. The two C-rich Miras are similar to those found in other dwarf irregulars and a comparison with models from Marigo *et al.* (2017) suggests initial masses $M_i \sim 3M_\odot$. The O-rich star is unusual; it has $M_{\rm bol} \sim -6.7$ and a comparison with the models suggests an initial mass $M_i \sim 5M_\odot$ and that it is in a short lived phase at the end of hot bottom burning. Figure 4 illustrates evolutionary tracks from Marigo *et al.* (2017) for both the C- and O-rich Miras.

3.2. *Miras in NGC 3109*

NGC 3109 is probably just outside the Local Group, has a low metallicity and is more massive than Sgr dIG. The IRSF JHK_s survey was shallow and will therefore detect only the most luminous of the presumably numerous Miras. Menzies *et al.* (2018) found eight Miras, seven probably O-rich and one probably C-rich. Five of the O-rich candidates are very similar to the HBB stars found in IC 1613, NGC 6822 and WLM. These are the brightest of the blue ($J - K_s < 1.2$) stars shown in Fig. 5, where the isochrones in the top panel suggest ages of around 100 to 160 Myr. The other two O-rich candidates are only slightly older, the isochrone in the central panel of Fig. 5 suggests an age of around 180 Myr. The reddest of these has a period of 1486 days. The C star candidate Mira has a period of 1109 days and an age around 500 Myr. If its C-rich nature is confirmed will be amongst the longest period C-Miras known. Miras with $P > 1000$ days are unusual in the Galaxy and in the Magellanic Clouds and these are among the most massive and or most evolved AGB variables known as is discussed further by Menzies *et al.* (2018).

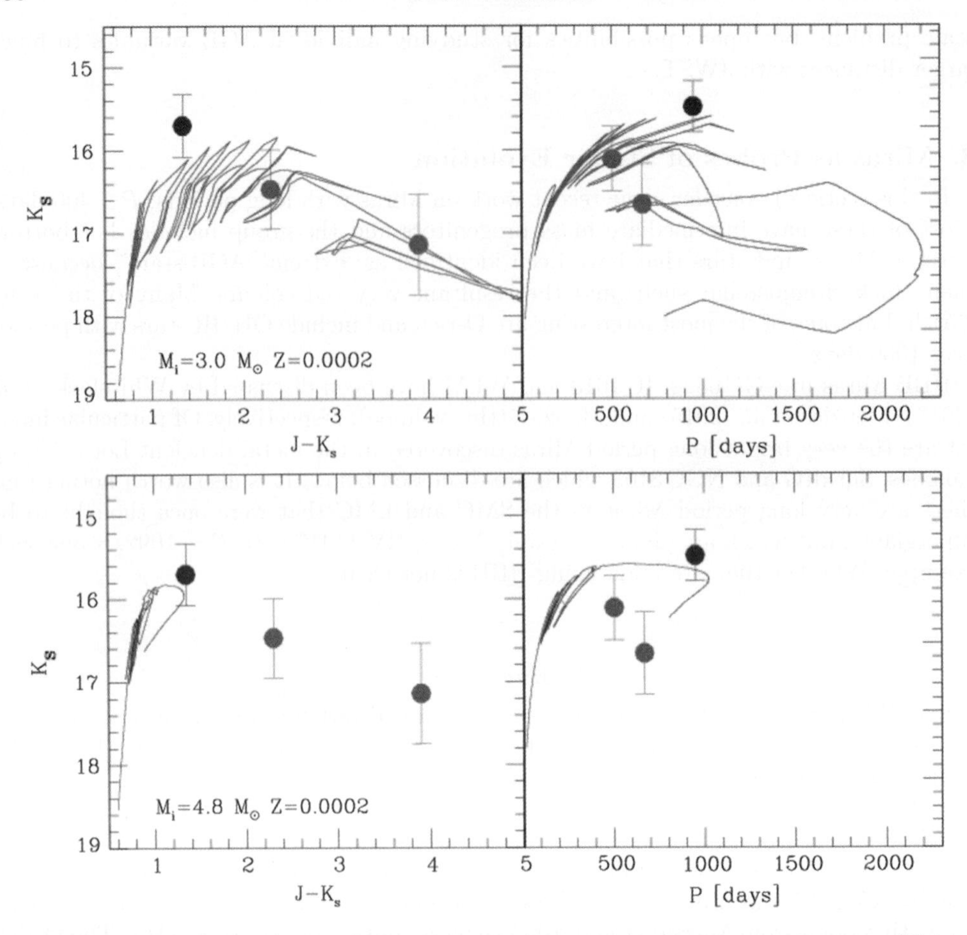

Figure 4. Evolutionary tracks in colour-magnitude and period-luminosity diagrams (C stars in red and the O-rich star in blue). Error bars show the variability range. AGB evolutionary tracks are shown for two choices of the initial mass as indicated and metallicity $Z = 0.0002$. Stages characterised by surface $C/O < 1$ and $C/O > 1$ are coloured in blue and red, respectively (from Whitelock *et al.* 2018).

3.3. *Miras in other galaxies*

The SPIRITS (Kasliwal *et al.* 2017; Karambelkar *et al.* 2018) and DUSTiNGS collaborations (Boyer *et al.* 2015; Boyer 2018; Goldman *et al.* 2018) have together obtained multiple observations of large numbers of galaxies with the Spitzer spacecraft at 3.6 and 4.5 μm. They reveal numerous red, large amplitude, long period variables most of which will be Miras. We can anticipate learning a great deal about mass-loss and AGB evolution as we start to understand these stars.

4. Conclusions

Short period Miras are showing great potential as distance indicators, and their importance is likely to increase as accurate measurements at mid-infrared wavelengths become commonplace. Long period Miras, particularly those with $P > 1000$ days are intriguing and there is need for more observations in different environments to understand the effects of mass and metellicity. There is also a great need for better understanding of pulsation and of evolution of these unusual stars.

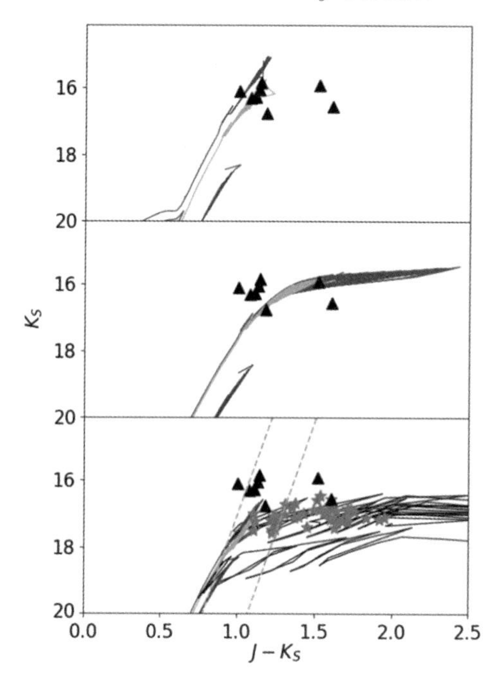

Figure 5. Comparison of the NGC 3109 AGB variables (black triangles) from Menzies *et al.* 2018 with Padova isochrones (Marigo *et al.* 2017). Top: Isochrones for Z = 0.001 and ages of 0.1, 0.158 and 1 Gyr (later phases omitted). Middle: Isochrones for Z = 0.003 and ages of 0.158, 0.178 and 1 Gyr (later phases omitted). Bottom: Isochrones for Z = 0.001 and ages of 0.398, 0.501 and 1 Gyr and non-Mira C-stars illustrated as red stars. In the bottom plot, the black lines show the evolutionary phases where the models predict stars to have C/O>1; note agreement with the observations. The parallel dashed lines show the region where according to Cioni *et al.* (2006) only O-rich stars with Z = 0.001 should be found (it obviously does not apply to this galaxy).

Acknowledgements

I am grateful to my collaborators particularly John Menzies and Michael Feast for their part in this work and to John Menzies for a critical reading of a draft of this paper. My thanks to the South African National Research Foundation for a research grant.

References

Blommaert, J.A.D.L., Groenewegen, M.A.T., Justtanont, K., & Decin, L. 2018, *MNRAS*, 479, 3545

Bonanos, A.Z., *et al.* 2006, *ApJ*, 652, 313

Boyer, M.L. 2018, Proc IAU Symp. 343, (Cambridge: Cambridge Univ. Press), these proceedings

Boyer, M.L., *et al.* 2015, *ApJ*, 800, 51

Cioni, M.-R.L., Girardi, L., Marigo P., & Habing H.J. 2006, *A&A*, 448, 77

Feast, M.W. 2009, in: AGB Stars and Related Phenomena, (eds.) T. Ueta, N. Matsunaga & Y. Ita, p. 48

Feast, M.W., Glass, I.S., Whitelock, P.A., & Catchpole, R.M. 1989, *MNRAS*, 241, 375

Gieren, W., *et al.* 2013, *ApJ*, 773, 69

Goldman, S.R., Boyer, M., & the DUSTiNGS team, 2018, Proc IAU Symp. 343, (Cambridge: Cambridge Univ. Press), these proceedings

He, S., Yuan, W., Huang, J.Z., Long, J., & Macri, L.M. 2016, *AJ*, 152, 164

Huang, C.D., *et al.* 2018, *ApJ*, 857, 67

Ita, Y., & Matsunaga, N. 2011, *MNRAS*, 412, 2345

Karakas, A.I., Lugaro, M., Carlos, M., Cseh, B., Kamath, D., & García-Hernández, D.A. 2018, *MNRAS*, 477, 421
Karambelkar, V., Adams, S., & the SPIRITS Collaboration, 2018, in preparation
Kasliwal, M.M., *et al.* 2017, *ApJ*, 839, 88
Marigo, P., *et al.* 2017, *ApJ*, 835, 77
Menzies, J.W., Whitelock, P.A., & Feast, M.W. 2015, *MNRAS*, 452, 910
Menzies, J.W., Whitelock, P.A., Feast, M.W., & Matsunaga, N. 2018, in preparation
Paczynski, B. 1970, *AcA*, 20, 47
Paladini, C., *et al.* 2017, *A&A*, 600, 136
Rejkuba, M. 2004, *A&A*, 413, 903
Riess, A.G., *et al.* 2016, *ApJ*, 826, 56
Sarajedini, A., Barker, M.K., Geisler, D., Harding, P., & Schommer, R. 2006, *AJ*, 132, 1361
Trabucchi, M., Wood, P.R., Montalbán, J., Marigo, P., Pastorelli, G., & Girardi, L. 2017, *ApJ*, 847, 139
van Langevelde, H.J. 2018, Proc IAU Symp. 348, (Cambridge: Cambridge Univ. Press), in press
Whitelock, P.A., Feast, M.W., van Loon, J.Th., & Zijlstra, A.A. 2003, *MNRAS*, 342, 86
Whitelock, P.A., Feast, M.W., & Van Leeuwen, F. 2008, *MNRAS*, 386, 313
Whitelock, P.A., Menzies, J.W., Feast, M.W., Nsengiyumva, F., & Matsunaga, N. 2013, *MNRAS*, 428, 2216
Whitelock, P.A., Menzies, J.W., Feast, M.W., & Marigo, P. 2018, *MNRAS*, 473, 173
Wood, P.R., *et al.* 1999, in IAU Symp. 191, Asymptotic Giant Branch Stars, (eds.) T. Le Bertre, A. Lebre, & C. Waelkens (Cambridge: Cambridge Univ. Press), 151
Yuan, W., Macri, L.M., He, S., Huang, J.Z., Kanbur, S.M., & Ngeow, C.-C. 2017, *AJ*, 154, 149
Yuan, W., Macri, L.M., Javadi, A. Lin, Z., & Huang, J.Z. 2018, *AJ*, 156, 112

Why Galaxies Care About AGB Stars:
A Continuing Challenge through Cosmic Time
Proceedings IAU Symposium No. 343, 2019
F. Kerschbaum, M. Groenewegen & H. Olofsson, eds.

doi:10.1017/S1743921318006671

AGB population as probes of galaxy structure and evolution

Atefeh Javadi[1] and Jacco Th. van Loon[2]

[1]School of Astronomy, Institute for Research in Fundamental Sciences (IPM), P.O. Box 19395-5531, Tehran, Iran
email: atefeh@ipm.ir

[2]Lennard-Jones Laboratories, Keele University, ST5 5BG, UK
email: j.t.van.loon@keele.ac.uk

Abstract. The evolution of galaxies is driven by the birth and death of stars. AGB stars are at the end points of their evolution and therefore their luminosities directly reflect their birth mass; this enables us to reconstruct the star formation history. These cool stars also produce dust grains that play an important role in the temperature regulation of the interstellar medium (ISM), chemistry, and the formation of planets. These stars can be resolved in all of the nearby galaxies. Therefore, the Local Group of galaxies offers us a superb near-field cosmology site. Here we can reconstruct the formation histories, and probe the structure and dynamics, of spiral galaxies, of the many dwarf satellite galaxies surrounding the Milky Way and Andromeda, and of isolated dwarf galaxies. It also offers a variety of environments in which to study the detailed processes of galaxy evolution through studying the mass-loss mechanism and dust production by cool evolved stars. In this paper, I will first review our recent efforts to identify mass-losing Asymptotic Giant Branch (AGB) stars and red supergiants (RSGs) in Local Group galaxies and to correlate spatial distributions of the AGB stars of different mass with galactic structures. Then, I will outline our methodology to reconstruct the star formation histories using variable pulsating AGB stars and RSGs and present the results for rates of mass–loss and dust production by pulsating AGB stars and their analysis in terms of stellar evolution and galaxy evolution.

Keywords. stars: luminosity function, mass function – stars: AGB and post–AGB – stars:mass–loss – galaxies: evolution – galaxies: individual: M 33 – galaxies: spiral – galaxies: dwarf – galaxies: stellar content – galaxies: structure

---

1. Introduction

Stars can be resolved in all of the Local Group galaxies. This allows the reconstruction of star formation histories (SFHs) by modelling the colour–magnitude diagram, or by using the luminosity distribution of specific stellar tracers. These galaxies have accurate distances based on the tip of the red giant branch (RGB), period–luminosity relation of relatively young Cepheids, or luminosities of old RR Lyrae. Cool evolved stars are among the most accessible probes of stellar populations due to their immense luminosity, from 2000 $L_\odot$ for tip–RGB stars, $\sim 10^4$ $L_\odot$ for asymptotic giant branch (AGB) stars, up to a few 10^5 $L_\odot$ for red supergiants. Their spectral energy distributions (SEDs) peak around 1 μm, so they stand out in the I–band (and reddening is reduced at long wavelengths). They have low surface gravity causing them to pulsate radially on timescales of months to years (Yuan *et al.* 2018). The most extreme examples among these long–period variables (LPVs) are Mira (AGB) variables, which can reach amplitudes of ten magnitudes at visual wavelengths. LPVs vary on timescales (not always strictly periodic) from $\sim$ 100 days for low mass AGB stars ($\sim$ 1 $M_\odot$; 10 Gyr old) to $\sim$ 1300 days for the dustiest massive AGB stars ($\sim$ 4–8 M$_\odot$; 30–200 Myr old); $\sim$ 600–900 days for red supergiants ($\sim$ 8–30 M$_\odot$; 10–30 Myr old). The variability helps identify these beacons; their luminosities can be used to reconstruct the star formation history; and their amplitudes pertain to the process by which they lose matter and ultimately terminate their evolution (van Loon *et al.* 2008; McDonald & Zijlstra 2016). The diagnostic power of LPVs has been demonstrated in M 33 (Javadi *et al.* 2013, 2017) and was illustrated once again by the discovery of a massive (5 M$_\odot$) LPV in the Sagittarius dwarf irregular galaxy (Whitelock *et al.* 2018).

In this project we aim: to construct the mass function of LPVs and derive from this the star formation history in different galaxy types; to correlate spatial distributions of the LPVs of different mass with galactic structures (spheroid, disc and spiral arm components); to measure the rate at which dust is produced and fed into the ISM; to establish correlations between the dust production rate, luminosity, and amplitude of LPVs; and to compare the *in situ* dust replenishment with the amount of pre–existing dust.

2. LPVs in nearby galaxies

Nearby galaxies in the Local Group provide excellent opportunities for studying dust–producing late stages of stellar evolution over a wide range of metallicity. This enables to study the detailed processes of galaxy evolution. Furthermore, we can investigate the formation histories, and probe the structure and dynamics, of spiral galaxies, of the many dwarf satellite galaxies surrounding the Milky Way and Andromeda, and of isolated dwarf galaxies.

2.1. *LPVs in M 33 galaxy*

The only spiral galaxies in the Local Group besides the Milky Way are M 31 and M 33. M 31 is highly inclines ($i \sim 77°$), and extinction therefore remains a problem. With inclination of $i = 56°$, and at d = 950 kpc only slightly more distant, M 33 is positioned much more favourably. Also, in contrast to M 31 the prominent disk of Sc galaxy M 33 bears evidence of recent star formation.

WFCAM and UIST on the UK InfraRed Telescope (UKIRT) was used to identify mass–losing AGB stars and red supergiants in M 33 galaxy from the central square kpc region to a square degree area (Fig. 1). K–band observations were complemented with occasional observations in J– and H–band to provide colour information. The photometric catalogue of the disc comprises 403 734 stars, among which 4643 stars display large–amplitude variability (Javadi *et al.* 2015). Likewise for the bulge we identified 18 398 stars among which 812 were identified as exhibiting large–amplitude variability (Javadi *et al.* 2010).

2.2. *LPVs in dwarf galaxies*

Among the pertinent questions regarding the dwarf galaxies are to what extent–and when–their star formation was quenched by gas removal mechanisms, be it as a result of internal feedback (e.g., supernova explosions) or external processes such as the interaction with massive haloes (Weisz *et al.* 2015). Can dwarf galaxies be rejuvenated, as some seem to harbour relatively young stars? Is gas only removed from dwarf galaxies by tidal or ram-pressure stripping, or can it also be (re–)accreted? To what extent is stellar death able to replenish (metals, dust) the interstellar medium, and to what extent does it heat it and drive galactic winds? Their star formation histories are among the clearest tracers of these processes having–or not having–occurred. Understanding the history of dwarf galaxies may also help us understand stellar streams and minor mergers, and massive globular clusters. A description of stellar mass–loss and dust production is of general

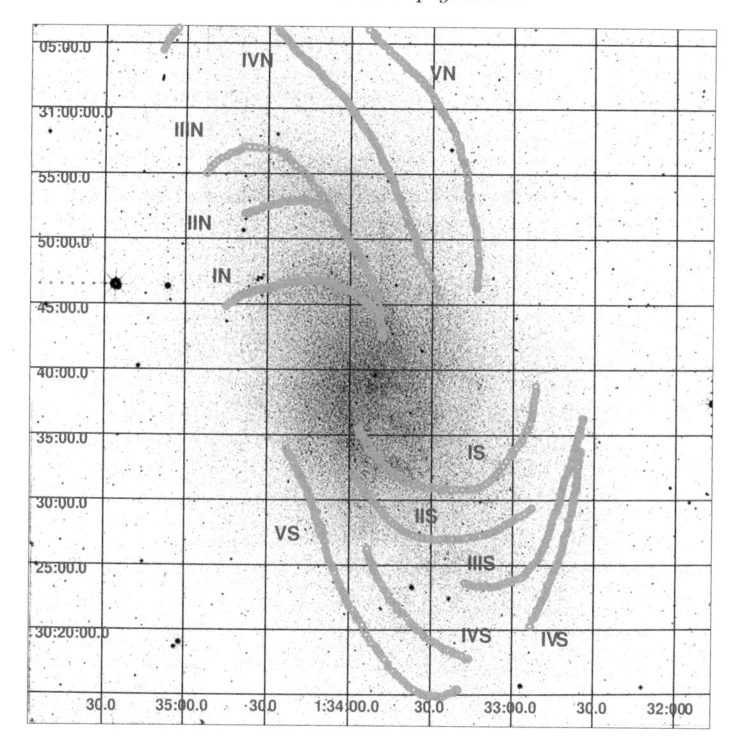

Figure 1. WFCAM K–band mosaic of M 33 with the system of five sets of spiral arms marked on it.

importance for understanding stellar and galaxy evolution. To answer the mentioned questions one the of robust ways is to identify the LPVs as tracers of galaxies star formation histories and chemical enrichment.

The LPVs are identified in some of the galaxies in the Local Group via long term monitoring surveys. In addition, a majority of dwarf galaxies have been observed with *Spitzer* at 3.6 and 4.5 μm via DUSTiNGS project (DUST in Nearby Galaxies with *Spitzer*; Boyer *et al.* 2015a,b). Due to lack of the monitoring survey of dwarf galaxies we started to monitor these galaxies with Isaac Newton Telescope (INT) with the purpose of identifying variable AGB stars (Saremi *et al.* 2017).

2.2.1. INT monitoring survey of dwarf galaxies in the Local Group

While the Milky Way satellite galaxies are spread all over the sky, a Northern hemisphere survey alone can be complete for the Andromeda system of satellite galaxies. Such a survey will benefit from the homogeneity in the distances and hence completeness and accuracy, and foreground populations and extinction are modest and similar between all Andromeda satellites. Surveying an entire satellite system enables us to determine variations among satellites due to their infall histories, cosmic reionization, and internal processes, and to examine how these variations depend on their structural properties such as total mass, gas mass, and distance to their galaxy host. For instance, the NGC 147 and NGC 185 pair are equal in mass but they differ in star formation history and gas content Weisz *et al.* 2015). We can also consider the system of satellites as a whole, and add sparse populations of stellar tracers within individual dwarf galaxies to mimic a much larger galaxy that has sufficient statistical value. About 20 Andromeda satellites are known to date, and their small number is the main limit on how clearly one can find trends and variance among them.

Individual Milky Way satellites observed as a comparison to the Andromeda system –is the Andromeda system a universal template for galaxy evolution, or just one particular case? Likewise, isolated galaxies such as Sextans A and B– or the massive and gas–rich IC 10 serve as references against which to assess the effects of galactic harassment affecting satellites. While a good few dwarf galaxies have been monitored over short campaigns to detect RR Lyrae and in some cases Cepheids, only a few (Southern) galaxies have been monitored (in the infrared) over sufficiently long periods of time (>year) to identify LPVs leaving a vast terrain unexplored. Looking ahead, the most luminous LPVs can be found as far away as the massive spiral galaxy M 101 (7 Mpc) to identify dusty supernova progenitors for spectroscopic follow–up with the James Webb Space Telescope.

We observed in I–band for identification of LPVs, as this is where the contrast between the LPVs and other stars is greatest, the bolometric corrections to determine luminosities are smallest, and the effects of attenuation by dust are minimal. However, we also monitored in the V–band. We prioritised the 62 targets, principally on the basis of their estimated number of AGB stars –populous galaxies include IC 10 ($> 10^4$ AGB stars) and Sextans A and B ($> 10^3$ AGB stars). We monitored the entire Andromeda system of satellites; next highest priority was given to isolated and/or gas–rich galaxies. We included distant globular clusters Pal 3 and 4, and NGC 2419 (Galactocentric distances 90–111 kpc) to investigate their connection to nucleated dwarf galaxies. Ultra–faint Milky Way satellites were given the lowest priority as they will have few (or no) LPVs. The face–on spiral galaxies M 101 and NGC 6946 are included to identify the red supergiant and super–AGB progenitors of imminent supernovae (9 SNe have been noticed over the past century in NGC 6946 (The Fireworks Galaxy). The $34'$ wide field of the INT camera covers each galaxy in one pointing, but dithering between repeat exposures is required to fill the gaps between the detectors. Even among the Andromeda system of satellites, none are near enough to one another to fit within one and the same field of view.

In this monitoring survey we aim to [1] reconstruct the SFHs, [2] perform accurate modelling of their SEDs, and [3] study the relation between pulsation amplitude and mass–loss (in conjunction with infrared measures of the dust, and theoretical models).

3. From LPVs counts to SFH

The LPVs are at the end points of their evolution and therefore their luminosity can be directly translated into their birth mass; this enables us to reconstruct the star formation history. The star formation history is described by the star formation rate, ξ, as a function of look-back time ("age"), t:

$$\xi(t) = \frac{f(K(M(t)))}{\Delta(M(t)) f_{\rm IMF}(M(t))}, \tag{1}$$

where $f(K)$ is the observed K-band distribution of pulsating giant stars, Δ is the duration of the evolutionary phase during which these stars display strong radial pulsation, and $f_{\rm IMF}$ is the Initial Mass Function describing the relative contribution to star formation by stars of different mass. Each of these functions depends on the stellar mass, M, and the mass of a pulsating star at the end of its evolution is directly related to its age (t) (Fig. 2). This new technique was successfully used in M 33 (Javadi *et al.* 2011a,b, 2017) the Magellanic Clouds (Rezaeikh *et al.* 2014), NGC 147 & NGC 185 (Golshan *et al.* 2017) and IC 1613 (Hashemi *et al.* 2018). The main results that we found in these galaxies are summarised as below:

- The disc of M 33 was built > 6 Gyr ago, when most stars in M 33 $\approx 73\%$ were formed. The second enhanced epoch of star formation in M 33 occurred ~ 250 Myr ago and

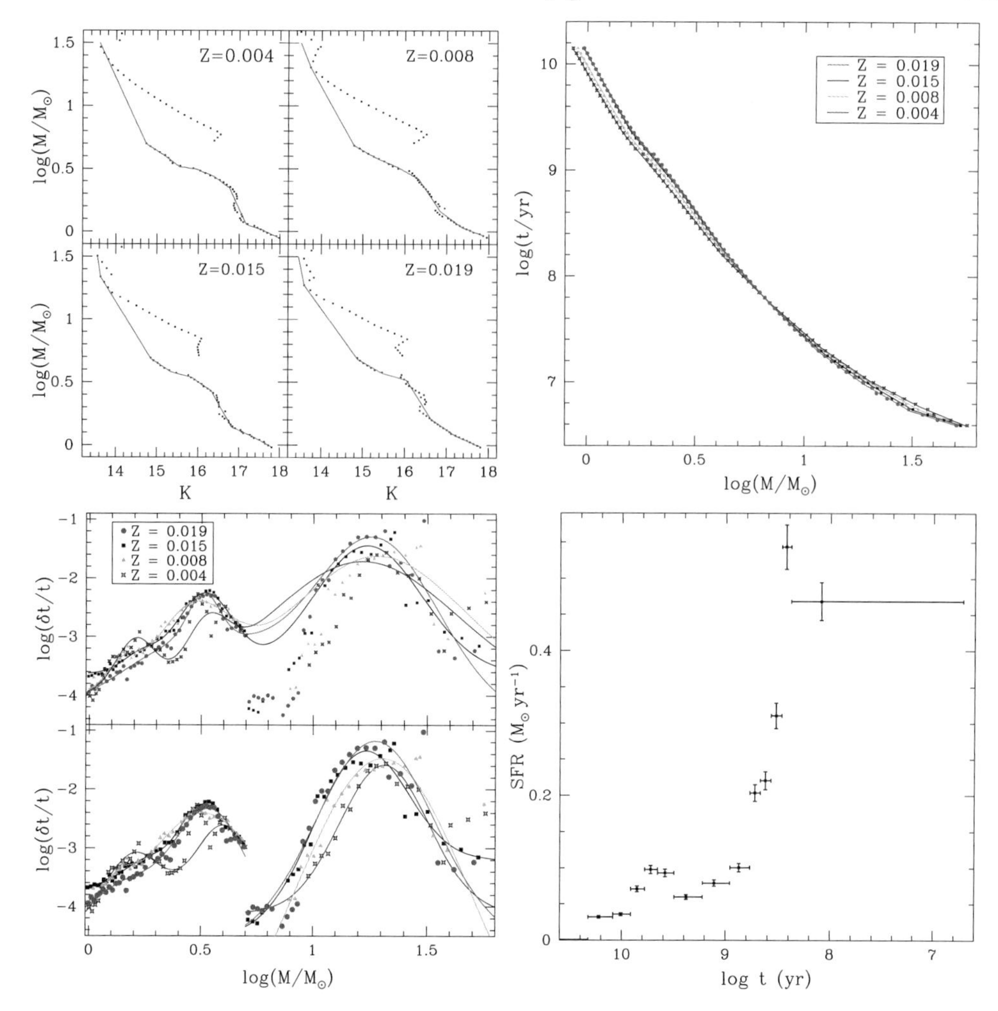

Figure 2. From LPVs counts to SFH (*Top left:*) Mass–Luminosity relation for $Z = 0.019$, 0.015, 0.008 and 0.004. The solid lines are the linear spline fits; (*Top right:*) (Birth) Mass–Age relation for AGB stars and red supergiants derived from the Marigo *et al.* (2008) isochrones; (*Bottom Left:*) Mass–Pulsation relation; (*Bottom right:*) The SFH across the entire disc of M 33.

contributed $\sim 6\%$ to M 33's historic star formation. Radial star formation history profiles suggest that the inner disc of M 33 was formed in an inside–out formation scenario.

• We found a significant difference in the ancient SFH of the LMC and the SMC. For the SMC the bulk of the stars formed a few Gyr later than the LMC. A secondary peak of SFH at ~ 700 Myr ago in the LMC and the SMC is possibly due to the tidal interaction between the Magellanic Clouds and their approach to the Milky Way.

• In spite of similar mass and morphological type, NGC 147 and NGC 185, which are two of the massive satellites of the Andromeda galaxy reveal completely different SFHs. NGC 185 formed earlier than NGC 147 but its star formation continued until recent times with almost constant rate while the star formation in NGC 147 quenched at least 300 Myr ago. These results are corroborated by strong tidal distortions of NGC 147 and the presence of gas in the centre of NGC 185.

• We do not find any enhanced period of star formation over the past 5 Gyr in IC 1613, which suggests that IC 1613 may have evolved in isolation for at least that long.

4. AGB stars as probes of galaxy structure

Populations of stars formed at different times may also reveal some of the galactic structures in M 33 galaxy. To this aim we separated the stars in our catalogue into massive stars, AGB stars, and RGB stars, on the basis of K-band magnitude and $J--K$ colour criteria. In the central parts, the AGB distribution shows clear signs of a double-component profile, with the break occurring around $r \sim 0.4$ kpc, so in this case we fitted a Sérsic profile, with $R_e = 0.30$ kpc and $n = 1.09$ (Javadi *et al.* 2011b). Possibly we are dealing with a bar–like feature, which is a disc-related structure and may be connected to the footpoints of the spiral arms. In addition, the spatial distributions of the massive stars, intermediate-age Asymptotic Giant Branch (AGB) stars and generally old Red Giant Branch (RGB) stars in this region suggest that young and intermediate-age stars were formed within the disc, while the oldest stars may inhabit a more dynamically-relaxed configuration. Interestingly, the massive stars concentrate in an area South of the nucleus, and the intermediate–age population shows signs of a "pseudo–bulge" that however may well be a bar–like feature. Furthermore, the distribution of stars with respect to five spiral arms in M 33, suggests that there is no evidence for a lag associated with the density wave having passed through the position of evolved stars, or any asymmetry at all. This means that spiral arms are transient features and not part of a global density wave potential. Based on these results, we concluded that dynamical mixing operates on timescales < 100 Myr (Javadi *et al.* 2017).

5. From mass–loss rates of LPVs to chemical enrichment of the galaxies

The LPVs are also important sources of dust and gas within galaxies. The variability of these cool evolved stars can be used to study their evolutionary state and mass loss. The pulsations are strongest when the mass loss also becomes strongest–it is likely that the pulsations help drive the outflow, assisted by dust formation inside the shocked elevated atmosphere (allowing radiation pressure to drive a wind) and possibly by mechanical or electro–magnetic waves. To estimate the mass–loss rate of variable stars we use the combination of near–IR and mid–IR data. In the case of M 33 almost 2000 variable stars have also been identified by *Spitzer* (Javadi *et al.* 2013, and in preparation). We modelled SEDs of variables stars with at least two measurements in near–IR bands and two mid–IR bands using the publicly available dust radiative transfer code DUSTY (Ivezić & Elitzur 1997). In addition 24–μm sources from Montiel *et al.* (2015) were modelled, because they contribute a large fraction to the total dust and mass return (Table 1, Fig. 3). Our results suggest that the mass–loss rate is approximately proportional to luminosity (and hence birth mass) with almost weaker dependence to pulsation period and/or amplitude (reflecting stellar evolution). In addition, the total mass lost by evolved stars ($\dot{M} \sim 0.1$ $M_\odot$ yr^{-1}) falls short by about a factor of four to sustain stars formation with a current rate, therefore requiring external sources of gas supply (Javadi *et al.* in preparation).

6. On–going works and conclusion

We are currently extending our M 33 study to the dwarf galaxies in the Local Group, to derive star formation history and dust production rate. We aim to identify all LPVs and obtain accurate time–averaged photometry and amplitudes of variability for all red giants and supergiants in the dwarf galaxies at Local Group.

In conclusion, this kind of research which is based on resolved AGB populations, is very important from both theoretical and observational perspectives; Firstly, it will give an unprecedented map of the temperature and radius variations as a function of luminosity and metallicity for mass-losing stars at the end of their evolution, which places important

Table 1. List of 24 μm variables in M 33, with UKIRT ID No. (Javadi *et al.* 2015). The luminosities and mass-loss rates are derived from SED fits, which should be appropriate for AGB stars and RSGs but not YSOs.

name	ID	RA(2000)	DEC (2000)	$\log L/L_\odot$	$\log \dot{M}$ ($M_\odot$ yr^{-1})	variable?	type
2	311369	01:34:22.85	+30:34:09.9	5.17	−2.79	no	YSO?
4	39836	01:33:32.64	+30:36:55.5	5.06	−3.37	no	YSO?
6	304069	01:33:29.70	+30:24:08.6	4.61	−3.56	yes	AGB
7	17077	01:34:12.95	+30:29:38.5	4.81	−3.55	no	AGB?
	160486	01:34:12.87	+30:29:40.1			no	AGB?
8	305279	01:33:28.38	+30:36:47.9	4.52	−3.75	yes	AGB
9	304597	01:34:27.85	+30:43:40.0	4.66	−3.65	yes	AGB
10	252686	01:33:50.06	+30:16:31.7	4.41	−3.66	probably	AGB
13	16033	01:33:26.65	+30:57:14.4	4.76	−3.80	yes	AGB?
14	453	01:34:12.25	+30:53:14.1	5.46	−4.60	no	RSG
16	8656	01:33:47.34	+30:16:32.4	4.52	−4.09	yes	AGB
17	24352	01:33:49.86	+30:52:41.3	5.06	−3.55	yes	RSG?
20	249854	01:33:19.68	+30:31:05.1	4.26	−4.04	probably	AGB
21	53418	01:33:41.54	+30:14:12.7	4.44	−4.03	yes	AGB
22	182878	01:33:37.43	+30:55:50.4	3.96	−3.46	no	YSO
23	43590	01:34:09.40	+30:55:18.2	4.41	−4.09	yes	AGB

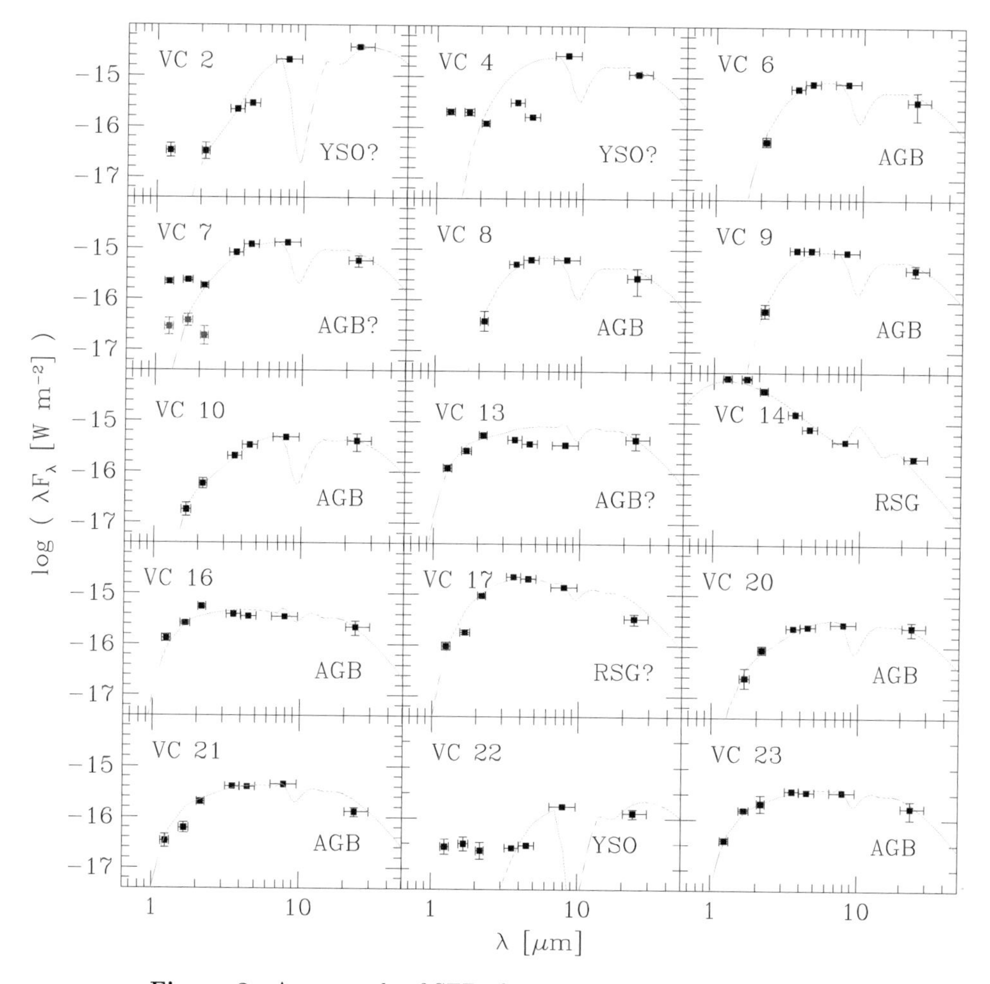

Figure 3. An example of SEDs for stars with 24–μm variability.

constraints on stellar evolution models and which is a vital ingredient in the much sought–after description of the mass–loss process. Secondly, from observational prospective, this research will gather independent diagnostics of the SFHs of different types of galaxies found in different environments, which help build a comprehensive picture of galaxy evolution in the Local Group. These two reasons together show the unprecedented success of AGB stars in investigating the galaxies formation and evolution.

Acknowledgements

AJ would like to thank the conference organisers for support. We are grateful for financial support by The Leverhulme Trust under grant No. RF/4/RFG/2007/0297, by the Royal Astronomical Society, and by the Royal Society under grant No. IE130487.

References

Boyer, M. L., McQuinn, K. B. W., Barmby, P., Bonanos, A. Z., Gehrz, R. D., Gordon, K. D., Groenewegen, M. A. T., *et al.* 2015a, *ApJS*, 216, 10

Boyer, M. L., McQuinn, K. B. W., Barmby, P., Bonanos, A. Z., Gehrz, R. D., Gordon, K. D., Groenewegen, M. A. T., *et al.* 2015b, *ApJS*, 800, 51

Hamedani Golshan, R., Javadi, A., van Loon, J.Th., Khosroshahi, H., & Saremi, E. 2017, *MNRAS*, 466, 1764

Hashemi, S. A., Javadi, A., & van Loon, J. Th. 2018, *MNRAS*, submitted

Ivezić, Ž, & Elitzur, M. 1997, *MNRAS*, 287, 799

Javadi, A., van Loon, J. Th., & Mirtorabi, M.T. 2010, *MNRAS*, 411, 263

Javadi, A., van Loon, J. Th., & Mirtorabi, M.T. 2011a, *ASPC*, 445, 497

Javadi, A., van Loon, J. Th., & Mirtorabi, M.T. 2011b, *MNRAS*, 414, 3394

Javadi, A., van Loon, J. Th., Khosroshahi, H., & Mirtorabi, M.T. 2013, *MNRAS*, 432, 2824

Javadi, A., Saberi, M., van Loon, J. Th., Khosroshahi, H., Golabatooni, N., & Mirtorabi, M.T. 2015, *MNRAS*, 447, 3973

Javadi, A., van Loon, J. Th., Khosroshahi, H., Tabatabaei, F., Hamedani Golshan, R., & Rashidi, M. 2017, *MNRAS*, 464, 2103

Marigo, P., Girardi, L., Bressan, A., Groenewegen, M. A. T., Silva, L., & Granato, G. L. 2008, *A&A*, 482, 883

McDonald, I., & Zijlstra, A. A. 2016, *ApJ*, 823, 38

Montiel, E. J., Srinivasan, S., Clayton, G. C., Engelbracht, C. W., & Johnson, C. B. 2015, *AJ*, 149, 57

Rezaeikh, S., Javadi, A., Khosroshahi, H., & van Loon, J.Th. 2014, *MNRAS*, 445, 2214

Saremi, E., Javadi, A., van Loon, J. Th., Khosroshahi, H., Abedi, A., Bamber, J., Hashemi, S. A., Nikzat, F., & Molaei Nezhad, A. 2017, *JPhCS*, 869, 2068

van Loon, J. Th., Cohen, M., Oliveira, J. M., Matsuura, M., McDonald, I., Sloan, G. C., Wood, P. R., & Zijlstra, A. A. 2008, *A&A*, 487, 1055

Weisz, D. R., Dolphin, A. E., Skillman, E. D., Holtzman, J., Gilbert, K. M., Dalcanton, J. J., & Williams, B. F. 2015, *ApJ*, 804, 136

Whitelock, P. A., Menzies, J. W., Feast, M. W., & Marigo, P. 2018, *MNRAS*, 473, 173

Yuan, W., Macri, L. M., Javadi, A., Lin, Z., & Huang, J. Z. 2018, *AJ*, 156, 112

Why Galaxies Care About AGB Stars: A Continuing Challenge through Cosmic Time
Proceedings IAU Symposium No. 343, 2019
F. Kerschbaum, M. Groenewegen & H. Olofsson, eds.

doi:10.1017/S1743921318006129

The role of AGB stars in the evolution of globular clusters

Paolo Ventura[1], Franca D'Antona[1], Marcella Di Criscienzo[1], Flavia Dell'Agli[2,3] and Marco Tailo[4]

[1]INAF, Observatory of Rome,
Via Frascati 33, 00077, Monte Porzio Catone (RM), Italy
email: paolo.ventura@inaf.it

[2]Instituto de Astrofísica de Canarias, E-38200 La Laguna, Tenerife, Spain

[3]Departamento de Astrofísica, Universidad de La Laguna,
E-38206 La Laguna, Tenerife, Spain

[4]Dipartimento di Fisica e Astronomia "Galileo Galilei", Universitá di Padova,
Vicolo dellOsservatorio 3, I-35122 Padova, Italy

Abstract. The results from high-resolution spectroscopy and accurate photometry have challenged the traditional paradigm that stars in globular clusters (GC) are simple stellar populations, rather suggesting that these structures harbor distinct groups of stars, differing in the chemical composition, particularly in the abundances of the light elements, from helium to silicon. Because this behavior is not shared by field stars, it is generally believed that some self-enrichment mechanism must have acted in GC, such that new stellar generations formed from the ashes of stars belonging to the original population. In this review, after presenting the state-of-the-art of the observations of GC stars, we discuss the possibility that the pollution of the intra-cluster medium was provided by the winds of AGB stars of initial mass above $\sim 3\ M_\odot$. These objects evolve with time scales of $40-100$ Myr and contaminate their surroundings with gas processed by p-capture nucleosynthesis, in agreement with the chemical patterns traced by GC stars.

Keywords. stars: abundances, stars: AGB and post-AGB, stars: evolution, globular clusters: general

1. Introduction

The observations collected in the last three decades have definitively challenged the traditional, long-standing belief, that GC stars provide the best example of a simple stellar population.

A series of studies, based on high-resolution spectroscopy, showed that star-to-star differences in the surface chemical composition exist in practically all the GC analyzed, which define well defined abundance patterns, the most clear being the C-N and O-Na anti correlations (Osborn 1971; Cottrell & Da Costa 1981; Kraft 1979; Gratton *et al.* 2001; Gratton *et al.* 2004; Carretta *et al.* 2009a; Carretta *et al.* 2009b). More recent results, based on high-resolution infrared spectroscopy, obtained by the Apache Point Observatory Galactic Evolution Experiment (APOGEE; Majewski *et al.* 2017), have further investigated the Mg-Al anti correlation patterns exhibited by stars in twelve GC, showing that the spread in the magnesium abundances detected is highly sensitive to the metallicity of the cluster (Mészáros *et al.* 2015).

On the photometric side, the first studies on the argument of multiple populations of GC were focused on the interpretation of the extremely peculiar morphology of the horizontal branch (hereafter HB) of some GC, such as NGC 2808 (D'Antona & Caloi 2004), M3 and M13 (Caloi & D'Antona 2005), NGC 6441 (Caloi & D'Antona 2007); the working hypothesis of all these investigations was that the morphology of the HB of each cluster is essentially determined by the distribution of the helium contents of the stars in the cluster, with the faintest and bluest objects being those most enriched in helium. This understanding was confirmed by independent studies, focused on the main sequence (hereafter MS) of some clusters with an extended and clumpy HB, which showed that a difference in the helium mass fractions among stars of the same cluster was required to reproduce the MS spread (or splitting, Bedin *et al.* 2004; Piotto *et al.* 2005, 2007; D'Antona *et al.* 2006).

The combination of the observational results given above confirm that the clusters showing the most extended chemical patterns are those hosting stars enriched in helium. This is a clear evidence that GC harbor one or more stellar components, formed from gas which was exposed to p-capture nucleosynthesis. Because this behavior is not shared by field stars, a self-enrichment mechanism, likely favored by the depth of the gravitational well in the central regions, was active in GC, so that new stellar generations formed from the ashes of the gas lost by stars belonging to the original, first generation (FG) of the cluster.

The recent years have witnessed a growing interest and a lively debate regarding the modalities with which such a self-enrichment occurred and the identification of the possible polluters.

In this review we will describe the "self-enrichment by AGB" scenario, according to which the actors of the pollution of the intra-cluster medium were stars of mass in the range $3.5 - 8\ M_{\odot}$, evolving through the AGB phase. In the context of this mechanism, originally proposed by D'Antona *et al.* (1983), later reconsidered on more solid grounds by Ventura *et al.* (2001), AGB stars belonging to the FG of the cluster evolved within ~ 100 Myr and contaminated their surrounding with gas exposed to p-capture activity, triggered by the hot temperatures which these stars attain at the base of the envelope, higher than ~ 40 MK, as a consequence of the ignition of hot bottom burning (HBB, Renzini & Voli 1981; Blöcker & Schönberner 1991). These stars can be considered as valuable polluters, because the gas which they eject is enriched in helium, as a consequence of the second dredge-up (SDU) episode (see e.g. Ventura 2010).

In the initial part of the present contribution we will provide a summary of the most relevant observational results from high-resolution spectroscopy and photometry of GC stars, thus fixing the main pre-requisites that the contaminant gas must fulfill. We will then discuss whether the ejecta from massive AGB stars can cope with such an observational evidence and provide a gross description of how stellar formation takes place in GC when self-enrichment by AGB stars is active.

2. Multiple populations in Globular Clusters: observational evidences

The possibility that GC stars harbor a significant fraction of stars contaminated by advanced p-capture nucleosynthesis was first deduced on the basis of results from high-resolution spectroscopy, which evidenced in all the GC examined the presence of stars with an anomalous chemical composition; initially, the most outstanding and confirmed chemical patterns were the O-Na and C-N anti-correlations, suggesting the exposure of the gas from which these stars formed to nuclear activity, based on p-capture reactions.

These studies were completed by investigations aimed at interpreting results from photometry, initially focused on understanding the peculiar morphology of the HB of a few clusters, then concentrated on the interpretation of the splitting of the main sequences and the giant branches of GC stars in the color-magnitude diagrams, depending on the combination of filters used. The outcome of these investigations was the identification of stellar populations enriched in helium, which further confirms the signature of p-capture processing in the gas from which part of the stars formed.

2.1. *Chemical patterns of globular cluster stars*

The detection of stars with anomalous chemistry dates back to a few decades ago, when it was clear the presence of star-to-star differences in the chemical composition of the sources observed (e.g. Kraft 1979). More recent studies outlined that the chemical patterns are present in all the galactic GC: Carretta *et al.* (2009a, 2009b) showed that the O-Na trend is a common feature of all GC, though with some dissimilarities from cluster to cluster, in the extension and the slope of the anti-correlation.

The recent results from APOGEE have shown that some GC exhibit a well defined Mg-Al anti-correlation: among twelve GC investigated, the study by Mészáros *et al.* (2015) showed that a significant depletion of Mg is present only in metal-poor clusters. These results, combined with the discovery of an extremely large Mg spread among the stars in the metal poor cluster NGC 2419 (Mucciarelli *et al.* 2012), indicate that the Mg-Al pattern, unlike the O-Na trend, is more sensitive to the metallicity, thus providing more information regarding the nucleosynthesis at which the contaminated material was exposed.

2.2. *The discovery of helium-rich stars in globular clusters*

In a seminal paper devoted to the analysis of the various factors which affect the distribution of stars along the HB, D'Antona *et al.* (2002) suggested that the peculiar morphology of the HB of some GC can be explained by hypothesizing a helium distribution among the stars in the cluster. A helium difference would have a minor impact on the morphology of the MS turn off, but would largely affect the distribution of the stars across the HB. This idea was successfully applied to interpret the HB of NGC 2808 (D'Antona & Caloi 2004), M13 (Caloi & D'Antona 2005), NGC 6441 (Caloi & D'Antona 2007), M3 (Caloi & D'Antona 2008), 47 Tuc (Di Criscienzo *et al.* 2010), NGC 2419 (Di Criscienzo *et al.* 2010, 2015). In some of these clusters, particularly NGC 2808 and NGC 2419, the bluest and faintest objects in the HB were interpreted as stars formed from gas enriched in helium, with a helium mass fraction $Y \sim 0.36 - 0.37$.

The hypothesis of differences among the helium abundances of GC stars was further reinforced by the discovery of MS splitting in NGC 2808 (D'Antona *et al.* 2005; Piotto *et al.* 2007) and from additional results, mainly based on HST photometry, which confirmed the complexity of the stellar population puzzle of some GC (Milone *et al.* 2012, 2013; Piotto *et al.* 2013; Milone *et al.* 2017).

2.3. *Spectroscopic and photometric evidences of multiple populations in GC*

To assemble the observational results from high-resolution spectroscopy and photometry, we discuss at the same time the helium enrichment of GC stars, deduced on the basis of the morphology of the HB and, when present, the MS broadening or splitting, and the extent of the p-capture nucleosynthesis. For the latter we use as a key indicator the magnesium spread among the stars of the same cluster; this is because, as stated previously, the activation of the Mg-Al-Si nucleosynthesis requires higher temperatures compared to the other nuclear channels involved (such as the CNO and the Ne-Na burning), thus it is more sensitive to the physical conditions of the gas from which the contaminated stars formed.

The results representing the state-of-the-art of the observations collected so far are shown in Fig. 1, where we report the helium enrichment as a function of the Mg spread, for clusters with different metallicity (see the color coding on the right side of the figure).

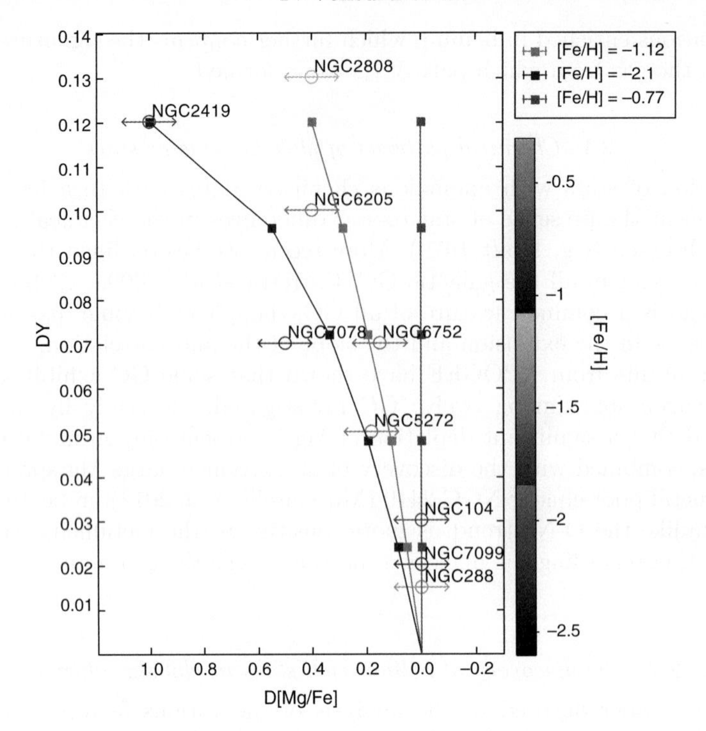

Figure 1. The magnesium and helium spread for some galactic globular clusters, for which the helium spread has been estimated.The color coding corresponds to different metallicities, as reported on the right, vertical axis. The dilution curves were obtained by mixing the AGB ejecta with various percentages of pristine gas for the metallicities [Fe/H] = −0.77 (red), [Fe/H] = −1.12 (green) and [Fe/H] = −2.1 (blue).

The most relevant results shown in Fig. 1 are the following:

(*a*) The helium enrichment is independent of metallicity. The two clusters exhibiting the largest helium spread, namely NGC 2419 and NGC 2808, have different metallicities, but the helium enhancement required to explain their very peculiar HB is practically the same ($\delta Y \sim 0.12 - 0.13$).

(*b*) The Mg spread, δMg, is sensitive to the metallicity of the cluster: the higher the metallicity, the lower δMg. No significant spread in magnesium has been observed so far in metal-rich clusters, even when a O-Na pattern is detected. For a given metallicity, δMg is correlated to the helium enrichment.

(*c*) At a given metallicity, some clusters show up the presence of a highly contaminated stellar population, whereas in other GC the surface chemistry is very homogeneous, quite similar to FG stars. This suggests that in the first case part of SG stars formed directly from the ejecta of the polluters, whereas in the latter case strong dilution with residual pristine gas in the cluster occurred.

2.4. *The indication from the observations to identify the polluters*

The observational results collected so far stimulated the search for a self-enrichment mechanism, which favored the formation of stars with a chemical composition contaminated by p-capture nucleosynthesis. The polluter stars of the FG must have evolved sufficiently fast, say below ~ 1 Gyr, because the observations of the MS of the clusters do no exhibit any trace of a significant age spread.

Based on the observational results described above, we may conclude that the polluting stars must fulfill the following pre-requisites:

(*a*) The gas ejected must be enriched in helium. The highest helium must not exceed $\sim 38\%$, corresponding to a net enrichment of $\delta Y = 0.13$. This is necessary to reproduce the peculiar morphology of the HB of clusters such as NGC 2419 and NGC 2808.

(*b*) The material lost by the polluters must show the imprinting of p-capture nucleosynthesis, able to reproduce the chemical patterns observed, particularly the C-N, O-Na and Mg-Al anti-correlations.

(*c*) While the helium enrichment is independent of metallicity, the extent of the p-capture experienced, identified with the spread in the Mg content of stars in the same cluster, must depend on the metallicity: the ejecta of the polluters must be Mg poor, but the Mg spread must decrease with metallicity and must vanish for the chemical composition of metal-rich clusters.

3. The ejecta of AGB stars

An appealing possibility regarding pollution in GC is that the gas from which new generations of stars formed was provided by stars evolving through the AGB phase. If this is the case, the attention must be focused on stars of initial mass above $\sim 3 - 4\ M_{\odot}$, because lower mass stars reach the C-star stage, thus the gas which they eject into the interstellar medium show a great enrichment in carbon (Karakas & Lattanzio 2014; Lattanzio, this volume), which is at odds with the observational evidences collected so far. We will consider only stars of mass in the range $3.5\ M_{\odot} \leq M \leq 8\ M_{\odot}$, to which we will refer in the following as "massive AGB stars".

3.1. *The helium enrichment in the gas from massive AGB stars*

Massive AGB stars produce helium-rich gas. This is shown in Fig. 2, showing the average helium in the ejecta of stars of different mass. The results refer to non rotating models of different metallicity, published by various research groups. The helium enrichment is mainly achieved during the second dredge-up (SDU) episode, which takes place after the end of the core helium burning phase, when the surface convection penetrates inwards, until reaching stellar layers previously touched by p-capture processing. The helium mass fraction increases with the mass of the star, because the higher the mass, the deeper is SDU. The helium enrichment is practically independent of metallicity. These findings are rather robust, as confirmed by the agreement among the results published by different groups. The reason for this is that SDU occurs before the beginning of the thermal pulses phase, thus the description of this phenomenon is only scarcely affected by the uncertainties related to AGB modelling, mainly related to convection modelling (Lattanzio, this volume).

Interestingly, the largest enrichment in helium, occurring in very massive AGB stars, is $\delta Y \sim 0.13$, which corresponds to mass fractions $Y \sim 0.37$. This is in agreement with the values required to fit the very peculiar HB of NGC 2808 and NGC 2419 (see Fig. 1). This upper limit in the helium enrichment would shift upwards if rotation was considered: for realistic rotation rates it is found that the helium abundance after the SDU reaches $Y \sim 0.40$ in the most massive AGB stars (A. Dotter, private communication).

3.2. *The signature of p-capture processing in the winds of AGB stars*

The ejecta from massive AGB stars definitively show the signature of p-capture nucleosynthesis. This is due to the activation of HBB at the base of the convective envelope,

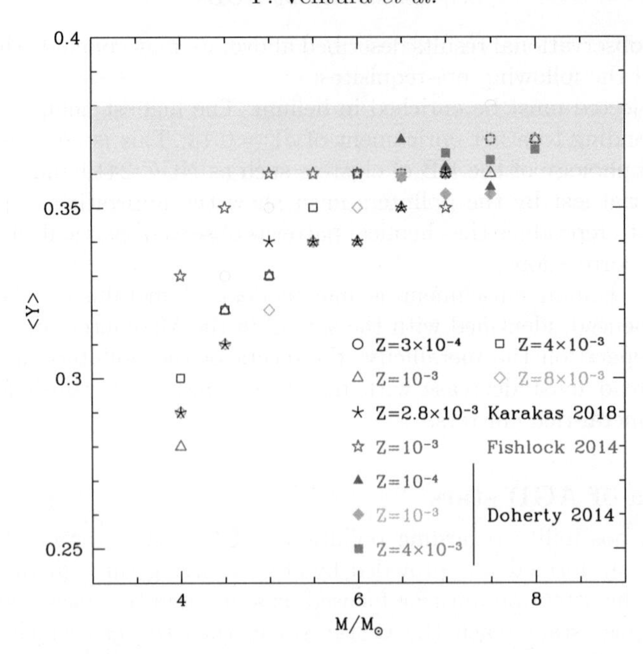

Figure 2. The helium abundance in the ejecta of AGB stars as a function of the initial mass. In the plot we show results from AGB models of various metallicities published in Ventura *et al.* (2013, 2014), low-metallicity models by Doherty *et al.* (2014), Doherty *et al.* (2014) and Karakas *et al.* (2018). The yellow shaded region indicates the helium values required to fit the faintest stars in the HB of NGC 2808 and NGC 2419.

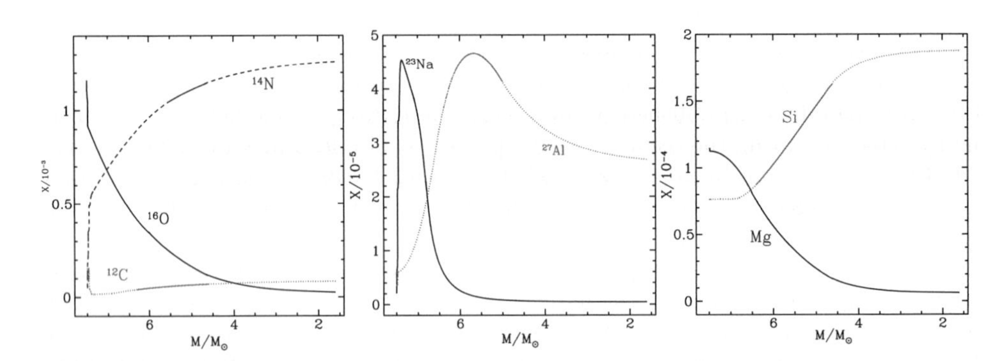

Figure 3. The variation of the surface mass fraction of the CNO elements (left panel), sodium and aluminum (middle), magnesium and silicon (right) during the AGB evolution of a star of initial mass 7.5 $M_\odot$, with the same chemical composition of NGC 2808 stars. The current mass of the star is reported on the abscissa.

as soon as the temperature in that region of the star, $T_{\rm bce}$, exceeds ~ 40 MK, which requires core masses $M_{\rm c} > 0.8\ M_\odot$.

The example shown in Fig. 3 refers to an AGB model of initial mass 7.5 $M_\odot$, recently used by Di Criscienzo *et al.* (2018) to study the multiple populations in NGC 2808. The figure reports the evolution of the surface chemical composition of the star, in terms of the mass fractions of the CNO elements (left panel), sodium and aluminum (middle panel) and magnesium and silicon (right panel). The abscissa shows the current mass of the star, to better understand the yields expected.

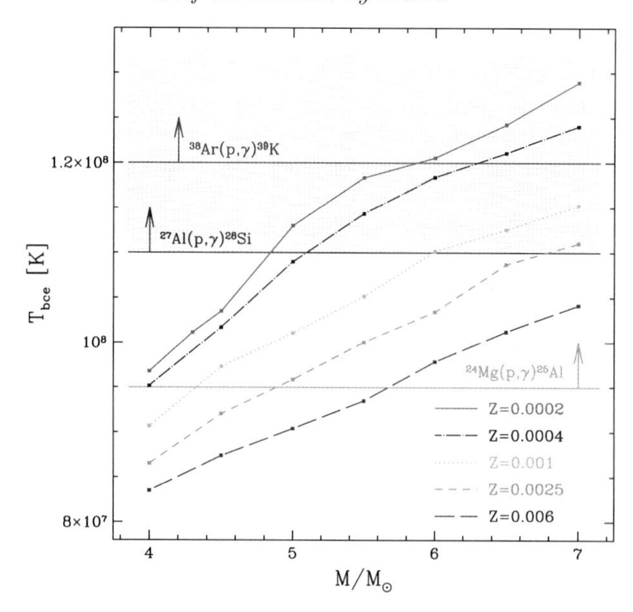

Figure 4. The temperature at the base of the convective envelope of AGB models of different mass, reported on the abscissa. The five lines correspond to different metallicities. The threshold temperatures required to activate proton capture reaction on ^{24}Mg, ^{27}Al and argon nuclei are also indicated

We notice in the left panel the effects of the full activation of the CNO cycle, with the synthesis of significant amounts of nitrogen at the expenses of carbon and oxygen. The extent of the oxygen depletion is extremely sensitive to $T_{\rm bce}$: while at temperatures in the range 40 MK $< T_{\rm bce} <$ 80 MK only CN cycling is activated, for temperatures above 80 MK oxygen is depleted efficiently.

The middle panel shows the behavior of sodium and aluminum, two species involved in both production and destruction channels, thus extremely sensitive to the temperature at which the nucleosynthesis takes place. In both cases we note an initial phase during which the two species are produced, owing to the ignition of ^{22}Ne and magnesium burning, followed by evolutionary stages when both Na and Al are destroyed by $^{23}\mathrm{Na}(p,\alpha)^{20}\mathrm{Ne}$ and $^{27}\mathrm{Al}(p,\alpha)^{24}\mathrm{Mg}$ reactions; the latter reactions are characterized by a steeper sensitivity to the temperature than the corresponding production rates, thus they become dominant when the temperature exceeds a given threshold.

The right panel of Fig. 3 shows the effects of a very advanced p-capture processing, with the depletion of the surface magnesium and the synthesis of silicon, the signature of the activation of the Mg-Al-Si nucleosynthesis, which requires temperatures $T_{\rm bce} \geq 90$ MK.

3.3. *The role of metallicity on the chemistry of AGB stars*

While the above arguments hold on general grounds, the overall situation is indeed extremely complex, because the degree of the nucleosynthesis experienced at the base of the envelope is sensitive to $T_{\rm bce}$, which, in turn, depends on both the core mass and the metallicity of the star.

This is described in Fig. 4, which shows the typical temperature at the base of the envelope of AGB stars of different initial mass (reported on the abscissa) and various metallicities, indicated with different lines and colors in the figure. The general trend with mass in that higher mass stars experience higher $T_{\rm bce}$, because they develop more massive cores at the beginning and during the AGB phase.

Besides the role of the initial mass it is clear in Fig. 4 the effect of metallicity, with lower Z stars reaching hotter temperatures at the base of the surface convective zone. This is consistent with the general behavior of stars supported by H-burning shells, where the gradients of the thermodynamic variables, primarily temperature, are steeper the lower the metallicity, which reflects into a higher efficiency of the shell in metal poor stars.

In Fig. 4 the temperatures attained by massive AGB stars are compared with the ignition temperatures of different nuclear channels. While in metal-poor stars the ignition of the Mg-Al-Si nucleosynthesis is activated in all the masses considered, in AGB stars with metallicity typical of metal-rich GC advanced nucleosynthesis can be ignited only in the most massive stars.

This has an important consequence on the yields from this class of objects and on the role that they might play in the self-enrichment of GC. In metal-poor clusters, provided that part of the SG formed from undiluted or scarcely diluted AGB ejecta, we expect to observe stars with the imprinting of advanced nucleosynthesis, showing up magnesium poor chemistry, and Al and Si enrichment. The Mg depletion is expected to become smaller and smaller as the metallicity increases, according to the results showed in Fig. 4. This trend with metallicity is not expected for the helium in the ejecta, because the latter depends on the extend of SDU, which is practically independent of metallicity.

We may therefore conclude that, at least on qualitative grounds, the yields from massive AGB stars share all the requisites, discussed in the previous section, required to reproduce the observational evidences collected so far, summarized in Fig. 1.

4. The role of AGB stars in the self-enrichment of Globular Clusters

The AGB yields described in the previous section have been successfully applied to explain the O-Na anti correlations observed in M3, M13, NGC 6752 (D'Antona & Ventura 2007), the O-Na and C-N patterns detected in M5, NGC 6388, NGC 6441 Ventura & D'Antona 2008, in NGC 6397 and M15 (Ventura & D'Antona 2009). Additional studies of this kind regard the interpretation of the extreme chemical composition of some stars in the metal-poor cluster NGC 2419 (Ventura *et al.* 2012), the chemical patterns exhibited by stars in 47 Tuc (Ventura *et al.* 2014), the O-Na, Mg-Al and Al-Si trends observed in NGC 2808 (Di Criscienzo *et al.* 2018). Dell'Agli *et al.* (2018) have recently interpreted the recent results from APOGEE, outlining how the trend with metallicity of the Mg-Al and Mg-Si trend can be nicely explained by invoking pollution from AGB stars.

In these studies the chemical patterns of the different clusters have been explained considering the pollution from AGB stars of the FG, using the yields of the same metallicity of each cluster, diluted at different extent with pristine gas. Dilution is an essential ingredient to explain most of the observations collected so far. These works were complemented by a series of studies, which tackled the argument of the formation of SG stars in GC with a dynamical approach.

In the seminal papers by D'Ercole *et al.* (2008, 2010) it is proposed that SG stars formed after the gas ejected from massive AGBs settled in the innermost regions of the cluster, under the effects of radiative cooling; we therefore expect that SG stars are initially more concentrated towards the center of the cluster in comparison with the FG. Such a model offers a valuable explanation of the significant fraction of SG stars observed, far in excess of what is expected considering the gas provided by massive AGBs (see e.g. Renzini *et al.* 2015): the gradual loss of stars from the outskirts of the cluster, which mainly involves FG stars, provokes a gradual increase in the SG/FG ratio, which poses the basis for a general context, where the clusters we observe nowadays are the relics of more massive structures, which lost $\sim 90\%$ of the stellar mass.

An exhaustive discussion on the factors affecting the temporal evolution of the radial distribution of the SG/FG ratio in GC is given in Vesperini *et al.* (2013), whereas the modalities with which pristine gas is re-accreted from the external regions of the clusters after the end of the type II SN explosions is described in D'Ercole *et al.* (2016). An interesting summary of the various modalities with which the formation of SG stars may occur in GC is given in the study by D'Antona *et al.* (2016), devoted at explaining the formation of 5 population in NGC 2808, as witnessed by the results from photometry, presented by Milone *et al.* (2015).

All the results obtained so far indicate that the general framework, which considers that massive AGB stars were the protagonists of self-enrichment in globular clusters, is robust and supported by the observational evidence. While additional investigations, possibly extended to extra-galactic GC, are required to draw definitive conclusions, it appears that pollution from AGB stars, for what concerns both the chemistry of the ejecta and the dynamics of formation of the SG, may account for the variety of situations observed, which make each cluster a unique case.

References

Bedin, L. R., Piotto, G., & Anderson, J., *et al.* 2004, *ApJ* (Letters), 605, L125
Blöcker, T., & Schönberner, D. 1991, *A&A*, 244, L43
Caloi, V., & D'Antona, F. 2005, *A&A*, 435, 987
Caloi, V., & D'Antona, F. 2007, *A&A*, 463, 949
Caloi, V., & D'Antona, F. 2008, *ApJ*, 673, 847
Carretta, E., *et al.* 2009a, *A&A*, 505, 117
Carretta, E., *et al.* 2009b, *ApJL*, 505, 139
Cottrell, P.L., & Da Costa, G.L. 1981, *ApJ* (Letters), 245, L79
D'Antona, F., Gratton, R., Chieffi, A. 1983, *MmSAI*, 54, 173
D'Antona, F., Caloi, V., Montalban, J., Ventura, P., & Gratton, R. 2002, *A&A*, 395, 69
D'Antona, F., & Caloi, V. 2004, *ApJ*, 611, 871
DAntona, F., Bellazzini, M., Caloi, V., Pecci, F. Fusi, Galleti, S., & Rood, R. T. 2006, *ApJ*, 631, 868
D'Antona, F., & Ventura, P. 2007, *MNRAS*, 379, 1431
D'Antona, F., Vesperini, E., D'Ercole, A., Ventura, P., Milone, A. P., Marino, A. F., & Tailo, M. 2016, *MNRAS*, 458, 2122
Dell'Agli, F., García-Hernández, D. A., Ventura, P., *et al.* 2018, *MNRAS*, 475, 3098
D'Ercole, A., Vesperini, E., D'Antona, F., McMillan, S. L. W., & Recchi, S. 2008, *MNRAS*, 391, 825
D'Ercole, A., D'Antona, F., Ventura, P., Vesperini, E., & McMillan, S. L. W. 2010, *MNRAS*, 407, 854
D'Ercole, A., D'Antona, F., & Vesperini, E. 2016, *MNRAS*, 461, 4088
Di Criscienzo, M., Ventura, P., D'Antona, F., Milone, A., & Piotto, G. 2010, *MNRAS*, 408, 999
Di Criscienzo, M., D'Antona, F., Milone, A. P., *et al.* 2011, *MNRAS*, 414, 3381
Di Criscienzo, M., Tailo, M., Milone, A. P., *et al.* 2015, *MNRAS*, 446, 1469
Di Criscienzo, M., Ventura, P., D'Antona, F., Dell'Agli, F., & Tailo, M. 2018, *MNRAS*, 479, 5325
Doherty, C. L., Gil-Pons, P., Lau, H. H. B., *et al.* 2014, *MNRAS*, 441, 582
Fishlock, C. K., Karakas, A. I., Lugaro, M., & Yong, D. 2014, *ApJ*, 797, 44
Gratton, R., *et al.* 2001, *A&A*, 369, 87
Gratton, R., Sneden, C., & Carretta E. 2004, *ARAA*, 42, 385
Karakas, A. I., & Lattanzio, J. C. 2014, *PASA*, 31, 30
Karakas, A. I., Lugaro, M., Carlos, M., *et al.* 2018, *MNRAS*, 477, 421
Kraft, R. P. 1979, *ARAA*, 17, 309
Majewski, S.R., *et al.* 2017, *AJ*, 154, 94
Mészáros, S., *et al.* 2015, *AJ*, 149, 153
Milone, A.P., *et al.* 2012, *ApJ*, 744, 58
Milone, A.P., *et al.* 2013, *ApJ*, 767, 120

Milone, A.P., *et al.* 2015, *ApJ*, 808, 51
Milone, A.P., Piotto, G., Renzini A., *et al.* 2017, *MNRAS*, 464, 3636
Mucciarelli, A., Bellazzini, M., Ibata, R., *et al.* 2012, *MNRAS*, 426, 2889
Osborn, W. 1971, *The Observatory*, 91, 223
Piotto, G., *et al.* 2005, *ApJ*, 621, 777
Piotto, G., *et al.* 2005, *ApJ* (Letters), 661, L53
Piotto, G., Milone, A. P., Marino, A. F., *et al.* 2013, *ApJ*, 775, 15
Renzini, A., & Voli, M. 1981, *A&A*, 94, 175
Renzini, A., D'Antona, F., Cassisi, S., *et al.* 2015, *MNRAS*, 454, 4197
Ventura, P., D'Antona, F., Mazzitelli, I., & Gratton, R. 2001, *ApJ* (Letters), 550, L65
Ventura, P., & D'Antona, F. 2008, *MNRAS*, 385, 2034
Ventura, P., & D'Antona, F. 2009, *A&A*, 499, 835
Ventura, P. 2010, *Light Elements in the Universe*, (SAO/NASA Astrophysics Data System), p. 147
Ventura, P., D'Antona, F., Di Criscienzo, M., *et al.* 2012 *ApJ* (Letters), 761, L30
Ventura, P., Di Criscienzo, M., Carini, R., & D'Antona, F. 2013, *MNRAS*, 431, 3642
Ventura, P., Di Criscienzo, M., D'Antona, F., *et al.* 2014, *MNRAS*, 437, 3274
Vesperini, E., McMillan, S. L., W., D'Antona, F., & D'Ercole, A. 2013, *MNRAS*, 429, 1913

Discussion

MARIGO: What is the current situation regarding star-to-star differences in the CNO abundances of globular cluster stars?

VENTURA: Initially, it was believed that the CNO is constant. More recent results suggest a possible spread of a factor of two in some globular clusters. This could be explained by a longer formation of the second generation, until times when carbon-rich gas by stars of mass 3-3.5 solar masses is ejected into the interstellar medium. On the observational side, a small increase in the overall CNO of second generation stars seems more compatible with the morphology of the horizontal branch of some metal poor clusters.

Why Galaxies Care About AGB Stars: A Continuing Challenge through Cosmic Time
Proceedings IAU Symposium No. 343, 2019
F. Kerschbaum, M. Groenewegen & H. Olofsson, eds.

doi:10.1017/S1743921318005409

Characterisation of long-period variables in the Magellanic Clouds

Michele Trabucchi[1], **Peter R. Wood**[2], **Josefina Montalbán**[1], **Paola Marigo**[1], **Giada Pastorelli**[1] **and Léo Girardi**[1,3]

[1]Dipartimento di Fisica e Astronomia Galileo Galilei, Universitá di Padova, Vicolo dell'Osservatorio 3, I-35122 Padova, Italy
email: `michele.trabucchi@unipd.it`

[2]Research School of Astronomy and Astrophysics, Australian National University, Canberra, ACT2611, Australia

[3]Astronomical Observatory of Padova–INAF, Vicolo dell'Osservatorio 3, I-35122 Padova, Italy

Abstract. Variability due to stellar pulsation on the Asymptotic Giant Branch (AGB) has a great potential for applications such as distance measurements, the study the evolution of stars and galaxies, and the estimate of global stellar parameters, as well as to constrain stellar evolutionary models. Given the importance of long-period variables (LPVs) in this sense, and given the lack of recent, updated sets of pulsation models, we computed an extended grid of pulsation models widely covering the space of AGB stellar parameters, including up-to-date opacities and accounting for the chemical evolution associated with third dredge-up events. We present the relevant properties of this grid and discuss the main results it allowed to obtain in terms of the interpretation of the observed properties of LPVs in the Large Magellanic Cloud (LMC).

Keywords. stars: AGB and post-AGB, stars: variables: other, Magellanic Clouds

1. Introduction

Long-Period Variables (LPVs) are evolved red giant stars exhibiting intrinsic variability due to stellar pulsations. They have periods of a few days to $\gtrsim 1000$ days and amplitudes from a milli-magnitude level to several magnitudes. They are often multi-periodic, although it is common practice to examine their properties in terms of a single pulsation mode, called "primary" period, *i.e.*, the one associated with the largest amplitude of variability in their light curve. It was first shown by Wood *et al.* (1999) using MACHO observations of the LMC (see also Wood (2000)) that LPVs follow several distinct period-luminosity (PL) relations, or "sequences", as displayed in the left panel of Fig. 1. LPVs are usually classified into the following three types, according to their variability properties: Miras (large-amplitude, fundamental mode pulsators with their primary periods on sequence C), semi-regular variables (SRVs, with lower amplitudes than Miras, and pulsating primarily in the first overtone (1O) and fundamental modes, with periods on sequences C$'$ and C, respectively), and OGLE Small Amplitude Red Giants (OSARGs, with several distinct radial and non radial modes and primary periods on sequences from A$'$ to C$'$). All of these sequences are the result of pulsation due to acoustic waves propagating in the stellar envelope. In addition to these modes, LPVs show another type of variability, the so-called long secondary periods (LSPs), the origin of which is still unknown. Despite of their name, they are often detected with amplitudes larger than regular modes, and thus classified as primary periods. This results in the formation of sequence D in the PL diagram.

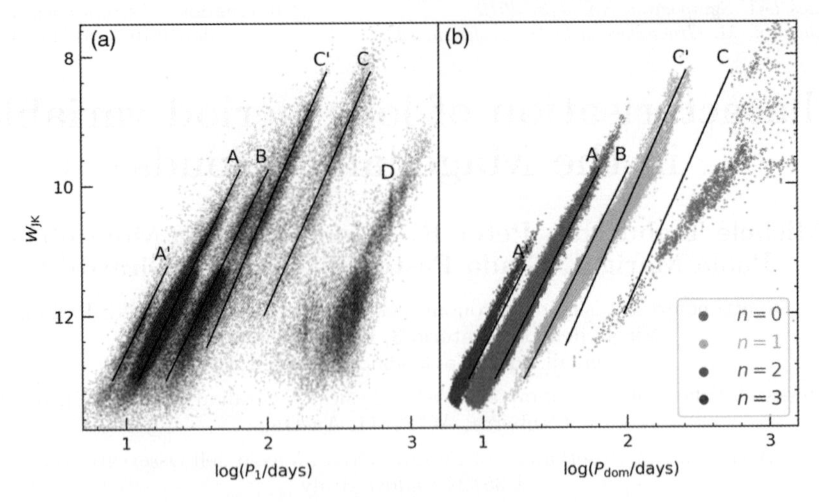

Figure 1. Left: primary periods of the LPVs in the LMC from OGLE-III data (Soszyński *et al.* (2009)). Right: periods of the dominant modes in a simulation of the population of LPVs in the LMC (Trabucchi *et al.* (2017)), labelled according to the radial order: $n=0$ for the fundamental mode, and $n=1,2,3$ for the first overtone (1O), second overtone (2O) and third overtone (3O) modes. The 4O mode is always stable in the models, and not required to reproduce observations. The Wesenheit index $W_{\rm JK}=K_{\rm s}-0.686\,(J-K_{\rm s})$ is an approximately reddening-free measure of luminosity.

LPVs are experiencing a growing interest from the scientific community for several reasons, the primary one being their great potential as distance indicators. This is especially true for Miras (*e.g.*, Whitelock *et al.* (2013), Menzies *et al.* (2015), Huang *et al.* (2018), see also Whitelock, these proceedings), owing to their high luminosity and large-amplitude pulsations, that allow for their detection and identification even in relatively distant extra-galactic systems. The potential of SRVs and OSARGs as standard candles is also being probed (Rau *et al.* (2018)).

Additionally, LPVs are of significant importance to study the evolution of stars and galaxies. Pulsation in AGB stars plays a key role in their mass loss by creating the conditions for the condensation of dust grains, that are accelerated by radiation pressure, resulting in strong stellar winds that contribute substantially to pollute the interstellar medium (*e.g.*, Höfner & Olofsson (2018)). Therefore, understanding LPVs is crucial both to improve our knowledge of complex processes related to advanced stellar evolution and to study the chemical evolution of galaxies.

Finally, observed periods and amplitudes of LPVs depend upon global stellar parameters, allowing for mass and radius estimates, as well as the calibration of the temperature scale, and in general provide an effective mean to constrain stellar models.

Such applications require a robust theoretical background, able to account for the interpretation of observations and for a consistent identification of the modes responsible for the PL sequences.

2. Pulsation Models

Extended grids of pulsation models of luminous red giants have long been missing from the scientific literature, since the works of, *e.g.*, Fox & Wood (1982); Ostlie & Cox (1986); Wood (1990). Important advancements were recently achieved using 3D models (Freytag *et al.* (2017) and references therein) and improving the treatment of convection (Xiong *et al.* (2018), and references therein). The time-consuming nature of such models, however, makes them unsuitable for the calculation of large grids. The main shortcomings of existing sets of models are (1) their limited sampling of the space of stellar parameters, especially in terms of chemical composition, and (2) the use of opacity data for scaled-solar mixtures only, not appropriate for the chemistry of TP-AGB stars (often drastically altered by the third dredge-up), in particular for C-stars.

Aiming to fill this gap, we computed a new grid of pulsation models, soon to become public (Trabucchi et al., in prep.), widely covering the variety of stellar parameters experienced during the TP-AGB, in accordance with recent results based on the `COLIBRI` evolutionary code (Marigo *et al.* (2017)). The grid provides full coverage of the mass-luminosity range of AGB evolution, as well as multiple values of core mass (to describe both quiescent evolution and thermal pulses) and of effective temperature. It also spans several values of metallicity, hydrogen mass fraction, and C/O ratio. All models are computed with opacity data (from the Opacity Project [Seaton (2005)], and the `ÆSOPUS` code [Marigo & Aringer (2009)]) consistent with the specific metal mixture assumed for the stellar envelope. We computed pulsation models with a linear, non-adiabatic code (Wood & Olivier (2014), and references therein), and provide periods and growth rates for five radial modes (from the fundamental to the 4O mode) for each model.

3. Discussion

We combined the grid of pulsation models with results from the population synthesis code `TRILEGAL` (Girardi *et al.* (2005))) to simulate the population of LPVs in the LMC (Trabucchi *et al.* 2017). We found the models to reproduce reasonably well the observed period-luminosity diagram (Fig. 1, right panel), with sequences A′ and A corresponding to dominant pulsation in the 3O and 2O modes, respectively, and the 1O mode being responsible for both sequences B and C′. The latter result is explained by a selection effect: stars developing a primary period on sequence D often have another period in the region between sequences B and C′. Since only primary periods are usually displayed in PL diagrams, the inclusion of LSPs on sequence D in the observed sample causes the exclusion of 1O periods between sequences B and C′, resulting in the appearance of a gap between those sequences. Models cannot describe sequence D variability, thus a consistent comparison requires LSPs to be removed from the sample. When this is done, primary periods are found to form a continuous distribution in place of the two sequences B and C′. These findings allow to bring into alignment the two main interpretations, incompatible with each other, previously suggested for the observed PL sequences (see, e.g. Wood (2015)).

Models are currently unable to reproduce correctly the observed sequence C, due to fundamental mode pulsation and associated with Mira variables. This is not only due to fundamental mode periods being overestimated at large luminosities, but also to the fact that theoretical growth rates, used to predict the edges of the PL sequences, depend on the treatment of convection and of its coupling with pulsation, and are thus uncertain. Preliminary tests suggest that a better agreement with observations can be obtained by assuming the mixing length parameter to decrease as a function of luminosity in evolutionary tracks, which effect is that of increasing the effective temperature, and thus of shortening the period, in the high-luminosity range. This is, however, insufficient to account for the discrepancy between theory and observations. We compared our results with periods obtained from a number of analytic prescriptions present in the literature (Fox & Wood (1982); Ostlie & Cox (1986); Wood (1990); Xiong & Deng (2007)), applied to the same synthetic stellar population model. The comparison shows that in all cases the fundamental mode periods are overestimated to some extent, and the models are unable to simultaneously reproduce the 1O and fundamental mode periods, as observed in SRVs on sequences C′ and C. This calls for an improvement of the models used to describe fundamental mode pulsation, especially desired for the interpretation of the observations from upcoming multi-epoch large sky surveys, such as LSST.

Acknowledgements

We acknowledge the support from the ERC Consolidator Grant funding scheme (*project STARKEY*, G.A. n. 615604).

References

Fox, M. W., & Wood, P. R. 1982, *ApJ*, 259, 198

Freytag, B., Liljegren, S., & Höfner, S. 2017, *A&A*, 600, A137

Girardi, L., Groenewegen, M. A. T., Hatziminaoglou E., & da Costa, L. 2005, *A&A*, 436, 895

Höfner, S., & Olofsson, H. 2018, *A&AR*, 26, 1

Huang, C. D., Riess, A. G., Hoffmann, S. L., Klein, C., Bloom, J., Yuan, W., Macri, L. M., Jones, D. O., Whitelock, P. A., Casertano, S., & Anderson, R. I. 2018, *ApJ*, 857, 67

Marigo, P., & B. Aringer 2009, *A&A*, 508, 1539

Marigo, P., Girardi, L., Bressan, A., Rosenfield, P., Aringer, B., Chen, Y., Dussin, M., Nanni, A., Pastorelli, G., Rodrigues, T. S., Trabucchi, M., Bladh, S., Dalcanton, J., Groenewegen, M. A. T., Montalbán, J., & Wood, P. R. 2017, *ApJ*, 835, 77.

Menzies, J. W., Whitelock P. A., & Feast, M. W. 2015, *MNRAS*, 452, 910

Ostlie, D. A., & Cox, A. N. 1986, *ApJ*, 311, 864

Rau, M. M., Koposov, S. E., Trac, H., & Mandelbaum, R. 2018, ArXiv e-print, arXiv:1806.02841.

Seaton, M. J. 2005, *MNRAS*, 362, L1

Soszyński, I., Udalski, A., Szymański, M. K., Kubiak, M., Pietrzyński, G., Wyrzykowski, Ł., Szewczyk, O., Ulaczyk, K., & Poleski, R. 2009, *AcA*, 59, 239

Trabucchi, M., Wood, P. R., Montalbán, J., Marigo, P., Pastorelli, G., & Girardi, L. 2017, *ApJ*, 847, 139

Whitelock, P. A., Menzies, J. W., Feast, M. W., Nsengiyumva, F., & Matsunaga, N. 2013, *MNRAS*, 428, 2216

Wood, P. R. 1990 in: M. O. Mennessier & A. Omont (eds.), *From Miras to Planetary Nebulae: Which Path for Stellar Evolution?*, Proc. International Colloquim, Montpellier, France, Sept. 4–7, 1989 (Gif-sur-Yvette, France: Éditions Frontières), p. 67

Wood, P. R., Alcock, C., Allsman, R. A., Alves, D., Axelrod, T. S., Becker, A. C., Bennett, D. P., Cook, K. H., Drake, A. J., Freeman, K. C., Griest, K., King, L. J., Lehner, M. J., Marshall, S. L., Minniti, D., Peterson, B. A., Pratt, M. R., Quinn, P. J., Stubbs, C. W., Sutherland, W., Tomaney, A., Vandehei, T., & Welch, D. L. 1999, in: T. Le Bertre, A. Lebre & C. Waelkens (eds.), *Asymptotic Giant Branch Stars*, Proc. IAU Symposium No. 191, p. 151

Wood, P. R. 2000, *PASA* 17, 18

Wood, P. R., & Olivier, E. A. 2014, *MNRAS*, 440, 2576

Wood, P. R. 2015, *MNRAS*, 448, 3829

Xiong, D. R., & Deng, L. 2007 *MNRAS* 378, 1270

Xiong, D. R., Deng, L., & Zhang, C. 2018, *MNRAS*, 480, 2698

Discussion

UTTENTHALER: In your 1.5 $M_\odot$ model, the fundamental mode only dominant for a relatively short time, during which the period is strongly changing. Do you think this is realistic? In that case most fundamental mode pulsators (Miras) should have strongly changing periods, while we know from observations that this is not the case.

TRABUCCHI: Current linear pulsation models tend to underpredict fundamental mode growth rates, resulting in a short duration of the phase in which the fundamental mode is dominant (the Mira phase). Fundamental mode periods are overestimated at large luminosities, likely causing the rather large changes in period during the last thermal pulses of the tracks. Non-linear models, more appropriate for large-amplitude pulsation, could help in this respect.

Why Galaxies Care About AGB Stars:
A Continuing Challenge through Cosmic Time
Proceedings IAU Symposium No. 343, 2019
F. Kerschbaum, M. Groenewegen & H. Olofsson, eds.

doi:10.1017/S1743921318005902

The End: Witnessing the Death of Extreme Carbon Stars

G. C. Sloan[1,2], K. E. Kraemer[3], I. McDonald[4] and A. A. Zijlstra[4]

[1]Space Telescope Science Institute, 3700 San Martin Dr., Baltimore, MD 21248, USA
email: `sloan@astro.cornell.edu`

[2]Univ. of North Carolina Chapel Hill, Chapel Hill, NC 27599-3255, USA

[3]Boston College, Chestnut Hill, MA 02467, USA

[4]Jodrell Bank Centre for Astrophysics, Alan Turing Building, Manchester, M13 9PL, UK

Abstract. A small number of the sample of 184 carbon stars in the Magellanic Clouds show signs that they are in the act of evolving off of the asymptotic giant branch. Most carbon stars grow progressively redder in all infrared colors and develop stronger pulsation amplitudes as their circumstellar dust shells become optically thicker. The reddest sources, however, have unexpectedly low pulsation amplitudes, and some even show blue excesses that could point to deviations from spherical symmetry as they eject the last of their envelopes. Previously, all dusty carbon-rich AGB stars have been labeled "extreme," but that term should be reserved for the truly extreme carbon stars. These objects may well hold the clues needed to disentangle what actually happens when a star ejects the last of its envelope and evolves off of the AGB.

Keywords. infrared: stars, stars: carbon stars: AGB and post-AGB, Magellanic Clouds

---

1. Introduction

Before the *Spitzer Space Telescope* ended the cryogenic portion of its mission, the Infrared Spectrograph (IRS) observed a total of 184 carbon stars in the Large and Small Magellanic Clouds (LMC and SMC). The spectra were obtained in several observing programs, each with its own scientific objectives. Despite these differences, the spectral and photometric properties of the sample reveal the multiple evolutionary phases through which carbon stars pass before they shed their envelopes and leave the asymptotic giant branch (AGB; Sloan *et al.* 2016).

Figure 1 shows the trends followed by the Magellanic carbon stars and how the reddest stars break those trends. Because carbon stars grow redder consistently in all near- and mid-infrared colors as the amount of circumstellar dust increases, they generally occupy a tight sequence in any infrared color-color space. However, some of the reddest carbon stars in the sample from the IRS (in the [5.8]−[8] color) have an excess at 3.6 μm which pulls them away from the sequence. Generally, more dust and redder colors are associated with stronger pulsation amplitudes, but in the IRS sample, the peak amplitude occurs at [5.8]−[8] $\sim$ 1.5. For redder objects, *more* dust is associated with *weaker* amplitudes. This behavior led Sloan *et al.* (2016) to hypothesize that the reddest Magellanic carbon stars observed by the IRS may have ejected their envelopes and be developing asymmetries in their circumstellar dust shells as they evolve off of the AGB.

2. The dusty AGB

Gruendl *et al.* (2008) first identified the reddest carbon stars in the LMC photometrically, initially describing them as *extremely red objects*, or EROs. When spectra from *Spitzer* revealed their carbon-rich nature, they could be more properly referred to as

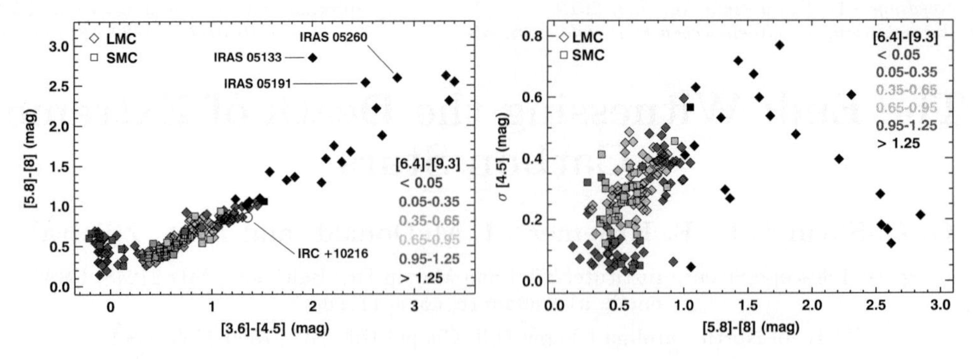

Figure 1. Left: A color-color plot of the IRS sample of Magellanic carbon stars using photometry from *Spitzer* and *WISE*. The data are coded by their [6.4]−[9.3] color, which is determined from their spectra and is a good proxy for the amount of dust. See the text for more on the labeled objects. **Right:** Variability, as measured by the standard deviation of *Spitzer* and *WISE* data at 4.5 and 4.6 μm, vs. [5.8]−[8] color. Both plots are adapted from Sloan *et al.* (2016).

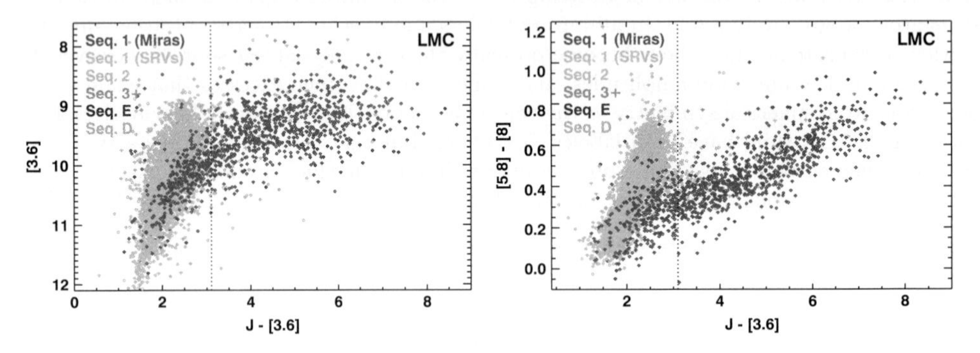

Figure 2. Left: A color-magnitude diagram of carbon stars in the OGLE-III survey (Soszyński *et al.* 2009), coded by their OGLE-III variability type and pulsation sequence. **Right:** The same carbon-rich sample in color-color space. The vertical line at $J-[3.6] = 3.1$ marks the boundary of the extreme AGB as defined by Blum *et al.* (2006).

extreme carbon stars. Their circumstellar dust shells are so optically thick, the SiC dust emission feature normally seen at ∼11.5 μm can appear in absorption, much like extreme carbon stars identified in the Galaxy (e.g. Speck *et al.* 1997, 2009, Pitman *et al.* 2007). Describing these objects as "extreme" reflects their rarity, with fewer than 20 or so known in either the LMC or the Galaxy, and none in the SMC.

However, another class of far more numerous objects has also been labeled "extreme," which confuses matters. Blum *et al.* (2006), noting a change in slope in the near-infrared color-magnitude diagram of the LMC at a $J-[3.6]$ color ∼ 3.1, identified all stars to the red as *extreme AGB*.

The two panels of Figure 2 plot all of the carbon-rich Miras and semi-regular variables identified in the LMC in the OGLE-III survey (Soszyński *et al.* 2009) for which we have found near- and mid-infrared photometry. The data are coded by their location on a period-luminosity diagram (see Fig. 1 by Sloan 2017), using the sequences defined by Fraser *et al.* (2005). Sequences 1, 2, and 3 refer to the fundamental mode, first overtone, and possibly higher overtones, respectively.

Both panels in Figure 2 show that a $J-[3.6]$ color of 3.1 forms a reasonable boundary between two sequences, with most of the semi-regular variables, no matter their pulsation mode, to the blue. Carbon stars with redder colors are almost always Miras. While both Miras and semi-regulars can pulsate in the fundamental mode, the Miras as defined by OGLE-III have stronger pulsation amplitudes. As explained by Sloan *et al.* (2015), different mechanisms produce the reddening along the two sequences. The sequence on the left can be described as the *molecular* sequence because increasing absorption from C_3 molecules at 5 μm leads to redder [5.8]−[8] colors. Stars on this sequence are pulsating at amplitudes too low to drive significant mass loss, and they have little circumstellar dust. The sequence to the right, previously referred to as the extreme AGB, is better described as the *dusty* sequence (or "D-AGB") because these stars have substantial mass-loss rates, and increasing amounts of circumstellar dust drives their reddening.

3. The dying carbon stars

We wish to draw attention to the population of objects more extreme than the dusty AGB. For comparison, Figure 1 (left) includes IRC +10216, in many ways the prototypical extreme carbon star in the Galaxy. This object was discovered in the Two-Micron Sky Survey (Neugebauer & Leighton 1969), but it did not stand out until it was found to be the brightest object outside the Solar System in the mid infrared (Becklin *et al.* 1969). Its thick dust shell gives it a $J-[3.6]$ color $\sim$ 10.7, which is off-scale in Figure 2, and yet the shell is still not thick enough to put SiC in absorption.

The location of IRC +10216 in the color-color diagram in Figure 1 is informative. Despite its optically thick dust shell, it is still a member of the main population of dusty carbon stars and to the blue of the break in the population density at [5.8]−[8] $\sim$ 1. Some caution is warranted with the population statistics of the IRS sample of Magellanic carbon stars, because it inherits the biases of the observing programs that contributed to it. Nonetheless, this break appears to be real. It would most likely be even more substantial in a less biased photometric sample, because Gruendl *et al.* (2008) targeted the majority of extreme carbon stars in the LMC. No comparable objects have been found in the SMC (Srinivasan *et al.* 2016). All of these arguments point to the extreme carbon stars, with their various indications of a highly evolved state and a possible departure from the AGB, as a special class of objects.

Given the many unanswered questions about the final moments of stellar evolution, these extreme carbon stars require further study. We have started that follow-up process. Figure 1 (left) marks the three objects with the greatest excess emission at 3.6 μm. IRAS 05133−6937 has the strongest excess. Our preliminary analysis of I-band imaging with the SOAR Adaptive Module (SAM) reveals a source with an apparent low-contrast halo, possibly as a result of scattering from light escaping from the dust shell, as Figure 3 illustrates.

The other two sources marked in Figure 1 tell a more complex story. Both IRAS 05260−7010 and IRAS 05191−6936 are accompanied by a naked star $\sim$0.9$''$ to the south, as previously noted by Gruendl *et al.* (2008). The I-band SOAR images show only the neighbors; the infrared sources are undetected. These neighbors raise the possibility that pollution in the beam of the *Spitzer* photometry could explain the excess at 3.6 μm. Our investigation of the *Spitzer* images at 3.6 and 4.5 μm reveals no indication that contamination at 3.6 μm is responsible, but we cannot yet draw a definitive conclusion.

These initial findings are the result of just one night observing with a 4-m telescope, and further work is needed to better understand the behavior of these extreme carbon stars. They are likely to be our window into the poorly understood processes driving the final moments of a star's lifetime on the AGB.

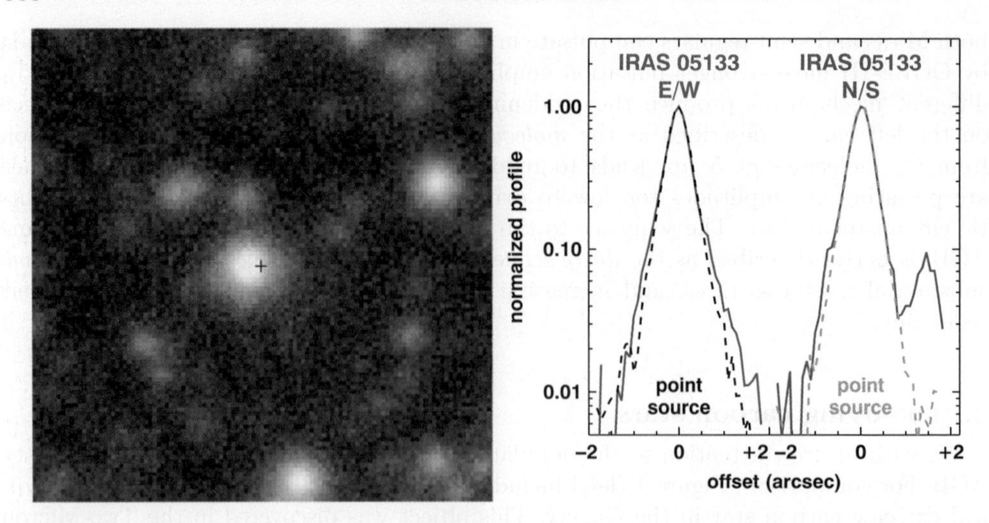

Figure 3. Left: An *I*-band image of IRAS 05133−6937 from the SOAR Adaptive Module (SAM) covering $10'' \times 10''$. The cross marks the infrared position of the source (Sloan *et al.* 2016), which is $0.26''$ from the center of the *I*-band source. **Right:** Profiles of IRAS 05133 showing a low-contrast halo (solid lines) compared to a nominal point source (dashed lines).

References

Becklin, E. E., *et al.* 1969 *ApJ*, 158, L133

Blum, R. D., *et al.* 2006, *AJ*, 132, 2034

Fraser, O. J., *et al.* *AJ*, 129, 768

Gruendl, R. A., *et al.* 2008, *ApJ*, 688, L9

Neugebauer, G., & Leighton, R. B. 1969, *Two-Micron Sky Survey—A Preliminary Catalog*, (NASA SP-3047) (Washington DC: Government Printing Office)

Pitman, K. M., Speck, A. K., & Hofmeister, A. M. 2006, *MNRAS* 371, 1744

Sloan, G. C., *et al.* 2015, in *Why Galaxies Care about AGB Stars III*, ed. F. Kerschbaum, *et al.*, *ASP Conf. Series*, 497, 429

Sloan, G. C., *et al.* 2016, *ApJ*, 826, 44

Sloan, G. C. 2017, *Plan. & Space Sci.*, 149, 32

Soszyński, I., *et al.* 2009 *Act. Astr.*, 59, 239

Speck, A. K., Barlow, M. J., & Skinner, C. J. 1997, *MNRAS*, 288, 431

Speck, A. K., *et al.* 2009, *ApJ*, 691, 1202

Srinivasan, S., *et al.* 2016, *MNRAS*, 457, 2814

Discussion

KWOK: One of the defining spectral characteristics of galactic extreme carbon stars is the presence of acetylene. However acetylene is not present in all your LMC examples. Would you care to comment on the presence of acetylene as a property of extreme carbon stars?

SLOAN: The acetylene bands at 7.5 and 13.7 μm appear in most spectra. The 13.7 μm band is almost always present, even in optically thick dust shells, which tells us that it does not originate in the stellar photosphere.

Why Galaxies Care About AGB Stars:
A Continuing Challenge through Cosmic Time
Proceedings IAU Symposium No. 343, 2019
F. Kerschbaum, M. Groenewegen & H. Olofsson, eds.

doi:10.1017/S1743921318006002

Oxygen-rich Long Period Variables in the X-Shooter Spectral Library

Ariane Lançon[1], Anaïs Gonneau[2], Scott C. Trager[3], Philippe Prugniel[4], Anke Arentsen[5], Yanping Chen[6], Matthijs Dries[3], Cécile Loup[1], Mariya Lyubenova[7], Reynier Peletier[3], Laure Telliez[1], Alexandre Vazdekis[8] and the XSL Collaboration

[1]Université de Strasbourg, CNRS, UMR7550, Observatoire astronomique de Strasbourg, 67000 Strasbourg, France
email: ariane.lancon@astro.unistra.fr

[2]Institute of Astronomy, University of Cambridge, United Kingdom

[3]Kapteyn Astronomical Institute, University of Groningen, The Netherlands

[4]Centre de Recherche Astrophysique de Lyon (UMR 5574), Université de Lyon, France

[5]Leibniz-Institut für Astrophysik Potsdam, Germany

[6]New York University in Abu Dhabi, United Arab Emirates

[7]European Southern Observatory, Garching, Germany

[8]Instituto de Astrofisica de Canarias, Santa Cruz de Tenerife, Spain

Abstract. The X-Shooter Spectral Library (XSL) contains more than 800 spectra of stars across the color-magnitude diagram, that extend from near-UV to near-IR wavelengths (320-2450 nm). We summarize properties of the spectra of O-rich Long Period Variables in the XSL, such as phase-related features, and we confront the data with synthetic spectra based on static and dynamical stellar atmosphere models. We discuss successes and remaining discrepancies, keeping in mind the applications to population synthesis modeling that XSL is designed for.

Keywords. stars: AGB, stars: pulsation, stars: atmospheres

---

1. The X-Shooter Spectral Library in context

Modern extragalactic surveys produce energy distributions and spectra of galaxies with an exquisite precision. Their interpretation calls for progress in the modeling of the integrated light of stellar populations (e.g. Powalka *et al.* 2016). With the shift from optical to near-infrared wavelengths that facilities such as the *James Webb Space Telescope* (JWST) or the Extremely Large Telescope (ELT) will bring about, it is more important than ever that stellar population synthesis models provide consistent predictions across the whole spectrum of stellar photospheric emission.

The X-Shooter Library project (XSL, Chen *et al.* 2014) was initiated in this context. The spectra extend from the near-ultraviolet (320 nm) to the near-infrared (2.45 μm), and this avoids some of the inter-connection issues between separate optical and near-infrared empirical spectral libraries that have existed before. Interpolation tools based on XSL will provide empirical spectra at a resolving power close to 10 000 over a broad range of stellar parameters, for direct usage in population synthesis modeling (e.g. Verro *et al.* this volume). In addition, a detailed comparison of the observed spectra with modern synthetic spectra will help us improve the agreement between the two, which is vital to validate the fundamental parameters of the stars observed (and hence the way they can

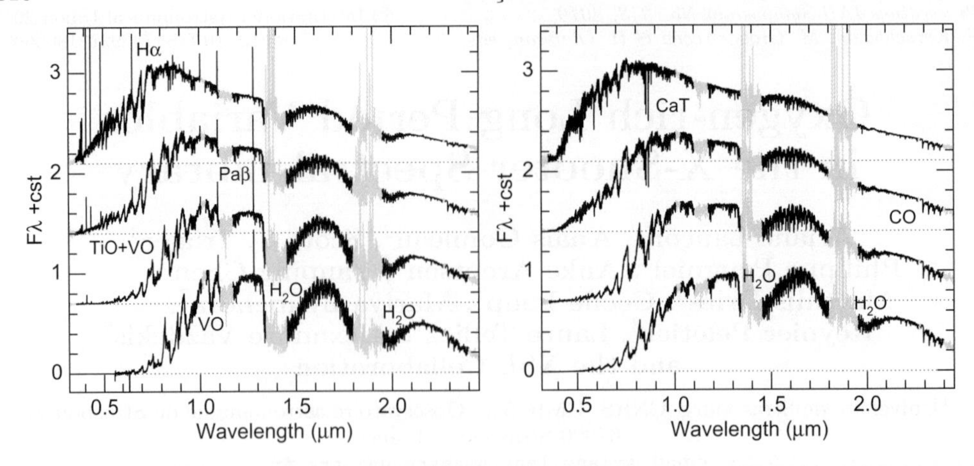

Figure 1. Average spectra of O-rich LPVs from the XSL project, in bins of similar $(R-K)$ colors (smoothed to $R = 3000$ for display). Left: spectra at phases that display emission lines. Right: spectra without emission lines. The color-bin definitions are the same in the two panels. Grey identifies ranges where the correction for telluric absorption leaves strong residuals.

be connected to loci on stellar evolution tracks), as well as to progress towards future population synthesis models based on synthetic spectra alone. Such purely theoretical models already exist, but the validity of their input synthetic stellar libraries has not been demonstrated over a wavelength range as broad as that of XSL.

XSL contains ~800 spectra of more than 650 stars. The second data release, with spectra from the three spectral arms of X-Shooter for all program stars, is nearing completion (Gonneau *et al.* in prep.). Because luminous cool stars are particularly important contributors to the red and near-infrared light of galaxies, XSL was designed to contain a large number of such objects. The spectra of 35 carbon stars were made available and compared to C-star models by Gonneau *et al.* (2016, 2017). Here, we focus on O-rich Long Period Variables (LPVs) with estimated temperatures lower than about 4200 K (from Arentsen *et al.* in prep). The sample used contains 160 spectra of 150 LPVs of the Milky Way and the Magellanic Clouds, with a range of periods, amplitudes and luminosities.

2. The spectra of O-rich LPVs in XSL

The spectra of O-rich LPVs in XSL display a range of colors and spectral features very similar to that observed by Lançon & Wood (2000) [hereafter LW2000], which indicates that each of these collections actually captures most of the natural variance in the range of colors sampled ($2 \leqslant R–K \leqslant 10$). The XSL spectra have a higher spectral resolution, especially at optical wavelengths. The existence of two data sets with such similar properties is useful in the investigation of features that are currently not explained by models, as it excludes with high confidence that these might be data artefacts. Such features include the molecular emission (most likely of TiO) sometimes seen near 1.24 μm and the blue slope of the pseudo-continuum around 1 μm, both seen preferably in spectra that also display emission lines and/or strong bands of VO. Note that some of these properties have been mentioned early-on in the literature on Miras (e.g. Wing 1974).

When averaging XSL spectra in bins of similar broad-band color, as was done with the LW2000 sample (Lançon & Mouhcine 2002), a very regular sequence of mean spectra is obtained. This will be useful for population synthesis purposes, but it hides the real variety of properties, and with the XSL sample we can now also attempt to exploit this variety. An example is shown in Fig. 1, where the left panel shows averages of spectra

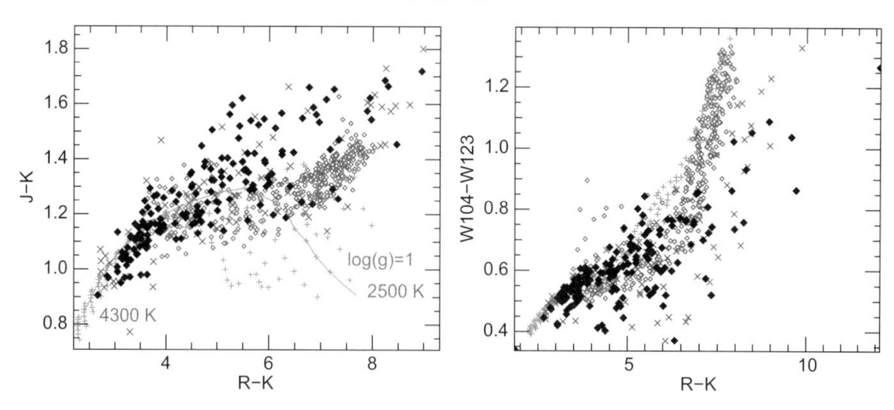

Figure 2. Photometric properties of the observations of O-rich LPVs (solid diamonds: XSL; crosses: LW2000), of the Phoenix models at solar metallicity (small plus-symbols; a line connects models at $\log(g) = 1$ in the left panel), and of the DARWIN models (small open diamonds).

of similar $(R-K)$ color at pulsation phases that display emission lines (typically, but not exclusively, phases near maximum light), while the right panel shows averages of spectra in exactly the same ranges of $(R-K)$ color, but selected not to display emission lines. The energy distributions differ, with more "triangular" shapes in the left panel; the strengths of molecular bands (TiO, VO, H_2O) are stronger in the left panel, while CO absorption band heads are more conspicuous in the right panel. The calcium triplet lines near 0.85 μm are strong in absorption in the right panel, while they disappear in the left panel (at the displayed resolution) as a consequence of stronger molecular absorption and line emission. The more rounded shapes in the redder stars of the right hand panel might indicate a stronger effect of extinction.

3. Comparison with models

We have compared the XSL spectra with two collections of synthetic spectra: (i) the static PHOENIX models of Husser *et al.* (2013), version 2, with solar-scaled abundances and a range of [Fe/H]; and (ii) the time series of DARWIN models at solar metallicity (Bladh *et al.*, 2015 and this volume). In total, about 2000 model spectra were considered. But these still represent a tiny fraction of the relevant space of stellar parameters, e.g. in terms of chemical composition and mass, and for the dynamical models in terms of pulsation properties. All models are spherically symmetric.

In traditional broad-band color-color diagrams, the loci of the observations tend to look compatible with the model collections, when allowing for extinction (Fig. 2, left panel). However, there are discrepancies in a number of diagrams that combine a broad-band color with narrow-band color-indices that target some of the features mentioned in Sect. 2. For instance, when narrow-band filters at 1.04 and 1.23 μm are used to measure the slope near 1 μm, and are plotted against $(R-K)$, it is seen that the models do not reach the blue (1.04-1.23) colors observed for LPVs at some phases (Fig. 2, right panel). The models also fail to reproduce the strong VO bands often associated with this property of the energy distribution (not shown). In these diagnostic diagrams, the dynamical models are displaced in the desired direction from the static models, but not quite enough. A few dynamical models display molecular bands in emission, but those of CN or CO rather than those of TiO.

When comparing spectra with models, excellent matches are readily obtained at temperatures near 4000 K. For small amplitude variables at those temperatures, a fit to the optical spectrum only, with a static model and a standard extinction law, often produces a very good representation of the whole XSL spectrum, with a reasonable match to the detailed spectral features (perfect matches of the metal line spectra are impossible with only solar-scaled abundances). Below 4000 K, one cannot expect a good representation of the near-IR spectrum when fitting only optical wavelengths. Decent representations of the energy distributions and the main molecular bands (except water and bands near 1 μm) may be obtained when constraining the fit with all available wavelengths. But for spectra near maximum light the effective temperatures thus obtained with static and dynamical models differ systematically, sometimes by more than 300 K. The most obvious discrepancies are found for the strongly peaked spectra that enter the averages in the left panels of Fig. 1. The models we have explored until now have a more “rounded” shape.

4. Conclusion

From this early work on the XSL spectra of O-rich LPVs, we have already learnt a few lessons. Without surprise, it remains difficult to match the energy distributions and spectral features of O-rich LPVs with synthetic spectra. Static models are impressively good matches to small amplitude LPV spectra at temperatures down to about 4000 K. In several aspects, dynamical models seem to provide improvements over static models for cool, larger amplitude LPVs (circumstellar dust, strength of the VO bands, shape of the SED near 1 μm), but the range of properties observed is not covered yet. It is clearly necessary to explore a wider range of model parameters in spherical symmetry, and to start considering 3D models (see Liljegren, Freytag, Chiavassa, this volume).

In population synthesis models that predict optical and near-infrared spectra, the warmest LPVs are most relevant (others have smaller contribution to the integrated light of all stars). Average spectra of LPVs remain a convenient practical choice to represent stars on the thermally pulsing AGB (TP-AGB). But we have identified the risk of a bias: the bluest stars in an empirical collection tend to include a larger fraction of stars caught near maximum light, and at a given broad band color the detailed spectra depend on phase. We hope that future dynamical models will help us obtain sufficiently good matches so we can relate instantaneous spectral properties to those of the parent static stars, in order to provide more meaningful averages.

We conclude with a tribute to Prof. Michael Scholz, whose dynamical models of have been a major contribution to our understanding of the relationships between atmospheric structure and spectrophotometric properties of Mira variables. They remain precious today.

Acknowledgments

AL, PP & AG thank the Programme National de Physique Stellaire (PNPS) and the Programme National Cosmologie & Galaxies (PNCG), France, for recurrent support. AG is supported by the EU-FP7 programme through grant number 320360.

References

Bladh, S., Höfner, S., Aringer, B., & Eriksson, K. 2015, *A&A*, 575, A105
Chen, Y.-P., Trager, S.C., Peletier, R.F. *et al.* 2014, *A&A*, 565, A117
Gonneau, A., Lançon, A., Trager, S.C., Aringer, B., *et al.* 2016, *A&A*, 589, A36
Gonneau, A., Lançon, A., Trager, S.C., Aringer, B., *et al.* 2017, *A&A*, 601, A141
Husser, T.-O., Wende-von Berg, S., Dreizler, S. *et al.* 2013, *A&A*, 553, A6
Lançon, A., & Wood, P. 2000, *A&AS*, 146, 217
Lançon, A., & Mouhcine, M. 2002, *A&A*, 393, 167
Powalka, M., Lançon, A., Puzia, T.H., *et al.* 2016, *ApJS*, 227, 12
Wing, R.F. 1974, *HiA*, 3, 285

Discussion

ARINGER: This work, that has just been started, is valuable to help us improve the models. At this time I do not know what might cause the "triangular" shapes seen at certain phases, but this is a big effect!

WITTKOWSKI: I am happy to see this new collection of spectra, as those of Lançon & Wood (2000) already proved to be highly useful. I would like to comment that differences between hydrostatic and dynamic models become much more obvious in spatially resolved visibility spectra. It could be useful to combine X-shooter spectra with interferometric data.

Why Galaxies Care About AGB Stars:
A Continuing Challenge through Cosmic Time
Proceedings IAU Symposium No. 343, 2019
F. Kerschbaum, M. Groenewegen & H. Olofsson, eds.

doi:10.1017/S174392131800563X

What Young Massive Clusters in the Magellanic Clouds teach us about Old Galactic Globular Clusters?

Francesca D'Antona[1], Paolo Ventura[1], Aaron Dotter[2], Sylvia Ekström[3] and Marco Tailo[4]

[1]INAF-OAR,
via di Frascati 33, I-00078 Monteporzio Catone, Italy
email: franca.dantona@gmail.com

[2]Harvard-Smithsonian Center for Astrophysics,
60 Garden Street, Cambridge, MA 02138, USA
email: aaron.dotter@gmail.com

[3]Geneva Observatory, University of Geneva,
Maillettes 51, CH-1290 Sauverny, Switzerland
email: sylvia.ekstrom@unige.ch

[4]Dipartimento di Fisica e Astronomia 'Galileo Galilei', Univ. di Padova,
Vicolo dell'Osservatorio 3, I-35122 Padova, Italy
email: mrctailo@gmail.com

Abstract. The Asymptotic Giant Branch (AGB) scenario ascribes the multiple populations in old Galactic Globular Clusters (GGC) to episodes of star formation in the gas contaminated by the ejecta of massive AGBs and super-AGBs of a first stellar population. The mass of these AGBs (4-8 $M_\odot$) today populate the Young Massive Clusters (YMC) of the Magellanic Clouds, where rapid rotation and its slowing down play an important role in shaping the color-magnitude diagram features. Consequently, we must reconsider whether the rotational evolution of these masses affects the yields, and whether the resulting abundances are compatible with the chemical patterns observed in GGC. We show the first results of a differential analysis, by computing the hot bottom burning evolution of non-rotating models with increased CNO-Na abundances at the second dredge-up, following the results of MESA rotational models.

Keywords. globular clusters: general, open clusters and associations: general, stars: abundances

1. Introduction

Mackey & Broby Nielsen (2007) discovered that the intermediate age (~1.5 Gyr) Globular Cluster (GC) NGC 1846 in the Large Magellanic Cloud (LMC) displayed a spread turnoff, signalling the presence of multiple populations, possibly similar to those revealed by chemical anomalies in the ancient Galactic GCs. A more or less extended turnoff region was revealed in many other YMC of the MCs (Milone *et al.* 2009), and was generally attributed to a stellar age spread. In 2015 this interpretation suddenly lost its appeal, when researchers considered results of the Geneva tracks and isochrones computed for a wide range of masses (ages) and rotation rates (e.g. Georgy *et al.* 2013)†. Brandt & Huang (2015) showed that the turnoff area covered between the non rotating and rotating isochrones increased with the cluster age up to about 1.5 Gyr, and then decreased, and the same behaviour was displayed by the YMC for increasing ages. In

† see the page web http://obswww.unige.ch/Recherche/evoldb/index/ created and maintained by C. Georgy and S. Ekström

terms of ages this would imply an increasing age spread (Niederhofer *et al.* 2015), if interpreted with non-rotating models only.

At the same time, Milone *et al.* (2015) found a different puzzling feature in the color magnitude diagram of the ~400 Myr cluster NGC 1856: a "split" main sequence. This feature could not be understood in terms of differences in age, metallicity or helium content, while splitting and the general features of the color-magnitude diagram were consistent with the superposition of two coeval populations, the first one including very rapidly rotating stars (the red main sequence and the upper part of the spread turnoff), the other one including slowly rotating stars (the blue main sequence and the lower luminosity turnoff stars) (D'Antona *et al.* 2015). Possibly the slowly rotating stars had also started as rapid rotators, but had been "braked" by dynamical tides (Zahn, 1977) due to interaction with a distant binary companion, as it occurs to the field binaries with orbital periods between 4 and 500 days (Abt & Boonyarak 2004). The role of braking was confirmed by D'Antona *et al.* (2017), who remarked that the slowly rotating 'blue' MSs in several young clusters required a ~10% younger age than the rapidly rotating upper 'red' turnoff stars. The alternative to a younger population is that the blue MS stars are in a less advanced nuclear burning stage with respect to the requirement of their slow rotation. In other words, these stars follow the evolution of their rapidly rotating counterparts, and feed fresh hydrogen in the H-burning core thanks to rotational mixing, until recent braking leaves them as slow rotators. Why the stars in these YMCs are born so rapidly rotating, and what are the precise mechanisms for braking which we see so well in the double MS and in the spread of the turnoff is to be worked out, but it is evident that rotation and its evolution *in the cluster ambient* is an important feature for these stars, and it is described quite well by the Geneva models.

Milone *et al.* (2018) have shown that the split MS phenomenon occurs for ages from ~30 to ~400 Myr, so it involves the range of masses from ~3 to ~9 $M_\odot$. This range includes the super-AGB and massive AGB masses whose envelopes, processed by Hot Bottom Burning (HBB), are responsible for the formation of second generation stars in the ancient GCs (Ventura *et al.* 2001), in the so called 'AGB scenario' (D'Ercole *et al.* 2008), so it is reasonable to assume that rotation, although subject to the quoted braking mechanisms, has been an important ingredient in the evolution of these stars too. On the other hand, Decressin *et al.* (2009) examined the consequences of rotation on the abundances of intermediate mass AGB and showed that rotational mixing increased so much the abundances of CNO at the second dredge up (2DU) that the AGBs could not be reasonable polluter sources for the multiple populations. As the AGB scenario is anyway the most adequate to deal with both the formation (D'Ercole *et al.* 2008), the chemistry (e.g. Ventura & D'Antona 2009, Ventura *et al.* 2013, D'Antona *et al.* 2016), the spatial distribution of the different populations (Vesperini *et al.* 2013), the role of binaries (e.g. Vesperini *et al.* 2011) —see also Ventura's review in these proceedings— in spite of its remaining problems (see, e.g. Renzini *et al.* 2015), it is now necessary to re-examine the problem of rotating evolution.

2. Abundances at the 2DU in recent rotating models

While new computations by one of us (Sylvia Ekström) with the Geneva code more or less reproduced the Decressin *et al.* (2009) results both at very low and intermediate metallicity, the results of the rotating MIST code (Choi *et al.* 2016), obtained from the MESA code (Paxton *et al.* 2011) give a much smaller CNO enhancement, it is at most a factor 4–5 for the abundances of the most metal poor GCs ([Fe/H]$\simeq$–2.2). The CNO increase at the metallicities typical of the bulk of the galactic GC population ([Fe/H]$\geqslant$–1) is negligible, so —if these models describe the rotational mixing correctly— we must worry about the difference in the yields only for the lowest metallicity clusters.

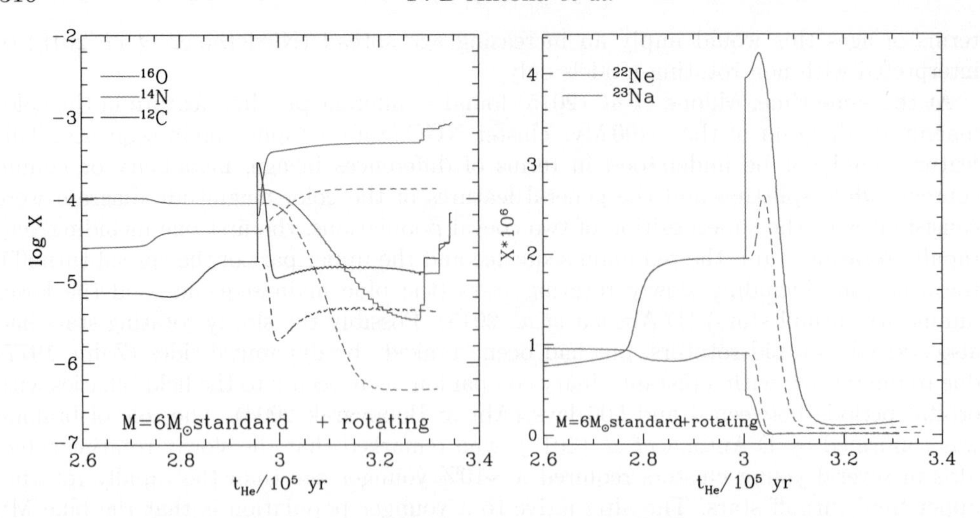

Figure 1. Evolution of the abundances of C, N and O (left panel) and Na and Ne (right panel) in non rotating model of $6\,M_\odot$ (dashed) and when the abundances are increased at the 2DU following rotating MESA models results (full lines). We see that sodium and oxygen survive longer in the rotating case.

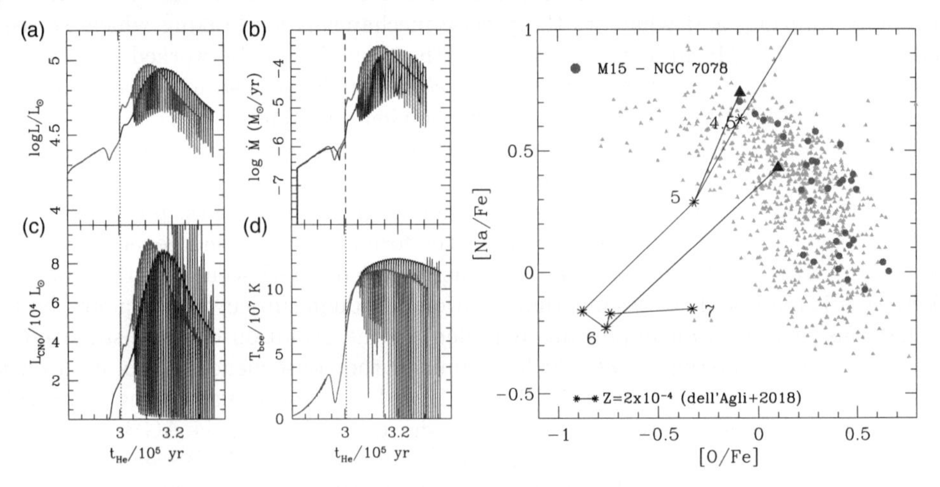

Figure 2. Left panel: time evolution of (a): log luminosity; (b): log mass-loss rate; (c): CNO luminosity; (d): log temperature at the base of the convective envelope. The blue lines refer to the standard track, the red ones refer to the track with increased CNO-Na abundances, simulating the result of rotational mixing. Right panel: in the [Na/Fe] versus [O/Fe], the green triangles refer to all GC data from Carretta *et al.* (2009), while the brown dots represent the stars in the low metallicity GC M15. The asterisks connected by black lines are the model yields from Dell'Agli *et al.* (2018), for the chemistry of M15, and the lines connecting to the blue triangles the yields of the 6 and $5\,M_\odot$ shift in the models with increased CNO-Na abundances.

As also ^{23}Na and ^{22}Ne abundances are important for the second generation pollution, we considered the whole results in MESA models computed by one of us (Aaron Dotter) for [Fe/H]=–2 and solar scaled abundances. We used the absolute values of ^{12}C, ^{14}N, ^{16}O, ^{23}Na and ^{22}Ne from these models and applied them to the standard (non rotating) models by Dell'Agli *et al.* (2018) which had the chemical abundances adequate to describe the detailed composition of the GC M 15 found in the APOGEE survey. By changing the abundances at the 2DU (Fig. 1), we may consider the differential effect of the different chemical abundances on the hot bottom burning phase and on the yields. Obviously, this is not a self-consistent result, but it gives us an interesting way of change in the abundance patterns.

We discovered that, for the 6 $M_{\odot}$ star, the ^{16}O and the ^{23}Na are ∼0.7 dex and ∼0.6 dex less depleted than in the standard models, and reproduce better the abundances in the cluster M 15, which lacks the extreme O depletions present in intermediate metallicity clusters (Fig. 2, right panel). The 5 $M_{\odot}$ shows a similar effect, by a smaller amount. In fact, the increased CNO mean a larger CNO luminosity at the 2DU (Figure 2, panel c) and thus a larger total luminosity (panel a), which in turn gives a larger mass-loss rate (panel b) during the first evolutionary phases. More mass is then lost when the ^{23}Na is high, and its yield is larger. Later on, however, the larger mass-loss rate provide a smaller temperature at the bottom of convection (panel d), and therefore a slower conversion of oxygen to nitrogen. Obviously, these results depend very much on the precise CNO and Na increase at the 2DU, and the total CNO increase assumed here are too large to be consistent with the CNO abundance determinations in the cluster, but the trends we find in this preliminary models deserve further investigation.

References

Abt, H. A., & Boonyarak, C. 2004, *ApJ*, 616, 562

Brandt, T. D., & Huang, C. X. 2015, *ApJ*, 807, 25

Carretta, E., Bragaglia, A., Gratton, R., & Lucatello, S. 2009, *A&A*, 505, 139

Choi, J., Dotter, A., Conroy, C., *et al.* 2016, *ApJ*, 823, 102

D'Antona, F., Di Criscienzo, M., Decressin, T., *et al.* 2015, *MNRAS*, 453, 2637

D'Antona F., Vesperini E., D'Ercole A., Ventura P., Milone A. P., Marino A. F., Tailo M. 2016, *MNRAS*, 458, 2122

D'Antona, F., Milone, A. P., Tailo, M., *et al.* 2017, *Nature Astronomy*, 1, 0186

Decressin, T., Charbonnel, C., Siess, L., *et al.* 2009, *A&A*, 505, 727

Dell'Agli, F., García-Hernández, D. A., Ventura, P., *et al.* 2018, *MNRAS*, 475, 3098

D'Ercole, A., Vesperini, E., D'Antona, F., McMillan, S. L. W., & Recchi, S. 2008, *MNRAS*, 391, 825

D'Ercole, A., D'Antona, F., Ventura, P., Vesperini, E., & McMillan, S. L. W. 2010, *MNRAS*, 407, 854

Georgy, C., Ekström, S., Granada, A., *et al.* 2013, *A&A*, 553, A24

Mackey, A. D., & Broby Nielsen, P. 2007, *MNRAS*, 379, 151

Milone, A. P., Bedin, L. R., Piotto, G., & Anderson, J. 2009, *A&A*, 497, 755

Milone, A. P., Bedin, L. R., Piotto, G., *et al.* 2015, *MNRAS*, 450, 3750

Milone, A. P., Marino, A. F., Di Criscienzo, M., *et al.* 2018, *MNRAS*, 477, 2640

Niederhofer, F., Georgy, C., Bastian, N., & Ekström, S. 2015, *MNRAS*, 453, 2070

Paxton, B., Bildsten, L., Dotter, A., *et al.* 2011, *ApJS*, 192, 3

Renzini A., D'Antona F., Cassisi S., *et al.* 2015, *MNRAS*, 454, 4197

Ventura, P., D'Antona, F., Mazzitelli, I., & Gratton, R. 2001, *ApJL*, 550, L65

Ventura P., & D'Antona F. 2009, *A&A*, 499, 835

Ventura P., Di Criscienzo M., Carini R., D'Antona F., 2013, *MNRAS*, 431, 3642

Vesperini, E., McMillan, S. L. W., D'Antona, F., & D'Ercole, A. 2011, *MNRAS*, 416, 355

Vesperini, E., McMillan, S. L. W., D'Antona, F., & D'Ercole, A. 2013, *MNRAS*, 429, 1913

Zahn, J.-P. 1977, *A&A*, 57, 383

Discussion

McSwain: Rapidly rotating massive stars, especially Be stars, are probably a population of stars that have been spun up by prior mass transfer in binary systems. Do you account for binarity among the rapidly rotating, evolved population as well?

D'Antona: If you refer to the MC models, they do not include mass transfer. In that case, actually, binarity is invoked as a mechanism to slow down the rapidly rotating stars by the resonant oscillations induced on the core by a far away companion.

Galaxy evolution, including the first AGB stars

Why Galaxies Care About AGB Stars: A Continuing Challenge through Cosmic Time
Proceedings IAU Symposium No. 343, 2019
F. Kerschbaum, M. Groenewegen & H. Olofsson, eds.

doi:10.1017/S1743921318005707

The Impact of AGB Stars on Galaxies

Martha L. Boyer

Space Telescope Science Institute,
3700 San Martin Dr., Baltimore, MD 21218 USA
email: mboyer@stsci.edu

Abstract. At the end of their evolution, asymptotic giant branch (AGB) stars undergo strong pulsation, mass loss, and dust production. Their mass loss results in substantial chemical and dust enrichment of the interstellar medium. Dust evolution models and isotope abundances in presolar grains suggest that AGB stars play a key role in both dust evolution and the star formation process. They are also the brightest stars in galaxies, potentially dominating in the near-infrared. As a result, AGB stars have a significant influence on the evolution and appearance of their host galaxies and thus must be accounted for when interpreting a galaxy's integrated light. I will highlight new results that describe the impact AGB stars have on galaxies, including how AGB stars are used to probe galaxy evolution.

Keywords. stars: AGB and post-AGB, stars: evolution, stars: winds, outflows, galaxies: fundamental parameters, infrared: stars

---

1. Introduction

At the end of their evolution, intermediate-mass stars ($\sim$1–8$M_\odot$) ascend the asymptotic giant branch (AGB). As they evolve, AGB stars develop some of the most complex aspects of stellar physics, including dredge up, thermal pulsation, radial pulsation, dust production, and strong mass loss. This phase ends with the star shedding its circumstellar envelope, exposing the stellar core (a white dwarf), and potentially developing a planetary nebula. AGB stars have several defining characteristics that have important impacts on galaxies, despite that fact that this phase is relatively short lived. I will review recent work that illustrates this impact, including the role AGB stars play in the enrichment of the interstellar medium (ISM; §2), the effect of their bright luminosities in the infrared (IR; §3), and their usefulness as probes of galaxy evolution (§4).

2. Enrichment of the ISM

AGB stars are important sources of ISM enrichment and galaxy chemical evolution, through both the synthesis of key elements and efficient dust production.

2.1. *Nucleosynthesis*

Solar abundances cannot be explained without nucleosynthesis processes in AGB stars. AGB stars are responsible for a significant fraction of some light elements: C, N, F, and possibly Li (e.g., Kobayashi *et al.* 2011a,b). They also produce neutron-rich isotopes of O, Ne, Mg, and Si and synthesize heavy elements via the *s*-process (e.g., Karakas *et al.* 2014). These elements are dredged up to the surface and distributed into the ISM via stellar winds, having a significant impact on the chemical evolution of their host galaxies. The signatures of AGB nucleosynthesis can be seen in surface abundances and can be used to estimate a star's initial mass by comparing ratios of different elements. For example, Rb is more likely to be present in more massive AGB stars (>4 $M_\odot$) and Ba is more likely present in low-mass stars (<3 $M_\odot$). Abundance ratios can therefore provide insight into the star formation histories as well as the nucleosynthesis processes in galaxies (Fishlock *et al.* 2014; Cristallo *et al.* 2015). For a more detailed review of AGB nucleosynthesis, see this conference's contribution from A. Karakas.

2.2. *Dust Production*

Thermally-pulsing (TP-)AGB stars are prolific dust producers and may be an important source of dust to the ISM. We see the signatures of AGB dust in the isotopic ratios of presolar grains (e.g., Nittler 2003; Zinner 2004; Hoppe 2008), and recent results suggest that some presolar grains may have originated in an electron-capture supernova from a super-AGB star (8–10 $M_\odot$; Nittler *et al.* 2018). These grains point to a possible significant role played by AGB stars in the dust evolution of galaxies, though their importance relative to supernova dust and grain growth in the ISM remains difficult to quantify.

Several works in the last 20 years have pointed to a dust budget "crisis", wherein the ISM dust masses observed in distant quasars is higher than the expected dust input from AGB stars and supernovae (e.g. Bertoldi *et al.* 2003; Robson *et al.* 2004; Beelen *et al.* 2006). However, recent observations in both the Small and Large Magellanic Clouds (SMC/LMC) are beginning to suggest that AGB stars may be a dominant source of dust. The Surveying the Agents of Galaxy Evolution *Spitzer* program (SAGE; Meixner *et al.* 2006; Gordon *et al.* 2011) was able to identify every dust-producing AGB star in both galaxies (Blum *et al.* 2006; Boyer *et al.* 2011) and quantify the dust input using several different techniques (Srinivasan *et al.* 2009, 2016; Matsuura *et al.* 2009; Riebel *et al.* 2012; Boyer *et al.* 2012). The total AGB dust input from these studies disagree by up to a factor of 5 owing primarily to the choice of dust optical constants. Recently, Nanni *et al.* (2016); Nanni *et al.* (2018) compared the SAGE TP-AGB infrared colors in the SMC to new AGB model spectra using a variety of different optical constants. They find that the best agreement is obtained when using optical constants that produce dust masses on the *high* end of the possible range estimated by the SAGE studies. Furthermore, Gordon *et al.* (2014) used *Herschel* data to revise the ISM dust masses in the SMC and LMC down by factors of 4–5, compared to previous dust mass estimates (Leroy *et al.* 2007; Bot *et al.* 2010). The decreased ISM dust mass and the increased AGB dust input from Nanni *et al.* both suggest than AGB stars can be a significant source of ISM dust. Dust evolution models from Zhukovska & Henning (2013) and Schneider *et al.* (2014) both imply that AGB dust can account for 15–70% of the total ISM dust even when including dust destruction, suggesting that the dust budget crisis observed in some quasars needs to be revisited.

Carbon stars can easily form their own condensation material and dredge it up to the surface for dust formation, but it is thought that oxygen-rich AGB stars require pre-existing condensation seeds. As a result, metal-poor oxygen-rich stars should have difficultly forming dust at low metallicity. This has consequences for the early Universe, suggesting that AGB dust input is delayed until the lower-mass carbon stars form ($t_{\rm age} > 100$–300 Myr). However, new evidence from the DUST in Nearby Galaxies with *Spitzer* (DUSTiNGS) survey suggests that oxygen-rich AGB stars can form significant dust masses even in very metal-poor galaxies. These stars were classified as O-rich (M type) using *HST*, and their [3.6]–[4.5] *Spitzer* colors suggest dust masses at least as high as those seen in their C type counterparts (Fig. 1; Boyer *et al.* 2017). The galaxies surveyed have very low gas-phase oxygen abundances ($12 + \log(\mathrm{O/H}) \lesssim 8$), implying that even the youngest stars are metal-poor. AGB stars with metallicities ~1 dex lower than the SMC may therefore contribute dust in the early Universe, as soon as 30 Myr after they form.

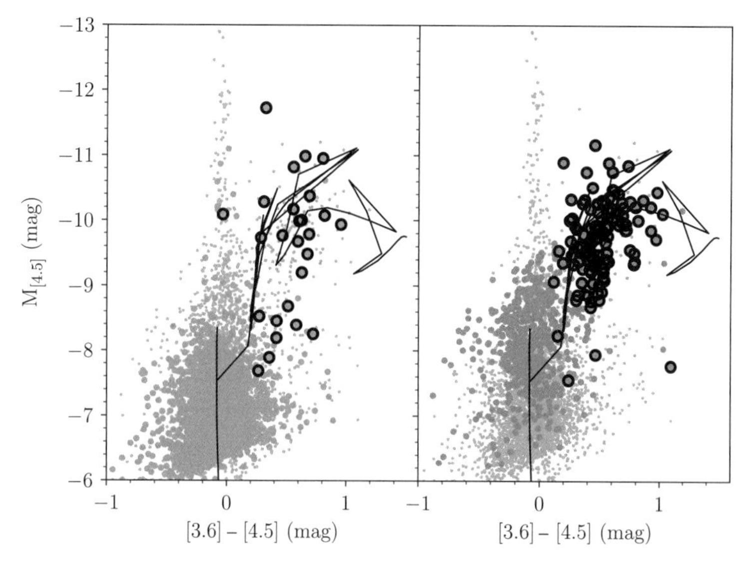

Figure 1. *Spitzer* and *HST* observations of metal-poor, star-forming dwarf galaxies show the presence of both M type (left) and C type (right) AGB stars that are producing dust (large, dark circles). Less-dusty AGB stars are marked with small blue (M) or red (C) symbols. These data suggest that, like C stars, primordial M type AGB stars can form substantial dust masses, similar to what is seen in the comparatively metal-rich Magellanic Clouds. This implies that massive AGB stars can contribute dust at very early times (∼30 Myr) in a galaxy's evolution. Produced using data from Boyer *et al.* (2017); isochrones from COLIBRI (Nanni *et al.* 2016).

3. Integrated Flux

Their high luminosities and cool temperatures place TP-AGB stars among the brightest objects in galaxies, especially in the near-IR. TP-AGB stars are short-lived and therefore rare, but the presence of even a small population has a strong impact on the integrated light of galaxies. For instance, population synthesis models from (Melbourne *et al.* 2012) show that TP-AGB stars are responsible for up to 70% of the 1.6 μm flux, with TP-AGB stars dominating over other bright stars by redshift ∼4–5.

3.1. *Near-IR*

Stellar population synthesis (SPS) models have difficulty reproducing the observed TP-AGB near-IR contribution, with model/data disagreements ranging from 0.2–3 in the 1–5 μm range (e.g., Johnson *et al.* 2013; Rosenfield *et al.* 2016). Uncertainties in the treatment of TP-AGB stars, particularly with metallicity, are responsible for these differences, with outcomes differing substantially depending on which models are applied in the population synthesis (e.g., Conroy *et al.* 2009). These uncertainties are thought to be due to the treatment of mass loss and dredge up, which both affect the stellar lifetime. Changes in the lifetime affect both the number of TP-AGB stars and their maximum luminosities, and therefore affect the integrated flux of the population. The ramifications of not accurately reproducing the TP-AGB luminosity contribution can be dramatic, since integrated near-IR light is often used to estimate galaxy stellar masses, mean ages, metallicities, and star formation histories (SFHs) using SPS models. For example, Conroy *et al.* (2009) find systematic biases in the derived galaxy stellar masses, with uncertainties up to 0.6 dex.

Baldwin *et al.* (2018) recently compared the effect that different models have on SFHs derived from optical and near-IR spectra of distant galaxies. They tested four different

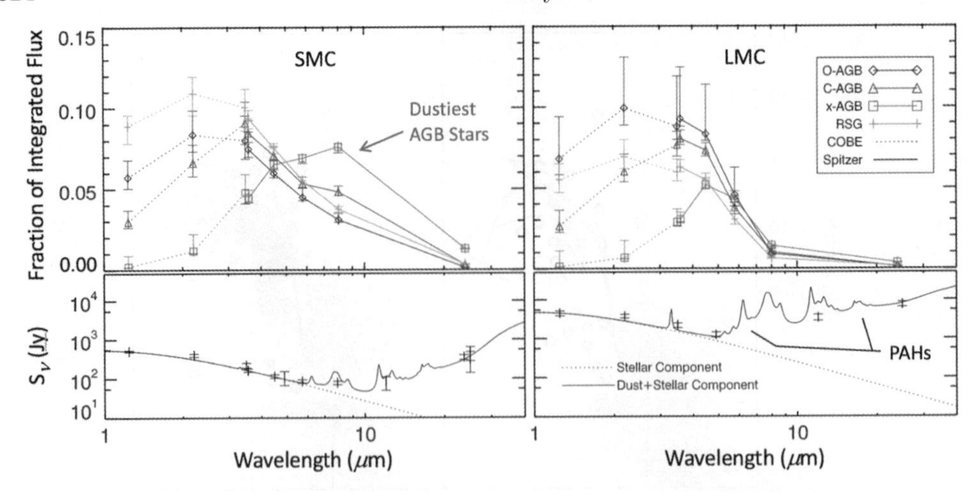

Figure 2. TP-AGB stars contribute a substantial fraction (up to 70%) of a galaxy's global flux, especially at 1–3 μm. Shown here is the measured contribution in the Magellanic Clouds, reproduced with permission from Melbourne & Boyer (2013). The lower panels show the global spectral energy distributions. In early-type galaxies or metal-poor galaxies with little ISM dust/PAHs, the contribution from dusty TP-AGB stars remains high out to 8 μm (e.g., in the SMC). In more metal-rich galaxies with strong star-formation, the mid-IR contribution is dominated by PAH emission (e.g., in the LMC).

SPS models (Bruzual & Charlot 2003; Conroy & Gunn 2010; Maraston & Strömbäck 2011; Vazdekis *et al.* 2016) and found that, while fits to optical spectra produced similar SFHs for all four models, the fits to the near-IR spectra produced SFHs that both disagree with each other and disagree with the optically-derived SFHs. All four SPS models produced reasonably similar optical SFHs for a given galaxy, but often produced four radically different SFHs in the IR. These results point to the importance of improving the TP-AGB models for SPS.

In addition, Baldwin *et al.* (2018) suggests that it may be the stellar spectral library, rather than the treatment of the TP-AGB phase, that drives the uncertainties. They tested this by inserting a new, high-resolution spectral library (C3K; see Conroy *et al.* 2013; Villaume *et al.* 2017) into the Flexible SPS models from Conroy & Gunn (2010) and testing the FSPS outcome using both the Padova (Marigo *et al.* 2008) and MIST (Choi *et al.* 2016) stellar evolution models. When the C3K library is used, the Padova and MIST models both produce reasonably good agreement between the optical and IR data and with each other.

3.2. *Mid-IR*

Circumstellar dust emission from TP-AGB stars can affect the integrated mid-IR flux of galaxies, particularly in early-type galaxies with little-to-no interstellar dust. Using the stellar catalogs from the Spitzer SAGE survey (Blum *et al.* 2006; Boyer *et al.* 2011), Melbourne & Boyer (2013) measured the contribution of each TP-AGB subclass to the light of the SMC and the LMC. They find that ≈350 stars in the superwind phase (the so-called extreme, or x-AGB stars) produce ∼10% of the 8-μm flux in the SMC, approximately equal to the contribution from the other ∼10,000 TP-AGB stars combined and suggesting a strong stochasticity in the mid-IR flux of low-mass, star-forming galaxies. However, the 8 μm TP-AGB contribution decreases to just 1–2% in the LMC, where emission from polycyclic aromatic hydrocarbons (PAHs) in star-forming regions has a much stronger contribution at these wavelengths (Fig. 2). Since PAH emission decreases precipitously at low metallicity (Sandstrom *et al.* 2010), we can expect the TP-AGB contribution to the mid-IR to increase in metal-poor galaxies. Beyond 10 μm, diffuse interstellar dust (if present) dominates the galaxy spectrum.

In quiescent, early-type galaxies that lacking cold interstellar dust, the signature of a TP-AGB population can be seen via the circumstellar 10 μm and 18 μm silicate features. Villaume *et al.* (2015) show models that can account for these features in early-type galaxy spectral energy distributions without invoking an interstellar dust component. These features are easily detectable in their sample of galaxy spectra, and their models suggest that the strength of the feature depends strongly on the age of the stellar population, with younger populations producing stronger silicate features. This particular AGB signature is therefore a useful tracer of population age in distant galaxies, particularly as high sensitivity mid-IR observations become possible with the launch of the *James Webb Space Telescope* (JWST). However, Simonian & Martini (2017) show that models of AGB dust may need some adjustments given their inability to reproduce the WISE colors ([3.4]–[12] and [3.4]–[22]) of a large sample of early-type galaxies.

In post-starburst galaxies, which have younger populations than their early-type counterparts, the mid-IR contribution from TP-AGB stars is less clear even in the case where there is no evidence for cold interstellar dust. For example, Alatalo *et al.* (2016) measured the WISE colors of several post-starburst galaxies and found them to be intermediate to the expected colors of TP-AGB stars and embedded active galactic nuclei (AGN). The mid-IR spectra may thus be a combination of the two, and inferences that rely on TP-AGB models in these systems must use caution.

3.3. *Long Period Variables*

AGB variability also produces a measurable signature in the integrated light of galaxies. Individual long-period variables (LPVs) can currently be identified out to distances of roughly 5 Mpc (See the P. Whitelock contribution). In distant and totally unresolved galaxies, the luminosity fluctuations of an LPV population averages out and is undetectable. However, there is an intermediate regime where these fluctuations are detectable both in semi-resolved galaxies and in crowded regions such as galactic bulges. Conroy *et al.* (2015) show that in the case where individual pixels contain $\lesssim$10 LPVs, the variability of this small number of stars can produce a pixel-shimmer effect across different observational epochs. This shimmer can be modeled, and tuning the LPV weight in the applied SPS models has a strong effect on the derived stellar age of the population, providing a potential age diagnostic for semi-resolved stellar populations.

3.4. *Recent Model Updates*

Modeling TP-AGB stars is notoriously difficult due to several complex processes: dredge up, mass loss, pulsation, etc. The clear effect of TP-AGB stars on galaxy integrated light is a strong motivation to improve the models, and significant progress has been made in recent years. Villaume *et al.* (2015), Dell'Agli *et al.* (2015a,b), and Nanni *et al.* (2018) have each presented new model results for circumstellar dust that provide good agreement to observed infrared colors, particularly in the Magellanic Clouds. The latest COLIBRI and MIST models both show significant improvements to both the expected number of TP-AGB stars and their expected near-IR flux, resulting in better agreement with observed TP-AGB luminosity functions in the Magellanic Clouds (Choi *et al.* 2016; Pastorelli *et al.* 2019) and in more distant resolved galaxies (Rosenfield *et al.* 2014, 2016).

4. Probes of Galaxy Evolution

Here, I describe examples of how AGB stars can trace galaxy evolution by providing population age diagnostics and morphological diagnostics. Their usefulness does not end at the termination of the TP-AGB phase; their end-products (post-AGB stars) can also shed light on the process of star-formation quenching in galaxies.

4.1. *Ionization*

When comparing the strength of galaxy emission line-intensity ratios (the BPT diagram), a class of galaxies with strong [N II] $\lambda 6583/H\alpha$ and weak [O III] $\lambda 5007/H\beta$ (e.g., Ho 2008) stands out. These galaxies account for a large fraction of massive galaxies (>30%) and have typically been classified as low-ionization nuclear emission line regions, or LINERs, with the ionization mechanism initially thought to be an AGN. However, there are problems with the AGN scenario that are difficult to explain, including an energy budget deficit (e.g., Eracleous *et al.* 2010), lack of correlation with radio emission (e.g., Best *et al.* 2005), and some hints that the emission line flux radial profile is extended beyond what is expected for a nuclear ionizing source (e.g., Sarzi *et al.* 2010). One possible alternative mechanism is photoionization by hot evolved stars distributed throughout the galaxy, especially post-AGB stars (de Serego Alighieri *et al.* 1990; Binette *et al.* 1994).

Recent spectral surveys that have high spatial resolution (SDSS-IV MaNGA, CALIFA) are able to investigate the spatial distribution of LINER emission in galaxies to test the AGN vs. post-AGB scenarios (e.g. Singh *et al.* 2013; Belfiore *et al.* 2016). Belfiore *et al.* (2016) analyze MaNGA data for 646 galaxies and show that LINER-like emission is present in many types of galaxies, sometimes in the outer halos and/or in the central regions of star-forming galaxies, and sometimes extended throughout the entire galaxy body. They find that the Hα surface brightness radial profiles are typically shallower than $1/r^2$ and that the ionization parameter of the gas shows a flat gradient, pointing to LINER emission originating from diffuse stellar sources rather than from a central nuclear source. Given the spatial information, they renamed LINERS to LIERS (omitting the 'nuclear' part), and noted cLIER (central) and eLIER (extended) subclasses.

The Belfiore *et al.* (2016) work points to post-AGB stars as the most likely candidate for the ionization mechanism, and SPS models indicate that post-AGB stars can easily produce LIER emission in galaxies, even when their luminosities are decreased by a factor of two (Byler *et al.* 2017). If post-AGB stars are indeed the source of the LIER emission, it suggests a possible galaxy evolutionary sequence that quenches star-formation from the inside-out, wherein star-forming galaxies evolve into green valley AGN, which evolve into quenched eLIERs (e.g. Sánchez *et al.* 2018). AGB evolution and their end-products (post-AGB stars) are clearly important to tracing the quenching process in galaxies.

4.2. *Spatial Distribution*

Most dwarf galaxies show radial age gradients, with the youngest stars concentrated at the centers and old red giant branch (RGB) stars present in the outskirts. This is often interpreted as evidence for outside-in growth, where star formation starts in galaxy outskirts and migrates to the center over time. El-Badry *et al.* (2016) proposed an alternate inside-out growth scenario, wherein short bursts in star formation lead to gas outflows (and subsequently, inflows as the gas cools). Stars inherit this motion, which leads to stellar migration. They show that in this scenario, most stars form in the galaxy center and migrate outwards on relatively short timescales. In this case, one would expect the

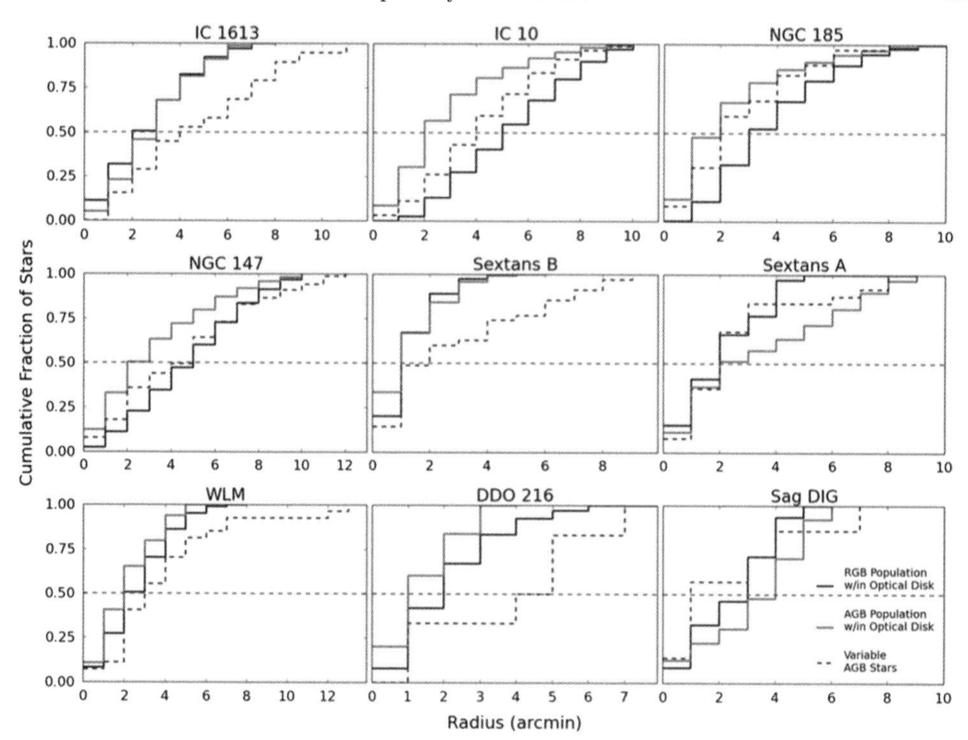

Figure 3. Radial profiles of evolved stars in nearby dwarf galaxies showing that TP-AGB stars (red; and blue indicating known LPVs) and RGB stars (black) are mixed at large radii, suggesting inside-out growth. Reproduced with permission from McQuinn *et al.* (2017).

intermediate-aged population (namely, AGB stars) to be mixed with the older population in dwarf galaxy outskirts, while the youngest stars remain concentrated in the centers.

Recently, the DUSTiNGS survey (Boyer *et al.* 2015a) obtained large 3–5 μm maps of 50 nearby dwarf galaxies that extend to well beyond the effective radii (R_e). We used this data to identify the AGB populations, using (1) statistical arguments that are able to isolate the TP-AGB population from unresolved AGN, which have similar IR colors and luminosities; and (2) multi-epoch data that leverage the long-period pulsation of TP-AGB stars (Boyer *et al.* 2015b). In McQuinn *et al.* (2017), we showed that the TP-AGB population is well mixed with the old RGB population out to $>3\times R_e$ (Fig. 3). These AGB radial profiles are evidence for inside-out growth, providing an additional diagnostic for understanding galaxy evolution.

5. Summary

AGB stars have a strong impact on galaxies, via the material they produce through nucleosynthesis and dust production and via their high luminosities, which peak at IR wavelengths. They also provide important insight into galaxy evolution.

References

Alatalo, K., Bitsakis, T., Lanz, L., *et al.* 2016, *ApJ*, 843, 9
Baldwin, C., McDermid, R. M., Kuntschner, H., *et al.* 2018, *MNRAS*, 473, 4698
Beelen, A., Cox, P., Benford, D., *et al.* 2006, *ApJ*, 642, 694
Belfiore, F., Maiolino, R., Maraston, C., *et al.* 2016, *MNRAS*, 461, 3111

Bertoldi, F., Carilli, C. L., Cox, P., *et al.* 2003, *A&A*, 406, 55
Best, P. N., Kauffmann, G., Heckman, T. M., *et al.* 2005, *MNRAS*, 362, 25
Binette, L., Magris, C. G., Stasińska, G., & Bruzual, A. G. 1994, *A&A*, 292, 13
Blum, R. D., Mould, J. R., Olsen, K. A., *et al.* 2006, *AJ*, 132, 2034
Bot, C., Ysard, N., Paradis, D., *et al.* 2010, *A&A*, 523, 20
Boyer, M. L., Srinivasan, S., van Loon, J. Th., *et al.* 2011, *AJ*, 142, 103
Boyer, M. L., Srinivasan, S., Riebel, D., *et al.* 2012, *ApJ*, 748, 40
Boyer, M. L., McQuinn, K. B. W., Barmby, P., *et al.* 2015a, *ApJS*, 216, 10
Boyer, M. L., McQuinn, K. B. W., Barmby, P., *et al.* 2015b, *ApJ*, 800, 51
Boyer, M. L., McQuinn, K. B. W., Groenewegen, M. A. T., *et al.* 2017, *ApJ*, 851, 152
Bruzual, G., & Charlot, S. 2003, *MNRAS*, 344, 1000
Byler, N., Dalcanton, J. J., Conroy, C., & Johnson, B. D. 2017, *ApJ*, 840, 44
Choi, J., Dotter, A., Conroy, C., *et al.* 2016, *ApJ*, 823, 102
Cristallo, S., Straniero, O., Piersanti, L., & Gobrecht, D. 2015, *ApJS*, 219, 40
Conroy, C., Gunn, J. E., & White, M. 2009, *ApJ*, 699, 486
Conroy, C., & Gunn, J. E. 2010, *ApJ*, 712, 833
Conroy, C., van Dokkum, P. G., & Choi, J. 2015, *Nature*, 527, 488
Conroy, C., van Dokkum, P. G., & Graves, G. J. 2013, *ApJL*, 763, L25
di Serego Alighieri, S., Trinchieri, G., & Brocato, E. 1990, *ASSL*, 160, 301
Dell'Agli, F., Ventura, P., Schneider, R., *et al.* 2015, *MNRAS*, 447, 2992
Dell'Agli, F., García-Hernández, D. A., Ventura, P., *et al.* 2015, *MNRAS*, 454, 4235
El-Badry, K., Wetzel, A. R., Geha, M., *et al.* 2016, *ApJ*, 820, 131
Eracleous, M., Hwang, J. A., & Flohic, H. M. L. G. 2010, *ApJ*, 711, 796
Fishlock, C. K., Karakas, A. I., Lugaro, M., & Yong, D. 2014, *ApJ*, 797, 44
Gordon, K. D., Meixner, M., Meade, M. R., *et al.* 2011, *AJ*, 142, 102
Gordon, K. D., Roman-Duval, J., Bot, C., *et al.* 2014, *ApJ*, 797, 85
Ho, L. C. 2008, *ARA&A*, 46, 475
Hoppe, P. 2008, *Space Sci. Revs*, 138, 43
Johnson, B. D., Weisz, D. R., Dalcanton, J., *et al.* 2013, *ApJ*, 772, 8
Karakas, A. I., & Lattanzio, J. C. 2014, *PASA*, 31, e030
Kobayashi, C., Izutani, N., Karakas, A. I., *et al.* 2011, *ApJL*, 739, L57
Kobayashi, C., Karakas, A. I., & Umeda, H. 2011, *MNRAS*, 414, 3231
Leroy, A., Bolatto, A., Stanimirovic, S., *et al.* 2007, *ApJ*, 658, 1027
Maraston, C., & Strömbäck, G. 2011, *MNRAS*, 418, 2785
Marigo, P., Girardi, L., Bressan, A., *et al.* 2008, *A&A*, 482, 883
Matsuura, M., Barlow, M. J., Zijlstra, A. A., *et al.* 2009, *MNRAS*, 396, 918
McQuinn, K. B. W., Boyer, M. L., Mitchell, M. B., *et al.* 2017, *ApJ*, 834, 78
Meixner, M., Gordon, K. D., Indebetouw, R., *et al.* 2006, *AJ*, 132, 2268
Melbourne, J., Williams, B. F., Dalcanton, J. J., *et al.* 2012, *ApJ*, 748, 47
Melbourne, J., & Boyer, M. L. 2013, *ApJ*, 764, 30
Nanni, A., Marigo, P., Groenewegen, M. A. T., *et al.* 2016, *MNRAS*, 462, 1215
Nanni, A., Marigo, P., Girardi, L., *et al.* 2018, *MNRAS*, 473, 5492
Nittler, L. R. 2003 *Earth Planet. Sci. Lett.*, 209, 259
Nittler, L. R., Alexander, C. M. O., Liu, N., & Wang, J. 2018, *ApJ*, 856, 24
Pastorelli, G., Girardi, L., Marigo, M., *et al.* 2019, *in preparation, presented at this conference*
Riebel, D., Srinivasan, S., Sargent, B., & Meixner, M. 2012, *ApJ*, 753, 71
Robson, I., Priddey, R. S., Isaak, K. G., & McMahon, R. G. 2004, *MNRAS*, 351, L29
Rosenfield, P., Marigo, P., Girardi, L., *et al.* 2014, *ApJ*, 790, 22
Rosenfield, P., Marigo, P., Girardi, L., *et al.* 2016, *ApJ*, 822, 73
Sánchez, S. F., Avila-Reese, V., Hernandez-Toledo, H., *et al.* 2018, *Rev. Mexicana AyA*, 54, 217
Sandstrom, K. M., Bolatto, A. D., Draine, B., Bot, C., & Stanimirovic, S. 2010, *ApJ*, 715, 701
Sarzi, M., Shields, J. C., Schawinski, K., *et al.* 2010, *MNRAS*, 402, 2187

Schneider, R., Valiante, R., Ventura, P., *et al.* 2014, *MNRAS*, 442, 1440
Simonian, G. V., & Martini, P. 2017, *MNRAS*, 464, 3920
Singh, R., van de Ven, G., Jahnke, K., *et al.* 2013, *A&A*, 558, A43
Srinivasan, S., Meixner, M., Leitherer, C., *et al.* 2009, *AJ*, 137, 4810
Srinivasan, S., Boyer, M. L., Kemper, F., *et al.* 2016, *MNRAS*, 457, 2814
Vazdekis, A., Koleva, M., Ricciardelli, E., *et al.* 2016, *MNRAS*, 463, 3409
Villaume, A., Conroy, C., & Johnson, B. 2015, *ApJ*, 806, 82
Villaume, A., Conroy, C., Johnson, B., *et al.* 2017, *ApJS*, 230, 23
Zhukovska, S., & Henning, T. 2013, *A&A*, 555, 99
Zinner, E. 2004, in: K.K. Turekian, H.D. Holland & A.M. Davis (eds.), *Treatise in Geochemistry 1 (Oxford and San Diego: Elsevier)*, p. 17

Discussion

SRINIVASAN: Dust opacities in the sub-mm are typically extrapolated from the mid-IR and are much lower than lab measurements. This contributes significantly to the ISM dust mass estimates. As a result, there may not be a high-redshift dust budget crisis.

Why Galaxies Care About AGB Stars:
A Continuing Challenge through Cosmic Time
Proceedings IAU Symposium No. 343, 2019
F. Kerschbaum, M. Groenewegen & H. Olofsson, eds.

doi:10.1017/S1743921318008116

On the origin of N in galaxies with galaxy evolution models

Fiorenzo Vincenzo and Chiaki Kobayashi

Centre for Astrophysics Research, University of Hertfordshire
College Lane, AL10 9AB, Hatfield, United Kingdom
emails: f.vincenzo@herts.ac.uk, c.kobayashi@herts.ac.uk

Abstract. Nitrogen is among the most abundant chemical elements in the cosmos, and asymptotic giant branch (AGB) stars are fundamental nucleosynthetic sources of N in galaxies. In this work, we show how the observed N/O versus O/H chemical abundance diagram, both in extragalactic systems and in our own Galaxy, can be used to constrain the nucleosynthetic origin of N in the cosmos. In particular, we review the results of our studies with chemical evolution models, embedded in full cosmological chemodynamical simulations.

Keywords. galaxies: abundances, galaxies: evolution, ISM: abundances, hydrodynamics, stars: abundances

---

In Fig. 1, we show the observed N/O–O/H chemical abundance diagram as observed in a sample of Milky Way (MW) thin and thick disc stars, and in a sample of MW open clusters Magrini *et al.* (2018), in unresolved star-forming regions within a sample of nearby disc galaxies as observed by the Sloan Digital Sky Survey IV Mapping Nearby Galaxies at Apache Point Observatory survey (MaNGA) Belfiore *et al.* (2017), in the HII regions of a sample of nearby spiral and irregular galaxies by Pilyugin *et al.* (2010), and nearby dwarf galaxies (Izotov *et al.* 2012; James *et al.* 2015; Berg *et al.* 2016; grey triangles with the error bars).

Interestingly, in Fig. 1, the observed stellar N/O–O/H diagram in the Galaxy qualitatively agrees with the observed gas-phase N/O–O/H diagram of the MaNGA survey, which collects chemical abundance measurements from a representative sample of 550 star-forming disc galaxies, with total stellar mass in the range $9.0 \leqslant \log(M_{\star}/\mathrm{M}_{\odot}) \leqslant 11.5$ dex. The offset in O/H between Belfiore *et al.* (2017) and Pilyugin *et al.* (2010) is because of the different assumed calibrations to measure the gas metallicity from the strong emission lines; in particular, Belfiore *et al.* (2017) assume as fiducial the calibration of Maiolino *et al.* (2008). Despite the large uncertainty in the assumed metallicity calibrations, the observations agree that N/O steeply increases when moving towards high O/H. At very low metallicity, N/O tends to remain flat around a mean value of ~ -1.5 dex, with a large scatter (see also the early work of Matteucci 1986).

The observed behaviour of N/O as a function of O/H depends on how N is produced by stars of different mass and metallicity. From a nucleosynthesis point of view, N is mostly produced as a *secondary* element; in particular, stars of larger metallicities produce – on average – larger amounts of N. The secondary N is synthesised during the CNO cycle of H-burning, at the expenses of the C and O nuclei already present in the gas mixture from which the star originated (see, for more details, Henry *et al.* 2000; Chiappini *et al.* 2005; Mollá *et al.* 2006; Gavilán *et al.* 2006; Vincenzo *et al.* 2016).

At very low metallicity, where massive stars are predominant in the chemical enrichment process, and the secondary N component is minimal, we observe a plateau of N/O

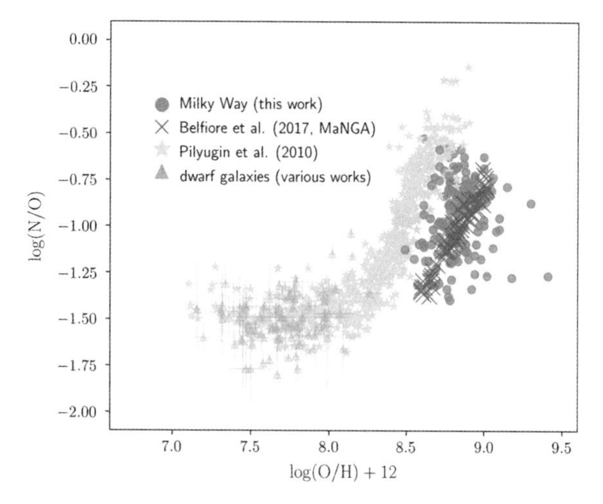

Figure 1. We compare the observed N/O–O/H relation in the MW stars and open cluster Magrini *et al.* (2018), in a sample of MaNGA survey galaxies (Belfiore *et al.* 2017; grey crosses), in the HII regions of nearby spiral and irregular galaxies (Pilyugin *et al.* 2010; pink stars), and in nearby dwarf galaxies (Izotov *et al.* 2012; James *et al.* 2015; Berg *et al.* 2016; grey triangles with the error bars).

versus O/H; to reproduce this plateau, an additional mechanism has been invoked by chemical evolution models of galaxies, requiring that N is mostly produced as a *primary* element by very metal-poor massive stars, namely its nucleosynthesis yields do not depend on the initial metal content of the stars (Matteucci 1986; Chiappini *et al.* 2005). Since stellar evolution models of massive stars typically fail in predicting the necessary amount of primary N to reproduce the observations, chemical evolution models assumed in the past a fixed, artificial amount of primary N from very metal-poor massive stars to reproduce N/O at very low metallicity.

We remark on the fact that also AGB stars can produce primary N during the so-called third dredge-up, when it occurs in conjunction with the hot-bottom burning, if nuclear burning at the base of the convective envelope is efficient (Renzini & Voli 1981; Ventura *et al.* 2013).

Cosmological chemodynamical simulations are nowadays among the best tools to study how chemical elements are produced within galaxies, to reconstruct also the spatial distribution of the chemical elements as a function of time within different galaxy environments. Our simulation code includes the main stellar nucleosynthetic sources in the cosmos (core-collapse and Type Ia supernovae, hypernovae, asymptotic giant branch stars, and stellar winds from stars of all masses and metallicities), and it is based on an updated version of the GADGET 3 code (Springel *et al.* 2001; Springel 2005; Kobayashi, 2004; Kobayashi *et al.* 2007; Kobayashi & Nakasato 2011).

In Figs. 2 and 3, we present the results of the cosmological chemodynamical simulation developed by Vincenzo & Kobayashi (2018a,b), including also the effect of failed supernovae (Kobayashi *et al.* in prep.). In Fig. 2 we show how the gas-phase N/O versus O/H abundance patterns vary within ten reference simulated galaxies, which have been selected because of their different star formation histories (SFHs); in particular, from galaxy 0 to galaxy 10, the SFH is concentrated towards later and later epochs (see Vincenzo & Kobayashi 2018b). The blue points in 2 correspond to the predictions of our simulation, while the pink crosses correspond to the MaNGA data of Belfiore *et al.* (2017); we also show the average observed relation as derived by Dopita *et al.* (2016; solid grey line). The black points with the error bars correspond to the predicted average N/O and O/H which have been computed by dividing our simulated galaxies in many

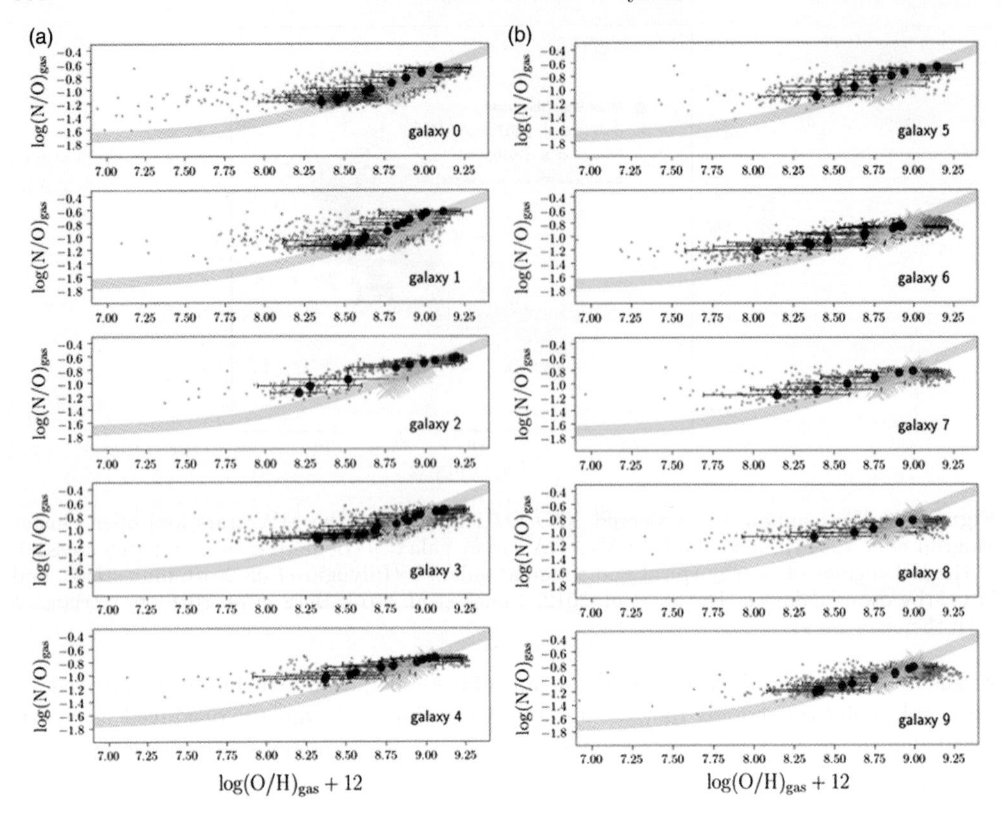

Figure 2. The predicted gas-phase N/O–O/H relation within our ten reference disc galaxies, where each blue point corresponds to a gas particle, and the black points with error bars have been computed, firstly by dividing in different annuli the simulated galaxies, and then by computing the average N/O and O/H within each annulus; the pink crosses correspond to the MaNGA survey data of Belfiore *et al.* (2017), while the solid grey line to the average observed relation of Dopita *et al.* (2016).

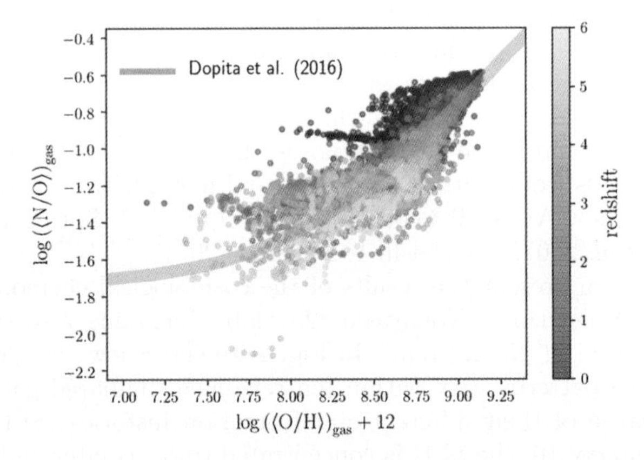

Figure 3. The predicted N/O–O/H relation, where the abundances correspond to SFR-weighted averages in the gas-phase of all 33 disc galaxies in our sample, which – in this case – are unresolved. See Vincenzo & Kobayashi (2018b) for more details about Figs. 2 and 3.

concentric annuli. We predict that – when we are able to spatially resolve galaxies – their N/O–O/H diagram tend to follow at the present time an average universal relation, which does not depend on the galaxy SFH, simply because the nucleosynthesis of N is universal, and it strictly depends – in the relatively high metallicity regime – on the metallicity of the stars. The trend of N/O versus O/H in Fig. 2 reflects chemical abundance gradients in our simulated galaxies, where the innermost metal-rich regions have higher N/O ratios than the outermost metal-poor regions.

In Fig. 3, we show how the average gas-phase N/O within 33 star-forming disc galaxies in our simulation volume vary as functions of the average gas-phase O/H. The colour-coding in the figure corresponds to the redshift. We predict that, even in the case of unresolved galaxies, their average N/O ratios tend to obey the same observed average relation as in the case of resolved galaxies. In this case, the relation between the average N/O and O/H is the consequence of a mass-metallicity relation that our simulated galaxies obey – on average – as they evolve across cosmic times, with the most massive galaxies having typically higher O/H (and hence higher N/O because of the main secondary origin of N) than the least massive systems.

Coming back to the original discussion on Fig. 1, our chemodynamical simulations have been able to predict and explain the reason why the gas-phase abundances of N/O versus O/H from the MaNGA survey are consistent with the abundances of the MW stars and open clusters. In conclusion, to respond also to the question of Letizia Stanghellini at the end of our presentation, we note that the stellar abundances in the MW stars can be used in the future to identify the best metallicity calibration for strong emission line diagnostics, as suggested in the original draft of Magrini *et al.* (2018).

References

Belfiore, F., Maiolino, R., Tremonti, C., *et al.* 2017, *MNRAS*, 469, 151
Berg, D. A., Skillman, E. D., Henry, R. B. C., Erb, D. K., & Carigi L., 2016, *ApJ*, 827, 126
Chiappini, C., Matteucci, F., & Ballero, S. K. 2005, *A&A*, 437, 429
Dopita, M. A., Kewley, L. J., Sutherland, R. S., & Nicholls D. C., 2016, *APSS*, 361, 61
Gavilán, M., Mollá, M., & Buell, J. F. 2006, *A&A*, 450, 509
Henry, R. B. C., Edmunds, M. G., & Köppen, J. 2000, *ApJ*, 541, 660
Izotov, Y. I., Thuan, T. X., & Guseva, N. G., 2012, *A&A*, 546, A122
Iwamoto, K., Brachwitz, F., Nomoto, K., *et al.* 1999, *ApJS*, 125, 439
James, B. L., Koposov, S., Stark D. P., *et al.*, 2015, *MNRAS*, 448, 2687
Karakas, A. I. 2010, *MNRAS*, 403, 1413
Kobayashi, C. 2004, *MNRAS*, 347, 740
Kobayashi, C., Springel, V., & White, S. D. M., 2007, *MNRAS*, 376, 1465
Kobayashi, C., & Nakasato, N., 2011, *ApJ*, 729, 16
Kroupa, P., Tout, C. A., & Gilmore, G., 1993, *MNRAS*, 262, 545
Magrini, L., Vincenzo, F., Randich, S., *et al.* 2018, *A&A*, 618, A102
Maiolino, R., Nagao, T., Grazian, A., *et al.* 2008, *ApJ*, 488, 463
Matteucci, F. 1986, *MNRAS*, 221, 911
Mollá, M., Vílchez, J. M., Gavilán, M., & Díaz, A. I. 2006, *MNRAS*, 372, 1069
Nomoto, K., Kobayashi, C., & Tominaga, N. 2013, *ARA&A*, 51, 457
Pilyugin, L. S., Vílchez, J. M., & Thuan, T. X. 2010, *ApJ*, 720, 1738
Renzini, A., & Voli, M. 1981, *A&A*, 94, 175
Springel, V., Yoshida, N., & White, S. D. M. 2001, *NA*, 6, 79
Springel, V. 2005, *MNRAS*, 364, 1105
Totani, T., Morokuma, T., Oda, T., Doi, M., & Yasuda, N. 2008, *PASJ*, 60, 1327
Ventura, P., Di Criscienzo, M., Carini, R., & D'Antona, F. 2013, *MNRAS*, 431, 3642
Vincenzo, F., Belfiore, F., Maiolino, R., Matteucci, F., & Ventura, P. 2016, *MNRAS*, 458, 3466
Vincenzo, F., & Kobayashi, C. 2018a, *A&A*, 610, L16
Vincenzo, F., & Kobayashi, C. 2018b, *MNRAS*, 478, 155

Why Galaxies Care About AGB Stars:
A Continuing Challenge through Cosmic Time
Proceedings IAU Symposium No. 343, 2019
F. Kerschbaum, M. Groenewegen & H. Olofsson, eds.

doi:10.1017/S1743921318005719

A Masing BAaDE's Window

Thousands of SiO Masers in Thousands of AGB Stars in the Galaxy

Lorant O. Sjouwerman[1], Ylva M. Pihlström[2], Adam C. Trapp[3], Michael C. Stroh[2], Luis Henry Quiroga-Nuñez[4], Megan O. Lewis[2], R. Michael Rich[3], Mark R. Morris[3], Huib Jan van Langevelde[4], Mark J Claussen[1] and the BAaDE collaboration

[1]National Radio Astronomy Observatory
P.O. Box O, Socorro, NM, U.S.A.
email: `lsjouwer@nrao.edu`

[2]University of New Mexico, Albuquerque NM, USA

[3]University of California, Los Angeles CA, USA

[4]Leiden Observatory, Leiden & Joint Institute for VLBI ERIC, Dwingeloo, The Netherlands

Abstract. We report on the Bulge Asymmetries and Dynamic Evolution (BAaDE) survey which has observed 19 000 *MSX* color selected red giant stars for SiO maser emission at 43 GHz with the VLA and is in the process of observing 9 000 of these stars with ALMA at 86 GHz in the Southern sky. Our setup covers the main maser transitions, as well as those of isotopologues and selected lines of carbon-bearing species. Observations of this set of lines allow a far-reaching catalog of line-of-sight velocities in the dust-obscured regions where optical surveys cannot reach. Our preliminary detection rate is close to 70%, predicting a wealth of new information on the distribution of metal rich stars, their kinematics as function of location in the Galaxy, as well as the occurrence of lines and line ratios between the different transitions in combination with the spectral energy distribution from about 1 to 100 μm. Similar to the OH/IR stars, a clear kinematic signature between disk and bulge stars can be seen. Furthermore, the SiO $J=1\rightarrow 0$ (v=3) line plays a prominent role in the derived maser properties.

Keywords. masers, surveys, stars: late-type, Galaxy: kinematics and dynamics, infrared: stars

---

1. Introduction

This symposium highlighted many reasons "Why Galaxies Care About AGB stars". The reasons relevant to the "Bulge Asymmetries and Dynamic Evolution (BAaDE)" survey, described here, can be summarized as:

– Asymptotic Giant Branch (AGB) stars are luminous and abundant; they are relatively easy to find and study in large quantities in the (near-)infrared where their spectral energy distributions (SEDs) peak.

– AGB stars are a mix of stellar populations: a range of stellar masses, ages and metallicities, for which detailed modeling provides a wealth of physical information as function of location in the Galaxy.

– AGB stars are variable stars and are oases of simple molecules; it is possible to determine a crude distance (with period-luminosity relations) and bolometric magnitude, and observing the molecular lines (thermal as well as maser lines) reveals the accurate stellar line-of-sight velocity instantly.

– Surveys of AGB stars can be combined with the physical and kinematic properties and outline the dynamics of the individual populations in the Galaxy.

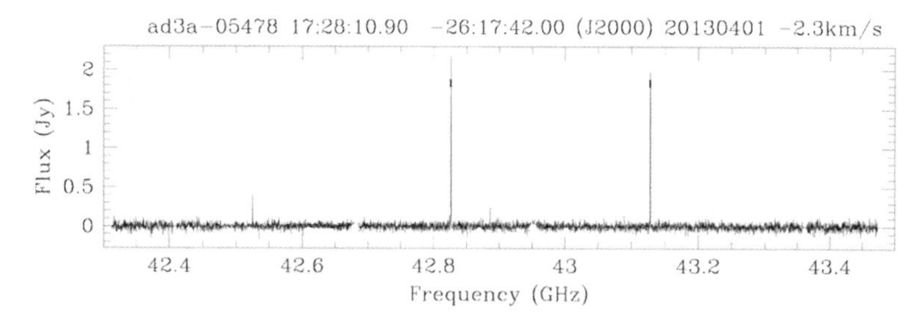

Figure 1. Typical VLA 43 GHz detection of bright SiO isotopologue $J = 1 \rightarrow 0$ maser lines in a source. From left to right: SiO (v=3), SiO (v=2), ^{29}SiO (v=0), and SiO (v=1).

AGB stars are excellent probes of the structure and evolution of the Galaxy!

The BAaDE survey aims to *significantly* improve the models of the dynamics and structure of the inner Galaxy, by probing into the regions of the Galactic Bulge and Galactic plane not reachable with optical surveys. This can be done by using stellar SiO masers as *radio* detected point masses. Similar surveys have been performed earlier, using the less abundant (~3000) OH masers found in thick-shell AGB stars (i.e., typically the OH/IR stars, e.g., Sevenster *et al.* 2000). However, with the release of the recent space-based (near-)infrared surveys in the Galactic Plane, in particular the Midcourse Space eXperiment (*MSX*) survey (Egan *et al.* 2003), many thousands of candidate thin-shell AGB stars (i.e., typically the Mira-type stars) have become available. With a judicious infrared color selection these stars can be surveyed for the SiO maser with a high detection rate (Sjouwerman *et al.* 2009).

BAaDE is surveying 28 000 *MSX* color-selected red giant stars with the Karl G. Jansky Very Large Array (VLA) north of Declination −35, and the Atacama Large Millimeter/submillimeter Array (ALMA) in the south for the range in Galactic Longitude that the VLA cannot observe ($-110 < l^\circ < -5$). Each individual detection provides a line-of-sight velocity for a given stellar position. Furthermore, for each source a comprehensive set of (near-)infrared photometry is compiled, in an attempt to characterize the central star and its surrounding circumstellar envelope (CSE) through modeling of the individual SEDs. Ultimately, it is the intent to obtain proper motion and parallax measurements from a subset of the sources with Very Long Baseline Interferometry (VLBI), to characterize the type of stellar orbits found in the inner Galaxy (Van Langevelde *et al.*, IAU Symp. 348).

The BAaDE survey is complementary to the optical *Gaia* survey that is obscured in the Bulge and plane (Gaia Collaboration *et al.* 2016). As BAaDE focuses on the inner Bulge area and and evolved stellar population, it is also complementary to the radio BeSSeL survey that uses methanol masers in star forming regions to model the spiral structure in the Galactic Plane (Brunthaler *et al.* 2011).

2. Results

During 2013 to 2017 ~19 000 sources were observed with the VLA covering four SiO $J = 1 \rightarrow 0$ transitions and three isotopologue lines at 43 GHz (Fig. 1). The preliminary detection rate, after analyzing about half of the data, is close to 70 % for both the VLA and ALMA data when we include the occasional detection of a line from a carbon-bearing species (i.e., from HC_5N and/or HC_7N) in the carbon-rich stars. Of the ~9 000 sources planned to be observed with ALMA covering three SiO $J = 2 \rightarrow 1$ transitions and one isotopologue line at 86 GHz and the CS line at 98 GHz, about 2 000 have been observed and analyzed in Cycles 2, 3 and 5. The results below are based on preliminary data sets.

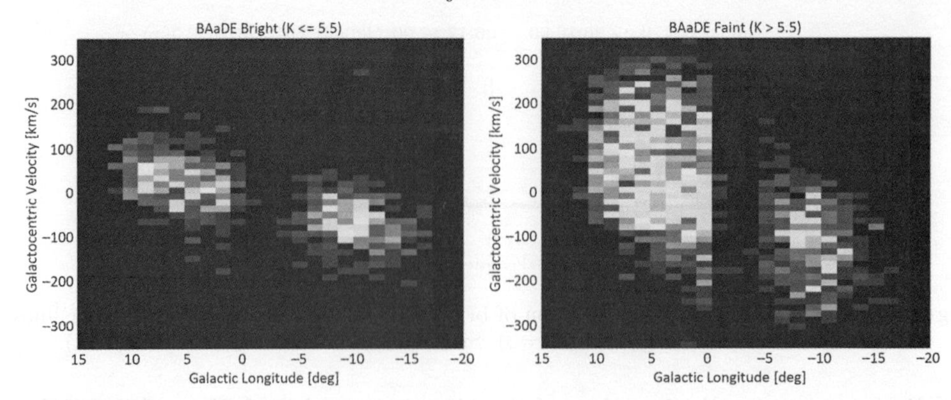

Figure 2. Longitude-velocity histogram plots of the first VLA ($0 < l^\circ < 10$) and ALMA ($l^\circ \approx -10$) observations. After splitting in a bright (left, K-magnitude = 5.5 or brighter) and a faint (right, $K > 5.5$) sample, the bright sample shows a tighter galactic rotation on either side of the Galactic Center ($l = 0$) than the more dispersed fainter sample. [Trapp *et al.* (2018)]

2.1. *Survey Sanity Checks*

A major bias in the BAaDE survey is that approximately 2/3 of the sample is observed in the 43 GHz $J = 1 \rightarrow 0$ transitions with the VLA and 1/3 in the 86 GHz $J = 2 \rightarrow 1$ transitions with ALMA. The assumption is that the detectability in either set of transitions is similar, and thus that a similar sensitivity in either set will sample a similar volume of the Galaxy. To check this assumption, we therefore used the Australia Telescope Compact Array (ATCA) to observe a (bright) subsample quasi-simultaneously at 43 and 86 GHz. The conclusions drawn by Stroh *et al.* (2018) are that the assumption in general holds, but it does depend on whether the 43 GHz $J = 1 \rightarrow 0$ (v=3) transition is detected or not.

In order to obtain proper SEDs for each source, it is important to identify the actual cross-matched source in the diverse (near-)infrared catalogs. As described by Pihlström *et al.* (2018), using the position of the parent *MSX* catalog directly makes it less reliable to securely find the photometry in other catalogs, compared to using the counterpart position from the 2MASS (Skrutskie *et al.* 2006), WISE (Wright *et al.* 2010) and *Gaia* DR2 (Gaia Collaboration *et al.* 2018). For follow-up work where the most accurate positions are needed, such as VLBI, the most useful initial astrometric positions are provided by 2MASS, or *Gaia* when available (i.e., if the source is not obscured in the optical).

With no effort to exclude C-rich CSEs (that would not necessarily be expected to emit in the SiO line), these sources may still contribute to the line-of-sight velocities of the entire sample by the detection of transitions of C-bearing molecules captured by the instrumental setup. Overall, a few percent of our detections are of C-rich sources and Lewis *et al.* (in prep.) are in the process of characterizing this sample.

2.2. *First Results on Galactic Dynamics*

Based on the first ∼2 700 SiO maser detections, Trapp *et al.* (2018) have analyzed the kinematics of the sample, which include a substantial number of line-of-sight velocities in high extinction regions within ±1° of the Galactic plane. They confirm that our radio-detected sample is consistent with Mira variables and mass-losing AGB stars, as anticipated using the *MSX* color selection. In Figs 2 and 3 they clearly distinguish two kinematic populations: a kinematic "cold" (small velocity dispersion) population proposed to be in the foreground Disk, and a kinematic "hot" (large velocity dispersion) candidate Bulge/Bar population. Only the kinematically hot giants include the reddest stars. Adopting 8.3 kpc to the Galactic center, and correcting for foreground

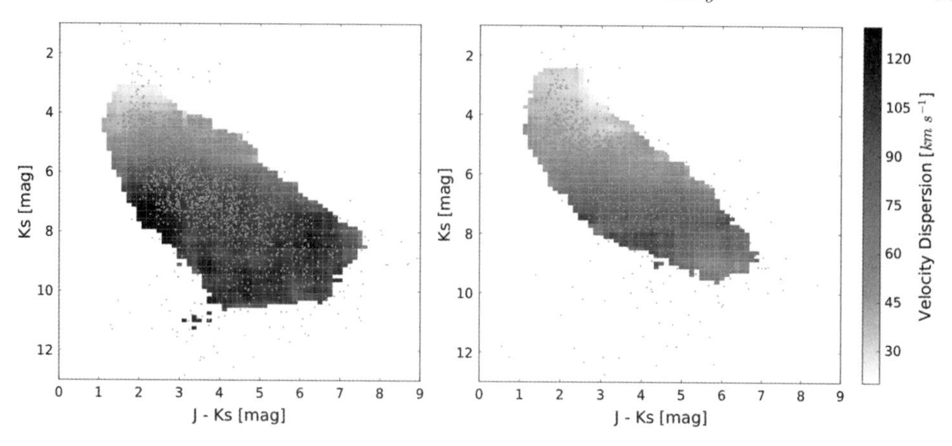

Figure 3. Color-magnitude dispersion diagrams of the VLA (left) and ALMA (right) samples. The clear transition from light to darker regions ($K > 5.5$ and $K > 6$, respectively) indicate a larger dispersion and probably is a signature of stars belonging to the Disk (bright and tight rotation) transitioning to stars making up the Bulge (fainter and dispersed). [Trapp *et al.* (2018)]

extinction, they find that most of the sources have $M_{\rm bol} \approx -5$, consistent with luminous intermediate-age AGB stars.

3. Conclusion

The BAaDE survey is well under way and promises excellent, new data for studying the dynamics and evolution of the Galactic Bulge by combining line-of-sight velocities measured from SiO maser emission in red giant stars with the infrared properties of their circumstellar environment.

Acknowledgements

The Bulge Asymmetries and Dynamic Evolution (BAaDE) project is funded by National Science Foundation Grants 1517970 (UNM) and 1518271 (UCLA).

References

Brunthaler, A., Reid, M.J., Menten, K.M., *et al.* 2004, *astro-ph*, 1102.5350

Egan, M.P., Price, S.D., *et al.* 2003, *"The Midcourse Space Experiment Point Source Catalog, Version 2.3"*, Air Force Research Laboratory Technical Report AFRL-VS-TR-2003-1589

Prusti, T., de Bruijne, J.H.J., Brown, A., *et al.* 2016, *A&A*, 595, A1

Brown, A.G.A., Vallenari, A., Prusti, *et al.* 2018, *A&A*, Special Issue on Gaia Data Release 2

Pihlström, Y.M., Sjouwerman, L.O., Morris, M.R., *et al.* 2018, *submitted*

Sevenster, M.N., Dejonghe, H., Van Caelenberg, K., Habing, H.J. 2000, *A&A*, 355, 537

Sjouwerman, L.O., Capen, S.M. & Claussen, M.J 2009, *ApJ*, 705, 1554

Skrutskie, M.F., Cutri, R.M., Weinberg, R., *et al.* 2006, *AJ*, 131, 1163

Stroh M.C., Pihlström, Y.M., Sjouwerman, L.O., *et al.* 2018, *ApJ*, 862, 153

Trapp, A.C., Rich, R.M., Morris, M.R., *et al.* 2018, *ApJ*, 861, 75

Wright, E.L., Eisenhardt, P.R.M., Mainzer, A.K., *et al.* 2010, *AJ*, 140, 1868

Posters

S343-Nakagawa
DivF-Mackebrandt
Astrometric observation of the Galactic LPVs with VERA; Miras and OH/IR stars

UNRAVELLING THE SULPHUR CHEMISTRY
OF (OXYGEN-RICH) AGB STARS
LOW MASS-LOSS RATE,
SEMI-REGULAR VARIABLE
HIGH MASS-LOSS RATE, MIRA VARIABLE

Imaging Red Supergiants with
VLT/SPHERE/ZIMPOL

Dust Structure Around Asymptotic Giant Branch Stars
Introduction
Objectives
Region of Interest
Inclination Angle
Methodology

WIN C-star model grid with new dust opacities
Kjell Eriksson, Susanne Höfner, Bernhard Aringer
Dust grain opacity
Binary interaction along the Red Giant Branch
The Barium Star perspective

Does 3DUP reduce AGB

star atmospheres modeling:
back to stellar evolutionary models
VLTI/GRAVITY
Exploring dust mass and dust
properties of nearby AGB stars
Introduction

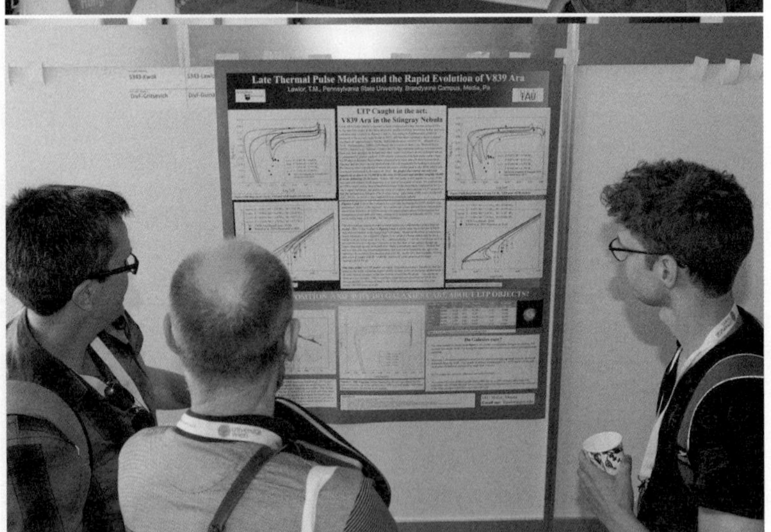
Late Thermal Pulse Models and the Rapid Evolution of V839 Ara
Lawlor, T.M., Pennsylvania State University, Brandywine Campus, Media, Pa
LTP Caught in the act:
V839 Ara in the Stingray Nebula

Why Galaxies Care About AGB Stars: A Continuing Challenge through Cosmic Time
Proceedings IAU Symposium No. 343, 2019
F. Kerschbaum, M. Groenewegen & H. Olofsson, eds.

doi:10.1017/S1743921318005021

M 1–92 revisited: the chemistry of a common envelope nebula?

Javier Alcolea[1], Marcelino Agúndez[2], Valentín Bujarrabal[1], Arancha Castro Carrizo[3], Jean-François Desmurs[1], Carmen Sánchez-Contreras[4] and Miguel Santander-García[1]

[1]Observatorio Astronómico Nacional (IGN/CNIG), Alfonso XII 3, E-28014 Madrid, Spain
[2]Group of Molecular Astrophysics (IFF, CSIC), Serrano 123, E-28006 Madrid, Spain
[3]IRAM, 300 Rue de Piscine, Domaine Universitaire, F-38406 S^{nt}Martin d'Héres, France
[4]Centro de Astrobiología (CSIC-INTA), ESAC, E-28691 Villafranca del Castillo, Spain

Abstract. We report on new molecular-line observations of the bipolar pre-planetary nebula M 1–92. The new IRAM 30 m MRT and NOEMA data shows the presence of shock induced chemistry in the nebula. From the derived [^{17}O]/[^{18}O] ratio, we suggest that the sudden mass loss event responsible for the formation of the nebula 1200 yr ago may also have resulted in the premature end of the AGB phase of the central star.

Keywords. stars: AGB and post-AGB, stars: evolution, stars: individual (M 1–92)

1. Introduction

M 1-92 is a $5'' \times 11''$ bipolar pre-planetary nebula with a 18,000 K (Sánchez-Contreras *et al.* 2008) central star and 10^4 L$_\odot$ at a distance of 2.5 kpc (Cohen & Kuhi 1977). The nebula is a bi-lobed structure divided by an equatorial flat disk, where most of the material is molecular gas, $\sim$ 0.9 M$_\odot$ (Bujarrabal *et al.* 1998a). ^{13}CO maps show that the nebula is dominated by a linear velocity gradient, most likely resulting from a sudden (common-envelope like) event occurred 1200 yr ago (Alcolea *et al.* 2007). Optical spectroscopy reveals the presence of a fast ionized wind very close to the star ($V_{\rm exp}$ up to 750 km s^{-1}; Sánchez-Contreras *et al.* 2008). Hα, H$_2$, OI, NII, OIII, SII are detected from compact knots located in the middle of the two lobes, revealing the existence of shocks along the symmetry axis, but amounting to just 10^{-3} M$_\odot$ (Bujarrabal *et al.* 1998b).

2. New observations

We have performed a full frequency scan of M 1–92 in the 3, 2, and 1.3 mm bands using the IRAM 30 m MRT, detecting for the first time C^{18}O, C^{17}O, HCO$^+$, H^{13}CO$^+$, HCN, H^{13}CN, CN, HNC, N$_2$H$^+$, SiO, SO, ^{34}SO, SO$_2$, ^{34}SO$_2$, SH$_2$, ^{34}SH$_2$, CS, NS and SO$^+$. Line profiles are of three kinds. CO and isotopologues show very similar broad profiles, –48 to +48 km s^{-1}, originating from the whole molecular nebula. Si– and S–bearing molecules only show emission from the central velocities, –20 to +20 km s^{-1}, suggesting that they arise from the equatorial component dividing the two lobes. Finally, HCO$^+$, HCN, HNC, N$_2$H$^+$, and CN show a triple peaked shape, only expected if the emission comes from the equator and tips of the nebula, but not from the lobe walls.

We have also conducted radio-interferometric observations, 8 GHz-wide, centered at 160, 176, 223, and 239 GHz, using the IRAM NOEMA interferometer, with resolutions of 0$''$.6–0$''$.8 and 2.5–3.5 km s^{-1}. We covered ^{13}CO, C^{18}O, C^{17}O, HCO$^+$, H^{13}CO$^+$, HCN,

$H^{13}CN$, and CN 2–1 lines, and several transitions of SO, SO_2, and SO^+. The maps confirmed the expected location of the different emissions according to their profile shapes, but with some surprises. S-bearing species trace two components, the outer parts of the equatorial disk and a very compact one close to the star. This compact component is more prominent in higher excitation lines, and shows a very low velocity dispersion of $5\,\mathrm{km\,s^{-1}}$. As expected, HCO^+, HCN, and CN are detected in the outer parts of the equatorial structure and at both axial tips. However, HCO^+ and HCN are also detected in the middle of the two lobes, at $\pm$ 2.″5 from the central star, just ahead of the compact knots seen in optical forbidden lines, a region devoid of CO. This component shows a large velocity dispersion, up to $60\,\mathrm{km\,s^{-1}}$, and a lower kinematic age of 600 yr. All this suggest that these species are the result of a shock-induced non-equilibrium chemistry.

3. Results

Combining 30 m MRT single-dish data and the maps obtained with NOEMA, we estimate excitation and abundances for several species in M 1–92, using rotational diagrams in the optically thin approximation. We divide the molecular line emission in two bins, the line core with $V_{\rm exp} \leqslant 20\,\mathrm{km\,s^{-1}}$, and the line wings with $V_{\rm exp}$ between 20 and $50\,\mathrm{km\,s^{-1}}$. All species show low excitation temperatures, $\sim$10–15 K, in agreement with previous results (Bujarrabal *et al.* 1998a); only SO_2 shows a component with $T_{\rm exc} \approx 30$ K or higher. For a $[^{13}CO/H_2]$ ratio of 10^{-5}, we derive a total mass of $0.9\,M_\odot$. For the low velocity emission we derive abundances (relative to H_2) of 6 and $4 \cdot 10^{-8}$ for SO_2 and SO, and 7 and $3 \cdot 10^{-9}$ for HCO^+ and HCN, respectively. For the shocked regions detected in HCN and HCO^+ we cannot derive relative abundances since ^{13}CO is not co-spatial, but HCN shows twice the abundance of HCO^+, contrarily to what happens in the equatorial parts of the nebula. The SO_2 warm component must arise from the central compact component detected in the NOEMA maps.

From the $C^{17}O$ and $C^{18}O$ J=1–0 and 2–1 observations, we derive a $[C^{17}O]/[C^{18}O]$ abundance ratio of 1.62. Since no important fractionation effects are expected, we conclude that the $[^{17}O]/[^{18}O]$ isotopic ratio is 1.6. This ratio is constant along the AGB evolution, and it can be used to determine the initial mass of the star (de Nutte *et al.* 2017). A value of 1.6 gives an initial mass of 1.7 $M_\odot$ (values above 4–5 $M_\odot$ are excluded because they imply a much faster post-AGB evolution Gesicki *et al.* 2014). For a $1.7\,M_\odot$ initial mass, it is expected that the 3$^{\rm rd}$ Dredge-Up results in the formation of a C-rich star. However, M 1–92 is O-rich (OH masers, strong SO and SO_2 lines, low $[^{12}C]/[^{13}C]$ ratio). One way to solve this apparent contradiction is assuming that the sudden $0.9\,M_\odot$ mass ejection that resulted in the formation of the nebula 1200 yr ago, prematurely ended the AGB evolution of the star, before it turned into C-rich.

Acknowledgements

This work is supported by the research grants AYA2016-78994-P and AYA2016-75066-C2-1-P (Spanish Ministry of Science, Innovation, and Universities).

References

Alcolea, J., Neri, R., & Bujarrabal, V. 2007, *A&A*, 468, L41
Bujarrabal, V., Alcolea, J., & Neri, R. 1998, *ApJ*, 504, 915
Bujarrabal, V., Alcolea, J., Sahai, R., Zamorano, J., & Zijlstra, A.A. 1998, *A&A*, 321, 361
Cohen, M. & Kuhi, L.V. 1977, *ApJ*, 213, 79
de Nutte, R., Decin, L., Olofsson, H., *et al.* 2017, *A&A*, 600, A71
Gesicki, K., Zijlstra, A.A., Hajduk, M., Szyszka, C. 2014, *A&A*, 566, A48
Sánchez-Contreras, C., Sahai, R., Gil de Paz, A., & Goodrich, R. 2008, *ApJS*, 179, 166

Why Galaxies Care About AGB Stars: A Continuing Challenge through Cosmic Time
Proceedings IAU Symposium No. 343, 2019
F. Kerschbaum, M. Groenewegen & H. Olofsson, eds.

doi:10.1017/S1743921318003812

The Evolutionary State of CEMP Stars

Johannes Andersen[1,2] **and Birgitta Nordström**[1,2]

[1]Dark Cosmology Centre, The Niels Bohr Institute, University of Copenhagen, Juliane Maries Vej 30, DK-2100 Copenhagen, Denmark.

[2]Stellar Astrophysics Centre, Aarhus University, Aarhus, Denmark. DK–2100 Copenhagen, Denmark
emails: ja@nbi.ku.dk, birgitta@nbi.ku.dk

Abstract. The standard scenario for the production of carbon-enhanced extremely metal-poor (CEMP) stars requires a more massive binary companion, which has evolved through the AGB stage and transferred carbon-rich material to the surface of the surviving, likewise extremely metal-poor (EMP) star. Evidently, the binary companion plays a key role in this process.

In order to characterise the polluting star, if any, the stage of evolution of the observed star (whether RGB or AGB), and whether pulsations exist, must be known. The Gaia DR2 parallaxes and photometry should contain the answer.

Keywords. Galaxy: halo, stars: Population II, stars: AGB and post-AGB, stars: carbon, stars: distances, stars: evolution.

---

1. Introduction

In 2015–2016, systematic long-term radial-velocity monitoring with high precision ($\sim$100 m s $^{-1}$) had surprisingly shown (Hansen *et al.* 2015, 2016a,b) that extremely metal-poor (EMP) stars are basically all single (i.e. have a Pop I normal binary fraction). Stars strongly enhanced in *r*-process elements, or in carbon without a similarly strong enhancement in *s*-process elements, all showed this. In stark contrast, EMP stars strongly enhanced in both carbon and *s*-process elements were nearly all binaries, interspersed with a small sprinkling of certified single stars. But no evidence was then available whether they were RGB or AGB stars. The second release of data from ESA's astrometric satellite Gaia has now changed that situation, providing individual distances for a subset of the brightest of those stars.

2. Previous Observations

The first step was a survey of the detailed composition of a sample of EMP stars. This requires an 8-metre-class telescope with an efficient high-resolution spectrograph, and we used the ESO VLT and UVES to identify stars in which a pronounced excess of carbon and/or of elements produced by neutron-capture processes might exist.

In contrast, precise and consistent binary identification by repeated determination of radial velocities over several years requires much less light for the individual observation by cross-correlation of low-S/N spectra, but a stable and repeatable instrumental setup over several years is essential. For this, we consequently used the 2.5-metre Nordic Optical Telescope (NOT) and its bench-mounted echelle spectrograph FIES in a separate, temperature-controlled underground vault.

Finally, determination of small stellar parallaxes with micro-arcsecond precision is only feasible in a space environment, free of atmospheric turbulence. For this, Gaia without competition–even in DR2, which is based on only two years of data.

3. Results of the Gaia Data

The Gaia DR2 trigonometric parallaxes and resulting distances for the 64 brightest stars show unambiguously that these stars are "normal" bright giants (although still EMP stars, with [Fe/H] ~ -2.5). They have $M_V \sim 0.5$ mag and are thus not true AGB stars ($M_V \sim -5$ mag). The 'standard' scenario for the origin of these stars thus basically still holds; they are intrinsically EMP giants. However, whereas the chemical anomalies in the single stars were imprinted on the material from which they formed (i.e., from birth), the binary CEMP-*s* stars acquired theirs by mass transfer from the companion to the surface layers of the star we see today. Thus, the mystery remains: How did the CEMP-no and the single CEMP-*s* stars acquire their marked carbon excesses in the absence of any AGB or other massive binary companions? Standard (local) chemical evolution models seem not to include this form of mass transfer across interstellar distances.

The Gaia DR2 data do include *G*, blue, and red photometry of micro-magnitude precision, but the DR2 time basis is still too short to allow conclusions on the possible presence and period of any pulsations in these stars (Hansen *et al.* 2016a,b).

References

Hansen, T.T., Andersen, J., Nordström, B. *et al.*, 2015, *The role of binaries in the enrichment of the early Galactic halo. I. The r-process enhanced metal-poor stars*, *A&A*, 583, A49

Hansen, T.T., Andersen, J., Nordström, B. *et al.*, 2016a, *The role of binaries in the enrichment of the early Galactic halo. II. Carbon-Enhanced Metal-Poor Stars - the CEMP-no stars*, *A&A*, 586, A160

Hansen, T.T., Andersen, J., Nordström, B. *et al.*, 2016b, *The role of binaries in the enrichment of the early Galactic halo. III. Carbon-Enhanced Metal-Poor Stars - the CEMP-s stars*, *A&A*, 588, A3

Why Galaxies Care About AGB Stars:
A Continuing Challenge through Cosmic Time
Proceedings IAU Symposium No. 343, 2019
F. Kerschbaum, M. Groenewegen & H. Olofsson, eds.

doi:10.1017/S1743921318007536

RAMSES II
Raman Search for Extragalactic Symbiotic Stars

Rodolfo Angeloni[1], Denise R. Gonçalves[2], Ruben J. Diaz[3] and the RAMSES II Team

[1]Instituto de Investigación Multidisciplinar en Ciencia y Tecnología, Universidad de La Serena, La Serena, Chile
email: `rangeloni@userena.cl`

[2]Observatório do Valongo, Universidade Federal do Rio de Janeiro, Rio de Janeiro, Brasil

[3]Gemini Observatory, Southern Operations Center, La Serena, Chile

Abstract. Symbiotic stars (SySts) are long-period interacting binaries composed of a hot compact star, an evolved giant star, and a tangled network of gas and dust nebulae. Presently, we know 252 SySts in the Milky Way and 62 in external galaxies. However, these numbers are still in striking contrast with the predicted population of SySts in our Galaxy. In this contribution, I present the concept and the early results from RAMSES II (Raman Search for Extragalactic Symbiotic Stars), a Gemini/GMOS Upgrade Project which makes use of the Raman OVI 6830Å band as a powerful photometric tool to identify new SySts, within and beyond the Galaxy.

Keywords. stars: AGB and post-AGB – binaries: symbiotic – techniques: photometric

---

1. Introduction

Symbiotic stars (hereafter SySts) are long-period interacting binaries composed of a hot compact star – generally a white dwarf (WD) – an evolved giant star, and a tangled network of gas and dust nebulae. SySts represent unique laboratories to study several important astrophysical phenomena and their reciprocal influence (Munari (2012)). Noteworthy, they are among the most promising candidates as progenitors of SNIa (e.g., Meng & Han 2016, Dimitriadis *et al.* 2014, Dilday *et al.* 2012).

Presently, we know 252 SySts in the Milky Way and 62 in external galaxies (Akras *et al.* 2018, submitted). However, the slowly growing number of known SySts is still in striking contrast with the predicted symbiotic population in our Galaxy that, according to different estimates, may oscillate between 10^3 (Allen 1984, Lü *et al.* 2012) and a few 10^5 (Magrini *et al.* (2003)). One of the reasons for this embarrassing discrepancy likely originates from the fact that, historically, the SySt group has been characterized on the basis of purely spectroscopic criteria (e.g., Belczyński *et al.* 2000).

Because of many other stellar sources that appear mimicking SySt colors (PNe, Be and T Tauri stars, CVs, Mira LPVs, etc. – see, e.g., Figs. 1 & 2 in Corradi *et al.* 2008), no photometric diagnostic tool has so far demonstrated the power to unambiguously identify a SySt, thus making the recourse to costly spectroscopic follow-up still inescapable.

The two intense Raman OVI bands at 6830Å and 7088Å, due to Raman scattering of the O VI $\lambda\lambda$1032, 1038 resonance doublet by neutral H (Schmid 1989), are nonetheless so unique to the symbiotic phenomenon that their presence has been used as a sufficient criterion for classifying a star as symbiotic, even in those cases where the cool companion appears to be hiding. Our team has thus submitted in 2016, and then been awarded, a

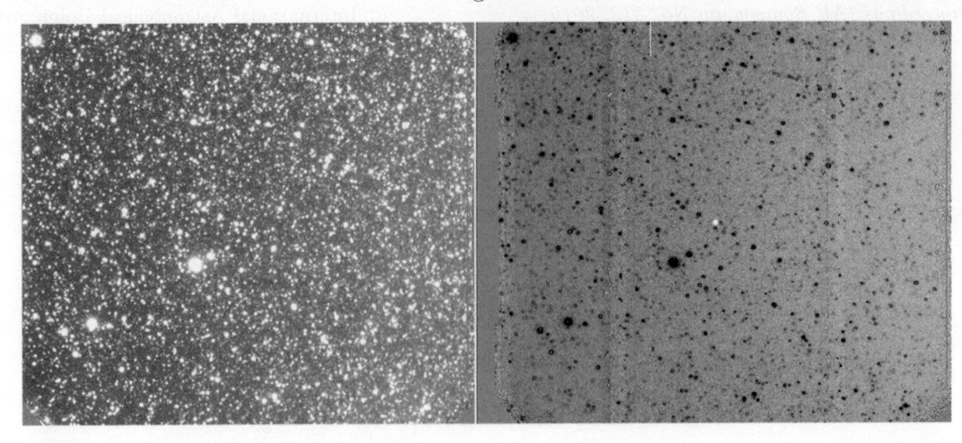

Figure 1. GMOS-South RAMSES II OVI (left) and OVI continuum-subtracted (right) images of Sanduleak's star, a Raman-emitter SySt in the Large Magellanic Cloud.

Gemini Instrument Upgrade Project with the aim of using Raman OVI emission as a new photometric diagnostic tool to systematically discover new SySts, within and beyond the Galaxy. Specifically, we have proposed to purchase two narrow-band filters for the GMOS at both Gemini telescopes: one filter centered on the Raman OVI 6830Å band, the other one centered in the adjacent continuum at 6780Å to be used as off-band filter.

2. Early results from the filter acceptance test at Gemini-South

After having gone through several tests at the optical labs of the Gemini Observatory at Cerro Pachón (Chile), the filters have been installed into GMOS-S at the beginning of 2018 for a series of day and on-sky Acceptance Tests (AT). In particular, a small sample of SySts with known Raman OVI bands have been observed in order to characterize the sensitivity of our method to Raman OVI emission of different intensities against different types of local continua. An example of GMOS-S/RAMSES II images of Sanduleak's star (Heo *et al.* 2016), a SySt in the Large Magellanic Cloud, is shown in Fig. 1. The Acceptance Tests will continue at Gemini North, and after a full characterization at both sites the full set of filters will be offered to the entire Gemini user community.

References

Allen, D. A. 1984, Ap&SS, 99, 101

Belczyński, K., Mikołajewska, J., Munari, U., Ivison, R. J., & Friedjung, M. 2000, A&AS, 146, 407

Corradi, R. L. M., Rodríguez-Flores, E. R., Mampaso, A., *et al.* 2008, A&A, 480, 409

Dilday, B., Howell, D. A., Cenko, S. B., *et al.* 2012, Science, 337, 942

Dimitriadis, G., Chiotellis, A., & Vink, J. 2014, MNRAS, 443, 1370

Heo, J.-E., Angeloni, R., Di Mille, F., Palma, T., & Lee, H.-W. 2016, ApJ, 833, 286

Lü, G.-L., Zhu, C.-H., Postnov, K. A., *et al.* 2012, MNRAS, 424, 2265

Magrini, L., Corradi, R. L. M., & Munari, U. 2003, Symbiotic Stars Probing Stellar Evolution, 303, 539

Meng, X., & Han, Z. 2016, A&A, 588, A88

Munari, U. 2012, Journal of the American Association of Variable Star Observers (JAAVSO), 40, 572

Schmid, H. M. 1989, A&A, 211, L31

Why Galaxies Care About AGB Stars:
A Continuing Challenge through Cosmic Time
Proceedings IAU Symposium No. 343, 2019
F. Kerschbaum, M. Groenewegen & H. Olofsson, eds.

doi:10.1017/S1743921319002291

Observational Properties of Miras in the KELT Survey

R. A. Arnold[1], M. Virginia McSwain[1], Joshua Pepper[1], Keivan G. Stassun[2] and the KELT Collaboration

[1]Dept. of Physics, Lehigh University, 16 Memorial Drive East, Bethlehem, PA 18015
emails: raa314@lehigh.edu, mcswain@lehigh.edu, joshua.pepper@lehigh.edu

[2]Dept. of Physics and Astronomy, Vanderbilt University,
2301 Vanderbilt Place, Nashville, TN 37235
email: keivan.stassun@vanderbilt.edu

Abstract. We present a catalog of the observed properties of Mira-type variable stars detected with the Kilodegree Extremely Little Telescope (KELT). Asymptotic giant branch (AGB) candidates were identified in KELT using a combination of photometric data from KELT and 2MASS colors. Of the 4 million objects with KELT photometry, 3332 Mira-like variables were identified. Here, we present their observed periods and luminosities which will place important constraints on future theoretical work on the effect convection has on pulsation periods and mode stability.

Keywords. stars: AGB and post-AGB, stars: variables: other, techniques: photometric

1. Introduction

Miras are large amplitude, long period, luminous variables near the tip of the asymptotic giant branch (AGB). They have the potential to serve as distance indicators as they follow a period-luminosity (PL) relation (Whitelock *et al.* 2008 and references therein). A catalog of galactic Miras may serve as an important tool in calibrating this PL relation and improving our understanding of the properties of Miras.

Our study uses photometric data from the Kilodegree Extremely Little Telescope (KELT), which is a robotic single-band photometric survey designed to detect exoplanet transits around stars with magnitudes of $8 < V < 13$. It has high photometric precision and cadence, and a long time baseline with the earliest observations dating from 2005.

2. Discussion

To find Miras we selected candidate AGB stars using $H - K_s$ and $J - H$ colors from 2MASS. Color criteria for selecting candidate AGBs were chosen using 2MASS colors of Miras from the GCVS and AAVSO catalogs. We tested light curves of candidate AGBs for coherent variability using the Stetson L (Stetson 1996) and the alarm variability statistic A (Tamuz *et al.* 2006). We adopted the criteria of $L > 1$ and $A > 50$ to select variable AGB candidates. For the selected stars, we then calculated the periodicity of the KELT light curves using the Lomb Scargle (LS) algorithm (Zechmeister & Kurster 2009; Press *et al.* 1992). We searched for periodic signals between 10 and 1000 days with a frequency step size of $\Delta f = 0.1/T$, where T is the length of observations. The period associated with the highest LS power was selected.

We selected Miras using their amplitudes and the strength of the periodic signal in light curves measured by the LS power. Miras are conventionally defined to have amplitudes of $\Delta V > 2.5$ mag, but at longer wavelengths such as those measured by KELT's R-band

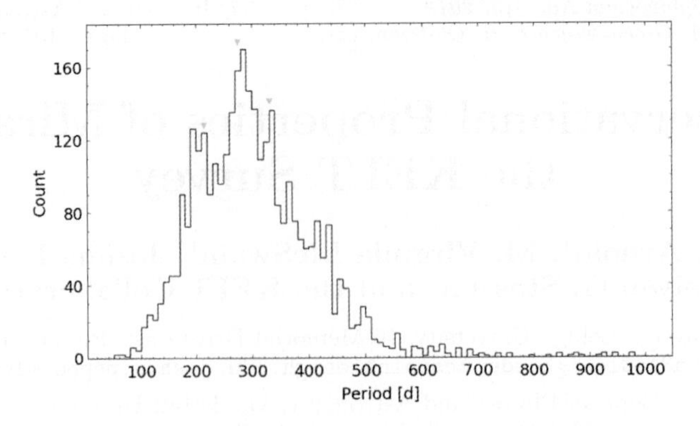

Figure 1. Period distribution of our Miras. The three red triangles mark period peaks found by Vogt *et al.* (2016)

filter their amplitudes decrease. We adopt the following criteria for identifying Miras: a LS power > 0.7, and a measured light curve amplitude > 1.5 mag.

We found 3332 Mira-like objects in the KELT survey. Of these, 258 Mira-like objects have not been included in previously published literature. The distribution of periods is shown in Fig. 1. For those stars also found by Vogt *et al.* (2016), we found good agreement between our periods and theirs. Of the 2875 Vogt *et al.* (2016) objects, 1415 were observed by KELT, and we identified 752 of them in our Mira-like catalog. Most of the failed matches were at the faint end of our magnitude range, but we recover 73% of the Vogt *et al.* (2016) targets with $V < 10$ mag.

We calculated absolute magnitudes of our Miras using 2MASS K_s magnitudes and preliminary distances derived from Gaia DR2 by Bailer-Jones *et al.* 2018. The absolute magnitudes were transformed to luminosities using color-dependent bolometric corrections from Houdashelt *et al.* (2000) and Kerschbaum *et al.* (2010). The typical error of the luminosities is a factor of 2 due to the imprecision of the Gaia DR2 parallaxes. This is mostly the result of their variability and their angular diameters being comparable to their parallaxes. In the future, we hope to compare the PL relation of our sample with theoretical results from MESA and GYRE. As this PL relation is better understood, Miras may serve as important extragalactic distance indicators in the future. We also plan to investigate the effects of convection on the theoretical PL relation through implementing turbulent convection in the open-source GYRE pulsation code, which can be used in conjunction with the MESA stellar evolution code.

References

Bailer-Jones, C. A. L., Rybizki, J., Fouesneau, M., *et al.* 2018, *AJ*, 156, 58
Green, G. M. *et al.* 2018, *MNRAS*, 478, 651
Houdashelt, M. L., Bell, R. A., Sweigart, A. V., & Wing, R. F. 2000, *AJ*, 119, 1424
Kerschbaum, F., Lebzelter, T. & Mekul, L. 2010, *A&A*, 524, 87
Paxton, B., *et al.* 2011, *ApJS*, 192, 3
Press, W.H., Teukolsky, S.A., Vetterling, W.T. & Flannery, B.P. 1992, Numerical Recipes in C, 2nd ed. (New York: Cambridge University Press)
Stetson, P. B. 1996, *PASP*, 108, 851
Tamuz, O., Mazeh T. & North P. 2006, *MNRAS*, 367, 1521
Townsend, R. H. D. & Teitler, S. A. 2013, *MNRAS* 435, 3406
Vogt, N., Contreras-Quijada, A., Fuentes, Morales, I., *et al.* 2016, *ApJS*, 227, 6
Whitelock, P. A. 2012, *Ap&SS*, 341, 123
Whitelock, P. A., Feast, M. W., & Van Leeuwen, F 2008, *MNRAS*, 386, 313
Zechmeister, M. & Kürster, M. 2009, *A&A*, 496, 577

Why Galaxies Care About AGB Stars: A Continuing Challenge through Cosmic Time
Proceedings IAU Symposium No. 343, 2019
F. Kerschbaum, M. Groenewegen & H. Olofsson, eds.

doi:10.1017/S1743921318006580

s-process abundances of Primary stars in the Sirius-like Systems: Constraints on pollution from AGB stars

Y. Bharat Kumar[1], X-M. Kong[2], G. Zhao[1] and J-K. Zhao[1]

[1]Key Laboratory of Optical Astronomy, National Astronomical Observatories, Chinese Academy of Sciences, Beijing 100101, China
email: bharat@bao.ac.cn

[2]School of Mechanical, Electrical and Information Engineering, Shandong University at Weihai, Weihai 264209, China

Abstract. We present the results from the abundance analysis of 21 primary stars in Sirius-like systems with various masses of white dwarf companions and orbital separation to understand the origin and nature of Ba stars. Three new Ba dwarfs are found for which masses are relatively low compared to Ba giants. Large fraction of the sample are found to be non-Ba stars, however, some of them have required WD mass and/or close orbital separation. Observed s-process abundances in Ba dwarfs are in good agreement with AGB models of respective WD companion mass, however, it required different pollution factors.

Keywords. stars: binaries (including multiple): close, stars: abundances, stars: AGB

1. Introduction

Sirius-like systems (SLSs) refer to white dwarfs (WDs) in binary or multiple star systems. They contain at least a less luminous companion of spectral type K or earlier (Holberg *et al.* 2013), which provide direct evidences to understand the origin and nature of Barium II (Ba) stars. Ba stars are G-K type giants and dwarfs that show excess of Ba and other neutron capture elements in their atmosphere. S-process elements are known to be present in the atmospheres of AGB stars, but Ba stars are not evolved enough to be self enriched like them. Therefore, s-process enrichment in Ba stars is proposed due to mass transfer within a binary system from its companion AGB star (now WD). The degree of chemical contamination in Ba stars is suggested to depend on the mass of the WD companion, orbital separation, and metallicity. However, the minimum WD mass limit of 0.51 $M_{\odot}$ is set to be its progenitor to evolve as AGB star. Other parameters, such as eccentricity, mass-loss mechanism, the efficiency of thermal pulses and dilution factors, also play a role in observed chemical peculiarities of Ba stars (see Kong *et al.* 2018b and references therein). Here, a systematic abundance analysis is performed for 21 FGK primary stars (15 dwarfs and 6 giants) of SLSs selected from Holberg *et al.* (2013).

2. Elemental abundances

Abundances are derived using measured equivalent widths of selected lines in high quality (SNR > 100) high resolution spectra, Kurucz model atmospheres, and ABONTEST8 program supplied by Dr. P. Magain (see Kong *et al.* (2018a) for details). Atmospheric parameters like effective temperature ($T_{\rm eff}$), surface gravity ($\log g$), metallicity ([Fe/H]) and microturbulence (ξ_t), are determined in standard procedure., e.g. imposing

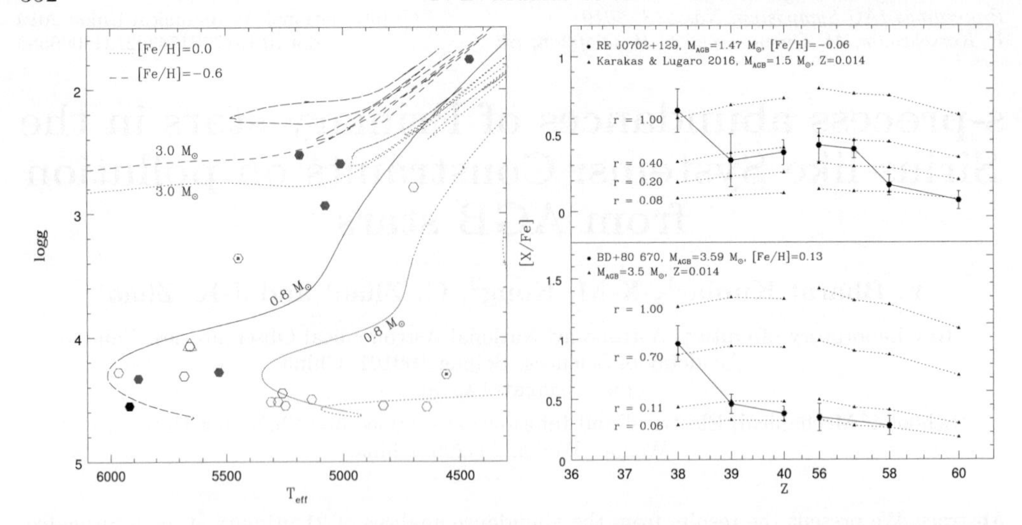

Figure 1. Left: Samples are on the HR-Diagram. Open and closed symbols are Ba ([s/Fe] > 0.25) and non-Ba stars. Red and blue represents WDs with mass less and greater than $0.51\,M_\odot$, respectively. Triangles represent stars with $M_{wd} > 0.51$ and close orbital separation. Right: Observed s-process enrichment of sample Ba dwarfs are compared with AGB models with different pollution factors.

excitation equilibrium from Fe I lines for T_{eff} and ionization equilibrium method from Fe I and Fe II lines for log g. Differential abundances ([X/Fe]) are obtained relative to solar values, derived from solar spectra observed along with programme stars (Kong *et al.* 2018b).

3. Results and Discussion

The abundance analysis for 21 primaries resulted in three new Ba dwarfs, 4 Ba giants, and 14 non-Ba stars. The location of these stars on the HR diagram suggests Ba dwarfs ($< 1.0\,M_\odot$) and relatively lower in mass compared to Ba giants ($> 1.5\,M_\odot$). Among the sample, two stars with WD mass $< 0.51\,M_\odot$ show the absence of s-process enrichment, which confirms the predictions and observations in the literature. Interestingly, 11 stars whose WD masses $> 0.51\,M_\odot$ show absence of s-process enrichment, suggesting the WD mass alone can't decide Ba nature. Also, the existence of a non-Ba star whose WD mass $> 0.51\,M_\odot$ and with an orbital separation of 58 AU suggests further investigation on pollution and accretion efficiency (Fig. 1, left). The observed s-process abundances for two Ba dwarfs are compared with AGB models (Karakas & Lugaro 2016) of respective companion masses which are estimated from their WD mass including different pollution factors (Kong *et al.* 2018a) (Fig. 1, right). This suggests further understanding of pollution factors which could constrain s-process enrichment.

Acknowledgements

We thank IAU for providing the partial financial support to present this work.

References

Holberg J. B., Oswalt T. D., Sion E. M., Barstow M. A., Burleigh M. R. 2013, *MNRAS*, 435, 2077

Karakas A. I., & Lugaro M., 2016 *ApJ*, 825, 26

Kong X. M., Bharat Kumar, Y., Zhao, G., Zhao, J. K., Fang, X. S., Shi, J. R., Wang, L., Zhang, J. B., Yan, H. L. 2018a, *MNRAS*, 474, 2129

Kong X. M., Zhao, G., Zhao, J. K., Shi, J. R., Bharat Kumar, Y.,Wang, L., Zhang, J. B., Zhou, Y. T. 2018a, *MNRAS*, 476, 724

Why Galaxies Care About AGB Stars:
A Continuing Challenge through Cosmic Time
Proceedings IAU Symposium No. 343, 2019
F. Kerschbaum, M. Groenewegen & H. Olofsson, eds.

doi:10.1017/S174392131800755X

On the nature and mass loss of Bulge OH/IR stars

Joris A. D. L. Blommaert[1], Martin A. T. Groenewegen[2], Kay Justtanont[3] and L. Decin,[4]

[1]Astronomy and Astrophysics Research Group, Department of Physics and Astrophysics, Vrije Universiteit Brussel, Pleinlaan 2, 1050 Brussels, Belgium

[2]Koninklijke Sterrenwacht van België, Ringlaan 3, 1180 Brussel, Belgium

[3]Chalmers University of Technology, Onsala Space Observatory, S-43992 Onsala, Sweden

[4]Instituut voor Sterrenkunde, K.U. Leuven, Celestijnenlaan 200D, 3001 Leuven, Belgium

Abstract. We report on the successful search for CO (2-1) and (3-2) emission associated with OH/IR stars in the Galactic Bulge. We observed a sample of eight extremely red AGB stars with the APEX telescope and detected seven. The sources were selected at sufficient high Galactic latitude to avoid interference by interstellar CO, which hampered previous studies of inner galaxy stars. We also collected photometric data and Spitzer IRS spectroscopy to construct the SEDs, which were analysed through radiative transfer modelling. We derived variability periods of our stars from the VVV and WISE surveys. Through dynamical modelling we then retrieve the total mass loss rates (MLR) and the gas-to-dust ratios. The luminosities range between approximately 4,000 and 5,500 $L_\odot$ and periods are below 700 days. The total MLR ranges between 10^{-5} and 10^{-4} $M_\odot$ yr^{-1}. The results are presented in Blommaert *et al.* 2018 and summarized below.

Keywords. Stars: AGB and post-AGB – Stars: mass-loss – circumstellar matter – dust – Galaxy: bulge – radio lines: stars

1. Introduction

Studying AGB stars in the Bulge gives the advantage of relatively well known distances (~ 8 kpc) and gives the opportunity to investigate not only the evolution on the Asymptotic Giant Branch but to shed light on Bulge stellar population and thus on the history of this central part of the Galaxy. So far, CO observations in this region had only limited success and concentrated on OH/IR stars near the Galactic centre. We selected a sample of stars at higher latitudes, ensuring their bulge membership and avoiding the strong interference from interstellar CO gas near the Galactic plane.

2. Results and conclusions

An example of the CO J= 2-1 and J= 3-2 transitions measurements with the APEX telescope is shown in Fig. 1.

On basis of our modelling of the observed SED and CO lines, we find that the stars have an average luminosity of 4729 ± 521 $L_\odot$ and the average MLR is (5.4 ± 3.0) 10^{-5} $M_\odot$ yr^{-1}. Such MLR is well above the classical limit, with a single scattering event per photon, for the luminosities in our sample. The gas-to-dust ratios are between 100 and 400 and are similar to what is found for OH/IR stars in the Galactic disk.

The variability periods of our OH/IR stars are below 700 days and do not follow the Mira-OH/IR PL relation (Whitelock *et al.* 1991).

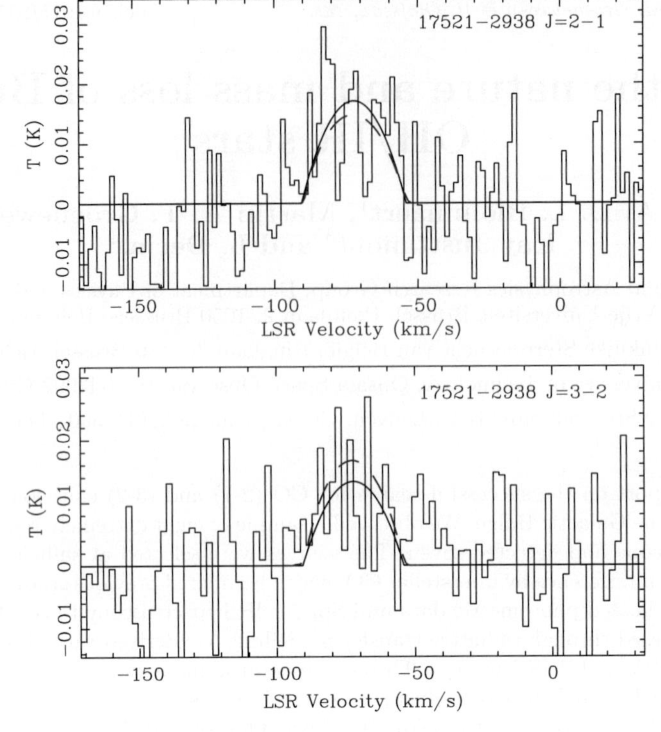

Figure 1. The APEX CO (2-1) and (3-2) line spectra, together with the line fits (in black) and the model predictions (dashed blue).

Comparison with the Vassiliadis & Wood (1993) evolutionary models shows that the progenitor mass of the bulge OH/IR stars is $\approx 1.5\,M_{\odot}$, similar to the bulge Miras and are of intermediate age (3 Gyr) (Groenewegen & Blommaert 2005). If more massive OH/IR stars are rare in the bulge this may explain the scarcity of bulge carbon stars.

Contrary to findings of bright OH/IR stars in the Galactic disk (Justtanont *et al.* 2013), our modeling does not impose a limit to the duration of the superwind below a thousand years.

One star, IRAS 17347-2319, has a very short period of approximately 300 days which may be decreasing further. It may belong to a class of Mira variables with a sudden change in period and may be connected to the occurance of a thermal pulse. It would be the first example of an OH/IR star in this class and deserves further follow-up observations.

References

Blommaert, J.A.D.L., Groenewegen, M.A.T., Justtanont, K., & Decin, L. 2018, *MNRAS*, 479, 3545

Groenewegen, M.A.T. & Blommaert, J.A.D.L. 2005, *A&A*, 443, 143

Justtanont, K., *et al.* 2013, *A&A*, 556, A101

Vassiliadis, E. & Wood, P. 1993, *ApJ*, 413, 641

Whitelock, P., Feast, M., Catchpole, R. 1991, *MNRAS*, 248, 276

Why Galaxies Care About AGB Stars: A Continuing Challenge through Cosmic Time
Proceedings IAU Symposium No. 343, 2019
F. Kerschbaum, M. Groenewegen & H. Olofsson, eds.

doi:10.1017/S1743921318006555

Understanding jets in post-AGB close binaries

Dylan Bollen[1,2], Devika Kamath[2,3], Hans Van Winckel[1] and Orsola De Marco[2,3]

[1]Instituut voor Sterrenkunde (IvS), KU Leuven, Celestijnenlaan 200D, B-3001 Leuven, Belgium
email: dylan.bollen@kuleuven.be

[2]Department of Physics & Astronomy, Macquarie University, Sydney, NSW 2109, Australia

[3]Astronomy, Astrophysics and Astrophotonics Research Centre, Macquarie University, Sydney, NSW 2109, Australia

Abstract. We have discovered jets in post-AGB binaries. The orbital motion allows us to carry out tomography of the jet as light from the primary star shines through the jet cone. Jets play a major role in many astrophysical environments, from young stellar objects to galaxies. They are also used to study the energetics of accretion phenomena in systems such as red transients and stellar mergers. We use high-resolution, optical, time-series spectra to constrain theories of jet launching, and the impact of jets on the evolution of these post-AGB binaries.

Keywords. Spectroscopy, post-AGB, jets and outflows, accretion, binaries

1. Introduction

Astrophysical jets are observed in many astronomical objects, ranging from young stellar objects to the extremely energetic jets from active galactic nuclei. Jets are key components in the shaping of non-spherical, collimated planetary nebulae and pre-planetary nebulae (Sahai & Trauger 1998). Although jets are a ubiquitous phenomenon in the Universe, the physical processes governing these energetic events remain poorly understood. Recently, we found that jets are commonly observed in post-AGB binary systems (Gorlova *et al.* 2012, Gorlova *et al.* 2015, Bollen *et al.* 2017).

Binaries with the primary on the asymptotic giant branch (AGB) can suffer mass transfer. Some enter a common envelope interaction, which shrinks the orbit and ejects the envelope, leaving a central compact binary and a collimated planetary nebula (De Marco 2009). Others seem to suffer less in-spiral for some unknown reasons and result in cooler post-AGB stars with companions in 100–1000-day orbits, dusty, circumbinary disks and, we now know, pervasive jets (Van Winckel 2003, Van Winckel *et al.* 2009). The jet originates from an accretion disk around the secondary, main sequence star component. In our previous work, we have analysed the observational data of post-AGB, jetting binary BD+46°442 (Bollen *et al.* 2017). We showed that these jets have wide opening angles (half-opening angles $< 40°$), with a fast, thin outflow along the jet axis and a slower, dense outflow at the jet walls. The source of the accretion might be from the circumbinary disk, not the primary, opening interpretation avenues for these systems. This study has revealed the potential of our spectroscopic data in deriving the jet structure, which gets to the heart of the launch mechanism.

Here, we aim at determining the spatio-kinematic structure of jets observed in post-AGB binary systems to get a more complete understanding of the accretion and jet launching processes.

2. Methodology and results

The high-resolution optical spectra (from the dedicated Mercator Telescope on La Palma; Raskin *et al.* 2011) show phase-dependent variations in the Balmer lines, which can be explained by the presence of a jet that emerges from an accretion disk around the companion. Due to the orbital motion of the binary system, the jet blocks the light coming from the primary component during each orbital period when it transits in front of the primary. The scattering of the continuum photons from the primary component by the hydrogen atoms in the jet gives us unique insight in the density and velocity structure in different parts of the jet. This tomography of the jet is a novel technique to determine its spatio-kinematic structure.

In this study, we focus on IRAS19135+3937, which is well-sampled throughout its orbital phase. We have written a code that creates a geometrical model of the jet configuration and computes a synthetic spectra from the model over the whole binary orbit, based on several parameters such as jet density, inner and outer velocity, jet opening angle, radius of the primary component, and inclination of the system. We implement two jet configurations in our model: an X-wind configuration (based on the X-wind model for protostellar outflows by Shu *et al.* 1994) and a disk wind configuration (based on the wind model by Blandford & Payne 1982). We fit these synthetic spectra to our observational data using an MCMC-fitting routine, in order to determine accurate parameters fully constraining the spatio-kinematic structure of the jets.

Both X-wind model and disk wind converge to the simplest jet configuration with a jet comprising a slow (10 km/s) dense cone with a fast (750–1050 km/s) low density core.

3. Conclusions

Theories of accretion and jet formation in general and their impact on post-AGB binary systems, planetary nebulae, and pre-planetary nebulae, in particular, are currently incomplete. Data of jets is hardly ever sufficient to fully constrain their parameters. Our data comprising more than 20 post-AGB objects with jets, thanks to the use of a dedicated telescope, has exquisite spectroscopic and temporal resolution. Eventually, we aim to determine the jet mass, mass-loss rate and disk accretion rates which will be key in a model of jet launching.

References

Blandford, R. D., & Payne, D. G. 1982, *MNRAS*, 199, 883
Bollen, D., Van Winckel, H., Kamath, D. 2017, *A&A*, 607, A60
De Marco, O. 2009, *PASP*, 12, 316
Gorlova, N., Van Winckel, H., Gielen, C., *et al.* 2012, *A&A*, 542, A27
Gorlova, N., Van Winckel, H., Ikonnikova, N.P., *et al.* 2015, *A&A*, 451, 2462
Raskin, G., Van Winckel, H., Hensberge, H., *et al.* 2011, *A&A* 526, A69
Van Winckel, H. 2003, *ARA&A*, 41, 391
Van Winckel, H., Lloyd Evans, T., Briquet, M., *et al.* 2009, *A&A*, 505, 1221
Sahai, R. & Trauger, J. T. 1998, *AJ*, 116, 1357S
Shu, F., Najita, J., Ostriker, E., *et al.* 1994, *ApJ*, 429, 781

Why Galaxies Care About AGB Stars:
A Continuing Challenge through Cosmic Time
Proceedings IAU Symposium No. 343, 2019
F. Kerschbaum, M. Groenewegen & H. Olofsson, eds.

doi:10.1017/S1743921318006348

To Be or Not to Be: EHB Stars and AGB Stars

David A. Brown

Specola Vaticana (Vatican Observatory),
V-00120, Cittá del Vaticano, Vatican City State, Europe
email: dbrown@specola.va

Abstract. The formation of EHB stars is linked to the lives of AGB stars by indications that such EHB/sdB stars might form in globular clusters with multiple populations linked to AGB evolution. Observations of massive globular clusters, such as ω-Centauri (Bedin *et al.* 2004, Piotto *et al.* 2005) suggest that single EHB stars might form from He-enhanced progenitors (D'Antona *et al.* 2005, D'Antona & Caloi 2008, Lee *et al.* 2005) in environments enriched by AGB ejecta. The studies conducted by Han *et al.* (2002), Han *et al.* (2003), and Han *et al.* (2007) have been able to provide a strong case for the binary formation of EHB/sdB stars in the Galactic field, though binary formation channels in globular clusters is uncertain. Simulations presented here are an extension of the simulations of Han *et al.* (2002) and Han *et al.* (2003), for low metallicities to examine the binary EHB population in globular clusters.

Keywords. EHB, AGB, helium-enrichment, star

1. Overview, Calculations, Results, and Conclusions

Various scenarios have been proposed by which EHBs may form. One compelling single star formation mechanism in globular clusters (GCs) for EHB (sdB) stars suggests that they are formed as the result of helium-enhanced ($Y > 0.25$, $(\Delta Y/\Delta Z) \sim 100$) progenitor stars (D'Antona *et al.* 2005, D'Antona & Caloi 2008), the result of earlier epochs of star formation in the cluster from gas enriched by the AGB ejecta, thought to be the case from observing multiple populations of stars in GCs such as ω-Centauri and NGC 2808 (Piotto *et al.* 2005), many displaying a Na-O anti-correlation (Carretta *et al.* 2006). Other single star formation scenarios invoke either high stellar winds during the RGB phase of evolution (D'Cruz *et al.* 1996) before the He-flash, or a delayed He-flash after rapid mass loss (Sweigart 1997). Then again, mass-loss scenarios because of binary interactions (Mengel, Morris & Gross 1976) could account for some EHBs through either common envelope (CE) ejection, or stable RLOF, or through the merger of two He WD stars in binaries as seen in studies done by Han *et al.* (2002) and Han *et al.* (2003). Here, detailed binary population synthesis calculations have been conducted by Brown following the methodology of Han *et al.* (2002) and Han *et al.* (2003) to examine EHB/sdB formation from binaries in GCs with multiple populations. A simple model GC is constructed with two populations having the same helium abundance but different metallicities $Z = 0.0001$ at 13 Gyr and $Z = 0.001$ for 1 and for 6 Gyr. In Fig. 2, the binary model produces EHB stars in the proper region of the CMD, but the fraction of binary EHBs is too high at $f \gtrsim 0.50$, even considering EHBs from the merger channel. This highlights the disparity in EHB/sdB binary fraction between the Galactic field ($f \sim 2/3$; Maxted *et al.* 2001) and in globular clusters ($f \lesssim 0.04 - 0.14$; Moni Bidin *et al.* 2011). In comparison, single EHBs produced via He-enrichment, seen in the model of Lee *et al.* (2005) (see Fig. 1),

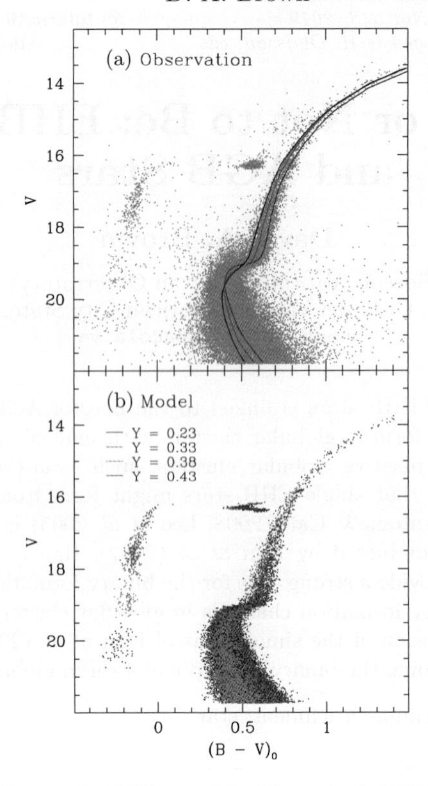

Figure 1. EHB model of Lee *et al.* (2005).

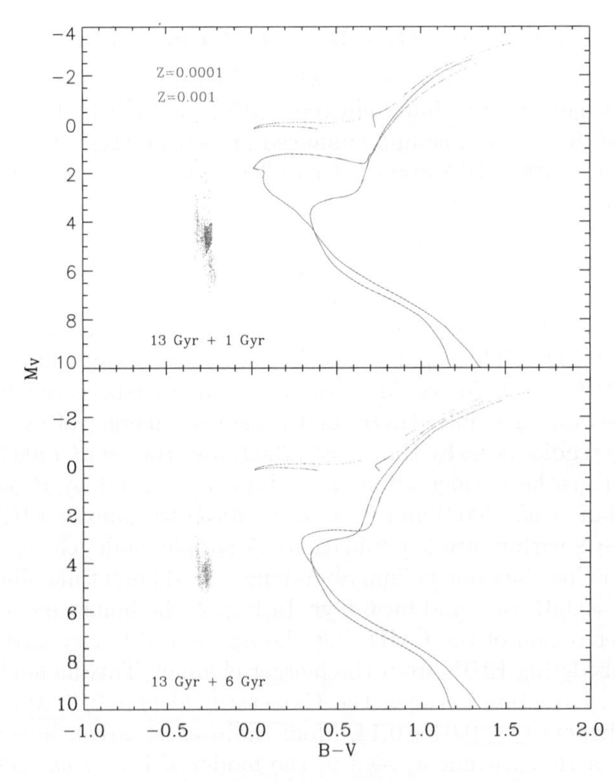

Figure 2. Binary EHB model.

make this a compelling scenario. It may be the case that diverse formation mechanisms for EHBs may indeed be operative in GCs.

References

Bedin, L.R., Piotto, G., Anderson, J., Cassisi, S., King, I.R., Momany, Y., & Carraro, G. 2004, *ApJ* (Letters), 605, L125

Carretta, E., Bragaglia, A., Gratton, R.G., Leone, F., Recio-Blanco, A., & Lucatello, S. 2006, *A& A*, 450, 523

D'Antona, F., Bellazzini, M., Caloi, V., Pecci, F.F., Galleti, S., & Rood, R.T. 2005, *ApJ*, 631, 868

D'Antona, F., & Caloi, V. 2008, *MNRAS*, 390, 693

D'Cruz N., Dorman B., Rood R., & OConnell, R. 1996, *ApJ*, 466, 359

Han, Z., Podsiadlowski, P., Maxted, P.F.L., Marsh, T.R., & Ivanova, N. 2002, *MNRAS*, 336, 449

Han, Z., Podsiadlowski, P., Maxted, P.F.L., & Marsh, T.R. 2003, *MNRAS*, 341, 669

Han, Z., Podsiadlowski, P., & Lynas-Gray, A.E. 2007, *MNRAS*, 380, 1098

Lee, Y.W., Joo, S.J., Han, S.I., Chung, C., Ree, C.H., Sohn, Y.J., Kim, Y.C., Yoon, S.J., Yi, S.K., & Demarque, P. 2005, *ApJ* (Letters), 621, L57

Maxted, P.F.L., Heber, U., Marsh, T.R., & North, R.C. 2001, *MNRAS*, 326, 1391

Mengel, J., Norris, J., & Gross, P. 1976, *ApJ*, 204, 488

Moni Bidin, C., Villanova, S., Piotto, G., & Momany, Y. 2011, *A&A*, 528, A127

Piotto G., Villanova S., Bedin L.R., Gratton R., Cassisi, S., Momany, Y., Recio-Blanco, A., Lucatello, S., Anderson, J., King, I.R., Pietrinferni, A., & Carraro, G. 2005, *ApJ*, 621, 777

Sweigart, A.V. 1997, *ApJS* 474, L23

Why Galaxies Care About AGB Stars:
A Continuing Challenge through Cosmic Time
Proceedings IAU Symposium No. 343, 2019
F. Kerschbaum, M. Groenewegen & H. Olofsson, eds.

doi:10.1017/S1743921318006312

The discovery of an asymmetric detached shell around the "fresh" carbon AGB star TX Psc

M. Brunner[1], M. Mečina[1], M. Maercker[2], E. A. Dorfi[1], F. Kerschbaum[1], H. Olofsson[2] and G. Rau[3,4]

[1]Department for Astrophysics, University of Vienna, Türkenschanzstraße 17, A-1180 Vienna
email: magdalena.brunner@univie.ac.at

[2]Department of Space, Earth & Environment, Chalmers Univ. of Tech., 43992 Onsala, Sweden

[3]NASA Goddard Space Flight Center, Code 667, Greenbelt, MD 20771, USA

[4]Department of Physics, The Catholic University of America, Washington, DC 20064, USA

Abstract. We present ALMA observations of the circumstellar envelope around the AGB carbon star TX Psc in molecular CO(2–1) emission, and detect a previously unknown detached shell with filamentary structure and elliptical shape. Up to now, all observed detached shells are found around carbon AGB stars and are of remarkable spherical symmetry. The elliptical shell around TX Psc is the first clear exception to that rule, with TX Psc being classified as rather "fresh" carbon star, that most likely has only experienced very few thermal pulses yet. We investigate and discuss the 3D structure of the CSE and its most likely formation scenarios, as well as the link of this peculiar detached shell to the AGB evolutionary status of TX Psc.

Keywords. stars: AGB and post-AGB – stars: carbon – stars: evolution – stars: mass-loss

---

TX Psc is a carbon-rich AGB star, with a relatively low C/O ratio (as summarised by Klotz *et al.* 2013). It has been observed in multiple wavelengths and spatial scales. At optical and infrared wavelengths, asymmetries and clumps have been detected close to the star (e.g. Cruzalebes *et al.* 1998, Hron *et al.* 2015). On much larger spatial scales, the interaction region between stellar wind and ISM has been mapped in thermal dust emission by *Herschel*/PACS observations during the Mass loss of Evolved StarS (MESS) observing program of AGB stars by Groenewegen *et al.* (2011). At these scales, Jorissen *et al.* (2011) have reported the detection of a ring-like structure around the star with a radius of $\sim 17''$ in addition to a well separated ISM interaction front. We present ALMA observations of the large scale structure of the circumstellar envelope (PI: M. Brunner, project ID 2015.1.00059.S), which show that the ring-like structure can also be seen in the submillimetre wavelengths in CO(2–1) emission, and in fact the previously observed structure is not spherically symmetric but represents an elliptical detached shell. This is the first detection of an elliptical detached shell, and we discuss the formation scenarios for such a structure in detail in Brunner *et al.* (2018), submitted.

Figure 1 shows an overlay of the ALMA ACA observations in CO(2–1) emission with the previously published *Herschel*/PACS dust emission. The elliptical CO(2–1) shell coincides with the ring-like dust structure, which we believe was formed through the high mass-loss event following a thermal pulse. A possible origin for the ellipticity of the detached shell is stellar rotation, increasing the wind velocity along the equatorial plane with respect to the poles of the mass-losing star. Following model calculations

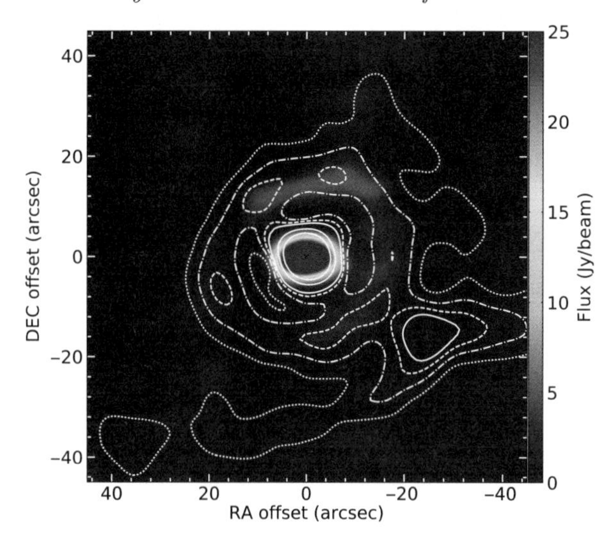

Figure 1. Integrated intensity ALMA CO(2–1) emission taken with the ACA in intermediate resolution (color scale) overlaid with *Herschel*/PACS dust emission at 70 μm (white contours).

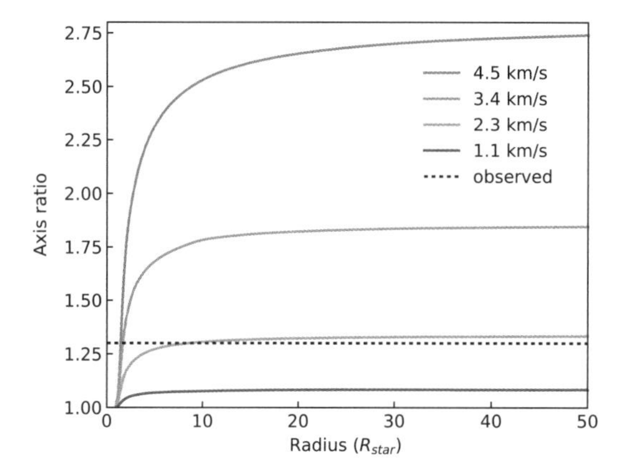

Figure 2. Model calculations of axis ratios for elliptical stellar winds generated through stellar rotation with different rotation rates (colored solid lines) compared to the observed axis ratio of 1.3 (dotted line).

by Dorfi & Höfner (1996), we calculate the stellar rotation rate that would create an elliptical shell with the observed axis ratio of 1.3 for a model star with the parameters of TX Psc. Figure 2 shows that a stellar rotation of ~2 km/s is sufficient to create the observed ellipticity.

References

Brunner, M., Mečina, M., Maercker, M., *et al.* 2018, submitted to *A&A*

Cruzalebes, P., Lopez, B., Bester, M., Gendron, E., & Sams, B. 1998, *A&A*, 338, 132

Dorfi, E. A. & Hoefner, S. 1996, *A&A*, 313, 605

Groenewegen, M. A. T., Waelkens, C., Barlow, M. J., *et al.* 2011, *A&A*, 526, A162

Hron, J., Uttenthaler, S., Aringer, B., *et al.* 2015, *A&A*, 584, A27

Jorissen, A., Mayer, A., van Eck, S., *et al.* 2011, *A&A*, 532, A135

Klotz, D., Paladini, C., Hron, J., *et al.* 2013, *A&A*, 550, A86

Why Galaxies Care About AGB Stars:
A Continuing Challenge through Cosmic Time
Proceedings IAU Symposium No. 343, 2019
F. Kerschbaum, M. Groenewegen & H. Olofsson, eds.

doi:10.1017/S1743921318006579

Imaging Red Supergiants with VLT/SPHERE/ZIMPOL

E. Cannon[1], M. Montargès[1], L. Decin[1] and A. de Koter[1,2]

[1]Instituut voor Sterrenkunde, KU Leuven, Leuven, Belgium
email: emily.cannon@kuleuven.be

[2]Anton Pannekoek Institute of Astronomy, University of Amsterdam, The Netherlands

Abstract. Using the VLT-SPHERE/ZIMPOL adaptive optics imaging polarimeter, images of a sample of nearby red supergiants (RSGs) were obtained in multiple filters. From these data, we obtain information on geometrical structures in the inner wind, the onset radius and spatial distribution of dust grains as well as dust properties such as grain size. As dust grains may play a role in initiating and/or driving the outflow, this could provide us with clues as to the wind driving mechanism.

Keywords. red supergiants, mass loss, winds, circumstellar matter, imaging

1. Introduction

During the red supergiant (RSG) phase, massive stars show powerful stellar winds, which strongly influence the supernova (progenitor) properties and control the nature of the compact object that is left behind. Furthermore, the material that is lost in the stellar wind, together with that ejected in the final core collapse, contributes to the chemical enrichment of the local interstellar medium. The mass-loss properties of RSGs are however poorly constrained. Moreover, little is known about the wind driving mechanism. To provide better constraints on both mass-loss rates and physics, high angular resolution observations are needed to unveil the inner regions of the circumstellar environment, where the mass loss is triggered.

2. The data

SPHERE/ZIMPOL is an adaptive optics imaging polarimeter (Beuzit, 2008) capable of achieving an angular resolution of 20 mas. Our sample consists of five RSGs (Antares, AH Sco, VX Sgr, UY Sct and T Cet) from two observing runs. Each of these RSGs, along with a corresponding calibrator star, were observed in four to six filters within the optical range. The reduced images were then deconvolved using the Lucy-Richardson deconvolution algorithm. A notable difference between Antares and the rest of the sample is that with an angular diameter of 37 mas (Ohnaka *et al.*, 2013), the photosphere is resolved by SPHERE. Though the photospheres of the other four stars, which have angular diameters ranging from 6 to 9 mas, remain unresolved we can still investigate their circumstellar environment.

3. Discussion

From the observations, we compute the degree of linear polarisation. This polarisation signal may be caused by anisotropic scattering on dust grains. Looking at Antares, we see a localised area with a significant signal in the degree of linear polarisation close to the surface of the star (Fig. 1). A similar observation has been obtained on Betelgeuse

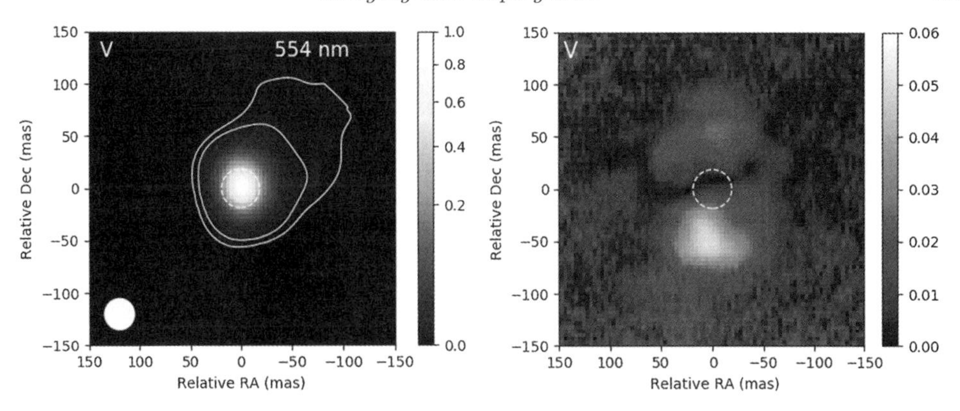

Figure 1. Left: Normalised intensity image of Antares in the V filter. The contours correspond to 20 and 50 times the noise. The dashed circle indicates the size of the photosphere and the filled circle is the beam size. Right: Degree of linear polarisation for Antares in the V filter.

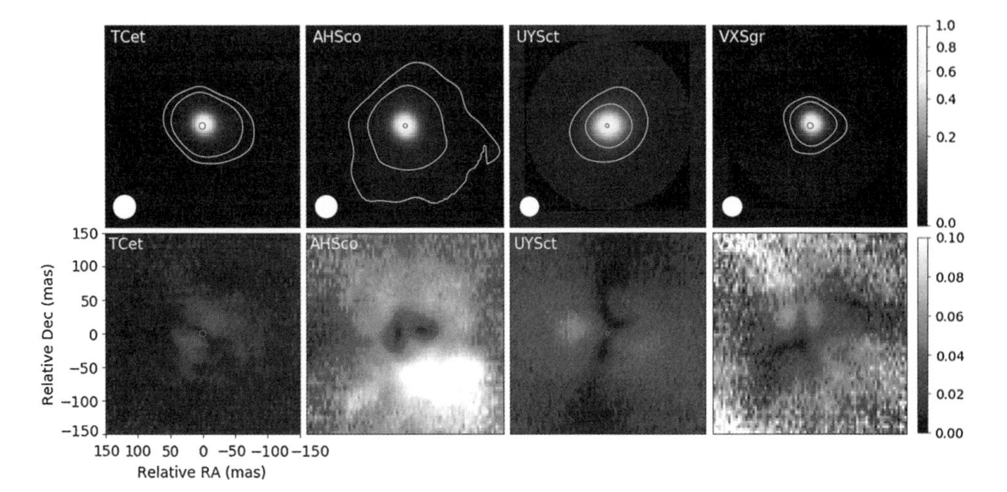

Figure 2. Top: Intensity images in the V filter. The dark circle corresponds to the photospheric diameters. Bottom: Degree of linear polarisation.

by Kervella *et al.* (2016). A significant linear polarisation signal is also seen in three out of four of the other RSGs in the sample (Fig. 2).

The asymmetries that are seen in the intensity images and the localised dust clump we see in Fig. 1 could be linked to processes on Antares' surface such as convection. Radiation pressure on dust grains that are this close to the star could have a significant influence on the mass loss mechanism, though whether this plays a role in driving the wind or perhaps makes a contribution to accelerating will require a detailed modelisation of the dust and its interaction with the radiation field.

4. Conclusions

SPHERE provides us with an unprecedented resolved view of the inner circumstellar environment of evolved stars. Thanks to linear polarisation, we are able to track dust very close to the star. This work will be continued by challenging numerical radiative dust models.

Based on observations made with ESO Telescopes at the La Silla Paranal Observatory under programme IDs 095.D-0458 and 099.D-0600

References

Beuzit, J.-L., Feldt, M., Dohlen, K., *et al.* 2008, *SPIE Conf. Ser*, 7014, 18

Kervella, P., Lagadec, E., Montargès, M., Ridgway, S.T., Chiavassa, A., Haubois, X., Schmid, H.-M, Langlois, M., Gallene, A., & Perrin, G. 2016, *A&A*, 585, A28

Ohnaka, K., Hofmann, K.-H., Schertl, D., Weigelt, G., Baffa, C., Chelli, A., Petrov, R., and Robbe-Dubois, S. 2013, *A&A*, 555, A24

*Why Galaxies Care About AGB Stars:
A Continuing Challenge through Cosmic Time
Proceedings IAU Symposium No. 343, 2019
F. Kerschbaum, M. Groenewegen & H. Olofsson, eds.*

doi:10.1017/S1743921318005872

The Impact of Dust/Gas Ratios on Chromospheric Activity in Red Giant and Supergiant Stars

Kenneth G. Carpenter[1] and Gioia Rau[1,2]

[1]NASA Goddard Space Flight Center, Code 667, Greenbelt, MD, 20771, USA
email: Kenneth.G.Carpenter@nasa.gov

[2]Dept. of Physics, Catholic University of America, USA
email: gioia.rau@nasa.gov

Abstract. Stencel *et al.* (1986) analyzed IUE spectra of a modest set of cool stars and found that they continue to produce chromospheres even in the presence of high dust levels in their outer atmospheres. This reversed the previous results of Jennings (1973) and Jennings & Dyck (1972). We describe an on-going extension of these studies to a sample of stars representing a broader range in dust/gas ratios, using archival IUE and archival and new HST data on both RGB and AGB stars. Surface fluxes in emission lines will be analyzed to assess the chromospheric activity and obscuration by dust in each star, as those fluxes will follow a different pattern for reduced activity (temperature/density dependent) vs. dust obscuration (wavelength dependent). Wind characteristics will be measured by modeling of wind-reversed chromospheric emission lines.

Keywords. stars: chromospheres, stars: winds, outflows, stars: late-type

---

1. Introduction

Stencel *et al.* (1986) examined UV chromospheric diagnostics in cool stars with a range of outer atmospheric dust levels. They found, as shown in Fig. 1a, that such objects continue to produce chromospheres even in the presence of high dust level in their outer atmospheres, although the surface fluxes were lower in the dustier stars (see Fig. 1b). This reversed the previous results of Jennings (1973) (cf. Jennings & Dyck (1972)).

2. HST Extension of IUE Studies

The goals of this new study are to: 1) verify the claim that cool stars still produce chromospheres even in highly dusty atmospheric environments, by studying a larger, more diverse (dust/gas content) sample of stars and 2) examine the characteristics of the winds, and how they change with outer atmospheric dust levels.

In this on-going study, we aim to resolve the discrepancy between the Stencel *et al.* (1986) and Jennings (1973) findings by examining the strength of the UV emission features in IUE and HST spectra. An overview of a subset of archival data are shown in Fig. 2a and a sample of wind diagnostic lines in Fig. 2b.

3. Conclusions and Future Work

Figures 1 and 2 show the feasibility of further observations of chromospheric lines using HST/STIS. By measuring their surface fluxes and studying the overlying wind absorptions (Fig. 2b) to map their winds, we will be able to assess the strength of their chromospheres and the characteristics of their winds and mass-loss rates.

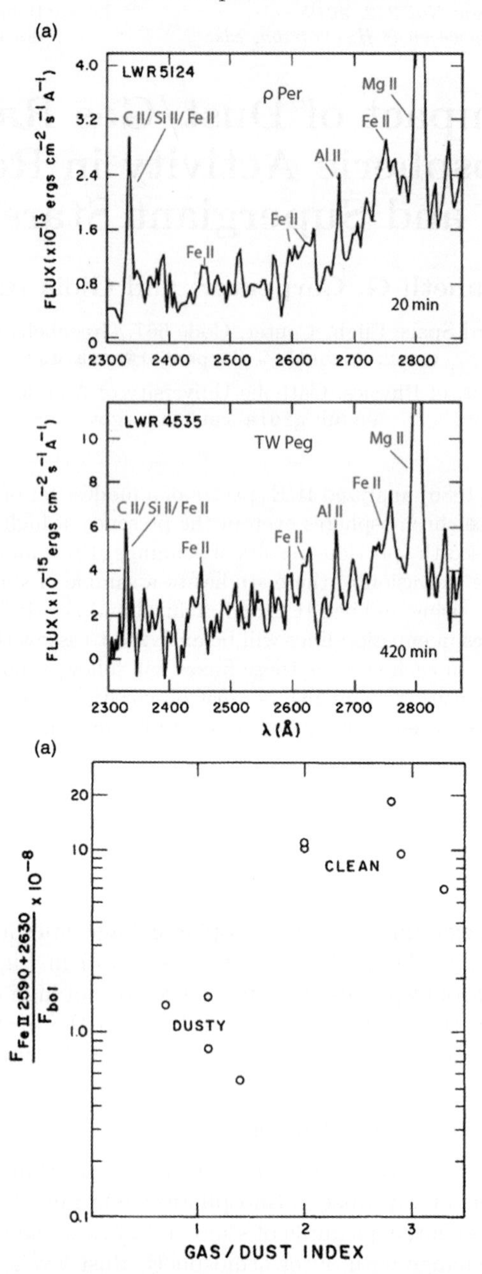

Figure 1. IUE Observations and Results: (a) IUE low-resolution spectra of two stars from the Stencel et al. study. Rho Per is a clean star (i.e. gas/dust>1) with strong optical Ca II emission, while TW Peg is a very dusty star (i.e. gas/dust<1) with no Ca II emission. All of the major UV chromospheric indicators in this region (Mg II, Al II, Fe II, and C II) are visible in both spectra. This demonstrates that the chromospheres in dusty stars are not fully quenched by high levels of dust. (b) total flux in two blends of Fe II lines near 2590 Å & 2630 Å, normalized by the stellar bolometric flux, vs. the gas/dust index of Hagen *et al.* (1983). The Fe II fluxes are substantially reduced in the dusty stars, but still measureable and significant.

(a)

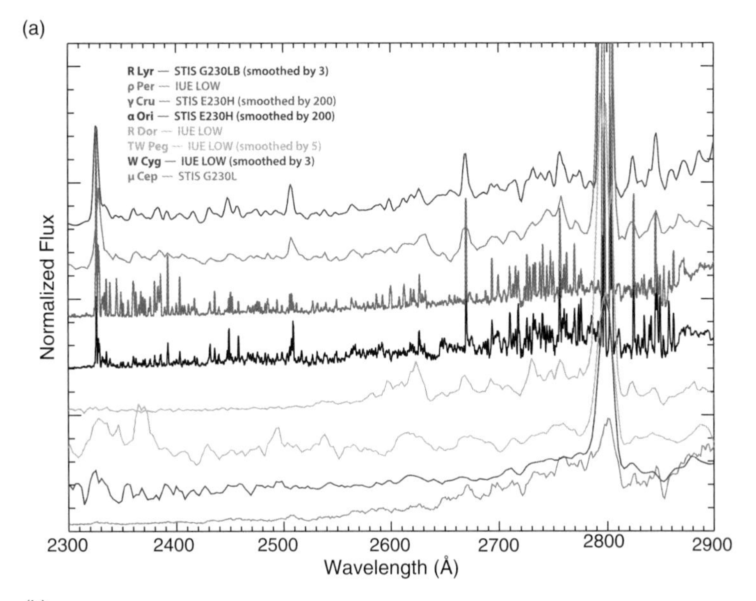

(b)

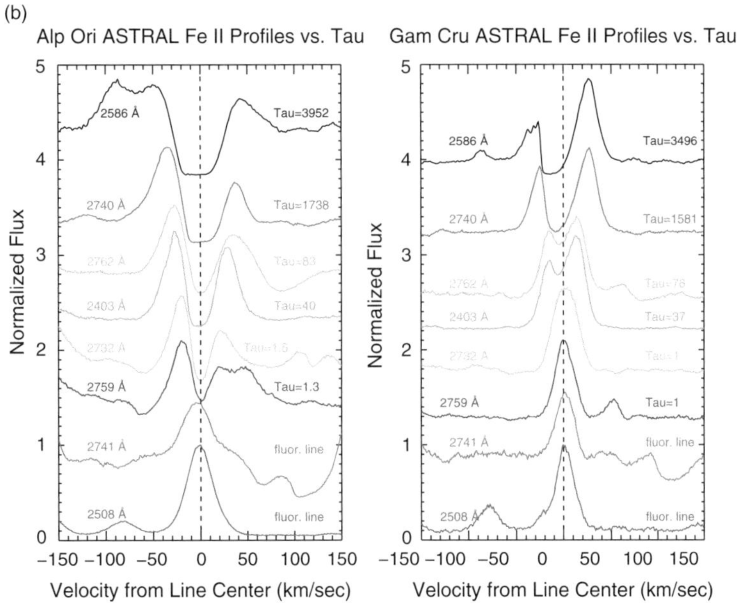

Figure 2. HST/IUE archival data and example wind lines: (a) Spectra of the overall UV region 2300-2900 Å for a subset of our stars. The spectra in this figure confirm that chromospheres persist in dusty stars, albeit at lower activity levels. We show IUE low resolution observations in this figure for clarity, when the IUE high resolution spectra are very noisy at shorter wavelengths. (b) For stars with sufficient S/N, we will also examine Fe II lines of various opacities (α Ori (left), γ Cru (right)), to map the wind acceleration and mass-loss, and characterize the impact of dust and chromosphere on these parameters.

References

Jennings, M.C. 1973, *ApJ*, 185, 197
Jennings, M.C. and Dyck, H.M. *ApJ*, 177, 427
Hagen, W., Stencel, R.E., & Dickenson, D. 1983, *ApJ*, 274, 286
Stencel, R.E., Carpenter, K.G., & Hagen, W. 1986, *ApJ*, 308, 859

Why Galaxies Care About AGB Stars: A Continuing Challenge through Cosmic Time
Proceedings IAU Symposium No. 343, 2019
F. Kerschbaum, M. Groenewegen & H. Olofsson, eds.

doi:10.1017/S1743921318006063

Metallic Line Doubling in the Spectra of the Variable Star R Scuti

K. Chafouai[1], A. Benhida[1], F. Sefyani[1], A. Ghout[1], Z. Benkhaldoun[1], P. Mathias[2], D. Gillet[3] and Y. El Jariri[1]

[1]Oukaïmeden Observatory LPHEA Cadi Ayyad University BP 2390 Marrakech, Morocco
email: chafouai.khadija@gmail.com

[2]CNRS UMR5277 Institut de Recherche en Astrophysique et Planétologie 14 Avenue Edouard Belin 31400 Toulouse, France

[3]Observatoire de Haute-Provence - CNRS/PYTHEAS/Université d'Aix-Marseille 04870 Saint Michel l'Observatoire, France

Abstract. In this work, we present spectroscopic results of the variable star R Scuti, obtained during the campaign of measures led in 2016 at the Oukaimeden observatory in Morrocco. High resolution spectra (R$\approx$12 000) were obtained between 4289 Å and 7125 Å. This intensive observing campaign spanned over 26 nights from June to November 2016.

Keywords. stars: variables: R Sct - shock waves - line: profiles - stars: individual

1. Introduction

The RV Tauri stars are pulsating variables characterized by alternating deep and shallow minima in their light curves. They are Population II variables of high luminosity ($10\,000 > L/L_{\odot} > 1\,500$) and pulsation period (30–150 d), as reported by Wallerstein & Cox (1984). In this study, we present the observation of line doubling absorption in the spectra of R Sct on the metal lines of Fe I and Ti I. The doubling of metal lines was observed in 1952 for the first time by Sanford on W Virginis. This phenomenon was interpreted by Schwarzschild (1952) on the basis of a two-layer atmosphere: during compression all the layers of the atmosphere move inwards, the front of the shock wave penetrates the lower layer of the atmosphere and a reversal of direction of motion occurs which results in rapid movement of the atmospheric layers from the inside to the outside.

2. Materials & methods

The instrument set up includes two telescopes mounted on each other. It is mounted at Oukaimeden Observatory (J43), a research entity belonging to the Cadi Ayyad University in Morocco. The instrument set up has already been presented in Benhida *et al.* (2018).

3. Discussion and Results

The dynamic atmosphere of R Sct is characterized by two shock waves during one luminosity period. These two shock waves were clearly observed by Lèbre and Gillet (1991a). They observed also the double absorption line of TiI (λ5866.46 Å) around $\varphi = 1.33 - 1.36$, $\varphi = 1.6$ and $\varphi = 1.83$. The interpretation of these features assumes that the first double absorption means that a weak infalling shock must certainly exist within the "Titanium layer" and is the consequence of the terminal infalling phase of the ballistic motion produced by the previous secondary shock, the second line absorption is the signature of the new ballistic motion of the "Titanium layer", the third line absorption

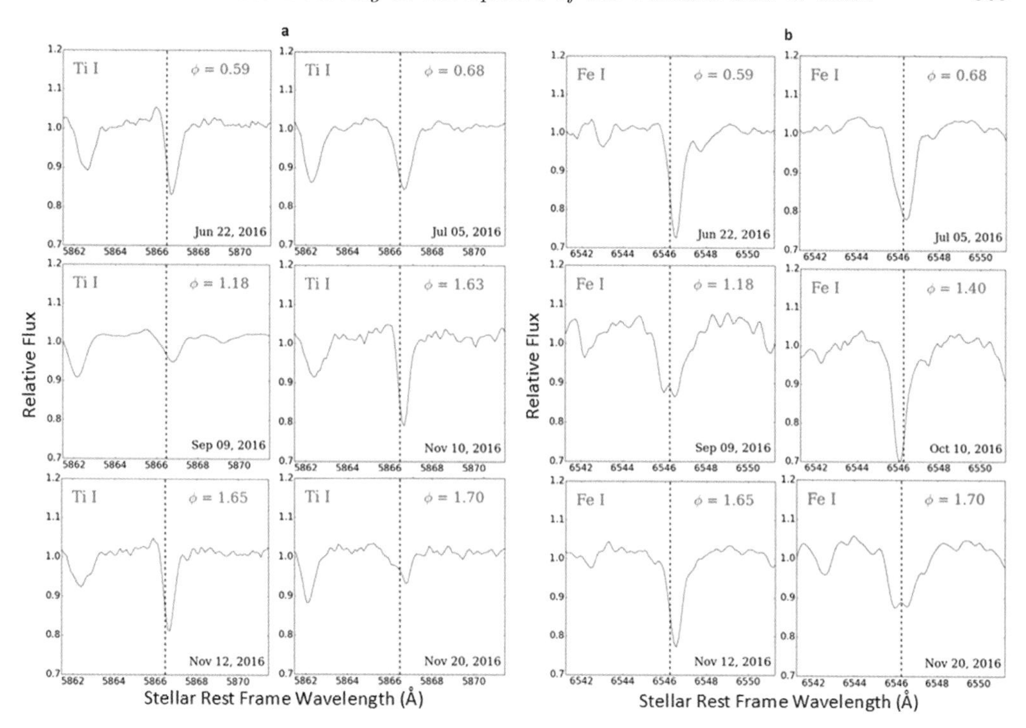

Figure 1. a) and b) The TiI λ5866.46 Å and FeI λ6546.245 Å profiles of R Sct in 2016 at Oukaimeden Observatory. The wavelengths are measured in the rest frame of R Sct (with systemic velocity equal to 43.8 km/s). The relative flux refers to the main level of the continuum. The vertical line represents the zero velocity in the stellar rest frame.

explains the existence of a weak infalling shock which would be the consequence of the acceleration of the external atmospheric layer caused by the stellar gravity. In this work, we present the observations of line doubling absorption in the spectra of R Sct on the metal lines of TiI(λ5866.46 Å) and FeI-TiI(λ6546.245 Å). In comparison with Lèbre and Gillet (1991a), we confirm the existence of two double absorption line of TiI(λ5866.46 Å). This double absorption, is well visible in Figure 1a, at phases φ= 1.18 and φ= 1.7, its blueshifted component weaker than its redshifted one. The third double line absorption is not observed in our results, supposedly the consequence of a low acceleration of the external atmospheric layers. Since 2016, the light curve in AAVSO observations of R Sct didn't show deep light minima. This disappearance may be the result of the decrease of the acceleration of the external atmospheric layers. On the other hand, for the first time, we observed a clear double absorption profile of the complex line FeI-TiI (λ6546.245 Å) in the same phases like TiI(λ5866.46 Å) profile at φ=1.18 and φ=1.70 (Fig. 1b).

4. Conclusion

Our work is based on R Sct spectroscopic observations in 2016. The light curve shape of the AAVSO data base does not allow us to distinguish the deep light minima in the 2016 cycle. For the first time, our spectroscopic results present two double absorption lines during one period for both titanium (λ5866.46 Å) and iron ($FeI\lambda$ 6546.245 Å). The third weak double absorption of the TiI (λ5866.46 Å) mentioned in the paper of Lèbre & Gillet doesn't appear in our results. These results allowed us to assume that the intensity of the main shock in R Sct has become low.

References

Benhida, A., *et al.* 2018, *Prossiding, Polish Astronomical Society*, 6, 279
Lèbre, A., Gillet, D. 1991, *A&A*, 246, 490
Lèbre, A., Gillet, D. 1991, *A&A*, 251, 549
Sanford, R.F. 1952, *ApJ*, 116, 331S
Schwarzschild, M. 1952, *Cambridge University Press*, 811
Wallerstein, G., Cox, A.N. 1984, *PASP*, 96, 677

Why Galaxies Care About AGB Stars:
A Continuing Challenge through Cosmic Time
Proceedings IAU Symposium No. 343, 2019
F. Kerschbaum, M. Groenewegen & H. Olofsson, eds.

doi:10.1017/S1743921318005070

Populations of accreting white dwarfs

Hai-Liang Chen[1,2,3], **Tyrone E. Woods**[4], **Lev Yungelson**[5], **Marat Gilfanov**[6,7] **and Zhanwen Han**[1,2,3]

[1]Yunnan Observatories, CAS, Kunming, 650011,China
email: chenhl@ynao.ac.cn

[2]Key Laboratory for the Structure and Evolution of Celestial Objects, CAS, Kunming 650011, China

[3]Center for Astronomical Mega-Science, Chinese Academy of Science, Beijing 100012, China

[4]Institute of Gravitational Wave Astronomy and School of Physics and Astronomy, University of Birmingham, Birmingham B15 2TT, UK

[5]Institute of Astronomy, RAS, 48 Pyatnitskaya Str., 119017 Moscow, Russia

[6]Max Planck Institute for Astrophysics, Karl-Schwarzschild-Str. 1, Garching b. München 85741, Germany

[7]Space Research Institute of Russian Academy of Sciences, Profsoyuznaya 84/32, 117997 Moscow, Russia

Abstract. Using a hybrid binary population synthesis approach, we modelled the formation and evolution of populations of accreting white dwarfs (WDs) for differing star formation histories. We found that the delay time distribution of SNe Ia in the single degenerate scenario is inconsistent with observations. Additionally, we found that our predicted X-ray and UV emission of populations of accreting WDs are consistent with the X-ray luminosities of early-type galaxies observed by Chandra and the HeII 4686 Å/Hβ line ratio measured in stacked SDSS spectra of passively evolving galaxies. Moreover, we found that the majority of current novae in elliptical-like galaxies have low-mass WDs, long decay times, long recurrence periods and are relatively faint. In contrast, the majority of current novae in spiral-like galaxies have massive WDs, short decay times, short recurrence periods and are relatively bright. Our predicted distribution of mass-loss timescales in an M31-like galaxy is consistent with observations for Andromeda.

Keywords. stars:novae, cataclysmic variables, white dwarfs

1. Introduction

It is now understood (e.g., Nomoto *et al.* (2007)) that for a narrow range of accretion rates, hydrogen will undergo steady nuclear-burning on the surface of an accreting white dwarf. If the accretion rate is below this range, hydrogen burns unstably, observed as novae. In this regime, accreting WDs produce little X-ray emission. In the stable burning regime, accreting WDs have a typical effective temperature of $10^5 - 10^6$K, and radiate predominantly in the soft X-ray and EUV bands. For accretion rates above the stable burning regime, the evolution of accreting WDs is still unclear. In one scenario, suggested by Hachisu *et al.* (1996), an optically thick wind is launched. In this case, the typical effective temperature of accreting WDs in the wind regime is $10^4 - 10^5$K and WDs emit prominently in the EUV. In this work, we modelled the formation and evolution of accreting WDs in different types of galaxies using a hybrid binary population synthesis approach. We study their X-ray and UV emission and properties of the nova population.

2. Hybrid Binary Population Synthesis Approach

Our calculations consist of two steps. First, we use the BSE code (Hurley *et al.* 2002) to obtain the population of WD binaries with non-degenerate companions at the onset of mass transfer. Then we use the MESA (Paxton *et al.* 2011) code to compute a grid of models describing the evolution of WD binaries with varying initial parameters. With the binary parameters at the onset of mass transfer from our BSE calculations, we can select the closest track in the grid of MESA calculations and follow the evolution of any WD binary. For more details, we refer to Chen *et al.* (2014).

3. Results

In Chen *et al.* (2014), we have computed the SN Ia rate in the single degenerate scenario in different types of galaxies, and compared these results with observational constraints. We found that the delay time distribution of SNe Ia in our calculation is inconsistent with observations (see Fig. 8 in Chen *et al.* (2014)), being more than 10 times smaller than the observationally inferred value.

In Chen *et al.* (2015), we computed the X-ray luminosity in the soft X-ray (0.3–0.7 keV) band for elliptical-like galaxies as a function of stellar age and found that it is comparable to the Chandra observational data of nearby elliptical galaxies (see Fig. 9 in Chen *et al.* (2015)). In addition, we computed the time evolution of the He II $\lambda4686/H\beta$ line ratio in passively evolving galaxies, and found that it is in good agreement with the line ratio measured in stacked SDSS spectra of retired galaxies (Johansson *et al.* 2014, 2016; Fig. 11 in Chen *et al.* (2015)).

In Chen *et al.* (2016), we have modelled the evolution of the nova population in different types of galaxies. We found that the current nova rate per unit mass in elliptical-like galaxies is 10–20 times smaller than that in spiral-like galaxies. Moreover, we found that the current nova population in elliptical-like galaxies has lower-mass WDs, longer decay times, relatively fainter absolute magnitudes and longer recurrence periods. The current nova population in spiral-like galaxies have massive WDs, short decay times, are relatively bright and have short recurrence periods. In addition, the predicted distribution of mass-loss timescale in a M31-like galaxy is in good agreement with observed statistics (see Fig. 3 in Chen *et al.* (2016)).

Acknowledgements

This work is partially supported by the National Natural Science Foundation of China (Grant no. 11703081,11521303,11733008), Yunnan Province (No. 2017HC018) and the CAS light of West China Program.

References

Johansson J., Woods T.E., Gilfanov M. *et al.* 2014 *MNRAS*, 442, 1079
Johansson J., Woods T.E., Gilfanov M. *et al.* 2016 *MNRAS*, 461, 4505
Hachisu, I., Kato, M., Nomoto, K. 1996 *ApJ*, 470, L97
Chen, H.-L., Woods, T. E., Yungelson, L. R., Gilfanov, M., Han, Z. 2014 *MNRAS*, 445, 1912
Chen, H.-L., Woods, T. E., Yungelson, L. R., Gilfanov, M., Han, Z. 2015 *MNRAS* 453, 3024
Chen, H.-L., Woods, T. E., Yungelson, L. R., Gilfanov, M., Han, Z. 2016 *MNRAS* 458, 2916
Hurley, J. R., Tout, C. A., Pols, O. R. 2002 *MNRSA* 329, 897
Nomoto, K., Saio, H., Kato, M., & Hachisu, I. 2007, *ApJ*, 663, 1269
Paxton, B., Bildsten, L., Dotter, A., Herwig, F., Lesaffre, P., Timmes, F. 2011 *ApJS* 192, 3

Why Galaxies Care About AGB Stars:
A Continuing Challenge through Cosmic Time
Proceedings IAU Symposium No. 343, 2019
F. Kerschbaum, M. Groenewegen & H. Olofsson, eds.

doi:10.1017/S1743921318003885

Using Gaia to measure the atmospheric dynamics in AGB stars

Andrea Chiavassa[1], Bernd Freytag[2] and Mathias Schultheis[1]

[1]Université Côte d'Azur, Observatoire de la Côte d'Azur, CNRS, Lagrange, CS 34229, Nice, France
email: andrea.chiavassa@oca.eu

[2]Department of Physics and Astronomy at Uppsala University, Regementsvägen 1, Box 516, SE-75120 Uppsala, Sweden

Abstract. We use 3D radiative-hydrodynamics simulations of convection with CO5BOLD and the post-processing radiative transfer code Optim3D to compute intensity maps in the Gaia *G* band [325–1030 nm]. We calculate the intensity-weighted mean of all emitting points tiling the visible stellar surface (i.e., the photo-center) and evaluate its motion as a function of time. We show that the convection-related variability accounts for a substantial part to the Gaia DR2 parallax error of our sample of semiregular variables. Finally, we denote that Gaia parallax variations could be exploited quantitatively to extract stellar parameters using appropriate RHD simulations corresponding to the observed star.

Keywords. stars: atmospheres – stars: AGB – astrometry – parallaxes – hydrodynamics

---

1. Introduction

AGBs are low- to intermediate-mass stars that evolve to the red giant and asymptotic giant branch increasing the mass-loss during this evolution. Their complex dynamics affect the measurements and amplify the uncertainties on stellar parameters. The convection-related variability, in the context of Gaia astrometric measurement, can be considered as "noise" that must be quantified in order to better characterize any resulting error on the parallax determination.

2. Methods

We employed the code Optim3D (Chiavassa *et al.* 2009) to compute intensity maps (Fig. 1, left panel) in the Gaia *G* photometric system (Evans *et al.* 2018). For this purpose, we based the calculation on snapshots from the radiation-hydrodynamics (RHD) simulations of AGB stars (Freytag *et al.* 2017) computed with CO^5BOLD (Freytag *et al.* 2012) code.

3. Comparison and predictions

First, we extracted the parallax error (σ_{ϖ}) from Gaia DR2 for a sample of semiregular variables (SRV) from Tabur *et al.* (2009), Glass & van Leeuwen (2007), and Jura *et al.* (1993) that match the theoretical luminosities of RHD simulations. It has to be noted that σ_{ϖ} may still vary in the following data releases because: (i) the mean number of measurements for each source amounts to 26 (Mowlavi *et al.* 2018) and it will be 70–80 times in total at the end of the nominal mission; (ii) and new solutions may be applied to adjust the imperfect chromaticity correction (Arenou *et al.* 2018).

We investigated if the parallax errors of our SRV sample can be explained by the resulting motion of the stellar photo-center seen in the RHD simulations. Fig. 1 (central panel)

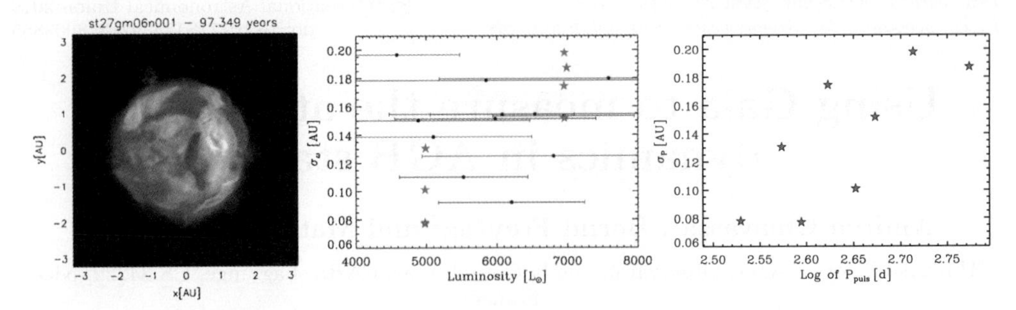

Figure 1. *Left:* Example of the squared root intensity map in the Gaia G photometric system (Evans *et al.* 2018). *Center:* luminosity against the parallax error of the observations (σ_{ϖ}, circle symbol in black) and the standard deviation of the photo-center displacement for the RHD simulations (σ_P, star symbol in red). *Right:* σ_P against logarithm of the stellar period for RHD simulations. Figures from Chiavassa *et al.* (2018).

displays that Gaia parallax errors are in good agreement with the standard deviations of the photo-center displacement in the simulations (in particular for higher luminosities). This attests that convection-related variability accounts for a substantial part to the parallax error in Gaia measurements. In addition to this, Fig. 1 (right panel) reveals the correlation between the photo-center displacement and the logarithm of the pulsation: larger values of σ_P correspond to longer pulsation periods.

4. Summary

The visible fluffy stellar surface (Fig. 1, left) is made of shock waves, that are produced in the interior and that are shaped by the top of the convection zone as they travel outward. The surface is characterized by the presence of few large and long-lived convective cells accompanied by short-lived and small scale structures. As a consequence, the position of the photo-center is affected by temporal fluctuations. We found a good agreement with observations probing that convection-related variability accounts for a substantial part to the parallax error. This result let us denote that parallax variations from Gaia measurements could be exploited quantitatively using appropriate RHD simulations corresponding to the observed star. More details can be found in Chiavassa *et al.* (2018).

Acknowledgements

This work has made use of data from the European Space Agency (ESA) mission Gaia (`https://www.cosmos.esa.int/gaia`), processed by the Gaia Data Processing and Analysis Consortium (DPAC, `https://www.cosmos.esa.int/web/`). Funding for the DPAC has been provided by national institutions, in particular the institutions participating in the Gaia Multilateral Agreement.

References

Arenou, F., Luri, X., Babusiaux, C., *et al.* 2018, arXiv:1804.09375
Chiavassa, A., Plez, B., Josselin, E., & Freytag, B. 2009, A&A, 506, 1351
Chiavassa, A., Freytag, B., & Schultheis, M. 2018, arXiv:1808.02548
Evans, D. W., Riello, M., De Angeli, F., *et al.* 2018, arXiv:1804.09368
Freytag, B., Steffen, M., Ludwig, H.-G., *et al.* 2012, Journal of Computational Physics, 231, 919
Freytag, B., Liljegren, S., & Höfner, S. 2017, A&A, 600, A137
Glass, I. S., & van Leeuwen, F. 2007, MNRAS, 378, 1543
Jura, M., Yamamoto, A., & Kleinmann, S. G. 1993, ApJ, 413, 298
Mowlavi, N., Lecoeur-Taïbi, I., Lebzelter, T., *et al.* 2018, arXiv:1805.02035
Tabur, V., Bedding, T. R., Kiss, L. L., *et al.* 2009, MNRAS, 400, 1945

Why Galaxies Care About AGB Stars: A Continuing Challenge through Cosmic Time
Proceedings IAU Symposium No. 343, 2019
F. Kerschbaum, M. Groenewegen & H. Olofsson, eds.

doi:10.1017/S1743921318006208

A critical test to disentangle the role of overshooting and rotation in stars

G. Costa[1], L. Girardi[2], A. Bressan[1], P. Marigo[3], Y. Chen[3], B. Kanniah[4], A. Lanza[1] and T. S. Rodrigues[2]

[1]SISSA, via Bonomea 365, Trieste, Italy
[2]OAPD–INAF, Vicolo dell'Osservatorio 5, Padova, Italy
[3]Dipartimento di Fisica e Astronomia Galileo Galilei, Università di Padova, Italy
[4]MIT, Cambridge, Massachusetts, USA

Abstract. We study the mixing in low-intermediate massive stars using eclipsing binaries. We compute stellar evolutionary models with a varying convective core overshooting parameter and different rotation rates. Using a Bayesian estimation method, we found that the coexistence of the two phenomena may be a reasonable explanation of the observed extra-mixing.

Keywords. stars: evolution, (stars:) binaries: eclipsing

---

1. Introduction and methodology

Detached double-lined eclipsing binaries (DLEBs) are an ideal laboratory to investigate the mixing processes that are at work in deep stellar interiors. For example, their position in the HR diagram at a given age strongly depends on the amount of internal mixing during the previous evolution. Claret & Torres (2018, 2017) analyzed 37 such binaries to calibrate the strength of core overshooting, one of the sources of extra mixing. They found a growing efficiency from $\sim$1 to $\sim$2 $M_\odot$ and a constant value thereafter. An important feature of their results is the presence of a significant star to star variation in the adopted Mixing Length parameter. Here we use the same sample to check the possible role of the additional mixing induced by rotation. To this purpose, we analyze the above DLEB sample with the code PARAM (Rodrigues *et al.* 2014, 2017). For each observed star, we have accurate determinations of mass M, radius R, effective temperature T_{eff} and metallicity Z. We explore two different hypotheses. In the first one, we assume that the extra mixing is only due to overshooting from the convective core while, in the second one, we ascribe the extra mixing to the interplay of overshooting and rotation. We computed evolutionary tracks with an updated version of the PARSEC code that account for rotation (V2.0, Costa et al. 2018, in prep.). For the first hypothesis, the model dataset consists of non rotating stellar models with an overshooting efficiency λ_{ov} from 0 to 0.8 H_p in steps of 0.1 (in our scheme the mean free path of the elements is across the border, i.e. $\sim$ twice the value commonly adopted). In the second case, we computed sets of models with a fixed overshooting λ_{ov}= 0.4 and different initial rotational velocities, with angular rotation rate over the critical one, ω, from 0.0 to 0.6 in steps of 0.1. In all cases, the metallicity varies from Z = 0.002 to 0.02, and the MLT parameter is fixed (1.74). With PARAM, we apply a Bayesian estimation method to compute the posterior probability density function (PDF) for the age and λ_{ov} parameters (age and ω in the second case). To derive the best age and λ_{ov} parameters (or age and ω in the second case), we finally constrain the binary components to be coeval. More details can be found in Costa et al. (2018 in prep).

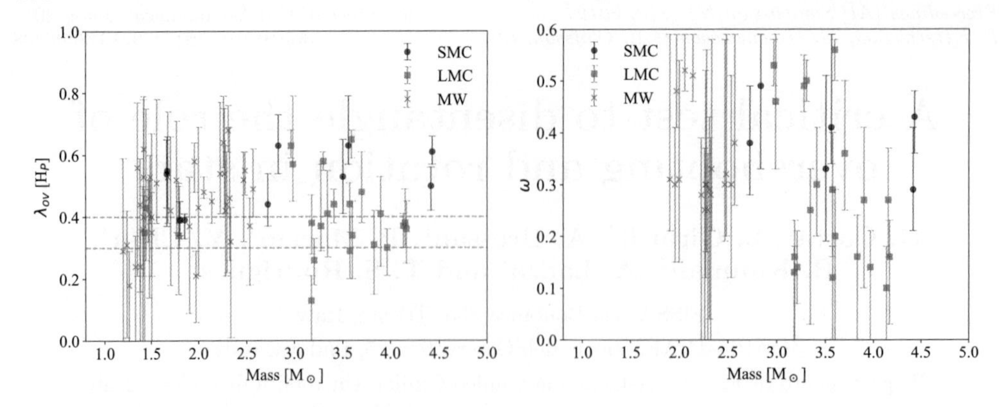

Figure 1. Left panel. λ_{ov} as a function of M_i for the 37 DLEBs. The best values and the error bars are coloured to divide stars of different galaxies. The gray dashed lines are drawn for an easier reading. Right panel. Angular velocity rate ω as a function of the M_i for $M_i \geq 2.0\ M_\odot$.

2. Results and conclusions

From the PDF maps, we extract the best values of the parameters using the median value of the marginalized distribution. These values are shown in Fig. 1 as a function of the initial mass. If we interpret the extra mixing only in terms of overshooting (left panel), we may draw the following conclusions:

1. There is a trend of λ_{ov} as a function of the mass, that grows from ~ 1 to $2\,M_\odot$, and remains flat ($\lambda_{ov} \sim 0.45$) at greater masses (see (Claret & Torres 2017, 2018)).
2. In the flat region, λ_{ov} shows a large scatter, even for similar initial masses.
3. In the flat region, there is an evident lack of points below λ_{ov}= 0.3–0.4.

A large scatter in λ_{ov} at the same initial mass is difficult to explain within the current convection theories that assume fixed values for the mixing parameters (including the MLT value). At the same time, point 3 suggests the existence of a minimum extra-mixing. We argue that the minimum extra-mixing is the result of convective overshooting and that the scatter above this value is caused by the additional mixing eventually induced by rotation. In this second scenario, we considered only DLEBs with $M_i > 2$ $M_\odot$ which populate the region of the unexpected large scatter in λ_{ov} and the results are shown in the right panel of Fig. 1. To conclude, our analysis provides insights into the relative role played by the overshooting and rotation in stellar interiors, and in particular, indicates that both may significantly contribute to internal mixing already from the H-burning stage. The overshooting seems responsible for only a fraction of the extra mixing, $\lambda_{ov} = 0.3$–0.4. Meanwhile, rotation provides a likely explanation of the additional mixing which shows a stochastic behaviour above the minimum threshold. These effects will propagate until the most advanced stellar phases and the impact on the evolved phases, in particular on the calibration of the AGB phase, will be soon investigated. Evolutionary tracks and isochrones will be soon available at: http://stev.oapd.inaf.it/cmd or http://starkey.astro.unipd.it/cgi-bin/cmd.

We acknowledge support from the ERC Consolidator Grant funding scheme (project STARKEY, G.A. n. 615604).

References

Claret, A. and Torres, G. 2017, *ApJ*, 849, 18
Claret, A. and Torres, G. 2018, *ApJ*, 859, 2
Rodrigues, T. S. and Girardi, L. and Miglio, A. *et al.* 2014, *MNRAS* 445, 3
Rodrigues, T.S. and Bossini, D. and Miglio, A. and Girardi, L. *et al.* 2017, *MNRAS* 467, 2

Why Galaxies Care About AGB Stars:
A Continuing Challenge through Cosmic Time
Proceedings IAU Symposium No. 343, 2019
F. Kerschbaum, M. Groenewegen & H. Olofsson, eds.

doi:10.1017/S1743921318003800

The role of shocks in the determination of empirical abundances for type-I PNe

Roberto D. D. Costa and Paulo J. A. Lago

Departamento de Astronomia - IAG - Universidade de São Paulo
Rua do Matão 1226, Cidade Universitária, 05508-090, São Paulo/SP, Brasil
email: roberto.costa@iag.usp.br

Abstract. We investigate, in the light of new diagnostic diagrams, the role of shocks in the ionization profile of type-I planetary nebulae, and their relation to the empirical derivation of chemical abundances. We apply our technique to two well-known type-I objects: NGC 2440 and NGC 6302. Our results indicate that shocks play a very important role in the spectra of both nebulae and, since the presence of shocks reinforces the flux of low ionization lines, this artificial reinforcement can lead to incorrect chemical abundances, when they are derived through Ionization Correction Factors, at least for type-I PNe.

Keywords. ISM: abundances, planetary nebulae: general, planetary nebulae: individual (NGC 2440, NGC 6302), techniques: spectroscopic

1. Introduction

Despite the growing use of photoionization models to derive chemical abundances of Planetary Nebulae (PNe), empirical abundances derived with the help of Ionization Correction Factors (ICFs) are still very useful for large samples when frequently no information on the central star is available for most nebulae.

This work aims to investigate the role of shocks in the excitation/ionization of type-I PNe, and their relation with the empirical derivation of chemical abundances, using the new diagnostic diagrams developed by Akras & Gonçalves (2016). To perform such diagnostics, an accurate description of the morpho-kinematic structure of PNe is required. This is challenging since these objects can have several morphological types, from the most regular ones, with spherical symmetry, to strongly asymmetrical, multipolar objects.

2. Targets and methodology

We applied the diagnostics to NGC 2440 and NGC 6302, both very well known; the first is multipolar with at least two bipolar components, and the second is bipolar. A whole kind of substructures known as LIS and FLIS (low ionization structures and their fast versions) were already studied in both objects, and were recently linked with shocks, that have a role in the ionization mechanisms. Line flux ratios and density maps from literature were used for both objects. For NGC 2440, velocity fields were obtained from L'opez *et al.* (1998), and from Lago & Costa (2016). For NGC 6302 we used the velocity fields from Szyszka *et al.* (2011), and our own data, derived using coudé spectroscopy.

Akras & Gonçalves (2016) (*loc. cit.*) developed diagnostic diagrams to distinguish photoionization from shocks in PNe, according to the ratio f_{shock}/f_{*}, where f_{shock} is the flux of ionizing photons due to shocks and f_{*} is the flux of ionizing photons from the central star. Using the definition of energy flux from shocks, F_{shock}, by

Dopita & Sutherland (1996) and converting it to photon flux f_{shock}, the nebula can be divided in three regions:

- $log(f_{shock}/f_*) > -1$: region excited by shocks
- $-2 < log(f_{shock}/f_*) < -1$: transition region
- $log(f_{shock}/f_*) < -2$: region excited by photoionization

For NGC 2440 we used narrow filter images from Cuesta & Phillips (2000) to build the diagnostic diagrams. For NGC 6302 we used the data from Rauber *et al.* (2014), combined with velocities derived from our own long slit spectra. The values of f_{shock} and f_* were calculated using the formulae by Akras & Gonçalves. Velocity fields are also required to estimate correctly the contribution of shocks. For NGC 2440 they were modelled by Lago & Costa (2016) (*loc. cit.*) and NGC 6302 had its velocity field estimated from our long slit spectra.

The environment where shocks play an important role are those at high velocity regimes ($\sim$100 km/s), far from the central star, and in the rims of PNe, where there is interaction with the ISM. One of the main consequences is the reinforcement of the low ionization lines in these regions.

3. Results and conclusions

For NGC 2440 we examined points at the peripheral region of the nebula, chosen by morphological criteria, in particular rims and knots, and for NGC 6302 we used points throughout the entire nebula in both lobes.

Results indicate clearly that shocks play a crucial role in the spectra of both nebulae. For NGC 2440, our analysis shows that all points in the peripheral zones lie in the shock regime, with $log(f_{shock}/f_*) > -1$. This raises the hypothesis of interaction of the main nebula with a halo. For NGC 6302 the whole nebula was analysed along the symmetry axis, in steps of 4 arcsecs, and the result shows that many points in the inner part of the nebula lie in the transition zone, with $-2 < log(f_{shock}/f_*) < -1$, while the outer parts are in the shock regime.

ICFs based on line ratios are commonly used to derive elemental abundances in PNe since for most atoms only a few ions are available in the spectra. The presence of shocks reinforces the flux of low ionization lines, and the results discussed above show that, for these nebulae, shocks are present indeed. This artificial reinforcement of low ionization lines can lead to incorrect chemical abundances, when derived through ICFs, at least for type-I PNe.

A comprehensive study of this effect, applied to a larger PNe sample and including all types, ideally a statistically complete sample, would be required to clarify this proposition.

References

Akras, S. & Gonçalves, D.R. 2016, *MNRAS*, 455, 930

Cuesta, L. & Phillips, J.P. *ApJ*, 453, 754

Dopita, M.A. & Sutherland, M.S., *ApJS*, 102, 161

Lago, P.J.A. & Costa, R.D.D. 2016, *Rev.Mex.A.&A*, 52, 329

López, J. A.; Meaburn, J.; Bryce, M. & Holloway, A. J. 1998, *ApJ*, 493, 803

Rauber, A.B., Copetti, M.V.F. & Krabbe, A.C. 2014, *A&A*, 563, A42

Szyszka, C., Zijlstra, A.A., & Walsh, J.A. 2011, *MNRAS*, 416, 715

Why Galaxies Care About AGB Stars:
A Continuing Challenge through Cosmic Time
Proceedings IAU Symposium No. 343, 2019
F. Kerschbaum, M. Groenewegen & H. Olofsson, eds.

doi:10.1017/S1743921318004969

Unravelling the sulphur chemistry of AGB stars

Taïssa Danilovich†

Department of Physics and Astronomy, Institute of Astronomy, KU Leuven,
Celestijnenlaan 200D, 3001 Leuven, Belgium
email: taissa.danilovich@kuleuven.be

Abstract. There are clear differences in what sulphur molecules form in AGB circumstellar envelopes (CSEs) across chemical types. CS forms more readily in the CSEs of carbon stars, while SO and SO_2 have only been detected towards oxygen-rich stars. However, we have also discovered differences in sulphur chemistry based on the density of the CSE, as traced by mass-loss rate divided by expansion velocity. For example, the radial distribution of SO is drastically different between AGB stars with lower and higher density CSEs. H_2S can be found in high abundances towards higher density oxygen-rich stars, whereas SiS accounts for a significant portion of the circumstellar sulphur for higher density carbon stars.

Keywords. stars: AGB and post-AGB, circumstellar matter, stars: evolution

1. Introduction

Sulphur is the tenth most abundant element in the universe and its behaviour in terms of what molecules it forms has been seen to vary for different types of AGB stars. Sulphur is not synthesised in AGB stars nor in their main sequence progenitors and studies of post-AGB stars show us that sulphur is not significantly depleted onto dust during the AGB phase (Reyniers & van Winckel 2007; Waelkens *et al.* 1991), making it a good tracer of changing circumstellar chemistry across the AGB. This also means we can study sulphur-bearing molecules in AGB circumstellar envelopes (CSEs) by estimating the total sulphur abundance, based on the solar abundance.

The most common sulphur-bearing molecules found in AGB CSEs are CS, SiS, SO, SO_2 and H_2S. We conducted an APEX survey of these molecules across a range of AGB chemical types and mass-loss rates, accompanied by radiative transfer modelling (Danilovich *et al.* 2016, 2017, 2018). Combining these results with some ALMA observations from Decin *et al.* (2018), we here summarise the overarching trends.

2. Carbon-rich and S-type stars

For carbon stars, up to half the available sulphur is in the form of CS, with the other half in SiS for the higher mass-loss rate stars. For low mass-loss rate stars, we didn't detect any SiS and expect a much lower abundance if any is present at all. Notably, no SiS was detected towards any semi-regular variables in our study. The only other sulphur-bearing molecules seen to date are H_2S and organic molecules such as C_2S, C_3S (Cernicharo *et al.* 1987), and H_2CS (Agúndez *et al.* 2008), which together account for only a small portion of the available sulphur.

For S-type stars, both CS and SiS abundances seem correlated with each other and with CSE density. No H_2S, SO, or SO_2 lines were detected in our survey.

† Postdoctoral Fellow of the Fund for Scientific Research (FWO), Flanders, Belgium

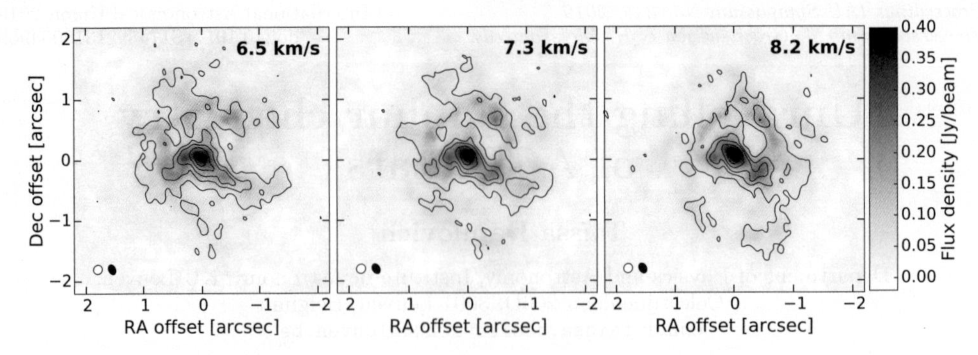

Figure 1. SO($8_8 \rightarrow 7_7$) (gradient, white beam) and SO$_2$($20_{1,19} \rightarrow 19_{2,18}$) (contours at 1, 3, 5, 10, and 20 times the rms noise, black beam) towards R Dor, close to $v_{\rm LSR} = 7.5$.

3. Oxygen-rich stars

3.1. *Low mass-loss rate, semi-regular variable*

Here the sulphur budget is accounted for by SO and SO_2, which are both formed close to the star and occupy a similar region of the CSE. This is particularly clear when comparing ALMA observations of the two molecules, shown in Fig. 1 for R Dor.

3.2. *Low mass-loss rate, Mira variable*

The sulphur budget for these stars is mostly accounted for by SO and SO_2, both formed close to the star. However, very sensitive ALMA observations reveal additional low abundances of CS and SiS, also formed (very) close to the star. For example, we find CS and SiS abundances a couple of orders of magnitude lower than for SO for W Hya.

3.3. *Intermediate mass-loss rate, Mira variable*

In such stars, the sulphur budget is shared in a more equal way between SO, SO_2, SiS and H_2S, with lower abundances of CS (which still requires shocks to form in the oxygen-rich CSE). Unlike the lower mass-loss rate stars, SO peaks in a shell close to the peak of OH abundance (which is produced from the photodissociation of H_2O), and with a lower abundance present close to the star. This behaviour is seen in IK Tau, towards which NS has also been detected, (Velilla Prieto *et al.* 2017) but not yet fully analysed.

3.4. *High mass-loss rate, Mira variable*

For the highest mass-loss rate AGB stars (probably including the extreme OH/IR stars), the sulphur budget is dominated by H_2S, with lower abundances of SO, SO_2, SiS and CS, which are still clearly seen. An example of such a star is V1300 Aql and a deeper analysis of the high mass-loss rate OH/IR stars is coming soon (APEX observations underway).

References

Agúndez, M., Fonfría, J. P., Cernicharo, J., Pardo, J. R., & Guélin, M. 2008, *A&A* 479, 493
Cernicharo, J., Kahane, C., Guelin, M., & Hein, H. 1987, *A&A* 181, L9
Danilovich, T., De Beck, E., Black, J. H., Olofsson, H., & Justtanont, K. 2016, *A&A* 588, A119
Danilovich, T., Ramstedt, S., Gobrecht, D., *et al.* 2018, *A&A* Forthcoming
Danilovich, T., Van de Sande, M., De Beck, E., *et al.* 2017, *A&A* 606, A124
Decin, L., Richards, A. M. S., Danilovich, T., Homan, W., & Nuth, J. A. 2018, *A&A* 615, A28
Reyniers, M. & van Winckel, H. 2007, *A&A* 463, L1
Velilla Prieto, L., Sánchez Contreras, C., Cernicharo, J., *et al.* 2017, *A&A* 597, A25
Waelkens, C., Van Winckel, H., Bogaert, E., & Trams, N. R. 1991, *A&A* 251, 495

Why Galaxies Care About AGB Stars:
A Continuing Challenge through Cosmic Time
Proceedings IAU Symposium No. 343, 2019
F. Kerschbaum, M. Groenewegen & H. Olofsson, eds.

doi:10.1017/S1743921318005392

MAGRITTE: a new multidimensional accelerated general-purpose radiative transfer code

Frederik De Ceuster[1,2], Jeremy Yates[1], Peter Boyle[3], Leen Decin[2,4] and James Hetherington[5]

[1]Dept. of Physics and Astronomy, University College London, London, WC1E 6BT, UK
email: frederik.deceuster@kuleuven.be

[2]Dept. of Physics and Astronomy, KU Leuven, Celestijnenlaan 200D, 3001 Leuven, Belgium

[3]School of Physics and Astronomy, The University of Edinburgh, Edinburgh EH9 3FD, UK

[4]School of Chemistry, University of Leeds, Leeds LS2 9JT, UK

[5]The Alan Turing Institute, Euston Road, London NW1 2DB, UK

Abstract. MAGRITTE is a new deterministic radiative transfer code. It is a ray-tracing code that computes the radiation field by solving the radiative transfer equation along a fixed set of rays for each grid cell. Its ray-tracing algorithm is independent of the type of input grid and thus can handle smoothed-particle hydrodynamics (SPH) particles, structured as well as unstructured grids. The radiative transfer solver is highly parallelized and optimized to have well scaling performance on several computer architectures. MAGRITTE also contains separate dedicated modules for chemistry and thermal balance. These enable it to self-consistently model the interdependence between the radiation field and the local thermal and chemical states. The source code for MAGRITTE will be made publically available at github.com/Magritte-code.

Keywords. radiative transfer, astrochemistry, methods: numerical

---

1. Introduction

Radiative transfer plays a key role in the dynamics, the chemistry and the energy balance of various astrophysical objects. Therefore, it is essential in astrophysical modelling to properly take into account all radiative processes and their interdependence. The ever growing size and complexity of these models requires fast and scalable methods to compute the radiation field. MAGRITTE is a new general-purpose radiative transfer solver written in modern C++. In contrast to popular (probabilistic) Monte Carlo codes, Magritte is a deterministic ray-tracer which computes the radiation field by solving the transfer equation along a fixed set of rays originating from each grid cell. Being a deterministic code allows for various optimizations and facilitates it to exploit the various layers of parallelism in the calculation.

2. Solving the transfer equation

MAGRITTE's ray-tracing algorithm only uses the locations of the cell centers and the nearest neighbor lists. Hence it can cope with SPH particle data as well as with structured or unstructured grids. The local emissivities and opacities account for contributions from both lines and continua. Scattering is taken into account iteratively, adding an extra source and opacity. MAGRITTE's solver is modular enough to cope with the most general anisotropic scattering formalisms. Once all emissivities and opacities are computed, the transfer equation is solved using a numerically more stable version

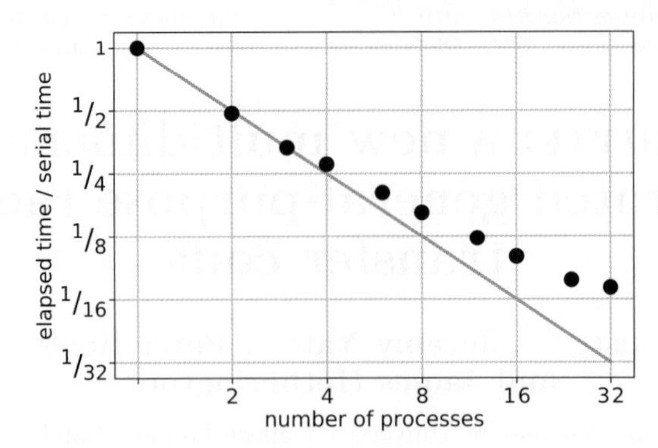

Figure 1. Plot of the (preliminary) strong scaling behavior of the MPI distributed parallelization over the rays for a test model containing 192 rays, 220 frequency bins and 12,133 grid cells. The total elapsed time when computed on a single core (serial time) was 221 seconds.

of Feautrier's second-order formulation (Feautrier 1964; Rybicki & Hummer 1991). MAGRITTE's transfer solver, the chemistry and thermal balance modules are designed to be modular to facilitate coupling it (as a whole or in parts) with other codes.

3. Parallelizing a deterministic ray-tracer

There are three common parallel programming paradigms: message passing, threading, and single instruction multiple data (SIMD) vectorization. MAGRITTE uses a combination of all three to ensure performance on both shared and distributed memory architectures. The computations for different rays in our algorithm are independent within an iteration. Therefore these can easily be distributed over different processes and the results communicated at the end of each iteration. This is done using the standard message passing interface (MPI). Figure 1 shows the (preliminary) scaling of this parallelization layer. Solving the transfer equation along a certain ray requires data from different grid cells and frequencies and thus can better be kept as local as possible in memory. Therefore, within each process, the computations for different cells are threaded using the OpenMP standard. The computations along a certain ray for different frequencies require exactly the same operations but with different values for the emissivities and opacities for each frequency. Hence these computations are ideally suited for SIMD vectorization. To achieve this in a portable way, Magritte uses the SIMD vector types provided in the GRID library (Boyle *et al.* 2016). In future versions, we will also explore the possibility of offloading the whole radiative transfer solver to graphics processing units (GPUs).

Acknowledgements

FDC is supported by the EPSRC iCASE studentship programme, Intel Corporation and Cray Inc. FDC and LD acknowledge support from the ERC consolidator grant 646758 AEROSOL. This work was performed using the Cambridge Service for Data Driven Discovery (CSD3), part of which is operated by the University of Cambridge Research Computing on behalf of the STFC DiRAC HPC Facility (www.dirac.ac.uk). The DiRAC component of CSD3 was funded by BEIS capital funding via STFC capital grants ST/P002307/1 and ST/R002452/1 and STFC operations grant ST/R00689X/1. DiRAC is part of the National e-Infrastructure.

References

Boyle, P., Yamaguchi, A., Cossu, G., & Portelli, A. 2016, *Proceedings of Science* 251 *Lattice* 2015
Feautrier, P. 1964, *Comptes Rendus Academie des Sciences (serie non specifiee)* 258
Rybicki, G. B. & Hummer, D. G. 1991, *Astron. Astrophys.* 245, 171

Why Galaxies Care About AGB Stars:
A Continuing Challenge through Cosmic Time
Proceedings IAU Symposium No. 343, 2019
F. Kerschbaum, M. Groenewegen & H. Olofsson, eds.

doi:10.1017/S1743921318006981

Stacking analysis of HERITAGE data to statistically study far-IR dust emission from evolved stars

Thavisha E. Dharmawardena[1,2], Francisca Kemper[1], Sundar Srinivasan[1], Sacha Hony[3], Olivia Jones[4] and Peter Scicluna[1]

[1]Academia Sinica Institute of Astronomy and Astrophysics,
11F of AS/NTU Astronomy-Mathematics Building,
No.1, Sect. 4, Roosevelt Rd, Taipei 10617, Taiwan, R.O.C.

[2]Graduate Institute of Astronomy, National Central University,
300 Zhongda Road, Zhongli 32001, Taoyuan, Taiwan, R.O.C.
email: tdharmawardena@asiaa.sinica.edu.tw

[3]Heidelberg University, Center for Astronomy,
Institute of Theoretical Astrophysics,
Albert-Ueberle-Str. 2, 69120 Heidelberg, Germany

[4]UK Astronomy Technology Centre, Royal Observatory,
Blackford Hill View, Edinburgh EH9 3HJ, UK

Abstract. We aim to analyse the co-added Herschel images of various categories of evolved stars in the LMC and SMC from the Herschel HERITAGE survey in order to identify, in a statistical sense, a cool historic dust mass component emitted by these sources. The fluxes derived from the co-added stacks can then be compared with those predicted by the GRAMS model grid in order to refine the DPRs estimated for the SMC and LMC.

Keywords. stars: AGB and post-AGB - galaxies: Magellanic Clouds - stars: mass-loss

---

1. Introduction

By fitting mid-IR SEDs of evolved stars in the Small Magellanic Cloud (SMC) and Large Magellanic Cloud (LMC) using the Grid of RSG and AGB ModelS (GRAMS) radiative transfer model grid, Riebel *et al.* (2012) and Srinivasan *et al.* (2016) estimated the dust budgets in the Magellanic clouds. However this method is primarily sensitive to the present day mass-loss rate and may not have taken into account a cooler older dust component visible only at longer, far-IR and sub-mm wavelengths.

Although emission from cool historic mass-loss will peak at Herschel wavelengths (Boyer *et al.* 2012), confusion caused by Interstellar Medium (ISM) background emission and limited spatial resolution makes it extremely difficult to detect it. Jones *et al.* (2015) only found 35 Herschel point source counter parts to the tens-of thousand evolved stars identified by Spitzer.

2. Stacking Analysis Methods

We revisit the Herschel HERITAGE data (Meixner *et al.* (2013)) in order to determine the presence of the cooler dust component in a statistical sense. In order to achieve this, we utilize a stacking and co-adding method of postage stamp size cutouts ($\sim 2' \times 2'$) of the Herschel HERITAGE maps of the SMC and LMC. By co-adding and averaging a stack

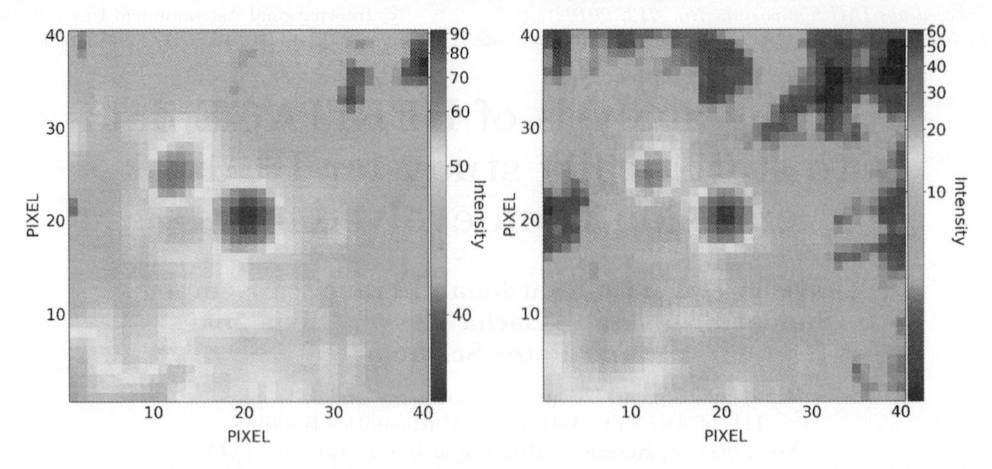

Figure 1. Left: PACS 160 μm stack of LMC Evolved stars from Jones *et al.* (2015); Right: same as left with a background annulus subtracted. The bright central source is our stacked emission. The bright source centred at pixel (12,25) to the top left of our source is a bright background emission source found in a few frames.

of sources we reduce the noise by $\sim \sqrt{No. of Observation}$, allowing us to detect fainter emission as the contrast improves. This method is more commonly used in extragalactic observations, and will allow us to improve the signal-to-noise of the cold dust emission. We divide the sample into sub-categories based on multiple parameters which can then be individually co-added and stacked. The stacks can be divided into subcategories, e.g. chemistry (C-rich, O-rich, S-type), mass-loss rates, initial mass (AGB, RSG), evolutionary state (E-AGB, TP-AGB, X-AGB).

The mean photometric fluxes derived from the co-added stack of each parameter will then be compared to the co-added fluxes predicted by GRAMS in order to identify the presence of the cold historic mass-loss component.

3. Preliminary Results and Future Work

Initial experiments have shown that confusion with background emission is currently the limiting part of the analysis. Co-addition of the 28 high mass-losing sources identified by Jones *et al.* (2015) show that while we enhance the source we also enhance the ISM background emission, which in the case of the LMC has significant structure (see Fig. 1, left). Subtracting using backgrounds estimated in an annulus centered at the source enhances the source significantly, however, there is still bright background structure, which affects the signal of the stacked source (see Fig. 1, right). Therefore we require a more complex background subtraction method such as polynomial subtraction to help remove the large scale structure.

Once we are able to successfully separate the stacked source from the background we will then be able to obtain accurate fluxes and hence refine the dust budget estimates determined for the sample by Riebel *et al.* (2012) and Srinivasan *et al.* (2016). The derived SED of the cold dust component can then be applied to analyse and adjust the average DPRs in the sample and thus the Magellanic clouds as a whole.

References

Boyer, M. L., Srinivasan, S., Riebel, D., *et al.* 2012, *ApJ*, 748, 40
Jones, O. C., Meixner, M., Sargent, *et al.* 2015, *ApJ*, 811, 145
Meixner, M., Panuzzo, P., Roman-Duval, J., *et al.* 2013, *AJ*, 146, 62
Riebel, D., Srinivasan, S., Sargent, B., and Meixner, M. 2012, *ApJ*, 753, 71
Srinivasan, S., Boyer, M. L., Kemper, F., *et al.* 2016, *MNRAS*, 457, 2814

Why Galaxies Care About AGB Stars: A Continuing Challenge through Cosmic Time
Proceedings IAU Symposium No. 343, 2019
F. Kerschbaum, M. Groenewegen & H. Olofsson, eds.

doi:10.1017/S1743921318005057

Observations of the Ultraviolet-Bright Star Barnard 29 in the Globular Cluster M13 (NGC 6205)

William V. Dixon[1], Pierre Chayer[1], I. N. Reid[1] and Marcelo Miguel Miller Bertolami[2]

[1]Space Telescope Science Institute,
3700 San Martin Drive,
Baltimore, MD 21218, USA
email: dixon@stci.edu

[2]Instituto de Astrofísica de La Plata, UNLP-CONICET,
Paseo del Bosque s/n, 1900 La Plata, Argentina

Abstract. We have analyzed *FUSE*, COS, GHRS, and Keck/HIRES spectra of the UV-bright star Barnard 29 in M13. Fits to the star's optical spectrum yield $T_{\rm eff} = 20,000 \pm 100$ K and $\log g = 3.00 \pm 0.01$. Using modern stellar-atmosphere models, we are able to reproduce the complex shape of the Balmer Hα feature. We derive photospheric abundances of He, C, N, O, Mg, Al, Si, P, S, Cl, Ar, Ti, Cr, Fe, Ni, and Ge. Barnard 29 exhibits an abundance pattern typical of the first-generation stars in M13, enhanced in oxygen and depleted in aluminum. We see no evidence of significant chemical evolution since the star left the RGB; in particular, it did not undergo third dredge-up. Previous workers found that the star's FUV spectra yield an iron abundance about 0.5 dex lower than its optical spectrum, but the iron abundances derived from all of our spectra are consistent with one another and with the cluster value. We attribute this difference to our use of model atmospheres without microturbulence. By comparing our best-fit model with the star's optical magnitudes, we derive a mass $M_*/M_\odot = 0.40 - 0.49$ and luminosity $\log L_*/L_\odot = 3.20 - 3.29$, depending on the cluster distance. Comparison with stellar-evolution models suggests that Barnard 29 evolved from a ZAHB star of mass $M_*/M_\odot \sim 0.50$, placing it near the boundary between the extreme and blue horizontal branches.

Keywords. stars: abundances — stars: atmospheres — stars: individual (NGC 6205 ZNG1) — ultraviolet: stars

---

The most luminous object in Messier 13 (NGC 6205) is the famous star Barnard 29. It is a UV-bright star, brighter than the horizontal branch and bluer than the red-giant branch. Some UV-bright stars are post-AGB stars, evolving from the asymptotic giant branch (AGB) to the white-dwarf cooling curve at high luminosity; others are AGB-manqué stars, evolving directly from the extreme horizontal branch (EHB) at lower luminosity.

We analyze archival spectra of Barnard 29 obtained with *FUSE*, with COS and GHRS aboard *HST*, and with the Keck HIRES. We derive stellar parameters $T_{\rm eff} = 20,000 \pm 100$ K, $\log g = 3.00 \pm 0.01$, and $\log N({\rm He})/\,N({\rm H}) = -0.89 \pm 0.01$. The star's photospheric abundances are plotted in Fig. 1. Our FUV results are roughly consistent with the optical values (where available) and with each other, though the scatter is sometimes greater than the error bars. Our optical results are consistent with those of Thompson *et al.* (2007), which are also shown. Preferred values are plotted as stars.

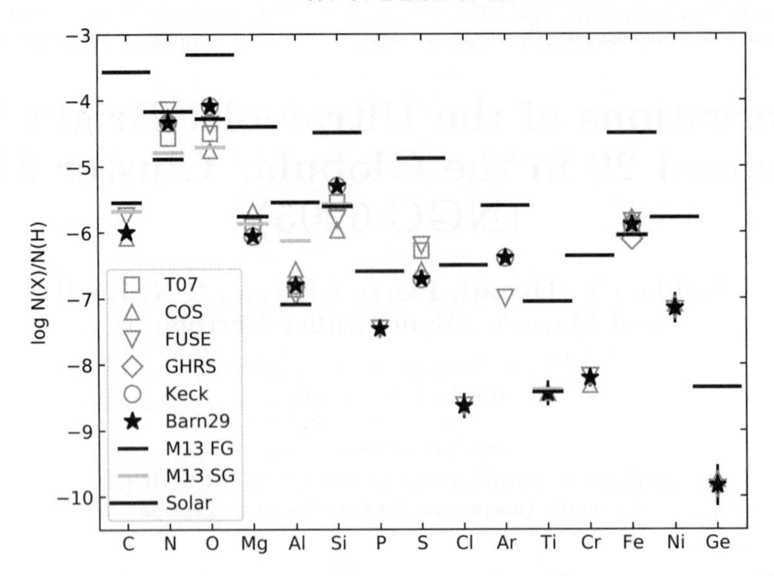

Figure 1. Photospheric abundances of Barnard 29 (stars), the solar photosphere (long black lines), and the first-generation (FG, short black lines) and second-generation (SG, short grey lines) stars in M13. Open squares are from fits to the star's optical spectrum by Thompson *et al.* 2007. Other symbols are from fits to the COS, *FUSE*, GHRS, and Keck spectra, as indicated.

We independently derive the iron abundance of Barnard 29 from its Keck, *FUSE*, COS, and GHRS spectra. As shown in Fig. 1, our values are consistent with one another and with the cluster iron abundance ($\log N(\mathrm{Fe})/N(\mathrm{H}) = -6.05$; Mészáros *et al.* 2015). Previous authors have found that the iron abundance derived from the star's far-ultraviolet spectrum is roughly 0.5 dex lower than that derived from optical data, which is consistent with the cluster mean (Moehler *et al.* 1998, Thompson *et al.* 2007). Moehler *et al.* adopt the microturbulent velocity $\xi = 10$ km s^{-1} derived by Conlon *et al.* (1994). All of our models were computed with a value $\xi = 0$ km s^{-1}, which yields line profiles that better reproduce the HIRES spectrum. Computing models with $\xi = 10$ km s^{-1} and re-fitting the GHRS spectrum yields a best-fit iron abundance $\log N(\mathrm{Fe})/N(\mathrm{H}) = -6.74 \pm 0.11$, matching the Moehler *et al.* result.

Galactic globular clusters host multiple stellar populations. First-generation (FG) stars display abundances typical of halo field stars, while second-generation (SG) stars, which may have multiple subpopulations, are enriched in Na and Al and depleted in O and Mg. Models suggest that the second generation is formed from gas polluted by material expelled by massive stars of the first generation. Mészáros *et al.* (2015) identified two stellar populations on the RGB of M13. FG stars (short black lines in Fig. 1) are richer in oxygen and poorer in aluminum than SG stars (short grey lines). Barnard 29 is clearly a FG star, richer in oxygen and poorer in aluminum. The star is slightly enhanced in nitrogen; otherwise, its abundances appear to have changed little since it left the RGB. In particular, its low carbon abundance ($N_C/N_O = 0.01$) indicates that the star did not undergo third dredge-up.

References

Conlon, E. S., Dufton, P. L., & Keenan, F. P. 1994, *A&A*, 290, 897
Mészáros, S., *et al.* 2015, *AJ*, 149, 153,
Moehler, S., Heber, U., Lemke, M., & Napiwotzki, R. 1998, *A&A*, 339, 537
Thompson, H. M. A., Keenan, F. P., Dufton, P. L., *et al.* 2007, *MNRAS*, 378, 1619

Why Galaxies Care About AGB Stars:
A Continuing Challenge through Cosmic Time
Proceedings IAU Symposium No. 343, 2019
F. Kerschbaum, M. Groenewegen & H. Olofsson, eds.

doi:10.1017/S1743921318006968

A systematic survey of grain growth in discs around post-AGB binaries with PACS and SPIRE photometry

K. Dsilva, H. Van Winckel and J. Kluska

Instituut voor Sterrenkunde, KU Leuven,
Celestijnenlaan 200D, 3001 Leuven, Belgium
email: karansingh.dsilva@student.kuleuven.be

Abstract. Post-AGB stars are the final stage of evolution of low-intermediate mass stars ($M < 8\,M_{\odot}$). Those in binary systems have stable circumbinary discs. Using data from Herschel (PACS/SPIRE), we extend the SEDs of 50 galactic post-AGB binary systems to sub-millimetre wavelengths and use the slope of the SED as a diagnostic tool to probe the presence of large grains. Using a Monte Carlo radiative transfer code (MCMax), we create a large grid of models to quantify the observed spectral indices, and use the presence of large grains in the disc as a proxy for evolution.

Keywords. stars: AGB and post-AGB - circumstellar matter - binaries: general - techniques: photometric - infrared: stars

---

1. Introduction

The presence of a binary companion drastically affects the evolution of an AGB star. Mass loss, mass transfer and tidal interaction strongly impact the outcome of this interaction (Pols 2004). Through long-term radial velocity monitoring, it is now well established that post-AGB binary systems are often surrounded by long-lived, stable circumbinary discs (van Winckel *et al.* 2009; van Winckel 2017). Interaction between the circumbinary disc and the central system might impact the evolution of the system, and hence it is important to understand the evolution of the disc. The formation and lifetime of the discs are unknown. Here, we present a systematic survey of 50 galactic post-AGB binaries and with the help of radiative transfer models, attempt to use the grain-sizes in the disc as a proxy for evolution.

2. Methods

The first step was to obtain accurate photometric fluxes from the PACS (Poglitsch *et al.* 2010) and SPIRE (Griffin *et al.* 2010) instruments on board the Herschel space telescope. This was done using the recommended software HIPE (Ott 2010).

Once the fluxes were obtained, they were added to the respective SEDs of the sources. In order to systematically study the sample, a comparison of the PACS and SPIRE slopes (called the spectral index) were made, called n_{160} and n_{500} respectively. Sources with missing fluxes or upper limits were excluded, and we proceeded with 37 targets.

Upon comparing n_{500} and n_{160}, the relative change δn in slope was measured and the sample was classified into three types:

- **Type 1**: Slopes with a knee like feature with $\delta n > 10\%$ (13/37 sources).
- **Type 2**: No change in slope with $\delta n < 10\%$ (12/37 sources).
- **Type 3**: An infrared excess with $\delta n > 10\%$ (12/37 sources).

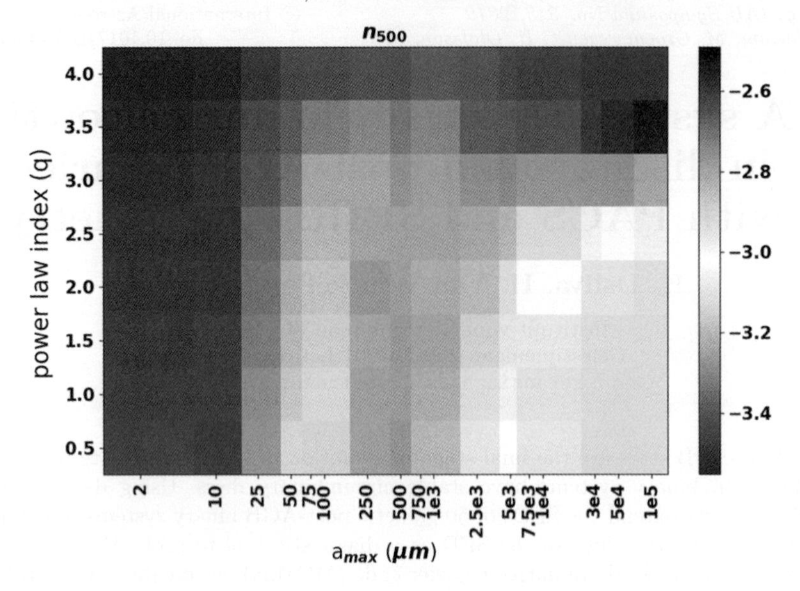

Figure 1. The synthetic spectral indices (n_{500})

After a parameter study using the radiative transfer code MCMax (Min *et al.* 2009), the maximum grain-size (a_{max}) and the grain-size distribution power-law index (q) were varied across a fixed input for a second generation disc with parameters adapted from Kluska *et al.* (2018). The reader is referred to Min *et al.* (2009) for an in-depth explanation of the parameters. The parameter space can be seen in Figure 1.

3. Results and discussion

Using Figure 1, we can compare the observed value of n_{500} to the synthetic spectral index and constrain the appropriate q and a_{max} value for each source.

We observe that the spectral index at 500 is the most sensitive to the presence of large grains and can reliably be used to q and a_{max}, which helps in reducing the degeneracies that plague disc modelling.

The maximum grain size in all sources is much larger than what is thought to be during the AGB wind phase ($a_{max} \gg 0.1\,\mu$m). The interpretation made here is that the timescale of grain-growth is relatively short, and hence grain-growth is omnipresent.

Many discs show grains of the order of 1 mm. Further follow-ups of 'pebble'-sized grains in these second-generation discs could help in understanding grain-growth in protoplanetary discs.

References

van Winckel, H. 2017, Planetary Nebulae: Multi-Wavelength Probes of Stellar and Galactic Evolution, 323, 231

van Winckel, H., Lloyd Evans, T., Briquet, M., *et al.* 2009, 505, 1221

Pols, O. R. 2004, 75, 749

Poglitsch, A., Waelkens, C., Geis, N., *et al.* 2010, 518, L2

Griffin, M. J., Abergel, A., Abreu, A., *et al.* 2010, 518, L3

Ott, S. 2010, Astronomical Data Analysis Software and Systems XIX, 434, 139

Min, M., Dullemond, C. P., Dominik, C., de Koter, A., & Hovenier, J. W. 2009, 497, 155

Kluska, J., Hillen, M., Van Winckel, H., *et al.* 2018, 616, A153

Why Galaxies Care About AGB Stars:
A Continuing Challenge through Cosmic Time
Proceedings IAU Symposium No. 343, 2019
F. Kerschbaum, M. Groenewegen & H. Olofsson, eds.

doi:10.1017/S1743921318006592

The loss of large amplitude pulsations at the end of AGB evolution

D. Engels[1], S. Etoka[2] and E. Gérard[3]

[1]Hamburger Sternwarte, Universität Hamburg, Germany,
email: dengels@hs.uni-hamburg.de

[2]Jodrell Bank Centre for Astrophysics, University of Manchester, UK,
email: sandra.etoka@googlemail.com

[3]GEPI, Observatoire de Paris, Meudon, France,
email: eric.gerard@obspm.fr

Abstract. Since 2013, we are performing with the Nancay Radio Telescope (NRT) a monitoring program of >100 Galactic disk OH/IR stars, having bright 1612-MHz OH maser emission. The variations of the maser emission are used to probe the underlying stellar variability. We wish to understand how the large-amplitude variations are lost during the AGB – post-AGB transition. The fading out of pulsations with steadily declining amplitudes seems to be a viable process.

Keywords. stars: AGB and post-AGB, masers, stars: evolution

Stars evolving on the thermal-pulsing Asymptotic Giant Branch (AGB) are in general observed as large-amplitude variables, but are almost non-variable in the post-AGB phase. In models covering the AGB –post-AGB transition, the evolutionary timescales depend on the assumptions of the change of the mass-loss rates. They must drop on short timescales from late AGB values of 10^{-5}–10^{-4} to post-AGB values of 10^{-7}–10^{-8} $M_{\odot}$ yr^{-1}. While the mass loss rates are parametrized on the AGB as a function of pulsation period, they are completely unconstrained starting with the time after which the pulsation ceased until the time that a radiation driven wind as observed in Planetary Nebulae takes over (Miller Bertolami 2016; MB16 hereafter).

Towards the end of AGB evolution, stars can develop very high mass loss rates, which enshrouds them completely by dust and gas. Among them are the OH/IR stars, which encompass large-amplitude variables on the AGB (L-AGB stars) with periods ~ 700–2000 days and almost non-variable stars (S-pAGB: small amplitude post-AGB stars, including 'non-variable' stars), which are thought to evolve in the early post-AGB phase. In both phases, the stars are still deeply embedded in their dusty circumstellar shell. H_2O and OH maser emissions are present in both phases. The association of the S-pAGB stars with the post-AGB phase is supported by observations that some of them already have diluted dust shells (Engels 2007), which indicate a recent decrease of the mass loss rates, and that others show prominent bipolar outflows (f.e. OH17.7–2.0 = IRAS 18276–1431, Sánchez-Contreras *et al.* 2007; OH 53.6–0.2 = IRAS 19292+1806, Sahai *et al.* 2007) including "water fountains" (f.e. W43A = OH 31.0+0.0 = IRAS 18450–0148, Chong *et al.* 2015). It is during the obscured phase that (at least in the more massive stars) the AGB – post-AGB evolutionary transition takes place and the stars stop pulsating.

Monitoring the stars via their bright and relatively stable OH maser emission is needed, because especially the S-pAGB candidates have very red spectral energy distributions, and cannot be monitored in the optical or the near-infrared. As a basic sample to study the transformation of the variability characteristics, we use the full sample of OH/IR stars of Baud *et al.* (1981), updated by Engels & Jiménez-Esteban (2007). This "Bright

OH/IR star sample" comprises 115 stars, with almost all located at $10 < l < 150°$, $|b| < 4°$ along the Galactic plane. It is quite complete for bright 1612-MHz OH masers ($F_\nu > 4$ Jy). The brighter part of the sample has been monitored by Herman & Habing (1985)(herafter HH85), who reported several sub-groups with different amplitudes and periodicity among S-pAGB stars. Objects, which are currently transiting from L-AGB to S-pAGB variability may hide in the sample. To find them, we are monitoring, since 2013, the 1612-MHz OH masers with the NRT, to probe the underlying stellar variability.

In our sample, the L-AGB and S-pAGB stars are almost of equal number. Assuming similar OH maser luminosities, this implies that the "pulsating" phase connected to relatively high mass-loss rates ($\dot{M} > 10^{-5}\,M_\odot$/yr) is of similar duration as the early post-AGB phase (Engels 2002). OH/IR stars must have experienced hot bottom burning on the AGB to avoid being converted to carbon-rich stars, and as such they must have had massive progenitors on the main sequence $M \geqslant 3\,M_\odot$. According to MB16, the predicted transition times τ_{tr} during the early (and obscured) post-AGB phase until the optical reappearance of the central stars last only < 1000 years. In the later post-AGB phase, the dust shells are dispersed, and, in general, maser emission disappears. Assuming a minimum lifetime of the OH maser emission in the "Bright OH/IR stars sample" of 2000 years (Engels & Jiménez-Esteban 2007), the time for massive AGB stars to appear as obscured OH/IR stars can last only a few thousand years.

As of May 2018, we have the variability characterizations for 52 stars (34 L-AGB, 18 S-pAGB). Another 28 stars are currently (2018/2019) monitored to obtain a characterization, while the remaining stars are planned to be monitored in 2020/2021. Monitoring of newly recognized L-AGB stars is continued until the period is determined. S-pAGB stars are re-observed occasionally to search for long-term trends, such as found by Wolak *et al.* (2014). They reported that the OH maser of one of the S-pAGB stars, OH 17.7–2.0, is continuously fading since its discovery and predict that the maser will fall below the detection limit around 2030. While the L-AGB stars in HH85 are confirmed, some of their S-pAGB stars had to be reclassified as L-AGB variables with periods $P > 1000$ days. Among 15 OH/IR stars not monitored by HH85, we found 9 L-AGB stars (60%), while the reminder shows at most irregular fluctuations qualifying them as S-pAGB stars.

No stars with short-period, small amplitude pulsations have been found, as assumed to exist as transition objects by Blöcker (1995). However, we found a couple of stars, which show periodic variations with periods similar to those of L-AGB stars but with significantly smaller amplitude (Engels *et al.* 2018). We consider them as the best candidates for transition objects. While an instantaneous cessation of the pulsation (Vassiliadis & Wood 1994; MB16) cannot be ruled out, we consider the fading out of pulsations with steadily declining amplitudes (damped oscillator) as a viable process.

References

Baud, B., Habing, H. J., Matthews, H. E., & Winnberg, A., 1981, *A&A*, 95, 156

Blöcker, T., 1995, *A&A*, 299, 755

Chong S.-N., Imai H., & Diamond, P.J., 2015, *ApJ*, 805, 53

Engels, D., 2002, *A&A*, 388, 252

Engels, D., 2007, "Asym. Planetary Neb. IV; http://www.iac.es/proyect/apn4, article #52"

Engels, D. & Jiménez-Esteban, F., 2007, *A&A*, 475, 941

Engels, D., Etoka, S., West, M., & Gérard, E., 2018, "Astrophysical Masers, IAUS 336", p. 389

Herman, J. & Habing, H. J., 1985, *A&AS*, 59, 523 (HH85)

Miller Bertolami, M. M., 2016 *A&A*, 588, A25 (MB16)

Sahai, R., Morris, M., Sánchez Contreras, C., & Claussen, M., 2007, *AJ*, 134, 2200

Sánchez Contreras, C., Le Mignant, D., Sahai, R., *et al.*, 2007, *ApJ*, 656, 1150

Vassiliadis, E. & Wood, P. R., 1994, *ApJS*, 92, 125

Wolak, P., Szymczak, M., Bartkiewicz, A., & Gérard, E., 2014, "EVN 2014; http://pos.sissa.it/cgi-bin/reader/conf.cgi?confid=230, id.116"

Why Galaxies Care About AGB Stars:
A Continuing Challenge through Cosmic Time
Proceedings IAU Symposium No. 343, 2019
F. Kerschbaum, M. Groenewegen & H. Olofsson, eds.

doi:10.1017/S1743921318007299

A DARWIN C-star model grid with new dust opacities

Kjell Eriksson[1], Susanne Höfner[1] and Bernhard Aringer[2]

[1]Dept. of Astronomy & Space Physics, Uppsala University,
Box 516, SE-75120 Uppsala, Sweden
email: kjell.eriksson@physics.uu.se

[2]Dipartimento di Fisica e Astronomia Galileo Galilei, Universita di Padova,
Vicolo dell'Osservatorio 3, I-35122 Padova, Italy

Abstract. We have improved the treatment of dust opacity from the small-particle limit approximation to size-dependent which leads to models with smaller grains, lower dust-to-gas ratios, but about the same mass-loss rates and outflow velocities. The K-magnitudes get brighter, whereas the V-magnitudes can be either brighter or dimmer depending on the wind properties.

Keywords. stars: AGB and post-AGB, stars: winds, outflows, stars: carbon

1. Introduction

Winds of AGB stars are presumably driven by a combination of pulsation-induced shock waves and radiation pressure on dust. We have computed a grid of RadiationHydroDynamic atmosphere+wind models for C-rich AGB stars using the DARWIN code. It calculates time-dependent radial structures by solving the hydrodynamical equations using non-gray radiation transport and time-dependent dust formation/destruction. The gas opacities come from the COMA11 code. The main difference to the Eriksson *et al.* (2014) grid is that we now use **size-dependent dust opacities (SDO)** instead of the **small-particle limit (SPL)** approximation for the amorphous carbon grains, see also Mattsson & Höfner (2011). For each model, synthetic spectra and photometry in the range $0.3 - 25\,\mu$m were computed *a posteriori* with the COMA11 code (Aringer *et al.* 2009) for selected time-steps (about 200 covering about six pulsation periods per model).

2. Grid parameters

The grid parameters are the same as in Eriksson *et al.* (2014): **Teff**: 2600, 2800, 3000, 3200 K, **log L**: 3.55, 3.70, 3.85, 4.00 $L_{\odot}$, and **Current mass**: 0.75, 1.0, 1.5, 2.0 $M_{\odot}$. For each of these, we use three carbon excesses: **log (C-O)**: 8.2, 8.5, 8.8, and three **pulsational amplitudes**: 2, 4, 6 km/s.

3. Results

As seen in Fig. 1, showing quantities for the SDO vs. the SPL models, we find that: the mass-loss rates and the outflow velocities are similar, while the carbon condensation degreees, the grain sizes, and the dust-to-gas mass ratios are smaller.

The dust opacity in the visual region with its high-momentum radiation is increased compared to the SPL case, especially for grains with radii of $0.1 - 0.4\,\mu$m. This gives a higher outward acceleration and the dust grains then move faster through the dust

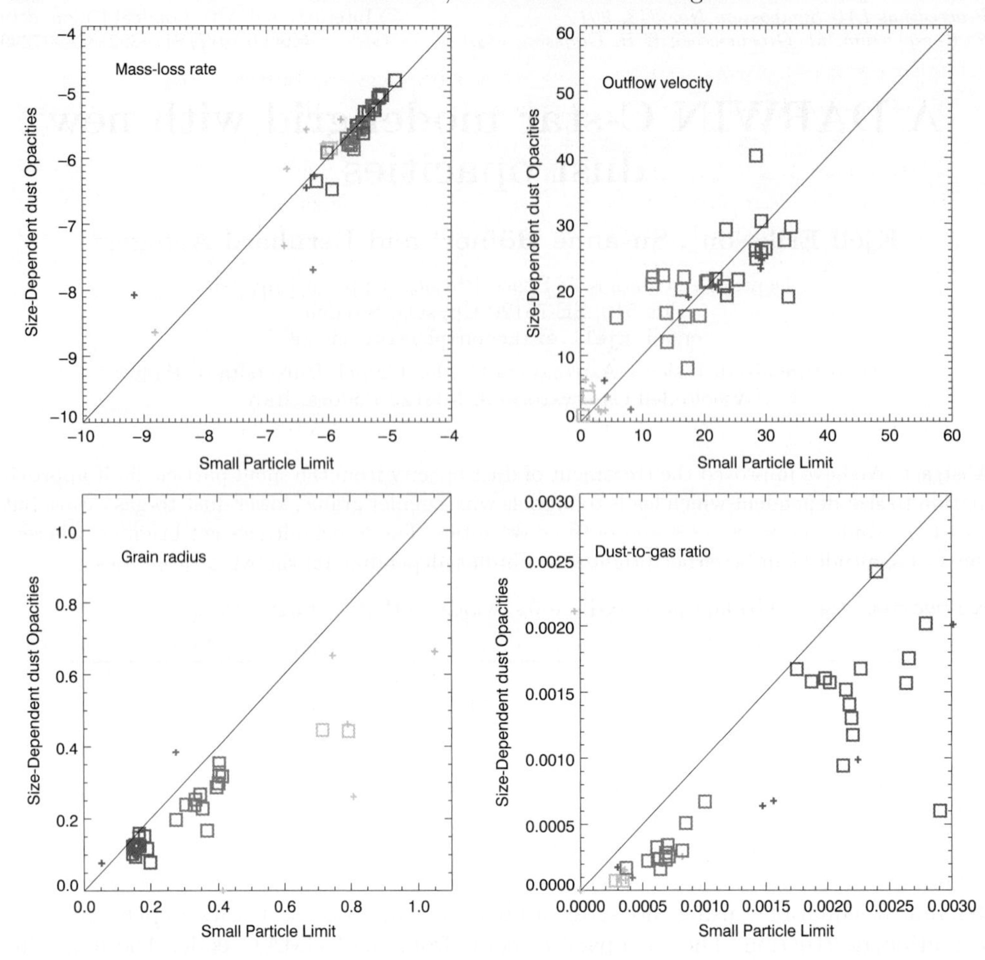

Figure 1. Comparing SDO results to the SPL ones. Squares: wind models, pluses: episodic models. Colours denote the carbon-excess: Green: 8.2, Blue: 8.5, Red: 8.8. Only results for 1 solar mass models are shown.

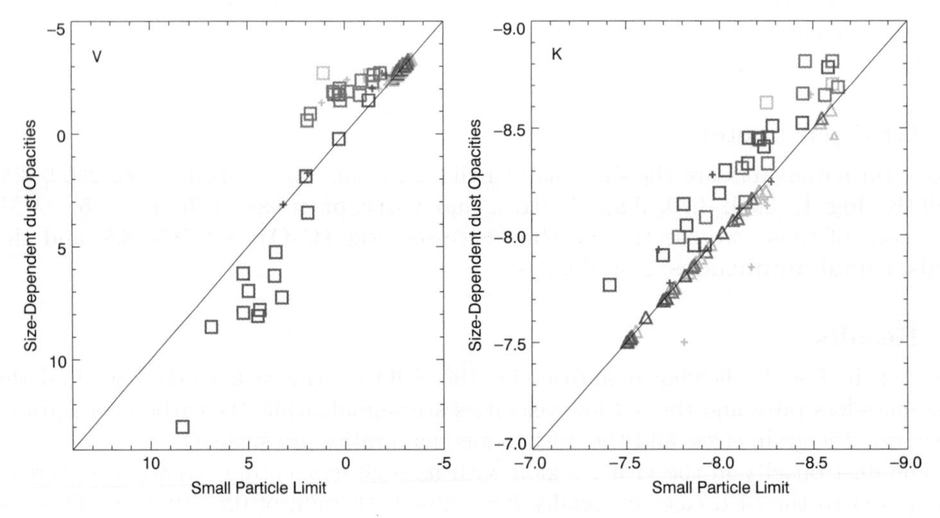

Figure 2. V and K photometry for SDO vs. SPL case. Symbols as in Fig. 1; in addition triangles denote no-wind models. Only 1 solar mass models are shown.

condensation zone. Hence, they will be smaller, like the condensation degree and the dust-to-gas mass ratio.

The different treatment of the dust opacity also affects the synthetic spectroscopy and photometry. For the V magnitude, we see in Fig. 2 that models with moderate winds (say, with $V < 0$) display brighter V magnitudes than in the SPL case: this is due to smaller dust condensation making the maxima brighter. Also the minima in the SPL case are usually significantly deeper. For the more massive winds in cool, luminous and carbon-rich models, the larger dust opacity makes V dimmer. For the K magnitudes, we see that for almost all models with a wind it is brighter than in the SPL case due to increased dust emission. More details will be given in Eriksson *et al.* (2018, in preparation).

References

Aringer, B., Girardi, L., Nowotny, W., Marigo, P., & Lederer, M.T. 2009, *A&A*, 503, 913
Eriksson, K., Nowotny, W., Höfner, S., Aringer, B. & Wachter, A. 2014, *A&A*, 566, A95
Mattsson, L., & Höfner, S. 2011, *A&A*, 533, A42

Why Galaxies Care About AGB Stars:
A Continuing Challenge through Cosmic Time
Proceedings IAU Symposium No. 343, 2019
F. Kerschbaum, M. Groenewegen & H. Olofsson, eds.

doi:10.1017/S1743921318005896

Binary interaction along the RGB: The Barium Star perspective

A. Escorza[1,2], L. Siess[2,3], D. Karinkuzhi[2], H. M. J. Boffin[4], A. Jorissen[2] and H. van Winckel[1]

[1]Institute of Astronomy, KU Leuven, Celestijnenlaan 200D, B-3001 Leuven, Belgium
email: ana.escorza@kuleuven.be

[2]Institut d'Astronomie et d'Astrophysique, Université Libre de Bruxelles, Campus Plaine C.P. 226, Boulevard du Triomphe, B-1050 Bruxelles, Belgium

[3]F.R.S.-FNRS, Belgium

[4]ESO, Garching bei München, Germany

Abstract. Barium (Ba) stars form via mass-transfer in binary systems, and can subsequently interact with their white dwarf companion in a second stage of binary interaction. We used observations of main-sequence Ba systems as input for our evolutionary models, and try to reproduce the orbits of the Ba giants. We show that to explain short and sometimes eccentric orbits, additional interaction mechanisms are needed along the RGB.

Keywords. stars: binaries, stars: evolution

---

1. Introduction

Stars with extended convective envelopes, like red or asymptotic giant branch (RGB or AGB) stars, in binary systems can fill a substantial fraction of their Roche Lobe and exchange angular momentum and possibly mass with their companion. In this study, we considered a family of chemically peculiar stars known as barium (Ba) stars to investigate binary evolution along the RGB.

Ba stars are main-sequence or giant stars formed due to binary interaction when a former AGB companion, which is now a dim white dwarf (WD), polluted them with heavy elements (McClure 1984). The interaction with the former AGB companion, i.e. the formation of the Ba star, is not well understood. However, we focused on the evolutionary link between dwarf and giant Ba stars which could be affected by a second phase of binary interaction between the Ba star ascending the RGB and its WD companion.

2. Methods

In Escorza *et al.* (2017), we presented a Hertzsprung-Russell diagram (HRD) of Ba stars using photometric stellar parameters and TGAS (Tycho-Gaia Astrometric Solution; Lindegren *et al.* 2016) parallaxes. Our database includes a sample of 90 objects which have a fully covered binary orbit (Jorrissen *et al.* 2016 and Escorza *et al.* in prep). We redetermined the stellar parameters of these from high-resolution spectra, and we used distances from Bailer-Jones *et al.* (2018) to compute new, more accurate luminosities. Finally, we determined the individual masses of the Ba stars by comparing their location on the HRD with a new grid of STAREVOL (Siess 2006) evolutionary models.

We used observations (masses, periods and eccentricities) of the main sequence Ba stars as input parameters for a grid of standard binary evolutionary models computed with the BINSTAR code (Siess *et al.* 2013). The grid of initial parameters covered four

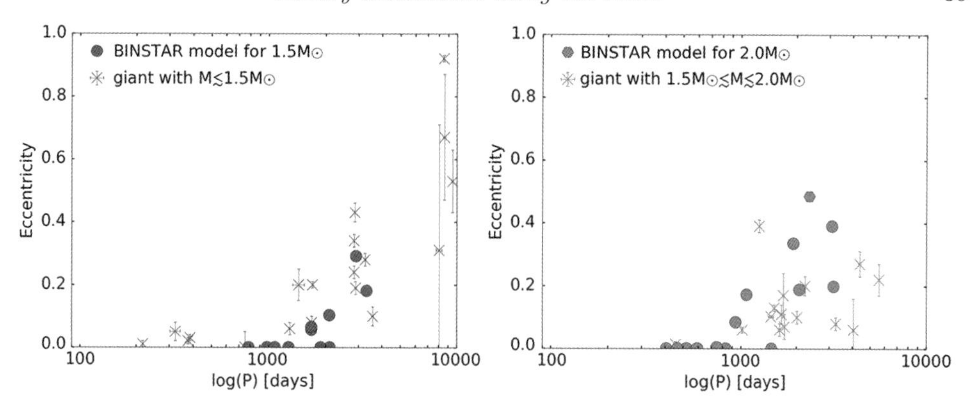

Figure 1. Observed and modelled orbits of Ba stars after the RGB phase.

initial masses: 1.5, 2.0, 2.5 and 3.0 $M_\odot$; six orbital periods: 100, 300, 600, 1000, 2000 and 3000 days; and three eccentricities: 0.2, 0.4 and 0.6. Since our goal was to investigate the evolution of the orbits during the RGB phase of the Ba star, the secondary was chosen to be a cool WD, and we let the systems evolve until the onset of core He-burning. Then we compared the final orbital parameters with observations of Ba giant systems.

3. Results and discussion

Figure 1 shows observed periods and eccentricities of Ba giants (crosses) and the final orbits of the models with 1.5 and 2.0 $M_\odot$ that reached the core He-burning phase (circles). Systems in which the interaction made the star leave the RGB phase or systems that did not significantly interact (those with 2.5 and 3.0 $M_\odot$) are not included in the figures.

Among the low-mass giants ($M \lesssim 1.5\,M_\odot$), there are several systems with periods shorter than those predicted by the models ($P < 700$ days). Other systems with 1000 days $\lesssim P \lesssim 3000$ days are significantly more eccentric than the models of the corresponding mass and period. This indicates that additional interaction mechanisms are operating during the RGB phase, independently of the past interaction with the former AGB primary that led to the pollution of the present Ba star.

We now plan to test several mechanisms that might help us reproduce the observed orbits better, for example, a tidally enhanced mass-loss (e.g. Tout & Eggleton 1988) during the RGB phase or a reduction of the tidal efficiency (e.g. Nie *et al.* 2017). Additionally, we plan to complement our observations with extrinsic S stars, which are thought to be the low-mass and more evolved counterparts of Ba stars, and might fall among the unpredicted orbits.

References

Bailer-Jones, C. A. L., Rybizki, J., Fouesneau, M., *et al.* 2018, AJ, 156, 58
Escorza, A., Boffin, H. M. J., Jorissen, A., *et al.* 2017, A&A, 608, A100
Jorissen, A., Van Eck, S., Van Winckel, H., *et al.* 2016, A&A, 586, A158
Lindegren, L., Lammers, U., Bastian, U., *et al.* 2016, A&A, 595, A4
McClure, R. D. 1984, PASP, 96, 117
Nie, J. D., Wood, P. R., & Nicholls, C. P. 2017, ApJ, 835, 209
Siess, L. 2006, A&A, 448, 717
Siess, L., Izzard, R. G., Davis, P. J., & Deschamps, R. 2013, A&A, 550, A100
Tout, C. A., & Eggleton, P. P. 1988, MNRAS, 231, 823

Why Galaxies Care About AGB Stars:
A Continuing Challenge through Cosmic Time
Proceedings IAU Symposium No. 343, 2019
F. Kerschbaum, M. Groenewegen & H. Olofsson, eds.

doi:10.1017/S1743921318006609

Ammonia in C-rich stars

Bartosz Etmański, Mirosław R. Schmidt, Bosco H. K. Yung and Ryszard Szczerba

Nicolaus Copernicus Astronomical Centre of the Polish Academy of Sciences,
Departament of Astrophysics, 87-140 Toruń, Rabiańska 8, Poland

Abstract. HIFI instrument onboard the Herschel satellite provided an unprecedented number of detections of rotational transitions of ammonia in circumstellar envelopes of evolved stars including massive red supergiants, Asymptotic Giant Branch (AGB), and post-AGB stars. The chemistry of ammonia formation in the circumstellar envelopes of evolved stars is poorly understood. The mechanisms proposed for its formation are processes behind the shock front, photochemistry in the inner part of the clumpy envelope, and formation on dust grains. We present results of the non−local thermodynamical equilibrium (non−LTE) radiative transfer modeling of ammonia transitions, mainly of the ground-state rotational one NH_3 $J_K = 1_0 - 0_0$ at 572.5 GHz, in selected AGB stars, aiming at the quantitative estimation of the NH_3 abundance. The model of ammonia includes IR radiative pumping via $v_2 = 1$ vibrational band at 10 μm.

Keywords. line: profiles - submilimeter: stars - stars: AGB and post-AGB - circumstellar matter

1. Introduction

Ammonia is the molecule which occurs in many different astrophysical objects, like dark clouds, merger remnants, massive star-forming regions, protoplanetary discs and in the circumstellar envelopes (CSE) around AGB stars. Its abundance found in CSE is much higher than predicted by models of stellar atmospheres of cool (super)giants, which is of order of 10^{-12}–10^{-10} relative to H_2 (Johnson & Sauval 1982). Abundances of ammonia in circumstellar environments of AGB stars are much higher ($\sim 10^{-6}$–10^{-7}). One of the explanations proposed to resolve this problem is dissociation of N_2 molecules by passage of shocks (Willacy & Cherchneff 1998). In C-rich stars, this process changes the abundance of ammonia only slightly (Gobrecht *et al.* 2016). Another explanation is that the clumpy CSE may be penetrated by the galactic background ultraviolet radiation even to the inner part of the envelope (Decin *et al.* 2010) producing atomic nitrogen. However, Li *et al.* (2016) have shown that N_2 may survive in inner regions due to self-shielding.

Recently, Schmidt *et al.* (2016) analysed all the lowest nine rotational transitions acquired with the HIFI instrument updated by observations of inversion lines of ammonia (Gong *et al.* 2015) in the very well studied C-rich AGB star CW Leo. Here, we follow their approach for the analysis of other C-rich objects.

2. Analysis and results

In Table 1, we present derived abundances of ortho-ammonia in five C-rich AGB stars together with their basic parameters. We estimate that the accuracy of determinations of ortho-ammonia abundances should be better than a factor of two.

For the modelling of molecular lines, we have applied the following procedure. The spectral energy distribution was modelled using MRT (Szczerba *et al.* 1997), and occasionally

Table 1. The basic parameters of analysed AGB stars and their CSE.

Object	P (days)	V_{LSR} (km s^{-1})	L_{star} (10^3 $L_\odot$)	T_{star} (K)	M_{loss} $M_\odot$ yr^{-1}	Distance (pc)	V_{exp} (km s^{-1})	f(o-NH_3)
V Cyg	421	15.0	6.6	1875	1.7×10^{-6}	458[1]	10.5	0.9×10^{-7}
CIT 6	640	−1.8	8.3	2470	5.2×10^{-6}	589[1]	16.5	1.8×10^{-7}
II Lup	580	−15.0	9.1	2000	1.5×10^{-5}	1917[1]	21.0	5.0×10^{-7}
LP And	614	−17.0	9.7	2040	2.2×10^{-5}	400	13.5	3.5×10^{-8}
V384 Per	535	−16.2	8.4	1820	4.2×10^{-6}	918[1]	14.5	2.2×10^{-7}

Notes:
[1]distance based on Gaia measurements (Brown *et al.* 2018)

DUSTY (Ivezic *et al.* 1999) codes. The thermal structure was derived using the code for computation of thermal structure THERMAL and modelling of CO transitions in the large velocity gradient approximation following Groenewegen *et al.* (1994). For modelling of molecular lines we have used the code for the solution of the multilevel radiative transfer problem in an expanding envelope, MOLEXCSE, applied earlier for analysis of emission lines of ammonia in CW Leo (Schmidt *et al.* 2016). The list of transitions and their strengths was extracted from the BYTe computations (Yurchenko *et al.* 2011) and compared with the list of lines in HITRAN database. The collisional data of rotational levels were adopted from the LAMDA database (up to 300 K). For more details of our approach, see Schmidt *et al.* (2016). The radius of NH_3 formation was fixed to the inner radius of the envelope. The photodissociation radius was calculated with CSENV (Mamon *et al.* 1988), while photodissociation rates were adopted from the website http://home.strw.leidenuniv.nl/ẽwine/photo/ (Heays *et al.* 2017).

Acknowledgements

The authors acknowledge support by the National Science Centre, Poland, under grant 2016/21/B/ST09/01626.

References

Brown, A.G.A., Vallenari, A., Prusti, T., de Bruijne, J.H.J, Babusiaux, C., & Bailer-Jones, C.A.L 2018, arXiv:1804.09365

Decin, L., Agndez, M., Barlow, M. J., Daniel, F., Cernicharo, J., Lombaert, R., De Beck, *et al.* 2010, *Nature* 467, 64

Gobrecht, D., Cherchneff, I, Sarangi, A., Plane, J.M.C., & Bromley, S.T. 2016, *A&A* 585, A6

Gong, Y., Henkel, C., Spezzano, S., *et al.* 2015, A&A, 574, A56

Heays, A. N., Bosman, A. D., & van Dishoeck, E. F. 2017, *A&A* 602, A 105

Ivezic, Z., Nenkova, M., & Elitzur, M. 1999, *http://www.pa.uky.edu/ moshe/dusty*

Johnson, H. R. & Sauval, A. J. 1982, *A&ASuppl.* 49, 77

Li, X., Millar, T.J., Heays, A. N., *et al.* 2016, *A&A* 588, A4

Mamon, G. A., Glassgold, A. E., & Huggins, P. J. 1988, *ApJ* 328, 797

Schmidt, M. R., He, J. H., Szczerba, R., Bujarrabal, V., Alcolea, J., Cernicharo, J., *et al.* 2016, *A&A* 592, A131

Szczerba, R., Omont, A., Volk, K., Cox, P., & Kwok, S. 1997, *A&A* 317, 859

Yurchenko, S. N., Barber, R. J., Tennyson, J. 2011, *New Astron. Revs* 46, 513

Willacy, K., & Cherchneff, I. 1998, *A&A* 330, 676

Why Galaxies Care About AGB Stars: A Continuing Challenge through Cosmic Time
Proceedings IAU Symposium No. 343, 2019
F. Kerschbaum, M. Groenewegen & H. Olofsson, eds.

doi:10.1017/S1743921318006178

The Maser-emitting Structure and Time Variability of the SiS Lines $J = 14 - 13$ and $15 - 14$ in IRC + 10216

J. P. Fonfría[1], M. Fernández-López[2], J. R. Pardo[1], M. Agúndez[1], C. Sánchez Contreras[3], L. Velilla-Prieto[1], J. Cernicharo[1], M. Santander-García[4], G. Quintana-Lacaci[1], A. Castro-Carrizo[5] and S. Curiel[6]

[1]Molecular Astrophysics Group, Instituto de Física Fundamental, CSIC, C/ Serrano, 123, 28006, Madrid (Spain)
email: jpablo.fonfria@csic.es

[2]Instituto Argentino de Radioastronomía, CCT-La Plata (CONICET), C.C.5, 1894, Villa Elisa (Argentina)

[3]Department of Astrophysics, Astrobiology Center (CSIC-INTA), Postal address: ESAC campus, P.O. Box 78, E-28691, Villanueva de la Cañada, Madrid (Spain)

[4]Observatorio Astronómico Nacional, OAN-IGN, Alfonso XII, 3, E-28014, Madrid (Spain)

[5]Institut de Radioastronomie Millimétrique, 300 Rue de la Piscine, 38406 Saint-Martin d'Hères (France)

[6]Departamento de Astrofísica Teórica, Instituto de Astronomía, Universidad Nacional Autónoma de México, Ciudad Universitaria, 04510, Mexico City (Mexico)

Abstract. AGB stars are important contributors of processed matter to the ISM. However, the physical and chemical mechanisms involved in its ejection are still poorly known. This process is expected to have remarkable effects in the innermost envelope, where the dust grains are formed, the gas is accelerated, the chemistry is active, and the radiative excitation becomes important. A good tracer of this region in C-rich stars is SiS, an abundant refractory molecule that can display maser lines, very sensitive to changes in the physical conditions. We present high angular resolution interferometer observations (HPBW $\gtrsim 0''.25$) of the $v = 0$ $J = 14 - 13$ and $15 - 14$ SiS maser lines towards the archetypal AGB star IRC+10216, carried out with CARMA and ALMA to explore the inner $1''$ region around the central star. We also present an ambitious monitoring of these lines along one single pulsation period carried out with the IRAM 30 m telescope.

Keywords. line: profiles, masers, radiative transfer, techniques: high angular resolution, techniques: interferometric, (stars:) circumstellar matter, stars: individual (IRC+10216)

---

1. SiS Maser emission and Spatial distribution

The SiS $J = 14 - 13$ and $15 - 14$ maser lines were observed toward the AGB star IRC+10216 with CARMA and ALMA (Fonfría *et al.* 2014; Cernicharo *et al.* 2013) with high spectral and angular resolutions (Fig. 1, upper left). SiS$(14 - 13)$ was modelled with the code developed by Fonfría *et al.* (2014). We assumed the spherically symmetric physical and chemical conditions for SiS derived by Fonfría *et al.* (2015) but allowing the rotational temperature, $T_{\rm rot}$, to be asymmetric if necessary. $T_{\rm rot}$ was varied only where needed to reproduce all the velocity-channel maps. It was chosen to be negative to describe population inversion only if there was no other way to reproduce the emission.

We found that the observed brightness distribution, comprising two bright spots, an extended component, and a bipolar structure, cannot be explained neither by thermal

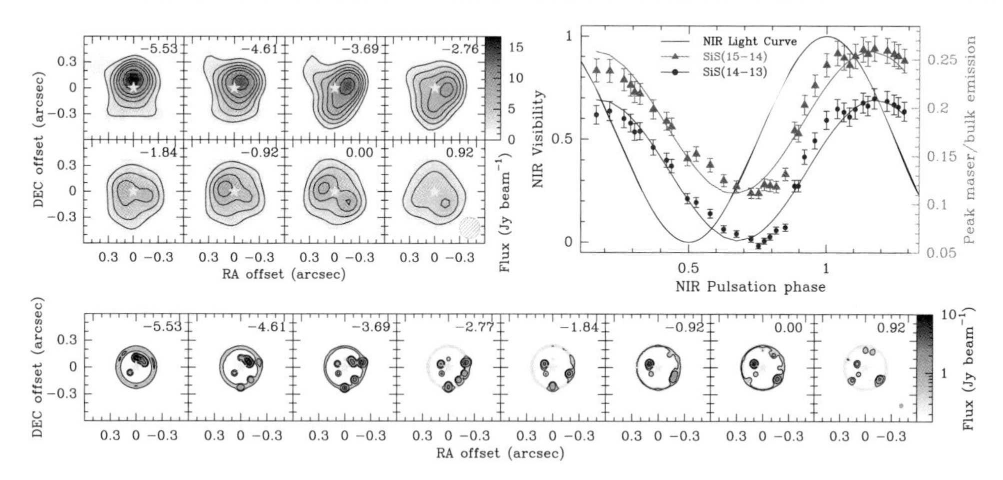

Figure 1. Several CARMA SiS(14 − 13) velocity-channel maps (HPBW $\simeq 0''.25$; *upper left*), synthetic emission in logarithmic scale (HPBW $= 0''.05$; *lower*), and time dependence of ratio of the strongest maser peak and the bulk emission in the same velocity interval (*upper right*).

emission nor by flux calibration errors, and maser emission is needed. The maser emission can be reproduced with a set of compact maser spots in an extended structure modelled as a shell (Fig. 1, lower). A large fraction of the total emission has a maser nature (75%). About 40% of the maser emission comes from the compact spots, mainly from the brightest ones located to the NW of the star, and 60% from the extended component.

2. Time Variability

We also monitored the SiS lines $J = 14 - 13$ and $15 - 14$ throughout a whole pulsation period with a sampling time of 16 days (Pardo *et al.* 2018). Some of the narrow maser components of these lines show evident time variations following the NIR light-curve of the star. Other spectral components display a milder time variability. The comparison of these SiS lines with thermally excited lines of other molecules suggests that the extended maser emission also varies over time. The time dependence of the ratio of the strongest maser peak and the bulk emission is offset by $\simeq 0.2$ pulsation periods with respect to the NIR light-curve (measured 10 years ago; Fig. 1, upper right), which could indicate a recent variation of the NIR light-curve or unexpected excitation effects.

The research leading to these results has received funding from the European Research Council under the European Union's Seventh Framework Programme (FP7/2007-2013)/ ERC grant agreement n° 610256: NANOCOSMOS

References

Cernicharo, J., Daniel, F., Castro-Carrizo, A., Agúndez, M., Marcelino, N., Joblin, C., Goicoechea, J. R., Guélin, M. 2013, *ApJ* (Letters), 778, L25

Fonfría, J. P., Fernández-López, M., Agúndez, M., Sánchez-Contreras, C., Curiel, S., Cernicharo, J. 2014, *MNRAS*, 445, 3289

Fonfría, J. P., Cernicharo, J., Richter, M. J., Fernández-López, M., Velilla-Prieto, L., Lacy, J. H. 2015, *MNRAS*, 453, 439

Fonfría, J. P., Fernández-López, M., Pardo, J. R., Agúndez, M., Sánchez Contreras, C., Velilla Prieto, L., Cernicharo, J., Santander-García, M., Quintana-Lacaci, G., Castro-Carrizo, A., Curiel, S. 2018, *ApJ*, 860, 162

Pardo, J. R., Cernicharo, J., Velilla Prieto, L., Fonfría, J. P., Agúndez, M., Quintana-Lacaci, G., Massalkhi, S., Tercero, B., Gómez-Garrido, M., de Vicente, P., Guélin, M., Kramer, C., Marka, C., Teyssier, D., Neufeld, D. 2018, *A&A*, 615, L4

Why Galaxies Care About AGB Stars:
A Continuing Challenge through Cosmic Time
Proceedings IAU Symposium No. 343, 2019
F. Kerschbaum, M. Groenewegen & H. Olofsson, eds.

doi:10.1017/S1743921318005641

Central Stars of Planetary Nebulae in Galactic Open Clusters: Providing additional data for the White Dwarf Initial-to-Final-Mass Relation

Vasiliki Fragkou[1], Quentin A. Parker[1], Albert Zijlstra[2], Richard Shaw[3] and Foteini Lykou[1]

[1]The University of Hong Kong, Department of Physics, Hong Kong SAR, China
emails: vfrag@hku.hk, quentinp@hku.hk, lykoufc@hku.hk

[2]The University of Manchester, Manchester, UK
email: a.zijlstra@manchester.ac.uk

[3] Space Telescope Science Institute, Maryland, USA
email: shaw@stsci.edu

Abstract. Accurate ($< 10\%$) distances of Galactic star clusters allow a precise estimation of the physical parameters of any physically associated Planetary Nebula (PN) and also that of its central star (CSPN) and its progenitor. The progenitor's mass can be related to the PN's chemical characteristics and, furthermore, provides additional data for the widely used white dwarf (WD) initial-to-final mass relation (IFMR) that is crucial for tracing the development of both carbon and nitrogen in entire galaxies. To date, there is only one PN (PHR1315- 6555) confirmed to be physically associated with a Galactic open cluster (ESO 96 -SC04) that has a turn-off mass ~ 2 $M_{\odot}$. Our deep HST photometry was used for the search of the CSPN of this currently unique PN. In this work, we present our results.

Keywords. Planetary Nebulae, Open Clusters

---

1. Introduction

CSPNe masses, crucial in understanding post-AGB evolution, provide additional data for the widely used WD IFMR. CSPNe studies are difficult as measurements of their masses require a precise determination of their distances. PNe members of Galactic star clusters allow the accurate determination of their distances from cluster Color-Magnitude Diagrams (CMDs), while photometric measurements of their CSPNe can constrain their intrinsic luminosity and mass and thus, these objects can be used as additional points for the IFMR. PHR 1315-6555 is a faint, bipolar, possible Type I PN, whose radial-velocity matches that of the distant Galactic open cluster ESO 96-SC04 (Parker *et al.* 2011).

In this work, we present our deep HST WFC3 F555W, F814W, F200LP and F350LP photometry of the cluster, the PN and its CSPN. Following, we create an improved cluster CMD and determine the physical properties of the nebular CSPN.

2. Methods and Results

The F555W and F814W HST filter exposures provide the deepest CMD of the cluster available to date. The theoretical Padova isochrone (Girardi *et al.* 2010) that fits best our CMD suggests that the distance to the cluster is 10.5 ± 0.4 kpc, it has an age of 0.72 ± 0.1 Gyrs, a reddening of 0.695 ± 0.06 and a turn-off mass of ~ 2 $M_{\odot}$.

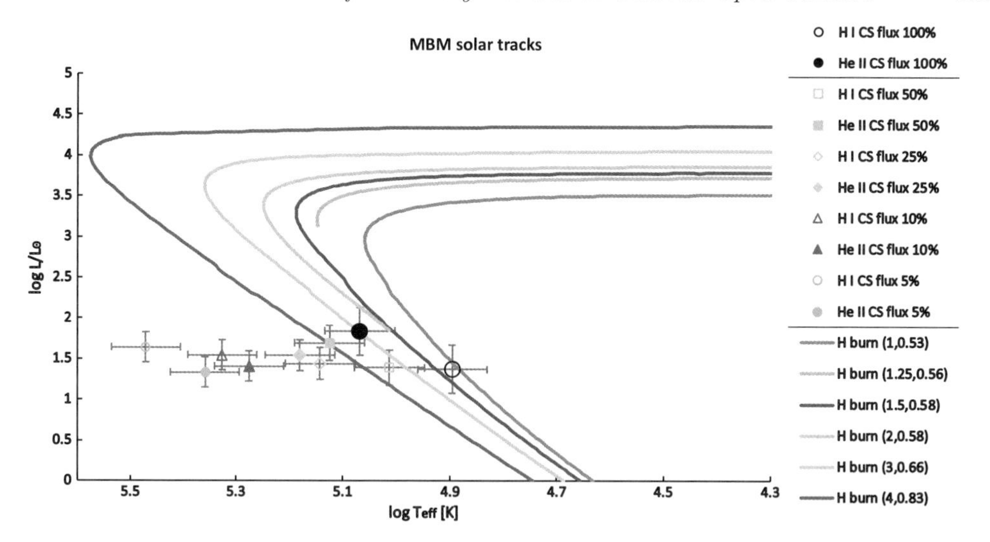

Figure 1. Derived CSPN temperatures and luminosities plotted along the MBM tracks.

The deep F200LP and F350LP HST filter exposures were used to find the CSPN since the F200LP-F350LP gives the near-UV continuum. This was the bluest star in the nebular field lying close to the projected nebula centre as might be expected. The CSPN temperature and luminosity were derived from the F555W CSPN measured magnitude using the Zanstra method (Zanstra 1931) and the nebular HI and HeII absolute fluxes (Parker *et al.* 2011). Previous steps were repeated assuming that the presence of an unresolved companion contributes to the measured CSPN visual flux in fractions from 50% to 95%. For all previous cases, the CSPN masses were determined by plotting the derived temperatures and luminosities in the logT-logL plane along with the Miller Bertolami (2016, MBM) post-AGB evolutionary tracks for solar metallicities (see Fig. 1). The CSPN mass was found to be 0.65 ± 0.09 and the presence of a companion contributing to the measured flux $> 50\%$ is deemed highly unlikely as this would place the CSPN too far from the evolutionary track expected from the estimated turn-off mass of the cluster. The measured F814W magnitude can provide any indications regarding the possible presence of an unresolved binary cool low mass companion. This would yield an IR excess (see Barker *et al.* 2018). $(F555W - F814W)_{\rm obs}$ - $(F555W - F814W)_{\rm mod} = 0.013 \pm 0.67$ and thus, the presence of a binary cool companion is not favored by the data.

The nebular log(N/O) abundance ratio has been measured to be around 0.87 (Parker *et al.* 2011). Stellar evolution models predict that such high N abundances in solar metallicity environments would be for stars with initial mass > 4 $M_\odot$ (Karakas & Lugaro 2016), which is not the case here. A much lower metallicity environment could explain the N yields, but is not supported from our CMD. If cluster metallicity is proven to be close to solar, the lower mass limits for hot bottom burning may need to be revised given our results.

References

Barker H., Zijlstra A., De Marco O., Frew D. J., Drew J. E., Corradi R. L. M., Eislöffel J., Parker Q. A. 2018, *MNRAS*, 475, 4504

Girardi L., Bressan A., Bertelli G., Chiosi C. 2000, *A&AS*, 141, 371

Karakas A. I., Lugaro M. 2016, *ApJ*, 825, 26

Miller Bertolami M. M. 2016, *A&A*, 588, A25

Parker Q. A., Frew D. J., Miszalski B., Kovacevic A. V., Frinchaboy P. M., Dobbie P. D., Köppen J. 2011, *MNRAS*, 413, 1835

Zanstra H. 1931, *ZA*, 2, 1

Why Galaxies Care About AGB Stars:
A Continuing Challenge through Cosmic Time
Proceedings IAU Symposium No. 343, 2019
F. Kerschbaum, M. Groenewegen & H. Olofsson, eds.

doi:10.1017/S1743921318005033

On cylindrically symmetric solutions of polarized radiative transfer equation

Juris Freimanis

Engineering Research Institute "Ventspils International Radio Astronomy Centre",
Ventspils University College, Inzenieru iela 101, Ventspils, LV-3600, Latvia
email: jurisf@venta.lv

Abstract. While the observed polarization maps of spatially resolved post-AGB objects usually require numerical modelling of radiative transfer, it is useful to have known analytical solutions of the polarized radiative transfer equation (PRTE) as benchmarks of computer codes in the simplest model cases. We consider two such solutions: cylindrically symmetric Green's function for an infinite medium and cylindrically symmetric inner eigenfunctions of PRTE.

Keywords. Radiative transfer, polarization, methods: analytical

1. Introduction

Strongly polarized radiation, with high spatial resolution mapping over the surface of often asymmetric gas-dust envelopes of AGB stars and post-AGB objects, has been observed in visible light (Ohnaka *et al.* 2017) and in the near infrared (Su *et al.* 2003). Physical interpretation of such observations includes numerical solution of the polarized radiative transfer equation (PRTE); however, it is desirable to test the numerical computer codes against the known analytic solutions of PRTE in the simplest model cases.

Here we review two options for an analytic or semi-analytic solution of PRTE in cylindrical symmetry, namely, Greens function for an infinite medium (Freimanis 2009) and cylindrically symmetric inner eigenfunctions.

2. The physical conditions and radiative transfer equation

Let us assume that the multiply scattering (polydisperse, dusty) medium within Euclidean space, and radiation field in it, obeys the following physical conditions:

• The conditions of validity of the radiative transfer equation in a polydisperse medium (Mishchenko, Travis & Lacis (2006)) are fulfilled;

• The medium is statistically homogeneous, isotropic and stationary. It is either the whole space or an infinitely long cylinder with nonreflecting surface;

• The effective refractive index of the medium is independent of polarization, and there is no circular birefringence nor circular dichroism;

• The radiation field is stationary and cylindrically symmetric, with the same symmetry axis as that of the medium.

Let us introduce a standard spatial cylindrical coordinate system ($\tau \equiv \alpha r, \Phi, \zeta$), with α being the scalar extinction coefficient, and let us characterize the direction of propagation of radiation by standard spherical angles ($\vartheta = \cos^{-1} \mu, \varphi$), with polar axis in the radial direction. Let us define the plane going through the radial direction and the direction of propagation of radiation as the linear polarization reference plane.

Describing Stokes vector $\mathbf{I}(\tau, \mu, \varphi)$ in circular polarization representation (Mishchenko 2006), the polarized radiative transfer equation is (Freimanis 2014)

$$\mu\frac{\partial\mathbf{I}(\tau;\mu,\varphi)}{\partial\tau}+\frac{(1-\mu^2)\sin^2\varphi}{\tau}\frac{\partial\mathbf{I}(\tau;\mu,\varphi)}{\partial\mu}-\frac{\mu\sin 2\varphi}{2\tau}\frac{\partial\mathbf{I}(\tau;\mu,\varphi)}{\partial\varphi}$$
$$+\frac{\sin 2\varphi}{2\tau}\mathbf{U}\mathbf{I}(\tau;\mu,\varphi)=-\mathbf{I}(\tau;\mu,\varphi)+\mathbf{B}_0(\tau;\mu,\varphi)$$
$$+\frac{1}{4\pi}\int_0^{2\pi}\sum_{s=-\infty}^{+\infty}\mathrm{e}^{-is(\varphi-\varphi')}d\varphi'\int_{-1}^{1}\mathbf{p}_s(\mu,\mu')\mathbf{I}(\tau;\mu',\varphi')d\mu', \tag{1}$$

where $\mathbf{U}$ is a constant diagonal 4×4 matrix, 4×4-matrices $\mathbf{p}_s(\mu,\mu')$ describe the scattering in the medium, and 4-vector $\mathbf{B}_0(\tau;\mu,\varphi)$ is the primary source function.

3. Green's function for infinite medium

If the medium is infinite, then the solution of equation (1) is given by its Greens function (Freimanis 2009). Its main advantage is that it is mathematically exact. Its main drawbacks are: i) it is applicable only for a homogeneous infinite medium, ii) it is extremely complex and hardly suitable for practical computations.

4. Cylindrically symmetric eigenfunctions

Cylindrically symmetric eigenfunctions of the homogeneous version of equation (1) were found heuristically earlier (Freimanis 2014), but without strict proof that they are eigenfunctions indeed. Now this has been proved, and the proof will be presented in Freimanis (2018). There are convergent inner eigenfunctions bounded in the vicinity of the cylindrical symmetry axis, and very strongly divergent outer eigenfunctions; more treatable expressions are to be found for them. We propose to use inner eigenfunctions for semi-analytic radiative transfer modelling in homogeneous cylinders of infinite length but finite radius, illuminated from outside. The advantage of this approach is that eigenfunctions are mathematically much simpler than the Green's function. The drawback of this method is that neither some orthogonality properties of eigenfunctions, nor the completeness of such eigenfunctions in some Banach space has been proved.

Acknowledgements

This work and participation of the author at IAUS 343 were financed by the ERDF project No. 1.1.1.1/16/A/213. Cofinancing by the basic budget of Ventspils International Radio Astronomy Centre and by Ventspils City Council was received as well. The author expresses his gratitude to all these entities.

References

Freimanis, J. 2009, *Journal of Quantitative Spectroscopy & Radiative Transfer*, 110, 1307
Freimanis, J. 2014, in: H.-W. Lee, Y.W. Kang & K.-C. Leung (eds.), *ASP-CS*, 482, p. 265
Freimanis, J., 2018, *Journal of Quantitative Spectroscopy & Radiative Transfer*, in preparation
Mishchenko, M.I., Travis, L.D., & Lacis, A.A. 2006, *Multiple Scattering of Light by Particles. Radiative Transfer and Coherent Backscattering.* Cambridge *et al.*, Cambridge University Press, 478 pp.
Ohnaka, K., Weigelt, G., & Hofmann, K.-H. 2017, *A&A*, 597, A20
Su, K.Y.L., Hrivnak, B.J., Kwok, S., & Sahai, R. 2003, *AJ*, 126, 848

Why Galaxies Care About AGB Stars:
A Continuing Challenge through Cosmic Time
Proceedings IAU Symposium No. 343, 2019
F. Kerschbaum, M. Groenewegen & H. Olofsson, eds.

doi:10.1017/S1743921318007056

GK Car and GZ Nor: Two low-luminous, depleted RV Tauri stars

I. Gezer[1,2], H. Van Winckel[3], R. Manick[3] and D. Kamath[4,5]

[1]Nicolaus Copernicus Astronomical Center, Rabiańska 8, 87-100 Toruń, Poland

[2]Astronomy and Space Science Department, Ege University, 35100 Bornova, Izmir, Turkey

[3]Institute of Astronomy, KU Leuven, Celestijnenlaan, 200D 3001 Leuven, Belgium

[4]Department of Physics and Astronomy, Macquarie University, Sydney, NSW 2100, Australia

[5]Australian Astronomical Observatory, PO Box 915, North Ryde, NSW 1670, Australia

Abstract. RV Tauri stars are luminous population II Cepheids which show a characteristic light curve of alternating deep and shallow minima. There are 126 RV Tauri variables in our Galaxy. Using WISE [3.4]-[4.6], [12]-[22] diagram we show that Galactic RV Tauri stars show three main types of IR properties in their SEDs; disc-type, non-IR and uncertain, which does not show a clear characteristic in the SED. We also show that there is a strong correlation between disc-type SED and binarity (Gezer *et al.* 2015). RV Tauri stars were linked to post AGB stars in early studies (Jura 1986), however, recent studies show that their evolutionary nature is more complex than previously thought (Kamath & Van Winckel 2014, and Manick *et al.* 2018). In this study, we intentionally selected two RV Tauri stars, GK Car (disc-type) and GZ Nor (uncertain), with different IR characteristics to compare their chemical and photometric properties.

Keywords. stars: AGB and post-AGB, stars: abundances, stars: variables: Cepheids.

---

1. Luminosity and Distance Estimates

Using ASAS photometry, the accurate pulsation periods have been obtained via period analysis. The obtained periods are used to derive luminosities and distances. The total extinction E(B-V) values are obtained from the SED fitting. For GK Car we obtained a total reddening of E(B-V)=0.41±0.1, while for GZ Nor E(B-V)=0.45±0.1 was found. We computed luminosities using three different methods. First, we derived the luminosities using the Period Luminosity Colour (PLC) relation given by Manick *et al.* (2017). Second, we obtained the bolometric luminosities, L_{SED}, for each star using the integrated flux below the dereddened SED model and the obtained distances. We also calculated distance and luminosity using GAIA parallax, however, GAIA parallax is available for only one of our stars (GK Car). Obtained distances and luminosities are given in Table 1.

2. Chemical Analysis

High-resolution, high signal-to-noise spectra for GK Car and GZ Nor were obtained with the Ultraviolet and Visual Echelle Spectrograph mounted on the 8m UT2 Kueyen Telescope of the VLT array at the Paranal Observatory of ESO in Chile. The abundances were calculated on the basis of LTE model atmospheres of Kurucz (Castelli & Kurucz 2003) and MARCS (Gustafsson *et al.* 2008) and the LTE chemical composition determination routine MOOG (version July 2009) (Sneden 1973). For GK Car we obtained $T_{\rm eff}$ = 5500 ±125 K, log g= 1.0 ±0.25 dex, microturbulent velocity ξ_t = 5.5 ±0.5 km/s, and [FeI/H] = −1.32 ±0.1. For GZ Nor the atmospheric parameters are as follow: $T_{\rm eff}$ = 4875 ±125 K, log g = 0.50 ±0.25 dex, ξ_t = 4.0 ±0.5 km/s, [FeI/H] = −2.05 ±0.1. With an

Table 1. The fundamental pulsation period (P_0) is given in Col. 2. Calculated distances and luminosities using PLC relation are shown in Cols. 3 and 4, respectively. Only for GK Car, the distance and luminosity is calculated using parallax and they are given in the last two columns.

Star	**P_0 (days)**	**Distance (kpc)**	**Luminosity (PLC) ($L_\odot$)**	**Luminosity (SED) ($L_\odot$)**	**Distance (plx) (kpc)**	**Luminosity (plx) ($L_\odot$)**
GK Car	27.6, 55.2	4.55±0.59	1762±450	1626±264	4.30±0.57	1455±390
GZ Nor	36.2, 72.4	8.42±1.0	1560±340	1425±234	–	–

[Fe/H]=−1.3 and a [Zn/Ti]=+1.2 for GK Car and a [Fe/H]=−2.0 and a [Zn/Ti]=+0.8 for GZ Nor, both stars show depletion of refractory elements in their photospheres. In a depleted photosphere, refractory elements, which have high dust condensation temperature, are underabundant, while volatiles, which have low condensation temperature, are more abundant (Van Winckel 2003). Waters, Trams, & Waelkens (1992) proposed that the most likely circumstance for the process to occur is the dust trapping in a circumstellar disc. All depleted atmospheres have been detected in binary post-AGB objects so far (Van Winckel, Waelkens, & Waters 1995, Gezer *et al.* 2015). The most characteristic chemical signatures of depleted photospheres are high [Zn/Fe], [Zn/Ti] and [S/Ti] ratios. GK Car and GZ Nor both show a depletion characteristic in their atmosphere. This would imply that they are likely binary objects.

3. Conclusions

In this study, we show that GZ Nor is RV Tauri variable with a disc. GK Car and GZ Nor are both depleted RV Tauri stars with disc hence we conclude that they are likely binary objects. All RV Tauri stars with the disc-type SED are likely binaries and they probably follow different evolutionary channels depending on the initial mass of their primaries. The luminosity of the tip of the Red Giant Branch (RGB-tip) for $1\,M_\odot$ is $2615\,L_\odot$ (Bertelli *et al.* 2008). The obtained luminosities for GK Car ($1762\,L_\odot$) and GZ Nor ($1560\,L_\odot$) are lower than the predicted RGB-tip luminosity of a $1\,M_\odot$ star. Thus, they very likely evolve off the RGB due to a strong binary interaction process, which occurs already on the RGB.

References

Bertelli G., Girardi L., Marigo P., Nasi E. 2008, *A&A*, 484, 815

Castelli F., Kurucz R. L. 2003, *IAUS*, 210, A20

Gezer I., Van Winckel H., Bozkurt Z., De Smedt K., Kamath D., Hillen M., Manick R. 2015, *MNRAS*, 453, 133

Gustafsson B., Edvardsson B., Eriksson K., Jørgensen U. G., Nordlund Å., Plez B. 2008, *A&A*, 486, 951

ApJ...309..732J Jura M. 1986, *ApJ*, 309, 732

Kamath D., Wood P. R., Van Winckel H. 2014, *MNRAS*, 439, 2211

Manick R., Van Winckel H., Kamath D., Hillen M., Escorza A. 2017, *A&A*, 597, A129

Manick R., Van Winckel H., Kamath D., Sekaran S., Kolenberg K. 2018, arXiv:1806.08210

Sneden C. A., 1973, PhDT

Van Winckel H., Waelkens C., Waters L. B. F. M. 1995, *A&A*, 293,

Van Winckel H. 2003, *ARAA*, 41, 391

Waters L. B. F. M., Trams N. R., Waelkens C. 1992, *A&A*, 262, L37

Why Galaxies Care About AGB Stars:
A Continuing Challenge through Cosmic Time
Proceedings IAU Symposium No. 343, 2019
F. Kerschbaum, M. Groenewegen & H. Olofsson, eds.

doi:10.1017/S174392131800707X

Infrared light curves of dusty & metal-poor AGB stars

Steven R. Goldman, Martha Boyer and the DUSTiNGS team

Space Telescope Science Institute, 3700 San Martin Drive,
Baltimore, MD 21218, USA
email: s.goldman@stsci.edu

Abstract. The effects of metallicity on both the dust production and mass loss of evolved stars have consequences for stellar masses, stellar lifetimes, progenitors of core-collapse SNe, and the origin of dust in the ISM. With the DUST in Nearby Galaxies with Spitzer (DUSTiNGS) survey, we have discovered samples of dusty evolved AGB stars out to the edge of the Local Group with metallicities down to 0.6% solar. This makes them the nearest analogs of AGB stars in high-redshift galaxies. We present new infrared light curves of the dustiest AGB stars in 10 galaxies from the DUSTiNGS survey and show how the infrared Period-Luminosity (PL) relation is affected by dust and metallicity. These results have implications for the efficiency of AGB dust production at high-redshift and for the use of the Mira PL relation as a distance indicator.

Keywords. stars: AGB and post-AGB, stars: variables: other, galaxies: dwarf, (galaxies:) Local Group, stars: carbon, infrared: stars

---

Understanding how the dust contribution of evolved stars is affected by changes in metallicity is critically important for understanding the origin of dust in the Universe, especially dust that has been seen at high redshift (e.g., Valiante *et al.* 2009; Dwek & Cherchneff 2011; Rowlands *et al.* 2014; Michałowski 2015). Mira variables are radial pulsators that pulsate primarily in the fundamental mode. As they levitate material out the large radii, the material cools, condenses into dust, and is driven out into the ISM by radiation pressure. We can probe this mechanism by measuring the pulsation periods of evolved stars in the infrared, and using their infrared colors to determine the dust content. We can then study the link between dust production and pulsation by investigating both the dust content and the pulsation in the same star. However, dusty objects are often faint or not detectable in optical and near-IR variability surveys. To detect the dustiest stars, we monitored DUSTiNGS targets at 3.6 and 4.5 microns (Fig. 1). By probing samples of nearby metal-poor dwarf galaxies, we can also study the effects of metallicities down to less than one hundredth solar and constrain the dust production rates of evolved stars in these environments

Recent observations including those from the *Spitzer Space Telescope* and the DUSTiNGS survey have already discovered samples of evolved stars producing dust in nearby metal-poor dwarf galaxies (Boyer *et al.* 2015a,b, 2017, Whitelock *et al.* 2018). Multi-epoch observations of each of these galaxies have highlighted the evolved stars through their variability, and with additional observations (using new and archival data) we have now been able to confirm the nature of these sources and study their pulsation behaviors (Fig. 2). Results have shown that while the infrared Mira PL relation is affected by changes in the *Spitzer* [3.6]−[4.5] color (shown to correlate well with the dust content) it is unaffected by changes in metallicity down to ∼0.6% solar. This is encouraging for the prospect of using the relation as a distance indicator, and for AGBs as dust producers at high redshift.

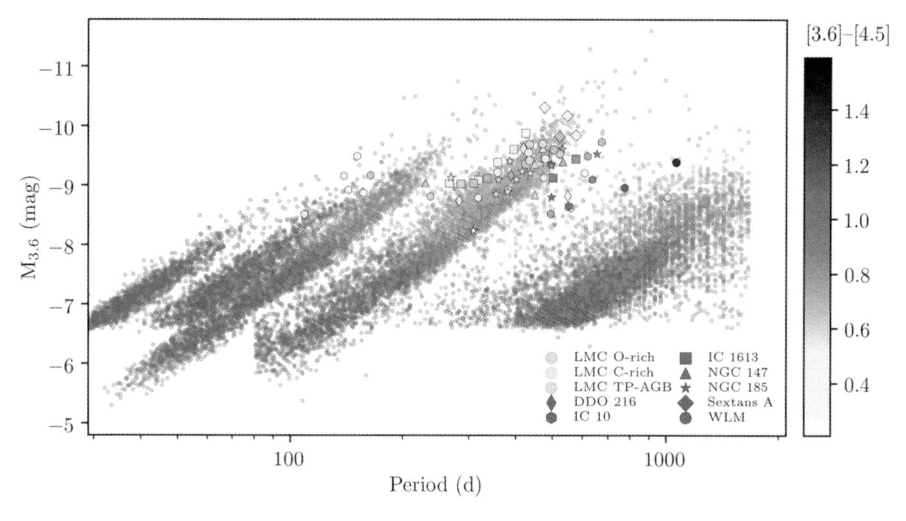

Figure 1. The P-L relation of the DUSTiNGS sample with the color showing the [3.6]−[4.5] color and galaxy membership shown using shapes. Also shown is the MACHO-SAGE sample from Riebel *et al.* (2015) containing oxygen- and carbon-rich AGB stars and more evolved and dusty TP-AGB stars of both types.

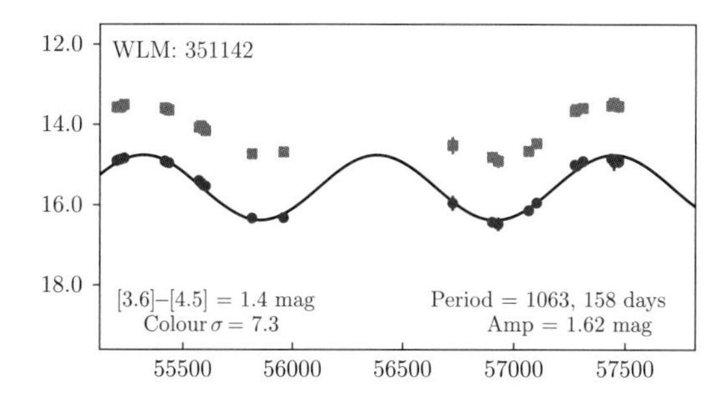

Figure 2. The lightcurve of one of our high-confidence long-period variables, with the Julian date vs. absolute 3.6 (blue) and 4.5 μm (red) magnitudes. Also shown are best-fit and second-best-fit pulsation period, best-fit amplitude, the [3.6]−[4.5] color and its standard deviation.

In addition to *Spitzer* data, we have *Hubble Space Telescope* medium-band photometry for each of the DUSTiNGS galaxies. The sensitivity of the observations reaches down below the tip of the red giant branch, ensuring adequate sensitivity for detecting all but the most dusty and obscured evolved stars. Through a method developed by Boyer *et al.* (2017), we have used the F127M, F139M, and F153M colors to clearly disentangle the carbon- and oxygen-rich evolved stars, which produce carbonaceous and silicate-rich dust, respectively. It is expected that the dust production of oxygen-rich evolved stars should be limited by the initial metallicity, as dust grain condensation should require s-process elements as seed nuclei. These results show that contrary to this, oxygen-rich AGBs have now been seen producing considerable dust at metallicities as low as 5% solar.

References

Boyer, M. L., McQuinn, K. B. W., Barmby, P., *et al.* 2015, *ApJ*, 800, 51
Boyer, M. L., McQuinn, K. B. W., Barmby, P., *et al.* 2015, *ApJS*, 216, 10
Boyer, M. L., McQuinn, K. B. W., Groenewegen, M. A. T., *et al.* 2017, *ApJ*, 851, 152

Dwek, E., and Cherchneff, I. 2011, *ApJ*, 727, 63
Michałowski, M. J. 2015, *A&A*, 577, 80
Riebel, D., Meixner, M., Fraser. O., *et al.* 2015, *ApJ*, 723, 1195
Rowlands K., Gomez, H. L., Dunne, L., *et al.* 2014, *MNRAS*, 441, 1040
Valiante, R., Schneider, R., Bianchi, S., and Andersen, A. C., 2009, *MNRAS*, 397, 1661
Whitelock, P. A., Menzies, J. W., Feast, M. W., and Marigo, P., 2018, *MNRAS*, 473, 173

Why Galaxies Care About AGB Stars: A Continuing Challenge through Cosmic Time
Proceedings IAU Symposium No. 343, 2019
F. Kerschbaum, M. Groenewegen & H. Olofsson, eds.

doi:10.1017/S1743921318007809

A step further on the physical, kinematic and excitation properties of PNe

Denise Rocha Gonçalves[1,2] and Stavros Akras[1]

[1]Valongo Observatory, Federal University of Rio de Janeiro, Ladeira Pedro Antonio 43, 20080-090, Rio de Janeiro, Brazil

[2]East Asian ALMA Regional Center, National Astronomical Observatory of Japan, 2-21-1 Osawa, Mitaka, Tokyo, 181-8588, Japan

emails: denise@astro.ufrj.br, akras@astro.ufrj.br

Abstract. PNe are known to be photoionized objects. However they also have low-ionization structures (LIS) with different excitation behavior. We are only now starting to answer why most LIS have lower electron densities than the PN shells hosting them, and whether or not their intense emission in low-ionization lines is the key to their main excitation mechanism. Can LIS line ratios, chemical abundances and kinematics enlight the interplay between the different excitation and formation processes in PNe? Based on the spectra of five PNe with LIS and using new diagnostic diagrams from shock models, we demonstrate that LIS's main excitation is due to shocks, whereas the other components are mainly photoionized. We propose new diagnostic diagrams involving a few emission lines ([N II], [O III], [S II]) and $f_{shocks}/f_{\star}$, where f_{shocks} and $f_{\star}$ are the ionization photon fluxes due to the shocks and to the central star ionizing continuum, respectively.

Keywords. ISM: kinematics and dynamics, (ISM:) planetary nebulae: general, ISM: jets and outflows, techniques: spectroscopic

1. Motivation and data

The generalized interacting stellar winds model (Balick & Frank 2002) can adequately explain the formation of the different PN nebular components – attached- (rim) and detached-shells and haloes – and their morphological classes: round, elliptical, bipolar or multiple bipolar. There are specific PN components, the low-ionization structures (LIS, Gonçalves *et al.* 2001) that need special attention, since we lack details on their formation and ionization mechanisms, though many observational constraints are available.

Long-slit optical spectroscopic data of five PNe with several LIS were obtained using the 2.5 m Isaac Newton Telescope (INT) and the 1.54 m Danish telescope. Three (NGC 6891, NGC 6572 and IC 4846) where observed on August 2001 and two (K 1-2 and Wray 17-1) on April 1997, with the INT and the Danish telescope, respectively.

2. Data interpretation and clues on the excitation of LIS

Spectra of the different nebular regions and LIS were extracted, then allowing us to carry out a study of LIS relative to their surrounding gas. The top-left panel of Fig. 1 illustrates such regions in Wray 17-1. Physical and chemical properties of all these spectra were obtained. We (Akras & Gonçalves 2016) thus confirm that LIS have the same chemical composition as the main nebular components (Fig. 1 top-right panel). The linear relation between log(N/O) and log(N/H) is consistent with the general picture for PNe (Perinotto *et al.* 2004) and the best fit agrees with that for PNe with [WR] and weak emission line central stars (Garcá-Rojas 2013).

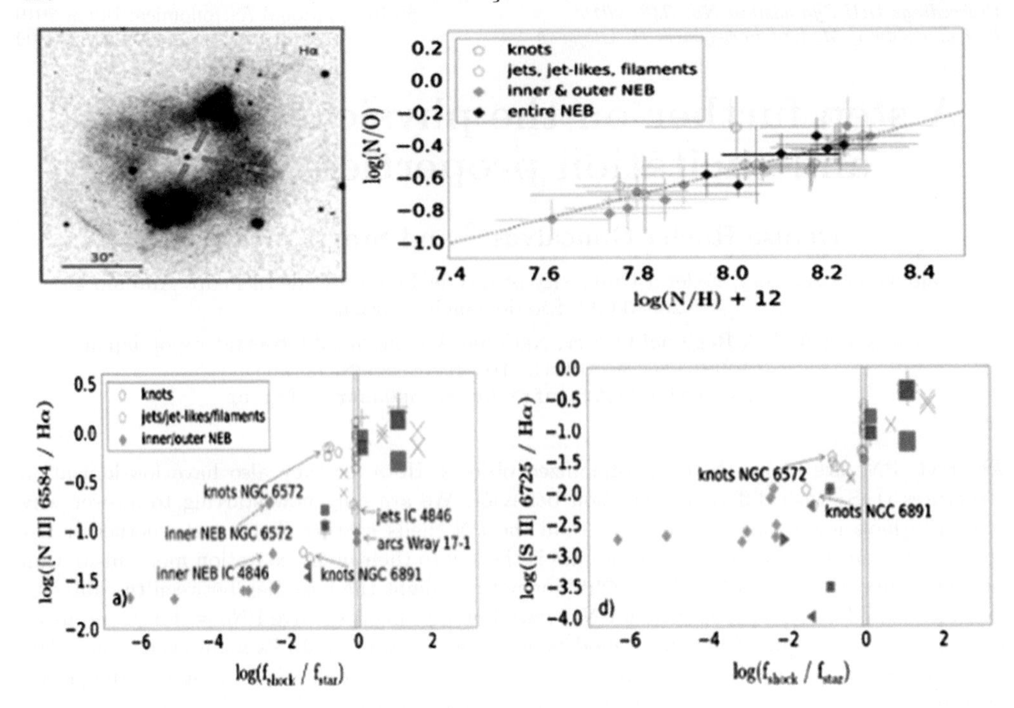

Figure 1. *Top-left*: the H_α image of Wray 17-1, highlighting the extracted spectra: LIS (dark blue and green) and inner shells and caps (red and cyan). *Top-right*: log(N/O) vs. log(N/H)+12 for all the structures of the whole sample. *Bottom:* new diagnostic diagrams to distinguish shock- from photo-excited PN components – [N II]/H_α (*left*), [S II]/H_α (*right*) vs. log(f_{shocks} /$f_\star$).

New diagnostic diagrams that apply LIS strong emission-line ratios and the $f_{shocks}/f_\star$ ratio – f_{shocks} and $f_\star$ are the ionization photon fluxes produced by the shocks and the central star's radiation field– are used to distinguish the photo-ionized from the shock-excited regions. All LIS exhibit systematically higher $f_{shocks}/f_\star$ ratio (>0.1) compared to the nebular gas (<0.01), thus indicating some shock activity in LIS (bottom panels of Fig. 1, Akras & Gonçalves 2016). For the construction of these new diagnostics both observations and models (Raga *et al.* 2008) were needed.

The presence of molecular hydrogen in several LIS of two PNe –K 4-47 and NGC 7662– has been recently confirmed (Akras *et al.* 2017). This should solve the problem of LIS's missing mass (they have systematically lower electron density than their surrounding nebula, which contradicts the models), at the same time that opens new challenges about their formation (Akras *et al.* 2017).

References

Akras, S., & Gonçalves, D. R. 2016, *MNRAS*, 455, 930
Akras, S., Gonçalves, D. R., Ramos-Larios, G. 2017, *MNRAS*, 465, 1289
Balick, B., & Frank, A. 2002 *ARA&A*, 40, 439
García-Rojas, J., *et al.* 2013, *A&A*, 558, 122
Gonçalves, D. R., Corradi, R. L. M., Mampaso, A., 2001, *ApJ*, 547, 302
Perinotto, M., Morbidelli, L., Scatarzi, A. 2004, *MNRAS*, 349, 793
Raga, A. C., Riera, A., Mellema, G., Esquivel, A., Velázquez, P. F. 2008, *A&A*, 489, 1141

Why Galaxies Care About AGB Stars:
A Continuing Challenge through Cosmic Time
Proceedings IAU Symposium No. 343, 2019
F. Kerschbaum, M. Groenewegen & H. Olofsson, eds.

doi:10.1017/S1743921318006932

Mid-IR colors and surface brightness fluctuations as tracers of stellar mass-loss in the TP-AGB

Rosa A. González-Lópezlira

Instituto de Radioastronomia y Astrofisica, UNAM
Campus Morelia, Michoacan, Mexico, C.P. 58089
email: r.gonzalez@irya.unam.mx

Abstract. I present integrated colors and surface brightness fluctuation magnitudes in the mid-IR, derived from stellar population synthesis models that include the effects of the dusty envelopes around TP-AGB stars. The models are based on the Bruzual & Charlot CB* isochrones; they are single-burst, range in age from a few Myr to 14 Gyr, and comprise metallicities between Z = 0.0001 and Z = 0.04. I compare these models to mid-IR data of AGB stars and star clusters in the Magellanic Clouds, and study the effects of varying self-consistently the mass-loss rate, the stellar parameters, and the output spectra of the stars plus their dusty envelopes.

Keywords. stars: AGB and post-AGB, stars: mass loss, stars: carbon, stars: evolution, Magellanic Clouds, infrared: stars, galaxies: stellar content, galaxies: star clusters

---

Asymptotic Giant Branch (AGB) stars are central to the chemical evolution of galaxies, and understanding the contribution of these evolved stars to the spectral energy distribution (SED) of galaxies is essential for the interpretation of galactic emission in the near and mid-infrared (IR). Thermally pulsing AGB (TP-AGB) evolution is very complex, however, on account of a large number of physical processes at work, and the difficulties to constrain them (see, for a brief recent summary, Marigo *et al.* 2013). While several processes and parameters —dredge-up efficiency, mixing-length, hot-bottom burning, pulsations— are degenerate on their effects on both TP-AGB lifetimes and luminosity functions, there is no doubt that mass-loss is the most important parameter determining the duration of the phase (e.g., Rosenfield *et al.* 2014; Rosenfield *et al.* 2016). On the other hand evolutionary synthesis models for the study of stellar populations have particular challenges: they are required to account for the mass-loss rate ($\dot{M}$) and the emission of the stellar dusty envelopes for all TP-AGB evolutionary stages and all metallicities. For this, they need an analytic approach, since there are no empirical spectra for all phases and metallicities. Furthermore, a good calibration of all the parameters involved in individual TP-AGB stars does not exist for supersolar Z.

I adopt the view (e.g., Willson 2000) that empirical relations between mass-loss and stellar parameters are the result of very strong selection effects, since stars with a low rate will not be detected as mass-losing, whereas stars with a high rate will be obscured by dust and/or extremely short-lived. In other words, regardless of the actual rate, mass-loss will appear to follow a Reimers' type relation† (Reimers 1975; Reimers 1977), and such relations give the properties of stars undergoing mass-loss, but do not describe how any

† $\dot{M} = \eta LR/M$, where M and L are, respectively, the stellar mass and luminosity, $R(L, M, Z)$ is the stellar radius, and η is a fitting parameter.

one star loses mass over time. Consequently, rather than, for example, varying η while leaving the stellar parameters unchanged, I vary together $\dot{M}$ and the stellar parameters, in a consistent fashion. The whole procedure has been described in detail in Appendix A1 of González-Lópezlira *et al.* (2010), and is quite iterative: a variation in $\dot{M}$ will imply changes in L, R, temperature, pulsation period, carbon to oxygen ratio, and wind expansion velocity $v_{\rm exp}$. For the calculation of the SEDs, the dust opacity τ is a function of $\dot{M}$, specific extinction coefficient κ_λ, gas-to-dust ratio Ψ, $v_{\rm exp}$, and L, but at the same time $\dot{M}$ is a function of L and τ, κ_λ is a function of τ, Ψ is a function of $\dot{M}$ and τ, and $v_{\rm exp}$ depends on L and $\dot{M}$. Lifetimes t and hence star numbers in a particular phase also change with $\dot{M}$, according to the fuel consumption theorem Renzini & Buzzoni (1986), i.e., assuming that the product Lt is constant (and equal to the value for fiducial $\dot{M}$).

I have compared the models to the integrated mid-infrared colors of individual AGB candidates (Srinivasan *et al.* 2009), and to integrated colors and surface brightness fluctuations (SBF) of eight artificial "superclusters, i.e., coadded data of Magellanic star clusters in bins with similar ages and metallicities, according to classes I - VII in the Searle *et al.* (1980) SWB categorization scheme, plus an ultra-young (pre-SWB) supercluster. The SWB types constitute a smooth, one-dimensional sequence of increasing age and decreasing Z; coaddition reduces the stochastic uncertainty produced by the inadequate sampling, in sparse clusters, of stars evolving through short evolutionary phases, of which the AGB is a prime example. If the numbers of stars in different evolutionary stages have a Poissonian distribution, then the theoretical relative errors scale as $M_{\rm tot}^{1/2}$, with $M_{\rm tot}$ the total mass of the cluster.

My conclusions are as follows (González-Lópezlira 2018): models with different mass-loss rates and metallicities differ significantly in their predicted mid-IR colors and SBF magnitudes; models with a higher than fiducial $\dot{M}$ are needed to fit the mid-IR colors of "extreme" single AGB stars in the LMC; the range of mid-IR colors of individual MC clusters is consistent with models with $Z = 0.008$, $\dot{M}$ between fiducial and 5 $\times$ fiducial, and the stochastic errors expected for a cluster population between 5 $\times 10^3$ and 5×10^4 M$_\odot$; in the case of artificial "superclusters", although models are compatible with the observations, integrated colors cannot strongly constrain $\dot{M}$, given the present data and theoretical uncertainties (the colors of the 3 Gyr old SWB VI cluster, however, suggest a higher than fiducial mass-loss rate); model SBF magnitudes are quite sensitive to metallicity for 4.5 μm and longer wavelengths, basically at all stellar population ages; fluctuation magnitudes are powerful diagnostics of mass-loss rate in the TP-AGB; the SBF measurements of the MC superclusters suggest a mass-loss rate close to fiducial.

References

González-Lópezlira, R. A. 2018, *ApJ*, 856, 170

González-Lópezlira, R. A., Bruzual-A., G., Charlot, S., Ballesteros-Paredes, J., & Loinard, L. 2010, *MNRAS*, 403, 1213

Marigo, P., Bressan, A., Nanni, A., Girardi, L., & Pumo, M. L. 2013, *MNRAS*, 434, 488

Reimers, D. 1975, *Memoires of the Societe Royale des Sciences de Liege*, 8, 369

Reimers, D. 1977, *A&A*, 61, 217

Renzini, A., & Buzzoni, A. 1986, *Spectral Evolution of Galaxies*, 122, 195

Rosenfield, P., Marigo, P., Girardi, L., *et al.* 2014, *ApJ*, 790, 22

Rosenfield, P., Marigo, P., Girardi, L., *et al.* 2016, *ApJ*, 822, 73

Searle, L., Wilkinson, A., & Bagnuolo, W. G. 1980, *ApJ*, 239, 803

Srinivasan, S., *et al.* 2009, *AJ*, 137, 4810

Willson, L. A. 2000, *ARAA*, 38, 573

Why Galaxies Care About AGB Stars:
A Continuing Challenge through Cosmic Time
Proceedings IAU Symposium No. 343, 2019
F. Kerschbaum, M. Groenewegen & H. Olofsson, eds.

doi:10.1017/S1743921318005598

Kepler K2: A Search for Very Red Stellar Objects

E. Hartig[1], K. H. Hinkle[2] and T. Lebzelter[3]

[1]Department of Astrophysics, University of Vienna, Austria,
email: erich.hartig@univie.ac.at,

[2]National Optical Astronomy Observatories, P.O. Box 26732, Tucson, AZ 85726 USA,

[3]Department of Astrophysics, University of Vienna, Austria

Abstract. Analyzing 41 targets data of the *Kepler* K2 Campaign 2 mission suspected to be Long Period Variables (LPVs), we developed a method for the prediction of periods longer than the observation period of 77.48d using the 3500 data points provided by K2. The 'Self-Flat-Field' method (K2SFF or SFF) of the '*Kepler* K2 High Level Science Product' (K2HLSP) corrected the instrumental effects best.

Keywords. stars: AGB and post-AGB, methods: data analysis, stars: variables: other

---

1. *Kepler* light curves of LPVs

The analysis of *Kepler* light curves of long period variables forms a particular challenge, because the period lengths typically exceed the observed time window significantly. On the other hand, the parts of the light change available is of such high sampling and accuracy that we attempt to derive reliable period estimates from these 2.5 months coverages. For this, we applied an RMS-error minimization using sinusoidal fits to the SFF data. An illustration from C0 is presented in Figure 1a, below.

2. Sinusoidal fits of the light curves

In the meantime, we have several candidates for testing our method, four of them are shown in Figure 2 with the light curve simulations on the left and the errors as a function of period on the right. For EP 202070273 we derive a long period of 763.3d in excellent agreement with the ASAS result of 730.0d. Example 4 shows a clear primary period of 24.9d, for the second period we get 276.3d or higher. Using a three terms sinusoidal function, as shown in Figure 1b, we constrain the second period to 353.2d.

3. *Kepler* C2 sinusoidal fits vs. ASAS

In our C2 sample we have nine stars with known ASAS periods >90d, Table 1. We use these objects to test and optimize our method. The plots of Figure 3 show the TPD and SFF LCs (left), the artificial LCs fit to the SFF (mid) and the error trend with the error bar and the ASAS period mark (right). Typically the ASAS period is close to the fit of the third local error period (marked S3), which is in 6 cases close to the period of the rms-error minimum. The ASAS period can be approximated by 86.5 % of the S3 period with a standard deviation σ of 0.83, except for AT Sco, EP 202913758.

Table 1. K2-C2, the properties of the 10 candidates, the sinusoidal rms-minimum, and the sinusoidal results compared with the ASAS periods.

C2-EP	Name	Type	Jmag	Kmag	Kpmag	SFF Sinusoidal		ASAS
						rms-min[d]	Result[d]	[d]
202913758	AT Sco	Mi	8.666	7.249	8.666	450.3	275.4	130.1
203529462	BQ Sco	Mi?	8.512	6.930	8.512	290.5	290.5	217.9
203748709	BW Sco	Mi	6.381	4.652	10.685	180.8	180.8	115.2
203763661	UZ Sco	Or	7.119	5.539	7.119	90.1	289.2	276.1
203785618	GSC 06797-00345	Star	6.648	5.209	12.039	754.6	286.5	335.
203795904	DI Sco	Mi?	8.679	7.227	8.679	367.8	272.9	197.5
204122147	TY Sco	MiCet	7.112	5.591	7.112	283.1	283.1	291.
204443100	TU Oph	Mi	8.464	6.985	12.044	293.2	293.2	246.
205087771	RR Oph	MiCet	4.533	2.510	8.829	366.9	366.9	299.
205240599	V1158 Sco	LP	8.254	6.815	8.254	449.7	449.7	LC only

(a) K2-C0, EP 202070273. sin.-fits.

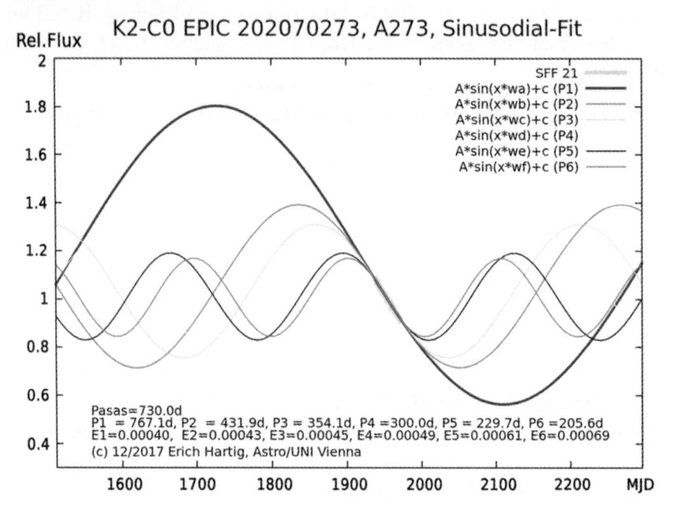

(b) K2-C3, three terms long period sinusoidal-fits.

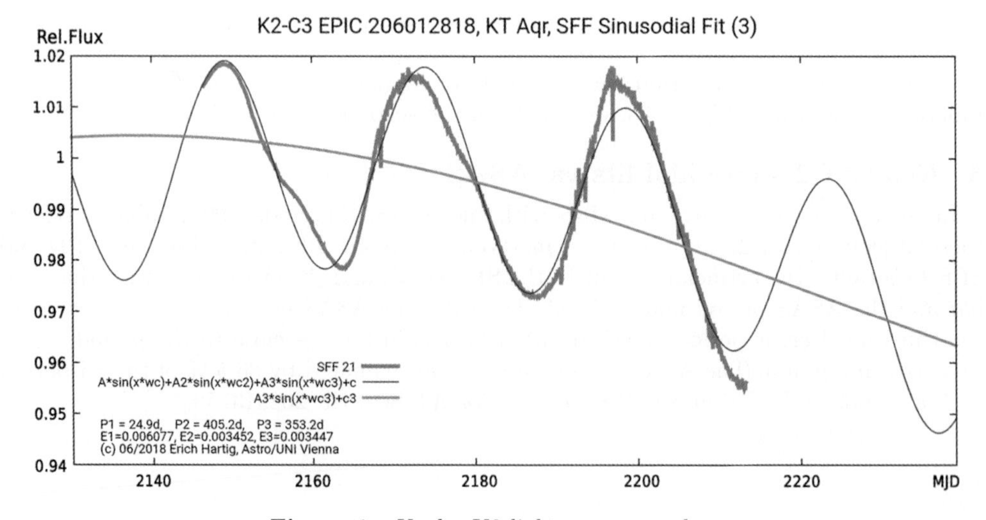

Figure 1. *Kepler*-K2 light curve samples.

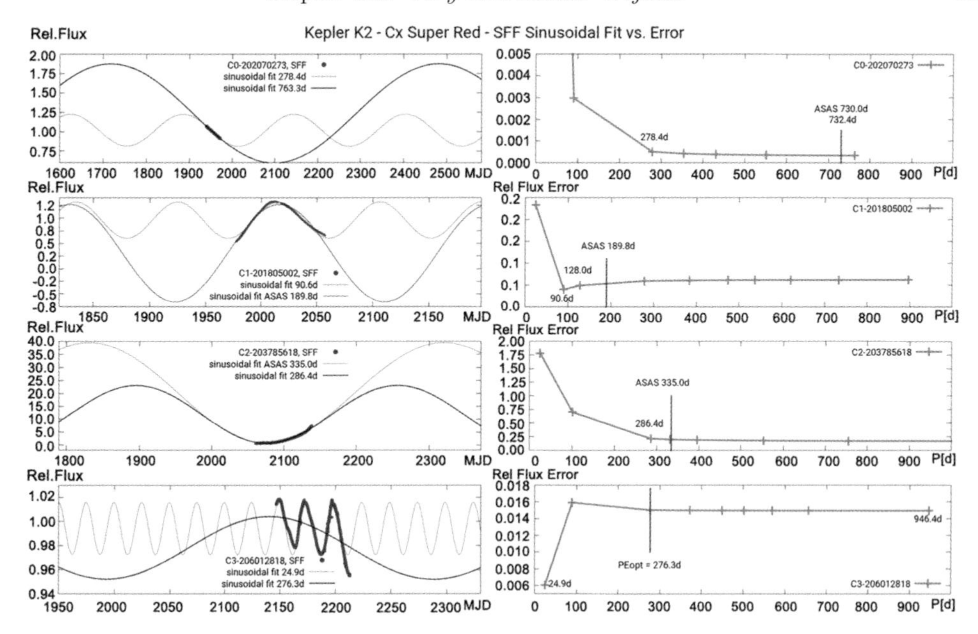

Figure 2. C0-C3, four samples of sinusoidal-fits and rms-errors.

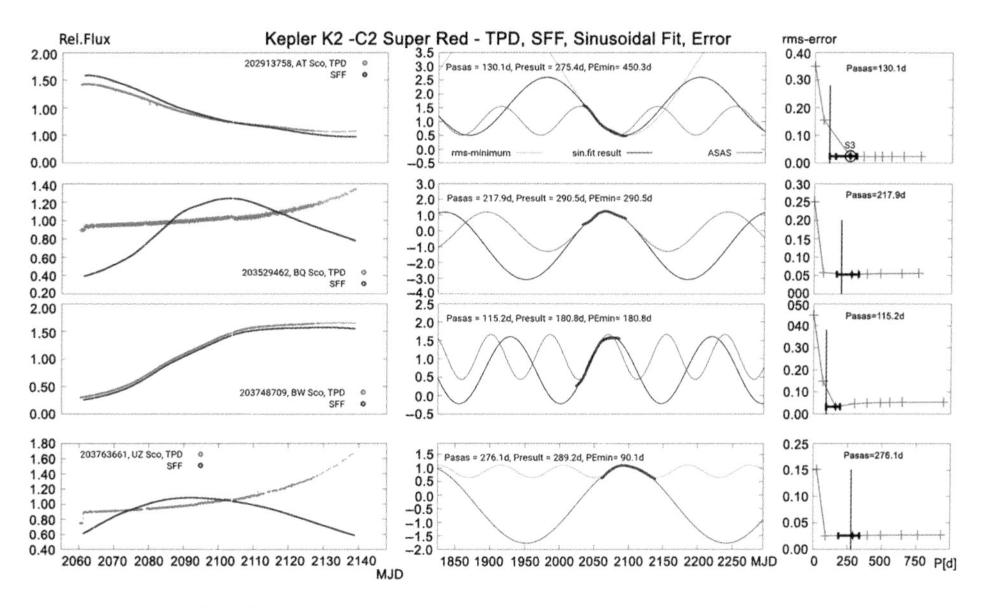

Figure 3. K2-C2, sinusoidal-fits samples, the rms-errors and the ASAS periods.

Why Galaxies Care About AGB Stars:
A Continuing Challenge through Cosmic Time
Proceedings IAU Symposium No. 343, 2019
F. Kerschbaum, M. Groenewegen & H. Olofsson, eds.

doi:10.1017/S1743921318005859

MIKE High Resolution Spectroscopy of Raman-scattered O VI and C II Lines in the Symbiotic Nova RR Telescopii

Jeong-Eun Heo[1], Hee-Won Lee[1], Rodolfo Angeloni[2], Tali Palma[3] and Francesco Di Mille[4]

[1]Department of Physics and Astronomy, Sejong University, Seoul, Korea
email: jeung6145@gmail.com

[2]Departamento de Física y Astronomía, Universidad de La Serena, La Serena, Chile

[3]Observatorio Astronómico, Universidad Nacional de Córdoba, Córdoba, Argentina

[4]Las Campanas Observatory, Carnegie Observatories, Casilla 601, La Serena, Chile

Abstract. We present a high-resolution optical spectrum of the symbiotic nova RR Tel obtained with *MIKE* at Magellan-Clay telescope. RR Tel is a wide binary system of a hot white dwarf and a Mira with an orbital period of a few decades, where the white dwarf is accreting through gravitational capture of some fraction of material shed by the Mira. We find broad emission features at 6825, 7082, 7023, and 7053 Å, which are formed through Raman scattering of far-UV O VI $\lambda\lambda$ 1032 and 1038 Å, C II $\lambda\lambda$ 1036 and 1037 Å with atomic hydrogen. Raman O VI features exhibit clear double-peak profiles indicative of an accretion flow with a characteristic speed ~ 30 km s^{-1}, whereas the Raman C II features have a single Gaussian profile. We perform a profile analysis of the Raman O VI by assuming that O VI emission traces the accretion flow around the white dwarf with a fiducial scale of 1 AU. A comparison of the restored fluxes of C II $\lambda\lambda$ 1036 and 1037 from Raman C II features with the observed C II λ1335 multiplet is consistent with the distance of RR Tel $\sim$ 2.6 kpc based on interstellar extinction of C II.

Keywords. binaries: symbiotic - stars: individual (RR Tel) - line: profiles - radiative transfer

1. Raman O VI Features and Accretion Flow

Unique spectral features at 6825 and 7082 Å with a broad width $\sim$ 20 Å are detected in a majority of symbiotic stars (Harries & Howarth 1996). The emission bands were identified as Raman-scattered O VI $\lambda\lambda$ 1032 and 1038 features by H I (Schmid 1989). Subsequent studies reveal that the Raman O VI features provide strong constraints on the kinematics of emission nebulae around the white dwarf and extend our understanding of the mass loss process in symbiotic systems (e.g. Heo *et al.* 2016).

In Fig. 1, we present the Raman-scattered O VI $\lambda\lambda$ 1032 and 1038 features at 6825 and 7082 Å of RR Tel in the left and right panels, respectively. The observations were performed on 2017 July 26 with *MIKE* spectrograph mounted on the Magellan-Clay telescope. The Raman O VI features of RR Tel are characterized by double-peak profiles with a peak separation of $\sim$ 60 km s^{-1}. The inelasticity of Raman-scattering requires that the Raman profiles reflect only the relative kinematics between the emission region and the scattering region, irrespective of the observer's sightline. The observed profiles of Raman O VI features are well fit with an O VI accretion disk around the white dwarf with a physical size of 1 AU. The disparate profiles of the two Raman O VI features are attributed to the local variation of the ratio $F(1032)/F(1038)$ in the O VI accretion disk

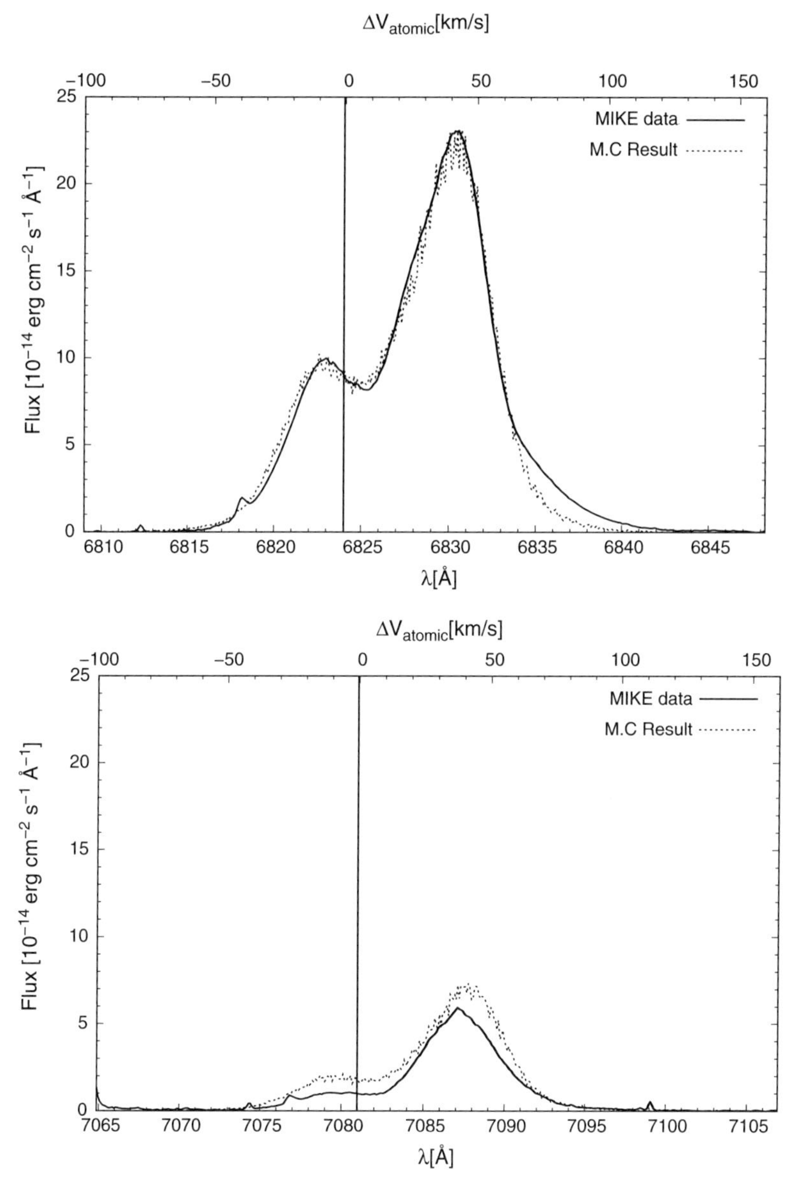

Figure 1. The Raman-scattered O VI $\lambda\lambda$1032 and 1038 at 6825 Å (left) and 7082 Å (right) in RR Tel. The solid line shows the *MIKE* observation, while the dotted line represents the result of our Monte Carlo simulations.

as discussed by Heo & Lee (2015). The red emission part, assumed to be of high density, is characterized by the flux ratio $F(1032)/F(1038) \sim 1$, whereas the blue emission region is much more sparse resulting in $F(1032)/F(1038) \sim 2$. Adopting the asymmetric O VI accretion disk model supplemented by the locally varying $F(1032)/F(1038)$, the best fit result is obtained for the mass loss rate $\dot{M} \sim 2\times\ 10^{-6}\ \mathrm{M}_\odot\ \mathrm{yr}^{-1}$ and the giant wind terminal velocity $v \sim 10\ \mathrm{km\ s}^{-1}$.

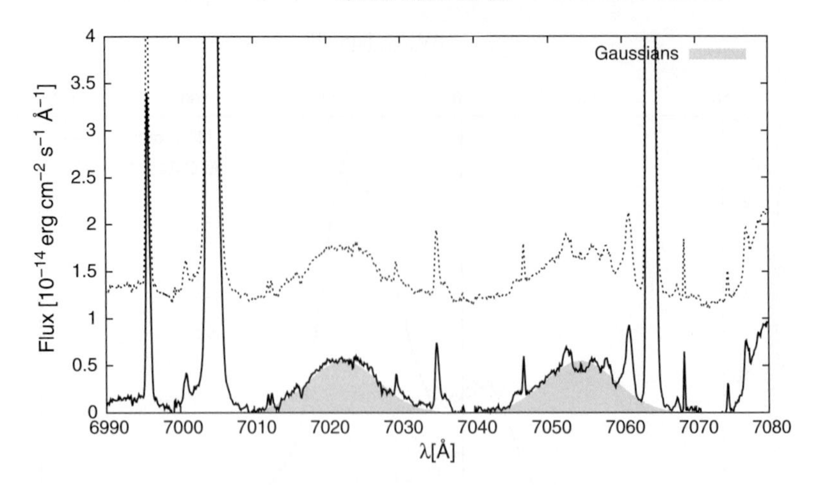

Figure 2. The Raman-scattered C II λλ1036 and 1037 at 7025 and 7052 Å. Our best fits with single Gaussian are shown by gray shaded regions.

2. Raman C II Features and Interstellar Extinction

Schild & Schmid (1996) detected two weak, broad emissions at 7023 and 7054 Å in the symbiotic nova V1016 Cyg. They identified the emission features as Raman-scattered lines of C II doublet λλ 1036 and 1037 Å by H I. Fig. 2 shows our *MIKE* spectrum of RR Tel in the range of the Raman C II features. The two features are well fitted with a single Gaussian profile, whose FWHM $\sim$ 11.3 Å and the center wavelengths of 7022.2 and 7054.3 Å, respectively. The total fluxes are obtained with $F(7023) = 6.56 \times 10^{-14}$ erg cm^{-2} s^{-1} and $F(7054) = 6.55 \times 10^{-14}$ erg cm^{-2} s^{-1}, which are about 6 times weaker than Raman O VI 1038 feature.

From our Monte Carlo analysis of the observed Raman scattered C II features, significant fluxes of C II λλ 1036 and 1037 $F(1036) = 6.06 \times 10^{-10}$ erg cm^{-2} s^{-1} and $F(1037) = 6.05 \times 10^{-10}$ erg cm^{-2} s^{-1} are expected, respectively. These far-UV lines are absent in the *FUSE* data, whereas the *IUE* data indicate the presence of C II 1335 triplet, implying the heavy interstellar extinction of C II λλ 1036 and 1037. The optical depths of C II λλ 1036 and 1037 and C II 1335 deduced from the comparison of the *FUSE* data, *IUE* data and Raman C II fluxes lead us to estimate the lower bound of the C II column density $N(CII) \sim 1.50{\times}10^{14}$ cm^{-2} toward RR Tel, which appears consistent with the presumed distance $D \sim 2.6$ kpc.

Acknowledgements

This research was supported by the Korea Astronomy and Space Science Institute under the R&D program (Project No. 2018-1-860-00) supervised by the Ministry of Science and ICT.

References

Harries, T. J., & Howarth, I. D. 1996, *A&AS*, 119, 61
Heo, J.-E., & Lee, H.-W. 2015, *JKAS*, 48, 105
Heo, J.-E., Angeloni, R., Di Mille, F., Palma, T. & Lee, H.-W. 2016, *ApJ*, 833, 286
Schild, H., & Schmid, H. M. 1996, *A&A*, 310, 211
Schmid, H. M. 1989, *A&A*, 211, L31

Why Galaxies Care About AGB Stars:
A Continuing Challenge through Cosmic Time
Proceedings IAU Symposium No. 343, 2019
F. Kerschbaum, M. Groenewegen & H. Olofsson, eds.

doi:10.1017/S1743921318005008

The Structure of the Inner Circumstellar Shell in Miras

Kenneth H. Hinkle[1] and Thomas Lebzelter[2]

[1]National Optical Astronomy Observatory, P.O. Box 26732, Tucson, Az, USA
email: hinkle@noao.edu

[2]Dept. of Astrophysics, Univ. of Vienna, Vienna, Austria
email: thomas.lebzelter@univie.ac.at

Abstract. We have measured CO line profiles in a time series of 42 high-resolution 1.6 - 2.5 μm spectra of R Cas. The low-excitation CO first overtone lines have a contribution from a ~1000 K region. We show that this region undergoes a periodic changes on time scales many times longer than the photospheric pulsation. Comparison with interferometry and models suggests that the ~1000 K region is at ~2 R_* and cospatial with the region of SiO masers and grain condensation. The CO lines are entirely in absorption requiring formation in a layer thin compared to the stellar diameter. The CO excitation temperature has been measured as low as 600 K suggesting that grains with a variety of compositions condense at ~2 R_*.

Keywords. stars: AGB and post-AGB, (stars:) circumstellar matter, stars: variables: other

---

Low-mass stars undergo most of their mass loss on the AGB with this accounting for ~75% of the mass returned to the Milky Way ISM. AGB mass loss results from the combined effects of stellar pulsation and radiation pressure on dust. We report here on observations of low-excitation CO first overtone lines in the mira R Cas. The lines have multiple velocity components one of which is formed in cool, infalling gas. The temperature and velocity of the cool, low-excitation component are significantly different than that of high-excitation CO lines formed in the stellar photosphere.

The presence of multiple velocity components in low-excitation 2-0 lines has been known for decades (Hinkle 1978, Hinkle *et al.* 1982). However, previous time series were insufficient to reveal the nature of the variability (Hinkle & Barnes 1979). We report here on part of an extensive time series of near-IR mira spectra. In most spectra the 2-0 lines are blends but on rare occasions the photospheric and ~1000 K gas have opposite velocities and the two components can be resolved.

Figure 1 shows velocities measured from low-excitation 2-0 CO lines. The CO probes a region that undergoes long term, aperiodic variations. These are very different from the periodic variations seen in the CO second overtone lines and tied to the light curve (Hinkle *et al.* 1984). For spectra where the velocity components are well separated we have measured excitation temperatures. The CO excitation temperature was measured dropping from ~1100 K to ~600 K over ~300 days. This velocity component then disappeared, probably as the result of transversing a shock. This kind of behavior appears in a number of recent models for regions at $\gtrsim$2 R_* (Ireland *et al.* 2011, Gail *et al.* 2016, Liljegren *et al.* 2017).

Direct evidence that the ~1000 K CO is at ~2 R_* comes from VLTI/AMBER R=12000 K-band spectral-spatial interferometry (Ohnaka *et al.* 2017). The AMBER data show CO extending to ~3 R_*. Assaf *et al.* (2011) has shown that the R Cas SiO masers have a maximum concentration at ~2 R_*. ALMA observations by Wong *et al.* (2016)

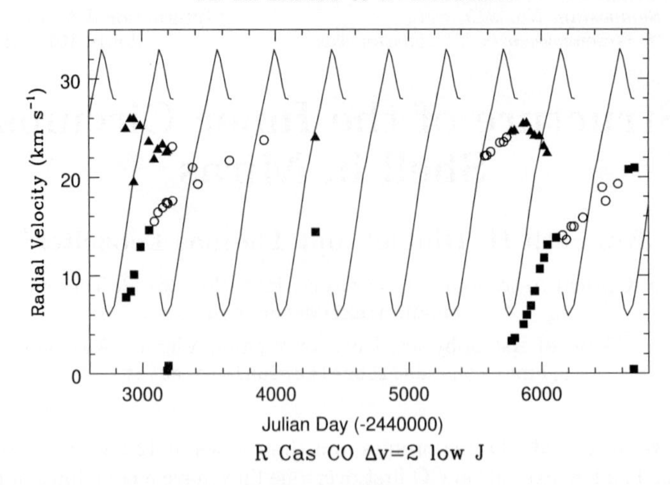

Figure 1. R Cas CO Δ v=2 low-excitation line velocities as a function of time. The points shown as filled squares have a high excitation temperature, the open circles are blends of high and low temperatures, and the filled triangles are at low-excitation temperature. The solid line is the mean velocity curve for the higher excitation CO Δ v=3 lines. The center of mass velocity of R Cas is 16.4 km s^{-1} so the cool gas is falling.

detect a ~1000 K molecular layer at ~2 R_*. The lack of P-Cyg type profiles in our observations show that the CO layer is thin. The line widths suggest the existence of large scale inhomogeneities that are also apparent in the SiO maser and ALMA results. An intriguing possibility is that the CO ~1000 K component is formed in clouds that also can produce SiO masers. The maser clouds are preferentially seen tangentially while the CO layer is seen in absorption against the stellar photosphere.

The CO observations reported here show temperatures low enough for SiO condensation on grains. Models by Gail *et al.* (2016) and Liljegren *et al.* (2017) show SiO dust formation in descending clouds. On the other hand, Wong *et al.* (2016) argues that silicon can not be significantly depleted on grains in the SiO maser region. Al_2O_3 (corundum) grain formation occurs at ~1000 K (Ohnaka *et al.* 2017). However, corundum is transparent to mira flux peak radiation (Woitke 2006) requiring the deposition of an opaque coating on the grain seeds if the grains are to be accelerated by radiation pressure (Bladh *et al.* 2015). Given the inhomogeneous nature of the 2 R_* region and our discovery of gas as cool as 600 K, grains with a variety of compositions should originate in this region.

References

Assaf, K.A., Diamond, P.J., Richards, A.M.S., *et al.* 2011 MNRAS 415 1083
Bladh, S., Höfner, S., Aringer, B., *et al.* 2015 A&A 557 A105
Gail, H.-P., Scholz, M., & Pucci, A. 2016 A&A 591 A17
Hinkle, K.H. 1978, ApJ, 220, 210
Hinkle, K.H. & Barnes, T.G. 1979, ApJ, 227, 923
Hinkle, K.H., Hall, D.N.B., & Ridgway, S.T. 1982, ApJ, 252, 697
Hinkle, K.H., Scharlach, W.W.G., & Hall, D.N.B. 1984, ApJS, 56, 1
Ireland, M.J., Scholz, M., & Wood, P.R. 2011 MNRAS 418 114
Liljegren, S., Höfner, S., Eriksson, K., *et al.* 2017 A&A 606 A6
Ohnaka, K., Weigelt, G., Hofmann, K.-H. 2017 A&A 597 A20
Woitke, P. 2006 A&A 460, L9
Wong, K.T., Kamiński, T., Menten, K.M., *et al.* 2016 A&A 590 A127

Why Galaxies Care About AGB Stars: A Continuing Challenge through Cosmic Time
Proceedings IAU Symposium No. 343, 2019
F. Kerschbaum, M. Groenewegen & H. Olofsson, eds.

doi:10.1017/S1743921318005884

Signs of rotating equatorial density enhancements around SRb pulsators

W. Homan[1], L. Decin[1,2], A. Richards[3] and P. Kervella[4]

[1]IvS, KU Leuven, Celestijnenlaan 200D, 3001 Leuven, Belgium
email: ward.homan@kuleuven.be

[2]University of Leeds, School of Chemistry, Leeds LS2 9JT, United Kingdom

[3]JBCA, Dept. Physics and Astronomy, University of Manchester, Manchester M13 9PL, UK

[4]LESIA (CNRS UMR 8109), Obs. de Paris, PSL, CNRS, UPMC, Univ. Paris-Diderot, France

Abstract. We observed the circumstellar environments (CSEs) of the semiregular AGB stars L2 Puppis, R Doradus and EP Aquarii with ALMA. (1) The molecular emission in the L_2 Pup nebula reveals an edge-on rotating disk. (2) PV diagrams of the ^{28}SiO emission in the inner CSE of R Dor expose a pattern pointing to an inclined rotating disk. (3) The CO emission in the CSE of EP Aqr reveals a nearly face-on equatorial density enhancement (EDE). The inner EDE strongly resembles theoretical wind-Roche-lobe-overflow models. The SiO emission points to a potential companion. The combination of (1), (2) and (3) suggests that a link may exist between the type of AGB pulsations and the morphological nature of the CSE.

Keywords. stars: AGB and post-AGB stars, circumstellar material, stars: mass loss, stars: imaging, submillimeter

1. Introduction

Asymptotic giant branch (AGB) stars are the evolutionary stage of low- and intermediate-mass stars before they turn into the planetary nebulae. These CSEs exhibit large-scale sphericity, but generally display a rich spectrum of smaller scale structural complexities. Hydrodynamic models show that perturbation of the AGB wind by the local gravity field of a non-giant binary companion may play a crucial role in explaining various deviations from spherical symmetry (Mastrodemos & Morris 1999). This should not not come as a surprise, as the multiplicity frequency of the main sequence (MS) predecessors of AGB stars is found to exceed 50% (Raghavan *et al.* 2010). Considering that recent studies show that on average every star in the Milky Way (Cassan *et al.* 2012) possesses one or more planets, this frequency can only be considered a lower limit. In this manuscript we briefly present high-resolution ALMA observations of three different M-type SRb AGB stars: L_2 Puppis, R Doradus and EP Aquarii.

2. Observations

L_2 Puppis (Homan *et al.* 2017) We obtained spatially resolved ALMA band 7 observations of the CSE of L_2 Pup. The data shows an edge-on differentially rotating gas disk (Kervella *et al.* 2016). The Keplerian motion of the gas allowed us to accurately determine the central mass of the L_2 Pup system. A significant asymmetry in the continuum emission of the disk points towards the presence of a binary companion. We modelled the observed emission of the ^{12}CO and ^{13}CO rotational transition J=3–2 with LIME (Brinch & Hogerheijde 2010) which allowed us to constrain the (thermo)dynamical and morphological properties of the circumstellar gas. This permitted us to deduce the mass of the disk. The angular momentum (AM) in the disk was found to be excessive compared to the host star's AM. We calculate that a binary companion of ~1 Jupiter mass would be sufficient to explain the AM contained in the disk.

R Doradus (Homan *et al.* 2018a) A spectral scan of the CSE of R Dor was acquired with ALMA at a spatial resolution of ~ 150 mas (Decin *et al.* 2018). Many molecular transitions show a distinct spatial offset between the blue and red shifted emission in the central 1" × 1". We constructed position-velocity diagrams of the compact ^{28}SiO v=1 emission, whose dynamical signature resembles that of a nearly edge-on rotating disk. We model the emission using the radiative transfer code LIME, and reproduce the morpho-kinematical emission features by assuming a compact disk. We calculate the mass of the disk and estimate that its AM exceeds the AM of the host star. An object with a mass of at least ~2.5 earth masses at a distance of 6 AU, the tentative disk inner rim location, could explain the AM contained within the disk, though its mass may be substantially higher.

EP Aquarii (Homan *et al.* 2018b) EP Aqr was observed with ALMA band 6 in cycle 4, probing the molecular emission of ^{12}CO , ^{28}SiO , and SO_2 emission. The spatially resolved CO emission reveals a bi-conical outflow with a bright central spiral structure-harbouring EDE. The EDE is extremely confined in velocity-space. The innermost regions of the CO emission exhibit a morphology that strongly resembles the wind Roche-lobe overflow (WRLOF) simulations of the Mira AB system. The SiO emission exhibits a localized emission void, located about 0.5" west of the continuum brightness peak in the red-shifted portion of the emission, located where the WRLOF simulations predict it to be. This feature may be a local environment caused by the presence of a companion, and estimate its to be at most ~0.1 $M_\odot$ based on its non-detection in the continuum. Finally, the SO_2 emission shows a clear sign of (co)rotation, with a maximum projected velocity amplitude of ~1 km s^{-1}. The odd nature of the spiral/EDE and the deprojected rotation speeds of the SO_2 gas favours the hypothesis that the equatorial matter is contained within a large face-on differentially rotating disk.

3. Concluding hypothesis

Observations/Modeling of the inner wind of these three oxygen-rich SRb-type AGB stars show strong indications for the presence of binary companions. Furthermore, it seems that the dominating velocity field of the equatorial matter in the inner CSE is tangential, resulting in the formation of a stable disk. High-resolution observation of mira-type carbon stars show that these typically possess spiral-shaped features in their winds, with no indications of rotation. We hypothesize that there may be a relation between the nature of the surface pulsations of the AGB star, and the resulting companion-induced wind morphology.

References

Mastrodemos N., & Morris M. 1999, *ApJ*, 523, 357
Raghavan D., McAlister H. A., Henry T. J, Latham D. W., , *et al.* 2010, *ApJS*, 190, 1
Cassan A., Kubas D., Beaulieu J.-P., Dominik M., Horne K., *et al.* 2012, *Nature*, 481, 167
Homan W., Richards A., Decin L., Kervella P., de Koter A., *et al.* 2017, *A&A*, 601A, 5H
Kervella P., Homan W., Richards A., Decin L., McDonald I., *et al.* 2016, *A&A*, 596A, 92K
Brinch C., & Hogerheijde M. 2010, *A&A*, 523A, 25B
Homan W., Danilovich T., Decin L., de Koter A., Nuth J., *et al.* 2018, *A&A*, 614A, 113H
Decin L., Rochards A., Danilovich T., Homan W., & Nuth J. 2018, *A&A*, 615A, 28D
Homan W., Richards A., Decin L., de Koter A., & Kervella P. 2018, *A&A*, 616A, 34H

Why Galaxies Care About AGB Stars:
A Continuing Challenge through Cosmic Time
Proceedings IAU Symposium No. 343, 2019
F. Kerschbaum, M. Groenewegen & H. Olofsson, eds.

doi:10.1017/S1743921318006956

Variability in Post-AGB Stars: Pulsation in Proto-Planetary Nebulae

Bruce Hrivnak[1], Gary Henson[2], Griet Van de Steene[3], Hans Van Winckel[4], Todd Hillwig[1] and Matthew Bremer[1]

[1]Dept. of Physics & Astronomy, Valparaiso University, Valparaiso, IN 46383, USA
emails: bruce.hrivnak@valpo.edu, todd.hillwig@valpo.edu, matthew.bremer@valpo.edu

[2]Dept. of Physics & Astronomy, East Tennessee State University, Johnson City, TN 37614, USA
email: hensong@mail.etsu.edu

[3]Royal Observatory of Belgium, Astronomy & Astrophysics, Ringlaan 3, Brussels, Belgium
email: g.vandesteene@oma.be

[4]Inst. voor Sterrenkunde, KU Leuven, Celestijnenlaan 200 D, B-3001, Leuven, Belgium
email: Hans.VanWinckel@kuleuven.ac.be

Abstract. We have been intensely monitoring photometric variability in proto-planetary nebulae (PPNe) over the past 25 years and radial velocity variability over the past ten years. Pulsational variability has been obvious, in both the light and velocity, although the resulting curves are complex, with multiple periods and varying amplitudes. Observed periods range from 25 to 160 days, and the periods and amplitudes reveal evolutionary trends. We will present our observational results to date for approximately 30 PPNe, and discuss these results, including the search for period changes that might help constrain post-AGB evolutionary timescales.

Keywords. AGB and post-AGB, stars: variable

1. Introduction

Proto-planetary nebulae (PPNe) are post-AGB objects in transition between the AGB and PN phases in the evolution of low- and intermediate-mass stars. They consist of a luminous central star of intermediate temperature (5000–30000 K), surrounded by a detached, expanding shell of gas and dust. It is a short-lived phase (few $\times$ 10^3 yr). Their identification and study began with the *IRAS* satellite ~30 years ago.

As post-AGB stars, it is not surprising that PPNe might vary due to pulsation. As they evolve in the post-AGB phase, they pass through the extension of the Cepheid instability strip (Kiss *et al.* 2007). There are several motivations for this study. (1) Pulsation provides an opportunity to determine fundamental properties of PPNe. The period (P) and amplitude, when combined with stellar pulsational models, can be used to find PPNe masses, which have not been measured directly (none are in known binaries). (2) Pulsation is a driver of mass loss, which is poorly understood physically, so it might lead to new insights. (3) The pulsation period is expected to change as the post-AGB star evolves, and this has the potential to help us measure the timescale of this evolution.

2. Light Curve and Radial Velocity Curve Studies

Light curve studies of PPNe have been carried out primarily by two groups (Arkhipova *et al.* 2010; Arkhipova *et al.* 2011; Hrivnak *et al.* 2010; Hrivnak *et al.* 2015) and examples of light curves are shown in their work. Our observations are carried out primarily at the Valparaiso Univ. Observatory by undergraduate students. About 20 objects have been observed annually for 25 years (1994–2018). Additional observations are from the SARA telescopes at KPNO and CTIO over 8–10 yr. We have also used ASAS and ASAS-SN survey data to find periods for additional stars, particularly in the southern hemisphere.

These light curves display cyclical behavior with varying amplitudes, with a few also showing longer-term increases or decreases in brightness. Periods range from 25 to 160 days (SpT=A–G), with hotter B types varying on shorter timescales of days or less. Amplitudes (ΔV) range from 0.1–0.7 mag (V, peak-to-peak), with cooler ones having larger ΔV. The stars are redder when fainter. The light curves are thus seen to be complex, with changing amplitude and multiple P values (P_1, P_2, ...). A period ratio of $P_2/P_1 = 0.90$–1.10 is found in most cases.

Closer examination reveals several trends in the light curve properties. Most significant is a linear correlation of P_1 with $T_{\rm eff}$; P is seen to decrease with increasing $T_{\rm eff}$ from 5000–8000 K. This is seen in both C-rich and O-rich stars. (See an early example in Hrivnak *et al.* 2010.) The amplitude of the pulsation also decreases with increasing $T_{\rm eff}$.

We have also begun in 2007 to monitor the radial velocity of a subset of seven bright PPNe using the 1.2-m telescope at the Dominion Astrophysical Observatory (Victoria, Canada) and the 1.2-m Mercator telescope with Hermes spectrograph at La Palma. This allows us to get contemporaneous light, color, and radial velocity curves. Similar periods are found in the radial velocity as in the light and color curves. Phase comparisons show that the light and color curves are in phase, while the radial velocity curves differ by a quarter of a cycle (Hrivnak *et al.* 2013).

3. Discussion and Conclusions

Theoretical models predict that for post-AGB stars, $T_{\rm eff}$ will increase with time. Using the current models of Miller Bertolami (2016), combined with our observed P–$T_{\rm eff}$ relationship, predicts a change of P with time of -0.05 day/yr or -1 day over 20 years. This is potentially observable! However, in a preliminary analysis we find it difficult to measure reliable period changes in these complex light curves.

The contemporaneous light, color, and radial velocity curves show that these stars are brightest when hottest and smallest. This agrees with the expectations from radial pulsations with the κ mechanism operating in the He II ionization zone. Note that this phase relationship is different from that found for Cepheids, in which the radial velocity is half a cycle out of phase with the light curve and the star is brightest when hottest and about average size and expanding (so called "phase lag"). New pulsation models are needed to update the early work and make use of these contemporaneous light and velocity curves to determine masses for these post-AGB objects.

Acknowledgements

BH acknowledges the support of the National Science Foundation (AST-1413660).

References

Arkhipova, V. P., Ikonnikova, N. P., & Komissarova, G.V. 2010, *Astron. Lett.*, 36, 269
Arkhipova, V. P., Ikonnikova, N. P., & Komissarova, G.V. 2011, *Astron. Lett.*, 37, 635
Hrivnak, B.J., Lu, W., Maupin, R.E., & Spitzbart, B.D. 2010, *ApJ*, 709, 1042
Hrivnak, B.J., Lu, W., & Nault, K.A. 2015a, *AJ*, 149, 184
Hrivnak, B.J., Lu, W., Sperauskas, J., *et al.* 2013, *ApJ*, 766, 116
Kiss, L.L., Derekas, A., Szabó, Gy.M., Bedding, T.R. & Szabados L. 2007, *MNRAS*, 375, 1338
Miller Bertolami, M.M. 2016, *A&A*, 588, A25

Why Galaxies Care About AGB Stars:
A Continuing Challenge through Cosmic Time
Proceedings IAU Symposium No. 343, 2019
F. Kerschbaum, M. Groenewegen & H. Olofsson, eds.

doi:10.1017/S174392131800697X

Are the silicate crystallinities of oxygen-rich evolved stars related to their mass loss rates?

Biwei Jiang[1], **Jiaming Liu**[2] **and Aigen Li**[3]

[1] Department of Astronomy, Beijing Normal University, Beijing 100875, China
email: bjiang@bnu.edu.cn

[2] National Astronomical Observatories, Chinese Academy of Sciences, Beijing 100012, China
email: jmliu@nao.cas.cn

[3] Department of Physics and Astronomy, University of Missouri, Columbia, MO 65211, USA
email: lia@missouri.edu

Abstract. A sample of 28 oxygen-rich evolved stars is selected based on the presence of crystalline silicate emission features in their ISO/SWS spectra. The crystallinity, measured as the flux fraction of crystalline silicate features, is found not to be related to mass loss rate that is derived from fitting the spectral energy distribution.

---

1. Introduction

Crystalline silicates are identified through a series of sharp spectral lines between 10-70 micron by the ISO and Spitzer space observations (Henning 2010). They are detected in various types of objects, from solar system objects – comets, to pre-main sequence stars, evolved stars and distant quasars. The spectral features of crystalline silicates are detected in every stage of evolved stars: AGB stars, post-AGB stars and planetary nebulae (e.g. Jiang *et al.* 2013). The degree of crystallinity, i.e. the mass percentage of crystalline silicate in all silicate dust, is found to range from a few percent to $>90\%$. What determines crystallinity has long been debated (Liu & Jiang 2014). Mass loss rate is thought to be an important factor. Theoretical calculations have shown that amorphous silicates cannot be crystallized in stars of low mass loss rate because the dust cannot be heated to temperatures high enough for crystallization, and that crystalline silicates can only be formed in stars undergoing substantial mass-loss with a critical value of $\dot{M} > 10^{-5}\,\mathrm{M_{\odot}\,yr^{-1}}$ and having high dust column densities (e.g. Gail *et al.* 1999). Jones *et al.* (2012) analyzed the Spitzer/IRS spectra of 315 evolved stars and found that the mass-loss rates of the stars exhibiting the crystalline silicate features at 23, 28 and 33 μm span over 3 dex, down to $10^{-9}\,\mathrm{M_{\odot}\,yr^{-1}}$. They investigated the possible correlation between $\dot{M}$ and the silicate crystallinity by examining the relation of $\dot{M}$ with the strengths of the 23, 28 and 33 μm features measured by their equivalent widths, but found no correlation. Kemper *et al.* (2001) performed an extensive radiative transfer calculation of the model IR emission spectra for O-rich AGB stars of mass-loss rates ranging from $10^{-7}\,\mathrm{M_{\odot}\,yr^{-1}}$ to $10^{-4}\,\mathrm{M_{\odot}\,yr^{-1}}$ and of a wide range of crystallinities up to 40 percent. They also found that crystallinity is not necessarily a function of mass-loss rate.

2. Data and method

We selected a sample of nearby 28 oxygen-rich evolved stars (mainly AGB stars and red supergiants) which show prominent spectral features of crystalline silicate as well as amorphous silicate in their ISO spectra. The mass loss rate is calculated by fitting the spectral energy distribution from (ultraviolet-)optical to far-infrared based on the

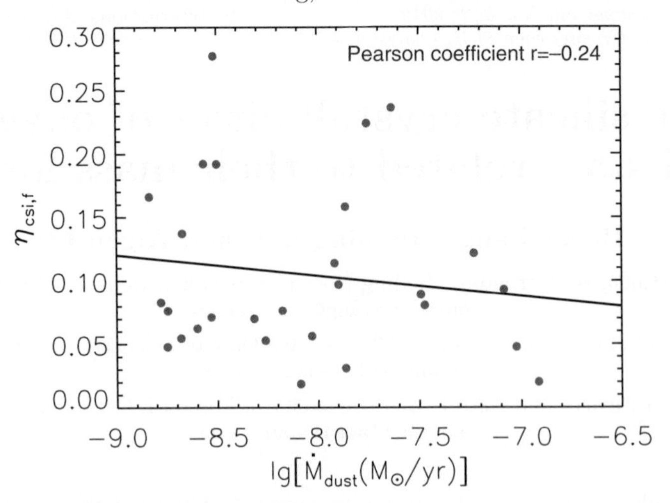

Figure 1. Relation between mass loss rates and silicate crystallinities of O-rich evolved stars.

photometry in the UBVRI, 2MASS/JHK, WISE and IRAS bands. The '2-DUST' radiative transfer code (Ueta & Meixner 2003) is used to derive the mass loss rate and other stellar and circumstellar parameters.

The silicate crystallinity ($\eta_{\rm csi,f}$) is characterized with the flux ratio of the emission features of crystalline silicates to that of all (crystalline + amorphous) silicates. This measure of crystallinity differs from the common measure of the mass ratio of crystalline to total silicate. The calculation of silicate mass suffers generally significant uncertainties from dust temperature in addition to the opacity since the species is not clearly identified. The flux ratio would represent the mass ratio if the UV/V/IR opacities of crystalline and amorphous silicates are comparable. The flux of crystalline and amorphous silicate is calculated with the PAHFIT code by decomposing the ISO spectrum into continuum and spectral features of amorphous and crystalline silicate.

3. Result

The derived mass loss rate and crystallinity have a Pearson correlation coefficient of -0.24 (Figure 1), which indicates that the silicate crystallinities and the mass-loss rates of these oxygen-rich evolved stars are not correlated (Liu *et al.* 2017). A further check also found no relation of crystallinity with stellar luminosity or effective temperature.

References

Gail H., & Sedlmayr, E. 1999, *A&A*, 347, 594
Henning T. 2010, *ARA&A*, 48, 21
Jiang, B. W., Zhang, K., Li, A., & Lisse, C. 2013, *ApJ*, 765, 72
Jones, O. *et al.* 2012, *MNRAS*, 427, 3209
Kemper, F., Waters, L., de Koter, A., & Tielens, A. 2001, *A&A*, 369, 132
Liu, J. M., & Jiang, B. W. 2014, *Progress in Astronomy*, 32, 2
Liu, J. M., Jiang, B. W., Li, A.G., & Gao, J. 2017, *MNRAS*, 466, 1963
Ueta, T., & Meixner, M. 2003, *ApJ*, 586, 1338

Why Galaxies Care About AGB Stars:
A Continuing Challenge through Cosmic Time
Proceedings IAU Symposium No. 343, 2019
F. Kerschbaum, M. Groenewegen & H. Olofsson, eds.

doi:10.1017/S1743921318005069

Binary evolution and double sequences of blue stragglers in globular clusters

Dengkai Jiang[1,2,3], Xuefei Chen[1,2,3], Lifang Li[1,2,3] and Zhanwen Han[1,2,3]

[1]Yunnan Observatories, Chinese Academy of Sciences,
396 Yangfangwang, Guandu District, Kunming, 650216, China
email: dengkai@ynao.ac.cn

[2]Center for Astronomical Mega-Science, Chinese Academy of Sciences,
20A Datun Road, Chaoyang District, Beijing, 100012, China

[3]Key Laboratory for the Structure and Evolution of Celestial Objects, Chinese Academy of Sciences, Kunming, 650011, China

Abstract. Binary evolution can produce different blue-straggler binaries, for example, blue stragglers with a bright, red component, or with a faint, blue component. In globular clusters, these blue-straggler binaries are generally observed as a single star, because two components can not be distinguished. Therefore, these blue-straggler binaries can be located in different regions of the color-magnitude diagram of globular clusters, e.g. blue sequence and red sequence observed in M30. We suggest that binary evolution can contribute to the blue stragglers in both of the sequences. Some blue stragglers in the blue sequence may have a faint white dwarf companion, while the red sequence includes some binaries experiencing mass transfer. It should be noted that the red sequence may also have other binaries, for example, the binaries just finished the mass transfer, and the binaries including a blue straggler (the accretors) that have evolved away from the blue sequence.

Keywords. binaries: general, blue stragglers, globular clusters, stars: evolution

---

1. Introduction

Blue stragglers are brighter and bluer than the main-sequence turn-off stars in the color-magnitude diagram in globular clusters. They are found in almost all Galactic globular clusters (Piotto *et al.* 2004). Their location in the color-magnitude diagram suggests that they are more massive than other main-sequence stars. According to stellar evolution theory, these stars should have evolved away from their location and become a white dwarf, if they are normal single stars in globular clusters. Two popular mechanisms can explain the existence of blue stragglers in globular cluster, direct stellar collision (Hills & Day 1976) and binary evolution (McCrea 1964). Binary evolution may play an important role in open clusters and in the field. However, it is still debatable which is the dominant formation mechanism in globular clusters. Two blue-straggler sequences in the color-magnitude diagram (Ferraro *et al.* 2009, Dalessandro *et al.* 2013, Simunovic *et al.* 2014), as one of the most important observations of blue stragglers, may have significant information about the origin of blue stragglers in globular clusters.

2. Binary evolution and two blue-straggler sequences

It should be noted that mass transfer in binary systems can produce two kinds of blue-straggler binaries (as shown in Figure 1): (1) blue stragglers with a bright, red component; (2) blue stragglers with a faint, blue component. For the binary experiencing mass transfer, the accretor becomes a blue straggler while the donor is still a bright and

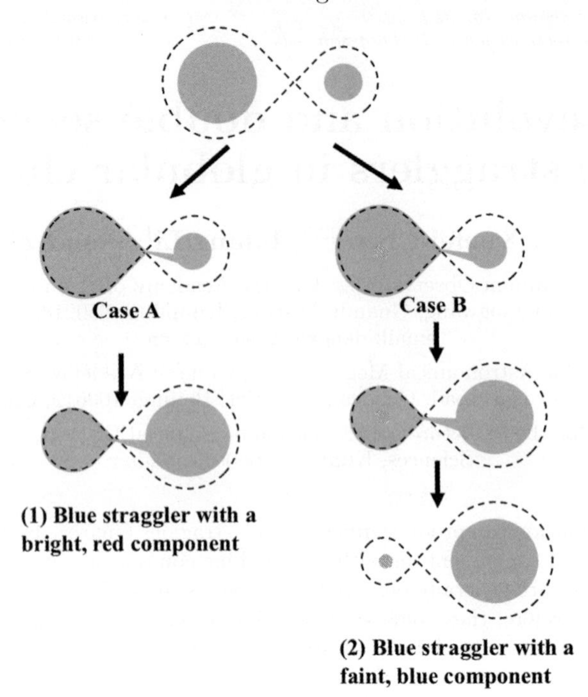

Figure 1. Path of different blue-straggler binaries from binary evolution.

red star. Therefore, they have a "low-luminosity boundary" (about 0.75 mag brighter than the ZAMS) given by Tian *et al.* (2006), and can match the observed red sequence in globular cluster M30 (Ferraro *et al.* 2009; Xin *et al.* 2015).

However, the location of the other blue-straggler binaries may be very different. If the blue stragglers have a faint, blue component (e.g. faint white dwarf), they would have similar location of single blue stragglers (e.g. blue stragglers from binary merger or stellar collision) in the color-magnitude diagram. When these blue stragglers with faint components are considered, binary evolution can produce the blue sequence as well as the red sequence observed in M30. Furthermore, they would go through the red sequence twice, for example, before they evolve into the blue sequence or after they have evolved away from the blue sequence.

References

Dalessandro, E., Ferraro, F. R., Massari, D., *et al.*. 2013, *ApJ*, 778, 135
Ferraro, F. R., Beccari, G., Dalessandro, E., *et al.*. 2009, *Natur*, 462, 1028
Hills, J. G., & Day, C. A. 1976, *ApL*, 17, 87
McCrea, W. H. 1964, *MNRAS*, 128, 147
Piotto, G., De Angeli, F., King, I. R., *et al.*. 2004, *ApJL*, 604, L109
Simunovic, M., Puzia, T. H., & Sills, A. 2014, *ApJL*, 795, L10
Tian, B., Deng, L., Han, Z., & Zhang, X. B. 2006, *A&A*, 455, 247
Xin, Y., Ferraro, F. R., Lu, P., *et al.*. 2015, *ApJ*, 801, 67

Why Galaxies Care About AGB Stars: A Continuing Challenge through Cosmic Time
Proceedings IAU Symposium No. 343, 2019
F. Kerschbaum, M. Groenewegen & H. Olofsson, eds.

doi:10.1017/S1743921318007342

Near-Infrared Stellar Populations in the metal-poor, Dwarf irregular Galaxies Sextans A and Leo A

Olivia C. Jones[1], Matthew T. Maclay[2], Martha L. Boyer[2], Margaret Meixner[2] and Iain McDonald[3]

[1]UK Astronomy Technology Centre, Royal Observatory, Blackford Hill, Edinburgh, EH9 3HJ
email: olivia.jones@stfc.ac.uk

[2]Space Telescope Science Institute, 3700 San Martin Drive, Baltimore, MD 21218, USA

[3]Jodrell Bank Centre for Astrophysics, Alan Turing Building, The University of Manchester, Oxford Road, Manchester, M13 9PL, UK

Abstract. We present JHK observations of the metal-poor ([Fe/H] $<$ -1.40) dwarf-irregular galaxies, Leo A and Sextans A, obtained with the WIYN High-resolution Infrared Camera. Their near-IR stellar populations are characterized by using a combination of color-magnitude diagrams and by identifying long-period variable (LPV) stars. We detected red giant and asymptotic giant branch (AGB) stars, consistent with membership of the galaxy's intermediate-age populations (2-8 Gyr old). We identify 32 dusty evolved stars in Leo A and 101 dusty stars in Sextans A, confirming that metal-poor stars can form substantial amounts of dust. We also find tentative evidence for oxygen-rich dust formation at low metallicity, contradicting previous models that suggest oxygen-rich dust production is inhibited in metal-poor environments. The majority of this dust is produced by a few very dusty evolved stars.

Keywords. stars: late-type – infrared: stars – circumstellar matter – stars: mass-loss

1. Introduction

Metal-poor ($-2.1 \lesssim$ [Fe/H] $\lesssim -1.1$) dwarf galaxies offer a fantastic opportunity to investigate evolved stellar populations over a wide range of environments, star-formation histories (SFH) and metallicity.

Ground-based observations of Sextans A and Leo A were obtained through the broadband JHK_s filters using the WIYN High-Resolution Infrared Camera (WHIRC; Meixner *et al.* 2015) mounted on the 3.5m WIYN telescope at the Kitt Peak National Observatory. For a source to be included in the full catalogue we require it to be detected in at least two bands or at two epochs with high-confidence. Over 750 sources are included in the final point-source catalogue for both of the galaxies (Jones *et al.* 2018).

2. Dusty Variable Stars

AGB stars pulsate, and are variable on timescales of hundreds of days. These pulsations levitate atmospheric material, leading to the formation of dust grains. Thus, stars with long period variations can show a significant IR excess. Our data examines source variability for three epochs, separated by 82, 267 and 363 days for Sextans A, and two epochs separated by 463 days for Leo A. Using the variability index defined by Vijh *et al.* (2009), we identify 30 large-amplitude variables in Sextans A and 50 variables in Leo A. We classify these variables using near-IR colours; sources are identified as AGB stars

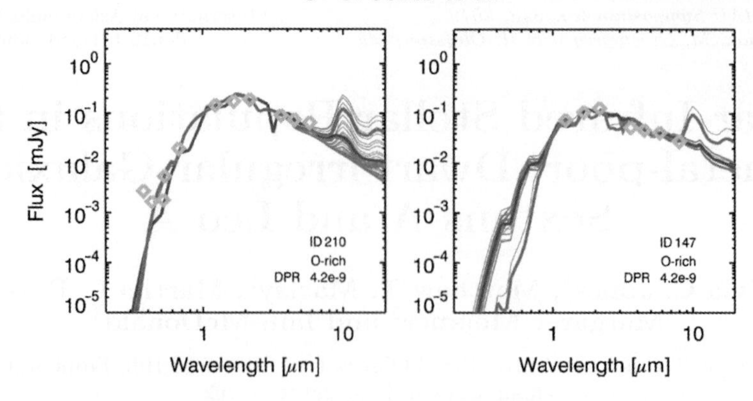

Figure 1. Spectral Energy Distributions (orange) of two dusty stars in Sextans A shown with the best-fitting O-rich model (blue) and carbon-rich model (red). The range of models which produce an acceptable fit are plotted in grey (Jones *et al.* 2018).

or RSGs candidates if they are brighter than the RGB tip and redder than foreground objects.

3. Dust Production in metal-poor galaxies

Dust-production rates (DPR) for every evolved star in Leo A and Sextans A can be computed by fitting their broadband spectral energy distributions (SEDs) with the GRAMS radiative transfer models (Sargent *et al.* 2011; Srinivasan *et al.* 2011), using the method described in Jones *et al.* (2014). This allows us to characterize the cumulative dust production in these galaxies, and constrain the evolutionary status and chemical type of the individual stars (see Fig. 1).

We identify 32 stars in Leo A and 101 stars in Sextans A with a dust production rate $> 10^{-11}$ $M_{\odot}$ yr^{-1} confirming that metal-poor stars can form substantial amounts of dust. In both galaxies, the total dust input into the ISM is dominated by one or two stars. Intriguingly, the star with the highest dust-production rate, SextansA-17, with a DPR of $(5.5 \pm 0.1) \times 10^{-7}$ $M_{\odot}$ yr^{-1}, is best fit with an oxygen-rich dust chemistry. Furthermore, it has been identified as variable x-AGB candidate by DUSTiNGS (Boyer *et al.* 2015). If this star is forming silicate dust then it may be more metal-rich, and hence younger and of higher mass, than the rest of the AGB population. Alternatively, this dust might be nucleating in a circumbinary disc, or that silicate dust may form via an alternate reaction mechanism to metal-rich stars. To verify these results, observations with *JWST* in the mid-IR would place significant constraints on both the dust chemistry and the stellar type (Jones *et al.* 2017a).

References

Boyer, M. L., McQuinn, K. B. W., Barmby, P., *et al.*, 2015, *ApJ*, 800, 51
Jones, O. C., Kemper, F., Srinivasan, S., *et al.* 2014, *MNRAS*, 440, 631
Jones, O. C., Meixner, M., Justtanont, K., & Glasse, A. 2017, *ApJ*, 841, 15
Jones, O. C., Maclay, M. T., Boyer, M. L., Meixner, M., McDonald, I., Meskhidze, H., 2018, *ApJ*, 854, 117
Meixner, M., Smee, S., Doering, R L., *et al.*, 2015, *PASP*, 122, 451
Sargent, B. A., Srinivasan, S., & Meixner, M. 2011, *ApJ*, 728, 93
Srinivasan, S., Sargent, B. A., & Meixner, M. 2011, *A&A*, 532, A54
Vijh, U. P., Meixner, M., Babler, B., *et al.*, 2009, *AJ*, 137, 3139

Why Galaxies Care About AGB Stars:
A Continuing Challenge through Cosmic Time
Proceedings IAU Symposium No. 343, 2019
F. Kerschbaum, M. Groenewegen & H. Olofsson, eds.

doi:10.1017/S1743921318007287

Spectroscopic binaries among AGB stars from HERMES/Mercator: the case of V Hya

Alain Jorissen[1], Sophie Van Eck[1], Thibault Merle[1] and Hans Van Winckel[2]

[1]Institut d'Astronomie et d'Astrophysique, Université libre de Bruxelles, CP 226, B-1050 Bruxelles, Belgium
email: ajorisse@ulb.ac.be

[2]Institute of Astronomy, KU Leuven, Celestijnenlaan 200D, B-3001 Leuven, Belgium
email: hans.vanwinckel@kuleuven.be

Abstract. We report on our search for spectroscopic binaries among a sample of AGB stars. Observations were carried out in the framework of the monitoring of radial velocities of (candidate) binary stars performed at the Mercator 1.2m telescope, using the HERMES spectrograph. We found evidence for duplicity in UV Cam, TU Tau, BL Ori, VZ Per, T Dra, and V Hya.

Keywords. binaries: spectroscopic, stars: carbon, stars: AGB and post-AGB

Several methods exist to find binaries: by detecting radial-velocity variations, eclipsing or ellipsoidal light curve, proper-motion anomalies, composite spectra, X-ray or UV flux excess, spiral arms in the circumstellar dust or molecular emission, binary signature in interferometric or lunar-occultation data (see the reviews by Jorissen 2004, and Jorissen & Frankowski 2008). In the case of binaries involving mass-losing AGB stars, the duplicity sometime produces symbiotic activity. However, the fraction of AGB binaries exhibiting symbiotic activity is unknown. We therefore started a radial-velocity survey of non-symbiotic AGB stars. Very few spectroscopic binaries are known so far among non-symbiotic AGB stars. One such rare case is the M5III SRb variable (with $P_{\rm GCVS} = 44.3$ d) RR UMi with $P_{\rm orb} = 749$ d (Batten & Fletcher 1986). Spectroscopic binaries are difficult to find among AGB stars because their radial velocities bear signatures from regular envelope pulsation, or in less extreme (but more frequent) cases, atmospheric jitter. Both are the sources of annoying noise which may possibly mask the presence of orbital variations in the radial-velocity curve. Therefore, the detection of spectroscopic binaries is only possible if it triggers orbital velocity variations larger than this intrinsic velocity scatter (represented by the dashed line in Fig. 1, calibrated on a comparison sample of M giants; Jorissen *et al.* 2009).

Observations were carried out in the framework of the monitoring of radial velocities of (candidate) binary stars performed at the Mercator 1.2m telescope at La Palma (Gorlova *et al.* 2014), using the HERMES high-resolution spectrograph (Raskin *et al.* 2011). The monitoring started in mid-2009, spans about 250 nights/yr, and involves AGB and carbon stars with suspicion of binarity (because of a composite spectrum, X-ray or UV flux, interferometric data, or lunar-occultation data; Jorissen 2004, Sahai *et al.* 2008). We found evidence for duplicity in UV Cam, TU Tau, BL Ori, VZ Per, T Dra, and V Hya (large filled circles above the dashed line in Fig. 1). The situation for the carbon Mira V Hya is displayed in Fig. 2. An orbital period of 8.5 yr is suspected by Sahai *et al.* (2016; also this volume) from the dynamics of 'bullets' in jets, but this value

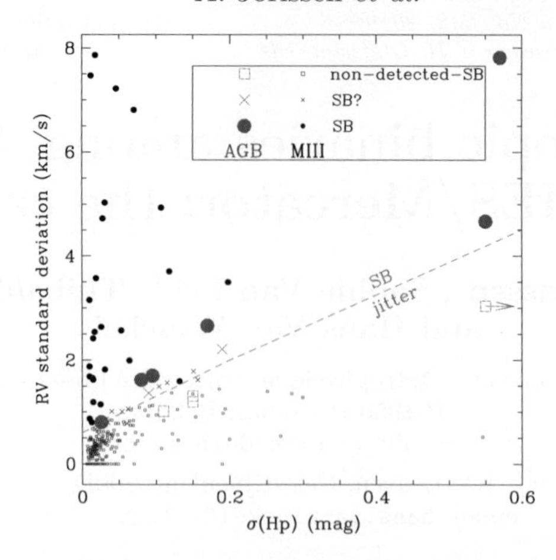

Figure 1. The samples of candidate SB among AGB and carbon stars (large symbols), as compared to the reference sample of M giants (small symbols), from Jorissen *et al.* (2009), who identified SBs (small filled circles), possible SBs (small crosses), and non-SBs (small open squares). Spectroscopic binaries should thus be located above the dashed magenta line. $\sigma(Hp)$ is the standard deviation of the Hipparcos Hp magnitude. V Hya is in the upper right corner.

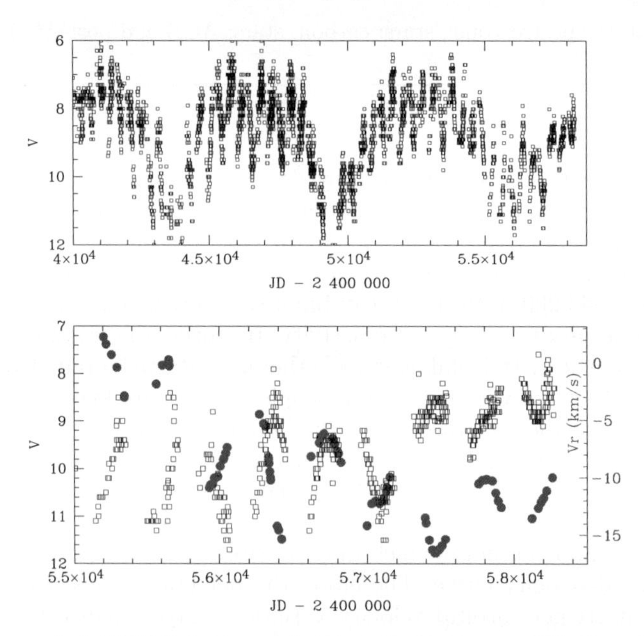

Figure 2. Top panel: AAVSO light curve showing the long-term orbital modulation superimposed on the Mira pulsations. Bottom panel: Same as top with radial velocities added (filled circles). Note the clear phase shift between velocity and light curves.

is not confirmed by our monitoring which now spans just over 8.5 yr. The orbital period looks instead similar to the long-term modulation observed in the AAVSO light curve (Knapp *et al.* 1999), and is therefore likely caused by obscuration from a circumbinary disc, a characteristic feature of binary RV Tau stars of the b subtype (Manick *et al.* 2018).

References

Batten, A.H., Fletcher, J.M. 1986, *PASP* 98, 647
Gorlova, N., Van Winckel, H., Vos, J., *et al.* 2014, arXiv:1403.2287

Jorissen, A. 2004, Chapter 9 (Binary AGB stars) in: H. Habing & H. Olofsson (eds.), *AGB stars*, A&A Library (New York: Springer Verlag)
Jorissen, A., Frankowski, A., Famaey, B., Van Eck, S. 2009, *A&A*, 498, 489
Jorissen, A., Frankowski, A. 2008, American Institute of Physics Conference Series, Vol. 1057, p. 1 (arXiv:0804.3720)
Knapp, G. R., Dobrovolsky, S. I., Ivezić, Z., *et al.* 1999, *A&A* 351, 97
Manick, R., Kamath, D., Van Winckel, H., *et al.* 2018, *A&A*, in preparation
Raskin, G., Van Winckel, H., Hensberge, H. *et al.* 2011, *A&A* 526, A69
Sahai, R., Findeisen, K., Gil de Paz, A., Sànchez Contreras, C. 2008, *ApJ* 689, 1274
Sahai, R., Scibelli, S., Morris, M.R. 2016, *ApJ*, 827, 92

Why Galaxies Care About AGB Stars:
A Continuing Challenge through Cosmic Time
Proceedings IAU Symposium No. 343, 2019
F. Kerschbaum, M. Groenewegen & H. Olofsson, eds.

doi:10.1017/S1743921318007998

KIC 5110739: A new Red Giant in NGC 6819

Edward Jurua, Otto Trust and Felix Kampindi

Department of Physics, Mbarara University of Science and Technology, P.O. Box 1410, Mbarara, Uganda
email: ejurua@must.ac.ug

Abstract. The *Kepler* Input Catalogue (KIC) misclassified a number of red giant stars as sub giants. This could have resulted from the large uncertainties in the KIC surface gravities. This resulted in 1523 stars which were recently classified as red giant stars. The cluster membership of the 1523 red giant stars was determined using age, distance modulus, and variation of colour magnitude with large frequency separation. We found that one star, KIC 5110739, is a member of NGC 6819.

Keywords. stars: fundamental parameters, stars: mass loss, stars: general

---

1. Introduction

The *Kepler* input catalog (KIC) misclassified some red giant stars as sub giants (Molenda-Żakowicz *et al.* 2011) as a result of incorrect KIC surface gravities (Yu *et al.* 2016). This could imply that the KIC stellar parameters of sub giant stars are significantly biased (Thygesen *et al.* 2012). This motivated Yu *et al.* (2016) to analyze 4758 solar-like oscillating sub giant stars in the *Kepler* field, coming up with 1523 red giants. Increase in the number of red giants in a cluster may influence the cluster parameters such as distance modulus, age and rate of mass loss. Since stars in the same cluster are formed from a common cloud, they are expected to possess approximately similar properties such as distance modulus and age.

2. Methods and Results

The cluster membership of the 1523 stars was determined. The data used were obtained from Multi-mission Archive at Space Telescope (MAST) website.† The stellar coordinates for the 1523 stars were plotted. Stars that lie in the regions that correspond to the cluster sizes were considered to be probable members of the respective clusters. Using this criterion, 21 candidate members of the *Kepler* open cluster were obtained.

However, there is a possibility of some stars lying in the field of view (FOV) of a given *Kepler* cluster, but if so they are, actually, relative to the observer, either farther or nearer than the cluster. The distance modulus for each of the 21 red giant stars was then determined using Equation (12) of Wu *et al.* (2014a).

Comparisons were made between the values of distance modulus for the 21 candidate members and the available confirmed members of the respective *Kepler* clusters. The stars with distance modulus within the range of the confirmed members were considered to be members of those clusters. This resulted in KIC: 2303304, 2438069 and 2439420 as the only candidate members of NGC 6791 (8.0 ± 0.4 Gyr) (Wu *et al.* 2014b) and 5110739 and 5113985 for NGC 6819 (2.5 ± 0.05 Gyr) (Balona *et al.* 2013).

† https://archive.stsci.edu/kepler/data_search/search.php

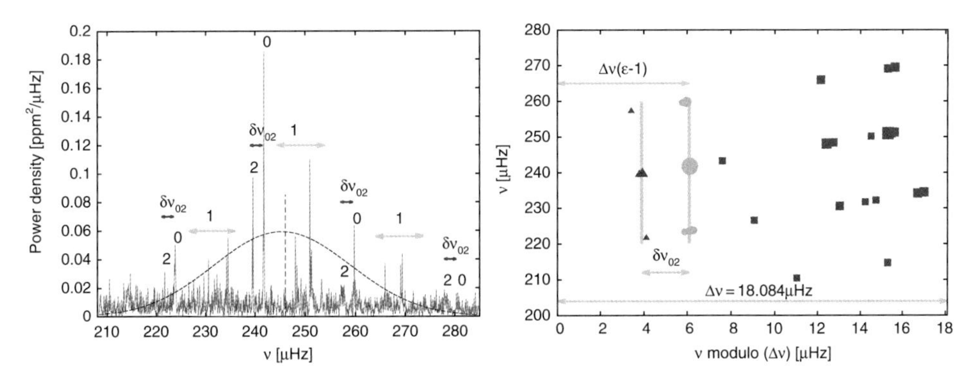

Figure 1. Left: the power density spectrum of KIC 5110739 indicating oscillations. The black dashed curve shows a heavily smoothed power spectrum indicating the power excess envelope of the solar-like oscillations. The black dashed vertical line represents $\nu_{\rm max} = 246.08$ νHz. The numbers indicate the degree, l, of the modes and $\delta\nu_{02}$ represent the small frequency separation (~ 2.4 νHz). Right: Echelle diagram of KIC 5110739 showing the frequencies (red squares, green points and blue triangles) as determined from the power spectrum: modes with $l = 0$, $l = 1$, and $l = 2$ are represented with circles, squares and triangles, respectively.

The 5 stars were over-plotted with the well known red giants on a colour magnitude-large frequency separation diagram. This technique eliminated KIC 2439420 from the group. In addition, using isochrone fitting, the ages of KIC: 2303304, 2438069, 5110739 and 5113985 were found to be 0.60, 0.77, 2.45 and 2.90 Gyr, respectively. This confirmed KIC 5110739 as the only star from the RG catalog by Yu *et al.* (2016), as a member of the *Kepler* open cluster NGC 6819.

With high quality continuous photometric *Kepler* data, we observed that KIC 5110739 pulsates stochastically and the solar-like oscillations are clearly manifested in its power density spectrum represented in Figure 1 (left).

With an intention of knowing whether KIC 5110739 is on the RGB or RGC, its echelle diagram was drawn (Figure 1, right). Using frequencies with $l = 1$, the period spacing (ΔP = 2.42 s) was calculated which confirms KIC 5110739 to be on the RGB (Balona 2010).

3. Conclusions

KIC 5110739 is confirmed as a cluster member (NGC 6819) by all the three criteria. Being the only star of the 1523 previously misclassified red giant stars, we think that its effect on the cluster parameters could be insignificant.

Acknowledgments

The International Science Program (ISP) from Uppsala University in Sweden for funding the study. The *Kepler* team for generously making data publicly available.

References

Balona, L. A. 2010, Challenges In Stellar Pulsation

Balona, L. A., Medupe, T., Abedigamba, O. P., *et al.* 2013, MNRAS, 430, 3472

Molenda-Żakowicz, J., Latham, D. W., Catanzaro, G., Frasca, A., & Quinn, S. N. 2011, MNRAS, 412, 1210

Thygesen, A. O., Frandsen, S., Bruntt, H., *et al.* 2012, A&A, 543, A160

Wu, T., Li, Y., & Hekker, S. 2014a, ApJ, 786, 10

—. 2014b, ApJ, 781, 44

Yu, J., Huber, D., Bedding, T. R., *et al.* 2016, MNRAS, 463, 1297

Why Galaxies Care About AGB Stars:
A Continuing Challenge through Cosmic Time
Proceedings IAU Symposium No. 343, 2019
F. Kerschbaum, M. Groenewegen & H. Olofsson, eds.

doi:10.1017/S1743921318007986

ALMA spectrum of the extreme OH/IR star OH 26.5+0.6

K. Justtanont[1], S. Muller[1], M. J. Barlow[2], D. Engels[3], D. A. García-Hernández[4,5], M. A. T. Groenewegen[6], M. Matsuura[7], H. Olofsson[1], D. Teyssier[8], I. Marti-Vidal[1], T. Khouri[1], M. Van de Sande[9], W. Homan[9], T. Danilovich[9], A. de Koter[10], L. Decin[9], L. B. F. M. Waters[11,10], R. Stancliffe[12], W. Vlemmings[1], P. Royer[9], F. Kerschbaum[13], C. Paladini[14], J. Blommaert[9] and R. de Nutte[9]

[1]Chalmers University of Technology, Onsala Space Observatory, S-439 92 Onsala, Sweden
[2]Univ. College London, Dept. of Physics & Astronomy, Gower Street, London, UK
[3]Hamburger Sternwarte, Gojenbergsweg 112, D-21029 Hamburg, Germany
[4]Instituto de Astrofísica de Canarias, E-38205 La Laguna, Tenerife, Spain
[5]Departamento de Astrofísica, Univ. de La Laguna (ULL), E-38206 La Laguna, Tenerife, Spain
[6]Koninklijke Sterrenwacht van België, Ringlaan 3, 1180 Brussels, Belgium
[7]School of Physics & Astronomy, Cardiff University, The Parade, Cardiff, UK
[8]Telespazio Vega UK Ltd for ESA/ESAC, Camino bajo del Castillo, s/n, Urbanizacion Villafranca del Castillo, Villanueva de la Cañada, E-28692 Madrid, Spain
[9]Institute of Astronomy, KU Leuven, Celestijnenlaan 200D, 3001 Leuven, Belgium
[10]Sterrenkundig Instituut "Anton Pannekoek", Science Park 904, 1098 XH Amsterdam, The Netherlands
[11]SRON Netherlands Institute for Space Research, PO Box 800, 9700 AV Groningen, The Netherlands
[12]Argelander-Inst. fr Astronomie, Univ. of Bonn, Auf dem Hügel 71, 53121 Bonn, Germany
[13]University of Vienna, Dept. of Astrophysics, Türkenschanzstrasse 17, 1180 Wien, Austria
[14]European Southern Observatory, Alonso de Córdova 3107, Vitacura, Santiago, Chile

Abstract. We present ALMA band 7 data of the extreme OH/IR star, OH 26.5+0.6. In addition to lines of CO and its isotopologues, the circumstellar envelope also exhibits a number of emission lines due to metal-containing molecules, e.g., NaCl and KCl. A lack of $C^{18}O$ is expected, but a non-detection of $C^{17}O$ is puzzling given the strengths of $H_2^{17}O$ in Herschel spectra of the star. However, a line associated with $Si^{17}O$ is detected. We also report a tentative detection of a gas-phase emission line of MgS. The ALMA spectrum of this object reveals intriguing features which may be used to investigate chemical processes and dust formation during a high mass-loss phase.

Keywords. stars: AGB and post-AGB, circumstellar matter, stars: individual (OH 26.5+0.6), stars: late-type, stars: abundances

---

1. Introduction

A number of intermediate-mass ($\sim$ 4–8 $M_\odot$) that evolve on the AGB are known to be undergoing hot-bottom burning (HBB) from observations of enhancement of ^{7}Li and other s-process elements (e.g., Garcia *et al.* 2013). The CNO cycle operates during this

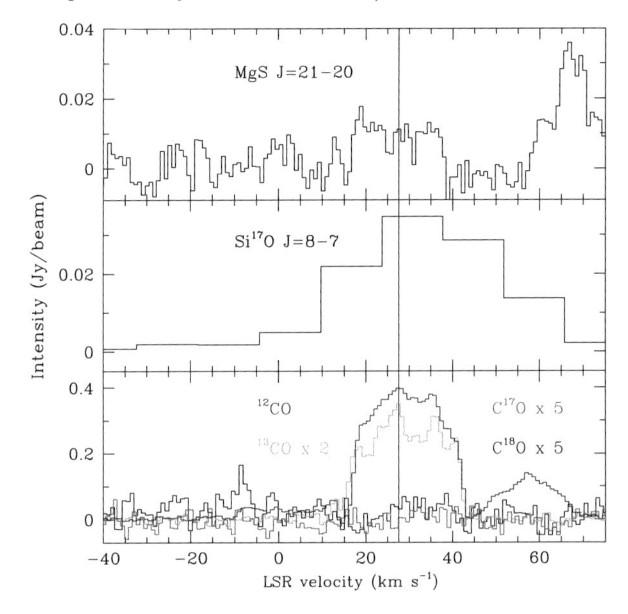

Figure 1. ALMA spectrum of CO J=3-2 and its isotopologues (bottom), $Si^{17}O$ and a tentative detection of MgS. The vertical line denotes the LSR velocity of OH 26.5+0.6.

evolutionary phase and drives the $^{12}C/^{13}C$ towards the equilibrium value of $\sim$4. The process shuts down when the envelope mass is reduced to 1 $M_{\odot}$ (Karakas & Lattanzio 2014). The Herschel spectrum OH 26.5+0.6 shows a lack of $H_2^{18}O$ while $H_2^{16}O$ and $H_2^{17}O$ are readily detected (Justtanont *et al.* 2013). HBB preferentially destroys ^{18}O (Karakas & Lattanzio 2014) thereby confirming that the progenitor of OH 26.5+0.6 is an intermediate-mass star. We subsequently observed the object with ALMA in band 7 in 2016 with spectral windows centered on the transition J=3-2 of CO, ^{13}CO, $C^{17}O$ and $C^{18}O$ (Justtanont *et al.* 2018, ADS/JAO.ALMA#2015.1.00054.S).

2. The ALMA spectrum

A total of about 60 emission lines have been detected in our ALMA observations. Fig. 1 shows the spectrum of CO isotopologues. The $C^{17}O$ J=3-2 is not detected above the noise which is unexpected considering that strong $H_2^{17}O$ lines have been detected in the Herschel spectrum of the star. However, we detected a line which can be attributed to $Si^{17}O$ J=8-7 at 334.3015 GHz. The resolution of this line is 14 km s^{-1} as it falls in a spectral window assigned to a continuum measurement. The ALMA spectrum indicates a possible chemical pathway of molecular formation of oxygen in a high density environment: ^{17}O is locked up in $H_2^{17}O$ and $Si^{17}O$ rather than $C^{17}O$.

A line at 335.9845 GHz may be assigned to a new circumstellar molecule. It corresponds to the MgS J=21-20 transition. Previously, a broad dust emission feature at 30 μm has been attributed to MgS dust, but this has been observed only towards C-rich circumstellar environments. A number of lines in the spectrum are due to the lines of SO and metal-containing molecules like NaCl, KCl and their isotopologues. Unlike the low-mass AGB stars, no SO_2 lines are detected within the spectral range covered by our observations.

References

García-Hernández, D. A., Zamora, O., Yagüe, A., *et al.* 2013, *A&A*, 555, L3
Justtanont, K., Muller, S., Barlow, M. J., *et al.* 2018, *A&A*, submitted
Justtanont, K., Teyssier, D., Barlow, M. J., *et al.* 2013, *A&A*, 556, A101
Karakas, A. I. & Lattanzio, J. C. 2014, *PASA*, 31, 30

Why Galaxies Care About AGB Stars: A Continuing Challenge through Cosmic Time
Proceedings IAU Symposium No. 343, 2019
F. Kerschbaum, M. Groenewegen & H. Olofsson, eds.

doi:10.1017/S1743921318006567

When binaries keep track of recent nucleosynthesis

D. Karinkuzhi[1,2], S. Van Eck[1], A. Jorissen[1], S. Goriely[1], L. Siess[1], T. Merle[1], A. Escorza[1,3], M. Van der Swaelmen[1], H. M. J. Boffin[4], T. Masseron[5,6], S. Shetye[1,3] and B. Plez[7]

[1]Institut d'Astronomie et d'Astrophysique, Université Libre de Bruxelles, ULB, Campus Plaine C.P. 226, Boulevard du Triomphe, B-1050 Bruxelles, Belgium

[2]Dept. of Physics, Bangalore University, Bangalore, India. 560056.

[3]Institute of Astronomy, KU Leuven, Celestijnenlaan 200D, 3001 Leuven, Belgium

[4]ESO, Karl-Schwarzschild-Straße 2, D-85748 Garching bei München, Germany

[5]Instituto de Astrofísica de Canarias, E-38205 La Laguna, Tenerife, Spain

[6]Departamento de Astrofísica, Universidad de La Laguna, E-38206 La Laguna, Tenerife, Spain

[7]Laboratoire Univers et Particules de Montpellier, Université Montpellier, CNRS, 34095, Montpellier Cedex 05, France

emails: dkarinku@ulb.ac.be, svaneck@astro.ulb.ac.be

Abstract. We determine Zr and Nb elemental abundances in barium stars to probe the operation temperature of the s-process that occurred in the companion asymptotic giant branch (AGB) stars. Along with Zr and Nb, we derive the abundances of a large number of heavy elements. They provide constraints on the s-process operation temperature and therefore on the s-process neutron source. The results are then compared with stellar evolution and nucleosynthesis models. We compare the nucleosynthetic profile of the present sample stars with those of CEMP-s, CEMP-rs and CEMP-r stars. One barium star of our sample is potentially identified as the highest-metallicity CEMP-rs star yet discovered.

Keywords. stars: AGB, stars: binaries, stars: abundances

---

1. Introduction

Barium stars are s-process enriched giants; they owe their chemical peculiarities to a past mass transfer, during which they were polluted by their binary companion, which at the time was an asymptotic giant branch (AGB) star, but now an extinct white dwarf. Hence barium stars are ideal targets to understand and constrain the s-process in AGB stars. Actually, since the ^{93}Zr/Zr isotopic ratio is a sensitive function of the s-process operation temperature (independently of stellar evolution models), and since, in extrinsic stars, ^{93}Zr has fully decayed into mono-isotopic ^{93}Nb , we can use the Nb/Zr abundance ratio to constrain the s-process operation temperature. Adopting a recent methodology (Neyskens *et al.* 2015), we analyze a sample of highly-enriched barium stars observed with the high-resolution HERMES spectrograph (Raskin *et al.* 2011) mounted on the MERCATOR telescope (La Palma). Atmospheric parameters for the majority of the programme stars are derived using the BACCHUS pipeline (Masseron *et al.* 2016). Abundances for all the elements are derived by comparing observed spectra with synthetic spectra generated by the Turbospectrum radiative transfer code using MARCS model atmospheres.

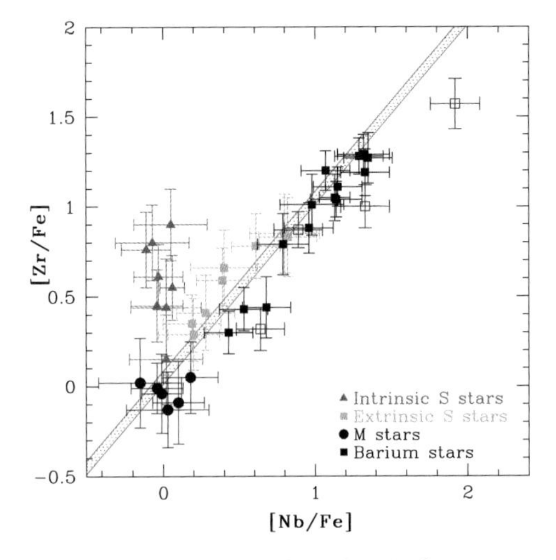

Figure 1. Distribution of [Zr/Fe] vs. [Nb/Fe] for our target stars.

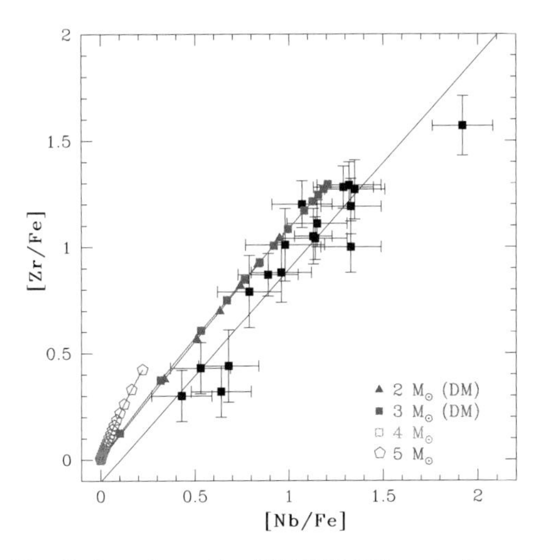

Figure 2. Predictions from the STAREVOL code for extrinsic stars.

2. Results and Discussions

Four objects in our sample are found to be nitrogen rich (blue squares in Fig. 1). Among these four objects, three are found to be Nb-rich (see Fig. 1) with high Nb/Zr ratio. Also, these three objects are found to be Na-rich indicating possible hot bottom burning (HBB) in their companion. We could not derive Mg abundances in all these objects, hence we were not able to test the HBB occuring in massive AGB stars. Most of the objects are falling in the magenta shaded band in Fig. 1, which indicates the expected location of stars polluted by material resulting from the s-process operating at temperatures between 10^8 K (upper line) and $3{\cdot}10^8$ K (lower line) for $^{13}C(\alpha,n)^{16}O$ neutron source. Moreover, Zr and Nb predictions from the STAREVOL code for 2, 3, 4, or $5\,M_\odot$ stars with [Fe/H] = -0.5 (Fig. 2) rule out the operation of a high temperature s-process in the former companion AGB stars that polluted our programme stars.

Possible i-process candidates

Among the three N- and Nb- rich objects, two objects, HD 100503 and HD 121447 are found to be also enriched with r-process elements and show high [hs/ls] ratios. HD 121447 is highly s-process enriched which is consistent with its short orbital period of 185 days. We have compared the abundance patterns in these objects with CEMP-s, CEMP-r and CEMP-rs stars from Masseron *et al.* (2010). HD 100503 is found to be in the CEMP-rs group and is identified as a potential analogue of CEMP-rs star at higher metallicity; if its abundance pattern originates from the i-process, it could help constraining this process at higher metallicities.

Acknowledgements

DK acknowledges the financial support from BELSPO and SERB -DST, Government of India (through PDF/2017/002338 and ITS/2018/003014). SVE thanks Fondation ULB for its support.

References

Masseron, T., Johnson, J. A., Plez, B., Van Eck, S., Primas, F., Goriely, S., & Jorissen, A. 2010, *A&A*, 509, A93

Masseron, T., Merle, T., & Hawkins, K., 2016, *ascl*, 1605.004

Neyskens, P., Van Eck, S., Jorissen, A., Goriely, S., Siess, S., Plez, B. 2015, *nature*, 517, 174

Raskin, G., van Winckel, H., Hensberge, H., Jorissen, A., Lehmann, H., Waelkens, Avila, G., de Cuyper, J.-P., *et al.* 2011, *A&A*, 526, A69

Why Galaxies Care About AGB Stars:
A Continuing Challenge through Cosmic Time
Proceedings IAU Symposium No. 343, 2019
F. Kerschbaum, M. Groenewegen & H. Olofsson, eds.

doi:10.1017/S1743921318005343

Tomography of the red supergiant star μ Cep

K. Kravchenko[1,2], A. Chiavassa[3], S. Van Eck[2], A. Jorissen[2], T. Merle[2] and B. Freytag[4]

[1]European Southern Observatory, Karl-Schwarzschild-Str. 2, 85748 Garching bei München, Germany
email: kateryna.kravchenko@eso.org

[2]Institut d'Astronomie et d'Astrophysique, Université Libre de Bruxelles, CP226, Boulevard du Triomphe, 1050 Bruxelles, Belgium

[3]Université Côte d'Azur, Observatoire de la Côte d'Azur, CNRS, Lagrange, CS 34229, 06304 Nice Cedex 4, France

[4]Department of Physics and Astronomy at Uppsala University, Regementsvägen 1, Box 516, 75120 Uppsala, Sweden

Abstract. A tomographic method, aiming at probing velocity fields at depth in stellar atmospheres, is applied to the red supergiant star μ Cep and to snapshots of 3D radiative-hydrodynamics simulation in order to constrain atmospheric motions and relate them to photometric variability.

Keywords. Techniques: spectroscopic, stars: atmospheres, supergiants.

---

1. Introduction

Red supergiant (RSG) stars show irregular photometric variations characterized by two main periods (Kiss *et al.* 2006): a short one (few hundred days) and a longer one (few thousand days). We focus on the short photometric period which was proposed to be due to either stellar pulsations or convection. We aim at constraining the atmospheric motions in the RSG star μ Cep and at relating them to photometic variability using a tomographic method (Kravchenko *et al.* 2018).

Tomographic method aims at recovering the line-of-sight velocity field as a function of the optical depth in the atmosphere. The method is based on sorting spectral lines according to their formation depth provided by the contribution function (CF) maximum at each wavelength. This allows to split the atmosphere into different layers and construct masks which contain lines forming in the corresponding ranges of optical depths. The cross-correlation of masks with stellar spectra provides the velocity in the different atmospheric layers.

2. Application to μ Cep and 3D radiative-hydrodynamics (RHD) simulations

μ Cep was observed with the HERMES spectrograph (Raskin *et al.* 2011, R $\sim$ 86000). 95 spectra were obtained between April 2011 and January 2018, corresponding to a time span of about 7 years. A set of 5 tomographic masks was constructed from a 1D MARCS (Gustafsson *et al.* 2008) model atmosphere with $T_{\rm eff} = 3400$ K and $\log g = -0.4$. The masks were cross-correlated with μ Cep spectra in order to provide RV in each atmospheric layer.

The RVs in masks C1-C5 are compared with the AAVSO light curve and the effective temperature ($T_{\rm eff}$) in Fig. 1. A phase shift of about 100 days is observed between RV

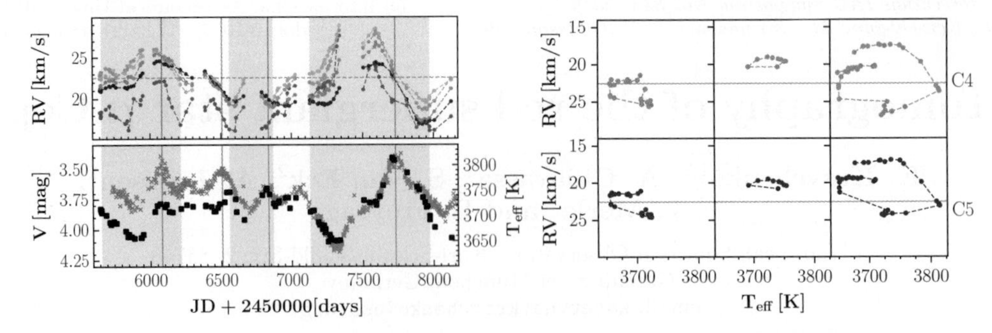

Figure 1. *Left panel:* The RVs in different masks compared to the AAVSO visual light curve (crosses) and $T_{\rm eff}$ (squares). Vertical lines indicate times of a maximum light. Shaded areas define pseudo light cycles which correspond to hysteresis loops. *Right panel:* Hysteresis loops for masks C4 and C5 (rows) corresponding to three pseudo-cycles (columns).

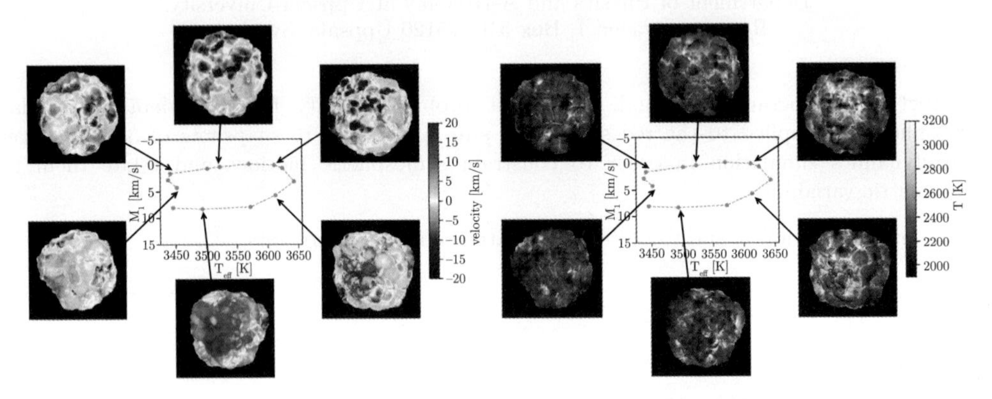

Figure 2. Velocity (left) and temperature (right) maps weighted by the CF of a line from mask C4 for snapshots along the hysteresis loop.

and V magnitude (and $T_{\rm eff}$) variations. This phase lag results in hysteresis loops in the $T_{\rm eff}$ – RV plane (Fig. 1) with timescales similar to the photometric ones. Similar hysteresis loops were observed by Gray (2008) for Betelgeuse and interpreted as the turn-over of the material through a large convective cell.

The tomographic method was applied to snapshots from the 3D RHD simulation of a RSG star performed with the CO5BOLD code (Freytag *et al.* 2012). Hysteresis loops in the $T_{\rm eff}$ – RV plane were detected and show timescales similar to the photometric ones. Velocity and temperature maps for snapshots along the hysteresis loop of mask C4 were weighted by the CF of a spectral line from the same mask. They are shown in Fig. 2 and reveal convective motions in the atmosphere. The consistency between the hysteresis loop timescales measured in μ Cep and in the 3D simulation indicates that convection might account for the short-period photometric variations in μ Cep.

References

Freytag, B., Steffen, M., Ludwig, H.-G., Wedemeyer-Böhm, S., Schaffenberger, W., Steiner, O. 2012, *J. Comput. Phys.*, 231, 919

Gray, D.F. 2008, *AJ*, 135, 1450

Gustafsson, B., Edvardsson, B., Eriksson, K., Jørgensen, U.G., *et al.* 2008, *A&A*, 486, 951

Kiss, L.L., Szab, G.M., Bedding, T.R. 2006, *MNRAS*, 372, 1721

Kravchenko, K., Van Eck, S., Chiavassa, A., Jorissen, A., *et al.* 2018, *A&A*, 610, A29

Raskin, G., van Winckel, H., Hensberge, H., Jorissen, A., *et al.* 2011, *A&A*, 526, A69

Why Galaxies Care About AGB Stars:
A Continuing Challenge through Cosmic Time
Proceedings IAU Symposium No. 343, 2019
F. Kerschbaum, M. Groenewegen & H. Olofsson, eds.

doi:10.1017/S1743921318007524

Chemical enrichment of galaxies as the result of organic synthesis in evolved stars

Sun Kwok[1,2], **SeyedAbdolreza Sadjadi**[2] **and Yong Zhang**[2,3]

[1]Dept. of Earth, Ocean and Atmospheric Sciences, University of British Columbia, Canada
emails: skwok@eoas.ubc.ca, sunkwok@hku.hk

[2]Laboratory for Space Research, The University of Hong Kong, Hong Kong, China

[3]School of Physics & Astronomy, Sun Yat Sen University, Zhuhai, China

Abstract. Infrared spectroscopic observations have shown that complex organics with mixed aromatic-aliphatic structures are synthesized in large quantities during the late stages of stellar evolution. These organics are ejected into the interstellar medium and spread across the Galaxy. Due to the sturdy structures of these organic particles, they can survive through long journeys across the Galaxy under strong UV background and shock conditions. The implications that stellar organics were embedded in the primordial solar nebula is discussed.

Keywords. stars: AGB and post-AGB, ISM: molecules, planetary nebulae: general

---

1. Introduction

It has been recognized since the 1950s that heavy elements synthesized in AGB stars contribute to the atomic chemical enrichment of the Galaxy. Advances in millimeter-wave and infrared spectroscopy have led to our current realization that AGB stars are major galactic sources of molecules and minerals. Over 80 gas-phase molecules and various inorganic solids such as silicates, silicon carbide, and refractory oxides have been detected in the circumstellar envelopes of AGB stars. These molecules and solids are formed in the stellar winds of AGB stars over very short ($\sim 10^3$ yr) time scales and are ejected into the interstellar medium via stellar winds (Kwok 2004).

2. Evolved stars as sources of complex organics

The unidentified infrared emission (UIE) bands, consisting of a family of emission bands and broad emission plateaus superimposed on an underlying continuum, was first detected in planetary nebulae, descendents of AGB stars. The UIE bands are now seen in reflection nebulae, diffuse interstellar medium, and external galaxies. In some active galaxies, the amount of energy emitted in the UIE bands can be as high as 20% of the total energy output of the galaxy. In the astronomical community, it is commonly believed that small (< 50 carbon atoms) polycyclic aromatic hydrocarbon (PAH) molecules are responsible for the UIE phenomenon. However, the PAH hypothesis has a number of problems (Kwok & Zhang 2013, Zhang & Kwok 2015). As an alternative, a mixed aromatic-aliphatic model (MAON) has been proposed (Kwok & Zhang 2011, Figure 1).

Mixed aromatic/aliphatic hydrocarbons with amorphous structures are natural products of combustion and are common products of energetic bomdbarments of simple hydrocarbon molecules in laboratory simulations. Experimental infrared spectra of such compounds show naturally broad emission features resembling the astronomical UIE bands (Dischler *et al.* 1983). We are currently undertaking quantum chemistry calculations to study the vibrational modes of MAON-like molecules with the goal of

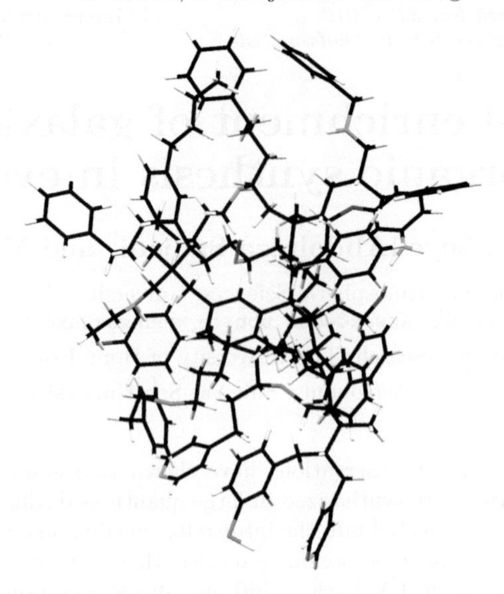

Figure 1. The MAON structure is characterized by a highly disorganized arrangement of small units of aromatic rings linked by aliphatic chains. This structure contains 169 C atoms (in black) and 225 H atoms (in white). Impurities such as O (in red), N (in blue), and S (in yellow) are also present. A typical MAON particle may consist of multiple structures similar to this one.

testing the hypothesis that MAONs are the chemical carriers of astronomical UIE bands (Sadjadi *et al.* 2015, 2017).

3. Implications

The large-scale production and distribution of complex organics by evolved stars suggests that the Galaxy has been heavily enriched by organic compounds. The detection of UIE bands in galaxies with redshift of as high as 2 suggests that organic synthesis already began as early as 10 billion years ago. Pre-planetary systems including the primordial solar nebula could have been enriched by stellar organics. The fact that macro-organics with MAON-like structures have been identified in meteorites, comets, interplanetary dust particles, and planetary satellites suggests that there may be a stellar-solar system connection (Kwok 2016). It is even possible that the early Earth may have inherited some of these stellar organics (Kwok 2017). The possible role of stellar organics on the origin of life on Earth represents a fascinating topic for future investigations (Kwok 2018).

Acknowledgment

This work was supported by a grant from the Natural Sciences and Engineering Research Council of Canada.

References

Dischler, B., Bubenzer, A., & Koidl, P. 1983 *Solid State Commun*, 48, 105
Kwok, S. 2004 *Nature*, 430, 985
Kwok, S. 2016 *A&AR*, 24, 8
Kwok, S. 2017 *Nature Astronomy*, 1, 642
Kwok, S., 2018 Chapter 4.2 in *Handbook of Astrobiology*, Vera Kolb (ed.), CRP Press
Kwok, S., & Zhang, Y. 2011, *Nature*, 479, 80
Kwok, S., & Zhang, Y. 2013, *ApJ*, 771, 5
Sadjadi, S., Zhang, Y., & Kwok, S. 2015 *ApJ*, 801, 34
Sadjadi, S., Zhang, Y., & Kwok , S. 2017 *ApJ*, 845, 123
Zhang, Y., & Kwok, S. 2015, *ApJ*, 798, 37

Why Galaxies Care About AGB Stars:
A Continuing Challenge through Cosmic Time
Proceedings IAU Symposium No. 343, 2019
F. Kerschbaum, M. Groenewegen & H. Olofsson, eds.

doi:10.1017/S1743921318003733

Late Thermal Pulse Models and the Rapid Evolution of V839 Ara

Timothy M. Lawlor

Pennsylvania State University - Brandywine,
Media, PA, United States
email: tlawlor@psu.edu

Abstract. We present evolution calculations from the Asymptotic Giant Branch (AGB) to the Planetary Nebula (PNe) phase for models of mass 1.0 to 2.0 $M_\odot$ over a range of metallicities. The understanding of these objects plays an important role in galactic evolution and composition. Here, we particularly focus on Late Thermal Pulse (LTP) models, which are models that experience an intense helium-shell pulse that occurs just following AGB departure and causes a rapid looping evolution between the AGB and PN phases. The transient phases only last decades and centuries while increasing and decreasing in temperature dramatically. We use our models to make comparisons to V839 Ara (SAO 244567). This star has been observed rapidly heating over more than 50 years. Observations have proven difficult to model because the central star has a small radius, high surface gravity, and low temperature compared to our models.

Keywords. stars: AGB and post-AGB, stars: evolution, stars: abundances

---

1. Late Thermal Pulse Caught in the Act: V839 Ara in the Stingray Nebula

V839 ARA (SAO 244567) appears to have experienced a late thermal pulse (LTP). It has been observed rapidly evolving, becoming hotter and less luminous. According to Parthasarathy (2000) it appeared to be an early type B star in 1971, and has been reported to have a central star mass of approximately 0.6 $M_\odot$. LTP objects are a cousin of Very Late Thermal Pulse (VLTP) objects (e.g. Sakurai's Object) and both undergo a helium shell flash that dredges up helium and other metals, and mixes down hydrogen which is consumed by proton capture. LTP objects experience a helium pulse earlier because they contain comparatively more mass in their helium shell. The star will eventually be hydrogen deficient. We (among others) predicted V839 Ara would ultimately evolve back to the AGB, which has now been confirmed by Reindl *et al.* (2016). We predict the central star will cool rapidly by as much as 15,000 K in the next 25–50 years and another 10,000 K in the following 25–50 years. Within 200–300 years, it will appear as a cool AGB star, and will retrace AGB departure over tens of thousands of years. We confirm that an LTP scenario is the most likely explanation for the stars rapid evolution. Another piece of evidence that supports a LTP scenario is that it is the youngest known planetary nebula.

The observations appear to reach a lower luminosity and smaller radius than our models. This is evident in Figure 1 (left side) which shows the evolution of V839 Ara and our models in the $\log(g) - \log(T_{\rm eff})$ plane. Based on the trend of reaching a lower luminosity for a higher Z we predict a better match may be for a model with an initial mass of 1.2 $M_\odot$ and metallicity $Z = 0.030$. Our choice of Z is an estimate based on the stars location in the thin disk of our Galaxy, though the major observed species in Reindl *et al.* imply a metallicity near 0.015. A model that is less massive may lower the

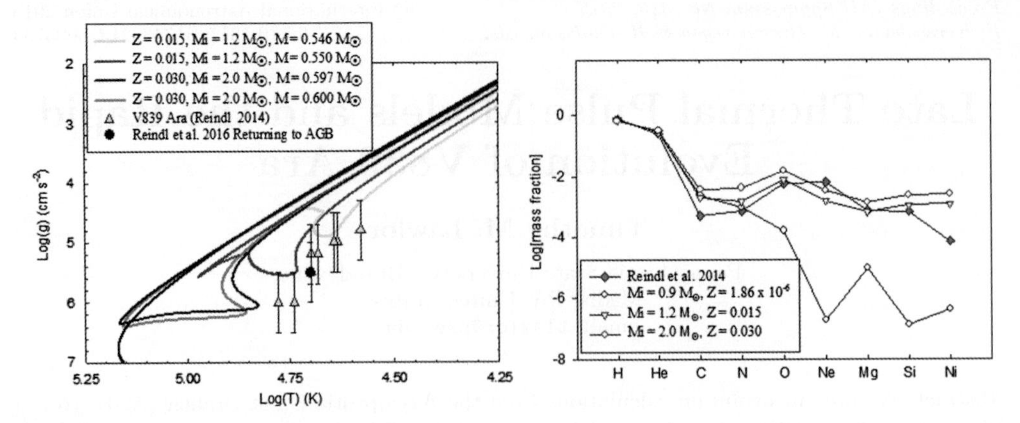

Figure 1. Left: LTP models in the $\log(g) - \log(T_{\rm eff})$ plane compared to V839 Ara from Reindl *et al.* 2014, Reindl *et al.* 2016. Right: log[mass fraction] for LTP models compared to V839 Ara taken from Reindl *et al.* (2014).

temperature, but how much less is limited by the age of the universe. It cannot be less massive than 0.85 $M_{\odot}$ on the MS.

Figure 1 (right) also shows surface abundances for V839 Ara from Reindl *et al.* (2014) compared to our LTP models taken during the same epoch. The most abundant elements are similar, which supports that the initial Z for this star is between 0.015 and 0.03, and final mass is close to 0.56 $M_{\odot}$. In general, our models produce too much heavy elements at this stage; however, this may just be due to how we set our initial values and our starting models may begin with too much helium, a crucial parameter in predicting how a LTP will evolve.

2. Take away: Do Galaxies care?

For models that do not undergo a LTP, dredge up is modest and so mass loss on the AGB and during the superwind phase returns mass that is not significantly enriched. However; LTP models experience a period of convective dredge up when it traces its second departure from the AGB. This results in surface abundances of $X = 0.55$ and $Y = 0.40$, and most other abundances enhanced by more than 10 times. VLTP models are similarly enhanced in metals and deficient in hydrogen, but more extremely so. We estimate (Lawlor & MacDonald 2003, Lawlor & MacDonald 2006) that up to 20% of lower mass stars will experience a thermal helium pulse, increasing the metallicity of material returned to the ISM, perhaps earlier than previously thought.

References

Lawlor, T. M., & MacDonald, J. 2003, *ApJ*, 583, 913

Lawlor, T. M., & MacDonald, J. 2006, *MNRAS*, 371, 263

Parthasarathy, M. 2000, *Bull. Astro. So. India*, 28, 217

Reindl, N., Rauch, T., Parthasarathy, M., Werner, K., Kruk, J. W., Hamann, W.-R., Sander, A., & Todt, H. 2014, *A&A*, 565, 40

Reindl, N., Rauch, T., Miller Bertolami, M. M., Todt, H., & Werner, K. 2016, *MNRAS*, 464, L51

Why Galaxies Care About AGB Stars: A Continuing Challenge through Cosmic Time
Proceedings IAU Symposium No. 343, 2019
F. Kerschbaum, M. Groenewegen & H. Olofsson, eds.

doi:10.1017/S1743921318007044

Carbon and oxygen isotopes in AGB stars. From the cores of AGB stars to presolar dust

Thomas Lebzelter[1], Kenneth Hinkle[2] and Oscar Straniero[3]

[1]Department of Astrophysics, University of Vienna, Austria
email: thomas.lebzelter@univie.ac.at

[2]National Optical Astronomy Observatory
950 North Cherry Ave, Tucson, AZ 85719, USA
email: khinkle@noao.edu

[3]INAF, Osservatorio Astronomico di Teramo, 64100, Teramo, Italy
email: oscar.straniero@inaf.it

Abstract. Isotopic ratios are a powerful tool for gaining insights into stellar evolution and nucleosynthesis. The isotopic ratios of the key elements carbon and oxygen are perfectly suited to investigate the pristine composition of red giants, the conditions in their interiors, and the mixing in their extended atmospheres. Of course the dust ejected from red giants in their final evolution also contains isotopically tagged material. This red giant dust is present in the solar system as presolar dust grains. We have measured isotopic ratios of carbon and oxygen in spectra from a large sample of AGB stars including both Miras and semiregular variables. We show how the derived ratios compare with expectations from stellar models and with measurements in presolar grains. Comparison of isotopes that are affected by different types of nucleosynthesis provides insights into galactic evolution.

Keywords. stars: abundances, stars: evolution, stars: AGB and post-AGB

1. Introduction

The surface carbon and oxygen isotopic ratios found in AGB stars are modified by nucleosynthesis and mixing processes in the stellar interior over the lifetime of the star. A network of processes during H- and He-burning lead to phases of enrichment and depletion of ^{12}C, ^{13}C, ^{17}O, and ^{18}O. The ashes of these fundamental nucleosynthesis processes can appear at the surface as a consequence of a deep mixing.

We present the continuation of a work obtaining the above mentioned isotopic abundances in long period variables (LPVs). Results on Miras and SRa have been published elsewhere (Hinkle *et al.* 2016). There we show that it is possible to estimate the initial masses of LPVs due to the sensitivity of oxygen and carbon isotopic ratios on stellar mass. Here we report on measurements of isotopic ratios for the smaller amplitude cousins of these variables, namely a sample of SRbs. The observational material for this project is drawn entirely from the archives of the Kitt Peak National Observatory (KPNO) 4m Mayall Telescope Fourier Transform Spectrometer (FTS). Isotopic ratios have been determined using a differential curve-of-growth technique described in detail in Hinkle *et al.* 2016.

2. Results

Preliminary results are shown in Fig. 1. Our sample consists of 35 M-type stars, 9 S-type stars and 3 C-type stars. A large fraction of SRb stars shows a carbon isotopic

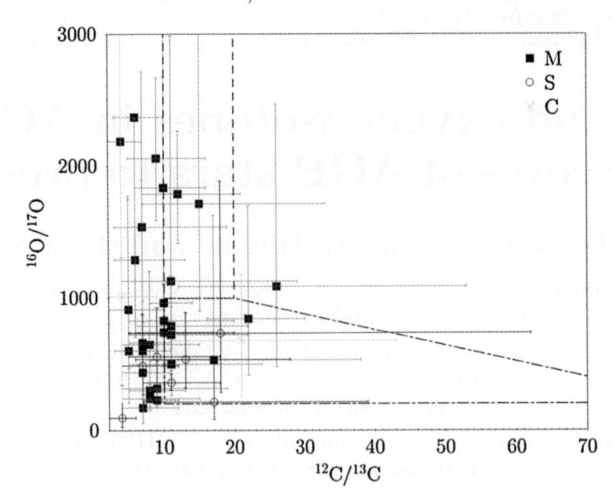

Figure 1. Carbon versus oxygen isotopic ratio measurements for our sample of SRb variables. M-stars are marked by filled squares, S-stars by open circles, and C-stars by crosses. Error bars give the maximum possible range for the respective isotopic ratios. Dashed lines mark the area of model predicted ratios for stars with $M \leqslant 1.5\ M_\odot$, while the dashed dotted lines indicate the range for $1.5 M_\odot < M < 3\ M_\odot$.

ratio below 10, while only very few Mira/SRa stars are found in that range. In contrast, only very few SRbs show $^{12}C/^{13}C$ values above 20. Such low values indicate an FDU composition without significant third dredge up changes. $^{12}C/^{13}C < 10$ implies extra-mixing on the RGB with $T_{max} = 23 - 24$ MK. Exposing the material to such temperatures should lead to a depletion of ^{18}O, and indeed we observe this effect in our data.

Their location in a $^{16}O/^{17}O$ vs. $^{12}C/^{13}C$ diagram suggests, by comparison with stellar evolution models (Hinkle *et al.* 2016), typical masses of M-type SRbs of 1.5 $M_\odot$ and below. All C- and S-type stars in our sample of SRbs show masses $> 1.5\,M_\odot$. There is no obvious distinction between intrinsic and extrinsic S-stars.

The isotopic composition measured in AGB stars will also go into the dust grains formed in the circumstellar environment of these objects. We can thus compare the isotopic ratios found in stars with the ratios in presolar dust grains found in meteorites (e.g. Nittler *et al.* 2008). In our previous paper, we made a comparison with the values for Miras and SRa and found an offset towards higher $^{18}O/^{16}O$ values relative to the solar value and the bulk of the presolar grains, which is likely due to galactic evolution. For SRbs, this offset is not seen suggesting an on average higher age of SRbs compared to SRas and Miras as is also indicated by the somewhat lower mass estimates from the $^{16}O/^{17}O$ ratios.

References

Hinkle, K.H., Lebzelter, T., & Straniero, O. 2016, *ApJ*, 825, 38

Nittler, L.R., Conel, M.O'D.A., Gallino, R., Hoppe, P., Nguyen, A.N., Stadermann, F.J., Zinner, E.K. 2008, *ApJ*, 682, 1450

Why Galaxies Care About AGB Stars:
A Continuing Challenge through Cosmic Time
Proceedings IAU Symposium No. 343, 2019
F. Kerschbaum, M. Groenewegen & H. Olofsson, eds.

doi:10.1017/S1743921318006130

Stellar Wind Accretion and Raman O VI Spectroscopy of the Symbiotic Star AG Draconis

Young-Min Lee[1], Jeong-Eun Heo[1], Hee-Won Lee[1], Ho-Gyu Lee[2], Rodolfo Angeloni[3], Francesco Di Mille[4] and Tali Palma[5]

[1]Department of Physics and Astronomy, Sejong University, Seoul, Korea
email: ymlee9211@gmail.com

[2]Korea Astronomy and Space Science Institute, Daejeon, Korea

[3]Departamento de Física y Astronomía, Universidad de La Serena, La Serena, Chile

[4]Las Campanas Observatory, Carnegie Observatories, Casilla 601, La Serena, Chile

[5]Observatorio Astronómico, Universidad Nacional de Córdoba, Córdoba, Argentina

Abstract. Raman scattered O VI features at 6825 Å and 7082 Å found in symbiotic stars are important spectroscopic tools to probe the mass transfer process. Adopting a Monte Carlo approach, we perform a profile analysis of Raman O VI features of the yellow SySt AG Draconis and make a comparison with the spectrum obtained with CFHT. It is assumed that the accretion flow is convergent on the entering side with enhanced O VI emission and the flux ratio $F(1032)/F(1038) \sim 1$, whereas on the opposite side the flow is divergent with low O VI emission and $F(1032/F(1038) \sim 2$. Our best fit to the spectrum is obtained from our model with a mass-loss rate of the giant $\sim 4 \times 10^{-7}$ $M_\odot$ yr^{-1}. A slight red wing excess in the spectrum suggests the presence of bipolar neutral components receding in the directions perpendicular to the binary orbital plane with a speed $\sim$ 70km s^{-1}

Keywords. binaries: symbiotic - stars: individual: (AG Dra) - accretion disks - radiative transfer

1. Raman O VI Spectroscopy of the Symbiotic Star AG Draconis

Symbiotic stars (SySts) are a wide binary system of a giant and a hot white dwarf, where various astrophysical activities are attributed to accretion of a fraction of slow stellar wind from the giant. Very broad features at 6825 Å and 7082 Å, unique to symbiotic stars, are formed through Raman scattering of O VI 1032 Å and 1038 Å with atomic hydrogen (Schmid 1989). They are characterized by complicated profiles that reflect the kinematics of the O VI emission region around the white dwarf. AG Dra, having a K type giant mass donor, is known as a yellow symbiotic star.

We present our high resolution spectrum of AG Dra obtained with the ESPaDOnS installed on the *Canada-France-Hawaii Telescope* on 2014 September 6. In Fig. 1, the upper and lower panels show the parts of the *CFHT* spectrum of AG Dra around Raman O VI 6825 Å and 7082 Å, respectively. The 7082 Å feature exhibits a clear double peak structure, whereas a prominent blue shoulder is found in the 6825 Å feature. The full width at zero intensity (FWZI) is measured to be 200 km s^{-1}.

2. Raman O VI Profile Analysis

Fig. 2 shows a schematic illustration of the model adopted in our profile analysis. The slow stellar wind region around the giant is identified with the Raman scattering site, on

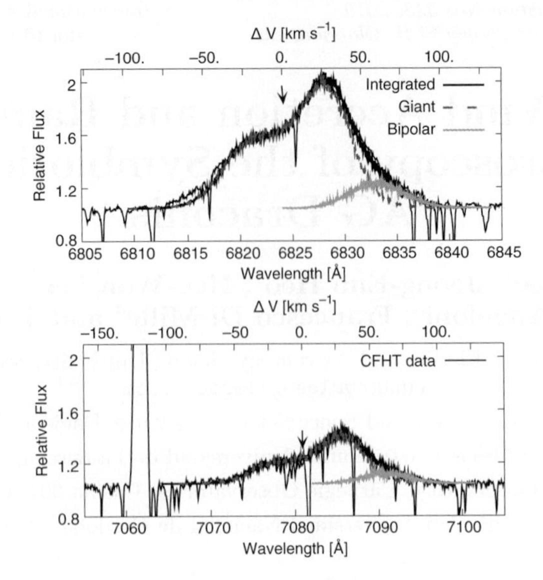

Figure 1. Our best fit result is consistent with the assumption of the asymmetric accretion flow around the white dwarf and additional receding components.

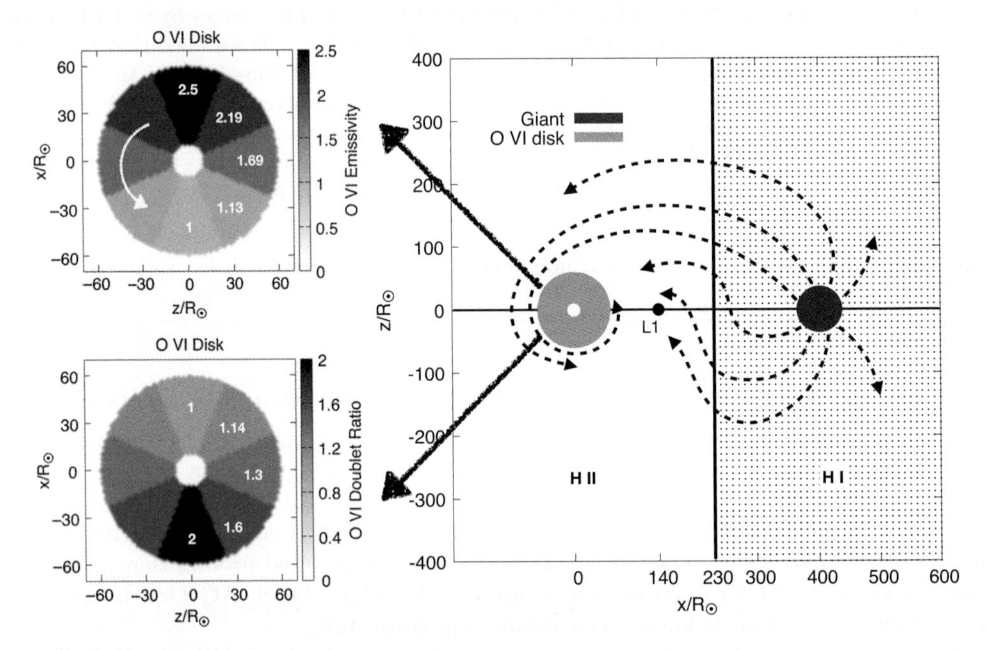

Figure 2. Schematic illustration of the ionization front of AG Dra. Inset figures are the emissivity of O VI (upper panel) and flux ratios (lower panel) as a function of the azimuthal angle.

which copious O VI 1032 Å and 1038 Å photons are incident. We also assume that the O VI emission region constitutes a part of the accretion flow around the white dwarf, where the entering side is of higher density than the opposite side (Heo & Lee 2015).

Based on the measured FWZI, we take the range of Keplerian velocities of the O VI emission region from 40 km s^{-1} to 70 km s^{-1} with the white dwarf mass $M_{WD}= 0.5\,M_\odot$ (Mikołajewska 2002). In view of the clear disparity of the 6825 Å and 7082 Å profiles, we also consider the flux ratio $F(1032)/F(1038)$ varying locally, where $F(1032)/F(1038) \sim 1$ near the entering region and increases to ~ 2 on the opposite side.

The best fit is obtained with the mass loss rate of the giant $\dot{M} = 4 \times 10^{-7}\,M_{\odot}\,yr^{-1}$, for which the Raman conversion efficiency ~ 0.1 is deduced (e.g. Birriel *et al.* 2000). We also find excess in the red wing part in both the 6825 Å and 7082 Å features, which can be explained by placing additional neutral regions moving away from the binary orbital plane in the perpendicular directions with a speed of $70\,km\,s^{-1}$. We propose that the additional neutral component may be associated with the collimated outflows in AG Dra reported in the radio study by Torbett & Campbell (1987).

Acknowledgements

This research was supported by the Korea Astronomy and Space Science Institute under the R&D program (Project No. 2018-1-860-00) supervised by the Ministry of Science and ICT.

References

Birriel, J. J., Espey, B. R., Schulte-Ladbeck, R. E., 2000, *ApJ*, 545, 1020
Heo, J.-E., Lee, H.-W., 2015, *JKAS*, 48, 105
Mikołajewska, J., 2002, *MNRAS*, 335, 33
Schmid, H. M., 1989, *A&A*, 211, L31
Torbett, M. V., Campbell, B., 1987, *ApJ*, 318, 29

Why Galaxies Care About AGB Stars:
A Continuing Challenge through Cosmic Time
Proceedings IAU Symposium No. 343, 2019
F. Kerschbaum, M. Groenewegen & H. Olofsson, eds.

doi:10.1017/S1743921318005616

S-process Elements in Binary Central Stars of Planetary Nebulae

Lisa Löbling[1,2] and Henri Boffin[1]

[1]European Southern Observatory,
Karl-Schwarzschild-Str. 2, 85748 Garching bei München, Germany
email: lloebling@eso.org

[2]Institute for Astronomy and Astrophysics, Kepler Center for Astro and Particle Physics,
Eberhard Karls University, Sand 1, 72076 Tübingen, Germany

Abstract. Low- and intermediate-mass stars experience a phase of carbon enrichment and slow neutron-capture nucleosynthesis (s-process) on the asymptotic giant branch. An interesting element is the radioactive technetium, whose presence is a clear indication that nucleosynthesis happened recently. Analysing the element abundances not only in the hot evolved stars at the center of planetary nebulae helps to derive constraints for the evolution of these stars. Doing so also in their companions if they are in a binary, provides information on the mass-transfer history.

Keywords. planetary nebulae: individual: Hen 2–39 – stars: abundances – stars: evolution – AGB and post-AGB stars – chemically peculiar stars: barium-stars

1. Introduction

Barium (Ba) stars are polluted with asymptotic giant branch (AGB) nucleosynthesis products like carbon (C) and s-process elements from an evolved companion via (wind-)Roche-lobe overflow (RLOF, Jones & Boffin 2017). Hen 2–39 hosts a binary with a red giant companion dominating in the visible. Miszalski *et al.* (2013) determined an enrichment in C and Ba placing this star in the small group of five known Ba central stars of planetary nebulae (Ba-CSPNe).

Technetium (Tc), the lightest element without stable isotopes, is found in evolved stars and must be synthesised in late stages of stellar evolution, since the half-life of ^{99}Tc of 210 000 years is much shorter than the time spent as a giant. The analysis of the Tc surface abundance of the Ba-CSPN may provide the mass-transfer link between the binary components. Taking into account typical post-AGB ages of CSPNe of some 10^3 years (Miller Bertolami 2016) and the short duration of dynamical mass transfer of only hundreds of years (Iben & Livio 1993), a large fraction of the transferred ^{99}Tc must still be present in the stellar atmosphere.

2. Analysis and Results

Observations, Analysis Techniques, and Atomic Data. We analysed optical spectra obtained with the Ultraviolet and Visual Echelle Spectrograph (UVES) at the Very Large Telescope (VLT). For synthetic spectra, we used SPECTRUM (Gray & Corbally 1994) based on ATLAS9 (Castelli & Kurucz 2003) model atmospheres. Tc data was taken from the atomic spectra database of the National Institute of Standards and Technology (NIST) and from Palmeri *et al.* (2007).

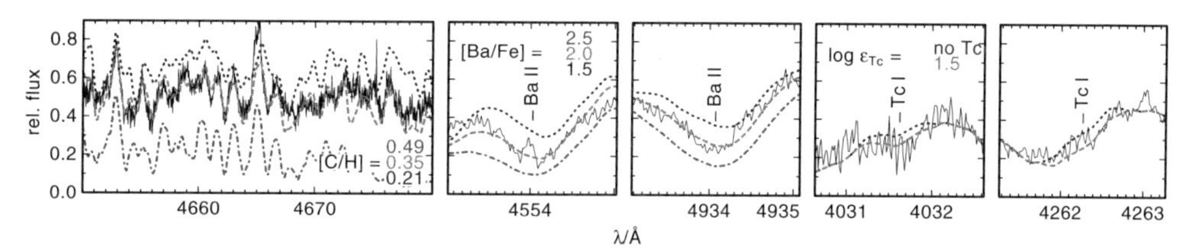

Figure 1. Observation (solid line) compared to model spectra for regions of absorption due to C with [C/H] = 0.49, 0.35, 0.21 dex (red dash-dotted, green dashed, blue dotted), Ba with [Ba/Fe] = 2.0, 1.5, 1.0 dex, and Tc with $\log \epsilon_{\mathrm{Tc}} = 1.5$ (green dashed) and without Tc (blue dotted).

Fundamental Parameters. We determined $T_{\mathrm{eff}} = (4260 \pm 170)$ K from characteristic lines of Fe I, Fe II, Ti I, Ti II, Sc I, and Sc II. The surface gravity $\log g$ could not be constrained by a fit of these lines and a spectroscopic determination is hampered by uncertain values for the distance and brightness. Thus, we adopted a typical value of $\log(\mathrm{g/cm/s^2}) = 2.5$ for the giant. This approach seems reasonable since a change of $\Delta \log g = 0.5$ does barely affect the derived abundance values compared to the statistical error limits.

Element Abundances. The region of strong C_2 absorption (4650 - 4737 Å) yields [C/H] = 0.35 ± 0.02 dex (Fig. 1). Using Ba II λλ 4554.0, 4934.1, 5853,7, 6141.7, 6496.9 Å, we determined [Ba/Fe] = 2.0 ± 0.3 dex (Fig. 1). Absorption features at Tc I λλ 4031.6, 4095.7, 4238.2, 4262.3, 4297.1 Å appear in the computed spectra but cannot be identified in the observation. We found an upper abundance limit of $\log \epsilon_{\mathrm{Tc}} = 1.5$ ($\log \varepsilon_{\mathrm{X}}$ are normalized to $\log \sum \mu_{\mathrm{X}} \varepsilon_{\mathrm{X}} = 12$, Fig. 1).

Comparison with AGB Models and Mass-Transfer History. We compared the determined abundances to the yields of nucleosynthesis models from Karakas *et al.* (2018) with a metallicity of $Z = 0.0028$. With the observed and calculated enrichment factors for C, we get $\Delta M_2 / M_{2,\mathrm{env}} = 0.4$ for the envelope $M_{2,\mathrm{env}}$ and the accreted mass ΔM_2. To reproduce the observed Ba enrichment, the required [Ba/Fe] = 2.5 dex for the donor is only almost reached for the model with an initial mass of 2.5 $M_\odot$. Assuming this mass for the AGB star, the companion should have $M_{\mathrm{ini}} = 1.5$ - 1.9 $M_\odot$ to be a red giant, which would have a convective envelope of about 1.2 - 1.6 $M_\odot$. Thus, it needs to have accreted 0.46 - 0.64 $M_\odot$ which equals the amount of mass lost by the AGB star in the last thermal pulse demanding quasi-conservative mass transfer. Model calculations of Iaconi *et al.* (2017) support that this can be obtained in RLOF. Alternatively, a quasi-conservative wind-RLOF (e.g., Chen *et al.* 2017) may be possible.

References

Castelli, F. & Kurucz, R. L. 2003, in IAU Symposium, Vol. 210, A20
Chen, Z., Frank, A., Blackman, E. G., Nordhaus, J., *et al.* 2017, *MNRAS*, 468, 4465
Jones, D. & Boffin, H. M. J. 2017, Nature Astronomy, 1, 0117
Gray, R. O. & Corbally, C. J. 1994, *AJ*, 107, 742
Iaconi, R., Reichardt, T., Staff, J., *et al.* 2017, *MNRAS*, 464, 4028
Iben, Jr., I. & Livio, M. 1993, *PASP*, 105, 1373
Karakas, A. I., Lugaro, M., Carlos, M., *et al.* 2018, *MNRAS*, 477, 421
Miller Bertolami, M. M. 2016, *A&A*, 588, A25
Miszalski, B., Boffin, H. M. J., Jones, D., *et al.* 2013, *MNRAS*, 436, 3068
Palmeri, P., Quinet, P., Biémont, É., *et al.* 2007, *MNRAS*, 374, 63

Why Galaxies Care About AGB Stars:
A Continuing Challenge through Cosmic Time
Proceedings IAU Symposium No. 343, 2019
F. Kerschbaum, M. Groenewegen & H. Olofsson, eds.

doi:10.1017/S1743921318007901

OH/IR stars versus YSOs in infrared photometric surveys

Cécile Loup[1], Mark Allen[1], Ariane Lançon[2] and Anais Oberto[1]

[1]Observatoire Astronomique de Strasbourg, CNRS, UMR7550,
Centre de Données astronomiques de Strasbourg,
11 rue de l'Université, 67000 Strasbourg, France
email: `cecile.loup@astro.unistra.fr`

[2]Université de Strasbourg, CNRS, UMR7550, Observatoire astronomique de Strasbourg,
67000 Strasbourg, France

Abstract. AGB stars play a major role in the chemical evolution of the galaxies. It thus is important to establish reliable photometric selection criteria to count them, especially AGB stars at the last stages of AGB evolution like OH/IR stars. Here, we have identified about 1500 OH/IR stars and 500 YSOs with methanol masers, in all major mid– and far–infrared surveys (IRAS, MSX, AKARI, WISE, GLIMPSE, and Hi–Gal). We show that AGB stars with high mass-loss rates cannot be disentagled from YSOs with only mid–infrared photometry; far–infrared photometry is essential. In the region observed by GLIMPSE, we show that the proportion of AGB stars has been severely underestimated in previous works: about 70% of "intrinsically" red objects in GLIMPSE are AGB stars rather than YSOs.

Keywords. stars: AGB and post-AGB, stars: pre–main-sequence

1. Samples and identifications

There are about 2000 OH/IR stars known in the Galaxy. They are characterised by a double-peaked OH profile, where the distance between the peaks is a direct measurement of the expansion velocity of the circumstellar envelope. OH/IR stars have been observed in various ways with different astrometric accuracies: in single-dish surveys (2 to 30 arcmin accuracy), colour-selected IRAS samples (10 to 90 arcsec accuracy), or in interferometric surveys (1 to 6 arcsec accuracy). Their identification in modern infrared surveys, and in crowded regions like the galactic plane, has thus been challenging. Mid–infrared IRAS sources were first identified in the AKARI and WISE surveys, taking into account the resolutions, the consistency between fluxes, and blending. This program has actually been applied to all IRAS sources in order to improve their identifications in the SIMBAD database. OH observations were cross-identified between themselves on the basis of the velocity profile. Finally, about 1500 OH/IR stars have a secure identification in MSX, AKARI, WISE, GLIMPSE, MIPSGAL (Gutermuth & Heyer 2015) or/and Hi–Gal (Molinari *et al.* 2016). The sample of YSOs contains some well known OH masers, but mostly methanol masers from the MMB survey (Caswell *et al.* 2010, astrometric accuracy 0.4"). About 500 could be identified in infrared surveys (MSX, AKARI, WISE, GLIMPSE, MIPSGAL, or/and Hi–Gal).

2. Results and comparison with previous works

Figure 1 shows the newly defined samples of OH/IR stars and YSOs on mid- and far-infrared colour-colour diagrams. At first view, it seems that YSOs are "redder" than

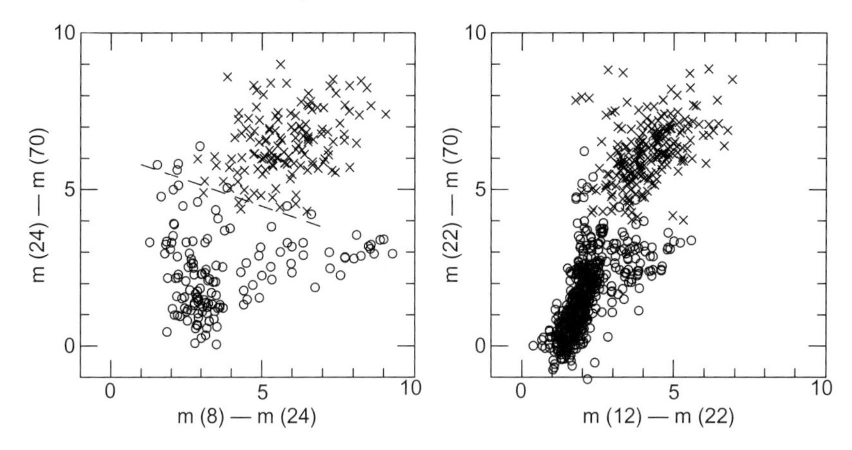

Figure 1. Infrared colour-colour diagrams of OH/IR stars (open circles) and YSOs (crosses). Left: GLIMPSE + MIPSGAL + Hi–Gal; the dashed line shows the separation between OH/IR stars and YSOs. Right: WISE + Hi-Gal.

OH/IR stars. This, however, is only true with the colours combining mid– and far-infrared. For mid–infrared colours, OH/IR stars can be as red as YSOs, up to $m(8) - m(24) = 10$ based on Spitzer data, or up to $m(12) - m(22) = 5$ using WISE photometry.

In the past years, the lack of available far–infrared photometric catalogues led many authors to set up selection criteria based on mid–infrared observations only. For instance, Robitaille *et al.* (2008) selected 16500 intrinsically red sources in the GLIMPSE and MIPSGAL surveys. Their selection criteria were based on a colour–magnitude diagram combining $m(4.5)$ and $m(8) - m(24)$. They concluded that their sample contained over 11000 YSOs and about 7000 AGB stars. Using combined mid– and far–infrared colour-colour diagrams we reach the opposite conclusion: 11300 sources are expected to be AGB stars, 3100 cannot be classified, and only 2100 are very likely YSOs.

3. Conclusions

AGB stars with optically thick circumstellar envelopes cannot be disentangled from YSOs with mid–infrared photometric observations only. Far–infrared observations are needed to separate them. Checking the environment on infrared images can also help as YSOs are usually located in star–forming regions. We have established reliable selection criteria to separate OH/IR stars and YSOs in combined mid– and far–infrared colour–colour diagrams. We conclude that the intrinsically red sources selected by Robitaille *et al.* (2008) towards the galactic plane are mostly AGB stars – about 70% – rather than YSOs.

References

Caswell, J.L., Fuller, G.A., Green, J.A., *et al.* 2010, *MNRAS*, 404, 1029
Gutermuth, R.A., & Heyer, M. 2015, *AJ*, 149, 64
Molinari, S., Schisano, E., Elia, D., *et al.* 2016, *A&A*, 591, 149
Robitaille, T.P., Meade, M.R., Babler, B.L., *et al.* 2008, *AJ*, 136, 2413

Why Galaxies Care About AGB Stars:
A Continuing Challenge through Cosmic Time
Proceedings IAU Symposium No. 343, 2019
F. Kerschbaum, M. Groenewegen & H. Olofsson, eds.

doi:10.1017/S1743921318006300

Zooming into the complex dusty envelopes of C-rich AGB stars

Foteini Lykou[1,2], Josef Hron[3] and Daniela Klotz[3]

[1]Department of Physics, The University of Hong Kong, Hong Kong S.A.R., China
[2]Laboratory for Space Research, The University of Hong Kong, Hong Kong S.A.R., China
[3]Institute for Astrophysics, University of Vienna, Vienna, Austria

Abstract. Recent advances in high-angular resolution instruments (VLT and VLTI, ALMA) have enabled us to delve deep into the circumstellar envelopes of AGB stars from the optical to the sub-mm wavelengths, thus allowing us to study in detail the gas and dust formation zones (e.g., their geometry, chemistry and kinematics). This work focuses on four (4) C-rich AGB stars observed with a high-angular resolution technique in the near-infrared: a multi-wavelength tomographic study of the dusty layers of the circumstellar envelopes of these C-rich stars, i.e. the variations in the morphology and temperature distribution.

Keywords. techniques: high angular resolution, techniques: interferometric, stars: AGB and post-AGB, infrared: stars

---

Our sample is composed of four C-rich AGB stars (namely R For, R Scl, R Lep and II Lup) selected for showing secondary periods known as "obscuration events" (Whitelock *et al.* 2006). Such events may be related to the orbital motion of a companion, and they have been found thus far only in C-rich AGB stars. All four targets were observed in the near-infrared using an aperture masking technique on the VLT which essentially converts a single-dish 8-m telescope into an interferometer (SAM/NACO; Tuthill *et al.* 2010). We report here on our preliminary results.

Aperture masking is proven to be an excellent tool in detecting companion stars. However, no such detection was found in R For, R Scl and R Lep data (a companion may have been detected in II Lup, Lykou *et al.* 2018). If such companions exist then they may be fainter than the SAM/NACO detection limit, or they are wider binaries and therefore the companion is located outside the field-of-view. Here, the dusty envelopes (pseudo-continuum) do not present morphological asymmetries, with the exception of II Lup in K, L and M (Lykou *et al.* 2018) and possibly R For in H (Lykou *et al.*, in prep.). The angular sizes of the dusty envelopes were derived from the azimuthally-averaged visibility data after fitting circular Gaussian functions (Fig. 1), and they are only indicative of the true scale of the circumstellar envelopes.

An attempt to estimate envelope temperatures was made using the relation of Bedding *et al.* (1997), the angular sizes derived from this work, and the bolometric magnitudes from the P–L relation of Whitelock *et al.* (2006) and the *Gaia* DR2 distances. We find that the envelope temperature of R Lep and R Scl is decreasing with increasing wavelength with values ranging from 1700 to 1000 K, which is expected taking into account that cooler layers of the envelope are probed at longer wavelengths. However, the envelope of R For appears to be much cooler in J. The difference in sizes compared to literature data (Fig. 1), and the apparent variation of the envelope radii and temperatures with wavelength, can be explained as:

(a) a true signature of dynamic variations within the envelope with pulsation phase and wavelength observed, or

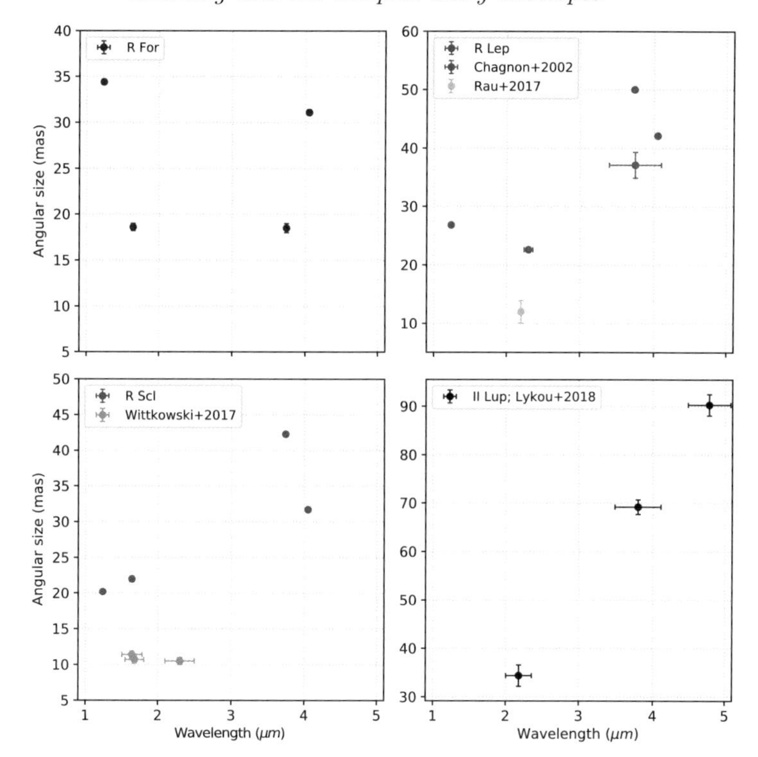

Figure 1. The angular sizes of the four AGB stars compared to literature values (Chagnon *et al.* 2002, Lykou *et al.* 2018, Rau *et al.* 2017, Wittkowski *et al.* 2017). Horizontal errorbars correspond to the filter bandwidths.

(b) a limiting case of detecting circumstellar envelopes with this technique, where the actual envelopes are smaller than the SAM/NACO detection limit.

Aperture masking is a time-effective technique when searching for temporal variations in the circumstellar envelopes of AGB stars. The technique can work in synergy with current spectro-interferometric instruments, such as GRAVITY and MATISSE on the VLTI, as it offers shorter baselines in similar wavelength ranges that can fill the gap between small (≤ 10 mas; VLTI) and intermediate spatial scales (20–400 mas; SAM). We intend to explore such a combination of techniques in the near future to achieve a better understanding on the variability of these four sources.

Acknowledgements

FL acknowledges with thanks support from the IAU and the SOC for the provision of a travel grant, as well as The University of Hong Kong for the URC-CRGC/201707170670 conference grant. This project has been supported by the FWF (AP23006, P.I. Josef Hron) and the HKU Postdoctoral Fellowship Scheme.

References

Bedding T.R., Zijlstra A.A., von der Lühe O., *et al.* 1997, *MNRAS*, 286, 957
Chagnon G., Mennesson B., Perrin G., *et al.* 2002, *AJ*, 124, 2821
Lykou, F., Zijlstra, A.A., Kluska, J., *et al.* 2018, *MNRAS*, 480, 1006
Rau G., Hron J, Paladini C., *et al.* 2017, *A&A*, 600, A92
Tuthill P.G., Lacour S., Amico P., *et al.* 2010, *Proceedings of SPIE*, 7735, 77351O
Whitelock, P.A., Feast, M.W., Marang, F., Groenewegen, M.A.T. 2006, *MNRAS*, 369, 751
Wittkowski M., Hoffman K.-H., Höfner S., *et al.* 2017, *A&A*, 601, A3

Why Galaxies Care About AGB Stars:
A Continuing Challenge through Cosmic Time
Proceedings IAU Symposium No. 343, 2019
F. Kerschbaum, M. Groenewegen & H. Olofsson, eds.

doi:10.1017/S1743921318005604

Mass loss rates of Li-rich AGB/RGB stars

Walter J. Maciel and Roberto D. D. Costa

Astronomy Department, University of São Paulo, São Paulo, Brazil
email: wjmaciel@iag.usp.br

Abstract. A sample of AGB/RGB stars with an excess of Li abundances is considered in order to estimate their mass loss rates. Our method is based on a correlation between the Li abundances and the stellar luminosity, using a modified version of the Reimers formula. We have adopted a calibration based on an empirical correlation between the mass loss rate and stellar parameters. We conclude that most Li-rich stars have lower mass loss rates compared with the majority of AGB/RGB stars, which show no evidences of Li enhancements, so that the Li enrichment process is probably not associated with an increased mass loss rate.

Keywords. stars: abundances, stars: mass loss, stars: AGB and post-AGB

---

1. Introduction

Several AGB/RGB stars present some Li enrichment, characterized by abundances $\epsilon(\text{Li}) = \log(\text{Li/H}) + 12 > 1.5$. Li enrichment has sometimes been associated with enhanced mass loss rates, and some Li-rich giants show evidences of mass loss and chromospheric activity. However, there are also suggestions in the literature that no important mass loss phenomena are associated with these stars. In this work, we estimate the mass loss rates of a sample of Li-rich AGB/RGB stars based on a correlation between the Li abundance and the stellar luminosity. We use a modified Reimers formula calibrated by an independently derived empirical correlation between the mass loss rate and stellar parameters. As a result, we estimate the mass loss rates of a large sample of AGB/RGB stars with well determined Li enhancements.

2. Determination of mass loss rates

We have adopted a sample containing 159 Li-rich stars for which reliable determinations are available for the Li abundances as well of some stellar parameters, such as the effective temperature T_{eff} and gravity $\log g$. For details, see Maciel & Costa (2016) and Maciel & Costa (2018). The full table containing the data and results can be retrieved from http://www.astro.iag.usp.br/~maciel/research/research.html. In the adopted correlation between the Li abundances and the stellar luminosity, the lithium enhancements show some dispersion for each selected luminosity, since for some stars Li may have been more strongly destroyed than for others. However, there is an upper envelope suggesting that the maximum Li enrichment increases with the stellar luminosity. Choosing the maximum contribution at each bin as representative of the Li enhancement process, and adopting 9 luminosity bins, we are able to derive a polynomial fit to the maximum abundance at each luminosity bin. From this relation, the luminosity can be obtained. Since the observed Li abundance may have any value lower or equal to the maximum value, the corresponding luminosity is generally a lower limit. The mass loss rates ($M_\odot$/yr) are obtained with a modified version of the Reimers formula given by

$$\frac{dM}{dt} = 4 \times 10^{-13}\, \eta\, \frac{(L/L_\odot)\,(R/R_\odot)}{(M/M_\odot)}. \tag{1}$$

The η parameter is taken as a free parameter, to be determined on the basis of an adequate calibration. The adopted procedure is as follows: from the Li abundance, we estimate the luminosity using the obtained fit. From the luminosity and the effective temperature, the stellar mass can be estimated from recent detailed evolutionary tracks for giant stars; using the stellar gravity, the radius can be obtained. We have calibrated the Reimers formula using an empirical formula derived by van Loon *et al.* (2005), which is based on the modelling of the spectral energy distributions of a sample of red giants in the Large Magellanic Cloud. The formula can be written as

$$\log \frac{dM}{dt} = \alpha + \beta \log\left(\frac{L}{10000\,L_{\odot}}\right) + \gamma \log\left(\frac{T_{\rm eff}}{3500\,{\rm K}}\right), \tag{2}$$

where the mass loss rates are given in $M_{\odot}$/yr, and the constants are $\alpha = -5.64 \pm 0.15$, $\beta = 1.05 \pm 0.14$ and $\gamma = -6.3 \pm 1.2$. This corresponds to an approximately linear logarithmic relation between the mass loss rate and the stellar luminosity, in agreement with predictions from dust radiative driven winds.

3. Results and discussion

Adopting the polynomial fit with luminosities derived from the $L(\epsilon)$ correlation, we derive the distribution of the mass loss rates of the sample stars. The same is done using Eq. (2.1), which is calibrated to get the same distribution from Eq. (2.2). We derive the best fit parameter, finding $\eta = 5.7$. The derived mass loss rates as a function of the luminosity for our stellar sample are then essentially the same both for the modified Reimers formula and the rates obtained by the empirical formula by van Loon.

There are many reliable determinations in the literature of the luminosities and mass loss rates of AGB/RGB stars with no evidences of Li enhancements. As an example, we have taken into account data from Gullieuszik *et al.* (2012), Groenewegen *et al.* (2009), and Groenewegen & Sloan (2018), selecting O-rich objects. We find that most of these objects have higher luminosities and mass loss rates compared with the Li-rich stars, with very few exceptions. It can then be concluded that the Li-rich objects are generally associated with mass loss rates much lower than in the case of the majority of AGB/RGB stars, which are Li-poor objects. In other words, Li enhancements seem to be a low-luminosity feature associated with lower mass loss rates compared with the majority of these stars.

Acknowledgements

This work was partially supported by CNPq (Process 302556/2015-0) and FAPESP (Process 2010/18835-3 and 2018/04562-7).

References

Groenewegen, M. A. T., Sloan, G. C., Soszyński, I., Peterson, E. A. 2009, *A&A*, 506, 1277
Groenewegen, M. A. T., & Sloan, G. C. 2018, *A&A*, 609, A114
Gullieuszik, M., Groenewegen, M. A. T., Cioni, M. R. L., *et al.* 2012, *A&A*, 537, A105
Maciel, W. J., & Costa, R. D. D. 2016, In: G. A. Feiden (ed.), *The 19th Cambridge Workshop on Cool Stars, Stellar Systems, and the Sun*, Zenodo. http://doi.org/10.5281/zenodo.59278
Maciel, W. J., & Costa, R. D. D. 2018, *AN*, 339, 168
van Loon, J. Th., Cioni, M. R. L., Zijlstra, A. A., Loup, C. 2005, *A&A*, 438, 273

Why Galaxies Care About AGB Stars: A Continuing Challenge through Cosmic Time
Proceedings IAU Symposium No. 343, 2019
F. Kerschbaum, M. Groenewegen & H. Olofsson, eds.

doi:10.1017/S1743921318007871

Abundance Estimates in Carbon Star Envelopes

S. Massalkhi, M. Agúndez and J. Cernicharo

Molecular Astrophysics Group, IFF, CSIC, C/ Serrano 123, E-28006, Madrid, Spain

Abstract. The synthesis of dust grains mostly takes place in the circumstellar envelopes (CSEs) of asymptotic giant branch (AGB) stars. What are the precursor seeds of condensation nuclei and how do these particles evolve toward the micrometer sized grains that populate the interstellar medium? These are key questions of the NANOCOSMOS project. In this study, we carried out an observational study to constrain what the main gas-phase precursors of dust in C-rich AGB stars are.

Keywords. astrochemistry, (stars:) circumstellar matter, AGB and post-AGB

1. Introduction

During the late stages of their evolution, AGB stars experience significant mass loss, which results in extended CSEs. These environments are efficient factories of molecules and dust grains and their chemical nature depends to a large extent on the C/O elemental abundance ratio at the stellar surface. Although much has been advanced recently, there is still much to understand about how dust grains form, what their main gas-phase seeds are and what is their role in the formation of dust around AGB stars. To investigate this, we carried out observations with the IRAM 30m telescope of 25 C-rich AGB stars of diverse mass-loss rates to search for emission of SiC_2, SiO, SiS and CS in the λ 2 mm band. We aim to model their emission and determine their abundances in each source to provide a global view of how abundant they are in CSEs around C-rich AGB stars.

2. Abundance Estimate

We adopted a common physical scenario consisting of a spherically symmetric envelope of gas and dust expanding with a constant velocity and mass-loss rate around a central AGB star. The spherical envelope is described by the radial profile of various physical quantities, such as the gas density, the kinetic temperature of the gas, the dust temperature, the expansion velocity, and the microturbulence velocity. We also consider a dusty component of the envelope. We performed non-LTE excitation and radiative transfer calculations to model the line emission of the molecules, based on the multi-shell LVG (Large Velocity Gradient) method. The circumstellar envelope is divided into a number of concentric shells, each of them with a characteristic set of physical properties, and statistical equilibrium equations are solved in each of them. In each shell, the contribution of the background radiation field (cosmic microwave background, stellar radiation, and thermal emission from surrounding dust) is included. We consider that the molecules are formed close to the star with a given fractional abundance, which remains constant across the envelope up to the outer regions, where they are photodissociated by the ambient ultraviolet radiation field of the local interstellar medium. To model the emission lines and determine the abundance, we varied the initial fractional abundance relative to H_2 of the molecule, f_0, until the calculated line profiles matched the observed ones. We choose as best-fit model the one which results in the best overall agreement

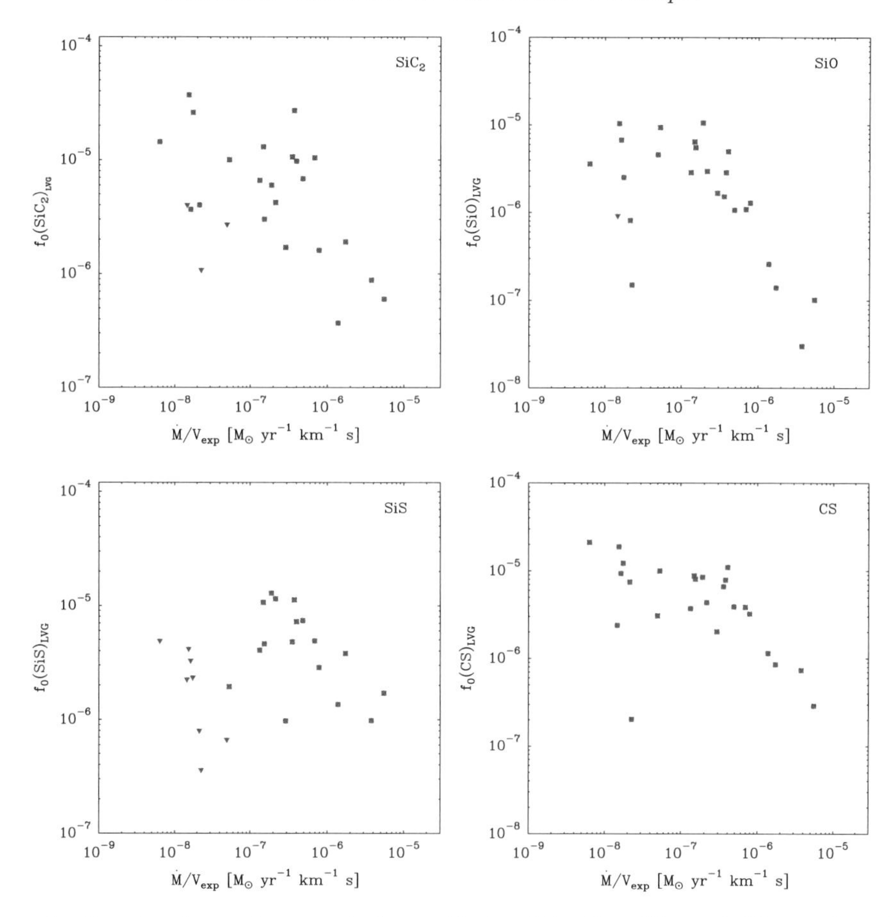

Figure 1. Fractional abundances relative to H_2 for SiC_2, SiO, SiS, and CS are plotted as a function of density measure ($\dot{M}/V_{\rm exp}$). Upper limits for the non-detections are denoted with downward triangles.

between calculated and observed line profiles for the entire set of lines observed. In those cases in which no lines are detected, we derive upper limits to the abundance by choosing the maximum abundance that results in line intensities compatible with the noise level of the observations.

3. Results

The fractional abundance of SiC_2 shows an interesting trend with the density in the envelope, evaluated through the quantity $\dot{M}/V_{\rm exp}$. As shown in Fig. 1, SiC_2 becomes less abundant as the density in the envelope increases. Our interpretation is that the SiC_2 molecules deplete from the gas phase to incorporate into solid dust grains, a process that is favored at higher densities owing to the higher rate at which collisions between particles occur. The ring molecule SiC_2 thus emerges as a very likely gas-phase precursor in the process of formation of SiC dust in envelopes around C-rich AGB stars (Massalkhi *et al.* 2018). We note a similar trend for SiO and CS, which we interpret as evidence of efficient adsorption of SiO and CS onto dust grains. In contrast to these results, SiS does not show any clear trend, suggesting that it is less likely to adsorb onto dust grains than SiC_2, SiO and CS (Massalkhi *et al.*, in prep.).

Reference

Massalkhi, S., Agúndez, M., Cernicharo, J., *et al.* 2018, *A&A*, 611, A29

Why Galaxies Care About AGB Stars:
A Continuing Challenge through Cosmic Time
Proceedings IAU Symposium No. 343, 2019
F. Kerschbaum, M. Groenewegen & H. Olofsson, eds.

doi:10.1017/S1743921318006294

Separation of gas and dust in the winds of AGB stars

Lars Mattsson[1], Christer Sandin[1] and Paolo Ventura[2]

[1]Nordita, Roslagstullsbacken 23, SE-106 91 Stockholm, Sweden
email: lars.mattsson@nordita.org

[2]INAF-Osservatorio Astronomico di Roma, Monteporzio, Italy

Abstract. We present first results from a project aiming at a better understanding of how gas and dust interact in dust-driven winds from Asymptotic Giant Branch (AGB) stars. We are at the final stage of developing a new parallelised radiation-hydrodynamics (RHD) code for AGB-wind modelling including a new generalised implementation of drift. We also discuss first results from high-resolution box simulations of forced turbulence intended to give quantitative "3D corrections" to dust-driven winds from AGB stars. It is argued that modelling of dust-driven winds of AGB stars is a problem that may need to be treated in a less holistic way, where some parts of the problem are treated separately in detailed simulations and are parameterised back into a less detailed (1D spherically symmetric) model describing the entire picture.

Keywords. stars: AGB and post-AGB, stars: winds, outflows, stars: atmospheres

1. Background

The current state-of-the-art codes used to model AGB winds include time-dependent dust-grain growth and frequency-dependent radiative transfer, but assume direct position coupling between gas and dust, i.e., there is no drift between the two phases in the model (see, e.g., Mattsson *et al.* 2010, Mattsson & Höfner 2011, Höfner *et al.* 2016). It is well established, though, that kinetic drag and drift is indeed an essential part of the wind-formation mechanism in AGB stars (Krüger *et al.* 1994, Sandin & Höfner 2003a, Sandin & Höfner 2003b, Sandin & Höfner 2004).

Three dimensional (3D) models for the inner (convective) part of AGB atmospheres are now being used instead of a simple piston boundary to account for pulsation (see, e.g., Liljegren *et al.* 2018). But 3D inhomogeneous models of the wind region is still an inhibiting computational challenge. However, effects of relaxing the spherical-symmetry assumption in the wind region may be greater than the effects of replacing the simplistic piston boundary with simulations of large-scale convection in 3D. The reason for this is that in reality, turbulence is expected to develop and in the presence of turbulence, inertial dust grains will cluster, which significantly alters the radiative balance.

For the reasons given above, we have developed a new simulation code for AGB winds, which includes a new generalised implementation of drift as well as corrections for 3D effects in the wind region.

2. Towards a new improved AGB wind model

We are currently finalising the development of a new RHD code built upon the work by Sandin (2008). In comparison to existing codes, our improved code is based on modern powerful features in Fortran 90/95. Additionally, it has drastically improved numerics; the radiative transfer is parallelised, and the code is structured in such a way that it can easily be extended with more accurate physical descriptions. Currently the code can do gas-dust drift, as well as convection, but more features are added continuously. Below, we detail the new physics considered at present.

Radial drift. By having $1 + N$ momentum equations for the gas and N dust sizes/species we can model drift with rather generic approach, which does not necessarily require dust dynamics to be treated in terms of a "mean component", but actually allow a multidisperse dust component. This, combined with high-resolution frequency-dependent radiative transfer, leads to different rates of dust condensation and sublimation compared to models with position coupling as well as compared to models that include drift and *grey* radiative transfer.

3D effects. How long does the dust spend in low- and in high-density regions, respectively? As we hinted above, decoupled grains are more susceptible to radiative heating and sublimation. They also grow slower as they are less exposed to their growth-species molecules. Separation of gas and dust occurs not only in the radial direction and effects on the radiative transfer through the wind region can be significant. High-resolution box simulations of moderately forced turbulence in a local co-moving region of the wind provide quantitative "3D corrections" to the 1D radiative transfer, which have bearing on the wind formation and the efficiency of momentum transfer from radiation to gas via dust grains.

Grain clustering. One type of "3D correction" is particularly important: grains in a turbulent medium will cluster on scales shorter than that of the gas, which leads to an uneven dust-drag on the gas as the dust is accelerated by radiation pressure. Moreover, grain do not necessarily cluster where the gas is; the dust grains are exposed to more radiation as they are not shielded by the gas. It also means that the effective dust opacity (or optical depth) is less than it would be assuming a homogeneous mix of gas and dust.

3. Concluding remarks

Test models for carbon-rich AGB stars show that the decoupling of gas and dust leads to complex two-fluid dynamics, which suggest reduced mass-loss rates. The preferential (size-dependent) concentration and gas-drag acceleration of grains also create nontrivial size distributions, which in turn suggest an evolving grain-temperature distribution. This may be an important parameter in observational estimates of dust around AGB stars, just as well as it is in supernova remnants (Mattsson *et al.* 2015).

Finally, a thought-provoking question: *if dust-driven stellar winds are formed, as dust drags gas along – can we say we understand the formation of such winds before the gas-to-dust coupling is fully understood?*

References

Krüger A., Gauger D. & Sedlmayr E. 1994, *A&A*, 290, 573
Höfner S., Bladh S., Aringer B. & Ahuja R. 2016, *A&A*, 594, A108
Liljegren S., Höfner S., Freytag B. & Bladh S. 2018, arXiv:1808.05043
Mattsson L., Whalin R., Höfner S. & Eriksson K. 2008, *A&A*, 484, L5
Mattsson L., Whalin R. & Höfner S. 2010, *A&A*, 509, A14
Mattsson L. & Höfner S. 2011, *A&A*, 533, A42
Mattsson L., Gomez H. L., Andersen A. C., & Matsuura M. 2015, *MNRAS*, 449, 4079
Sandin C. 2008, *MNRAS*, 385, 215
Sandin C. & Höfner S. 2003a, *A&A*, 398, 253
Sandin C. & Höfner S. 2003b, *A&A*, 404, 789
Sandin C. & Höfner S. 2004, *A&A*, 413, 789

Why Galaxies Care About AGB Stars:
A Continuing Challenge through Cosmic Time
Proceedings IAU Symposium No. 343, 2019
F. Kerschbaum, M. Groenewegen & H. Olofsson, eds.

doi:10.1017/S1743921318005860

The onset of mass loss in AGB stars

Iain McDonald

Jodrell Bank Centre for Astrophysics, University of Manchester,
Manchester, M13 9PL, UK
email: iain.mcdonald-2@manchester.ac.uk

Abstract. The factors controlling strong mass loss from evolved stars remain elusive, frustrating efforts to parameterise mass loss in models of evolved stars. We herein describe evidence we have collected to show that the mass-loss rate of stars is controlled by stellar pulsations, and that we are close to providing improved prescriptions for mass-loss rates from many kinds of evolved stars.

Keywords. stars: mass loss, stars: AGB and post-AGB, stars: variables: other, circumstellar matter, stars: evolution, stars: late-type, stars: chromospheres, stars: magnetic fields, stars: Population II, stars: winds, outflows

---

Strong mass loss from asymptotic giant branch (AGB) stars is thought to occur via a combination of atmospheric levitation by long-period stellar pulsations, and radiation pressure on dust forming in that levitated atmosphere. The conditions required to initiate that mass loss, however, remain a problematic source of uncertainty for models of stellar structure and evolution (Höfner & Olofsson 2018). While we observe the effects of dust formation and wind acceleration, it is not clear what determines the rate at which mass is lost, meaning it cannot be described by stellar evolution modellers (J. Lattanzio, this proceedings). The distinction becomes important when one applies these rules to stars of differing metallicity: if pulsation sets the mass-loss rate, mass loss should be largely independent of metallicity; if radiation pressure on dust sets the mass-loss rate, mass loss should be strongly metallicity dependent. A third option also exists: the weaker mass loss promoted by magnetically driven winds may continue to play a role on the AGB (e.g. Dupree, Hartmann & Avrett 1984).

Mass-loss rates from evolved stars have historically been estimated via empirical laws. These notably include Reimers (1975), a function of luminosity (L), radius (R) and mass (M) described by $\dot{M} = 4 \times 10^{-13} \eta LR/M$ M$_\odot$ yr^{-1}, where $\eta \approx 0.477$ (McDonald & Ziljstra 2015) is a scaling constant, normally calibrated on globular clusters. This is well-known to under-predict mass-loss rates from intermediate-mass evolved stars, which are undergoing this dust-driven "superwind" (Renzini & Voli 1981). However, it also *over*-estimates mass-loss rates of many open cluster and field stars that are not yet undergoing a superwind (Miglio *et al.* 2012, Handberg *et al.* 2017, Groenewegen 2014): Reimers' law (with $\eta \approx 0.4$) predicts a mass-loss rate of $\sim 10^{-7}$ M$_\odot$ yr^{-1} for RGB-tip stars; yet where such mass-loss rates are observed, stars invariably show infrared excess indicative of strong dust production, which is not typically observed for RGB-tip stars in clusters or the field (e.g. McDonald *et al.* 2011, McDonald *et al.* 2014, McDonald & Ziljstra 2016).

Infrared excess, indicating strong mass loss ($\gtrsim 10^{-7}$ M$_\odot$ yr^{-1}), seems confined to the AGB. It typically begins close to the RGB tip, and is present only in variable stars (McDonald, Zijlstra & Boyer 2012; McDonald, Zijlstra & Watson 2017). A critical point appears to be when stars reach a 60-day period, when $K_{\rm s} - [22]$ colour increases from ~ 0

mag to 1–2 mag: this corresponds approximately to the point at which low-mass stars transit onto the C' sequence in the period–luminosity diagram (McDonald & Ziljstra 2016), and is associated with an increase in pulsation amplitude (Trabucchi *et al.* 2017 & these proceedings).

The cause of this increase in excess infrared flux has been unclear. It may represent a real increase in mass-loss rate driven by radiation or pulsation, or simply an increase in dust-production efficiency in a magnetically supported wind. To investigate, we obtained spectra of carbon monoxide J=3–2 lines towards 11 nearby, oxygen-rich stars that straddle the 60-day boundary, using the Atacama Pathfinder Experiment (APEX) telescope (McDonald *et al.*, in press). Stars with infrared excess ($K_s - [22] > 0.55$ mag) were detected with APEX (estimated average $\dot{M} \sim 3.7 \times 10^{-7}$ $M_\odot$ yr^{-1}); stars without infrared excess were not detected ($\dot{M} \lesssim 1 \times 10^{-7}$ $M_\odot$ yr^{-1}), and $K_s - [22]$ colour is shown to correlate with mass-loss rate within a factor of ~6 between $0 \leqslant K_s - [22] \lesssim 5$ mag. Wind velocity shows little change, maybe slowing slightly across this transition, from ~10–12 km s^{-1} (Groenewegen 2014; McDonald *et al.* 2016) to ~3–11 km s^{-1} for stars above the $K_s - [22] = 0.55$ mag transition. The transition is near-instantaneous in evolutionary terms, so is unlikely to be tied to stellar luminosity, hence radiation pressure on dust. Thus, the transition around 60 days corresponds to a real increase in mass-loss rate, tied to pulsations and not either more efficient dust condensation, or a change in the star's magnetic wind driving. We estimate the increase to be by a factor of ~100.

The substantial increase in mass-loss rate when stars begin strong pulsation strongly implies that mass-loss rate is set predominantly by the pulsation properties of the star, not radiation pressure on dust. In turn, this implies that radiation pressure on dust should not greatly influence the mass-loss rate, setting only the terminal velocity of the wind. To investigate this, we must observe outflow velocities of stars in metal-poor systems. Observations in the Magellanic Clouds are possible but difficult, due to the distance (e.g. Groenewegen *et al.* 2016), as are observations in globular clusters, due to the strong irradiation (e.g. McDonald *et al.* 2015). Recently, we have obtained detections of a wind from a star in a globular cluster and a Galactic Halo star, using the Atacama Large Millimetre Array. Initial analysis of these observations support the idea that the winds of these metal-poor stars are still driven by radiation pressure on dust (McDonald *et al.*, in prep.).

References

Dupree, A.K., Hartmann, L., & Avrett, E.H. 1984, *ApJ*, 281, L37
Groenewegen, M.A.T. 2014, *A&A*, 561, L11
Groenewegen, M.A.T., *et al.* 2016, *A&A*, 596, A50
Handberg, R., *et al.* 2017, *MNRAS*, 472, 979
Höfner, S., & Olofsson, H. 2018, *AARv*, 26, 1
McDonald, I., *et al.* 2011, *ApJS*, 193, 23
McDonald, I., Zijlstra A.A., & Boyer, M.L. 2012, *MNRAS*, 427, 343
McDonald, I., *et al.* 2014, *MNRAS*, 439, 2618
McDonald, I., & Zijlstra, A.A. 2015, *MNRAS*, 448, 502
McDonald, I., *et al.* 2015, *MNRAS*, 453, 4324
McDonald, I., & Zijlstra, A.A. 2016, *ApJ*, 823, L38
McDonald, I., *et al.* 2016, *MNRAS*, 456, 4542
McDonald, I., Zijlstra A.A., & Watson, R.A. 2017, *MNRAS*, 471, 770
Miglio, A., *et al.* 2012, *MNRAS*, 419, 2077
Reimers, D. 1975, *Mem. Soc. Royale Sci. Liège*, 8, 369
Renzini, A., & Voli, M. 1975, *A&A*, 94, 175
Trabucchi, M., *et al.* 2017, *ApJ*, 847, 139

Why Galaxies Care About AGB Stars:
A Continuing Challenge through Cosmic Time
Proceedings IAU Symposium No. 343, 2019
F. Kerschbaum, M. Groenewegen & H. Olofsson, eds.

doi:10.1017/S1743921318005082

Dust properties in the circumstellar environment of carbon stars

M. Mečina[1], B. Aringer[2], M. Brunner[1], F. Kerschbaum[1], M. A. T. Groenewegen[3] and W. Nowotny[1]

[1]Institut für Astrophysik, Universität Wien, Türkenschanzstraße 17, 1180 Wien, Austria
email: marko.mecina@univie.ac.at

[2]Dipart. di Fisica e Astronomia, Universit di Padova, via Marzolo 8, 35131 Padova, Italy

[3]Royal Observatory of Belgium, Ringlaan 3, 1180 Brussels, Belgium

Abstract. Herschel PACS imaging observations of carbon stars show well-resolved spherically symmetric detached shells around several objects. In the case of U Hya the shell is additionally detected in scattered visible light and in the far UV. The remarkable spherical symmetry justifies a straightforward application of 1D models to constrain the properties of the dust envelope, whose modulation in density is a consequence of short epochs of highly increased mass loss and/or wind-wind interaction between outflows of different velocity. We perform dust radiative transfer calculations, first based on a parametrised density distribution, and in a more sophisticated approach on a combination of stationary wind models. The impact of dust properties, particularly grain geometry, on the results is highlighted.

Keywords. stars: AGB and post-AGB, stars: carbon, stars: mass loss

---

U Hya, a semi-regular carbon star, is surrounded by a thin detached dust shell with a radius of 114″ (0.12 pc) that has been resolved in detail by Herschel/PACS. The spherically symmetric dust structure also shows up, and very well resembles the far IR data, in optical scattered light observations by the Pan-STARRS programme. In addition, Sanchez *et al.* (2015) found clumped far UV emission in GALEX observations, cospatial with Herschel and Pan-STARRS data. Since in the near-UV band corresponding signatures could not be clearly identified, they concluded that shocks in the shell due to the stellar motion through the ISM, rather than scattering of the (inter-)stellar radiation field are causing the high-energy emission. Such interaction is also indicated by minor deviations from spherical symmetry seen in scattered light and far IR. There is no observed gaseous counterpart to the dust, likely due to photodissociation and low densities ten thousands of AUs from the star.

All detached shell objects in our sample for which ISO spectra are available show a very distinct C_3 feature at 5 μm. This strongly suggests that they are carbon-rich objects with a fairly high C/O and thus must have undergone several dredge up events. Hence one can assume that the dust in the detached shell is also already dominated by carbonaceous species. However, inconsistencies in JHK colours between observations and stationary wind models indicate that additional factors such as the role of C_2H_2 must be considered (Aringer, priv. comm.). Models using more complete molecular line data may shed new light on this discrepancy.

We use *More of DUSTY* (Groenewegen 2012) to calculate the radiative transfer through the circumstellar dust envelope. Adopting a dust mixture of amorphous carbon and SiC, a best fit model is evaluated regarding photometric data, IR spectra, and PACS imaging data. The results suggest a rather gradual decrease in mass loss after a high mass loss period that presumably caused the detached shell.

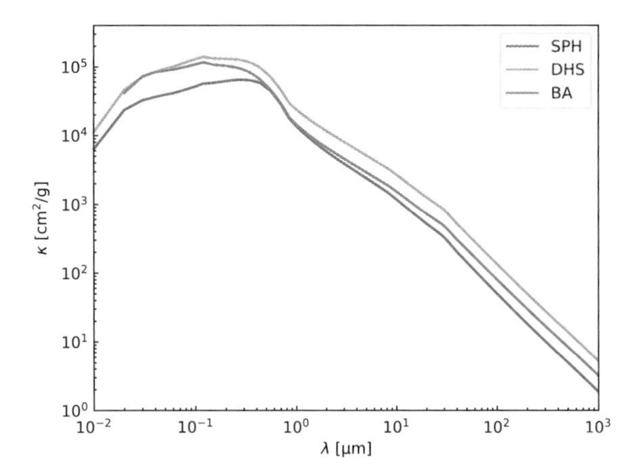

Figure 1. Opacities calculated for solid spheres, a distribution of hollow spheres, and ballistic agglomeration models with the same effective size, using optical properties from Suh (2000).

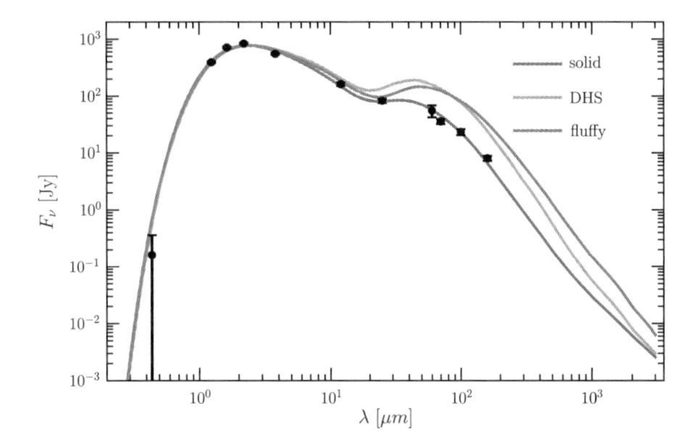

Figure 2. SED model fits to a carbon star with detached dust shell. All models contain the same amount of carbon dust, but differ in the used opacities (see Fig. 1).

Grain geometry plays a crucial role in dust opacity calculations (Fig. 1). We compare solid spheres, distributions of hollow spheres (DHS), and "fluffy" grains formed by ballistic agglomeration (BA, see Draine & Flatau 1994). DHS and BA mimic the porous structure of, e.g., amorphous carbon, thus requiring models to assume significantly less dust mass to match the observations (Fig. 2). In addition, the shape of narrow spectral features as well as the slope of the SED are affected.

Acknowledgements

This work was supported by the Austrian Science Fund FWF under project number AP23586 and the FFG ASAP HIL project. BA acknowledges the support from the project STARKEY funded by the ERC Consolidator Grant, G.A. n.615604. MB acknowledges funding through the Abschlussstipendium fellowship of the University of Vienna.

References

Draine, B.T. & Flatau, P.J. 1994, *Journal of the Optical Society of America A*, 11, 1491
Groenewegen, M.A.T. 2012, *A&A*, 543, A36
Sanchez, E., Montez, R., Ramstedt, S., *et al.* 2015, *ApJL*, 798, 39
Suh, K.-W. 2000, *MNRAS*, 315, 740

Why Galaxies Care About AGB Stars: A Continuing Challenge through Cosmic Time
Proceedings IAU Symposium No. 343, 2019
F. Kerschbaum, M. Groenewegen & H. Olofsson, eds.

doi:10.1017/S1743921318006051

Rotating and magnetic stellar models of intermediate-mass stars up to the AGB

Luiz T. S. Mendes[1], **Natália R. Landin**[2] **and Paolo Ventura**[3]

[1]Depto. Engenharia Eletrônica, Universidade Federal de Minas Gerais, Brazil
email: luizt@cpdee.ufmg.br

[2]Campus UVF Florestal, Universidade Federal de Viçosa, Brazil
email: nlandin@ufv.br

[3]INAF - Osservatorio Astronomico di Roma, Monte Porzio, Rome, Italy
email: paolo.ventura@oa-roma.inaf.it

Abstract. Aiming at investigating the roles of rotation and magnetic fields on AGB stars, the rotating version of the ATON stellar evolution code is being extended in order to account for intermediate--mass stars and their later evolutionary stages. Here we report some preliminary results on the effects of rotation and of a large-scale magnetic field on the structure and evolution of 3 and 5 $M_{\odot}$ stellar models from the pre-main sequence up to the AGB.

Keywords. stars: rotation, stars: magnetic fields, stars: AGB and post-AGB

1. Introduction and Methods

Rotation and magnetic fields can play an important role on evolved low- and intermediate-mass stars. The impact of rotation-induced mixing on the nucleosynthesis, and the driving of the cool bottom processing episode in asymptotic giant branch (AGB) stars are some examples of topics respectively related to those physical properties. To address these issues, the rotating version of the ATON stellar evolution code (Ventura *et al.* 1998) is being extended to deal with intermediate-mass stars and their later evolutionary stages, and also to incorporate the effects of a large-scale magnetic field.

Rotation is implemented according to the Kippenhahn & Thomas (1970) method, which essentially replaces the spherical surfaces of non-rotating models by equipotential, non-spherical ones. From these equipotential surfaces one can calculate suitable correction factors f_p and f_t that enter the stellar structure equations of hydrostatic equilibrium and of energy transport, respectively. The ATON code currently supports three different rotation laws: solid body rotation, differential rotation at each mass shell, and combined solid body rotation in convective regions and differential rotation in radiative ones.

As for the magnetic fields, we use the method by Lydon & Sofia (1995, hereafter LS95) which treats the magnetic field as a perturbation on the stellar structure equations, by means of a new state variable $\chi = B^2/(8\pi\rho)$ representing the magnetic energy density from which the magnetic pressure $P_\chi = (\gamma - 1)\chi\rho$ can be computed and then introduced in the equation of state. The numerical factor γ crudely commands the transition from the intrinsic 3-D magnetic field geometry to a 1-D approximation. The reader is referred to the LS95 paper for full details regarding this technique.

2. Results

We computed solar metallicity models starting from the pre-main sequence (PMS) and with $\alpha = 1.5$ for the mixing length. The initial rotation rate was taken from Kawaler's

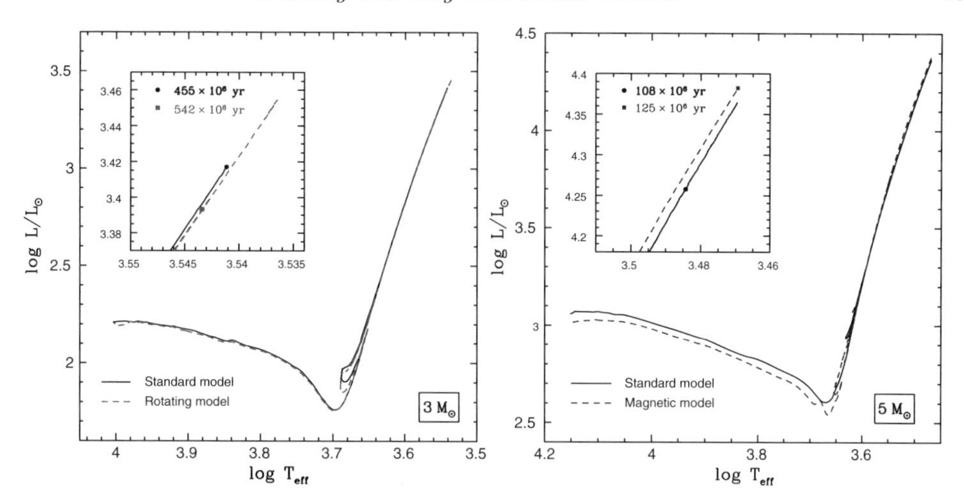

Figure 1. Rotating (left) and magnetic (right) evolutionary tracks of 3 and 5 M$_\odot$ stellar models, respectively, from the main sequence turn-off to the beginning of the AGB.

(1987) mass-radius and mass-moment of inertia relations for low- and intermediate--mass stars. Fig. 1 shows that rotation has a little effect on the evolutionary track of a 3 M$_\odot$ model; the impact of rotation in this case is even smaller than for the 1 M$_\odot$ case, discussed by Mendes *et al.* (2013). In any case, rotating models always reach the AGB at later ages and lower luminosities than standard models, independently of the chosen rotation law.

For models with a magnetic field we considered a field topology whose magnetic force component is perpendicular to the field lines ($\gamma = 2$), and with no magnetic flux associated to the convective motions. The initial surface magnetic field strength $|\mathbf{B}|_{\rm surf}$ is of order 20 G, which is sort of an "average" between the mean solar dipole field of 1 G and the kG values observed in T Tauri stars. The scaling of $|\mathbf{B}|_{\rm surf}$ throughout the stellar interior was treated according to the *constant ratio scaling* (D'Antona *et al.* 2000). As also shown in Fig. 1, the magnetic models have lower luminosities from the PMS up to the middle of the Red Giant Branch (RGB), when this situation reverses, and reach the AGB at later ages than non-magnetic ones. The value of $|\mathbf{B}|_{\rm surf}$ at this point is of order of a few tens of Gauss, which is compatible with the range of 2.2-115 G obtained for some AGB stars through H_2O maser observations (Leal-Ferreira *et al.* 2013). However, an issue with these magnetic models that needs further analysis is the too high, apparently unrealistic interior magnetic field strength (e.g. $\approx 10^{11}$ G in the case of the 5 M$_\odot$ model).

Acknowledgments

L. T. S. Mendes gratefully acknowledges a travel grant from the IAU.

References

D'Antona, F., Ventura, P., & Mazzitelli, I. 2000, *ApJL*, 543, L77

Kawaler, S. 1987, *PASP*, 99, 1322

Kippenhahn, R., & Thomas, H.-C. 1970, in *Stellar Rotation*, ed. A. Sletteback (D. Reidel), 20

Leal-Ferreira, M. L., Vlemmings, W. H. T., Kemball, A., & Amiri, N. 2013, *A&A*, 554, A134

Lydon, T. J., & Sofia, S. 1995, *ApJS*, 101, 357

Mendes, L. T. S., Landin, N. R., & Vaz, L. P. R. 2013, in *Magnetic Fields throughout stellar evolution*, ed. P. Petit, M. Jardine, & H. C. Spruit (Cambridge University Press), 112

Ventura, P., Zeppieri, A., Mazzitelli, I., & D'Antona, F. 1998, *A&A*, 334, 953

Why Galaxies Care About AGB Stars:
A Continuing Challenge through Cosmic Time
Proceedings IAU Symposium No. 343, 2019
F. Kerschbaum, M. Groenewegen & H. Olofsson, eds.

doi:10.1017/S1743921318006543

The common-envelope wind model for type Ia supernovae

Xiangcun Meng[1] and Philipp. Podsiadlowski[2]

[1]Yunnan Observatories, Chinese Academy of Sciences, 650216 Kunming, PR China
email: `xiangcunmeng@ynao.ac.cn`

[2]Department of Astronomy, Oxford University, Oxford OX1 3RH, UK

Abstract. We have developed a new version of the SD model for type Ia supernovae (SNe Ia) in which a common envelope (CE) is assumed to form if the mass-transfer rate between a carbon/oxygen white dwarf (CO WD) and its companion exceeds a critical accretion rate. Based on this model, we found that both SN 2002cx-like and SN Ia-CSM objects may share a similar origin, i.e. these peculiar objects may originate from the explosion of hybrid carbon/oxygen/neon white dwarfs (CONe WDs) in SD systems, where SNe Ia-CSM explode in systems with a massive CE of $\sim 1\,M_{\odot}$, while SN 2002cx-like events correspond to events without a massive CE.

Keywords. binaries: close, stars: evolution, supernovae: general, white dwarfs.

---

1. Introduction

Although type Ia supernovae (SNe Ia) are important in many areas of astrophysics, e.g. as distance indicators to measure cosmological parameters, the nature of their progenitor systems has remained unclear (Riess *et al.* 1998; Perlmutter *et al.* 1999; Maoz *et al.* 2014). SNe Ia originate from the thermonuclear explosion of a carbon/oxygen white dwarf (CO WD) in a binary system, where the companion may be a normal non-degenerate star (the SD model). The most popular SD model is the optically thick wind (OTW) model in which, if the mass-transfer rate between the CO WD and its companion exceeds a critical mass-accretion rate, the WD accretes material at the critical accretion rate, while the remainder is lost from the system as an OTW (Hachisu *et al.* 1996). However, some of the predictions of the model are in conflict with observations. For example, it predicts that there should be no SNe Ia in low-metallicity or high-redshift environments, but many SNe Ia have been found in such environments (Rodney *et al.* 2015).

2. Model

To resolve the shortcomings of the OTW model, we constructed a new version of the SD model. We assume that, when the mass-transfer rate exceeds the critical accretion rate in a WD+MS system, the WD becomes a red-giant-like object and fills and ultimately overfills its Roche lobe, rather than developing an OTW: this leads to the formation of a common envelope (CE), in which the inner region of the CE corotates with the binary, while the angular velocity of the outer region drops off as a power law. The WD gradually increases its mass at the base of CE similarly to how the core grows in a thermally pulsing AGB star. The large nuclear luminosity for stable hydrogen burning drives a CE wind (CEW) from the surface of the CE. Because of the low CE density, the binary system can avoid a fast spiral-in phase and finally re-emerges from the CE as a detached system. The SN Ia may explode in the CE, a stable or weakly unstable hydrogen burning phase.

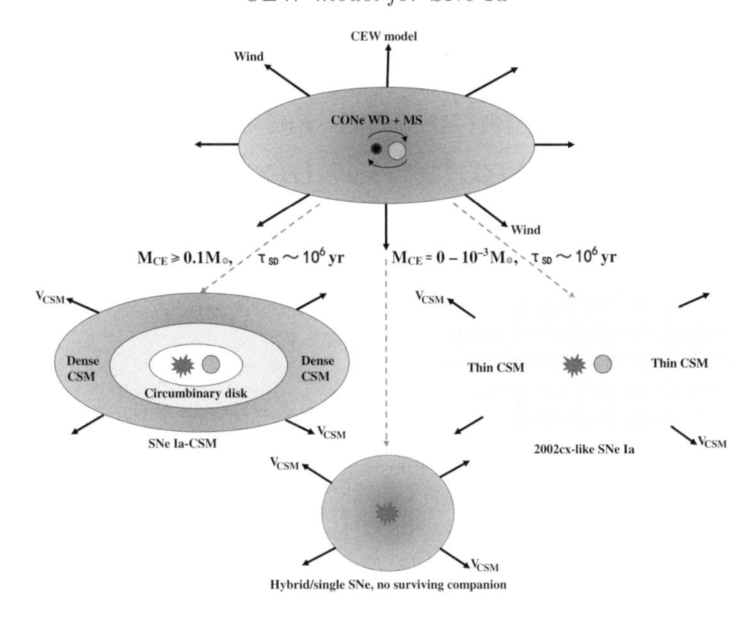

Figure 1. Schematic diagram illustrating the different channels for forming SNe Ia-CSM, SN 2002cx-like, and hybrid/single SNe (the figure is from Meng & Podsiadlowski 2018)

3. Virtues of the CEW model

The model (1) does not depend strongly on metallicity, (2) predicts a low-wind velocity, (3) allows for the self-adjustment of helium and hydrogen burning, (4) may explain the low number of observed supersoft X-ray sources and (5) the properties of recurrent novae, (6) leads to a higher SN Ia rate than the OTW model, (7) may produce 2002ic-like SNe, naturally explains (8) the high-velocity features of SNe Ia, (9) associations with planetary nebulae, and (10) the shape of some supernova remnants. Also (11) the CEW model is quite robust to various input assumptions (Meng & Podsiadlowski 2017).

4. SNe Ia-CSM and 2002cx-like

Based on the CEW model, we found that both SNe Ia-CSM and 2002cx-like objects may originate from systems with a hybrid CONe WD + MS, where those exploding in massive CEs show the properties of SNe Ia-CSM, while those without massive CEs are the progenitor of 2002cx-like SNe Ia (Fig. 1). The model predicts a number ratio of SNe Ia-CSM to SN 2002cx-like objects between 1/3 and 2/3, consistent with observations. Depending on the stage when the explosion occurs, our model produces a sequence of SNe Ia with a range of CSM densities: from SNe Ia-CSM to SN 2002cx-like events and normal SNe Ia, consistent with existing radio constraints. We also find a new subclass of hybrid SNe that share the properties of Type II and Type Ia SNe, without a surviving companion (Meng & Podsiadlowski 2018).

References

Hachisu I., Kato M., Nomoto K. 1996, *ApJ*, 470, L97
Maoz D., Mannucci F., Nelemans G. 2014, *ARAA*, 52, 107
Meng, X. & Podsiadlowski, Ph. 2017, *MNRAS*, 469, 4763
Meng, X. & Podsiadlowski, Ph. 2018, *ApJ*, 861,127
Perlmutter, S., Aldering, G., Goldhaber, G., *et al.* 1999, *ApJ*, 517, 565
Riess, A., Filippenko, A. V., Challis, P., *et al.* 1998, *AJ*, 116, 1009
Rodney S. A. *et al.* 2015, *AJ*, 150, 156

Why Galaxies Care About AGB Stars:
A Continuing Challenge through Cosmic Time
Proceedings IAU Symposium No. 343, 2019
F. Kerschbaum, M. Groenewegen & H. Olofsson, eds.

doi:10.1017/S1743921318005653

Long Period Variables in Local Group dwarf Irregular Galaxies

John Menzies

South African Astronomical Observatory,
PO Box 9, Observatory, 7935, South Africa
email: jwm@saao.ac.za

Abstract. The long-term SAAO survey of Local Group galaxies in the near-infrared (JHK_s) has included five dwarf irregulars (dIrr), namely, NGC 6822, IC 1613, WLM, Sgr dIG and NGC 3109. We have found long-period (Mira) variables in all of them. Most of the Miras, which follow a linear LMC period-luminosity (PL) relation well, are carbon-rich. A small group of oxygen-rich Miras are brighter than the linear PL relation predicts, presumably because they are undergoing hot-bottom burning (HBB).

Keywords. galaxies: Local Group, stars: AGB, stars: variables: Mira

1. Introduction

A survey of Local Group galaxies has been conducted at SAAO using the IRSF 1.4-m telescope and the SIRIUS camera that produces simultaneous JHK_s images. Five dwarf irregular galaxies have been observed, for three of which results have been published (NGC 6822: Whitelock *et al.* 2013; IC 1613: Menzies *et al.* 2015; Sgr dIG: Whitelock *et al.* 2018). Data for the other two (WLM and NGC 3109) are being prepared for publication. Because of the relatively small size of the telescope, the limiting magnitude is typically 18.5 mag at Ks. This limits the population of stars that we can observe in a given galaxy. Thus in NGC 6822 we reach well below the tip of the RGB (TRGB), while in NGC 3109 the limit is about 1 mag brighter than the TRGB. There are repercussions for very red stars, which might be well above the Ks limit but are too faint for J; this effect should be borne in mind when interpreting relative numbers of variables in different galaxies.

2. WLM and NGC 3109

WLM is an isolated galaxy in the Local Group, with a moderately low metallicity, [Fe/H] = −1.28, in which about 30% of the stars were formed by 10 Gyr ago, while there was a recent burst of star formation starting about 3 Gyr ago. Six Mira variables with periods in the range 184 to 520 days have been found. This is likely to be an underestimate of the full complement of Miras in the galaxy. Five of these are known or suspected C stars, while the other is O-rich. The colour magnitude diagram is shown in Fig. 1.

NGC 3109 is part of a small group of galaxies just outside the Local Group. The bulk of the stars are very metal poor, [Fe/H] = −1.84, though the young blue supergiants have [Fe/H] = −0.67. About 80% of the stars were formed before 10 Gyr ago, and there has been more recent star formation starting about 2 Gyr ago. A probably incomplete census of Mira variables reveals eight O-rich and one probably C-rich Miras. One of the O-rich stars has a period of about 1480 days while the probable C-rich variable has a period of about 1100 days; these are unusually long periods for such variables.

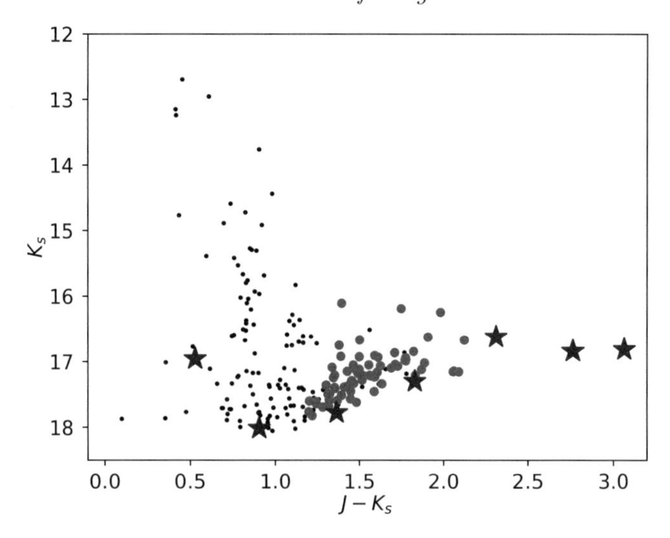

Figure 1. Colour-magnitude diagram of WLM. Large star symbols (blue) show 6 Mira variables and one Cepheid, large filled circles (red) show the known carbon stars, and small dots show field stars and WLM AGB stars.

3. Discussion

All the five galaxies have a complement of Miras, ranging from over 50 in NGC 6822 to at least three in Sgr dIG. These counts are almost certain to be underestimates in the case of the three most distant of the galaxies (Sgr dIG, WLM and NGC 3109). Most of these are C stars, which follow the LMC linear PL relation (Whitelock *et al.* 2008), particularly once bolometric corrections are applied, in the same way as for the dwarf spheroidals. However, unlike in the dwarf spheroidals, all except WLM have O-rich Miras with periods greater than about 400 days. These stars are brighter than expected from the linear PL relation, presumably because they are undergoing hot-bottom burning. The dwarf spheroidals generally comprise an old to intermediate age metal-poor population with a moderate range of metal abundances. The situation in dwarf Irregulars is more complex. In these galaxies there tends to be a wide range of ages and metallicities and in the case of NGC 3109, at least some of the AGB stars, including Miras, are quite young.

References

Whitelock, P. A., Feast, M. W. & Van Leeuwen, F. *MNRAS*, 386, 313

Whitelock, P. A., Menzies, J. W., Feast, M. W., Nsengiyumva, F. & Matsunaga, N. *MNRAS*, 428, 2216

Whitelock, P. A., Menzies, J. W., Feast, M. W. & Marigo, P. *MNRAS*, 473, 173

Menzies, J. W., Whitelock, P. A., & Feast, M. W. 2015, *MNRAS*, 452, 910

Why Galaxies Care About AGB Stars: A Continuing Challenge through Cosmic Time
Proceedings IAU Symposium No. 343, 2019
F. Kerschbaum, M. Groenewegen & H. Olofsson, eds.

doi:10.1017/S1743921318007160

Updates on the Ultraviolet Emission from Asymptotic Giant Branch Stars

Rodolfo Montez Jr.[1], Sofia Ramstedt[2], Joel H. Kastner[3] and Wouter Vlemmings[4]

[1]Smithsonian Astrophysical Observatory, Cambridge, MA USA
email: rodolfo.montez.jr@gmail.com
[2]Uppsala University, Uppsala, Sweden
[3]Rochester Institute of Technology, Rochester, NY USA
[4]Chalmers University of Technology, Onsala Space Observatory, Onsala, Sweden

Abstract. A comprehensive study of UV emission from asymptotic giant branch (AGB) stars with the Galaxy Evolution Explorer (GALEX) revealed that out of the 316 observed AGB stars, 57% were detected in the near-UV (NUV) bandpass and 12% were detected in the far-UV (FUV) bandpass (Montez *et al.* 2017). A cross-match between our sample and Gaia DR2 results in parallax estimates for 90% of the sample of AGB stars, compared to only 30% from Hipparcos. This increase allowed us to further probe trends and conclusions of our initial study. Specifically, that the detection of UV emission from AGB stars is subject to proximity and favorable lines of sight in our Galaxy. These improved results support the notion that some of the GALEX-detected UV emission is intrinsic to AGB stars, likely due to a combination of photospheric and chromospheric emission.

Keywords. stars: AGB and post-AGB, ultraviolet: stars, stars: evolution

---

Asymptotic Giant Branch (AGB) stars are luminous sources from optical to radio wavelengths, however, their ultraviolet (UV) emission is poorly-characterized. In Montez *et al.* 2017), we presented a comprehensive study of the UV emission in the near and far UV emission (NUV and FUV, respectively) based on observations of 316 AGB stars by the Galaxy Evolution Explorer (GALEX). The sample included carbon, M-type, and S-type AGB stars and both photometric and spectroscopic observations. 179 of the AGB stars were detected and 137 were not detected. We reported that the NUV emission from AGB stars is correlated with the optical to the near-infrared emission and is often found to vary in phase with phased visible light curves. Our study also found evidence for anti-correlation between the circumstellar envelope density and the NUV – and possibly FUV – emission. Including Hipparcos parallax estimates, we found that the detections and non-detections indicated higher detection fractions from the closest AGB stars, as well as the influence of galactic extinction on the detectability of UV emission from AGB stars.

We cross-correlated our GALEX AGB catalog with the 2nd Data Release of Gaia (Gaia DR2; Gaia Collaboration 2016, 2017) using a search radius of $6''$. The resulting sample increases the number of AGB stars with parallax estimates compared to the Hipparcos-based results. However, given the variability and projected sizes of the nearest AGB stars, the Gaia DR2 parallax estimates for the closest AGB stars have lower precision due to potential brightness variations on their surfaces that can lead to photocenter shifts (e.g., Chiavassa *et al.* 2018). Indeed, this problem is evident in the Gaia DR2 Astrometric

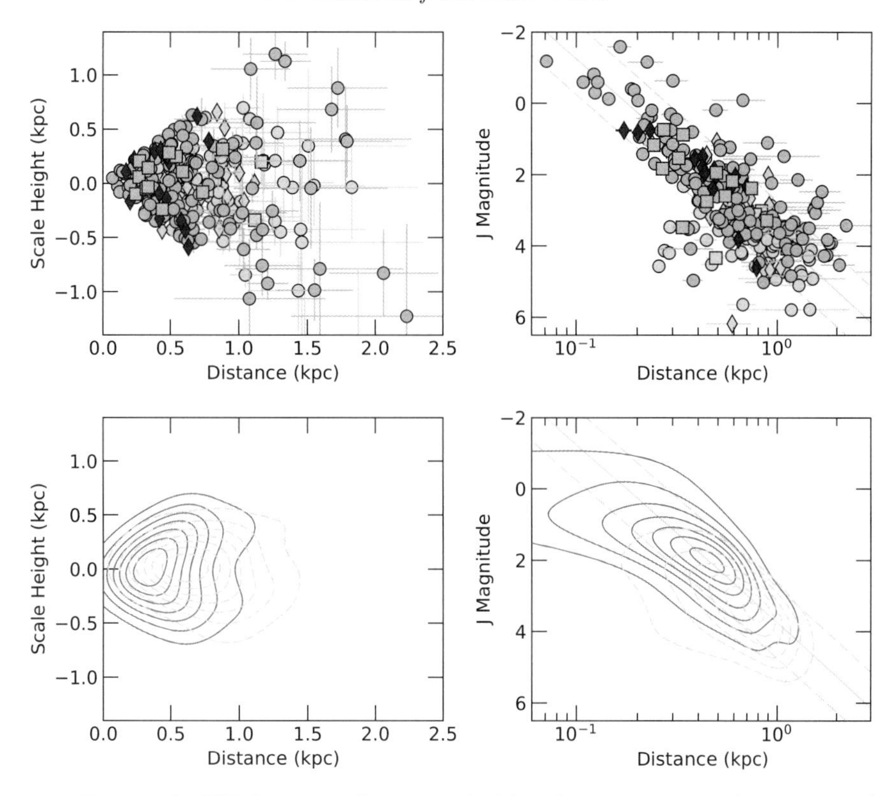

Figure 1. Top panels: UV detections (green symbols) and non-detections (gray symbols) of the AGB star sample with Gaia parallax measurements and their scale heights (left) and 2MASS J band magnitudes (right). Bottom panels show the same data after applying a kernel density estimator to better represent the detection and non-detection distributions. Symbols shapes represent the various types of AGB stars (circles for M-type, diamonds for carbon, and squares for S-type).

Goodness of Fit in the Along-Scan direction (`GOF AL`) metric for the sample. The `GOF AL` values are worse for the brightest/closest AGB stars and flattens to acceptable values for the fainter/farther AGB stars. As a result, reported parallax estimates for nearby ($J < 2$ mag) individual AGB stars are suspect, but when treated as a population, as done here, the influence of the potential photocenter shifts is reduced.

The trends determined from Hipparcos parallax estimates for our AGB sample are further strengthened by the Gaia parallax estimates. We more readily detect UV emission from the brighter and closer AGB stars and can detect AGB stars farther away when they are at higher galactic scale heights. These trends further support the notions that UV emission is an inherent (and hence most likely intrinsic) characteristic of AGB stars, and that galactic extinction hampers UV detection of AGB stars.

References

Chiavassa, A., Freytag, B., & Schultheis, M. 2018, *A&A*, 617, L1
Gaia Collaboration 2017, *A&A*, 605, A79
Gaia Collaboration 2016, *A&A*, 595, A1
Montez, R., Jr., Ramstedt, S., *et al.* 2017, ApJ, 841, 33

Why Galaxies Care About AGB Stars:
A Continuing Challenge through Cosmic Time
Proceedings IAU Symposium No. 343, 2019
F. Kerschbaum, M. Groenewegen & H. Olofsson, eds.

doi:10.1017/S174392131800635X

Astrometric observation of the Galactic LPVs with VERA; Mira and OH/IR stars

Akiharu Nakagawa[1], Tomoharu Kurayama[2], Gabor Orosz[1], Tomoaki Oyama[3], Takumi Nagayama[3] and Toshihiro Omodaka[1]

[1]Kagoshima University, 1-21-35, Korimoto, Kagoshima-shi, Kagoshima, Japan
email: nakagawa@sci.kagoshima-u.ac.jp

[2]Teikyo University of Science, 2-2-1 Sakuragi, Senjyu, Adachi-ku, Tokyo, Japan

[3]National Astronomical Observatory of Japan, Mizusawa VLBI Observatory, 2-12 Hoshigaoka-cho, Mizusawa-ku, Oshu, Iwate, Japan

Abstract. We present studies of Long Period Variables (LPVs) in our Galaxy based on astrometric VLBI observations of H_2O and SiO masers. The Galactic Miras and OH/IR stars are our main targets. For Miras, we present the distribution of the LPVs on the $M_K - \log P$ plane. Galactic Miras show consistency with PLR in the LMC except for some fainter sources. Parallaxes of the LPVs determined from VLBI and Gaia are compared. There seems to be some offset.

Keywords. astrometry, interferometric, variables, AGB and post-AGB

1. Introduction

The LPVs are 1 to 8 $M_\odot$ AGB stars pulsating with a typical period range of 100 to 1000 days. They are in the late stage of their life time, and show a high mass loss rate (10^{-7} $M_\odot$ yr^{-1}) before they evolve to planetary nebulae. LPVs in the LMC show some period-luminosity relations (PLR), and the PLR is used as distance indicator. Since there is a metallicity difference between the LMC and our Galaxy, it is also important to explore PLRs of Miras in our Galaxy. The PLR of Galactic Miras determined from our study is reported to be $M_K = -3.52 \log P + 1.09 \pm 0.14$ (solid line in Fig. 1, Nakagawa *et al.* 2016). There are some OH/IR stars showing quite long periods, longer than 1000 days (extreme-OH/IR). They are thought to have initial masses of $\sim 4\,M_\odot$ and ages of $\sim 10^8$ yr. Recent studies predict galactic spiral arms bifurcating/merging on time scales of 10^8 yr. So, the extreme-OH/IR stars can be used as a new probe to survey spiral arms structure and its evolution. We started 43 GHz astrometric VLBI observations of two extreme-OH/IR stars, OH127.8+0.0 and NSV25875 from Nov. 2017. Since the evolutional relation between Miras and OH/IR stars is still an open question, sequential studies of LPVs along wide period axis are also crucial.

2. Observation

Miras and OH/IR stars often represent H_2O, SiO, and OH masers, so they are good targets of VLBI. To derive parallaxes, we observe the target maser source and position reference QSO using a phase referencing VLBI technique. Part of the source names of our program are given in Table 1 in Nakagawa *et al.* (2018) together with sources from other studies. Recently, we selected a few samples of extreme-OH/IR stars from the Database of Circumstellar Masers (Engels & Bunzel 2015), and started 43 GHz (SiO maser) astrometric VLBI observations from Nov. 2017.

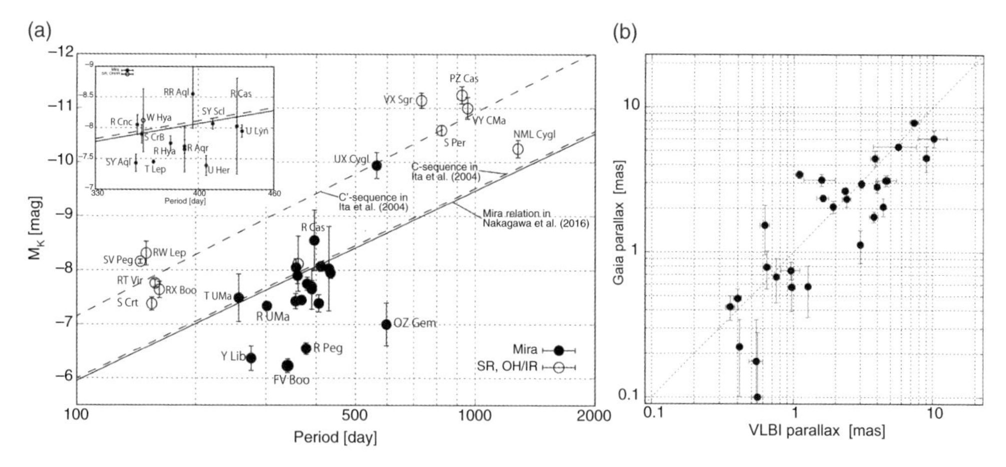

Figure 1. (a) M_K of Galactic LPVs obtained from apparent magnitude (m_K) and VLBI parallaxes. Filled and open circles are results of Miras and other type of variables. Dashed lines are C and C' sequences in Ita *et al.* (2004). (b) Annual parallaxes determined from VLBI (horizontal axis) and Gaia (vertical axis).

3. Results and Discussion

Figure 1(a) shows the $M_K - \log P$ diagram. Filled and open circles indicate Miras and other types of variables (SR, OH/IR stars). We find that many Galactic Miras coincide with the C-sequence of the LMC in Ita *et al.* (2004), but some Miras are fainter than the LMC sequence. Though the discrepancy should carefully be investigated, this can indicate different properties of Miras in the LMC and our Galaxy. A 43 GHz phase referencing observation of an extreme-OH/IR star NSV25875 and position reference source J2231+5922 gives a parallax of 0.38 ± 0.13 mas, corresponding to a distance of 2.60 ± 0.85 kpc. The distance was consistent with that estimated from two different methods of kinematic distance and OH maser phase-lag distance.

On 25 April 2018, parallaxes of more than 1.3×10^9 sources were released as Gaia DR2 (https://www.cosmos.esa.int/web/gaia/dr2). Since the larger part of VLBI parallaxes are determined for star forming regions which are deeply obscured by heavy dust and molecular clouds, it is difficult to find counterparts of the VLBI targets in the Gaia DR2 catalog. On the other hand, counterparts of the Galactic LPVs can easily be identified in the Gaia DR2. They are good samples for comparison of parallax from VLBI and Gaia. We compiled VLBI parallaxes from various studies (e.g. Vlemmings *et al.* 2003), and compared them with Gaia parallaxes in Figure 1(b). We see a trend that Gaia parallaxes give slightly smaller values than VLBI parallaxes. There seems to be a small offset between VLBI and Gaia parallaxes, and a more detailed study is needed for better understanding. Since the Galactic LPVs with longer period tend to be fainter in visible band, VLBI astrometry still plays a promising and complementary role even in Gaia era.

References

Engels, D., & Bunzel, F. 2015, *A&A*, 582, A68

Ita, Y., Tanabé, T., Matsunaga, N., *et al.* 2004, *MNRAS*, 353, 705

Nakagawa, A., Kurayama, T., Matsui, M., *et al.* 2016, *PASJ*, 68, 78

Nakagawa, A., Kurayama, T., Orosz, G., *et al.* 2018, Astrophysical Masers: Unlocking the Mysteries of the Universe, 336, 365

Vlemmings, W. H. T., van Langevelde, H. J., Diamond, P. J., Habing, H. J., & Schilizzi, R. T. 2003, *A&A*, 407, 213

Why Galaxies Care About AGB Stars:
A Continuing Challenge through Cosmic Time
Proceedings IAU Symposium No. 343, 2019
F. Kerschbaum, M. Groenewegen & H. Olofsson, eds.

doi:10.1017/S1743921318005677

The dust production rate of carbon-rich stars in the Magellanic Clouds

Ambra Nanni[1], Martin A. T. Groenewegen[2], Bernhard Aringer[1], Paola Marigo[1], Stefano Rubele[1] and Alessandro Bressan[3]

[1]Dipartimento di Fisica e Astronomia Galileo Galilei, Università di Padova, Vicolo dell'Osservatorio 3, I-35122 Padova, Italy
email: ambra.nanni@unipd.it

[2]Koninklijke Sterrenwacht van België, Ringlaan 3, B-1180 Brussel, Belgium

[3]SISSA, via Bonomea 265, I-34136 Trieste, Italy

Abstract. We present our new investigation aimed to estimate the mass-loss and dust production rates of carbon-rich stars (C-stars) in the Magellanic Clouds (MCs). We compute dust growth and radiative transfer in circumstellar envelopes of C-stars for a grid of stellar parameters and for selected optical constants that simultaneously reproduce the main colour–colour diagrams in the infrared. We employ these grids of spectra to fit the spectral energy distribution of C-stars in the MCs. We find that our estimates can be significantly different from the other ones in the literature.

Keywords. stars: AGB and post-AGB– stars: carbon – circumstellar matter – stars: mass loss – stars: winds, outflows – Magellanic Clouds

1. Introduction

The spectra and colours of thermally pulsing asymptotic giant branch (TP-AGB) stars are deeply affected by the properties of dust grains condensed in their circumstellar envelopes (Nanni *et al.* 2016). For carbon (C-) stars, amorphous carbon (amC) dust is particularly relevant. In radiative transfer calculations, the grain size distribution and optical constants of amC dust are chosen among several data sets (Hanner 1988; Rouleau & Martin 1991; Zubko *et al.* 1996; Jager *et al.* 1998). Such a choice affects the estimate of the dust production rate (DPR) and mass-loss rates of C-stars (Nanni *et al.* 2018; Groenewegen & Sloan 2018).

2. Model

In our description, grain growth is coupled with a spherical symmetric, stationary wind, as described in Nanni *et al.* (2013, 2014), based on Ferrarotti & Gail (2006). This scheme is coupled with a radiative transfer code (MoD; Groenewegen (2012) Ivezic & Elitzur 1997), and allows one to compute large grids of spectra reprocessed by dust. The input quantities of our code are the stellar parameters and the photospheric spectra (COMARCS; Aringer *et al.* 2016). The outflow expansion velocity ($v_{\rm exp}$) and the gas-to-dust ratio are predicted by our calculations. We employ the combinations of optical constants and grain sizes of amC dust that reproduce the main colour–colour diagrams (CCDs) in the infrared (Nanni *et al.* 2016). We then estimate the DPR and mass-loss rates of C-stars in the Magellanic Clouds (MCs), by fitting their spectral energy distribution (SED) over the grids.

Table 1. Total DPR for the MCs computed with the ACAR sample by Zubko *et al.* (1996), and compared with the results in the literature. The C- and extreme (x-) stars are classified according to Cioni *et al.* (2006) and Blum *et al.* (2006).

LMC			
This work, Zubko *et al.* 1996	C-stars $(2.60 \pm 0.52) \times 10^{-6}$	X-stars $(1.72 \pm 0.43) \times 10^{-5}$	Total $(1.98 \pm 0.49) \times 10^{-5}$
Riebel *et al.* 2012	$(1.34 \pm 0.40) \times 10^{-6}$	$(1.29 \pm 0.27) \times 10^{-5}$	$(1.43 \pm 0.31) \times 10^{-5}$
Srinivasan *et al.* 2016	7.6×10^{-7}	1.2×10^{-5}	1.2×10^{-5}
Dell'Agli *et al.* 2015	–	–	$\approx 4 \times 10^{-5}$
SMC			
This work, Zubko *et al.* 1996	$(4.4 \pm 1.2) \times 10^{-7}$	$(2.6 \pm 1.1) \times 10^{-6}$	$(3.0 \pm 1.2) \times 10^{-6}$
Srinivasan *et al.* 2016	$\approx 1.2 \times 10^{-7}$	$\approx 6.8 \times 10^{-7}$	$\approx 8.0 \times 10^{-7}$
Boyer *et al.* 2012	$\approx 1.2 \times 10^{-7}$	$\approx 6.3 \times 10^{-7}$	$\approx 7.5 \times 10^{-7}$
Matsuura *et al.* 2013	–	–	$\approx 4 \times 10^{-6}$

3. Results

In Table 1, we provide the DPR derived with our method for one selected optical data set for amC dust (ACAR sample; Zubko *et al.* 1996) with grain size between 0.06 and 0.09 μm. The DPRs from the literature are also listed. For the SMC, our DPR is, within the uncertainty, in agreement with the one by Matsuura *et al.* (2013), while the DPRs of Boyer *et al.* (2012) and of Srinivasan *et al.* (2016) are about four times lower than our estimate. Such a discrepancy is due to the different grain size distributions and $v_{\rm exp}$.

For the LMC, our DPR is in agreement with the one derived by Riebel *et al.* (2012) and about 1.7 times larger than the one by Srinivasan *et al.* (2016). On the other hand, our DPR is about two times lower than the one from Dell'Agli *et al.* (2015). A complete analysis performed with all the well-performing optical constants will be soon published.

References

Aringer, B., Girardi, L., Nowotny, W., Marigo, P., & Bressan, A. 2016, *MNRAS*, 457, 3611
Blum, R. D., Mould, J. R., Olsen, K. A., *et al.* 2006, *AJ*, 132, 2034
Boyer, M. L., Srinivasan, S., Riebel, D., *et al.* 2012, *ApJ*, 748, 40
Cioni, M.-R. L., Girardi, L., Marigo, P., & Habing, H. J. 2006, *A&A*, 448, 77
Dell'Agli, F., Ventura, P., Schneider, R., *et al.* 2015, *MNRAS*, 447, 2992
Ferrarotti, A. S., & Gail, H.-P. 2006, *A&A*, 447, 553
Groenewegen, M. A. T. 2012, *A&A*, 543, A36
Groenewegen, M. A. T., & Sloan, G. C. 2018, *A&A*, 609, A114
Hanner, M. S. 1988, Infrared Observations of Comets Halley and Wilson and Properties of the Grains
Ivezic, Z., & Elitzur, M. 1997, *MNRAS*, 287, 799
Jager, C., Mutschke, H., & Henning, T. 1998, *A&A*, 332, 291
Matsuura, M., Woods, P. M., & Owen, P. J. 2013, *MNRAS*, 429, 2527
Nanni, A., Bressan, A., Marigo, P., & Girardi, L. 2013, *MNRAS*, 434, 2390
Nanni, A., Bressan, A., Marigo, P., & Girardi, L. 2014, *MNRAS*, 438, 2328
Nanni, A., Marigo, P., Groenewegen, M. A. T., *et al.* 2016, *MNRAS*, 462, 1215
Nanni, A., Marigo, P., Girardi, L., *et al.* 2018, *MNRAS*, 473, 5492
Riebel, D., Srinivasan, S., Sargent, B., & Meixner, M. 2012, *ApJ*, 753, 71
Rouleau, F., & Martin, P. G. 1991, *ApJ*, 377, 526
Srinivasan, S., Boyer, M. L., Kemper, F., *et al.* 2016, *MNRAS*, 457, 2814
Zubko, V. G., Mennella, V., Colangeli, L., & Bussoletti, E. 1996, *MNRAS*, 282, 1321

Why Galaxies Care About AGB Stars: A Continuing Challenge through Cosmic Time
Proceedings IAU Symposium No. 343, 2019
F. Kerschbaum, M. Groenewegen & H. Olofsson, eds.

doi:10.1017/S1743921318007937

Near IR and visual polarimetry of the Planetary Nebula M2-9

Silvana G. Navarro[1], Omar Serrano[2], Abraham Luna[2], Rangaswami Devaraj[2], Luis J. Corral[1], Julio Ramírez Vélez[3] and David Hiriart[3]

[1]Instituto de Astronomía y Meteorología, Universidad de Guadalajara, 44130 Guadalajara, Jal., México
emails: silvana@astro.iam.udg.mx, lcorral@astro.iam.udg.mx

[2]Instituto Nacional de Astrofísica Óptica y Electrónica, 72840 Puebla, Pue., México
emails: arthas@inaoep.mx, aluna@inaoep.mx

[3]Instituto de Astronomía, sede Ensenada, Universidad Nacional Autónoma de México, Ensenada Baja California, México
emails: jramirez@astro.unam.mx, hiriart@astro.unam.mx

Abstract. Bipolar and more complex morphologies observed in planetary nebulae have been explained by two principal hypotheses: by the existence of a companion producing a circumstellar disk, by the effects of a magnetic field, or by a combination of both. The polarimetric analysis of these objects could give information about the presence of dust grains aligned with any preferential direction, due to a magnetic field or to the action of radiative torques (RAT). We performed polarimetric observations of some planetary nebulae in order to detect linear polarization and (in the best scenario) to detect the signature of an accretion disk in these objects. We observed in the visual region with POLIMA at the San Pedro Mártir observatory, and with POLICAN the NIR polarimeter in the Guillermo Haro observatory. We present the result of these observations in one of these objects: the PN M2-9.

Keywords. Polarization, stars: magnetic fields, ISM: Planetary Nebulae: individual: M2-9

1. Introduction

Polarization due to dust scattering is a useful tool to determine the geometry of extended envelopes in symbiotic systems and proto-planetary nebulae (Scarrott & Scarrott 1995, Gledhill 2005). It is also used to trace the circumstellar disk and jets/outflows present in such objects. M2-9 is a widely studied bipolar PN, it shows two coaxial shells and a series of bright filaments and knots, symmetrically located on both lobes, which have been explained by a lighthouse effect. Their apparent motion has been shown by Corradi *et al.* (2011). Recently, Castro-Carrizo *et al.* (2012), resolved two ring-shaped structures at the center of this object *giving evidence of their binary nature.*

2. Observations

M2-9 was included in an observation program of PNe using POLIMA polarimeter on the 84 cm telescope located in San Pedro Mártir observatory on July 2012. We used two narrow filters: Hα ($\lambda = 6564$ Å, $\delta\lambda = 72$ Å) and H6819 ($\lambda = 6819$ Å, $\delta\lambda = 86$ Å). The spatial resolution, with a 2x2 binning, is 0.528 arcsec/pix. We present the Hα polarization map to compare with those obtained at the IR. At these wavelengths, we use POLICAN, the IR polarimeter at the Guillermo Haro observatory at Cananea (Devaraj *et al.* 2018).

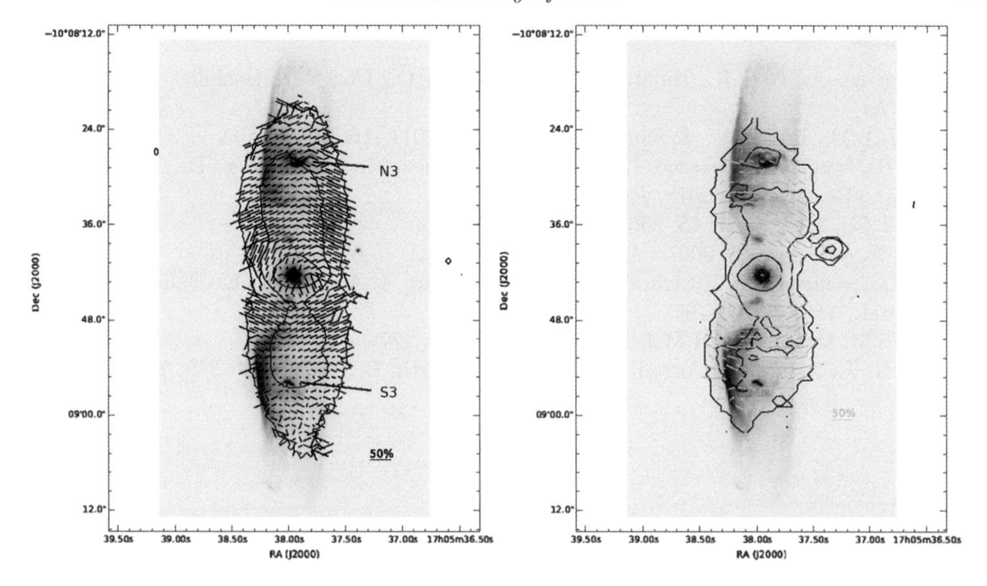

Figure 1. Left: Polarization vectors in Hα superimposed over the HST image of M2-9 at Hα. Right: Polarization vectors in the H band superimposed over the same HST image.

We obtained polarimetric images in the J and H bands in March 2017. The polarimeter, attached to the IR camera CANICA provides a spatial resolution of 0.32 arcsec/pix (0.64 arcsec/pix with a binning of 2x2). The observations were made in commissioning time with a seeing of 1.2 arcsec.

3. Results and analysis

3.1. *Polarization of M2-9 in Hα*

In this object, a high degree of polarization is detected, especially in the regions near the border, as we expect from scattered light coming from the lobes. In general, the direction of the polarization vectors is nearly perpendicular to the radiation from the central object, in accord with the grain alignment produced by radiative torques (RAT, Lazarian & Hoang 2007). Although at the position of the bright knots observed in the north and south lobes (N3 and S3 in Fig. 1), the percentage of polarization is lower. In these regions, there are indications of shocked material that produce such bright knots (Schwarz *et al.* 1997) and dominate over the scattered light in these regions. In the Hα polarization map (Fig. 1) at the central waist of the object, a thin region with higher polarization is observed. This could be an indication of the presence of dust in the core of the object as was shown by the observations of Lykou *et al.* (2011) in the NIR and Castro-Carrizo *et al.* (2012) in the ^{12}CO J=2-1 line.

3.2. *NIR Polarimetry of M2-9*

In the H band, M2-9 shows considerable polarization levels. As in Hα, the observed polarization angle is, in general, perpendicular to the radiation from the central object. In this case, only in one of the knots (at the north from the central object) the polarization is notoriously less important. At the central region, the polarization is minimum, and the thin region with a greater polarization level is not detected.

References

Castro-Carrizo A., Neri R., Bujabarral V., Chesneau O., Cox P. & Bachiller R. 2012, *A&A*, 545, A1

Corradi, R.L.M., Balick, B., & Santander-García M. 2011 *A&A*, 529, A43

Devaraj, R., Luna, A., Carrasco, L., Vazquez-Rodríguez, M. A., Maya, Y. D., Tánori, J. G. & Serrano-Bernal E. O. 2018, *PASP* 130, 987

Gledhill, T.M., 2005, *MNRAS*, 356, 883

Lazarian, A. & Hoang, T. 2007, *MNRAS*, 378, 910

Lykou, F., Chesneau, O., Zijlstra, A. A., Castro-Carrizo, A. A., Legdec, E., Balick, B. & Smith, N. 2011, *A&A*, 527, A105

Scarrott, S.M. & Scarrott, R.M.J. 1995, *MNRAS*, 277, 277

Schwarz, H. E., Aspin, C., Corradi, R. L. M. & Reipurth, B. 1997, *A&A*, 319, 267

Why Galaxies Care About AGB Stars:
A Continuing Challenge through Cosmic Time
Proceedings IAU Symposium No. 343, 2019
F. Kerschbaum, M. Groenewegen & H. Olofsson, eds.

doi:10.1017/S1743921318008104

SMA Spectral Line Survey of the Proto-Planetary Nebula CRL 618

Nimesh A. Patel[1], Carl Gottlieb[1], Ken Young[1], Tomasz Kaminski[1], Michael McCarthy[1], Karl Menten[2], Chin-Fei Lee[3] and Harshal Gupta[4]

[1]Harvard-Smithsonian Center for Astrophysics, Cambridge MA, USA

[2]Max Planck Institute for Radio Astronomy, Bonn, Germany

[3]Academia Sinica Institute of Astronomy and Astrophysics, Taipei, Taiwan

[4]National Science Foundation, Washington DC, USA

Abstract. Carbon-rich Asymptotic Giant Branch (AGB) stars are major sources of gas and dust in the interstellar medium. During the brief (~1000 yr) period in the evolution from AGB to the Planetary Nebula (PN) stage, the molecular composition evolves from mainly diatomic and small polyatomic species to more complex molecules. Using the Submillimeter Array (SMA), we have carried out a spectral line survey of CRL 618, covering a frequency range of 281.9 to 359.4 GHz. More than 1000 lines were detected in the ~60 GHz range, most of them assigned to HC_3N and c-C_3H_2, and their isotopologues. About 200 lines are unassigned. Lines of CO, HCO^+, and CS show the fast outflow wings, while the majority of line emission arises from a compact region of ~1" diameter. We have analyzed the lines of HC_3N, c-C_3H_2, CH_3CN, and their isotopologues with rotation temperature diagrams.

Keywords. (stars:) circumstellar matter, molecular processes, stars: AGB and post-AGB, submillimeter, stars: individual (CRL 618)

Earlier studies of the chemical evolution of carbon rich AGB and post-AGB sources with the IRAM 30m, CSO 10m, and ISO telescopes concluded that CO, ^{13}CO, and HCN emission are produced in shocks caused by the fast wind in CRL 618; H_2O and OH form in the inner parts of the envelope; more complex organic molecules appear; and finally as the star reaches the PN stage, molecules are converted to atoms (Herpin *et al.* 2002). By referring to chemical models, it was found that HCN is quickly photodissociated to produce CN; and HCCCN is subsequently produced by reactions of CN with C_2H_2 and C_2H (Cernicharo 2004). Mapping the spatial location of various molecular species is important for comparison with chemical model predictions. Single-dish observations do not have adequate angular resolution (e.g., Pardo *et al.* 2007a). Interferometric observations of CRL 618 are presented in a few studies, but the spectral coverage is limited (Remijan *et al.* 2005, Lee *et al.* 2013, Sanchez-Contreras 2004).

CRL 618 was observed with the SMA with an angular resolution of $0.''5 \sim 2.''5$. The frequency range was 282.0 to 360.0 GHz, with a spectral resolution of ~0.8 MHz, and rms line sensitivity of ~0.1 Jy/beam. A sample spectrum covering about 1/4 of the full range is shown in Fig. 1. A total of 1075 lines were detected. About 250 lines remain unassigned. Most of the lines are attributed to c-C_3H_2, and HC_3N and its isotopologues. CS, CN, HCN and CCH are detected in the fast outflow of CRL 618. There are no Si bearing molecules detected except SiO, in contrast to IRC+10216 where molecules such as SiCC are in much greater abundance. Eight hydrogen and two helium recombination lines are detected in the ionized region at the peak continuum emission which remains unresolved even at

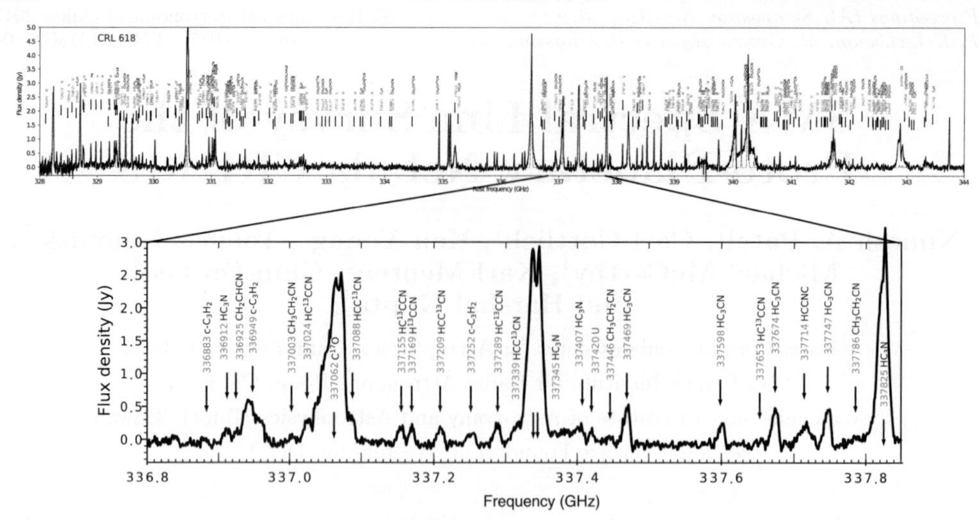

Figure 1. *Top:* Part of the spectrum from the line survey of CRL 618 covering 328 to 344 GHz. The red curve is an LTE model based on identified molecules and isotopologues. *Bottom:* Zoomed spectrum from above, to show details and some of the identifications.

Table 1. Summary of molecules and isotopologues detected in CRL 618

c-C_3H_2	~150	CH_3CN	21	CN	5	Recombination	
HC_3N	~ 170	CH_3C_2H	43	^{13}CN	2	lines	
		c-C_3H	11	$^{13}C^{15}N$	2		
$H^{13}CCCN$,		C_2H	3	HCN	4		
$HC^{13}CCN$,	144			$H^{13}CN$	3	H26α	1
$HCC^{13}CN$		H_2CO	6			H27α	1
		$H_2^{13}CO$	3	MgNC	1	H37α	1
$HCCC^{15}N$	1	$H_2C^{18}O$	1			H39α	1
HCCNC	4	CO	1			H33β	1
HC_5N	27	CS	2	^{13}CO	1	H34β	1
		^{13}CS	1	$C^{18}O$	1	H41δ	1
SiO	1			$C^{17}O$	1	H38γ	1
		HCO^+	1				
HCP	1	HCS^+	1	PN	1	He26α	1
CH_2NH	4						
$^{13}CH_2NH$	4						
CH_2CHCN	6						
CH_3CH_2CN	30						

0.″5 resolution. Table 1 summarizes the detected molecules and isotopologues in the line survey. LTE modeling is carried out for several of these molecules and isotopologues (Patel *et al.* 2018, in prep.).

References

Bachiller, R., Forveille, T., Huggins, P. J., *et al.* 1997, *A&A*, 324, 1123

Cernicharo, J. 2004, *ApJ* 608, L41

Herpin, F., Goicoechea, J. R., Pardo, J. R., & Cernicharo, J. 2002, *ApJ*, 577, 961

Knapp, G. R., Sandell, G., & Robson, E. I. 1993, *ApJSS*, 88, 173

Lee, C.-F., Yang, C.-H., Sahai, R., & Sánchez Contreras, C. 2013, *ApJ*, 770, 153

Pardo, J. R., Cernicharo, J., Goicoechea, J. R., Guélin, M., & Asensio Ramos, A. 2007, *ApJ*, 661, 250

Patel, N. A., Young, K. H., Gottlieb, C. A., *et al.* 2011, *ApJSS*, 193, 17

Remijan, A. J., Wyrowski, F., Friedel, D. N., Meier, D. S., & Snyder, L. E. 2005, *ApJ*, 626, 233

Sánchez Contreras, C., Sahai, R., & Gil de Paz, A. 2002, *ApJ*, 578, 269

Tafoya, D., Loinard, L., Fonfría, J. P., *et al.* 2013, *A&A*, 556, A35

Why Galaxies Care About AGB Stars:
A Continuing Challenge through Cosmic Time
Proceedings IAU Symposium No. 343, 2019
F. Kerschbaum, M. Groenewegen & H. Olofsson, eds.

doi:10.1017/S1743921318007810

SWAG: Distribution and Kinematics of an Obscured AGB Population toward the Galactic Center

Jürgen Ott[1,2], David S. Meier[2,1], Adam Ginsburg[1], Farhad Yusef-Zadeh[3], Nico Krieger[4] and Cornelia Jäschke[4]

[1]National Radio Astronomy Observatory, 1003 Lopezville Road, Socorro, NM 87801, USA
emails: jott@nrao.edu, aginsbur@nrao.edu

[2]New Mexico Institute of Mining and Technology, 801 Leroy Place, Socorro, NM 87801, USA
email: david.meier@nmt.edu

[3]Department of Physics and Astronomy and CIERA, Northwestern University, Evanston, IL 60208, USA
email: zadeh@northwestern.edu

[4]Max-Planck-Institut für Astronomie, Königstuhl 17, 69120 Heidelberg, Germany
emails: krieger@mpia.de, jaeschke@mpia.de

Abstract. Outflows from AGB stars enrich the Galactic environment with metals and inject mechanical energy into the ISM. Radio spectroscopy can recover both properties through observations of molecular lines. We present results from SWAG: "Survey of Water and Ammonia in the Galactic Center". The survey covers the entire Central Molecular Zone (CMZ), the inner $3.35^\circ \times 0.9^\circ$ ($\sim$480$\times$130 pc) of the Milky Way that contains $\sim 5\times10^7$ $M_\odot$ of molecular gas. Although our survey primarily targets the CMZ, we observe across the entire sightline through the Milky Way. AGB stars are revealed by their signature of double peaked 22 GHz water maser lines. They are distinguished by their spectral signatures and their luminosities, which reach up to 10^{-7} $L_\odot$. Higher luminosities are usually associated with Young Stellar Objects located in CMZ star forming regions. We detect a population of $\sim$600 new water masers that can likely be associated with AGB outflows.

Keywords. stars: AGB and post-AGB, ISM: molecules, Galaxy: center, radio lines: stars

1. SWAG

SWAG ("Survey of Water and Ammonia in the Galactic center") is a Large Project ($\sim$460 h on-source) to observe the entire CMZ (inner 500 pc) of the Milky Way with the Australia Telescope Compact Array (Krieger *et al.* 2017, Jäschke 2018, Ott *et al.* 2018 in prep.). We target 42 molecular lines and wideband continuum in the 21.2-25.4 GHz range across $\sim$6500 individual pointings. The spatial resolution is $\sim$20" ($\sim$0.8 pc) with up to 0.4 km s^{-1} spectral resolution. The line list contains typical shock, photon-dominated region, density, and temperature molecular tracers, as well as radio recombination lines. The list includes the $6_{16} - 5_{23}$ 22 GHz water (H_2O) maser and multiple ammonia lines.

2. Water Masers toward the CMZ

H_2O masers are typically associated with shocked regions, where density and temperatures are high enough to pump and invert the water level populations by collision, and path lengths long enough for amplification. 22 GHz is one of the brightest and best accessible water masers. Maser conditions are typically met in outflows, frequently in Young Stellar Object (YSOs) jets or envelopes of Asymptotic Giant Branch (AGB) stars.

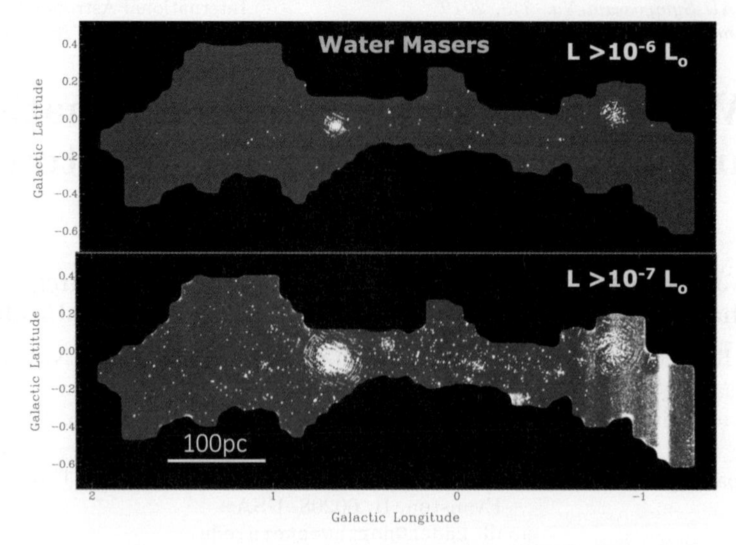

Figure 1. 22 GHz water masers toward the CMZ. **Top panel:** Masers with luminosities $> 10^{-6}\,L_\odot$; **Bottom panel:** Masers with $L > 10^{-7}\,L_\odot$. The distribution considerably broadens on the faint end.

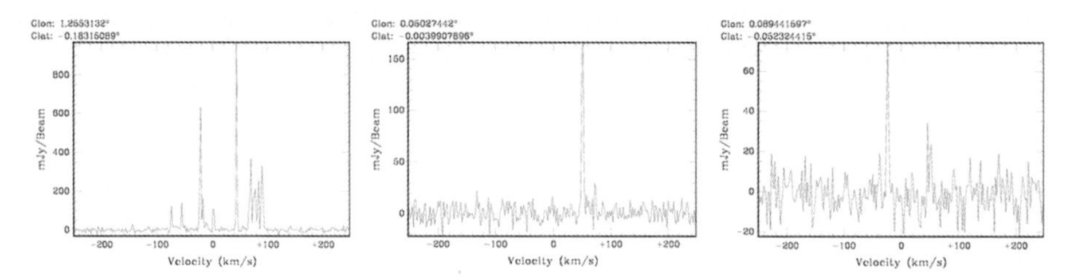

Figure 2. Sample spectra for different maser luminosities. Faint masers tend to show a second, even weaker velocity component, separated by typically $\sim$80 km s^{-1}.

The CMZ contains star forming regions and indeed we detect bright water masers that we associate with YSOs (Rickert 2017). In the upper panel of Fig. 1, we show masers with luminosities $> 10^{-6}\,L_\odot$ (assuming Galactic Center distance of 8.5 kpc for all sources). These masers appear to be frequently associated with dense molecular gas.

When the luminosity threshold is lowered by an order of magnitude to $\sim 10^{-7}\,L_\odot$, the distribution changes and widens in Galactic Latitude and Longitude. Overall, the masers follow more the Galactic potential than the distribution of dense gas in the CMZ. In addition, the weaker masers frequently show a second velocity component, with an average separation of $\sim$80 km s^{-1}. High luminosity masers, in contrast, typically show many more individual components at a greater velocity variation. Example spectra are shown in Fig. 2. The spatial distribution and spectral signatures of the newly detected, faint population of $\sim$600 masers (density of $\sim$200 deg^{-2}) is in agreement with a population of predominantly AGB stars across the entire Milky Way. The velocity separation of the maser components suggests that the traced AGB stars show expansion velocities with typical values around $\sim$40 km s^{-1}.

References

Jäschke, C. 2018, BS Thesis, University of Heidelberg, Germany
Krieger, N. *et al.* *ApJ*, 2017, 850, 77
Rickert, M. 2017, PhD Thesis, University of Illinois, USA

Why Galaxies Care About AGB Stars:
A Continuing Challenge through Cosmic Time
Proceedings IAU Symposium No. 343, 2019
F. Kerschbaum, M. Groenewegen & H. Olofsson, eds.

doi:10.1017/S1743921318004957

AGB stars of the Magellanic Clouds as seen within the Δa photometric system

Ernst Paunzen[1], Jan Janík[1], Petr Kurfürst[1], Jiří Liška[1], Martin Netopil[2], Marek Skarka[1,3] and Miloslav Zejda[1]

[1]Department of Theoretical Physics and Astrophysics, Masaryk University, Kotlářská 2, CZ-611 37 Brno, Czech Republic
email: epaunzen@physics.muni.cz

[2]Department of Astrophysics, University of Vienna, Austria

[3]Astronomical Institute, Czech Academy of Sciences, Czech Republic

Abstract. The a-index samples the flux of the 5200 Å region by comparing the flux at the center with the adjacent regions. The final intrinsic peculiarity index Δa was defined as the difference between the individual a-values and the a-values of normal stars of the same colour (spectral type). Here we present, for the first time, a case study to detect and analyse Asymptotic Giant Branch (AGB) stars in the Magellanic Clouds. For this, we use our photometric survey of the Magellanic Clouds within the a-index. We find that AGB stars can be easily detected on the basis of their Δa index in an efficient way.

Keywords. Magellanic Clouds, stars: AGB and post-AGB, techniques: photometric

1. Introduction

Asymptotic Giant Branch (AGB) stars trace the intermediate age population of galaxies (ages between one and several Gyr). In near-infrared (near-IR) light they are often the brightest isolated objects in a galaxy and can be studied at distances beyond 1 Mpc. There are two main kinds of AGB stars: oxygen-rich stars (spectral type M or K) and carbon-rich stars (spectral type N or C). When the number of O-atoms equals that of C-atoms, the AGB star is of type S. It is not straightforward to identify S stars either photometrically or spectroscopically, and this introduces as well intermediate classes of AGB stars (e.g. MS and SC). Single stars enter the AGB phase as oxygen-rich, but may be converted into carbon-rich stars after several short episodes in which matter that has been enriched in carbon nuclei by nuclear fusion is dredged-up to the stellar surface.

2. AGB stars in the Magellanic Clouds

Cioni & Habing (2003) showed that the ratio between C-rich and O-rich AGB stars (C/M ratio) varies within the Magellanic Clouds (MCs). Interpreting C/O changes as variations in the mean metallicity, they concluded that in the Large Magellanic Cloud (LMC) a metallicity gradient is present. Later, this result was revised (Cioni *et al.* 2006a) using new stellar models and taking into account different age distributions. It was suggested that a fit to the K_s-band magnitude distribution of both C- and O-rich AGB stars should be particularly useful to detect variations on the mean age and metallicity across the surface of nearby galaxies. For some well-defined LMC regions, both the mean age and metallicity were found to span the whole range of grid parameters. The Small Magellanic Cloud (SMC) stellar population was found to be on average 7 to 9 Gyr old, but older stars are present at its periphery and younger stars are present towards the

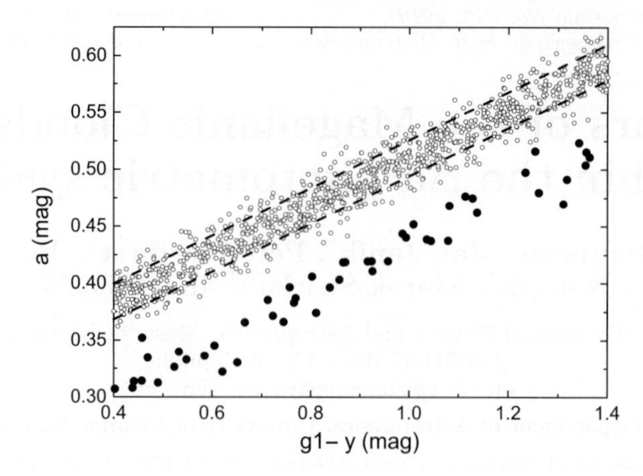

Figure 1. Δa photometry of a selected field in the LMC. The known AGB stars (filled circles) are clearly distinct from the normal MS stars (open circles). The dashed lines denote the 3σ level due to observational errors.

LMC (Cioni *et al.* 2006b). The metallicity distribution traces a ring-like structure that is more metal rich than the inner region of the LMC.

3. The Δa photometric system and results

Basically, the Δa photometric system consists of one filter which measures the 5200 Å region ($g2$) and an additional information about the continuum flux of the same object. This can be either achieved by measuring the flux at the adjacent spectral regions (using $g1$ and y) or by any other effective temperature sensitive colour.

During the last years, we have performed a survey of almost the complete MCs in the $g2$ filter. The CCD photometric measurements were obtained with the 1.54m Danish and 2.2m MPG telescope at the La Silla Observatory in Chile. The high accurate measurements yielded a mean error of a few mmags for the individual stars.

We have selected the stars published by Cioni & Habing (2003) and Cioni *et al.* 2006a,b) for our investigation. The stars on the main sequence are already quite faint in the $g2$ filter. From our analysis we conclude that

- 99% of all AGB stars in the MCs can be unambiguously detected (Figure 1).
- The C/M ratio can be directly deduced from the Δa value.
- This method works in the optical region without any pre-selection of targets.

In a forthcoming study, we publish all our observations in the MCs of AGB and post-AGB stars.

This study is supported by the grants of Ministry of Education of the Czech Republic LH14300 and 7AMB14AT030 as well as the GAČR international grant 17-01752J and the Postdoc@MUNI project CZ.02.2.69/0.0/0.0/16_027/0008360.

References

Cioni, M.-R.L., & Habing, H.J. 2003, *A&A*, 402, 133
Cioni, M.-R.L., Girardi, L., Marigo, P., & Habing, H.J. 2006a, *A&A*, 448, 77
Cioni, M.-R.L., Girardi, L., Marigo, P., & Habing, H.J. 2006b, *A&A*, 452, 195
Paunzen, E., Stütz, Ch., & Maitzen, H.M. 2005, *A&A*, 441, 631

Why Galaxies Care About AGB Stars: A Continuing Challenge through Cosmic Time
Proceedings IAU Symposium No. 343, 2019
F. Kerschbaum, M. Groenewegen & H. Olofsson, eds.

doi:10.1017/S174392131800501X

On the circumstellar effects on the Li and Ca abundances in massive Galactic O-rich AGB stars

V. Pérez-Mesa[1,2], O. Zamora[1,2], D. A. García-Hernández[1,2], Y. Ossorio[1,2], T. Masseron[1,2], B. Plez[3], A. Manchado[1,2,4], A. I. Karakas[5] and M. Lugaro[6]

[1]Instituto de Astrofísica de Canarias (IAC), E-38206 La Laguna, Tenerife, Spain
email: vperezme@iac.es

[2]Universidad de La Laguna (ULL), Departamento de Astrofísica, E-38206 La Laguna, Tenerife, Spain

[3]Laboratoire Univers et Particules de Montpellier, Université de Montpellier2, CNRS, 34095 Montpellier, France

[4]Consejo Superior de Investigaciones Científicas (CSIC), E-28006 Madrid, Spain

[5]Monash Centre for Astrophysics, School of Physics and Astronomy, Monash University, VIC3800, Australia

[6]Konkoly Observatory, Research Centre for Astronomy and Earth Sciences, Hungarian Academy of Sciences, 1121 Budapest, Hungary

Abstract. We explore the circumstellar effects on the Li and Ca abundances determination in a complete sample of massive Galactic AGB stars. The Li abundance is an indicator of the hot bottom burning (HBB) activation, while the total Ca abundance could be affected by overproduction of the short-lived radionuclide ^{41}Ca by the *s*-process. Li abundances were previously studied with hydrostatic models, while Ca abundances are determined here for the first time. The pseudo-dynamical abundances of Li and Ca are very similar to the hydrostatic ones, indicating that circumstellar effects are almost negligible. The new Li abundances confirm the (super-)Li-rich character of the sample Li-detected stars, supporting the HBB activation in massive Galactic AGB stars. Most sample stars display nearly solar Ca abundances that are consistent with predictions from the s-process nucleosynthesis models. A minority of the sample stars show a significant Ca depletion. Possible reasons for their (unexpected) low Ca content are given.

Keywords. stars, AGB, abundances, evolution, reactions, nucleosynthesis, atmospheres, late-type.

---

1. Observational data and models

We have used high-resolution (R $\sim$ 50,000) optical echelle spectra for the sample of massive Galactic AGB stars by García-Hernández *et al.* (2006, 2007) and the spectra of the super Li-rich AGBs reported by García-Hernández *et al.* (2013). By using a modified version of the spectral synthesis code Turbospectrum, we have obtained the Li and Ca abundances (from the 6708 Å Li I and 6463 Å resonance lines, respectively), applying new pseudo-dynamical model atmospheres developed by us (Zamora *et al.* 2014; Pérez-Mesa *et al.* 2017).

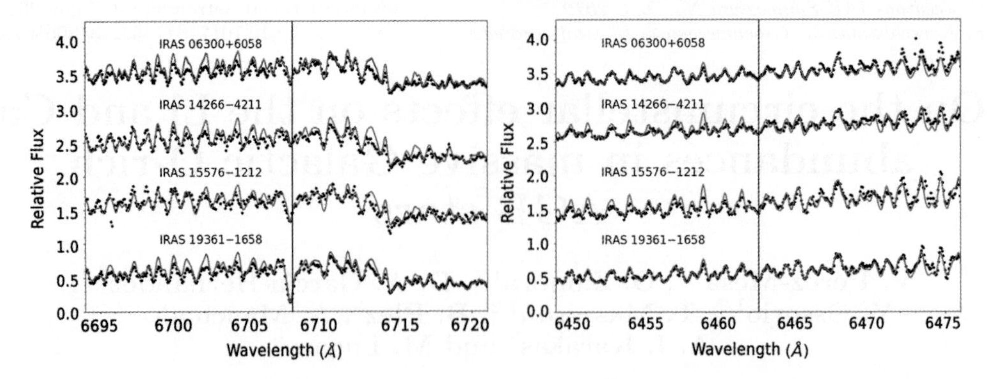

Figure 1. The Li I 6708 Å (*left*) and Ca I 6463 Å (*right*) spectral regions in four massive Galactic AGB stars. The pseudo-dynamical models (red lines) that best fit the observations (black dots) are shown. Note that the best hydrostatic models (blue lines) are indistinguishable.

2. Lithium

The new Li abundances derived from our pseudo-dynamical models are very similar (typically within 0.1 dex) to those obtained with the hydrostatic ones. Fig. 1 displays examples of the observed spectra and the best synthetic fits from hydrostatic and pseudo-dynamical models in the spectral region around Li I 6708 Å.

The new Li abundances obtained from extended atmosphere models confirm the Li-rich (and super Li-rich, in some cases) character of the sample stars with detected Li I lines, supporting strong activation of the HBB process in massive Galactic AGB stars. This is in good agreement with theoretical predictions for solar metallicity massive AGBs from nucleosynthesis models such as ATON (Ventura & D'Antona 2009), Monash (Karakas & Lugaro 2016) and NuGrid/MESA (Pignatari *et al.* 2016), but at odds with the FRUITY database (Cristallo *et al.* 2015) which predicts no Li production by HBB in such stars.

3. Calcium

The new pseudo-dynamical Ca abundances are identical to the hydrostatic ones (see Fig. 1). Most sample stars display nearly solar (within the estimated errors or considering possible NLTE effects) Ca abundances that are consistent with the available *s*-process nucleosynthesis models for solar metallicity massive AGB stars (e.g., Karakas & Lugaro 2016, Pignatari *et al.* 2016) which predict an important ^{41}Ca production but no change in the total Ca abundance. A minority of the sample stars seem to show, however, a significant Ca depletion (by $\sim$ -1.0 dex). Possible reasons to explain their apparent and unexpected Ca depletion are missed opacities in the stellar atmosphere models or Ca depletion into dust (see Pérez-Mesa *et al.* 2018 for more details).

References

Cristallo, S., Straniero, O., Piersanti, L., & Gobrecht, D., 2015, ApJs, 219, 40
García-Hernández, D. A., García-Lario, P., Plez, B., *et al.*, 2006, Science, 314, 1751
García-Hernández, D. A., García-Lario, P., Plez, B., *et al.*, 2007, *A&A*, 462, 711
García-Hernández, D. A., Zamora, O., Yagüe, A., *et al.*, 2013, *A&A*, 555, L3
Karakas, A. I., & Lugaro, M., 2016, ApJ, 825, 26
Pérez-Mesa, V., Zamora, O., García-Hernández, D. A., *et al.*, 2017, *A&A*, 606, A20
Pérez-Mesa, V., Zamora, O., García-Hernández, D. A., *et al.*, 2018, in prep.
Ventura, P., & D'Antona, F., 2009, *A&A*, 499, 835
Zamora, O., García-Hernández, D. A., Plez, B., & Manchado, A., 2014, *A&A*, 564, L4

Why Galaxies Care About AGB Stars:
A Continuing Challenge through Cosmic Time
Proceedings IAU Symposium No. 343, 2019
F. Kerschbaum, M. Groenewegen & H. Olofsson, eds.

doi:10.1017/S1743921318005628

AGB star atmospheres modeling as feedback to stellar evolutionary and galaxy models

Gioia Rau[1,2], M. Wittkowski[3], A. Chiavassa[4], K. Carpenter[1], K. Nielsen[2] and V. S. Airapetian[1,2]

[1]NASA Goddard Space Flight Center, Code 667, Greenbelt, MD 20771, USA,
email: gioia.rau@nasa.gov.

[2]Dept. of Physics, Catholic University of America.

[3]European Southern Observatory (ESO).

[4]Observatoire Côte d'Azur (OCA)

Abstract. The chemical enrichment of the Universe is considerably affected by the contribution of cool evolved stars. We studied the O-rich star R Peg and the C-rich star V Oph, using respectively the VLTI/GRAVITY and VLTI/MIDI instruments. We interpret the data using grids of 1-D and 3-D dynamic model atmospheres.

Keywords. AGB and post-AGB – atmospheres – mass-loss – stars: carbon – circumstellar matter – techniques: interferometric – techniques: high angular resolution –

1. Introduction

Toward the end of their lives, stars on the Asymptotic Giant Branch (AGB) produce in their atmospheres heavy chemical elements, molecules, and dust. Their mass loss produced by stellar winds will place this material into the interstellar medium. Even though many efforts have been carried out to-date (see e.g. Hron 1998), we still lack observational constraints and a consistent comparison with models, to retrieve fundamental information on this evolutionary stage. Ground-based interferometric measurements with high-angular resolution instruments from VLTI, such as GRAVITY and MIDI, can help to test dynamical models describing the behavior of the outer AGB atmospheres at various spatial scales.

2. Overview, observations, and goals

We present two separate projects. The first one consists of monitoring the photosphere and extended atmosphere of the O-rich AGB star R Peg with VLTI/GRAVITY, and studying its molecular composition using 1-D and 3-D model atmospheres. This work has been already published by Wittkowski *et al.* (2018). Hence, we focus here on the second ongoing project: modeling the C-rich star V Oph, observed with VLTI/MIDI in the N-band by Ohnaka *et al.* (2007), using 1-D dynamic model atmospheres. V Oph is indeed one of the few C-rich AGB stars showing interferometric variability in the N-band, probably related to its pulsation activity, and for this reason is a particularly interesting case-study for understanding the atmospheric dynamics in these type of stars. In this ongoing work we aim to: (1) better understand the physical processes responsible for molecule and dust formation close to the star; (2) verify if the extended molecular and dust layers can be explained by dust-driven winds triggered by large-amplitude stellar pulsation alone, or if some other contributing mechanism is required.

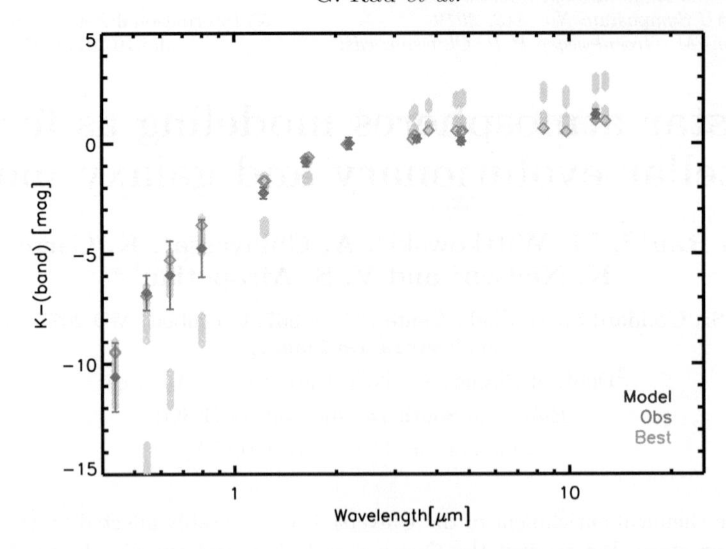

Figure 1. Photometric observations of V Oph. Observations (purple circles) are compared to the synthetic photometry derived from the dynamic models (gray diamonds). Orange diamonds show the best fitting time-steps of the star.

3. Preliminary results

We compare V Oph photometric data from the literature, and interferometric MIDI observations, with a grid of 1-D dynamic model atmospheres for C-rich stars by Eriksson *et al.* (2014) (based on Höfner *et al.* 2003, 2016), as similarly done by Rau *et al.* (2015, 2016, 2017). We derive the atmospheric stratifications and fundamental stellar parameters, and provide constraints for dynamic atmospheres, the mass loss process for carbon stars, and stellar evolutionary models. Figure 1 shows that the grid of C-rich stars dynamic model atmospheres fits the photometric data well at all wavelengths ($\chi^2 = 1.51$). We refer to a future paper (Rau *et al.* in prep.) for details on this analysis.

4. Future plans

Further investigation is upcoming concerning the fit to the interferometric data. We also need further studies to interpret how the reflection and dissipation of Alfvén waves (see also Airapetian *et al.* 2010) could explain the interferometric variability of V Oph. Radiative transfer modeling studies with RADMC3D, will deepen our knowledge on the causes of interferometric variability, and on the dust stratification around the star. Our team is planning to observe V Oph also with the second generation VLTI instrument in the *LM*- and *N*-bands: MATISSE. These observations will help to disentangle the interferometric variability, if due to pulsation as speculated by Ohnaka *et al.* (2007), or to yet unobserved surface structures.

References

Airapetian, V., Carpenter, K. G., & Ofman, L. 2010, ApJ, 723, 1210
Eriksson, K., Nowotny, W., Höfner, S., Aringer, B., & Wachter, A. 2014, *A&A*, 566, A95
Höfner, S., Gautschy-Loidl, R., Aringer, B., & Jørgensen, U. G. 2003, *A&A*, 399, 589
Höfner, S., Bladh, S., Aringer, B., & Ahuja, R. 2016, *A&A*, 594,108
Hron, J., Loidl, R., Höfner, S., *et al.* 1998, *A&A*, 335, L69
Rau, G. 2016, PhD thesis, University of Vienna, doi:10.5281/zenodo.1407903
Rau, G., Hron, J., Paladini, C., *et al.* 2017, *A&A*, 600, A92
Rau, G., Paladini, C., Hron, J., *et al.* 2015, *A&A*, 583, A106
Wittkowski, M., Rau, G., Chiavassa, A., *et al.* 2018, *A&A*, 613, L7
Ohnaka, K., Driebe, T., Weigelt, G., & Wittkowski, M. 2007, *A&A*, 466, 1099

Why Galaxies Care About AGB Stars:
A Continuing Challenge through Cosmic Time
Proceedings IAU Symposium No. 343, 2019
F. Kerschbaum, M. Groenewegen & H. Olofsson, eds.

doi:10.1017/S174392131800786X

Circumstellar molecular maser emission of AGB and post-AGB stars

Georgij Rudnitskij[1], Nuriya Ashimbaeva[1], Pierre Colom[2], Evgeny Lekht[1], Mikhail Pashchenko[1] and Alexander Tolmachev[3]

[1]Lomonosov Moscow State University, Sternberg Astronomical Institute, Moscow, 119234 Russia
email: gmr@sai.msu.ru

[2]LESIA, Observatoire de Paris-Meudon, UPMC, Université Paris–Diderot, 92195 Meudon CEDEX, France

[3]Pushchino Radio Astronomy Observatory, Astrospace Center, Lebedev Institute of Physics, Russian Academy of Sciences, Pushchino, 142290 Russia

Abstract. Results of long-term studies of circumstellar molecular maser emission of late-type giant and supergiant variable stars are reported. In the 1.35-cm H_2O line, the peak flux density correlates with the optical brightness lagging behind it by 0.3–0.4 P (P is the stellar period). "Superperiods" of 10 to 15 P are visible in several stars, demonstrated as high maxima in the visible light curve and associated flares in the H_2O maser line. In the 18-cm OH lines, full polarization of the maser emission has been measured. Variable Zeeman patterns suggesting a changing magnetic field of a few milligauss have been detected.

Keywords. stars: late-type, stars: winds, outflows, stars: mass loss, stars: magnetic fields, radio lines: stars.

Results of long-term studies of circumstellar molecular maser emission of late-type giant and supergiant stars are reported. The observations have been carried out during several decades in the H_2O line at a wavelength of 1.35 cm (22-meter telescope in Pushchino, Russia) and in the hydroxyl (OH) lines at 18 cm (Nançay radio telescope, France). A sample of ∼70 AGB long-period variable stars has been monitored in the 1.35-cm H_2O line in 1980–2018. It includes Mira-type stars (U Ori, RS Vir, U Her, Y Cas, R Cas, ...) and semiregular variables (R Crt, RT Vir, W Hya, VX Sgr, ...).

We have traced H_2O maser variations and found them to follow the optical brightness variations of the stars with a time delay of 0.3–0.4 P. Some strong H_2O flares were observed, which occur every 10 to 15 periods, probably due to long-term changes in the mass-loss rate; good examples are U Ori (Rudnitskij *et al.* 2000), R Cas (Pashchenko *et al.* 2004), and RS Vir (Lekht *et al.* 2001). In the star W Hya, the H_2O peak flux density reached several thousand janskys, while on the average it did not exceed 50–100 Jy (Rudnitskij *et al.* 1999). We consider the model of the flares with shock excitation of the H_2O maser in combination with stellar radiation pumping.

A sample of 70 AGB stars, mostly overlapping with the H_2O line program, was observed in the 18-cm OH lines. For 53 of them, the emission was detected in at least one of three OH lines (1612, 1665, or 1667 MHz). Appreciable polarization (nonzero Stokes parameters) was present in 41 stars (Rudnitskij *et al.* 2010). Circular and linear polarization of the maser emission was measured, yielding all four Stokes parameters. Features probably due to Zeeman splitting were detected in the OH line profiles of several stars. Estimated magnetic-field strengths in the maser sources are a few milligauss. In particular, we

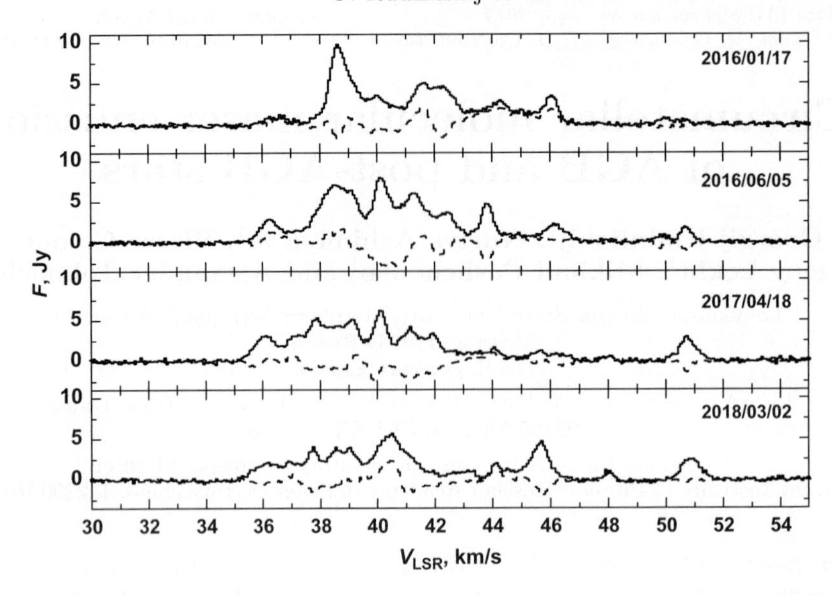

Figure 1. Variation of the semiregular variable star HU Puppis in the 1612-MHz OH line in 2016–2018. Solid curves: total flux density; dashed curves: Stokes parameter V (difference of flux densities in the right- and left-hand circular polarizations).

discuss the data on the 1612-MHz OH emission of the semiregular variable star HU Pup displaying maser emission in three ground-state OH lines 1612, 1665, and 1667 MHz. Figure 1 presents radial-velocity profiles of the 1612-MHz satellite OH line of this star measured on the Nançay radio telescope in 2016–2018. Earlier data on the maser emission of HU Pup were published by Colom *et al.* (2014). The total flux density profile (I) consisting of several peaks spread in velocity between 32 and 52 km/s demonstrates rather small changes, in contrast to the V Stokes parameter shown with dashed curves. Some alternating positive/negative V structures appearing occasionally in the V profile can be interpreted as Zeeman splitting in a few-milligauss magnetic filed in the OH masering region of the circumstellar envelope. The behaviour of the V profile suggests important variations in the pattern of the star's magnetic field producing polarization of the OH emission.

The complete version of this poster in the electronic form can be found at: http://comet.sai.msu.ru/~gmr/IAUS343_23554.pdf

References

Colom, P., Lekht, E.E., Pashchenko, M.I., Rudnitskij, G.M., & Tolmachev, A.M. 2014, *Astron. Lett.*, 40, 212

Lekht, E.E., Mendoza-Torres, J.E., Rudnitskij, G.M., & Tolmachev, A.M. 2001, *A&A*, 376, 928

Pashchenko, M.I., & Rudnitskij, G.M. 2004, *Astron. Rep.*, 48, 380

Rudnitskij, G.M., Lekht E.E., & Berulis, I.I. 1999, *Astron. Lett.*, 25, 398

Rudnitskij, G.M., Lekht, E.E., Mendoza-Torres, J.E., Pashchenko, M.I., & Berulis, I.I. 2000, *A&AS*, 146, 385

Rudnitskij, G.M., Pashchenko, M.I., & Colom, P. 2010, *Astron. Rep.*, 54, 400

Why Galaxies Care About AGB Stars: A Continuing Challenge through Cosmic Time
Proceedings IAU Symposium No. 343, 2019
F. Kerschbaum, M. Groenewegen & H. Olofsson, eds.

doi:10.1017/S1743921318007172

High angular-resolution infrared imaging and spectra of the carbon-rich AGB star V Hya

Raghvendra Sahai[1], Jayadev Rajagopal[2], Kenneth Hinkle[2], Richard Joyce[2] and Mark Morris[3]

[1]Jet Propulsion Laboratory, MS 183-900, California Institue of Technology, Pasadena, CA 91109, USA
email: raghvendra.sahai@jpl.nasa.gov

[2]National Optical Astronomy Observatory, 950 N. Cherry Ave, Tucson, AZ 85719, USA

[3]Department of Physics and Astronomy, University of California, Los Angeles, CA 90095, USA

Abstract. The carbon-rich AGB star V Hya is believed to be in the very brief transition phase between the AGB and a planetary nebula (PN). Using HST/STIS, we previously found a high-velocity ($> 200\,\mathrm{kms}^{-1}$) jet or blob of gas ejected only a few years ago from near (< 0.3 arcsec or 150 AU) the star (Sahai *et al.* 2003, Sahai *et al.* 2016). From multi-epoch high-resolution spectroscopy we found time-variable high-velocity absorption features in the CO 4.6 μm vibration-rotation lines of V Hya (Sahai *et al.* 2009). Modeling shows that these are produced in compact clumps of outflowing gas with significant radial temperature gradients consistent with strong shocks. Here, we present very high resolution ($\sim$100 milliarcsecond) imaging of the central region of V Hya using the coronagraphic mode of the Gemini Planet Imager (GPI) in the 1 μm band and spectral-spatial imaging of 4.6 μm CO 1-0 transitions using the Phoenix spectrometer. We report the detection of a compact central dust disk from GPI, and molecular emission from the Phoenix observations at relatively larger scales. We discuss models for the central structures in V Hya, in particular disks and outflows, using these and complementary images in the optical and radio.

Keywords. stars: AGB and post-AGB, instrumentation: High angular resolution, ISM: planetary nebulae: general

1. Mid-IR Imaging and Spectra

Fig. 1 shows Y band coronagraphic images of V Hya using GPI: the total intensity, polarized intensity and polarized fraction. The white circle marks the extent of a 0.57" (285 AU at a distance of 478 pc to V Hya) disk inferred by Townes *et al.* (2011) through mid-IR interferometry. The disk is clearly detected in the GPI polarized (scattered) light image. In our model the star is located at the center of a flared thick disk structure tilted such that its west side opens towards us (Fig. 2). In the GPI image, we are seeing light scattered from the surface of this disk, but only (the more distant) half of this disk is seen, from where the starlight is back-scattered towards us. As the disk is optically thick, and because it has a highly flared geometry, the illuminated part of its near side is not exposed to us directly. This explains why the disk-light is seen only on one side of the star.

Fig. 2 shows a schematic for the inner region of V Hya, with an equatorial, slowly expanding CO region, compact central dust disk and fast, clumpy bipolar outflows. The fast "bullets" close to the star have been detected in SII with HST/STIS (Sahai *et al.*

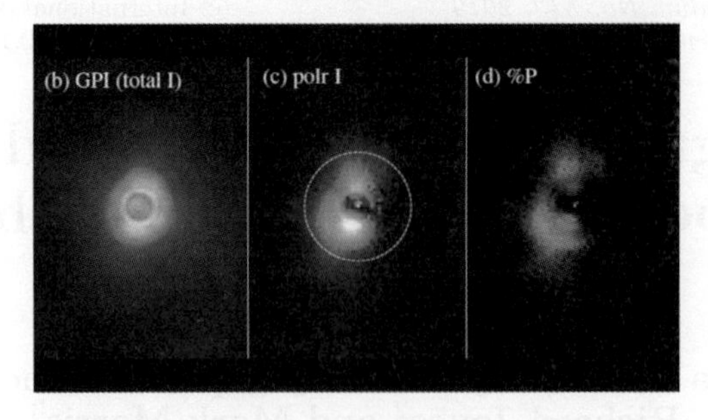

Figure 1. Y band coronagraphic images of V Hya with GPI. North is up, east to the left.

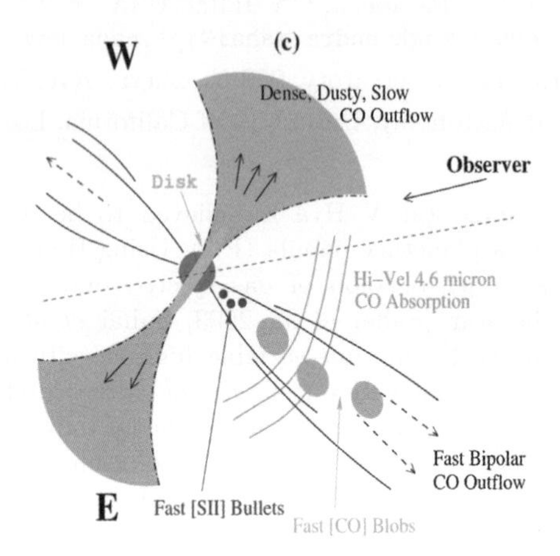

Figure 2. Schematic view of the central structures and outflows.

2003, Sahai *et al.* 2016, Scibelli *et al.* 2018) and the clumpy bipolar CO outflow with the SMA and IRAM 30m (Hirano *et al.* 2004, Kahane *et al.* 1995). For scale, the "bullets" are < 0.3" or 150 AU offset from the star, and the dense slow outflow (grey) is about 10" in extent on each side (not shown in full).

We detect both slow and fast outflows in spectra of the 4.6 μm CO band using Phoenix at Gemini with the slit offset by 0.5" south from the star and after subtracting the stellar continuum. An average of the 1-0 R1, R2, and R3 lines in emission shows a low velocity peak consistent with the normal, slowly expanding circumstellar envelope (CSE) and a high velocity peak at $\sim$ -110 kms^{-1}. This flow is seen to varying degrees in other offsets as well.

2. Summary

We show here high resolution coronagraphic images and spectra of the carbon-rich AGB star V Hya. These are part of a multiwavelength study of V Hya that builds a picture of the CSE of this star caught in transition from the AGB phase to a Planetary Nebula. There is now convincing evidence that V Hya has a slowly expanding equatorial torus (in CO and dust), a compact central disk (dust) and fast blobs or "bullets" (seen in SII and FeII) close to the star aligned with a fast, clumpy, bipolar flow seen at larger distances in CO. Sahai *et al.* (2016) propose an eccentric, 8.5 year period binary system as the engine for the outflow. Periastron passages by the companion through the outer envelope of the primary drive instabilities in the companion accretion disk and eject the collimated "bullet" outflow. These outflows are seen in other similar systems, and the V

Hya model is likely a common path from a symmetric AGB CSE to a bipolar (perhaps even multipolar) Planetary Nebula.

References

Sahai, R., Scibelli, S., & Morris, M. 2016, *ApJ*, 827, 92
Shai, R., Sugerman, Ben, E.K., Hinkle, K. 2009, *ApJ*, 699, 1015
Sahai, R., Morris, M., Knapp, G.R., Young, K., Barnbaum, C. 2003, *Nature*, 426, 261
Scibelli, S., Sahai, R., & Morris, M. 2018, *ApJ*, submitted
Hirano, N., Shinnaga, H., Dinh-V-Trung, *et al.* 2004 *ApJ*, 616, L43
Townes, C.H., Wishnow, E.H., & Ravi, V. 2011, *PASP*, 123, 1370
Kahane, C., Audinos, P., Barnbaum, C., & Morris, M. 1995, *A&A*, 314, 871

Why Galaxies Care About AGB Stars: A Continuing Challenge through Cosmic Time
Proceedings IAU Symposium No. 343, 2019
F. Kerschbaum, M. Groenewegen & H. Olofsson, eds.

doi:10.1017/S1743921318007068

Infrared Studies of the Variability and Mass Loss of Some of the Dustiest Asymptotic Giant Branch Stars in the Magellanic Clouds

B. Sargent[1], S. Srinivasan[2], M. Boyer[1], M. Feast[3,4], P. Whitelock[3,4], M. Marengo[5], M. A. T. Groenewegen[6], M. Meixner[1], J. L. Hora[7] and M. Otsuka[2,8]

[1]Space Telescope Science Institute, 3700 San Martin Drive, Baltimore, MD 21218, USA; email: sargent@stsci.edu

[2]Academia Sinica, Institute of Astronomy and Astrophysics, 11F of AS/NTU Astronomy-Mathematics Building, No. 1, Sec. 4, Roosevelt Rd., Taipei 10617, Taiwan, R.O.C.

[3]South African Astronomical Observatory, P.O. Box 9, Observatory, 7935, South Africa

[4]University of Cape Town, Private Bag X3, Rondebosch 7701, Republic of South Africa

[5]Iowa State University, A313E Zaffarano, Ames, IA 50010, USA

[6]Koninklijke Sterrenwacht van België, Ringlaan 3, B-1180 Brussels, Belgium

[7]Harvard-Smithsonian Center for Astrophysics, 60 Garden Street, MS-65, Cambridge, MA 02138-1516, USA

[8]Kyoto University, Okayama Observatory, Honjo, Kamogata, Asakuchi, Okayama, 719-0232, Japan

Abstract. The asymptotic giant branch (AGB) stars with the reddest colors have the largest amounts of circumstellar dust. AGB stars vary in their brightness, and studies show that the reddest AGB stars tend to have longer periods than other AGB stars and are more likely to be fundamental mode pulsators than other AGB stars. Such stars are difficult to study, as they are often not detected at optical wavelengths. Therefore, they must be observed at infrared wavelengths. Using the *Spitzer Space Telescope*, we have observed a sample of very dusty AGB stars in the Large Magellanic Cloud (LMC) and Small Magellanic Cloud (SMC) over Cycles 9 through 12 during the Warm Spitzer mission. For each cycle's program, we typically observed a set of AGB stars at both 3.6 and 4.5 μm wavelength approximately monthly for most of a year. We present results from our analysis of the data from these programs.

Keywords. infrared: stars, stars: AGB and post-AGB

The asymptotic giant branch (AGB) phase is the phase of a star's life when the star expels its own material, and dust forms in the outflow (Sloan *et al.* 2016). The Large Magellanic Cloud (LMC; 50 kpc; Schaefer *et al.* 2008) and Small Magellanic Cloud (SMC; 60 kpc; Szewczyk *et al.* 2009) are nearby galaxies with many AGB stars and low foreground extinction. Their mean metallicities are 0.5×solar (Dufour *et al.* 1982) and 0.2×solar (Asplund *et al.* 2004), respectively. Thus, they are ideal laboratories to study AGB stars.

Spitzer-IRAC measurements were obtained about once per month in Cycles 9-12. Cycle 9 (pid 90219) targets were from Gruendl *et al.* (2008) and Vijh *et al.* (2009), and Cycle 10 (pid 10154) targets were Gruendl *et al.* (2008) and SMC follow-up targets. Cycle 11 (pid 11163) targets were Far-infrared (FIR) bright SMC AGB stars from Polsdofer *et al.* (2015), 1 SMC globular cluster AGB star (Tanabé *et al.* 1997; Nishida *et al.* 2000), and

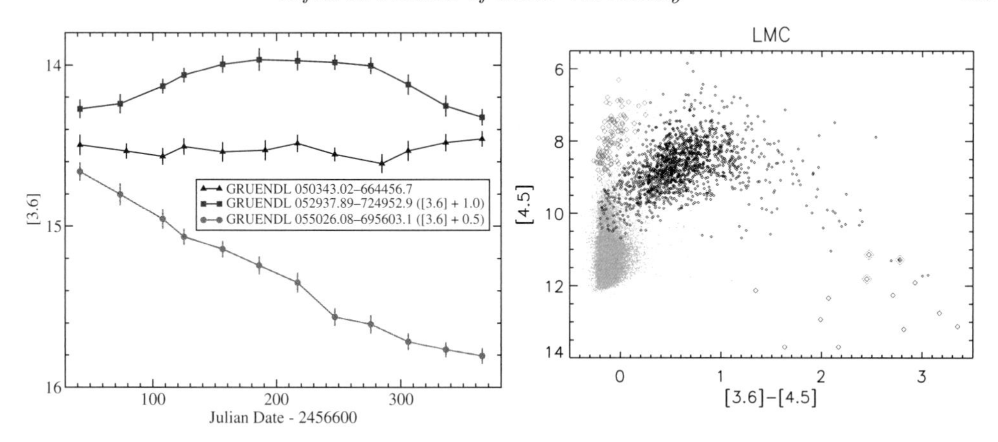

Figure 1. **Left**, light curves for 3 sources from the Gruendl *et al.* (2008) sample. **Right**, a CMD similar to a plot shown by Sloan *et al.* (2016). Green, cyan, black, and magenta points are oxygen-rich (O-rich), carbon-rich (C-rich), and extreme AGB candidate stars (Srinivasan *et al.* 2009), and red supergiants (Bonanos *et al.* 2009). The 3 red and blue points are the Gruendl *et al.* (2008) sources showing variability, while the 10 blue points are Gruendl *et al.* (2008) sources showing little to no significant variability.

20 SMC AGB stars from Ita (2005). Cycle 12 (pid 12097) targets included AGB and RSG stars detected by Herschel (Jones *et al.* 2015), an SMC OH/IR candidate (Polsdofer *et al.* 2015), optically-obscured AGB stars missing from optical surveys (Gruendl *et al.* 2008; Riebel *et al.* 2010; Srinivasan *et al.* 2016) and others from Vijh *et al.* (2009).

We expected most of our sample would show significant variability - high amplitudes of variation and longer periods. For the most part, this was true, except for the Gruendl *et al.* (2008) extremely red object (ERO) sample, constituting the reddest and dustiest AGB stars in the LMC. This sample was hardly variable at all (though see Figure 1). This, and their positions on the color-magnitude diagram (CMD; Figure 1), suggest they may be near or at the end of the AGB phase of their lives (see also Sloan *et al.* 2016). Future plans include monitoring a larger sample of the reddest extreme AGB candidate population, to look for additional non-variable extremely dusty stars. This work is based in part on observations made with the *Spitzer Space Telescope*, which is operated by the Jet Propulsion Laboratory, California Institute of Technology under NASA contract 1407. B.A.S. acknowledges funding from Spitzer-JPL contract RSA #1561703.

References

Asplund, M., Grevesse, N., and Sauval, A. J., *et al.* 2004, *A&A*, 417, 751
Bonanos, A. Z., Massa, D. L., Sewilo, M., *et al.* 2009, *AJ*, 138, 1003
Dufour, R. J., Shields, G. A., & Talbot, Jr., R. J. 1982, *ApJ*, 252, 461
Gruendl, R. A., Chu, Y.-H., Seale, J. P., *et al.* 2008, *ApJ*, 688, L9
Ita, Y. 2005, *The Astronomical Herald*, 98, 163
Jones, O., Meixner, M., Sargent, B. A., *et al.* 2015, *ApJ*, 811, 145
Nishida, S., Tanabé, T., Nakada, Y., *et al.* 2000, *MNRAS*, 313, 136
Polsdofer, E., Seale, J., Sewiło, M., *et al.* 2015, *AJ*, 149, 78
Riebel, D., Meixner, M., Fraser, O., *et al.* 2010, *ApJ*, 723, 1195
Schaefer, B. E., Simon, M., Prato, L., & Barman, T. 2008, *AJ*, 135, 112
Sloan, G. C., Kraemer, K. E., McDonald, I., *et al.* 2016, *ApJ*, 826, 44
Srinivasan, S., Meixner, M., Leitherer, C., *et al.* 2009, *AJ*, 137, 4810
Srinivasan, S., Boyer, M. L., Kemper, F., *et al.* 2016, *MNRAS* 457, 2814
Szewczyk, O., Pietrzyski, G., Gieren, W., *et al.* 2009, *AJ*, 138, 1661
Tanabé, T., Nishida, S., Matsumoto, S., *et al.* 1997, *Nature*, 385, 509
Vijh, U., Meixner, M., Babler, B., *et al.* 2009, *AJ*, 137, 3139

Why Galaxies Care About AGB Stars:
A Continuing Challenge through Cosmic Time
Proceedings IAU Symposium No. 343, 2019
F. Kerschbaum, M. Groenewegen & H. Olofsson, eds.

doi:10.1017/S1743921318007421

Observing the mass-loss of nearby red supergiants through high-contrast imaging

Peter Scicluna[1], **R. Siebenmorgen**[2], **J. A. D. L. Blommaert**[3], **F. Kemper**[1,2], **R. Wesson**[4] **and S. Wolf**[5]

[1]Academia Sinica Institute of Astronomy and Astrophysics (ASIAA), 11F Astronomy-Mathematics Building, No. 1, Sec. 4, Roosevelt Road, Taipei 10617, Taiwan
email: peterscicluna@asiaa.sinica.edu.tw

[2]ESO, Karl-Schwarzschild-Str. 2, 85748 Garching b. München, Germany

[3]Astronomy and Astrophysics Research Group, Dept. of Physics and Astrophysics, V.U. Brussel, Pleinlaan 2, 1050, Brussels, Belgium

[4]Dept. of Physics & Astronomy, University College London, Gower Street, London WC1E 6BT, UK

[5]ITAP, Universität zu Kiel, Leibnizstr. 15, 24118 Kiel, Germany

Abstract. We present observations of nearby red supergiants with SPHERE.

Keywords. stars: mass loss, circumstellar matter, stars: winds, outflows

---

1. Introduction

After massive stars ($M > 8\,M_{\odot}$) leave the main sequence, they undergo periods of enhanced mass loss before exploding as supernovae. This mass loss is a key factor in determining the further evolution of the stars (Georgy *et al.* 2013, Smith 2014, Meynet *et al.* 2015, Georgy & Ekström 2015). For stars with initial masses $\lesssim 30\,M_{\odot}$, a large fraction of the mass loss occurs during the RSG phase. The mechanisms driving mass-loss in this phase remain a matter of debate (van Loon *et al.* 2005, Harper *et al.* 2009), in particular the origins of mass-loss asymmetries, variability and eruptions. Answering outstanding questions regarding mass-loss from evolved massive stars requires a systematic approach, in which a large sample of supernova progenitors are homogeneously analysed. Hence, we are conducting a high-contrast imaging and polarimetry survey of nearby southern and equatorial evolved massive stars, primarily red supergiants, exploiting the capabilities of SPHERE. Some initial results of this survey are shown below.

2. Results

Optical polarimetry with ZIMPOL allows high-contrast imaging of extended emission, particularly dust scattering. Two examples of ZIMPOL observations are shown here. VY CMa (see Fig. 1) has clear extended emission in V- and I-bands which is strongly polarised. Fitting the polarisation data with a dust model can reveal the grain size; we find an average size in one region of $\langle a \rangle = 0.56\,\mu$m, and $\langle a \rangle = 0.42\,\mu$m in a second region. This provides direct confirmation of previous suggestions of $\sim\mu$m size grains in RSGs.

On the other hand, VX Sgr is less extended in intensity (Fig. 2). However, clear extension is seen in polarisation, with a centrosymmetric pattern characteristic of a shell or torus. Like VY CMa, the polarisation is higher in I band, suggesting similar grain sizes.

As RSGs emit the bulk of their radiation in the near-IR, sub-micron dust grains can receive a significant amount of radiation pressure by scattering the stellar emission rather

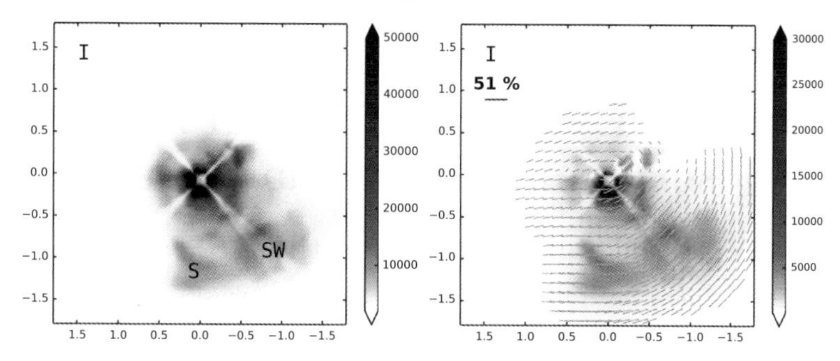

Figure 1. ZIMPOL observations of VY CMa based on Norris *et al.* (2012), with offsets in arc seconds and intensity scale in arbitrary units. *Left*: Total intensity. The locations of the South Knot and Southwest Clump are marked with 'S' and 'SW' respectively. *Right*: Polarised intensity. The overlaid vectors show the polarisation fraction and direction.

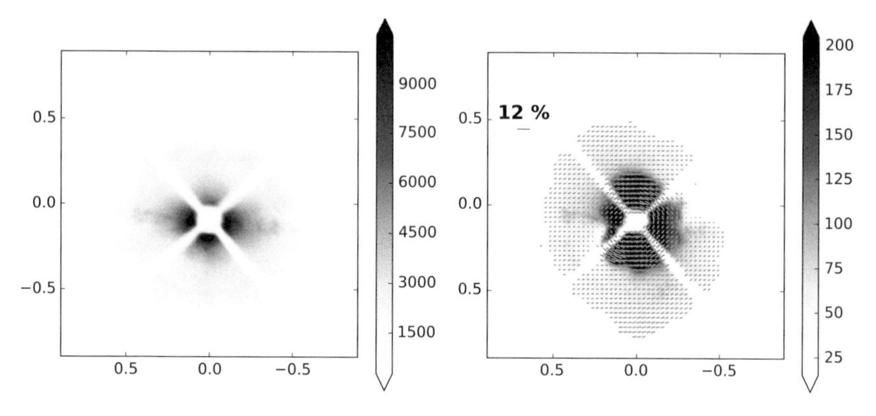

Figure 2. As above, for VX Sgr in the continuum 820 μm filter. The absolute orientation of the instrument has not been corrected for, but both images are given in the same orientation.

than absorbing it (Höfner 2008, Bladh & Höfner 2012). This has been found to be an effective mechanism for driving mass loss in oxygen-rich AGB stars (Norris *et al.* 2012).

3. Conclusions

SPHERE is a powerful tool for detecting extended dust-scattered light around bright evolved stars, including red supergiants. The full sample contains a further 18 red supergiants - the analysis of this data is ongoing, and will place strong constraints on the detectability of extended emission in sources with lower mass-loss rates. In addition, this dataset will reveal distant main-sequence companions over a large mass range.

References

Bladh, S., Höfner, S. 2012, *A&A*, 546, A76
Georgy, C. *et al.* 2013, *EAS Publications Series*, 60, 43
Georgy, C., Ekström, S. 2015, *EAS Publications Series*, 71, 41
Harper, G. M. *et al.* 2009, *ApJ*, 701, 1464
Höfner, S. 2008, *A&A*, 491, L1
Meynet, G. *et al.* 2015, *A&A*, 575, A60
Norris, B. R. M., *et al.* 2012, *Nature*, 484, 220
Scicluna, P. *et al.* 2015, *A&A*, 584, L10
Smith, N. 2014, *ARA&A*, 52, 487
van Loon, J. T., Cioni, M.-R. L., Zijlstra, A. A., Loup, C. 2005, *A&A*, 438, 273

Why Galaxies Care About AGB Stars:
A Continuing Challenge through Cosmic Time
Proceedings IAU Symposium No. 343, 2019
F. Kerschbaum, M. Groenewegen & H. Olofsson, eds.

doi:10.1017/S174392131800741X

The Nearby Evolved Stars Survey: Project description and initial results

Peter Scicluna, on behalf of the *NESS* team

Academia Sinica Institute of Astronomy and Astrophysics (ASIAA), 11F
Astronomy-Mathematics Building, No. 1, Sec. 4, Roosevelt Road, Taipei 10617, Taiwan
email: `peterscicluna@asiaa.sinica.edu.tw`

Abstract. The Nearby Evolved Stars Survey aims to observe over 400 evolved stars within 2 kpc, to determine why, and how much, *our Galaxy* cares about AGB stars. This contribution presents a brief introduction to the survey and data. NESS is an open project. Anyone is welcome to get involved and we aim to make as much data and code available to the community as possible.

Keywords. stars: AGB and post-AGB, stars: mass loss, circumstellar matter, stars: winds, outflows, surveys

1. Introduction

The Nearby Evolved Stars Survey (NESS) is a survey of galactic evolved stars in the sub-mm using the JCMT and APEX. We aim to observe over 400 evolved stars within 2 kpc of the Sun (see Fig. 1). Our primary objective is to determine the total gas and dust return to local ISM, but we also expect to explore global gas-to-dust ratios for individual sources, constraining sub-mm dust properties and the incidence of cold dust, and examine Galactic evolved stars as a population, amongst other goals.

2. Initial results

NESS data is being reduced and analysed by automated pipelines, some preliminary results of which are shown here (see Fig. 2 & 3). The pipelines automatically retrieve and reduce JCMT data, to give uniform data products. They also perform some simple analysis, for example fitting a simple function to the CO spectra to locate the line, and azimuthally-averaging SCUBA-2 observations to search for extended emission.

Further results can be seen in Dharmawardena *et al.* (this volume) and Wallström *et al.* (this volume).

3. Open Science

NESS is an open project - anyone is welcome to get involved. In addition, we plan to release raw, reduced and advanced products to the community through a catalogue hosted on the NESS website. We also aim to make all the code we write available as well as the data, via Github, so that our results are reproducible and the community can use the tools we develop. This includes automated pipelines, radiative-transfer fitting tools, machine-learning classifiers, and even the code used to generate plots.

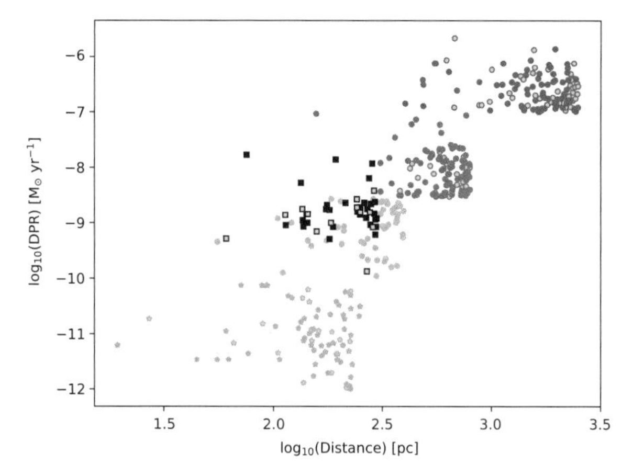

Figure 1. The distribution of NESS sources in distance–dust-production rate space. Sources with higher DPRs are selected out to larger distances as they are assumed to be intrinsically larger and brighter sources, and hence easier to detect. We aim to have a roughly similar number of sources in each of the 4 bins of DPR (each shown in a different colour). A smaller number of bright, nearby sources are selected for mapping in detail (black squares). Filled symbols indicate sources being observed with the JCMT, and outline symbols are being observed with APEX.

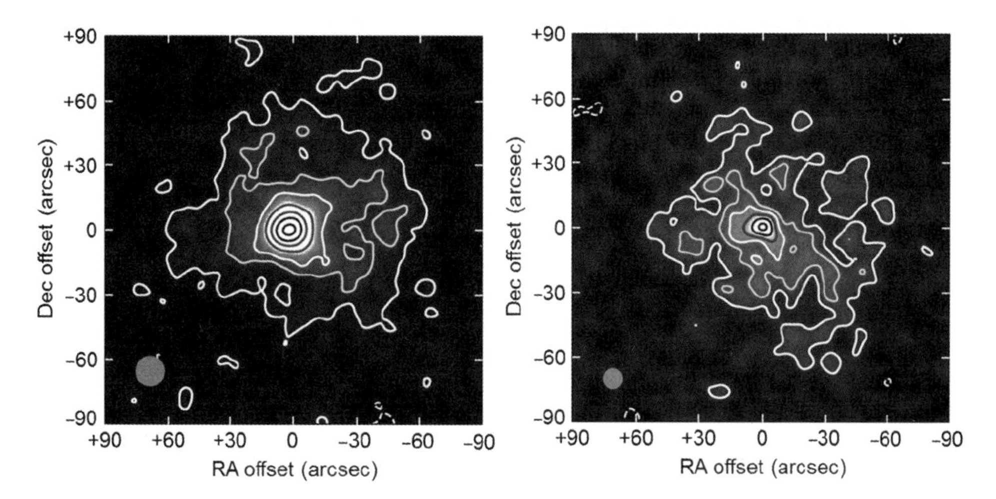

Figure 2. Example SCUBA-2 observations of T Cas (*left*: 850 μm, *right*: 450 μm). In both cases, the lowest contour is 3σ, and each successive contour is $+4\sigma$ thereafter.

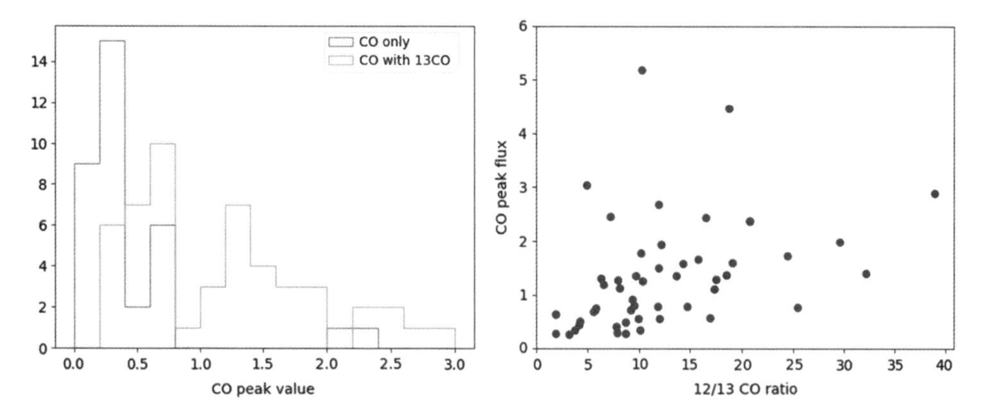

Figure 3. Distribution of pipeline-reduced CO(2–1) fluxes (left) and 12/13 CO ratios (right).

Why Galaxies Care About AGB Stars: A Continuing Challenge through Cosmic Time
Proceedings IAU Symposium No. 343, 2019
F. Kerschbaum, M. Groenewegen & H. Olofsson, eds.

doi:10.1017/S1743921318006944

Modelling gas and dust around carbon stars in the Large Magellanic Cloud

Sundar Srinivasan[1]†, I.-K. Chen[2], P. Scicluna[1], J. Cami[3] and F. Kemper[1],

[1]Academia Sinica Institute of Astronomy & Astrophysics (ASIAA)
11F Astronomy-Mathematics Building, No. 1, Sec. 4, Roosevelt Road, Taipei 10617, Taiwan
[2]College of Arts & Sciences, Cornell University, KG17 Klarman Hall, Ithaca, NY 14853, USA
[3] Department of Physics & Astronomy, The University of Western Ontario, London, ON N6A 3K7, Canada
email: s.srinivasan@irya.unam.mx

Abstract. In order to investigate the effect of dust production on the molecular absorption, we model the dust continuum and the 7.5 and 13.7 μm acetylene absorption features in the *Spitzer* IRS spectra of 148 carbon stars in the Large Magellanic Cloud (LMC). Our preliminary investigation does not find a strong correlation between the dust-production rate and the column density of acetylene for the LMC sample. However, we will construct more models at high optical depths and probe a larger range of dust properties for more robust results.

Keywords. stars: AGB and post-AGB, stars: mass loss, (stars:) circumstellar matter, radiative transfer

1. Background

Acetylene (C_2H_2) is among the most abundant molecules formed around C-rich asymptotic giant branch (AGB) stars (Olofsson 2005). It is a building block for more complex carbonaceous molecules, and may be an important precursor for dust formation. In the mid-infrared, the most obvious signatures of C_2H_2 are its absorption bands at $\sim$7 and $\sim$14 μm. Observations indicate that the absorption features persist in heavily dust-obscured carbon stars, the C_2H_2 abundance is independent of metallicity, and that the absorption strength correlates with the mass-loss rate (*e.g.*, Yamamura *et al.* 1999; Matsuura *et al.* 2006). A self-consistent radiative transfer model for the circumstellar gas and dust is required to fully address this problem. In this work, we present an initial step towards such modelling, by investigating correlations between the DPR and the gas column density.

2. Data and analysis

We model the 148 LMC carbon-star spectra obtained by the SAGE-Spec program (Kemper *et al.* 2010; Woods *et al.* 2011; Jones *et al.* 2017). We use the GRAMS carbon-star grid (Srinivasan *et al.* 2011; 12 243 models) to model the dust, and the latest HITRAN line lists for C_2H_2 (Gordon *et al.* 2017; models for 615 pairs of column density and excitation temperature values) to generate the normalised flux from the C_2H_2 slab, using the method described in Cami *et al.* (2010). By convolving the synthetic spectra from the gas and dust, we generate over 7.5 million models. We then compute χ^2 fits to each

† Present address: Instituto de Radioastronomía y Astrofísica, Universidad Nacional Autonóma de México, Antigua Carretera a Pátzcuaro #8701, Ex-Hda. San José de La Huerta, Morelia, Michoacán, 58089, México.

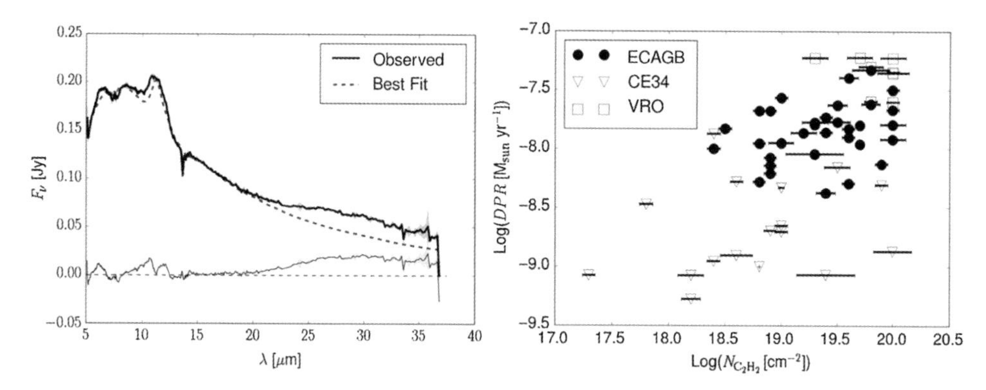

Figure 1. *Left*: best-fit model (thick dashed) to the IRS spectrum of IRAS 04473−6829 (thick solid), along with the fit residual (thin solid). *Right*: DPR as a function of C_2H_2 model column density for optically thin carbon stars (ECAGB; circles), stars with strong SiC emission (CE34; triangles), and for very red objects (VROs; squares).

observed spectrum. The GRAMS models only include amorphous carbon and SiC dust species, and do not account for the 30 μm feature, so we mask out this wavelength range when computing the fits.

3. Results and discussion

Fig. 1 compares the SAGE-Spec spectrum for the LMC carbon star IRAS 04473−6829 with its best-fit model. The molecular features are reproduced, but the shape of the SiC feature isn't. As the long-wavelength edge of this feature blends with the 13.7 μm C_2H_2 feature, it is important that we include a proper treatment in our modelling. The GRAMS grid consists of a dust mixture with a fixed SiC fraction (10% by mass), which affects the fit quality in many cases. We will vary the composition in future modelling.

Fig. 1 also shows the variation of DPR as a function of the best-fit C_2H_2 column density for three types of carbon stars – those lacking strong dust features (ECAGB), those with strong SiC emission (CE34), and the very red objects (VROs). The plot shows no strong trend in the DPR. However, this is partly due to the fact that we are only able to fit a small number of VROs, owing to the sparsity of very optically thick GRAMS models.

In the future, we will extend the GRAMS grid to higher optical depths and also explore a larger range of dust properties (including the 30 μm feature) in order to improve our analysis.

References

Cami, J., van Malderen R., Markwick, A. J. 2010, *ApJS*, 187, 409
Gordon, I. E., Rothman, L. S., Hill, C. *et al.* 2017, *JQSRT*, 203, 3
Jones, O. C., Woods, P. M., Kemper, F. *et al.* 2017, *MNRAS*, 420, 3250
Kemper, F., Woods, Paul M., Antoniou, V. *et al.*, 2010, *PASP*, 122, 683
Matsuura, M., Wood, P. R., Sloan, G. C. *et al.* 2006, *MNRAS*, 371, 415
Olofsson, H. 2005, in: Lis, D. C., Blake, G. A., Herbst, E. (eds.), *Astrochemistry: Recent Successes and Current Challenges*, Proc. IAU Symposium No. 231, p. 499
Srinivasan, S., Sargent, B. A., Meixner, M. 2011, *A&A*, 532, 54
Woods, Paul M., Oliveira, J. M., Kemper, F. *et al.* 2011, *MNRAS*, 411, 1597
Yamamura, I., de Jong, T., Waters, L. B. F. M. *et al.* 1999, in: Le Bertre, T, Lebre, A., Waelkens, C. (eds.), *Asymptotic Giant Branch Stars*, Proc. IAU Symposium No. 191, p. 267

Why Galaxies Care About AGB Stars:
A Continuing Challenge through Cosmic Time
Proceedings IAU Symposium No. 343, 2019
F. Kerschbaum, M. Groenewegen & H. Olofsson, eds.

doi:10.1017/S1743921318006361

Modelling dust around Nearby Evolved Stars Survey (NESS) Targets

Sundar Srinivasan†, T. Dharmawardena, F. Kemper, P. Scicluna and The NESS Collaboration

Academia Sinica Institute of Astronomy & Astrophysics (ASIAA)
11F Astronomy-Mathematics Building, No. 1, Sec. 4, Roosevelt Road, Taipei 10617, Taiwan
email: s.srinivasan@irya.unam.mx

Abstract. We present radiative transfer modelling of the dust around U Ant, a well-studied detached-shell source. U Ant is among the >400 sources targeted by the Nearby Evolved Stars Survey (NESS; PI: P. Scicluna), and the procedure used to model this source will be applied to the rest of the AGB sample in NESS.

Keywords. stars: AGB and post-AGB, stars: mass loss, (stars:) circumstellar matter, radiative transfer

1. Background

The Nearby Evolved Stars Survey (NESS; Scicluna *et al. in prep.*) is a large collaboration targeting a volume-limited ($d < 2$ kpc) sample of evolved stars in both CO line emission as well as continuum using multiple sub-mm facilities. The large sample size (over 400 objects) will allow NESS to derive robust estimates of the stellar gas and dust return to the Galactic interstellar medium. A pilot study for NESS imaged 14 AGB stars and one RSG with the SCUBA-2 instrument on the James Clerk Maxwell Telescope (JCMT) at 450 and 850 μm. Dust density profiles derived from these data showed a non-constant mass loss in all cases (Dharmawardena *et al.* 2018). In addition, multiple sources showed detached shells and asymmetric dust distributions. A systematic analysis of the NESS data requires us to develop a general framework for the radiative transfer (RT) modelling. We begin by modelling dust around the C-rich AGB star U Ant, a well-known detached shell source (*e.g.*, González Delgado *et al.* 2003; Kerschbaum *et al.* 2010; Maercker *et al.* 2010; Arimatsu *et al.* 2011; Maercker *et al.* 2018).

2. Analysis

The U Ant observations analysed here were obtained as part of the NESS project (see Dharmawardena *et al. in prep.* for details). Surface brightness profiles were derived using the methods described in Dharmawardena *et al.* (2018). The PSF-subtracted 850 μm residual profile is dominated by a broad peak in the $20'' - 40''$ range. We fit these data with synthetic intensity profiles generated from radiative transfer models for the U Ant detached shell. Following a procedure similar to that of Kerschbaum *et al.* (2010) and Maercker *et al.* (2018), we first constrain the properties of the central star and the recent mass loss (the "attached" shell). For this purpose, we fit the spectral energy distribution (SED) with models from the GRAMS carbon-star grid (Srinivasan *et al.* 2011). We then

† Present address: Instituto de Radioastronomía y Astrofísica, Universidad Nacional Autonóma de México, Antigua Carretera a Pátzcuaro #8701, Ex-Hda. San José de La Huerta, Morelia, Michoacán, 58089, México.

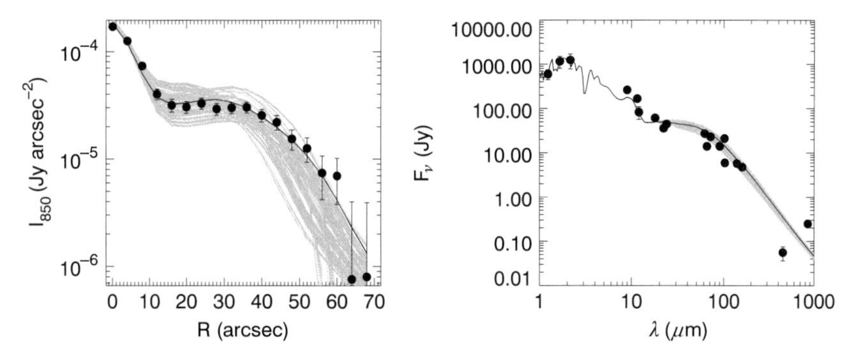

Figure 1. *Left*: best-fit models (black and grey curves) to the observed U Ant intensity profile (circles) at 850 μm. *Right*: the same models overlaid on the observed SED (circles).

construct a grid of models for the detached shell, exploring three parameters – the inner and outer radii $R_{\rm in}$ and $R_{\rm out}$ of the detached shell, and the density enhancement at $R_{\rm in}$ compared to the attached shell.

3. Results and discussion

Figure 1 shows the best-fit model along with the top 100 acceptable fits, which are used to estimate the parameter errors. These models also fit the overall SED (Figure 1). We find a dust-production rate of $(3.6 \pm 0.5) \times 10^{-9}$ $M_\odot$ yr^{-1} for the detached shell, corresponding to a dust mass of $(1.5 \pm 0.2) \times 10^{-5}$ $M_\odot$. Assuming a gas:dust ratio of 200, we obtain a gas mass-loss rate of $(7.2 \pm 1.1) \times 10^{-7}$ $M_\odot$ yr^{-1}. These estimates are consistent with those of Kerschbaum *et al.* (2010) and Maercker *et al.* (2018). While the modelled intensity profile and SED show good overall agreement with those derived from the data, our models seem to predict a lower 850 μm flux than observed (Figure 1). This could be due to the fact that we have ignored the complex geometry of the dust shell in our model, or possibly due to a variation in the dust properties in the shell. We will explore more free parameters in future modelling to address these issues, by combining our data with interferometric observations to constrain the properties in the inner ($R < 12''$) regions of the shell.

References

Arimatsu, K. Izumiura, H., Ueta, T., *et al.* 2011, *ApJ*, 729, 19

Dharmawardena, T., Kemper, F., Scicluna, P., *et al.* 2018, *MNRAS*, 479, 536

González Delgado, D., Olofsson, H., Schwarz, H. E., *et al.* 2001, *A&A*, 372, 885

González Delgado, D., Olofsson, H., Schwarz, H. E., *et al.* 2003, *A&A*, 399, 1021

Kerschbaum, F., Ladjal, D., Ottensamer, R., *et al.* 2010, *A&A* 518, 140

Maercker, M., Olofsson, H., Eriksson, K., *et al.* 2010, *A&A*, 511A, 37

Maercker, M., Khouri Silva, T., De Beck, E., *et al.* 2018, *A&A*, https://arxiv.org/abs/1807.11305

Why Galaxies Care About AGB Stars:
A Continuing Challenge through Cosmic Time
Proceedings IAU Symposium No. 343, 2019
F. Kerschbaum, M. Groenewegen & H. Olofsson, eds.

doi:10.1017/S1743921318006622

Population of AGB stars in the outer Galaxy

Ryszard Szczerba[1,2], Ilknur Gezer[1,3], Bosco H. K. Yung[1,2] and Marta Sewiło[4]

[1]Nicolaus Copernicus Astronomical Center, PAS, Rabiańska 8, 87-100 Toruń
email: szczerba@ncac.torun.pl

[2]Astronomical Observatory of the Jagiellonian University, Orla 171, 30-244 Kraków, Poland

[3]Astronomy and Space Science Department, Ege University, 35100 Bornova, Izmir, Turkey

[4]NASA Goddard Space Flight Center, 8800 Greenbelt Rd., Greenbelt, MD 20771, USA

Abstract. We present preliminary results of a study aimed at identifying and characterizing the Asymptotic Giant Branch (AGB) stars in the outer Galaxy using the color-color diagram (CCD) that combines the *Spitzer Space Telescope* and 2MASS photometry: $K_s - [8.0]$ vs. $K_s - [24]$. Our initial study concentrates on a region in the outer Galactic plane around a galactic longitude l of 105°, where we identified 777 O-rich and 200 C-rich AGB star candidates.

Keywords. stars: AGB and post-AGB, infrared: stars, catalogs

1. Introduction

We are conducting a systematic search for young and evolved stellar objects in the $\sim$24 deg^2 region in the outer Galaxy dubbed 'l105': $l = (102°, 109°)$ and $b = (-0.2°, 3.2°)$, covered by the "*Spitzer* Mapping of the Outer Galaxy" survey (SMOG, IRAC 3.6–8.0 μm and MIPS 24 μm; PI: Sean Carey). Szczerba *et al.* (2016) identified regions in the $K_s - [8.0]$ vs. $K_s - [24]$ CCD where Young Stellar Objects (YSO) and post-AGB stars are located. They also showed that the location of the hydrodynamical models computed by Steffen *et al.* (1998) reproduce the distribution of the O-rich and C-rich AGB stars from the Magellanic Clouds in this CCD very well. The $K_s - [8.0]$ vs. $K_s - [24]$ CCD allows us to separate C-rich and O-rich AGB stars quite effectively (Matsuura *et al.* 2014). Using this property in combination with hydrodynamical computations of Steffen *et al.* (1998), we selected O- and C-rich AGB star candidates in the l105 region.

2. Selection of the AGB candidates

Figure 1 shows the distribution of *Spitzer* sources in l105 in the $K_s - [8.0]$ vs. $K_s - [24]$ CCD displayed as a Hess diagram. Only sources with highly reliable *Spitzer* 8.0 μm and 24 μm, and 2MASS K_s-band photometry are included (15 311 objects). Most of the objects concentrate around ($K_s - [8.0]$, $K_s - [24]$) $\sim$ (0,0). Black lines and labels indicate regions in the CCD where different types of AGB stars are typically located; see also Fig. 10 in Matsuura *et al.* (2014). The bottom thick solid line and the top dotted line are exactly the same as the corresponding bottom and top lines from Matsuura *et al.* for $x \equiv K_s - [8.0] > 1$. Based on the position of our hydrodynamical tracks in the CCD (red points for C-rich and blue points for O-rich hydrodynamical tracks - see Szczerba *et al.* (2016) for details), we slightly redefined the boundary between O- and C-rich AGB stars, and significantly lowered the top boundary of Matsuura *et al.* for O-rich stars. The equations for these lines, assuming that $y \equiv K_s - [24]$, from bottom to top are: $y - 0.5 = 16/13(x - 1)$; $y - 2 = 14.5/11(x - 1)$; $y - 3.5 = 3.5/2.15(x - 1)$; and $y - 4 = 2(x - 1)$. Our

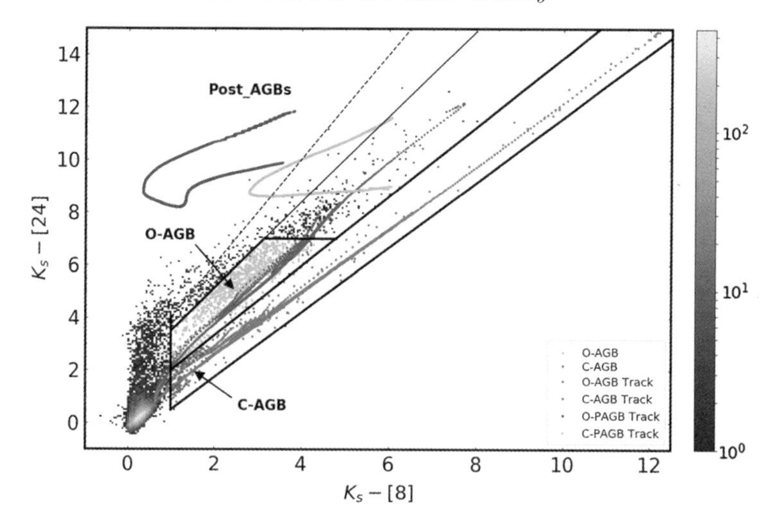

Figure 1. The $K_s - [8.0]$ vs. $K_s - [24]$ CCD showing the distribution of sources from the *Spitzer*/SMOG survey ('*l*105'). The regions occupied primarily by O-rich and C-rich AGB star candidates are delimited by the solid thick lines and labeled. The O-rich and C-rich AGB star candidates are indicated with cyan and green circles, respectively. See text for more details.

analysis excludes sources with $x < 1$ as it is difficult to separate O-rich from C-rich AGB stars in this region in the CCD; this region may also be populated by Red Giant Branch stars characterized by small mass loss rate. As shown by Szczerba *et al.* (2016), O-rich AGB stars with $y > 7$ are mixed with YSOs and post-AGB stars.

The regions within the thick solid lines in the $K_s - [8.0]$ vs. $K_s - [24]$ CCD are selected as areas mostly populated by C-rich and O-rich AGB stars. In total, we selected 200 C-rich and somewhat unexpectedly as much as 777 O-rich AGB star candidates. We used the SIMBAD Astronomical Database - CDS in Strasbourg to search for counterparts of *Spitzer* sources; we used a search radius of 2″. We found 99 entries for C-rich and only 44 for O-rich AGB star candidates. The probable reason is that selected O-rich candidates are redder and were less frequently observed than C-rich ones. Out of 99 SIMBAD sources matching *Spitzer*/SMOG sources selected based on the $K_s - [8.0]$, $K_s - [24]$ CCD, 38 are classified as 'C*' or 'C*?', while in O-rich sample only three were classified as 'variable' (possibly AGB), whereas eleven as young objects ('Be', 'YSO', or 'WR*'). We plan to further investigate the selected sample.

Acknowledgements

This work was supported by the Polish National Science Centre grant 2014/15/B/ST9/02111.

References

Matsuura, M., Bernard-Salas, J., Lloyd Evans, T., *et al.* 2014 *MNRAS*, 439, 1472

Szczerba, R., Bosco, H.K.Y., Sewiło, M., Siódmiak, N., & Karska, A. 2016, *Journal of Physics: Conference Series*, Volume 728, Issue 4, article id. 042004

Steffen, M., Szczerba, R., & Schönberner, D. 1998, *A&A*, 337, 149

Why Galaxies Care About AGB Stars:
A Continuing Challenge through Cosmic Time
Proceedings IAU Symposium No. 343, 2019
F. Kerschbaum, M. Groenewegen & H. Olofsson, eds.

doi:10.1017/S1743921318005665

The role of asymptotic giant branch stars in the chemical evolution of the Galaxy

G. Tautvaišienė[1], C. Viscasillas Vázquez[1], V. Bagdonas[1], R. Smiljanic[2], A. Drazdauskas[1], Š. Mikolaitis[1], R. Minkevičiūtėė[1] and E. Stonkutė[1]

[1]Institute of Theoretical Physics and Astronomy, Vilnius University,
Sauletekio av. 3, 10257, Vilnius, Lithuania
email: grazina.tautvaisiene@tfai.vu.lt

[2]Nicolaus Copernicus Astronomical Center, Polish Academy of Sciences,
Bartycka 18, 00-716, Warsaw, Poland

Abstract. Asymptotic giant branch stars play an important role in enriching galaxies by s-process elements. Recent studies have shown that their role in producing s-process elements in the Galactic disc was underestimated and should be reconsidered. Based on high-resolution spectra we have determined abundances of neutron-capture elements in a sample of 310 stars located in the field and open clusters and investigated elemental enrichment patterns according to their age and mean galactocentric distances.

Keywords. stars: abundances, Galaxy: evolution

1. Introduction

Young open clusters seem to have larger abundances of s-process dominated chemical elements compared to the older ones, however, it is still debatable, whether this phenomenon is similar in all s-process elements (D'Orazi *et al.* 2017 and references therein). We derived yttrium, zirconium, barium, lanthanum, cerium, praseodymium, neodymium, and europium abundances for 37 red giant branch (RGB) stars, which are probable members of 6 open clusters (NGC 5460, NGC 5822, NGC 6709, NGC 3680, NGC 6940, IC 4651). In addition, we observed all FGK dwarfs brighter than $V < 8$ mag (more than 400 stars) in two Galactic fields with the radii of 20° centered at $\alpha(2000) = 161^\circ.03552/\delta(2000) = 86^\circ.60225$ and at $\alpha(2000) = 265^\circ.08003/\delta(2000) = 39^\circ.58370$.

2. Observations and method of analysis

Spectra for the cluster stars were obtained with the 2.2 m MPG/ESO telescope at La Silla using the FEROS echelle spectrograph which provided a spectral resolving power of $R = 48\ 000$ in a spectral region of 3700–8600 Å. Galactic field stars were observed with the Vilnius University Echelle Spectrograph (VUES), mounted on the f/12 1.65 m Ritchey-Chretien telescope at the Molėtai Astronomical Observatory. With the VUES, we observed spectra in the 4000 to 8800 Å wavelength range with two spectral resolution modes ($R = 30\,000$ and $60\,000$).

For the determination of atmospheric parameters, the mean galactocentric distances $R_{\rm mean}$, and ages, we applied methods described by Mikolaitis *et al.* (2018).

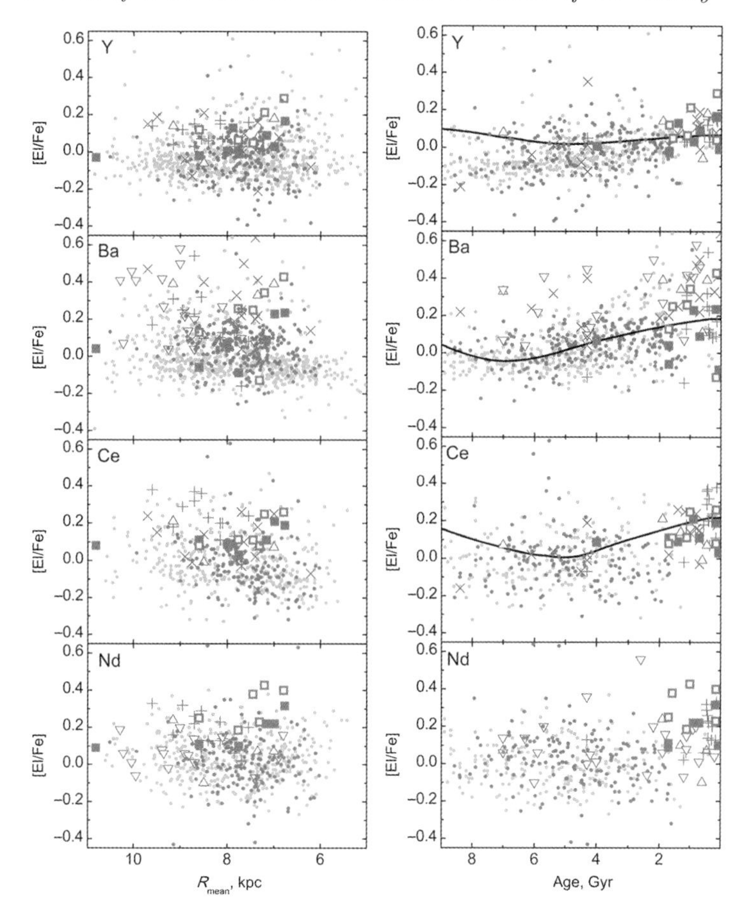

Figure 1. Elemental abundances in field stars and open clusters relatively to $R_{\rm mean}$ and ages. The dots indicate data for the Galactic field stars (the red – our results, the grey – by Bensby *et al.* 2014 and Battistini & Bensby 2016), the empty and filled red squares – our new and published data for open clusters, respectively (see Bagdonas *et al.* 2018 for references). The continuous lines indicate a chemical evolution model by Maiorca *et al.* (2012) at the solar radius.

3. Results

The available data (Fig. 1) do not show obvious s-process dominated element to iron abundance ratio gradients in respect to the $R_{\rm mean}$. A rise of abundances is visible with decreasing age. At young ages, the observations follow the chemical evolution models by Maiorca *et al.* (2012), however, we do not observe the raise of abundances in stars older than 6 Gyr predicted by the models.

Acknowledgements

We thank the Research Council of Lithuania (LAT-08/2016).

References

Bagdonas, V., Drazdauskas, A., Tautvaišienė, G., Smiljanic, R., & Chorniy, Y. 2018, *A&A*, 615, A165

Battistini, C., & Bensby, T. 2016, *A&A*, 586, A49

Bensby, T., Feltzing, S., & Oey, M. S. 2014, *A&A*, 562, A71

D'Orazi, V., De Silva, G. M., & Melo, C. F. H. 2017, *A&A*, 598, A86

Maiorca, E., Magrini, L., Busso, M., *et al.* 2012, *ApJ*, 747, 53

Mikolaitis, Š., Tautvaišienė, G., Drazdauskas, A., *et al.* 2018, *PASP*, 130, 074202

Why Galaxies Care About AGB Stars:
A Continuing Challenge through Cosmic Time
Proceedings IAU Symposium No. 343, 2019
F. Kerschbaum, M. Groenewegen & H. Olofsson, eds.

doi:10.1017/S1743921318007275

The star formation history of the M31 galaxy derived from Long-Period-Variable star counts

Maryam Torki[1,2], Atefeh Javadi[2], Jacco Th. van Loon[3] and Hossein Safari[1]

[1]Physics Department, Faculty of Science, University of Zanjan, Zanjan 45371-38791, Iran
email: Maryamtorki84@gmail.com

[2] School of Astronomy, Institute for Research in Fundamental Sciences (IPM), Tehran, 19395-5531, Iran

[3]Lennard-Jones Laboratories, Keele University, ST5 5BG, UK

Abstract. The determination of the star formation history is a key goal for understanding galaxies. In this regard, nearby galaxies in the Local Group offer us a complete suite of galactic environment that is perfect for studying the connection between stellar populations and galaxy evolution. In this paper, we present the star formation history of M31 using long period variable stars that are prime targets for studying the galaxy formation and evolution because of their evolutionary phase. In this method, at first, we convert the near-infrared K-band magnitude of evolved stars to mass and age and from this we reconstruct the star formation and evolution of the galaxy.

Keywords. stars: AGB and post-AGB - stars: luminosity function, mass function - galaxies: evolution - galaxies: formation - galaxies: individual: M31 - galaxies: stellar content - galaxies: structure

---

1. Star Formation History in the Andromeda Galaxy M31

M31 is the nearest spiral galaxy to the Milky Way and this is a great chance for us to study the formation and evolution of this galaxy. Our approach to investigate the star formation history (SFH) is based on employing long period variable stars (LPVs) which we have successfully applied in other galaxies in the Local Group such as M33 (Javadi *et al.* 2011a; Javadi *et al.* 2011b; Javadi *et al.* 2017), Magellanic Clouds (Rezaeikh *et al.* 2014), NGC147 and NGC185 (Golshan *et al.* 2017). For this research, we use the catalogue of LPVs in M31 from Mould *et al.* (2004). LPVs are mostly Asymptotic Giant Branch (AGB) stars as well as red supergiants (RSGs) in the final stage of evolution and this point of evolution is characterised by strong radial pulsations. AGB stars are luminous ($\sim 10^4\,L_\odot$) and cool ($< 4000\,K$) and hence stand out at near-infrared wavelengths. AGB stars will lose up to 85 percent of their initial mass at a rate of up to $10^{-4}\,M_\odot\,yr^{-1}$ (Javadi *et al.* 2013), so they have great influence on the chemical mixture and the rate of star formation in the universe. Actually, the SFH of a galaxy is a measure of the rate at which the gas mass was converted into stars over a time interval in the past. The star formation rate (SFR), ξ (in $M_\odot$ yr^{-1}) as a function of time is estimated by:

$$\xi(t) = \frac{\int_{\min}^{\max} f_{\rm IMF}(m) m\, dm}{\int_{m(t)}^{m(t+dt)} f_{\rm IMF}(m)\, dm} \frac{dn'(t)}{\delta t}, \tag{1.1}$$

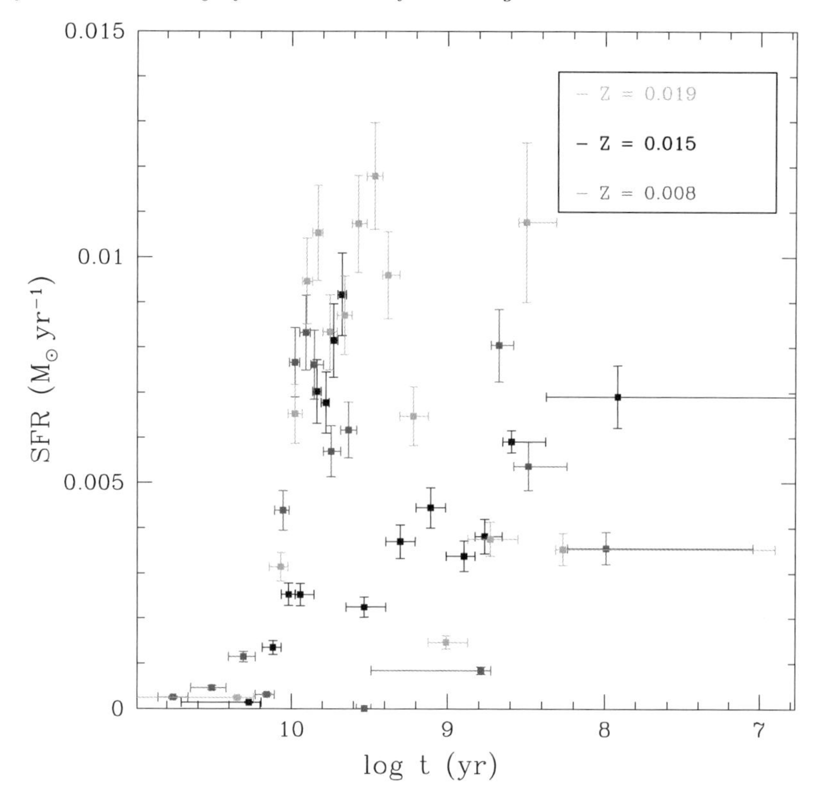

Figure 1. The SFH in M31 for a region displayed in Mould *et al.* (2004) for three choices of metallicity: Z = 0.019 (green), Z = 0.015 (black), Z = 0.008 (red).

where n' is the number of variables that we have identified, $f_{\rm IMF}$ is the initial mass function describing the relative contribution to star formation by stars of different mass and δt is the duration of variability of these stars displaying strong radial pulsation. The SFH in M31 is shown in Figure 1 for different metallicities. The horizontal errorbars represent the age bins and vertical lines demonstrate statistical errorbars.

2. Conclusions

We have applied the novel method of Javadi *et al.* (2011) using long period variable stars to derive the SFH in M31. With this method, we estimated the SFR in M31 during the broad time interval from 30 Myr to 10 Gyr ago.

References

Golshan, R. H., Javadi, A., van Loon, J. Th., *et al.*, 2017, MNRAS, 466, 1764

Javadi, A., van Loon, J. Th., and Mirtorabi, M. T., 2011a, ASPC, 445, 497

Javadi, A., van Loon, J. Th., and Mirtorabi, M. T., 2011b, MNRAS, 414, 3394

Javadi, A., van Loon, J. Th., Khosroshahi, H. G., and Mirtorabi, M. T. 2013, MNRAS, 432, 2824

Javadi, A., van Loon, J. Th., Khosroshahi, H. G., *et al.*, 2017, MNRAS, 464, 2103

Mould, J., Saha, A., Hughes, S., 2004, ApJ, 154, 623

Rezaeikh, S., Javadi, A., Khosroshahi, H. G., and van Loon, J. Th., 2014, MNRAS, 445, 2214

Why Galaxies Care About AGB Stars:
A Continuing Challenge through Cosmic Time
Proceedings IAU Symposium No. 343, 2019
F. Kerschbaum, M. Groenewegen & H. Olofsson, eds.

doi:10.1017/S1743921318006440

Comprehensive Panchromatic Data Analyses and Photoionization Modeling of NGC 6781

Toshiya Ueta[1], Masaaki Otsuka[2] and the HerPlaNS consortium

[1]University of Denver, Denver, CO 80112, U.S.A.,
email: toshiya.ueta@du.edu

[2]Okayama Observatory, Kyoto University, Okayama, Japan

Abstract. We characterized the dusty circumstellar nebula and central star of the C-rich bipolar planetary nebula (PN) NGC 6781 using our own Herschel data augmented with the archival data from UV to radio and constructed one of the most comprehensive photoionization PN models ever produced consisting of the ionized, atomic and molecular gas components as well as the dust component. We reproduced the observed spectral energy distribution (SED), constrained by 136 observational data points. The total nebula mass was estimated to be 0.41 $M_\odot$, with a significant fraction (about 70 %) of it existing in the photo-dissociation region (PDR) surrounding the ionized nebula. This finding demonstrates the critical importance of the PDR in PNe, which are typically recognized as the hallmark of ionized/H^+ region. It is therefore essential to characterize the PDR of the circumstellar nebula to understand material recycling in the Milky Way and other galaxies.

Keywords. planetary nebulae: individual (NGC 6781), circumstellar matter, stars: mass loss, stars: abundances

---

1. Plasma Diagnostics and Ionic/Elemental Abundance Analyses

We performed a comprehensive analysis of the panchromatic data obtained from the PN NGC 6781, spanning from UV to radio (consisting of photometric and spectroscopic data from a dozen facilities; Table 1 of Otsuka *et al.* 2017) to investigate the evolutionary status of the central star and the physical conditions of each of the ionized, atomic, and molecular gas components as well as the dust component in the nebula.

First, we carried out detailed plasma diagnostics to determine the spatially-resolved electron density and temperature along the radial direction in the equatorial plane of this nearly pole-on bipolar nebula. We found the following three distinct spatial regions at least: the ionized nebula ($n_{\rm H} = 300\,{\rm cm}^{-3}$, $T_{\rm e} = 24$ to 10 kK), the high density pile-up wall, which delineates the characteristic "ring" structure of the nebula ($n_{\rm H} = 960\,{\rm cm}^{-3}$, $T_{\rm e} = 10$ to 3 kK), and the even higher density PDR around the "ring" structure ($n_{\rm H} = 10^4\,{\rm cm}^{-3}$, $T_{\rm e} = 3$ to 1 kK).

We then derived elemental abundances in the nebula for the following nine species: He, C, N, O, Ne, Si, S, Cl, and Ar. By comparing the derived abundances with the abundance pattern of AGB nucleosynthesis models of Karakas (2010), we determined that the progenitor star of NGC 6781 was of $2.25 - 3.0\,M_\odot$ initially. We also derived the distance of 0.46 kpc by fitting the stellar luminosity as a function of the distance and effective temperature of the central star constrained by the adopted post-AGB evolutionary tracks. The detailed account of these comprehensive analyses are fully documented in our recent publication (Otsuka *et al.* 2017).

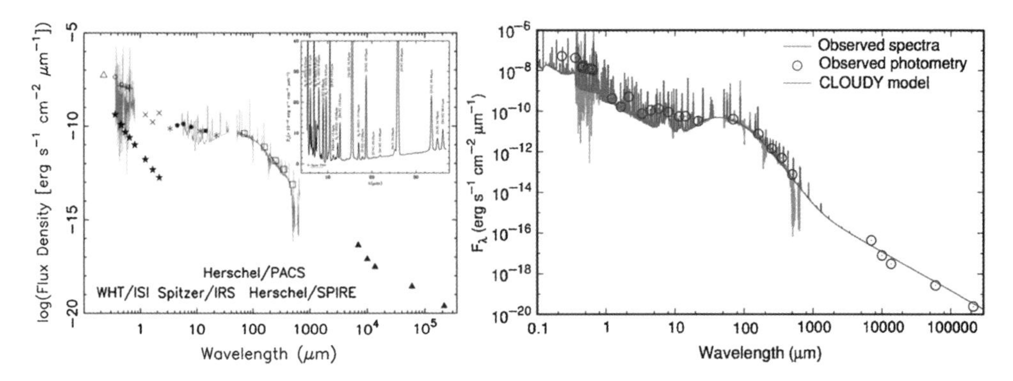

Figure 1. **[Left]** The panchromatic data of NGC 6781 adopted in the present study. Broadband photometry: GALEX (open triangle), ING/INT (open circles), ESO/NTT (pluses), UKIRT (crosses), WISE (asterisks), Spitzer (filled circles), ISO (filled square), Herschel (open squares), Radio (filled triangles), and HST/WFPC2 (filled stars for the central star). Spectra (grey lines): WHT/ISIS, Spitzer/IRS, and Herschel/PACS and SPIRE as indicated. The inset displays the Spitzer/IRS spectra in the mid-IR. **[Right]** The SED of the best-fit CLOUDY model of NGC 6781 (red line; spectral resolution $R = 300$), compared with the observational constraints: photometry data (blue circles) and spectroscopy data (grey line). For full details, refer to Otsuka *et al.* (2017).

2. Dusty Photoionization Modeling with CLOUDY

Next, by adopting the results of the above analyses as input parameters and model constraints, we constructed the best-fit photoionization model using CLOUDY (Ferland *et al.* 2013). The total gas mass was found to be $0.41\,M_\odot$, with only $0.09\,M_\odot$ ionized, $0.20\,M_\odot$ atomic, and $0.11\,M_\odot$ molecular gases. The total dust mass was found to be $1.53 \times 10^{-3}\,M_\odot$, which would make the gas-to-dust mass ratio to be 268. Hence, the total nebula mass accounted for purely from the available data turned out to be roughly 60 % of the amount of mass predicted to have been ejected during the last thermal pulse episode from a star of $2.5\,M_\odot$ initial mass (Karakas 2010). The fact that the ionized gas was found to be only about 20 % of the total mass in a PN emphasizes that the colder dusty PDR that surrounds the ionized nebula carries greater significance in terms of the progenitor's mass-loss history and cannot be neglected to account for the full energetics of the nebula, even though PNe are generally known as ionized gas and H^+ regions.

The present work demonstrated that PNe could indeed serve as (1) empirical constraints for stellar evolutionary models, because empirically derived central star and nebula parameters could now confront theoretical predictions (and the present AGB models are shown to be consistent), and (2) important probes of mass recycling and chemical evolution in galaxies because PNe would permit thorough mass accounting of the mass-loss ejecta in the circumstellar environments.

Acknowledgement

TU was partially supported by the Research Support Agreement (RSA) 1428128 issued through JPL/Caltech and Grant NNX15AF24G issued through the NASA Science Mission Directorate. MO was supported by the research fund 104-2811-M-001-138 and 104-2112-M-001-041-MY3 from the Ministry of Science and Technology (MOST), R.O.C.

References

Ferland, G. J., Porter, R. L., van Hoof, P. A. M., *et al.* 2013, *RMxAA*, 49, 137
Karakas, A. I. 2010, *MNRAS*, 403, 1413
Otsuka, M., *et al.* 2017, *ApJS*, 231, 22

Why Galaxies Care About AGB Stars:
A Continuing Challenge through Cosmic Time
Proceedings IAU Symposium No. 343, 2019
F. Kerschbaum, M. Groenewegen & H. Olofsson, eds.

doi:10.1017/S1743921318006397

Mass Loss History of Evolved Stars (MLHES) Excavated by AKARI

Toshiya Ueta[1], Andrew J. Torres[1], Hideyuki Izumiura[2] and Issei Yamamura[3]

[1]University of Denver, Denver, CO 80112, U.S.A.
email: toshiya.ueta@du.edu (TU)

[2]Okayama Branch Office, Subaru Telescope, NAOJ, Okayama, Japan

[3]Institute of Space and Aeronautical Science, JAXA, Kanagawa, Japan

Abstract. We performed a far-IR imaging survey of the circumstellar dust shells of 144 evolved stars as a mission program of the AKARI infrared astronomical satellite. Our objectives were to characterize the far-IR surface brightness distributions of the cold dust component in the circumstellar dust shells. We found that (1) far-IR emission was detected from all but one object, (2) roughly 60–70 % of the target sources showed some extension, (3) 29 sources were newly resolved in the far-IR in the vicinity of the target sources, (4) the results of photometry measurements were reasonable with respect to the entries in the AKARI/FIS Bright Source Catalogue, and (5) an IR two-color diagram would place the target sources in a roughly linear distribution that may correlate with the age of the circumstellar dust shell.

Keywords. stars: AGB and post-AGB, planetary nebulae: general, circumstellar matter, stars: mass loss, infrared: stars, surveys

---

1. The AKARI MLHES Mission Program

The *AKARI* Infrared Astronomical Satellite (*AKARI*; Murakami *et al.* 2007) was launched on February 21, 2006 (UT) to carry out its primary mission to map out the entire sky in the far-IR until the cryogen was exhausted on August 26, 2007. During all-sky survey scan observations that lasted for 1.5 years, some telescope time was allocated to perform pointed observations as *AKARI* mission programs. We carried out an imaging survey of the circumstellar dust shells (CDS) of evolved stars, at various stages of evolution from the tip of the first-ascent giant branch to planetary nebula phase, dubbed the "Mass Loss History of Evolved Stars" (MLHES), as one of such mission programs, using the Far-Infrared Surveyor (FIS; Kawada *et al.* 2007) and Infrared Camera (IRC; Onaka *et al.* 2007).

With MLHES, we aimed at observationally establishing the AGB mass loss history and understanding the mechanism(s) of mass loss and CDS formation by spatially resolving the CDS structures. Some of the MLHES data were already presented in the context of the interaction between the interstellar medium (ISM) and CDS (Ueta *et al.* 2008, 2010). In the context of MLHES, Izumiura *et al.* (2011) analyzed the detached CDS of the AGB star U Hya to conclude that (1) the temporal enhancement of mass loss due to thermal pulse and the subsequent two-wind interactions, or (2) the reverse/termination shock of the stellar wind bounced back from the wind-ISM interface. However, the marginal image quality at the preliminary stage of data reduction did not allow us to unequivocally conclude one of the two possibilities.

2. Surface Brightness Recalibration of AKARI/FIS Slow-Scan Maps

Far-IR Ge:Ga detectors such as FIS is known to exhibit reduced sensitivity to bright point sources that is attributed to the delay in response to the incoming flux (e.g., Kaneda *et al.* 2012). Such slow transient responses in scan observations of bright objects would typically manifest themselves as underestimated surface brightnesses, resulting in reduced fluxes. Hence, while FIS data were absolutely calibrated against large-scale diffuse background emission based on the *COBE*/DIRBE data (Matsuura *et al.* 2011), the surface brightness of slow-scan maps would have to be recalibrated if target sources were more compact than diffuse background cirrus clouds (i.e., just like CDSs). We have recently established a general method to recalibrate AKARI/FIS slow-scan maps with the inverse power-law FIS response function based on the scale invariance characteristic of the FIS maps (i.e., the PSF shape remains the same irrespective of the source flux; Ueta *et al.* 2017). With this correction, surface brightnesses of the MLHES maps that were underestimated depending on the actual signal strength of the emission distribution have now been properly corrected for.

3. Present Status

The recalibrated MLHES data set will be presented soon by Ueta *et al.* (2008) with bare photometry completed. With *AKARI*'s better spatial resolution in comparison with its predecessors (*IRAS* and *ISO*), 29 new nearby sources are resolved in the vicinity of the MLHES targets, several of which have not been identified previously. The photometry measurements are reasonable with respect to the corresponding entries in the AKARI/FIS Bright Source Catalogue (Yamamura *et al.* 2009), with a reasonable amount of discrepancies because of the fact that many MLHES sources are genuinely extended. An *AKARI*-WISE two-color diagram, showing the MLHES sources in a roughly linear distribution that may correlate with the age of the circumstellar dust shell, could potentially be used to identify which targets were more extended than others. The detailed analyses of structures of the MLHES CDSs are presently being conducted by removing the PSF effects using a scaled super-PSF map constructed from images of reference point sources. The results of the structural analysis of the evolved star circumstellar shell for the entire MLHES sample will be presented elsewhere (Torres *et al. in preparation*).

Acknowledgement

This work is based on observations with AKARI, a JAXA project with the participation of ESA. Ueta was partially supported by the Japan Society for the Promotion of Science via a long-term invitation fellowship program in FY2013.

References

Izumiura, H., *et al* 2011, *A&A*, 528, A29
Kaneda, H., *et al.* 2002, *Adv. Space Res.*, 30, 2105
Kawada, M., *et al.* 2007, *PASJ*, 59, S389
Matsuura, S., *et al.* 2011, *ApJ*, 737, 2
Murakami, H., *et al.* 2007, *PASJ*, 59, S369
Onaka, T., *et al.* 2007, *PASJ*, 59, 401
Ueta, T., *et al.* 2008, *PASJ*, 60, S407
Ueta, T., *et al.* 2010, *A&A*, 514, A16
Ueta, T., *et al.* 2017, *PASJ*, 69, 11
Ueta, T., *et al.* 2018, *PASJ*, in review
Yamamura, I., *et al.* 2009, in ASP Conf. Ser. Vol. 418, AKARI, a Light to Illuminate the Misty Universe, eds. Onaka, T., White, G. J., Nakagawa, T., & Yamamura, I., p.3

Why Galaxies Care About AGB Stars:
A Continuing Challenge through Cosmic Time
Proceedings IAU Symposium No. 343, 2019
F. Kerschbaum, M. Groenewegen & H. Olofsson, eds.

doi:10.1017/S1743921318006385

Herschel Planetary Nebula Survey Plus (HerPlaNS+)

Toshiya Ueta[1], Isabel Aleman[2], Masaaki Otsuka[3], Katrina Exter[4] and the HerPlaNS consortium

[1]University of Denver, Denver, CO 80112, U.S.A.
email: toshiya.ueta@du.edu (TU)

[2]IAG-USP, University of São Paulo, Brazil

[3]Okayama Observatory, Kyoto University, Japan

[4]KU Leuven, Belgium

Abstract. We present the current status update of the Herschel Planetary Nebula Survey Plus project (HerPlaNS+) based on the original General Observer HerPlaNS survey program during the OT1 cycle and the follow-up exhaustive archival search of PN observations using the PACS and SPIRE instruments on-board the Herschel Space Observatory.

Keywords. planetary nebulae: general, circumstellar matter, stars: mass loss, infrared: stars, surveys

---

1. HerPlaNS

The original Herschel Planetary Nebula Survey (HerPlaNS) was performed using all available observational modes of the PACS and SPIRE instruments aboard Herschel to investigate the far-IR characteristics of both the dust and gas components of the circumstellar nebulae for a set of 11 high-excitation PNe (Ueta *et al.* 2014). We obtained (1) broadband maps of the target sources at five far-IR bands, 70, 160, 250, 350, and 500 μm, with rms sensitivities of 0.01–0.1 mJy arcsec^{-2} (0.4–4 MJy sr^{-1}); (2) 5×5 IFU spectral cubes of 51–220 μm covering roughly a $50'' \times 50''$ field at multiple positions in the target sources, with rms sensitivities of 0.1–1 mJy arcsec^{-2} (4–40 MJy sr^{-1}) per wavelength bin; and (3) a sparsely sampled spectral array of 194–672 μm covering roughly a $3'$ field at multiple positions in the target sources, with rms sensitivities of 0.001–0.1 mJy arcsec^{-2} (0.04–4 MJy sr^{-1}) per wavelength bin.

A quick demonstration of the intended analyses using NGC 6781 as an example yielded the following: (1) the spectral fitting of the broadband images indicated that dust grains are composed mostly of amorphous-carbon based material (i.e., the power-law emissivity index of $\beta \approx 1$) of the temperature between 26 and 40 K; (2) the spatially-resolved plasma diagnostics using line ratios such as [O III] 52/88 μm and [N II] 122/205 μm resulted in the electron density and temperature (n_e, T_e) and ionic/elemental/relative abundance profiles exhibiting variations along the radial direction in the equatorial plane of this nearly pole-on bipolar nebula; and (3) the direct comparison between the above results allowed us to derive an empirical gas-to-dust mass ratio distribution projected roughly to the equatorial plane of this nebula, showing variation of the ratio from 550 at the inner edge of the nebula waist to 100 at the detected outer edge with the average ratio being 195 ± 110.

This demonstration signified the importance of direct and purely empirical comparison between the gas and dust components with PNe, especially in a spatially-resolved manner, to account for the amount of matter and energetics in these nebulae.

2. Detection of OH^+ and H I Laser Emission from PNe

Probing PNe in the less explored far-IR wavelength range allowed us to detect various exotic emission. First, we made the first detections of OH^+ in PNe (Aleman *et al.* 2014). The emission was detected in both PACS and SPIRE far-IR spectra of three of the 11 HerPlaNS PN sample (NGC 6445, 6720, and 6781), with the simultaneous and independent discovery in two other PNe (NGC 6853 and 7293) reported by Etxaluze *et al.* (2014). All five OH^+ PNe are molecule rich, with ring-like or torus-like structures and hot central stars ($T_{\rm eff} > 100\,000$ K). The OH^+ emission is most likely due to excitation in a photo-dissociation region (PDR). Although other factors such as high density and low C/O ratio may also play a role in the enhancement of the OH^+ emission, the fact that we do not detect OH^+ in objects with $T_{\rm eff} < 100\,000$ K suggests that the hardness of the ionizing central star spectra (i.e. the production of soft X-rays, at around 100–300 eV) could be an important factor in the production of OH^+ emission in PNe.

We then reported the detection of hydrogen recombination laser (HRL) lines in the far-IR to sub-mm spectrum of Mz 3 observed as part of HerPlaNS (Aleman *et al.* 2018). Comparison of optical to sub-mm HRL lines to theoretical calculations indicated that there was an enhancement in the far-IR to sub-mm HRLs. The likely explanation for this enhancement is the occurrence of a laser effect, and indicated was the presence of a dense and ionized gas ($n_{\rm H} > 10^8$ cm^{-3}) in the core of Mz 3, while the empirical analysis of forbidden lines suggested densities around 4 500 cm^{-3} in the surrounding lobes.

3. Comprehensive Panchromatic Data Analyses and Photoionization Modeling of NGC 6781

By augmenting the HerPLANS data with existing data from UV to radio, we performed one of the most thorough plasma diagnostics and ionic/elemental abundance analyses and constructed one of the most comprehensive dusty photo-ionization models ever produced for NGC 6781 (Otsuka *et al.* 2017), which is separately presented in this volume.

4. HerPlaNS+

Presently, we are compiling and analyzing all PACS and SPIRE data obtained for PNe found in the Herschel Science Archive as the expanded HerPlaNS+ survey. More than 1 000 PNe were observed in the photometry-mapping mode and a few dozens in the spectroscopy mode. These results will be published as the subsequent installments of the HerPlaNS series (Ueta *et al. in prep.*).

Acknowledgement

This work is based on observations made with the Herschel Space Observatory, a European Space Agency (ESA) Cornerstone Mission with significant participation by NASA. Support for this work was provided by the Research Support Agreement (RSA) 1428128 issued through JPL/Caltech and Grant NNX15AF24G issued through the NASA Science Mission Directorate.

References

Aleman, I., *et al.* 2014, *A&A*, 566, A79
Aleman, I., *et al.* 2018, *MNRAS*, 477, 4499
Etxaluze, M., *et al.* 2014, *A&A*, 566, A78
Otsuka, M., *et al.* 2017, *ApJS*, 231, 22
Ueta, T., *et al.* 2014, *A&A*, 565, A36

Why Galaxies Care About AGB Stars:
A Continuing Challenge through Cosmic Time
Proceedings IAU Symposium No. 343, 2019
F. Kerschbaum, M. Groenewegen & H. Olofsson, eds.

doi:10.1017/S1743921318006373

Morpho-Kinematics of the Circumstellar Environments around Post-AGB Stars

Toshiya Ueta[1], Hideyuki Izumiura[2], Issei Yamamura[3] and Masaaki Otsuka[4]

[1]University of Denver, Denver, CO 80112, U.S.A.
email: toshiya.ueta@du.edu (TU)

[2]Okayama Branch Office, Subaru Telescope, NAOJ, Okayama, Japan

[3]Institute of Space and Aeronautical Science, JAXA, Kanagawa, Japan

[4]Okayama Observatory, Kyoto University, Okayama, Japan

Abstract. We observed two proto-planetary nebulae, HD 56126 representing a source with an elliptical circumstellar shell, and IRAS 16594−4656 representing a source with a bipolar circumstellar shell, with ALMA in the ^{12}CO and ^{13}CO J=3−2 lines and neighboring continuum to see how the morpho-kinematics of CO gas and dust emission properties in their circumstellar environments differ.

Keywords. stars: AGB and post-AGB, planetary nebulae: general, circumstellar matter, stars: mass loss, radio lines: stars

---

1. Dual Morphologies of Proto-Planetary Nebulae

Proto-planetary nebulae (PPNe) are post-asymptotic giant branch (post-AGB) stars surrounded by a physically detached circumstellar envelope (CSE), which resulted from mass loss during the AGB phase up to about $10^{-4}\,M_{\odot}\,yr^{-1}$ (Kwok 1993, Van Winckel 2003). PPNe are known to possess either the elliptical or bipolar morphology that appears to be developed by the beginning of the PPN phase, which is punctuated by the cessation of mass loss, supposedly because of the equatorially-enhanced mass loss during the final epochs of mass loss along the AGB (Meixner *et al.* 2002, Ueta & Meixner 2003).

2. ALMA Observations of Proto-Planetary Nebulae

We observed two PPNe, HD 56126 and IRAS 16594−4656, each of which represents one of these two morphological archetypes, elliptical and bipolar, respectively, with the ALMA 12-m array in Band 7 during Cycle 3. These PPNe were observed in the ^{12}CO and ^{13}CO J=3-2 lines at the effective velocity resolution of 26 and 885 m s^{-1}, respectively, in 3840 channels. We also observed these sources in the neighboring continuum bands centered at 333.0 and 343.3 GHz at the velocity resolution of 27.2 km s^{-1} in 128 channels. The spatial resolution in terms of the beam size was $0.4'' \times 0.3''$ and $0.5'' \times 0.4''$, respectively. As for data reduction, we adopted the pipeline calibration performed with the Common Astronomy Software Application (CASA; ver. 4.5.3) as delivered and made use of the newer CASA (ver 5.1.1-rel5) to use the new TCLEAN function with the 3-σ threshold masking to clean each velocity channel map during the final image reconstruction. The use of the new TCLEAN function was necessary to clean algorithmically 800 and 3000 channels at which spatially-variable extended CO emission was detected from HD 56126 and IRAS 16594−4656, respectively. Fig. 1 shows the integrated maps of HD 56126 (left panel) and IRAS 16594−4656 (right panel) in the ^{12}CO, ^{13}CO, neighboring radio continuum, and continuum in the IR for comparison (from top-left to bottom-right), respectively.

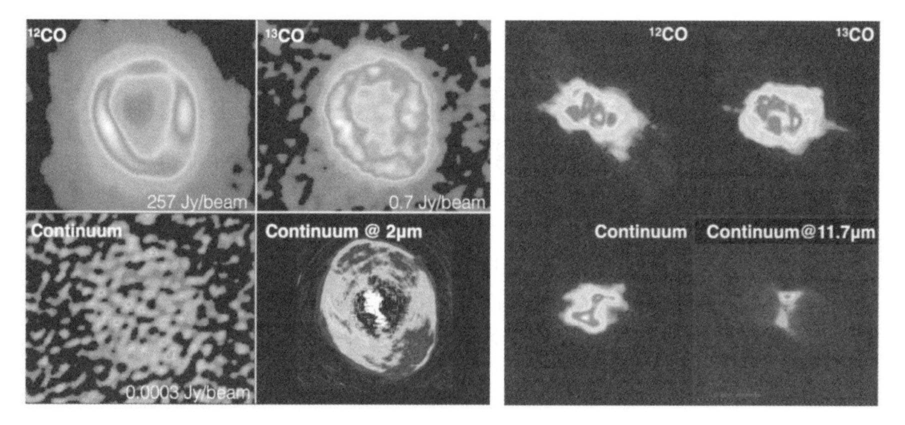

Figure 1. ALMA integrated maps in the ^{12}CO, ^{13}CO, and continuum, with a comparison continuum map in the IR for HD 56126 (left panels; the shell diameter is $\sim 4''$; the comparison $2\,\mu$m map by Ueta (2015)) and IRAS 16594−4656 (right panels; the FoV is roughly $8'' \times 8''$).

3. Morpho-Kinematics of Proto-Planetary Nebulae

Thermal dust emission at 330–340 GHz is almost non-existent for HD 56126, but shows some concentration at the bipolar waist (i.e., dust torus) seen edge-on which resembles what was detected at $11.7\,\mu$m (Volk *et al.* 2006) for IRAS 16594−4656, corroborating the optically-thin and thick nature of each of these dust shells, respectively.

HD 56126 is essentially a hollow shell expanding at $\sim 8\,\mathrm{km\,s^{-1}}$ with a slight velocity enhancement along the pole/long axis (at $\sim 12\,\mathrm{km\,s^{-1}}$). There are two velocity components at the pole at ~ 8 and $\sim 12\,\mathrm{km\,s^{-1}}$, indicating the presence of a separate, expanding cavity at the tips of the pole. IRAS 16584−4656 exhibits much more complex structures consisting of the central main bipolar structure surrounding the dust torus and at least four pairs of highly elongated but much fainter lobes at $\sim 25\,\mathrm{km\,s^{-1}}$ (three in the plane of the sky and one almost perpendicular). The morpho-kinematics is clearly distinct in these PPNe and is being analyzed to constrain the driving mechanism(s).

Acknowledgement

This paper makes use of the following ALMA data: ADS/JAO.ALMA#2015.1.00441.S. ALMA is a partnership of ESO (representing its member states), NSF (USA) and NINS (Japan), together with NRC (Canada), MOST and ASIAA (Taiwan), and KASI (Republic of Korea), in cooperation with the Republic of Chile. The Joint ALMA Observatory is operated by ESO, AUI/NRAO and NAOJ. The National Radio Astronomy Observatory is a facility of the National Science Foundation operated under cooperative agreement by Associated Universities, Inc. TU acknowledges technical assistance from the NA ARC at the NAASC. Dr. Daniel Tafoya at the EA ARC is especially thanked for his assistance in data reduction. In addition, TU appreciates fruitful discussions on the interpretation of the data with Dr. Griet Van de Steene at the Royal Observatory of Belgium.

References

Kwok, S. 1993, *ARA&A*, 31, 63
Meixner, M., Ueta, T., Bobrowsky, M., & Speck, A. 2002, *ApJ*, 571, 936
Ueta, T. 2015, *EAS Pub. Ser.*, 71-72, pp.265
Ueta, T., & Meixner, M. 2003, *ApJ*, 586, 1338
Van Winckel, H. 2003, *ARA&A*, 41, 391
Volk, K., Hrivnak, B. J., Su, K. Y. L., & Kwok, S. 2006, *ApJ*, 651, 294

Why Galaxies Care About AGB Stars:
A Continuing Challenge through Cosmic Time
Proceedings IAU Symposium No. 343, 2019
F. Kerschbaum, M. Groenewegen & H. Olofsson, eds.

doi:10.1017/S1743921318006452

Planetary Nebulae Detected in the AKARI Far-IR All-Sky Survey Maps

Toshiya Ueta[1], Ryszard Szczerba[2], Andrew G. Fullard[1] and Satoshi Takita[3]

[1]University of Denver, Denver, CO 80112, U.S.A.
email: toshiya.ueta@du.edu (TU)

[2]Nicolaus Copernicus Astronomical Centre, Toruń, Poland,

[3]National Astronomical Observatory of Japan, Tokyo, Japan

Abstract. The *AKARI* Far-IR All-Sky Survey (AFASS) maps produced by the *AKARI* Infrared Astronomical Satellite enabled us to probe the far-IR sky for objects having surface brightnesses greater than a few to a couple of dozen MJy sr^{-1}. Recently, we have verified that, if AFASS-measured fluxes are properly corrected for using the aperture correction method based on the empirical point-spread-function templates derived directly from the AFASS maps, point-source photometry measured from the AFASS maps reproduces fluxes in the *AKARI* bright source catalogue (BSC). We have surveyed the far-IR sky in the AFASS for Galactic planetary nebulae (PNe) based on the University of Hong Kong/Australian Astronomical Observatory/Strasbourg Observatory Hα Planetary Nebula database (HASHPNDB), preliminarily yielding far-IR fluxes for roughly 1000 Galactic PNe including a few hundreds of PNe not listed in the *AKARI*/BSC.

Keywords. planetary nebulae: general, circumstellar matter, stars: mass loss, infrared: stars, surveys

1. AKARI and AFASS

The *AKARI* Infrared Astronomical Satellite (*AKARI*; Murakami *et al.* 2007) was launched on February 21, 2006 (UT), carrying out its 550-day cryogen mission until it exhausted liquid Helium on August 26, 2007 and continuing its post-cryogen mission in the near-IR until the satellite was finally switched off on November 24, 2011. *AKARI* was outfitted with a cryogenically-cooled telescope of a diameter of 68.5 cm and two instruments, the Far-Infrared Surveyor (FIS; Kawada *et al.* 2007) and the Infrared Camera (IRC; Onaka *et al.* 2007), covering a wavelength range of 2–180 μm.

For the AKARI Far-IR All-Sky Survey (AFASS), the natural 100-min sun-synchronous orbit was used to scan the entire sky at $3.6' s^{-1}$ at four far-IR bands (65, 90, 140, and 160 μm) during the survey period from April 2006 to August 2007, achieving 99 % sky coverage (Doi *et al.* 2015; Takita *et al.* 2015). The presently archived AFASS map data (the Public Release Ver. 1, AFASSv1 hereafter)[†] are absolutely calibrated against large-scale background emission detected by *COBE*/DIRBE (Matsuura *et al.* 2011). Ueta *et al.* (2018) have recently verified that point-source photometry directly measured from the AFASS maps would be consistent with fluxes in the *AKARI* bright source catalogue ver. 2 (BSCv2, hereafter; Yamamura *et al.* 2009), for which photometry was done directly from time-series detector signal readouts of the scan observations.

[†] Maintained by ISAS/JAXA at the Data ARchives and Transmission System (DARTS; http://www.darts.isas.ac.jp/astro/akari/).

Table 1. Preliminary detections of PNe from HASHPNDB in the AFASSv1 maps

	BSC					AFASS				
PN Type	**N60**	**WIDE-S**	**WIDE-L**	**N160**	**All**	**N60**	**WIDE-S**	**WIDE-L**	**N160**	**All**
True	407	768	195	54	30	636	891	368	277	107
Likely	32	70	22	8	4	58	76	31	36	7
Possible	76	122	56	28	18	93	78	57	43	12
All	515	960	273	90	52	787	1045	456	356	126

2. Galactic PNe Detected in the BSC and AFASS Maps

We have used both the BSCv2 and AFASSv1 to obtain far-IR fluxes of PNe based on the University of Hong Kong/Australian Astronomical Observatory/Strasbourg Observatory Hα Planetary Nebula database ((HASHPNDB; Parker *et al.* 2017). This work is motivated by recent suggestions that the colder dusty photo-dissociation regions (PDRs) are important in understanding the global energetics of objects such as PNe that are known as the hallmark of ionized/H^+ regions (e.g., Otsuka *et al.* 2017).

First, by crossmatching HASHPNDB and BSCv2 with a $30''$ search radius, we find 515, 960, 273, and 90 detections with FQUAL = 3 (i.e., detection validated) in the N60, WIDE-S, WIDE-L, and N160 bands at 65, 90, 140, and 160 μm, respectively. There are 52 PNe for which the flux is measured in all four bands.

Then, we search AFASSv1 at each of the HASHPNDB coordinates in each band to see if there is any source detection. One notable potential complication here is that PNe are not necessarily always point sources. Our method essentially starts with the peak of the target PN at the HASHPNDB coordinates and continues to enlarge the extent of the target PN defined by the thresholding surface brightness by lowering the thresholding surface brightness until the presence of another object is recognized. We preliminary measure far-IR fluxes of 787, 1045, 456, and 356 PNe in AFASSv1 in the N60, WIDE-S, WIDE-L, and N160 bands, respectively (272, 85, 183, and 266 more detections than BSCv2), for the total of 1321 PNe, as summarized in Table 1.

Presently, we are hashing out false positives included in these preliminary detections. We will present the results of analyses of legitimate far-IR PN detections in due course.

Acknowledgements

This research is based on observations with AKARI, a JAXA project with the participation of ESA. The present research also has made use of the HASH PN database (http://hashpn.space) maintained at the University of Hong Kong.

References

Doi, Y., *et al.* 2015, *PASJ*, 1503, 6421
Kawada, M., *et al.* 2007, *PASJ*, 59, S389
Matsuura, S., *et al.* 2011, *ApJ*, 737, 2
Murakami, H., *et al.* 2007, *PASJ*, 59, S369
Neugebauer, G., *et al.* 1984, *ApJ*, 278, L1
Onaka, T., *et al.* 2007, *PASJ*, 59, 401
Otsuka, M., *et al.* 2017, *ApJS*, 231, 22

Parker, Q. A., *et al.* 2017, in IAU Symp. 323, Planetary Nebulae: Multi-Wavelength Probes of Stellar and Galactic Evolution, eds. Liu, X.-W., Stanghellini, L., & Karakas, A., p.36
Ueta, T., *et al.* 2018, *PASJ*, in press (arXiv:1808.00730)
Takita, S., *et al.* 2015, *PASJ*, 67, 51
Yamamura, I., *et al.* 2009, in ASP Conf. Ser. Vol. 418, AKARI, a Light to Illuminate the Misty Universe, eds. Onaka, T., White, G. J., Nakagawa, T., & Yamamura, I., p.3

Why Galaxies Care About AGB Stars:
A Continuing Challenge through Cosmic Time
Proceedings IAU Symposium No. 343, 2019
F. Kerschbaum, M. Groenewegen & H. Olofsson, eds.

doi:10.1017/S1743921318007858

Dust Structure Around Asymptotic Giant Branch Stars

Devendra Raj Upadhyay[1,2], Lochan Khanal[1], Priyanka Hamal[1] and Binil Aryal[1]

[1]Amrit Campus, Tribhuvan University, G.P.O. Box: 102, Lainchour, Kathmandu, Nepal
email: mnadphy03@gmail.com

[2]Central Department of Physics, Tribhuvan University, Kathmandu, Nepal
email: aryalbinil@gmail.com

Abstract. This paper presents mass, temperature profile, and the variation of Planck's function in different regions around asymptotic giant branch (AGB) stars. The physics of the interstellar medium (ISM) is extremely complex because the medium is very inhomogeneous and is made of regions with fairly diverse physical conditions. We studied the dust environment such as flux, temperature, mass, and inclination angle of the cavity structure around C-rich asymptotic giant branch stars in 60 μm and 100 μm wavelengths band using Infrared Astronomical Survey. We observed the data of AGB stars named IRAS 01142+6306 and IRAS 04369+4501. Flexible image transport system image was downloaded from Sky View Observatory; we obtained the surrounding flux density using software Aladin v2.5. The average dust color temperature and mass are found to be 25.08 K, 23.20 K and 4.73×10^{26} kg (0.00024 $M_\odot$), 2.58×10^{28} kg (0.013 $M_\odot$), respectively. The dust color temperature ranges from 18.76 K $\pm$ 3.16 K to 33.21 K $\pm$ 4.07 K and 22.84 K $\pm$ 0.18 K to 24.48 K $\pm$ 0.63 K. The isolated cavity like structure around the AGB stars has an extension of 45.67 pc $\times$ 17.02 pc and 42.25 pc $\times$ 17.76 pc, respectively. The core region is found to be edge-on having an inclination angle of 79.46° and 73.99°, respectively.

Keywords. Stars: AGB and post-AGB – mass-loss – circumstellar matter – dust – ISM: dust

1. Introduction

The interstellar medium (ISM) is the low density matter space between stars. It contains by mass 99% gas and 1% dust particles. The percentage of gas particles is 91% hydrogen, 9% helium and 0.1% are the atoms of elements heavier than hydrogen and helium. In this research work, we discuss how mass is distributed around the AGB stars, the temperature distribution and the structure of regions of dust around the AGB stars which forms near the cavity like structure and gives a wide range of area for the interaction between ISM and an AGB star. Aryal *et al.* (2006) studied nebulae and Jha *et al.* (2017) have done research work about KK loops and ATNF pulsars. Here, we have presented work related to asymptotic giant branch stars by using similar methods. We have investigated the surrounding temperature, mass and cavity around the C-rich AGB stars at a distance 4.49 kpc and 4.92 kpc using the paper by Guandalini *et al.* (2006). This research has made use of the SIMBAD (http://simbad.u-strasbg.fr/simbad/sim-fcoo (2018)) database and cross checked from SkyView Virtual Observatory (http://skyview.gsfc.nasa.gov/current/cgi/query.pl (2018)) respectively, using data reduction software like Aladin v2.5, Aladin v10 and other software, Origin 5.0 and 8.0, etc. and we obtained different phenomena regarding AGB stars. This work has been done using a formulation for dust color estimation by Schnee *et al.* (2005) and Dupac *et al.* (2005) whereas for dust mass and for inclination angle we have extensively used Young *et al.* (1993) and Holmberg *et al.* (1946) theoretical work, respectively.

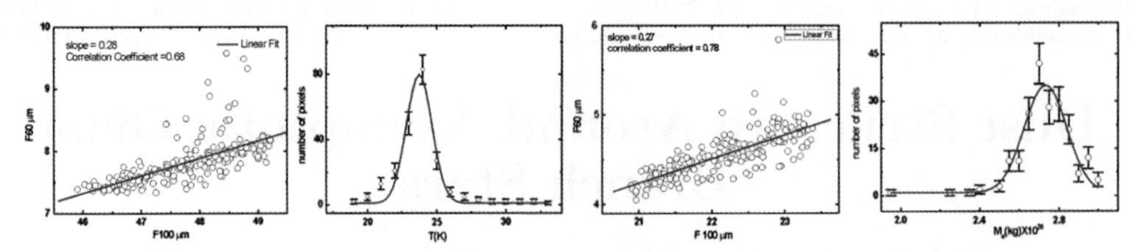

Figure 1. From left to right: flux at 60μm vs 100μm, Gaussian distribution of dust color temperature of star IRAS 01142+6306, flux at 60μm vs 100μm, and dust mass distribution of IRAS 04369+4501.

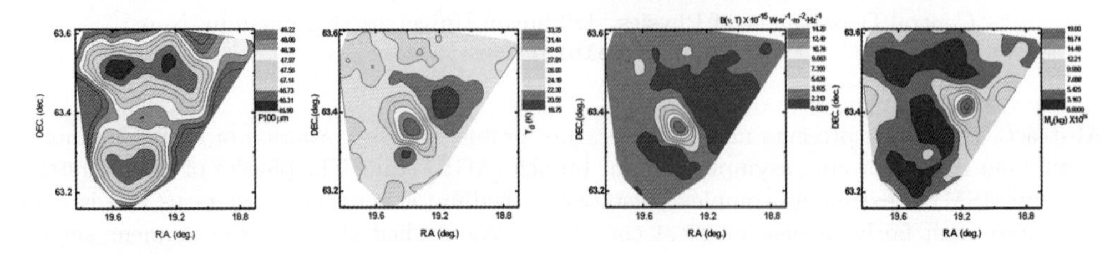

Figure 2. From left to right: flux, dust color temperature, intensity of radiation, and dust mass distribution contour plot, which follows expected trends i.e. mass is nearly inversely to the temperature distribution whereas the intensity of radiation is following temperature.

2. Results & Discussion

The maximum, minimum and average temperature of the ambient ISM around the AGB stars are found to be 33.21 K ± 4.06 K, 18.76 K ± 3.16 K, 25.08 K and 22.84 K ± 0.18 K to 24.48 K ± 0.63 K and 23.20 K, respectively. From more and less difference in temperature we realized that the stars are in the late and early AGB phase, respectively, whereas total mass and mass per pixel around AGB stars are found to be 1.04×10^{29} kg ($0.52\,M_\odot$), 4.73×10^{26} kg ($0.00024\,M_\odot$) and 5.156×10^{30} kg ($2.58\ M_\odot$), 2.58×10^{28} kg ($0.013\,M_\odot$), respectively. The inclination angle of the core region of IRAS01142+6306 and IRAS04369+450 are found to be 79.46° and 73.99°, suggesting edge-on appearance whereas sizes of the structures are found to be 45.67 pc × 17.02 pc and 42.25 pc × 17.76 pc, respectively.

Acknowledgement

We acknowledge CDS, Sky View Virtual Observatory, International Astronomical Union, Central Department of Physics, Tribhuvan University, Nepal, University Grant Commission, Nepal and Amrit Campus, Tribhuvan University, Nepal, for their different supports during this research work.

References

Aryal, B. & Weinberger, R., 2006, *Astronomy & Astrophysics* 448, 213.

Jha, A. K., Aryal, B. & Weinberger, R., 2017, *Revista Mexicana de Astronoma y Astrofísica* 53(2) 467-476

Guandalini, R., Busso, M., Ciprini, S., Silvestro, G., and P. Persi, 2006, *A&A* 445, 1069-1080

Schnee, S. L., Ridge, N.A., Goodman, A.A. & Jason, G.L., 2005, *ApJ*, 634, 442-450.

Dupac, X., Bernard, J.P., Boudet, N., Giard, M., Lamarre, J.M., Mny, C., Pajot, F., Ristorcelli, I., Serra, G., B. Stepnik, J.P. Torre, 2003, *A & A* 404 .

Young, K., Philips, T.G., Knapp, G.R., 1993, *ApJ*, 409, 725.

Holmberg, E. *Medd. Lund Astron. Obs. Ser. VI*, 117.

Why Galaxies Care About AGB Stars: A Continuing Challenge through Cosmic Time
Proceedings IAU Symposium No. 343, 2019
F. Kerschbaum, M. Groenewegen & H. Olofsson, eds.

doi:10.1017/S1743921318007354

Looking for new water-fountain stars

L. Uscanga[1], J. F. Gómez[2], B. H. K. Yung[3], H. Imai[4], J. R. Rizzo[5], O. Suárez[6], L. F. Miranda[2], M. A. Trinidad[1], G. Anglada[2] and J. M. Torrelles[7]

[1]University of Guanajuato, A.P. 144, 36000 Guanajuato, Gto., Mexico
email: lucero@astro.ugto.mx

[2]IAA−CSIC, Glorieta de la Astronomía s/n, 18008 Granada, Spain

[3]NCAC, ul. Rabiańska 8, 87-100 Toruń, Poland

[4]Kagoshima University, Korimoto 1-21-24, Kagoshima 890-8580, Japan

[5]CAB, INTA-CSIC, Ctra de Torrejón a Ajalvir, km 4, 28850 Torrejón de Ardoz, Madrid, Spain

[6]Observatoire de la Côte dAzur, CNRS, Laboratoire Lagrange, France

[7]ICE, CSIC, Carrer de Can Magrans, s/n 08193 Barcelona, Spain

Abstract. We carried out simultaneous observations of H_2O and OH masers, and radio continuum at 1.3 cm with the Karl G. Jansky Very Large Array (VLA) towards 4 water-fountain candidates. Water fountains (WFs) are evolved stars, in the AGB and post-AGB phase, with collimated jets traced by high-velocity H_2O masers. Up to now, only 15 sources have been confirmed as WFs through interferometric observations. We are interested in the discovery and study of new WFs. A higher number of these sources is important to understand their properties as a group, because they may represent one of the first manifestations of collimated mass-loss in evolved stars. These observations will provide information about the role of magnetic fields in the launching of jets in WFs. Our aim is to ascertain the WF nature of these candidates, and investigate the spatial distribution of the H_2O and OH masers.

Keywords. masers –stars: AGB and post-AGB

1. Introduction

The H_2O maser emission in WFs traces motions that are significantly faster than the typical expansion velocities of circumstellar envelopes during the AGB phase ($10-30\,\mathrm{km\,s^{-1}}$; Sevenster *et al.* 1997). These expansion velocities are traced by the double-peaked OH spectra typically seen in AGB and post-AGB stars, and thus, WF candidates are identified by H_2O maser spectra whose velocity spread is larger than that of the OH (te Lintel Hekkert *et al.* 1989; Imai *et al.* 2008).

With our recent confirmation of IRAS 17291−2147 and IRAS 18596+0315 as WFs (Gómez *et al.* 2017), there is a total of 15 WFs. Interferometric observations (with angular resolutions better than a few arcsec) are necessary, given the relatively low angular resolution of the single-dish observations (around 1 arcmin).

There remain several sources that are considered to be WF candidates based on single-dish observations (Yung *et al.* 2013, 2014, Gómez *et al.* 2015). Despite showing larger velocities in their H_2O maser spectrum than in OH, the velocity spread of some of these candidates is relatively low ($\simeq 40\,\mathrm{km\,s^{-1}}$), compared with the case of some WFs ($>100\,\mathrm{km\,s^{-1}}$). This may be a projection effect, if jets are moving close to the plane of the sky, but they might also represent a different kind of source (e.g., with less massive progenitors, or being relatively younger sources).

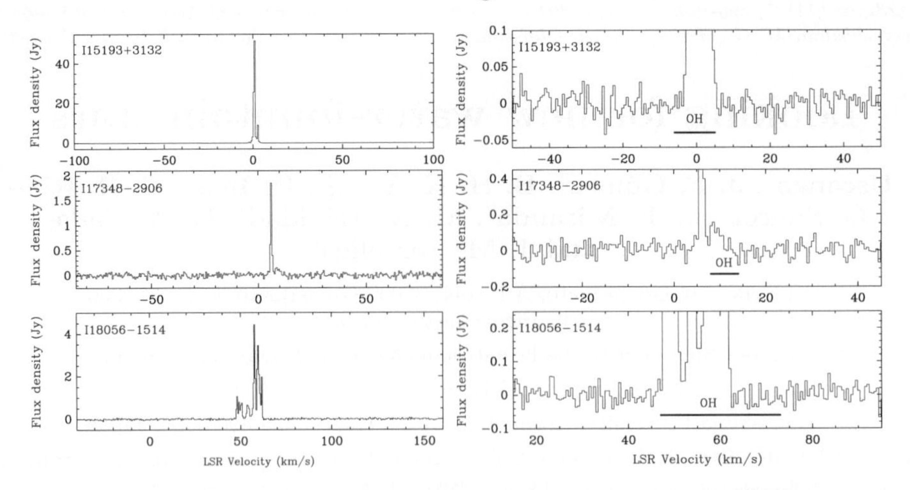

Figure 1. H_2O maser spectra toward the WF candidates, shown with different velocity and flux density ranges. The velocity range between the two OH maser peaks are shown by the thick lines.

2. Observations and Results

Our observations were carried out with the VLA of the National Radio Astronomy Observatory (NRAO), in its C configuration, on July and August 2017. We calibrated and processed the data using the CASA package. H_2O maser spectra was smoothed to a velocity resolution of 0.5 km s^{-1}.

Our main targets were IRAS 15193+3132, IRAS 17021−3109, IRAS 17348−2906 and IRAS 18056−1514. We confirm through interferometric observations that H_2O maser emission is associated with all sources, except in IRAS 17021−3109. Here, we present preliminary results of the H_2O maser spectra (Fig. 1). From these new data, more maser features are identified, showing variability in the flux density compared with the previous data. Unfortunately, some H_2O features with the higher velocity difference respect to the OH features, detected in the previous single-dish observations are absent in the new interferometric observations, so it is not possible to confirm whether they were associated with the target sources. Therefore, with the current data, we cannot confirm the nature of these candidates as WFs.

3. Summary

These WF candidates presented relatively high-velocity features in previous single-dish observations that have disappeared in the new interferometric observations. These sources could be different from the classical WFs, whose high-velocity emission is more stable over the time. Monitoring these candidates is crucial to ascertain their nature as WFs, and determine whether WF characteristics occur during a very limited time in stellar evolution, or whether they are recurrent episodes during longer periods of time.

References

Gómez, J. F., Rizzo, J. R., Suárez, O., Palau, A., *et al.* 2015, *A&A*, 578, A119

Gómez, J. F., Suárez, O., Rizzo, J. R., Uscanga, L., *et al.* 2017, *MNRAS*, 468, 2081

Imai, H., Diamond, P., Nakashima, J.-I., Kwok, S., *et al.* 2008, in *The role of VLBI in the Golden Age for Radio Astronomy*, Proc. 9th EVN Symposium (Proceedings of Science), p. 60

Sevenster, M. N., Chapman, J. M., Habing, H. J., Killeen, N. E. B., *et al.* 1997, *A&AS*, 122, 79

te Lintel Hekkert, P., Versteege-Hensel, H. A., Habing, H. J., Wiertz, M. 1989, *A&AS*, 78, 399

Yung, B. H. K., Nakashima, J.-I., & Henkel, C. 2014, *ApJ*, 794, 81

Yung, B. H. K., Nakashima, J.-I., Imai, H., Deguchi, S., *et al.* 2013, *ApJ*, 769, 20

Why Galaxies Care About AGB Stars: A Continuing Challenge through Cosmic Time
Proceedings IAU Symposium No. 343, 2019
F. Kerschbaum, M. Groenewegen & H. Olofsson, eds.

doi:10.1017/S1743921318006610

Does 3rd dredge-up reduce AGB mass-loss?

Stefan Uttenthaler[1], Iain McDonald[2], Klaus Bernhard[3,4], Sergio Cristallo[5,6] and David Gobrecht[7]

[1]Kuffner Observatory, Johann-Staud-Straße 10, 1160 Vienna, Austria
email: stefan.uttenthaler@gmail.com

[2]Jodrell Bank Centre for Astrophysics, Alan Turing Building, Manchester, M13 9PL, UK

[3]Bundesdeutsche Arbeitsgemeinschaft für Veränderliche Sterne e.V. (BAV), Berlin, Germany

[4]American Association of Variable Star Observers (AAVSO), Cambridge, MA, USA

[5]INAF – Osservatorio Astronomico, 64100 Italy

[6]INFN – Sezione di Perugia, Italy

[7]Instituut voor Sterrenkunde, Celestijnenlaan 200D, bus 2401, 3001 Leuven, Belgium

Abstract. We follow up on a previous finding that Miras containing the third dredge-up (3DUP) indicator technetium (Tc) in their atmosphere form a different sequence of $K - [22]$ colour as a function of pulsation period than Miras without Tc. A near-to-mid-infrared colour such as $K - [22]$ is a good probe for the dust mass-loss rate (MLR) of these AGB stars. Contrary to what one might naïvely expect, Tc-poor Miras show *redder* $K - [22]$ colours (i.e. higher dust MLRs) than Tc-rich Miras at a given period. In the follow-up work, the previous sample is extended and the analysis is expanded towards other colours and *ISO* dust spectra to check if the previous finding is due to a specific dust feature in the 22 μm band. We also investigate if the same two sequences can be revealed in the gas MLR. Different hypotheses to explain the observation of two sequences in the P vs. $K - [22]$ diagram are discussed and tested, but so far none of them convincingly explains the observations.

Uttenthaler (2013) demonstrated that two separate sequences of Miras exist in the P vs. $K - [22]$ diagram, if a distinction is made for the presence of Tc. Surprisingly, at a given period, Tc-poor Miras show *redder* $K - [22]$ colours (i.e. higher dust MLRs) than Tc-rich ones. Here, we report about the follow-up work Uttenthaler *et al.* (2018). Fig. 1 shows the updated P vs. $K - [22]$ diagram of our sample Miras. The two sequences have a tight correlation between $K - [22]$ and period. The $K - [22]$ sequence of Tc-rich Miras extends to longer periods and is much steeper. Also, the two groups are separated remarkably well in this diagram, most of the outliers can be explained by binarity or hot bottom burning.

To check if the two sequences are caused by a specific dust feature in the *WISE* 22 μm band, we expanded the analysis to other near-to-mid-IR colours and *ISO* dust spectra. Indeed, the two sequences can clearly be revealed with colours including bands as red as *Akari* 90 μm. Also *ISO* dust spectra reveal that Tc-rich Miras have much less dust emission than their Tc-poor siblings with similar periods. The previous finding based on the $K - [22]$ colour is thus not due to a specific dust feature. This could mean that, at a given period, Tc-poor Miras have a higher dust MLR than Tc-rich (post-3DUP) Miras.

An important question that arises from this is, if the two sequences can also be found in the gas MLRs? It would be important to know if only $K - [22]$ is affected by 3DUP, or also the gas MLR. This could help to identify the underlying process(es). We collected from the literature MLR data measured from CO radio lines to address this question. We confirm a clear increase of the gas MLR with increasing period, but unfortunately, the

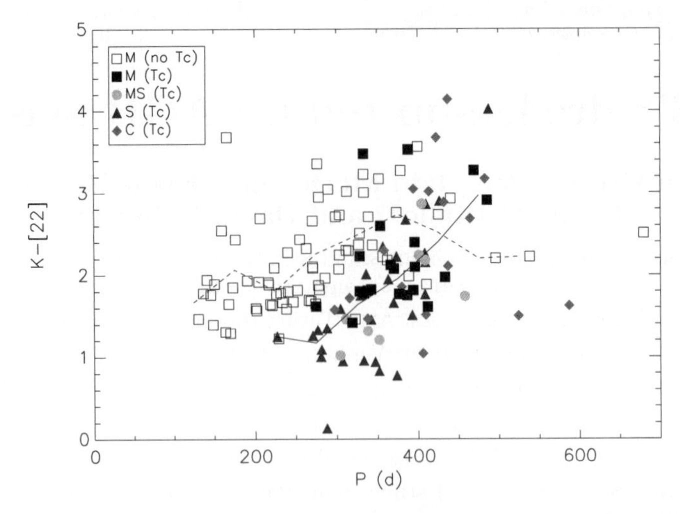

Figure 1. P vs. $K - [22]$ diagram of our sample Miras with and without Tc, see legend. The solid and dashed lines show a running mean of $K - [22]$ in 50 d period bins.

data are inconclusive on the existence of two sequences. Too few Miras with $P \lesssim 400$ d have been observed, there is little overlap between the Tc sample and CO observations, and large uncertainties attached to gas MLR measurements cause scatter. We encourage observers to target more short-period Miras with information on their Tc content in radio CO observations.

We put forward hypotheses to explain the surprising and unexpected observations:

H 1: *3DUP influences the pulsation period.* This is unlikely both on observational (Riebel *et al.* 2015) and theoretical (Scholz, Ireland & Wood 2014) grounds. This hypothesis can thus be discarded.

H 2: *The stars are groups of different masses that evolve differently in the P vs. $K - [22]$ plane.* We inspected the distance to the Galactic mid-plane and the radial velocity dispersion as indicators of age and hence mass (Feast 1963). None of them reveals a difference between Tc-poor and Tc-rich Miras at a given period.

H 3: *3DUP reduces MLR.* The decay of radioactively unstable isotopes could somehow reduce dust formation and thus MLR. Glassgold (1995) investigates the formation of ions in circumstellar shells of AGB stars by the decay of isotopes such as ^{26}Al. The effect on MLR is unclear, but ions could play a role in dust formation.

H 4: *3DUP reduces dust emissivity.* Similar to H 3, but only the dust emissivity (thus $K - [22]$) is reduced upon 3DUP, not MLR. Also here, the physical mechanism is unclear.

H 5: *Tc lines are unobservable in some stars.* This could be the case, if Tc line formation strongly depends on stellar parameters. Tc chemistry and line formation are indeed not well understood.

At the moment, none of the hypotheses satisfyingly explains the observations. Nevertheless, some of them deserve deeper investigation.

References

Feast, M.W. 1963, *MNRAS*, 125, 367
Glassgold, A.E. 1995, *ApJ*, 438, L111
Riebel, D., Boyer, M.L., Srinivasan, S., *et al.* 2015, *ApJ*, 807, 1
Scholz, M., Ireland, M.J., & Wood, P.R. 2014, *A&A*, 565, A119
Uttenthaler, S. 2013, *A&A*, 556, A38
Uttenthaler, S., McDonald, I., Bernhard, K., *et al.* 2018, submitted to *A&A*

Why Galaxies Care About AGB Stars:
A Continuing Challenge through Cosmic Time
Proceedings IAU Symposium No. 343, 2019
F. Kerschbaum, M. Groenewegen & H. Olofsson, eds.

doi:10.1017/S1743921318005434

The chemistry in clumpy AGB outflows

M. Van de Sande[1], J. O. Sundqvist[1], T. J. Millar[2], D. Keller[1] and L. Decin[1,3]

[1]Department of Physics and Astronomy, Institute of Astronomy, KU Leuven, Celestijnenlaan 200D, 3001 Leuven, Belgium
email: marie.vandesande@kuleuven.be

[2]Astrophysics Research Centre, School of Mathematics and Physics, Queen's University Belfast,
University Road, Belfast BT7 1NN, UK

[3]School of Physics and Astronomy University of Leeds, Leeds LS2 9JT, UK

Abstract. The chemistry within the outflow of an AGB star is determined by its elemental C/O abundance ratio. Thanks to the advent of high angular resolution observations, it is clear that most outflows do not have a smooth density distribution, but are inhomogeneous or "clumpy". We have developed a chemical model that takes into account the effect of a clumpy outflow on its gas-phase chemistry by using a theoretical porosity formalism. The clumpiness of the model increases the inner wind abundances of all so-called unexpected species, i.e. species that are not predicted to be present assuming an initial thermodynamic equilibrium chemistry. By applying the model to the distribution of cyanopolyynes and hydrocarbon radicals within the outflow of IRC+10216, we find that the chemistry traces the underlying density distribution.

Keywords. astrochemistry, molecular processes, stars: AGB and post-AGB stars, circumstellar material, stars: mass loss, stars: individual (IRC+10216)

1. Introduction

By using a theoretical porosity formalism (e.g., Owocki *et al.* 2006; Sundqvist *et al.* 2014), we have developed a model of the gas-phase chemistry that takes into account the effects of clumpiness throughout the outflow. This model is the first model that includes the effect of both the increased penetration rate of interstellar UV photons (porosity) and the relative over-density within clumps (clumping). Porosity is taken into account by modifying the optical depth of the outflow. A crucial parameter for this is the porosity length h_* at the stellar surface, which can be seen as the local mean free path between two clumps. Clumping is achieved by splitting the outflow into an overdense clump component and a rarefied inter-clump component (Van de Sande *et al.* 2018).

2. Effect on the inner wind abundances

In the inner regions of AGB outflows, several molecules have been detected with abundances much higher than expected from thermodynamic equilibrium chemical models. Non-equilibrium chemical models that take into account the effect of shocks, induced by the pulsating AGB star, can account for the presence of most, but not all, of these so-called unexpected species, but not all. We found that a clumpy outflow influences the abundances of the unexpected species. They can all form close to the star, including species not formed by shock-induced non-equilibrium chemical models (e.g., NH_3). The largest increase is seen in outflows with a large clump overdensity and large porosity (Van de Sande *et al.* 2018). The effect of porosity on the abundances is illustrated in

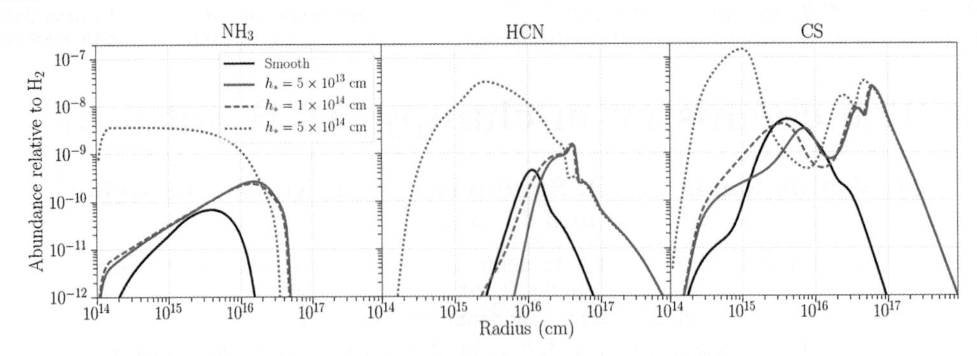

Figure 1. Abundance of NH_3, HCN, and CS relative to H_2 throughout an O-rich outflow with a mass-loss rate of 10^{-5} $M_\odot$ yr^{-1}. In the outflows shown here, all material is located inside the clumps (void inter-clump component) and the clumps take up 10% of the total volume.

Fig. 1. However, not all observed abundances can be reproduced. This can potentially be solved by combining shock-induced non-equilibrium chemistry and a clumpy outflow.

3. Cyanopolyynes and hydrocarbon radicals in IRC+10216

Cyanopolyynes, $HC_{2n+1}N$, and hydrocarbon radicals, $C_{2n}H$ ($n = 1, 2, 3...$), are located in molecular shells around IRC+10216. The $HC_{2n+1}N$ shells show a radial sequence, which is expected from a bottom-up formation route. The $C_{2n}H$ shells, however, are observed to be cospatial. The outflow of IRC+10216 is not smooth. Density-enhanced shells are present, which are moreover distinctly clumpy (e.g., Agúndez *et al.* 2017; Keller, *et al.* 2018). Previous chemical models have aimed to explain the discrepancy between the locations of the $HC_{2n+1}N$ and the $C_{2n}H$ families by either taking into account the effect of the density-enhanced shells (i.e. Cordiner & Millar 2009) or the increased UV radiation rate due to their clumpy nature (i.e. Agúndez *et al.* 2017). Both types of models found a cospatial distribution of both the $HC_{2n+1}N$ and the $C_{2n}H$.

We used our clumpy model to quantify the effects of clumping and porosity on the locations of the molecular shells. We find that one model cannot reproduce the observed locations of both the $HC_{2n+1}N$ and the $C_{2n}H$. Rather, we find two independent families of models. The family of solutions that best reproduce the locations of the $HC_{2n+1}N$ are characterised by a larger mass-loss rate (corresponding to a larger overall density) and a lower porosity compared to the family of solutions of the $C_{2n}H$ shells. This corresponds to the observed structure in the outflow: the $HC_{2n+1}N$ are located inside a density-enhanced shell, whereas the $C_{2n}H$ are located at the outer edge of the shell. The density distribution within the outflow can hence be traced by simple chemical models. However, to reproduce the locations of both the $HC_{2n+1}N$ and the $C_{2n}H$, a more comprehensive model is needed. It is clear that such a model must also include the significant effects of clumping and porosity on the chemistry.

References

Agúndez, M., Cernicharo, J., Quintana-Lacaci, G. *et al.* 2017, *A&A*, 601, A4
Cordiner, M.A. & Millar, T.J. 2009, *ApJ*, 697, 68
Keller, D., de Koter, A., Decin, L. *et al.* 2018, *A&A*, under review
Owocki, S. P. & Cohen, D. H. 2006, *ApJ*, 648, 565
Sundqvist, J. O., Puls, J. & Owocki, S. P. 2014, *A&A*, 568, A59
Van de Sande, M., Sundqvist, J. O., Millar, T. J. *et al.* 2018, *A&A*, 616, A106

Why Galaxies Care About AGB Stars:
A Continuing Challenge through Cosmic Time
Proceedings IAU Symposium No. 343, 2019
F. Kerschbaum, M. Groenewegen & H. Olofsson, eds.

doi:10.1017/S1743921318007214

Radial velocity variability in post-AGB stars: V448 Lac

G. C. Van de Steene[1], B. J. Hrivnak[2] and H. Van Winckel[3]

[1]Royal Observatory of Belgium, Astronomy and Astrophysics, Ringlaan 3, Brussels, Belgium
email: g.vandesteene@oma.be

[2]Department of Physics and Astronomy, Valparaiso University, Valparaiso, IN 46383, USA
email: bruce.hrivnak@valpo.edu

[3]Instituut voor Sterrenkunde, K.U. Leuven University, Celestijnenlaan 200 D, B-3001 Leuven, Belgium
email: hans.vanwinckel@kuleuven.ac.be

Abstract. To investigate the binary hypothesis in the formation of planetary nebulae, we have been doing long-term photometry and radial velocity (RV) monitoring of bright post-AGB stars which possess bipolar or ellipsoidal nebulae but no indication of a disk in their spectral energy distribution, indicative of a binary companion. RV's are determined by cross correlating high-resolution spectra with a line mask. Stellar variability and companions both deform the cross correlation function (CCF) and induce periodic variations in the RV. To uniformly quantify the asymmetry of the CCF from a Gaussian, we propose to fit the CCF profile with a Gauss-Hermite series and determine all CCF parameters (RV, skewness, FWHM, and depth) in one single fit. We analyze the correlation and time series of these CCF parameters for V448 Lac and conclude that its RV variability is most likely due to stellar pulsation and not to an orbiting body.

Keywords. line:profiles, methods:data analysis, techniques: radial velocities, stars: AGB and post-AGB, individual (V448 Lac)

---

1. Introduction

We have been doing long-term photometry and radial velocity study of seven bright post-AGB stars which possess bipolar or ellipsoidal nebulae but no indication of a disk in their spectral energy distribution, to investigate the binary hypothesis for shaping planetary nebulae (Hrivnak *et al.* 2013, Hrivnak *et al.* 2017).

V448 Lac (IRAS 22223+4327) is a metal-poor, C-rich, F-type post-AGB star. Its spectral energy distribution is double peaked, with a peak in the visible arising from the (reddened) photosphere and a second peak in the mid-infrared arising from reradiation from cool dust, but has no near-infrared excess which would be an indication for the presence of a disk. The disk type post-AGB stars have been shown to be binaries (Manick *et al.* 2017).

2. Observations and analysis

We obtained 107 high-resolution spectra of V448 Lac with the HERMES spectrograph (Raskin *et al.* 2011) mounted on the 1.2-m Mercator telescope at La Palma from 2009 to 2017 in the framework of a large program on evolved binaries (Van Winckel *et al.* 2010). The individual RV's are obtained by cross-correlating the reduced HERMES spectra with a software mask for F0 type star containing a large set of distinct spectral lines typical for the spectral type. The CCF profile is a weighted mean of all photospheric lines included in the mask. Spectral lines of pulsating, active, and binary stars are notoriously asymmetric. To determine the RV, the CCF is usually fit by a Gaussian and several authors have tried to quantify the CCF asymmetry with an extra parameter related to the bisector (Queloz *et al.* 2001, Boissé *et al.* 2011) or by fitting a double Gaussian to the CCF (Figueira *et al.* 2013). To uniformly quantify the asymmetry of the CCF from a Gaussian, we propose to fit the CCF profile with a Gauss-Hermite series and determine all CCF parameters (RV, skewness, FWHM, and depth) consistently in one single fit for each observation. The lowest order term of the series is a Gaussian, higher orders quantify the (a)symmetry of the Gaussian profile. We fit the CCF using the Kapteyn package (Terlouw & Vogelaar 2015). The skewness parameter is equivalent to the bisector inverse span as defined by Queloz *et al.* (2001).

For V448 Lac we determined the correlation between the CCF shape indicators (depth, FWHM, and skewness) with RV. We find that the depth is not correlated with RV, the FWHM is anti-correlated with RV, and the skewness parameter is correlated with RV. The latter warns us that the signal is related to the stellar surface and not to an orbiting body. However, it is peculiar that the sign of the slope is opposite to what is usually found in the literature (Delgado *et al.* 2018).

3. Period determination

Period determination of the CCF parameters was done with the generalized Lomb Scargle method (GLS; Zechmeister & Kürster 2009). We find a period of 86.6 days for the RV and a strong period of 86.4 days for the CCF depth, both in agreement with the photometric period of 86.8 days. The RV is about −0.25 period out of phase with the light curve, while the depth of the CCF is in phase with the photometry. The star is brightest when hottest and smallest (Hrivnak *et al.* 2018), which is when the CCF is shallowest. For the skewness parameter and FWHM we don't find the periodicity of 86.6 days as found in the RV and photometry. When phased to 86.6 days they show more erratic variations.

Further analysis of the spectral lines in V448 Lac aims to investigate the phenomena of stellar variability causing these characteristics of the CCF parameters.

References

Boissé, I., Bouchy, F., Hébrard, G., *et al.* 2011, *A&A* 528, 4
Delgado Mena, E., Lovis, C., Santos, N. C., *et al.* 2018, arXiv:180709608
Figueira, P. ,Santos, N.C., Pepe, F., *et al.* 2013, *A&A* 557, A93
Hrivnak, B. J., Lu, W., Sperauskas, J., *et al.* 2013, *ApJ* 766, 116
Hrivnak, B.J.,Van de Steene, G.C., Van Winckel, *et al.* 2017, *ApJ* 846, 96
Hrivnak, B.J.,Van de Steene, G.C., Van Winckel, *et al.* 2018, *this volume*
Kochanek, C. S., Shappee, B. J., Stanek, K. Z., *et al.* 2017, *PASP* 129,4502
Manick, R., Van Winckel, H., Kamath, *et al.* 2017, *A&A* 597, A129
Raskin, G., Van Winckel, H., Hensberge, *et al.*, 2011, *A&A* 526A,69
Terlouw, J. P. & Vogelaar, M. G. R. 2015, *Kapteyn package*
Queloz, D., Henry, G.W., Sivan, J. P., *et al.* 2001, *A&A* 379, 279
Van Winckel, H., Jorissen, A., Gorlova, N., *et al.* 2010, *MmSAI* 81, 1022
Zechmeister, M. & Kürster, M. 2009, *A&A* 496, 577

*Why Galaxies Care About AGB Stars:
A Continuing Challenge through Cosmic Time
Proceedings IAU Symposium No. 343, 2019
F. Kerschbaum, M. Groenewegen & H. Olofsson, eds.*

doi:10.1017/S1743921318005410

Circumstellar chemistry of Si-C bearing molecules in the C-rich AGB star IRC+10216

L. Velilla-Prieto[1,2], J. Cernicharo[1], M. Agúndez[1], J. P. Fonfría[1], A. Castro-Carrizo[3], G. Quintana-Lacaci[1], N. Marcelino[1], M. C. McCarthy[4], C. A. Gottlieb[4], C. Sánchez Contreras[5], K. H. Young[4], N. A. Patel[4], C. Joblin[6] and J. A. Martín-Gago[7]

[1]Instituto de Física Fundamental (IFF-CSIC),
Serrano, 123, CP 28006, Madrid, Spain

[2]Dept. of Space, Earth and Environment, Chalmers Univ. of Technology,
Onsala Space Observatory, 43992 Onsala, Sweden
email: luis.velilla@chalmers.se

[3]Institut de Radioastronomie Millimétrique (IRAM),
38406 Saint-Martin-d'Hères, France

[4]Harvard Smithsonian Center for Astrophysics,
60 Garden Street, Cambridge, MA 02138, USA

[5]Centro de Astrobiologia, (CSIC-INTA),
28691 Villanueva de la Cañada, Spain

[6]IRAP - Université de Toulouse, CNRS, UPS, CNES,
BP 44346, 31028 Toulouse Cedex 4, France

[7]Instituto de Ciencia de Materiales de Madrid (ICMM-CSIC),
Cantoblanco, 28049 Madrid, Spain

Abstract. Silicon carbide together with amorphous carbon are the main components of dust grains in the atmospheres of C-rich AGB stars. Small gaseous Si-C bearing molecules (such as SiC, SiCSi, and SiC_2) are efficiently formed close to the stellar photosphere. They likely condense onto dust seeds owing to their highly refractory nature at the lower temperatures (i.e., below about 2500 K) in the dust growth zone which extends a few stellar radii from the photosphere. Beyond this region, the abundances of Si-C bearing molecules are expected to decrease until they are eventually reformed in the outer shells of the circumstellar envelope, owing to the interaction between the gas and the interstellar UV radiation field. Our goal is to understand the time-dependent chemical evolution of Si-C bond carriers probed by molecular spectral line emission in the circumstellar envelope of IRC+10216 at millimeter wavelengths.

Keywords. stars: AGB & post-AGB, (stars:) circumstellar matter, stars: individual (IRC+10216).

1. Introduction

Dust formation and growth is far from being well understood because there are many unknowns in the formation pathways, the condensation sequences of refractory species, and the dependence on stellar and circumstellar properties. Dust grains in the atmospheres of C-rich AGB stars are mainly composed of silicon carbide and amorphous carbon (Savage & Mathis 1979, and references therein). Therefore, the study of gas phase carriers of Si-C bonds in the envelopes of C-rich stars is a promising approach for

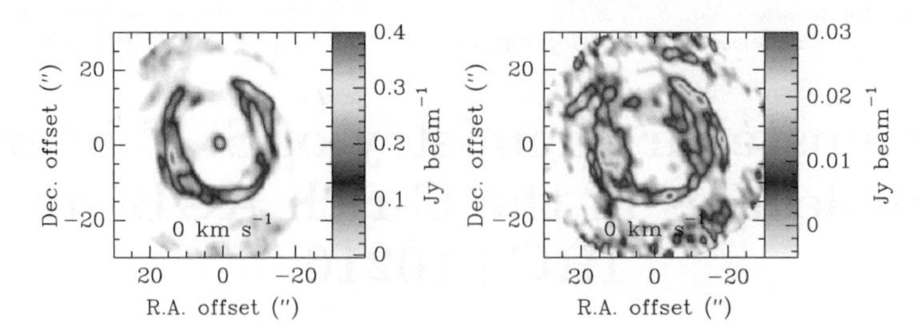

Figure 1. NOEMA maps of the SiC_2 $J_{Ka,Kc}$=$7_{0,7}$–$6_{0,6}$ (left) and the SiC Ω=2 J=4–3 e, f (right) emission lines toward IRC+10216 with v_{LSR}=$v - v_{sys}$ (v_{sys}=-26.5 km s^{-1}). The spectral resolution is $\Delta v \sim$4 km s^{-1}, and the spatial resolution is $\sim$2.″5.

shedding light on the formation of SiC dust. Here, we present our latest results derived from the analysis of single-dish (IRAM-30m) and interferometric data (SMA, NOEMA, and ALMA) of Si-C bearing, and other Si-bearing molecules in the circumstellar envelope of the prototypical AGB star IRC+10216.

2. Observations and analysis

We have carried out a comprehensive observational study of most of the Si-bearing molecules detected in the circumstellar envelope of IRC+10216. We have used ALMA, NOEMA, SMA, and the IRAM-30m telescope to observe several emission lines corresponding to SiC, SiC_2, SiCSi, SiS, and SiO. Calibration, reduction and merging standard procedures followed using CASA and GILDAS. The best spatial resolution achieved corresponds to the ALMA observations, with angular resolution as high as 0.″6.

Our analysis is based on radiative transfer analysis (MADEX, Cernicharo 2012) and chemical modelling of the molecules detected. The physical model of the CSE is essentially an updated revision of the models in Agúndez *et al.* (2012) and Cernicharo *et al.* (2013). Updated parameters are the kinetic temperature radial profile and mass-loss rate (Guélin *et al.* 2018). For more details about the chemical model see e.g. Velilla Prieto *et al.* (2015).

3. Results

SiC_2 and SiCSi are the main carriers of Si-C bonds in the innermost regions of the CSE with similar abundances in this zone ($\sim 10^{-7}$). This result is consistent with predictions of thermodynamical equilibrium (Cernicharo *et al.* 2015). The lack of SiC emission arising from the innermost regions of the CSE might be evidence that SiC is condensed onto the dust grains as one of the main building blocks of dust grains in C-rich CSEs (Treffers & Cohen 1974). In the outermost parts of the CSE, SiC and SiC_2 are reincorporated into the gas phase at $r \sim 15''$ (see Fig. 1).

We have derived the radial profiles of the fractional abundances of SiO and SiS, which present oscillations that probably reflect the mass-loss episodicity of the star. On average, the fractional abundances estimated are f(SiO)$\sim 10^{-7}$ and f(SiS)$\sim 10^{-6}$. The observations and analysis we have done pose strong evidences of a variable and episodic mass loss process acting on IRC+10216 creating dense shells with timescales of hundreds of years.

Acknowledgments

We thank funding support through ERC-2013-SyG-610256 (NANOCOSMOS), AYA2012-32032 and AYA2016-75066-C2-1-P grants.

References

Agúndez, M., Fonfría, J. P., Cernicharo, J., *et al.* 2012, *A&A*, 543, A48
Cernicharo, J. 2012, *EAS Publications Series*, 58, 251
Cernicharo, J., Daniel, F., Castro-Carrizo, A., *et al.* 2013, *ApJ* (Letter), 778, L25
Cernicharo, J., McCarthy, M. C., Gottlieb, C. A., *et al.* 2015, *ApJ* (Letter), 806, L3
Guélin, M., Patel, N. A., Bremer, M., *et al.* 2018, *A&A*, 610, A4
Savage, B. D., & Mathis, J. S. 1979, *ARAA*, 17, 73
Treffers, R., & Cohen, M. 1974, *ApJ*, 188, 545
Velilla Prieto, L., Sánchez Contreras, C., Cernicharo, J., *et al.* 2015, *A&A*, 575, A84

Why Galaxies Care About AGB Stars:
A Continuing Challenge through Cosmic Time
Proceedings IAU Symposium No. 343, 2019
F. Kerschbaum, M. Groenewegen & H. Olofsson, eds.

doi:10.1017/S1743921318004994

Measuring spatially resolved gas-to-dust ratios in AGB stars

Sofia Wallström[1], T. Dharmawardena[1], B. Rodríguez Marquina[2], P. Scicluna[1], S. Srinivasan[1], F. Kemper[1] and The NESS Collaboration

[1]Academia Sinica Institute of Astronomy and Astrophysics (ASIAA)
11F Astronomy-Mathematics Bldg., No.1, Sec. 4, Roosevelt Rd, Taipei 10617, Taiwan

[2]Sección Física, Departamento de Ciencias, Pontificia Universidad Católica del Perú
Av. Universitaria 1801 San Miguel, Lima 32, Perú

Abstract. Gas-to-dust ratios in Asymptotic Giant Branch (AGB) stars are used to calculate gas masses from measured dust masses and vice versa, but can vary widely and are rarely directly measured. In this work, we present spatially resolved gas and dust masses for a sample of 8 nearby AGB stars, using JCMT CO-line and continuum observations, and compare them. This serves as a pilot study for the Nearby Evolved Stars Survey (NESS; PI: P. Scicluna) project which will provide similar observations of $\sim$ 400 AGB stars in a volume-limited sample.

Keywords. stars: AGB and post-AGB, stars: mass loss, (stars:) circumstellar matter

---

1. Background

AGB stars form significant circumstellar envelopes of gas and dust, and the gas-to-dust ratio across the envelope carries vital information about, e.g., the dust-condensation efficiency. A canonical value of 100 or 200 is often used for Galactic stars. Knapp (1985) measured gas-to-dust ratios in Galactic evolved stars – finding ratios around 160 for O-rich stars, and 400 for C-rich stars – which remains the only direct measurement of this ratio for evolved stars in the Solar Neighbourhood. The Nearby Evolved Stars Survey (NESS; PI: P. Scicluna) project will provide CO and continuum observations of $\sim$ 400 AGB stars within 2 kpc. The statistical robustness of a volume-limited sample will allow us to address a range of questions about mass return to the Galactic ISM and the physics of mass loss from evolved stars, including direct determination of the gas-to-dust ratio. Mapping of a subsample of the closest sources allows the determination of spatially resolved gas-to-dust ratios and an investigation of changes in mass-loss over time.

2. Data analysis

We have used JCMT jiggle maps of CO 3–2 (with HARP) and 850 μm continuum (with SCUBA-2) emission on an initial sample of 8 nearby AGB stars from the NESS sample. Azimuthal averaging of the surface brightness was used to create radial profiles, assuming the circumstellar envelopes are spherically symmetric. From these radial profiles we subtracted the telescope PSF (beam $\sim 14''$) profiles to quantify the amount of extended emission. The continuum results are taken from Dharmawardena *et al.* (2018) who use SED fitting (70 – 850 μm) at each radial point to find a dust surface density and then integrate over a given annulus to find a dust mass. We get a crude estimate of the gas masses by integrating the CO emission within the PSF and using the empirical formula from Ramstedt *et al.* (2008) to calculate a central mass-loss rate, which is then multiplied by the spatial extent of the PSF and the expansion velocity to get a central

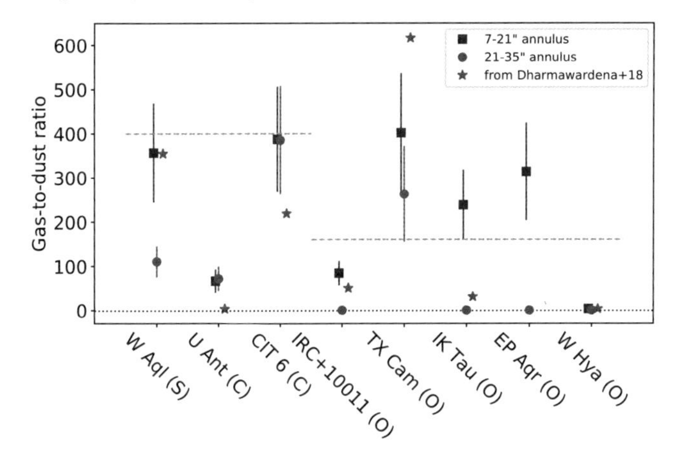

Figure 1. Gas-to-dust ratios integrated over radial annuli for each source, with uncertainties from the dust masses from Dharmawardena *et al.* (2018) plus 30%. Canonical values from Knapp (1985) are shown with grey dashed lines.

gas mass. This I_{CO} – M_{gas} relation for each source is applied to the extended emission, integrated in beam-width annuli, to estimate the extended gas masses.

3. Results

Across our sample, the median dust extent is 1.5 times larger than the CO extent, and a median of 50% of the 850 μm flux and 30% of the CO flux is in the extended component. The gas-to-dust ratios are shown in Fig. 1, calculated in integrated annuli from 7–21$''$ and 21–35$''$. Most sources have smaller gas-to-dust ratios in the outer annuli, consistent with the smaller gas extent which is likely due to a decrease in temperature leading to less efficient CO 3–2 emission, or CO dissociation by the interstellar radiation field. However, we see some distinction between chemical types: the two C-rich AGB stars, U Ant and CIT 6, show no decline in the gas-to-dust ratio. Furthermore, our sources do not seem to follow the overall gas-to-dust ratios found by Knapp (1985), which are indicated by grey dashed lines in Fig. 1. Also plotted are the gas-to-dust ratios calculated for the entire envelope by Dharmawardena *et al.* (2018), based on their continuum observations and CO modeling from De Beck *et al.* (2010). These total gas-to-dust ratios can vary significantly from our measurements, showing the importance of spatially resolving the emission and measuring both gas and dust.

4. Further work

We will model the CO maps with the 1D radiative transfer code MLINE (Kemper *et al.* 2003) to better constrain gas masses, and eventually extend our analysis to the full NESS sample. Proposed follow-up interferometric observations will resolve both CO and continuum emission, revealing detailed variations in the historic gas and dust mass-loss and gas-to-dust ratio for both individual sources and a statistical sample of AGB stars.

References

De Beck, E., Decin, L., de Koter, A., *et al.* 2010, *A&A*, 523, 18
Dharmawardena, T. E., Kemper, F., Scicluna, P., *et al.* 2018, *MNRAS*, 479, 536
Kemper, F., Stark, R., Justtanont, K., *et al.* 2003, *A&A*, 407, 609
Knapp, G. R. 1985, *ApJ*, 293, 273
Ramstedt, S., Schöier, F. L., Olofsson, H. & Lundgren, A. A. 2008, *A&A*, 487, 645

Why Galaxies Care About AGB Stars: A Continuing Challenge through Cosmic Time
Proceedings IAU Symposium No. 343, 2019
F. Kerschbaum, M. Groenewegen & H. Olofsson, eds.

doi:10.1017/S1743921318005689

WD+AGB star systems as the progenitors of type Ia supernovae

Bo Wang

Yunnan Observatories, Chinese Academy of Sciences, Kunming 650216, China
email: wangbo@ynao.ac.cn

Key Laboratory for the Structure and Evolution of Celestial Objects, Chinese Academy of Sciences, Kunming 650216, China

Abstract. WD+AGB star systems have been suggested as an alternative way for producing type Ia supernovae (SNe Ia), known as the core-degenerate (CD) scenario. In the CD scenario, SNe Ia are produced at the final phase during the evolution of common-envelope through a merger between a carbon-oxygen (CO) WD and the CO core of an AGB secondary. However, the rates of SNe Ia from this scenario are still uncertain. In this work, I carried out a detailed investigation on the CD scenario based on a binary population synthesis approach. I found that the Galactic rates of SNe Ia from this scenario are not more than 20% of total SNe Ia due to more careful treatment of mass transfer, and that their delay times are in the range of $\sim 90-2500\,\mathrm{Myr}$, mainly contributing to the observed SNe Ia with short and intermediate delay times.

Keywords. binaries: close – supernovae: general – white dwarfs

1. Introduction

Type Ia supernovae (SNe Ia) have high scientific values in the cosmic evolution. They are thought to be arised from thermonuclear explosions of carbon-oxygen white dwarfs (CO WDs) in binaries, although their progenitor systems are still unclear. Over the past few years, two classic progenitor models of SNe Ia have been suggested, i.e., the single-degenerate model and the double-degenerate model (e.g., Hachisu *et al.* 1996; Han & Podsiadlowski 2004; Wang *et al.* 2009; Toonen *et al.* 2012). Some variants of these two progenitor models are proposed to reproduce the observed diversity of SNe Ia (e.g., Wang & Han 2012; Maoz *et al.* 2014; Wang 2018).

Previous simulations on the double-degenerate model are mainly related to the merger of two cold CO WDs (e.g., Toonen *et al.* 2012). However, a CO WD can also merge with the hot CO core of an asymptotic giant branch (AGB) star, and then produce a SN Ia, known as the core-degenerate (CD) scenario (e.g., Soker 2013). In the CD scenario, a SN Ia explosion might occur shortly or a long time after the common-envelope (CE) phase. Although the CD scenario may explain some properties of SN Ia diversity, SN Ia rates from this scenario are still uncertain (e.g., Ilkov & Soker 2013). Ilkov & Soker (2013) argued that the CD scenario can explain the observed rates of total SNe Ia based on a simplified binary population synthesis (BPS) code. The purpose of this study is to investigate SN Ia rates and delay times for the CD scenario using a detailed Monte Carlo BPS approach.

2. Model and Results

In the CD scenario, a WD with Chandrasekhar or super-Chandrasekhar mass could be produced through the merger of a cold CO WD with the hot CO core of an AGB star.

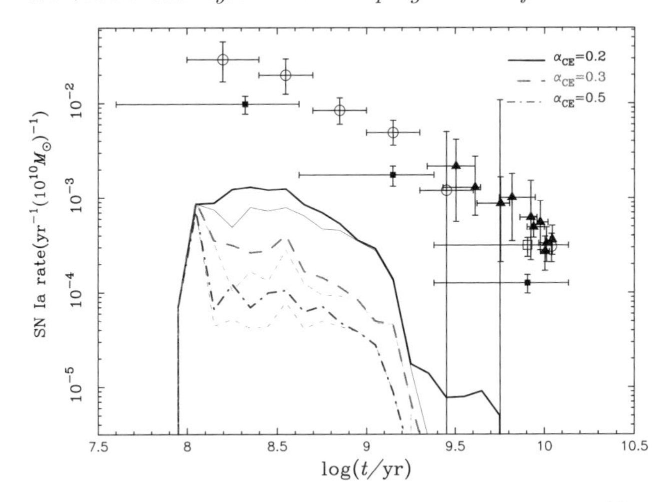

Figure 1. SN Ia delay time distributions based on a single starburst of 10^{10} M$_\odot$. The thick lines are for all SNe Ia from the CD scenario, and the thin lines are only for ones with circumstellar material like PTF 11kx. Source: From Wang *et al.* (2017).

A series of Monte Carlo BPS simulations for the CD scenario are performed using the Hurley binary evolution code (e.g., Hurley *et al.* 2002). Here, I consider three different values (e.g., 0.2, 0.3 and 0.5) of the CE ejection efficiency $\alpha_{\rm CE}$ to examine its influence on the final results. I adopted the following assumptions as the criteria for producing SNe Ia through the CD scenario (e.g., Soker 2013; Wang *et al.* 2017): (1) The WD and the AGB core merge during the final stage of CE evolution. (2) The combined mass of the CO WD ($\rm M_{WD}$) and the AGB core ($\rm M_{core}$) during the final phase of CE evolution is larger than or equal to the Chandrasekhar limit. (3) If $\rm M_{WD} > M_{core}$, I assume that the SN explosion occurs shortly after the CE stage, resulting in a SN Ia inside a planetary nebula shell like PTF 11kx.

Fig. 1 presents the SN Ia delay time distributions for the CD scenario based on a single starburst of 10^{10} M$_\odot$. The delay times of SNe Ia for this scenario are mainly in the range of $\sim 90 - 2500$ Myr after the starburst, which may contribute to the SNe Ia with short and intermediate delay times. I also found that the theoretical Galactic rates from this scenario are no more than 20% of the total SNe Ia in the observations. Especially, SNe Ia with circumstellar material from this scenario contribute to 0.7−10% of total SNe Ia, which indicates that the CD scenario can reproduce the observed rates of SNe Ia like PTF 11kx.

References

Hachisu, I., Kato, M., & Nomoto, K. 1996, *ApJ*, 470, L97
Han, Z., & Podsiadlowski, Ph. 2004, *MNRAS*, 350, 1301
Hurley, J. R., Tout, C. A., & Pols, O. R. 2002, *MNRAS*, 329, 897
Ilkov, M., & Soker, N. 2013, *MNRAS*, 428, 579
Maoz, D., Mannucci, F., & Nelemans, G. 2014, *ARA&A*, 52, 107
Soker, N. 2013, *IAUS*, 281, 72
Toonen, S., Nelemans, G., & Portegies Zwart, S. 2012, *A&A*, 546, A70
Wang, B. 2018, *Res. Astron. Astrophys.*, 18, 49
Wang, B., & Han, Z. 2012, *New Astron. Rev.*, 56, 122
Wang, B., Meng, X., Chen, X., & Han, Z. 2009, *MNRAS*, 395, 847
Wang, B., *et al.* 2017, *MNRAS*, 464, 3965

Why Galaxies Care About AGB Stars:
A Continuing Challenge through Cosmic Time
Proceedings IAU Symposium No. 343, 2019
F. Kerschbaum, M. Groenewegen & H. Olofsson, eds.

doi:10.1017/S1743921318004982

Exploring dust mass and dust properties of nearby AGB stars

J. Wiegert, M. A. T. Groenewegen and the STARLAB team

Royal Observatory of Belgium, Ringlaan 3, B-1180 Brussels, Belgium
email: joachim.wiegert@oma.be

Abstract. Low and intermediate mass stars evolve to the asymptotic red giant branch (AGB) late in their lives. These are surrounded by a circumstellar envelope (CSE) filled with gas and dust. The dust is formed close to the star at sublimation radii and is pushed away by the stellar wind. The dust in turn pushes gases from the envelope into the interstellar medium, thus enriching it with metals. This poster summary is a general description of the next piece of a larger project, whereas the first half has been published by Nicolaes *et al.* (2018). We now aim to use radiative transfer simulations to model spectral energy distributions (SED) of dust and fit them to far-infrared observations for the same 40 sources. We will use 2D and 3D simulations and models containing several dust species simultaneously.

Keywords. stars: AGB and post-AGB, (stars:) circumstellar matter, stars: mass loss, stars: winds, outflows, infrared: stars, submillimeter

1. Introduction

Low and intermediate mass stars, up to about 8 $M_{\odot}$, enter the AGB in the late stages of their lives. AGB stars are surrounded by a CSE consisting of a rich mix of molecules and dust. The dust, which is formed relatively near the star at certain condensation radii (a few up to $\sim$ 10 stellar radii), is pushed outwards by intense stellar winds through the CSE. This outflow pulls with it gases from the CSE which enriches the interstellar medium (ISM) with metals, see e.g. review by Höfner & Olofsson (2018).

Our aim is to use more detailed models for SED fitting than what is commonly used, and to estimate the morphology and mass of the dust in the envelopes of a sample of 37 AGB stars and 3 red super giants (Nicolaes *et al.* 2018) for which *Herschel*/PACS (Pilbratt *et al.* 2010, Poglitsch *et al.* 2010) and, in some cases SPIRE (Griffin *et al.* 2010) spectra are available. A large portion of these sources were also in the Mass loss of Evolved StarS *Herschel* Key Programme (MESS, Groenewegen *et al.* 2011). Archived observational photometric and spectral data will also be used, mainly to determine the stars' SEDs. Models of the stellar photospheric SEDs are extracted from the MARCS grid (Gustafsson *et al.* 2008)†.

We will simulate dust SEDs with MoD (More of Dusty, Groenewegen 2012)‡ and the monte-carlo based program RADMC-3D (Dullemon 2012)§ for non-standard morphologies and dust distributions. This way we may obtain estimates on grain sizes and constituents, which gives dust masses and also dust mass losses. In the cases where the gas mass is known we may estimate dust-to-gas ratios.

† http://marcs.astro.uu.se
‡ http://homepage.oma.be/marting/codes.html
§ http://www.ita.uni-heidelberg.de/~dullemond/software/radmc-3d/index.html

2. Project description

We employ SED fitting to estimate e.g. total dust masses, dust properties, and to explore the spatial distribution of different dust species around nearby AGB stars. The sources are within 1 kpc with few exceptions. The most distant source is V1365 Aql at 4.30 kpc (Justtanont *et al.* 2006).

We will use 2D and 3D radiative transfer. The aforementioned RADMC-3D is still under development, but is already applicable and has been proven useful in a variety of publications†. Some of the strengths of RADMC-3D are that we can assume any dust distribution of several arbitrary species and any number of stellar sources.

It is common to assume spherical symmetric dust distributions of one species when studying dust in CSEs. We will initially assume spherical dust distributions also, however, we will consider several dust species simultaneously, and will vary the density distributions (assumed to be a powerlaw, $\rho \propto R^{-k}$) and size of the dust cloud. The inner radii of each species is constrained by their sublimation temperatures and we will also consider interferometric observations, when available, to better model the inner part of the dust shell.

The risk is an ever-growing parameter tree while SED fitting relies on only one dimension. This issue can be addressed with spectral data from which the constituents and the size range of the dust grains can be constrained. We can also use theory to constrain some of the parametres as e.g. the inner radii, the density distribution, and a range of realistic grain size distributions. The grain sizes will initially be monosize, however, later we may adopt e.g. an MRN-like distribution as in the ISM (Mathis *et al.* 1977).

We limit the dust species to two cases. For O-rich stars, we will initially assume that iron poor magnesium-silicates are formed at ~ 2 to $3\,R_{\rm star}$ from the stellar centre (we also note that aluminium-oxides and titanium-oxides may be important as grain seeds). Iron will be added further out as $MgFeSiO_4$-dust ($\sim$ 4–5 to $9\,R_{\rm star}$). For C-rich stars the model is similar, but we will assume amorphous carbon grains as initial grains instead.

Acknowledgements

This research is funded by the Belgian Science Policy Office under contract BR/143/A2/STARLAB.

References

Dullemond, C. P., 2012, RADMC-3D: A multi-purpose radiative transfer tool, astrophysics Source Code Library

Gustafsson B., Edvardsson B., Eriksson K., *et al.* 2008, *A&A*, 486, 951

Groenewegen, M. A. T., Waelkens, C., Barlow, M. J., *et al.* 2011, *A&A*, 526, A162

Groenewegen, M. A. T. 2012, *A&A*, 543, A36

Griffin, M.J., Abergel, A., Abreu, A., *et al.* 2010, *A&A*, 518, L3

Höfner, S., & Olofsson, H. 2018, *A&AR*, 26, 1

Justtanont, K., Olofsson, G., Dijkstra, C., & Meyer, A. W. 2006, *A&A*, 450, 1051

Mathis, J. S., Rumpl, W., & Nordsieck, K. H. 1977, *ApJ*, 217, 425

Nicolaes, D., Groenewegen, M. A. T., Royer, P., *et al.* 2018, preprint (arXiv:1808.03467)

Pilbratt, G.L., Riedinger, J.R., Passvogel, T., *et al.* 2010, *A&A*, 518, L1

Poglitsch, A., Waelkens, C., Geis, N., *et al.* 2010, *A&A*, 518, L2

† An incomplete list of publications at:
`http://www.ita.uni-heidelberg.de/~dullemond/software/radmc-3d/publications.html`

Why Galaxies Care About AGB Stars:
A Continuing Challenge through Cosmic Time
Proceedings IAU Symposium No. 343, 2019
F. Kerschbaum, M. Groenewegen & H. Olofsson, eds.

doi:10.1017/S1743921318006993

K–Type Supergiants in the Large Magellanic Cloud

Robert F. Wing

Astronomy Department, Ohio State University
140 West 18th Avenue, Columbus, Ohio 43210, USA
email: wing.1@osu.edu

Abstract. Two-dimensional spectral classifications, on a narrow-band photometric system that measures near-infrared bands of TiO and CN, are being obtained for several hundred previously unclassified "suspected late-type supergiants" in the Large Magellanic Cloud. The objective is to identify supergiants of spectral type K, which are known to be plentiful in the Small Magellanic Cloud but were thought to be rare in the LMC. In the fields examined to date, 35 % of the targets are found to be K-type supergiants, while 25 % are early-M supergiants and 40 % are foreground stars of lower luminosity.

Keywords. Magellanic Clouds, stars: late-type, supergiants

---

1. The Problem

The brightest stars of the Large Magellanic Cloud are supergiants with magnitudes in the range $V = 11 - 14$ and colors that are either very blue (spectral types O, B, and A) or very red (type M). The apparent absence of supergiants of intermediate types (F, G, K) is particularly noteworthy because the Small Magellanic Cloud is known to contain large numbers of K-type supergiants.

Are K-type supergiants really absent from the LMC, or have they simply been missed by the survey techniques employed to date? Near-infrared objective-prism surveys of the LMC by Westerlund *et al.* (1981) and Blanco (unpublished) have identified several hundred M supergiants on the basis of their TiO absorption near 7100 Å. These TiO-based surveys, which had no way of identifying K stars, were subsequently used by Humphreys (1979) in her spectroscopic study of the red supergiants in the LMC. By contrast, when Elias, Frogel, & Humphreys (1985) studied the red supergiants of the SMC, no such infrared TiO-based survey of the SMC was available, and they selected their targets from a survey by Prévot *et al.* (1983) in which the main criterion had been the slope of the spectrum on their blue/visual objective-prism plates.

In an attempt to avoid the selection effects that may have artificially removed K-type supergiants from the LMC sample, I have gone back to the blue objective-prism survey by Sanduleak & Philip (1977) who reported 609 stars suspected of being late-type supergiants on the basis of their spectral slope and a magnitude in the range $V = 11.0 - 13.5$. Interestingly, the great majority of these had not been picked up in Westerlund's earlier TiO-based surveys. Are these the "missing" K-type supergiants? They must be of spectral types G, K, or early M, but are they LMC supergiants or foreground stars of lower luminosity?

2. Method

Sanduleak & Philip (1977) marked their suspected LMC late-type supergiants, as well as carbon stars, on charts of the Hodge-Wright Atlas (Hodge & Wright 1967). In the central 5×5 array of Atlas fields, which includes nearly all of the visible Cloud and some outlying regions, they marked 589 suspected late-type supergiants, and these comprise my target list.

Stars are assigned two-dimensional (temperature/luminosity) classifications from photometric measurements in 5 narrow-band filters, which include bands of TiO and CN and continuum points in the region 7100–10400 Å. The great majority of stars in this sample can be given unambiguous two-dimensional spectral types by this method, although a few peculiar stars have also been encountered (including two very red, weak-banded stars which may be halo giants of extremely low metallicity).

All observations for this project have been obtained at the Cerro Tololo Inter-American Observatory. The work began in the 1990s with a photoelectric photometer at the 1.0-m telescope, but was discontinued when photoelectric photometry stopped being supported. After a hiatus, observations have resumed with a CCD photometer at the 0.9-m telescope, now operated by the SMARTS consortium. With the latter arrangement, observations of the LMC target stars consist of 60-second integrations at each of the first four filters centered at 7120, 7540, 7810, and 8120 Å, while a 600-second integration is required for the long-wavelength (10350 Å) continuum point.

3. Preliminary Results

Since the work is still in progress, useful statistics can be given only for the 6 fields which have complete observations (Hodge-Wright fields 37, 38, 46, 54, 62, and 63). Results for the 133 suspected late-type supergiants in these fields break down as follows:

Confirmed K-type supergiants :	46	(35 %)
Early-M supergiants :	34	(25 %)
Probable foreground stars :	53	(40 %)

In short, the majority (60 %) of the target stars in these fields are classified as supergiants of luminosity classes Ib, Iab, or Ia, and are thus spectroscopically confirmed as members of the LMC. Their spectral types are nearly all K4, K5, M0, and M1 – and their distribution is similar to that found by Elias *et al.* (1985) in the SMC.

Individual types will, of course, be published when the observations are complete.

References

Elias, J.H., Frogel, J.A., Humphreys, R.M. 1985 *ApJ Suppl.*, 57, 91

Hodge, P.W., Wright, F.W. 1967, The Large Magellanic Cloud, *Smithsonian Press*, Washington D.C.

Humphreys, R.M. 1979, *ApJ Suppl.*, 39, 1979

Prévot, L., Martin, N., Maurice, E., Rebeirot, E., Rousseau, J. 1983, *Astron. Astrophys. Suppl. Ser.*, 53, 255

Sanduleak, N., Philip, A.G.D. 1977 *Publ. Warner and Swasey Obs.*, 2, No. 5

Westerlund, B.E., Olander, N., Hedin, B. 1981, *Astron. Astrophys. Suppl. Ser.*, 43, 267

Why Galaxies Care About AGB Stars: A Continuing Challenge through Cosmic Time
Proceedings IAU Symposium No. 343, 2019
F. Kerschbaum, M. Groenewegen & H. Olofsson, eds.

doi:10.1017/S1743921318004970

The carbon star R Sculptoris sheds its skin

Markus Wittkowski

ESO, Karl-Schwarzschild-Str. 2, 85748 Garching, Germany,
email: mwittkow@eso.org

Abstract. We describe near-IR H-band VLTI-PIONIER aperture synthesis images of the carbon AGB star R Sculptoris with an angular resolution of 2.5 mas. The data show a stellar disc of diameter ~ 9 mas exhibiting a complex substructure including one dominant bright spot with a peak intensity of 40% to 60% above the average intensity. We interpret the complex structure as caused by giant convection cells, resulting in large-scale shock fronts, and their effects on clumpy molecule and dust formation seen against the photosphere at distances of 2–3 stellar radii. Moreover, we derive fundamental parameters of R Scl, which match evolutionary tracks of initial mass 1.5 ± 0.5 $M_{\odot}$. Our visibility data are best fit by a dynamic model without a wind, which may point to problems with current wind models at low mass-loss rates.

1. Introduction

Mass loss becomes increasingly important during the AGB phase, both for the stellar evolution, and for the return of material to the interstellar medium. Some carbon-rich AGB stars are known to exhibit a clumpiness of their circumstellar environment (e.g., Weigelt *et al.* 1998). R Scl is a carbon-rich semi-regular pulsating AGB star with a period of 370 days at a distance of 370 ± 100 pc. ALMA observations in CO revealed a spiral structure, indicating the presence of a previously unknown companion (Maercker *et al.* 2012).

The near-IR imaging results of the stellar disc presented here are based on Wittkowski *et al.* (2017). Further descriptions of our results are available in an ESO blog† and an ESO picture of the week‡.

2. Observations and results

We obtained VLTI-PIONIER data of R Scl at three spectral channels in the near-IR H-band with baselines between 11 m and 140 m, providing an angular resolution of 2.5 mas. We reconstructed images with the `IRBis` package (Hofmann *et al.* 2014), using a best-fit model as a start image, a flat prior, and smoothness as regularisation. We investigated different start images, regularisations, priors, and image reconstruction packages, and obtained very similar reconstructions in all cases. The resulting images (1.68 μm example in Fig. 1) show a complex structure within the stellar disc, including a dominant bright spot with a peak intensity of 40-60% above the average intensity.

We interpret the features in our images as dust clumps at radii of 2–3 $R_{\rm star}$ seen against the photosphere. Such dust clumps may be caused by giant convection cells resulting in large-scale shock fronts and leading to clumpy molecule and dust formation, as modeled by Freytag & Höfner (2008) and Freytag *et al.* (2017).

We compared the VLTI-PIONIER, and VLTI-AMBER, data to a grid of dynamic atmosphere and wind models by Eriksson *et al.* (2014). We obtained a best fit with

† Available at http://www.eso.org/public/blog/how-stars-die/
‡ Available at http://www.eso.org/public/images/potw1807a/

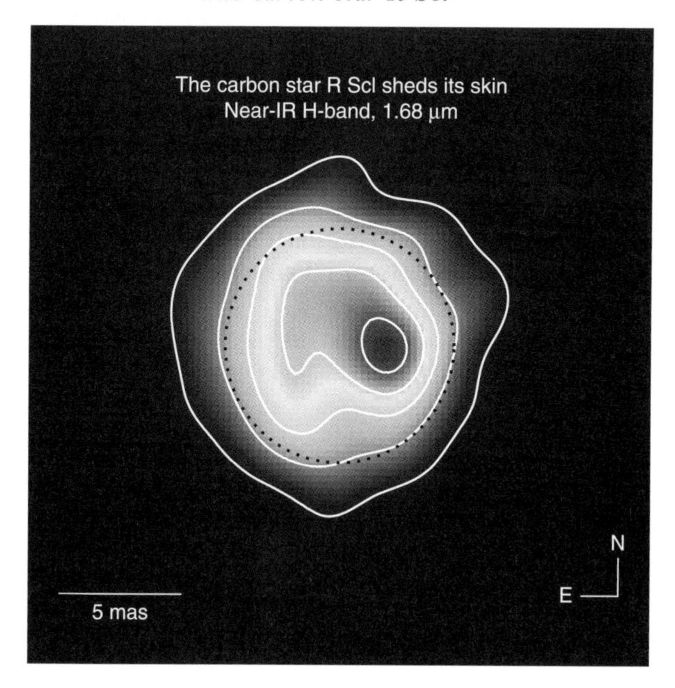

Figure 1. Image of R Scl at 1.68 μm, reconstructed with the `IRBis` package. The estimated Rosseland angular diameter is indicated by the dashed black circle. Contours are drawn at levels of 0.9, 0.7, 0.5, 0.3, 0.1. From Wittkowski *et al.* (2017).

a model without a wind, which may point to problems with current wind models at low mass-loss rates. We estimated an angular Rosseland diameter of 8.9 ± 0.3 mas, and derived further fundamental parameters of R Scl, which match evolutionary tracks (Lagarde *et al.* 2012, Marigo *et al.* 2013) of initial mass $1.5 \pm 0.5\,M_{\odot}$.

References

Eriksson, K., Nowotny, W., Höfner, S., *et al.* 2014, *A&A*, 566, A95
Freytag, B., & Höfner, S. 2008, *A&A*, 483, 571
Freytag, B., Liljegren, S., & Höfner, S.. 2017, *A&A*, 600, A137
Hofmann, K.-H., Weigelt, G., & Schertl, D. 2014, *A&A*, 565, A48
Lagarde, N., Decressin, T., Charbonnel, C., *et al.* 2012, *A&A*, 543, A108
Maercker, M., Mohmaned, S., Vlemmings, W., *et al.* 2012, *Nature*, 490, 232
Marigo, P., Bressan, A., Nanni, A., *et al.* 2013, *MNRAS*, 434, 488
Weigelt, G., Balega, Y., Blöcker, T., *et al.* 1998, *A&A*, 333, L51
Wittkowski, M., Hofmann, K.-H., Höfner, S., *et al.* 2017, *A&A*, 601, A3

Why Galaxies Care About AGB Stars: A Continuing Challenge through Cosmic Time
Proceedings IAU Symposium No. 343, 2019
F. Kerschbaum, M. Groenewegen & H. Olofsson, eds.

doi:10.1017/S1743921318007718

TIG*vival*: High-resolution spectroscopic monitoring of LPV stars

Uwe Wolter[1], **Dieter Engels**[1], **Bernhard Aringer**[2] **and Bernd Freytag**[3]

[1]Hamburger Sternwarte, Universität Hamburg, Germany
email: uwolter@hs.uni-hamburg.de

[2]Dipartimento di Fisica e Astronomia Galileo Galilei, Università di Padova, Italy

[3]Institutionen för fysik och astronomi, Uppsala Universitet, Sweden

Abstract. TIG*vival* is a spectroscopic monitoring program of long-period variables (LPV) using our robotic telescope TIGRE. Since 2013, we obtain low-noise, high-resolution spectra (R= 20 000) that cover the optical regime (3800 Å to 8800 Å). We are now continuously monitoring 7 LPVs with different periods and chemical properties. Our 350+ spectra evenly sample the target cycles, as far as ground-based observations allow. Analyzing the TIG*vival* spectra of Mira as a sample case, our measurements indicate that the strength of the TiO-absorption is phase-shifted with respect to the visual light curve.

Keywords. stars: AGB and post-AGB, molecular processes, techniques: spectroscopic

1. Introduction

Our long-term monitoring program TIG*vival* provides spectroscopic time series of currently seven long-period variables: G Her (89 days), BK Vir (150 d), U Hya (183 d), *o* Cet (=Mira, 332 d), W Hya (361 d), R Hya (380 d) and R Lep (445 d) – the corresponding cycle periods are given in brackets. The acronym TIG*vival* illustrates that we plan to continue our monitoring for several more years: TIGRE *vigila variables de largo periodo a largo plazo* = TIGRE long-term monitoring of LPVs. All our time series currently cover almost two continuous years, with a total of more than 350 spectra. They evenly sample the target cycles, as far as visibility from the TIGRE site in central Mexico allows.

TIGRE is a 1.2 meter robotic telescope. It uses a two-armed echelle spectrograph to cover much of the optical range from the near UV to the near IR with a spectral resolution of 20 000. With the exception of R Lep, essentially all TIG*vival* spectra have a signal-to-noise ratio that exceeds e.g. 100 at 7000 Å.

2. Mira's TiO molecular bands: illustrating the potential of TIG*vival*

The formation of molecules (TiO, VO) in the extended atmospheres of Mira variables dominates their visual variations (Reid & Goldstone 2002). Furthermore, molecular absorption also causes the different stellar diameters measured in different optical narrow-band filters (Tuthill *et al.* 1995) – as well as their cyclic variations (Young *et al.* 2000).

The bolometric emission of LPVs is difficult to measure directly. However, the H_2O and SiO maser emission of LPVs are assumed to be directly correlated with the bolometric emission. These maser emissions follow the visual light curve, although with a lag of approximately 0.2 in phase (Pardo *et al.* 2004, Brand *et al.* 2019 in prep.).

As a first step to study the TiO absorption in LPV, we performed a simple analysis: we integrated Mira's spectral flux inside two wavelength bands, each several dozen

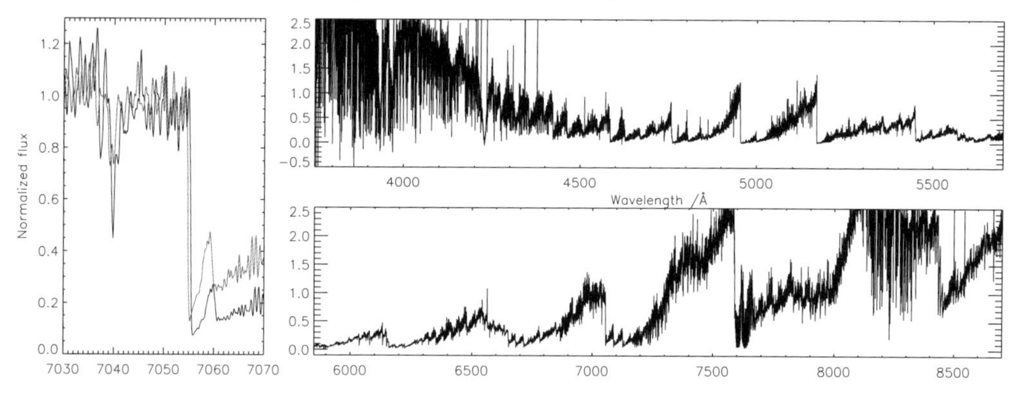

Figure 1. Representative TIG*vival* spectrum of Mira (*o* Ceti) taken near visual phase 0.3 (black graph, see also Fig. 2). The panels on the right show the entire spectrum, while the left panel zooms in on the same spectrum as well as on a second spectrum taken at about visual phase 0.16 (red). Note that above a wavelength of ∼4500 Å no signal noise is visible in this figure.

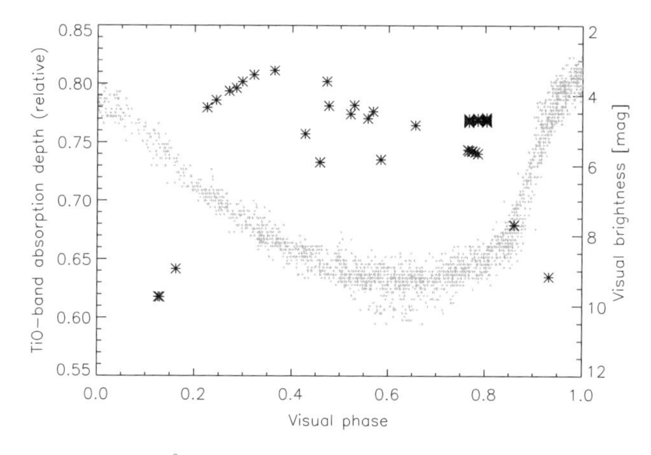

Figure 2. Depth of the 7062 Å TiO absorption for Mira (asterisks), in units of the nearby pseudo-continuum. Phase zero represents maximum visual brightness, the AAVSO magnitudes (January 2014 to September 2018) are shown in gray for comparison.

Ångströms wide, situated near the TiO bandhead at ~ 7055 Å (Fig. 1). We measure the strength of the TiO absorption, using the ratio of the flux redward of this bandhead (TB) divided by the pseudo-continuum TC directly blueward of the bandhead as a proxy. Here we define the *arbsorption depth* as $(1 - TB/TC)$. The phases we use here were estimated from visual AAVSO measurements spanning more than a decade.

Our result is shown in Fig. 2. We find the weakest TiO absorption close to visual phase zero. While the TiO band depth largely seems to follow the visual light curve, it shows a substantial lag. It reaches its maximum later, at approximately phase 0.35. We will continue our TIG*vival* monitoring of Mira to check the reality of this delay.

References

Brand, J., Winnberg A., Engels, D., 2018, A&A, in preparation
Pardo *et al.* 2004, A&A, 424 145
Reid, M.J. & Goldstone, J.E., 2002, ApJ, 568, 931
Tuthill, P.G., Haniff, C.A., & Baldwin, J.E., 1995, MNRAS, 277, 1541
Young *et al.*, 2000, MNRAS, 318, 381

Author index